Physical Chemistry:
A Comprehensive Approach

Physical Chemistry: A Comprehensive Approach

Edited by
Finn Miller

WILLFORD PRESS

www.willfordpress.com

Published by Willford Press,
118-35 Queens Blvd., Suite 400,
Forest Hills, NY 11375, USA

ISBN: 978-1-68285-375-7

Cataloging-in-Publication Data

Physical chemistry : a comprehensive approach / edited by Finn Miller.
    p. cm.
Includes bibliographical references and index.
ISBN 978-1-68285-375-7
1. Chemistry, Physical and theoretical. 2. Chemistry. I. Miller, Finn.
QD453.3 .P49 2017
541--dc23

For information on all Willford Press publications
visit our website at www.willfordpress.com

WILLFORD PRESS

Printed in the United States of America.

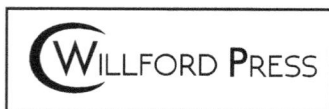

Contents

Preface...IX

Chapter 1 **Vortex flows impart chirality-specific lift forces**..1
Thomas M. Hermans, Kyle J.M. Bishop, Peter S. Stewart, Stephen H. Davis and
Bartosz A. Grzybowski

Chapter 2 **Mapping multidimensional electronic structure and ultrafast dynamics with
single-element detection and compressive sensing**...9
Austin P. Spencer, Boris Spokoyny, Supratim Ray, Fahad Sarvari and Elad Harel

Chapter 3 **Pulse-density modulation control of chemical oscillation far from equilibrium
in a droplet open-reactor system**..15
Haruka Sugiura, Manami Ito, Tomoya Okuaki, Yoshihito Mori,
Hiroyuki Kitahata and Masahiro Takinoue

Chapter 4 **Native characterization of nucleic acid motif thermodynamics via non-covalent
catalysis**..24
Chunyan Wang, Jin H. Bae and David Yu Zhang

Chapter 5 **Electropolymerization on wireless electrodes towards conducting polymer
microfibre networks**...35
Yuki Koizumi, Naoki Shida, Masato Ohira, Hiroki Nishiyama,
Ikuyoshi Tomita and Shinsuke Inagi

Chapter 6 **A simple and versatile design concept for fluorophore derivatives with
intramolecular photostabilization**..41
Jasper H. M. van der Velde, Jens Oelerich, Jingyi Huang, Jochem H. Smit,
Atieh Aminian Jazi, Silvia Galiani, Kirill Kolmakov, Giorgos Guoridis,
Christian Eggeling, Andreas Herrmann, Gerard Roelfes and Thorben Cordes

Chapter 7 **Design of crystal-like aperiodic solids with selective disorder–phonon coupling**............56
Alistair R. Overy, Andrew B. Cairns, Matthew J. Cliffe, Arkadiy Simonov,
Matthew G. Tucker and Andrew L. Goodwin

Chapter 8 **Ionic polarization-induced current–voltage hysteresis in $CH_3NH_3PbX_3$ perovskite
solar cells**...64
Simone Meloni, Thomas Moehl, Wolfgang Tress, Marius Franckevičius,
Michael Saliba, Yong Hui Lee, Peng Gao, Mohammad Khaja Nazeeruddin,
Shaik Mohammed Zakeeruddin, Ursula Rothlisberger and Michael Graetzel

Chapter 9 **High-contrast and fast electrochromic switching enabled by plasmonics**............................73
Ting Xu, Erich C. Walter, Amit Agrawal, Christopher Bohn, Jeyavel Velmurugan,
Wenqi Zhu, Henri J. Lezec and A. Alec Talin

Chapter 10   **Transition state theory demonstrated at the micron scale with out-of-equilibrium transport in a confined environment**...79
Christian L. Vestergaard, Morten Bo Mikkelsen, Walter Reisner,
Anders Kristensen and Henrik Flyvbjerg

Chapter 11   **Sub-10 nm rutile titanium dioxide nanoparticles for efficient visible-light-driven photocatalytic hydrogen production**...88
Landong Li, Junqing Yan, Tuo Wang, Zhi-Jian Zhao, Jian Zhang,
Jinlong Gong and Naijia Guan

Chapter 12   **Visualizing the orientational dependence of an intermolecular potential**...........................98
Adam Sweetman, Mohammad A. Rashid, Samuel P. Jarvis, Janette L. Dunn,
Philipp Rahe and Philip Moriarty

Chapter 13   **Three-dimensional controlled growth of monodisperse sub-50 nm heterogeneous nanocrystals**...105
Deming Liu, Xiaoxue Xu, Yi Du, Xian Qin, Yuhai Zhang, Chenshuo Ma, Shihui Wen,
Wei Ren, Ewa M. Goldys, James A. Piper, Shixue Dou, Xiaogang Liu and Dayong Jin

Chapter 14   **Electrochemically driven mechanical energy harvesting**...113
Sangtae Kim, Soon Ju Choi, Kejie Zhao, Hui Yang, Giorgia Gobbi, Sulin Zhang and Ju Li

Chapter 15   **Optimal metal domain size for photocatalysis with hybrid semiconductor-metal nanorods**..120
Yuval Ben-Shahar, Francesco Scotognella, Ilka Kriegel, Luca Moretti, Giulio Cerullo,
Eran Rabani and Uri Banin

Chapter 16   **The hydrogen-bond network of water supports propagating optical phonon-like modes**...127
Daniel C. Elton and Marivi Fernández-Serra

Chapter 17   **Two distinctive energy migration pathways of monolayer molecules on metal nanoparticle surfaces**..135
Jiebo Li, Huifeng Qian, Hailong Chen, Zhun Zhao, Kaijun Yuan, Guangxu Chen,
Andrea Miranda, Xunmin Guo, Yajing Chen, Nanfeng Zheng,
Michael S. Wong and Junrong Zheng

Chapter 18   **Hierarchy of bond stiffnesses within icosahedral-based gold clusters protected by thiolates**..143
Seiji Yamazoe, Shinjiro Takano, Wataru Kurashige, Toshihiko Yokoyama,
Kiyofumi Nitta, Yuichi Negishi and Tatsuya Tsukuda

Chapter 19   **Silicon oxycarbide glass-graphene composite paper electrode for long-cycle lithium-ion batteries**...150
Lamuel David, Romil Bhandavat, Uriel Barrera and Gurpreet Singh

Chapter 20   **Electron–phonon coupling in hybrid lead halide perovskites**......................................160
Adam D. Wright, Carla Verdi, Rebecca L. Milot, Giles E. Eperon,
Miguel A. Pérez-Osorio, Henry J. Snaith, Feliciano Giustino,
Michael B. Johnston and Laura M. Herz

Chapter 21   **Insertion compounds and composites made by ball milling for advanced sodium-ion batteries**...169
Biao Zhang, Romain Dugas, Gwenaelle Rousse, Patrick Rozier,
Artem M. Abakumov and Jean-Marie Tarascon

Chapter 22   **Putting pressure on aromaticity along with *in situ* experimental electron density of a molecular crystal**......178
Nicola Casati, Annette Kleppe, Andrew P. Jephcoat and Piero Macchi

Chapter 23   **Structural complexity of simple Fe₂O₃ at high pressures and temperatures**......186
E. Bykova, L. Dubrovinsky, N. Dubrovinskaia, M. Bykov, C. McCammon,
S.V. Ovsyannikov, H.-P. Liermann, I. Kupenko, A.I. Chumakov, R. Rüffer,
M. Hanfland and V. Prakapenka

Chapter 24   **Direct observation of mineral–organic composite formation reveals occlusion mechanism**......192
Kang Rae Cho, Yi-Yeoun Kim, Pengcheng Yang, Wei Cai, Haihua Pan,
Alexander N. Kulak, Jolene L. Lau, Prashant Kulshreshtha, Steven P. Armes,
Fiona C. Meldrum and James J. De Yoreo

Chapter 25   **Water electrolysis on La₁₋ₓSrₓCoO₃₋δ perovskite electrocatalysts**......199
J. Tyler Mefford, Xi Rong, Artem M. Abakumov, William G. Hardin, Sheng Dai6,
Alexie M. Kolpak, Keith P. Johnston and Keith J. Stevenson

Chapter 26   **Light-enhanced liquid-phase exfoliation and current photoswitching in graphene–azobenzene composites**......210
Markus Döbbelin, Artur Ciesielski, Sébastien Haar, Silvio Osella, Matteo Bruna,
Andrea Minoia, Luca Grisanti, Thomas Mosciatti, Fanny Richard,
Eko Adi Prasetyanto, Luisa De Cola, Vincenzo Palermo, Raffaello Mazzaro,
Vittorio Morandi, Roberto Lazzaroni, Andrea C. Ferrari, David Beljonne and
Paolo Samorı

Chapter 27   **Creating single-atom Pt-ceria catalysts by surface step decoration**......220
Filip Dvořák, Matteo Farnesi Camellone, Andrii Tovt, Nguyen-Dung Tran,
Fabio R. Negreiros, Mykhailo Vorokhta, Tomáš Skála, Iva Matolínová,
Josef Mysliveček, Vladimír Matolín and Stefano Fabris

Permissions

List of Contributors

Index

Preface

Physical chemistry is an interdisciplinary field integrating the principles of physics and chemistry. Physical chemistry focuses on understanding the physical properties of atoms and molecules. It involves analyzing materials to discover the potential use of the materials. Physical chemistry requires the use of analytical instruments like lasers, electron microscopes, mass spectrometers, etc. This book discusses the fundamentals as well as modern approaches of physical chemistry to help develop a comprehensive understanding of this field. From theories to research to practical applications, case studies related to all contemporary topics of relevance to this field have been included in this book. It is appropriate for students seeking detailed information in physical chemistry as well as for experts.

This book has been the outcome of endless efforts put in by authors and researchers on various issues and topics within the field. The book is a comprehensive collection of significant researches that are addressed in a variety of chapters. It will surely enhance the knowledge of the field among readers across the globe.

It gives us an immense pleasure to thank our researchers and authors for their efforts to submit their piece of writing before the deadlines. Finally in the end, I would like to thank my family and colleagues who have been a great source of inspiration and support.

Editor

Vortex flows impart chirality-specific lift forces

Thomas M. Hermans[1,†], Kyle J.M. Bishop[2], Peter S. Stewart[3,†], Stephen H. Davis[3] & Bartosz A. Grzybowski[1]

Recent reports that macroscopic vortex flows can discriminate between chiral molecules or their assemblies sparked considerable scientific interest both for their implications to separations technologies and for their relevance to the origins of biological homochirality. However, these earlier results are inconclusive due to questions arising from instrumental artifacts and/or insufficient experimental control. After a decade of controversy, the question remains unresolved—how do vortex flows interact with different stereoisomers? Here, we implement a model experimental system to show that chiral objects in a Taylor–Couette cell experience a chirality-specific lift force. This force is directed parallel to the shear plane in contrast to previous studies in which helices, bacteria and chiral cubes experience chirality-specific forces perpendicular to the shear plane. We present a quantitative hydrodynamic model that explains how chirality-specific motions arise in non-linear shear flows through the interplay between the shear-induced rotation of the particle and its orbital translation. The scaling laws derived here suggest that rotating flows can be used to achieve chiral separation at the micro- and nanoscales.

[1] Department of Chemical and Biological Engineering and Department of Chemistry, Northwestern University, 2145 Sheridan Road, Evanston, Illinois 60208, USA. [2] Department of Chemical Engineering, The Pennsylvania State University, 132C Fenske Lab, University Park, Pennsylvania 16802, USA. [3] Engineering Sciences and Applied Mathematics, Northwestern University, 2145 Sheridan Road, Evanston, Illinois 60208, USA. † Present address: Institut de Science et d'Ingénierie Supramoléculaires (ISIS), UMR7006, 8 allée Gaspard Monge, 67000 Strasbourg, France (T.M.H.); School of Mathematics and Statistics, University of Glasgow, University Gardens, Glasgow G12 0RB, UK (P.S.S.). Correspondence and requests for materials should be addressed to B.A.G. (email: grzybor@northwestern.edu).

The idea that fluid flows can discriminate between chiral objects of opposite handedness was initially proposed by Howard *et al.*[1] and has since been examined theoretically in considerable detail[2–10]. Achieving separation of enantiomers[11,12] without the use of a chiral stationary phase—the most costly component in the separation process—would revolutionize the pharmaceutical industry. Surprisingly, however, there is still no agreement as to the magnitudes or even directions of forces affecting chiral objects in fluid flows[7,8,13]. Although several experimental studies report chirality-specific flow effects—on scales from molecular (porphyrin aggregation during rotary evaporation)[14,15], through microscopic (helical bacteria[4] and particles[16]), to macroscopic[13,17,18]—they remain largely phenomenological. Part of the problem is the possibly ambiguous interpretation of data[19–21] (for example, from circular dichroism) and the experimental design in which vortex flows are implemented with imprecise systems such as rotary evaporators or magnetic stirrers[22], for which the flow structure on different length scales is largely unknown.

In this study, we use the well-controlled flow generated in a Couette cell containing macroscopic chiral particles (confined to the liquid/air interface) to clarify and explain the chirality-selective motions of solid objects in non-linear shear flows. Importantly, by confining the motion of the particles to a two-dimensional (2D) interface, we are able to capture the detailed rotational and translational motion of the particles in time—not just their average drifting motion as in previous studies[4,16,18]. Such detailed experimental data allow for quantitative comparisons with predictions of the hydrodynamic model that we develop. Building on these experimental results, we develop quantitative scaling laws that describe how in-plane chiral-selective motions depend on the size and shape of the particle and on the geometry of the non-linear flow field. These scaling laws predict that rotating flows can provide physical means of separating chiral objects at the micro- and nanoscales.

Results

Experimental design. Our experiments are performed in a custom-machined Couette cell (Fig. 1a,b) with the radius of the stationary inner cylinder $R_i = 25.4 \pm 0.01$ mm and that of the rotating outer cylinder $R_o = 55.9 \pm 0.01$ mm. The cylinders are aligned coaxially with variations in the gap width, $G = R_o - R_i$, <0.5 mm. The cell is partially filled with paraffin oil (dynamic viscosity, $\eta = 0.17 \pm 0.06$ Pa·s; density, $\rho = 0.86 \pm 0.03$ g ml^{-1}) and in each experiment, a single millimetre-sized particle (7.70 × 2.46 × 0.20 mm, fabricated in SU-8 by photolithography; density 1.19 g ml^{-1}) is placed onto the oil/air interface such that it is fully immersed in the oil save for its top face (Fig. 1c). The particles are s-shaped (D$_{2h}$ symmetry) such that when placed onto the interface, they have two non-superimposable—that is, chiral—orientations, which we denote 'R' and 'S' (Fig. 1b) by analogy to chemical convention. A shear flow is generated in the

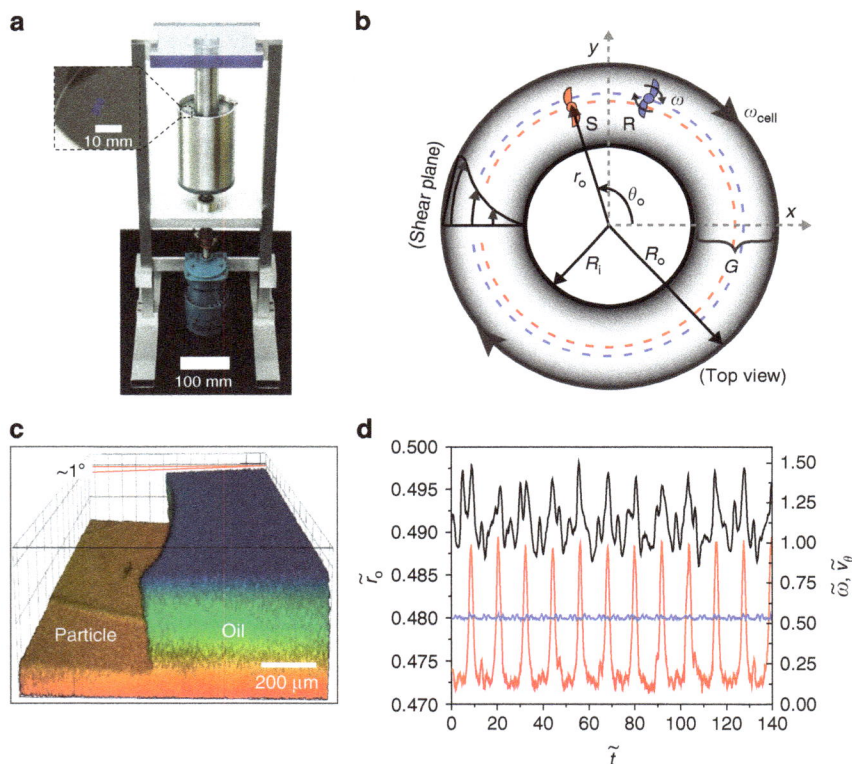

Figure 1 | Experimental arrangement. (a) Optical image of the Couette cell with a millimetre-sized chiral particle placed at the oil/air interface. **(b)** Top-view scheme of the system with an inner cylinder stationary and an outer cylinder rotating—here, in the CW direction—at an angular velocity ω_{cell}. Two chiral particles (denoted R and S) are illustrated; these particles revolve around the cell with linear velocity v_θ and also rotate around their centres with an angular velocity ω. The shear plane is indicated by the curved arrows. **(c)** Three-dimensional confocal microscopy of a chiral particle floating stationary on top of paraffin oil (containing 1 μM diketopyrrolopyrrole to make the oil fluorescent). The particle is not fluorescent and therefore appears as a void space on the left of the image. No significant meniscus was observed; however, as the particle is slightly denser than the fluid, the interface slopes downward at ~1° near the particle edge (see Supplementary Note 1). The grid size is 50 × 50 μm. **(d)** Typical experimental data showing the radial position $\tilde{r}_o = (r_o - R_i)/G$ (black), linear velocity $\tilde{v}_\theta = v_\theta/R_o\omega_{cell}$ (blue), and angular velocity $\tilde{\omega} = \omega/\omega_{cell}$ (red) of an S particle as a function of time $\tilde{t} = t\omega_{cell}$ (after the particle has reached a stable orbit); here, the cell velocity is $\omega_{cell} = 3.02 \pm 0.13$ rad s^{-1}.

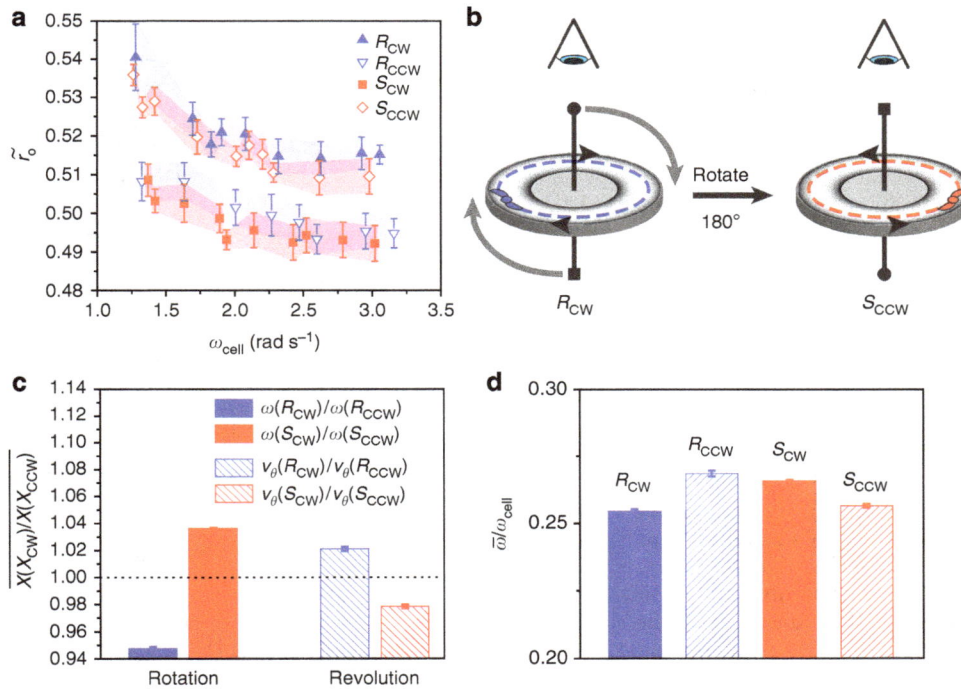

Figure 2 | Chirality effects in a vortex flow. (**a**) The mean orbit radii $\tilde{r}_o = (\bar{r}_o - R_i)/G$ observed in experiments depends on the chirality of the particles and on the direction of flow (solid markers = CW cell rotation; open markers = CCW rotation). For all cell velocities ω_{cell}, $\tilde{r}_o(R_{CW}) > \tilde{r}_o(S_{CW})$ and $\tilde{r}_o(S_{CCW}) > \tilde{r}_o(R_{CCW})$. Upon 'flow reversal' (without any manipulation of particles), $\tilde{r}_o(R_{CW}) \approx \tilde{r}_o(S_{CCW})$ and $\tilde{r}_o(R_{CCW}) \approx \tilde{r}_o(S_{CW})$. Error bars correspond to standard deviations in $\tilde{r}_o$, which are based on at least 10 independent experiments (12,500 data points per run recorded) for each value of ω_{cell} (note: four particles of each type were tested). (**b**) Scheme illustrating why the behaviour of a chiral particle in a CW flow is expected to be equivalent to its 'mirror image' in a CCW flow. Rotation of the entire system by 180° about a horizontal axis (denoted by grey arrows) changes the particle chirality and the flow direction (with respect to an observer looking down the vertical axis as denoted by the eye cartoon) while the orbit radius remains unchanged. These symmetry arguments are in agreement with the experimental observations in **a**. (**c**) When the direction of flow reverses from CW to CCW, the linear and angular velocities, ω and v_θ, also change as illustrated by the bar graph showing the ratios of the particle velocities for CW and CCW flows averaged over all experiments. Error bars represent s.d. based on at least 200 experiments for each bar. (**d**) The average angular velocity of the particles $\bar{\omega}$ is consistently smaller than that of the Couette cell ω_{cell}. Error bars represent s.d. based on at least 200 experiments for each bar.

cell by rotating the outer cylinder (a configuration for which Taylor instabilities cannot occur) at an angular frequency $\omega_{cell} = 1$–$3\,\text{rad}\,\text{s}^{-1}$. For the typical particle dimensions and velocity, the particle Reynolds number is $Re = \rho L R_o \omega_{cell}/\eta \sim 1$ where $L \approx 4\,\text{mm}$ is a characteristic linear dimension of the particle.

Dynamics of chiral particles. As the circular flow in the cell is established, the particles evolve within $\sim 30\,\text{min}$ to stable orbits. Importantly, the mean radii of these orbits depends on the particles' chirality (R,S) but not on the initial particle position. All particle trajectories are recorded and analysed using house-written ImageJ and MATLAB scripts. The output of these scripts contain the linear velocity v_θ of the particle in the θ-direction, the angular velocity ω of the particle about its centre, and the radial position r_o of the particle centre measured from the axis of the Couette cell (Fig. 1d). Both R and S chiral particles revolve around the cell at a nearly constant speed v_θ with variations of only $\sim 3\%$ about the mean $\bar{v}_\theta$ (Fig. 1d, blue line). On the other hand, the R and S particles exhibit large periodic variations in their angular velocities ω (Fig. 1d, red line)—the so-called Jeffery orbits[23] characteristic of elongated particles in shear flows.

The key experimental results of the present study are illustrated in Fig. 2a for R and S particles subject to clockwise (CW) and counter-clockwise (CCW) directions of cell rotation. For CW rotation, the mean radius of the orbit traced by the R particles is always greater than that of the S particles in the same flow,

$\tilde{r}_o(R_{CW}) > \tilde{r}_o(S_{CW})$. In contrast, in a CCW flow, the opposite is true and $\tilde{r}_o(S_{CCW}) > \tilde{r}_o(R_{CCW})$. Moreover, $\tilde{r}_o(R_{CW}) \approx \tilde{r}_o(S_{CCW})$ and $\tilde{r}_o(R_{CCW}) \approx \tilde{r}_o(S_{CW})$, which means that when the flow is reversed (from CW to CCW or from CCW to CW), the orbit of the R particle becomes that of S and vice versa. These results explicitly eliminate any influence of particle shape imperfections, centrifugal forces and capillarity on relative motions of R and S particles. The orbit radii change when only the flow is reversed and the particles are not manipulated in any way. Additional evidence that capillary effects are negligible comes from confocal microscopy imaging (Fig. 1c), which shows no appreciable meniscus between the immersed particle and the surrounding fluid (see also Supplementary Note 1). Furthermore, the above results indicate that the system can be treated as 2D whereby a hypothetical rotation of the Couette cell by 180 degrees around a horizontal axis (Fig. 2b) acts to reverse the apparent chirality of the particles and the direction of flow (for example, R→S and CW→CCW in Fig. 2b) but does not alter the particle orbits. These symmetries require that $\bar{r}_o(R_{CW}) = \bar{r}_o(S_{CCW})$ and $\bar{r}_o(R_{CCW}) = \bar{r}_o(S_{CW})$ as observed in experiments.

As expected for Taylor–Couette flow, in which flow velocity increases with radial position, particles revolving around larger orbits have higher velocities v_θ—that is, $\bar{v}_\theta(R_{CW}) \approx \bar{v}_\theta(S_{CCW}) > \bar{v}_\theta(R_{CCW}) \approx \bar{v}_\theta(S_{CW})$ (hatched bars in Fig. 2c). At the same time, the average angular velocity $\bar{\omega}$ of a particle decreases with increasing radial position—that is, $\bar{\omega}(R_{CW}) \approx \bar{\omega}(S_{CCW}) < \bar{\omega}(R_{CCW}) \approx \bar{\omega}(S_{CW})$ (solid bars in Fig. 2c,d)—because $\bar{\omega}$ depends linearly on the shear rate $\dot{\gamma}$ (ref. 23), which decreases with increasing r. Importantly, for achiral particles (disks or ellipsoids), no difference is observed in any of the measured variables (Supplementary Fig. 1). Taken together, the above results indicate that the differences in the orbits of the R and S particles are due to their chirality. Because stable orbits correspond to the loci where all radial forces on the particles are balanced, we conclude that the R and S particles experience a lift force that is parallel to the shear plane (unlike in refs 2,4,7,18) and whose direction depends on the chirality of the object with respect to the flow.

Hydrodynamic model of chirality-specific motions. To gain further insight into the origin of this chiral lift, we consider the motion of a single rigid particle moving in the Taylor–Couette flow

$$\mathbf{u}(\mathbf{r}) = \frac{\omega_{cell}R_iR_o^2}{R_o^2 - R_i^2}\left(\frac{r}{R_i} - \frac{R_i}{r}\right)\mathbf{e}_\theta, \qquad (1)$$

where $\mathbf{e}_\theta$ is the unit vector in the azimuthal direction. The particle is approximated by a collection of N small spheres of radius a that move together as a single object with linear velocity $\mathbf{U}$ and angular velocity $\boldsymbol{\omega}$. We first consider the limit of small Reynolds numbers ($Re \ll 1$) in which fluid inertia is negligible and the hydrodynamic force on sphere α can be approximated as

$$\mathbf{F}_\alpha = \sum_\beta \mathcal{R}_{\alpha\beta} \cdot \left(\mathbf{u}(\mathbf{r}_\beta) - \mathbf{v}_\beta\right), \qquad (2)$$

where $\mathcal{R}$ is the grand resistance tensor (see Methods), $\mathbf{u}(\mathbf{r}_\beta)$ is the flow velocity at the centre of sphere β, and $\mathbf{v}_\beta = d\mathbf{r}_\beta/dt$ is the velocity of sphere β. Physically, this force originates from the motion of the sphere relative to that of the fluid and from hydrodynamic interactions with other spheres that make up the composite particle. The rigid body motion of the particle implies that the velocity of sphere β is related to that of the particle as $\mathbf{v}_\beta = \mathbf{U} + \boldsymbol{\omega} \times \mathbf{R}_\beta$ where $\mathbf{R}_\beta$ denotes the position of sphere β relative to the moving particle origin O (that is, $\mathbf{R}_\beta = \mathbf{r}_\beta - \mathbf{r}_o$). To maintain the relative distances between the spheres, the hydrodynamic forces are exactly balanced by intraparticle forces between the spheres such that the net force $\mathbf{F}_H$ and torque $\mathbf{T}_H$ on the particle are identically zero

$$\mathbf{F}_H = \sum_\alpha \mathbf{F}_\alpha = 0 \text{ and } \mathbf{T}_H = \sum_\alpha \mathbf{R}_\alpha \times \mathbf{F}_\alpha = 0. \qquad (3)$$

Together with the kinematic conditions for rigid body motion, Equations (2) and (3) govern the dynamics of the particle position $\mathbf{r}_o$ and its angular orientation (see Methods for details). This model is similar to that of Hänggi[6] and co-workers but includes hydrodynamic interactions between the spheres, which we found to have a significant impact on the magnitude of the chiral lift force and its scaling with particle size (Supplementary Fig. 3).

Numerical integration of the above equations provides several insights into the experimental observations (Fig. 3). First, the linear velocity of the particle is roughly constant and equal to that of the fluid evaluated at the particle's centre, $v_\theta \approx u_\theta(r_o)$ (compare Figs 1d and 3b). The angular velocity of the particle oscillates in time with a period equal to that of the particle's rotation (Fig. 3b); the period of these Jeffery orbits is well approximated as $T \approx \pi(\chi + \chi^{-1})/\dot{\gamma}$, where $\chi = 3.1$ is the aspect ratio of the particles and $\dot{\gamma} = R_i^2R_o^2\omega_{cell}/r_o^2(R_o^2 - R_i^2)$ is the shear rate. Despite the simplifications involved, the hydrodynamic model predicts the particle rotation period in quantitative agreement with experiment—for example, for $\omega_{cell} = 3 \text{ rad s}^{-1}$, $T = 7.9 \text{ s}$ in experiment as compared with $T = 7.2 \text{ s}$ in the model using no tunable parameters.

Importantly, the radial position of the particle also oscillates in time but with a steady drifting motion that depends on both the

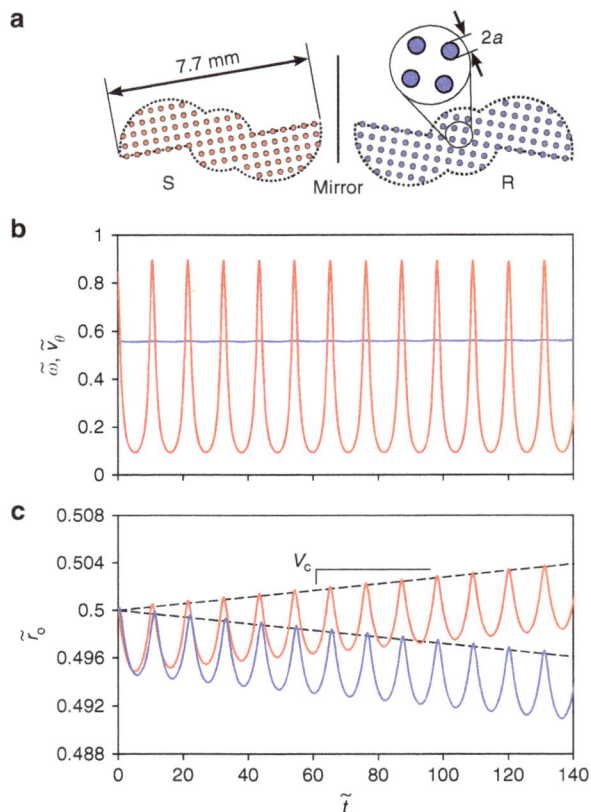

Figure 3 | Modelling of particle dynamics in Taylor–Couette flow. (**a**) Chiral particles (R and S) like those used in experiment are comprised of $N = 91$ spheres of radius $a = 0.1$ mm corresponding to the particle thickness; the spacing between neighbouring spheres is $4a$. The characteristic linear dimension of the particle (assumed equal to the square root of its area) is $L = 4$ mm; the dimensionless chiral factor is $\kappa = 1.9$; and the aspect ratio is $\chi = 3.1$ (see Methods). (**b**) Numerical solution of the linear velocity $\tilde{v}_\theta = v_\theta/R_o\omega_{cell}$ and the angular velocity $\tilde{\omega} = \omega/\omega_{cell}$ of an S particle moving in a CCW flow as a function of time $\tilde{t} = t\omega_{cell}$. (**c**) The radial position of the particle's centre $\tilde{r}_o = (r_o - R_i)/G$ oscillates in time with a steady drift velocity v_c whose direction depends on the chirality of the particle.

chirality of the particle and the direction of the flow. For CCW rotation of the outer cylinder ($\omega_{cell} > 0$), S particles drift away from the origin while R particles drift toward the origin in agreement with experimental observations; the opposite behaviour is observed for CW rotation. A detailed scaling analysis of the hydrodynamic model reveals that the chiral drift velocity v_c is well approximated as $v_c \approx \kappa aL/Tr_o$, where κ is a dimensionless 'chiral factor' that characterizes the direction and magnitude of the chiral drift and depends only on the shape of the particle (see Methods); for the model particles used in the simulations, $\kappa(S) = -\kappa(R) = 1.9$.

In the absence of inertial effects or hydrodynamic interactions with the walls of the Couette cell, chiral particles drift monotonically towards or away from the centre of the rotating flow. To understand the formation of the stable orbits observed in experiment, it is necessary to consider the role of fluid inertia, which acts to centre the particle within the gap separating the concentric cylinders. Near the midline position, $r_m = \frac{1}{2}(R_o + R_i)$, the inertial lift velocity v_i has been calculated previously[24] for spherical particles to be $v_i(r) \approx (r_m - r)/\tau$, where $\tau \approx 36\eta G^3/\rho\omega_{cell}^2R_o^2L^3$ is a characteristic time for the particle to relax to the steady-state orbit r_m. For chiral particles, the chiral drift velocity is expected to perturb the orbit of the particle away from r_m; the resulting steady-state orbit r_{ss} corresponds to that at which the chiral drift velocity balances the restoring lift velocity, $v_c(r_{ss}) \approx v_i(r_{ss})$. Using the above estimates, the difference in the steady-state orbits between R and S particles is estimated as $\Delta r_{ss} \approx \tau v_c(r_m) \sim 1$ mm and the relaxation time as $\tau \sim 10$ min, both of which are in quantitative agreement with experiment. Interestingly, this mechanism also predicts that larger separations between chiral particles can be achieved at lower Reynolds numbers. This prediction is supported by experiments that show that the chiral separation Δr_{ss} increases from 2.3 to 3.0 mm as the rotation speed is reduced by a factor of two (Fig. 2a).

Physical origins of chirality-specfic motions. Analysis of the hydrodynamic model reveals that chiral drift parallel to the shear plane occurs through a previously unreported mechanism requiring the presence of non-linear flow fields (that is, $\nabla\nabla\mathbf{u} \neq 0$). In general, the hydrodynamic force/torque $\mathcal{F}_H$ on a rigid particle depends both on its linear/angular velocity $\mathcal{U}$ and on the fluid velocity $\mathbf{u}$ as

$$\mathcal{F}_H = -\mathcal{R}_{FU} \cdot \mathcal{U} + \mathcal{C}_2 \cdot \mathbf{u}_o + \mathcal{C}_3 : (\nabla\mathbf{u})_o + \mathcal{C}_4 \vdots (\nabla\nabla\mathbf{u})_o + \ldots \quad (4)$$

where $\mathcal{R}_{FU}$ is the particle resistance tensor, $\mathcal{C}_n$ are nth order tensors that depend on the shape and size of the particle, and the velocity gradients are evaluated at the particle origin O (see Methods)[25–28]. To first approximation, a force and torque-free particle ($\mathcal{F}_H = 0$) moves at the local fluid velocity ($\mathbf{u}_o$) and rotates at a rate equal to one-half the local fluid vorticity ($\frac{1}{2}\nabla \times \mathbf{u}_o$). Deviations of the particle's translational and rotational velocity from that of the fluid are caused by straining flows which are incompatible with the rigid body motion of the particle. To describe these effects, it is most common to truncate the Taylor expansion of equation (4) at the level of velocity gradients and neglect higher-order derivatives of the velocity field. This approximation has proven effective in describing the chiral drift of helical particles in linear shear flows along the direction normal to the shear plane[4,7]. By contrast, we find that chirality-dependent drifting motions in two dimensions require consideration of second-order derivatives of the velocity field along with fourth-order tensors ($\mathcal{C}_4$) to describe the detailed shape of the particle.

Physically, fluid shear within the Couette cell acts to rotate the particle about its centre as it translates along a circular orbit. The interplay between rotation and orbital translation cause the particle to move back and forth along the radial direction perpendicular to the direction of flow (Fig. 3c). Each full rotation of the particle contributes a net displacement that depends on the particle's chirality as quantified by the chiral factor κ. Although these drifting motions are typically small, their steady accumulation can lead to large displacements over many particle rotations and may, therefore, provide a basis for the hydrodynamic separation of chiral objects.

Effects of Brownian motion. To assess the possibility of separating smaller micro- and nanoscale objects, we investigate the effect of Brownian motion on the chiral drift velocity through a series of stochastic simulations (see Methods). Briefly, we add a fluctuating force/torque $\mathcal{F}_B$ with zero mean and covariance $\langle \mathcal{F}_B(0)\mathcal{F}_B(t)\rangle = 2k_B T \mathcal{R}_{FU}\delta(t)$ to the otherwise deterministic equations described above. The addition of these Brownian contributions introduces one additional parameter—namely, the thermal energy $k_B T$—which controls the strength of thermal motion relative to the deterministic hydrodynamic motion of the particle. This parameter is conveniently expressed as the inverse of the Péclet number $Pe = v_c G/D$, where v_c is the chiral drift velocity in the absence of Brownian motion, $D = k_B T/K$ is a

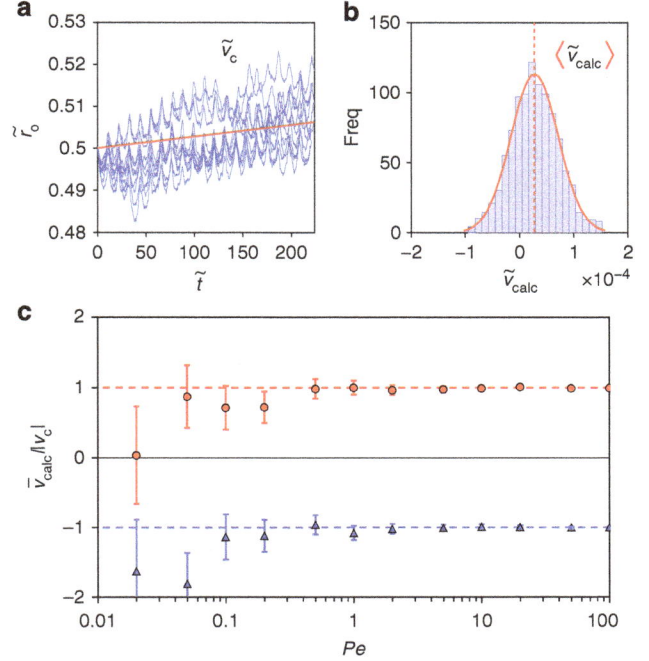

Figure 4 | Stochastic particles simulations. (a) Ten (of 1,000) stochastic particle trajectories of an R particle (Fig. 3a) for Péclet number, $Pe = 1$. **(b)** Histogram showing the corresponding distribution of drift velocities $\tilde{v}_{calc} = v_{calc}/G\omega_{cell}$ for $Pe = 1$. **(c)** Average drift velocity $\bar{v}_{calc}$ as a function of Pe for chiral particles (R and S) shown in Fig. 3a; the calculated (calc) drift velocity is scaled by the magnitude of chiral drift $|v_c|$ in the absence of Brownian motion. The error bars represent 95% confidence intervals on the mean drift velocity based on 1,000 realizations of the stochastic process.

characteristic diffusion coefficient ($K = |\mathbf{K}|$ is the magnitude of the resistance tensor $\mathbf{K}$; for example, $K = 6\pi\eta a$ for a sphere of radius a in an unbounded medium) and G is a length scale over which convection and diffusion are compared (for example, the gap width of a Taylor–Couette cell).

For each value of the Péclet number, we generate 1,000 stochastic particle trajectories of duration $t_s = 10T$ (that is, 10 particle rotations) and calculate the resulting drift velocity, $v_{calc} = [r_o(t_s) - r_o(0)]/t_s$ (Fig. 4). For weak thermal agitation ($Pe \gg 1$), the average drift velocity $\bar{v}_{calc}$ agrees closely with the deterministic result, $\bar{v}_{calc} \approx v_c$. By contrast, for strong thermal agitation ($Pe \ll 1$), the particle trajectories are increasingly erratic such that displacement due to chiral drift is increasingly insignificant compared with that of Brownian motion. In the simulations, the length $G = v_c t_s$ was kept small for computational reasons (simulating the dynamics of microscopic particles over macroscopic lengths was computationally prohibitive); however, the choice of G is arbitrary. Chiral drift (like any convective motion) will ultimately beat out diffusion over large length scales as displacement due to drift scales linearly with time whereas that due to diffusion scales as $t^{1/2}$. In other words, the Péclet number can be made increasingly large by increasing the length G over which the chiral separation is to occur. For a given length G, the separation of particles of opposite chirality should indeed be feasible provided that $Pe \gg 1$ such that the effects of Brownian motion are small.

Feasibility of chiral separations. To separate objects over the gap width G of the Couette cell, transport due to chiral drift must exceed that due to diffusion: $v_c \gg D/G$. Approximating the chiral drift velocity as $v_c \approx \kappa aL/Tr_o$ and the diffusivity as $D \approx k_B T/3\pi\eta L$, this condition implies that the particle size must be $L \gg (k_B T/$

$\eta\omega_{\text{cell}})^{1/3}$ assuming the gap is small (that is, $G \ll R_{\text{o}}$). In water ($\eta = 10^{-3}\,\text{Pa}\cdot\text{s}$) at room temperature ($k_{\text{B}}T = 4 \times 10^{-21}\,\text{J}$), a Couette cell rotating at a rate of $\omega_{\text{cell}} = 10^3\,\text{rad s}^{-1}$ should be capable of separating chiral objects provided $L > 100\,\text{nm}$. Separating even smaller chiral objects by this mechanism may also be possible in highly viscous fluids—for example, in glycerol ($\eta = 1\,\text{Pa}\cdot\text{s}$), the minimum size is $L > 10\,\text{nm}$.

Discussion

Interestingly, the chiral drift mechanism described here does not require the alignment of the particle along particular orientations (beyond the initial confinement to the interface) and is, therefore, insensitive to Brownian rotation. The flow-induced rotation of the particle leads—on average—to more rotations in one direction than the other resulting in steady drifting motions. Over sufficiently long times (again corresponding to the condition, $Pe \gg 1$), this steady rotational motion will always dominate that of Brownian motion. This behaviour is in sharp contrast with previously reported mechanisms of chiral drift in three dimensions perpendicular to the shear plane[4,7], which require the flow-induced alignment of the particles to achieve steady drifting motions. In these mechanisms, chiral drift vanishes at small scales due to Brownian rotation, which acts to randomize the particle's orientation. Thus, while the present mechanism is typically weaker than those described previously (for example, v_{c} scales as L^2 as opposed to L), it may be more robust to Brownian motion at small scales.

In summary, we provide conclusive evidence that particles in 2D vortex flows experience in-plane lift forces that depend on particle chirality. This lift force does not vanish at the nanoscale, and should allow for the separation of chiral objects tens of nanometres across even in the presence of Brownian motion. Although the present study focuses exclusively on 2D particles, in-plane lift forces may also act on three-dimensional (3D) chiral objects if and only if they are confined at or near a liquid interface parallel to the shear plane. In the absence of such an interface, rotation of the Couette cell by 180° (cf. Fig. 2b) reverses the apparent flow rotation direction but not the chirality of a 3D chiral particle; consequently, in-plane chiral lift forces are prohibited in extended Couette flows by symmetry considerations (Supplementary Fig. 5). To test these predictions, the present experiments should be scaled down and extended to neutrally buoyant particles that can migrate and rotate in 3D; however, the implementation and imaging of such a system is expected to be significantly more challenging. Future models of chirality-selective motions in 3D should incorporate the effects of higher-order velocity gradients ($\nabla\nabla\mathbf{u}$) that arise in non-linear flows as these contributions may enable potentially important mechanisms for the hydrodynamic separation of chiral objects.

Methods

Achiral control experiments.
Achiral control particles are made using SU-8 photolithography (Supplementary Fig. 1a). Both disks (D) and ellipses (E) are made with the same area (in contact with the liquid) and mass as the chiral R and S particles. Experiments are performed in which the direction of rotation of the Couette cell is reversed during the experiment (see dashed vertical line at $\tilde{t} = 400$ in Supplementary Fig. 1b). Upon reversal of the direction, the S particle starts moving towards its steady-state position (see Fig. 2a), both for the CW to CCW and CCW to CW case (Supplementary Fig. 1b). The ellipse E, however, shows a small perturbation in the radial position (due to the sudden reversal of the flow), but quickly returns to the same radial position $\tilde{r}_{\text{o}}$ (Supplementary Fig. 1b, black and magenta lines). In Supplementary Fig. 1c, we compare the CW versus CCW behaviour of all measured particles (that is, D, E, S and R). No significant differences in angular velocity ω, linear velocity v_{θ}, or radial position r are observed for the achiral particles D and E, whereas R and S chiral particles show significantly different behaviours (as described in the text). These achiral control experiments demonstrate that particle chirality is necessary to obtain

changes in ω, v_{θ} and r, when comparing CW versus CCW rotation in Taylor–Couette flows.

General hydrodynamic model.
Equations (2) and (3) describe the motion of a single rigid particle of arbitrary shape moving within an unbounded Stokes flow, $\mathbf{u}(\mathbf{r})$, in the absence of external forces and torques. To approximate the grand resistance tensor $\mathcal{R}$, we first compute the grand mobility tensor $\mathcal{M}$ though pairwise summation of the far field hydrodynamic interactions between the N spheres that make up the composite particle. This tensor is then inverted to obtain $\mathcal{R} = \mathcal{M}^{-1}$, which includes the long range, many-body interactions between the spheres. Assuming that the distance between neighbouring spheres is large compared with their radius a, the components of the mobility tensor are well approximated as

$$\mathcal{M}^{\alpha\alpha} = \frac{\delta}{6\pi\eta a} \quad \text{and} \quad \mathcal{M}^{\alpha\beta} = \frac{1}{8\pi\eta}\left(\frac{\delta}{r_{\alpha\beta}} + \frac{\mathbf{r}_{\alpha\beta}\mathbf{r}_{\alpha\beta}}{r_{\alpha\beta}^3}\right) \qquad (5)$$

where δ is the identity tensor, η is the fluid viscosity, and $\mathbf{r}_{\alpha\beta} = \mathbf{r}_\beta - \mathbf{r}_\alpha$ is the vector drawn from sphere α to sphere β. This approach is akin to the Stokesian dynamics method[29,30]; however, it neglects (i) higher-order moments of the force density on the particles' surface (for example, torques, stresslets, and so on) and (ii) near field (lubrication) interactions between the spheres.

Substituting equation (2) for the force on sphere α into the equation (3) for the net force and torque, we obtain

$$\begin{pmatrix} \mathbf{F}_{\text{H}} \\ \mathbf{T}_{\text{H}} \end{pmatrix} = -\begin{pmatrix} \mathbf{K} & \mathbf{C}^{\dagger} \\ \mathbf{C} & \Omega \end{pmatrix} \cdot \begin{pmatrix} \mathbf{U} \\ \omega \end{pmatrix} + \sum_{\alpha,\beta}\begin{pmatrix} \mathcal{R}_{\alpha\beta} \\ \mathcal{C}_{\alpha\beta} \end{pmatrix} \cdot \mathbf{u}(\mathbf{r}_\beta), \qquad (6)$$

where $\mathbf{K}$ is the symmetric resistance tensor[25], Ω is the symmetric rotation tensor[25], $\mathbf{C}$ is the coupling tensor[26] (not necessarily symmetric, $\mathbf{C} \neq \mathbf{C}^{\dagger}$), and $\mathcal{C}_{\alpha\beta} \equiv -(\mathbf{R}_\alpha \cdot \varepsilon) \cdot \mathcal{R}_{\alpha\beta}$ where ε is the alternating unit tensor ($\mathbf{a} \times \mathbf{b} = -\mathbf{a} \cdot \varepsilon \cdot \mathbf{b}$). For the composite particles described here, the resistance tensors are defined explicitly as

$$\mathbf{K} = \sum_{\alpha,\beta}\mathcal{R}_{\alpha\beta}, \quad \mathbf{C} = \sum_{\alpha,\beta}\mathcal{C}_{\alpha\beta}, \quad \text{and} \quad \Omega = \sum_{\alpha,\beta}\mathcal{C}_{\alpha\beta} \cdot (\mathbf{R}_\beta \cdot \varepsilon). \qquad (7)$$

Equation (6) is directly analogous to that derived by Brenner[28] for the motion of a solid particle and fully specifies the linear and angular velocities of the composite particle.

If the particle is small relative to spatial variations in the external flow field, we can approximate the velocity at the position of sphere β by a Taylor series about the particle origin

$$\mathbf{u}(\mathbf{r}_\beta) = \mathbf{u}_{\text{o}} + \mathbf{R}_\beta \cdot (\nabla\mathbf{u})_{\text{o}} + \tfrac{1}{2}\mathbf{R}_\beta\mathbf{R}_\beta : (\nabla\nabla\mathbf{u})_{\text{o}} + \cdots. \qquad (8)$$

Substituting this approximation into equation (6), we obtain

$$\begin{pmatrix} \mathbf{F}_{\text{H}} \\ \mathbf{T}_{\text{H}} \end{pmatrix} = -\begin{pmatrix} \mathbf{K} & \mathbf{C}^{\dagger} \\ \mathbf{C} & \Omega \end{pmatrix} \cdot \begin{pmatrix} \mathbf{U} \\ \omega \end{pmatrix} + \begin{pmatrix} \mathbf{K} \\ \mathbf{C} \end{pmatrix} \cdot \mathbf{u}_{\text{o}} + \begin{pmatrix} {}^{3}\Phi \\ {}^{3}\Gamma \end{pmatrix} : (\nabla\mathbf{u})_{\text{o}} + \begin{pmatrix} {}^{4}\Phi \\ {}^{4}\Gamma \end{pmatrix} \vdots (\nabla\nabla\mathbf{u})_{\text{o}} + \cdots,$$
$$(9)$$

where the tensors ${}^{n}\Phi$ and ${}^{n}\Gamma$ are constant n-adic intrinsic resistance coefficients[28] that depend only on the external shape of the particle

$$^{3}\Phi = \sum_{\alpha,\beta}\mathcal{R}_{\alpha\beta}\mathbf{R}_\beta \quad \text{and} \quad {}^{3}\Gamma = \sum_{\alpha,\beta}\mathcal{C}_{\alpha\beta}\mathbf{R}_\beta, \qquad (10)$$

$$^{4}\Phi = \frac{1}{2}\sum_{\alpha,\beta}\mathcal{R}_{\alpha\beta}\mathbf{R}_\beta\mathbf{R}_\beta \quad \text{and} \quad {}^{4}\Gamma = \frac{1}{2}\sum_{\alpha,\beta}\mathcal{C}_{\alpha\beta}\mathbf{R}_\beta\mathbf{R}_\beta. \qquad (11)$$

The level of approximation of equation (9) is precisely that which is required to describe the chiral drift of planar particles in 2D flows (cf. below).

Dynamics of planar particles.
The dynamical equation (6) is quite general and applies to 3D particles of arbitrary shape moving in any 3D Stokes flow. We now focus our attention on 'planar' particles, which are symmetric with respect to reflection about the xy plane, and their movement in 2D flows for which $u_z = 0$. In particular, we limit our analysis to unidirectional circular flows of the form

$$\mathbf{u} = \left(\frac{b}{r} + cr\right)\mathbf{e}_{\theta}, \qquad (12)$$

as permitted by the incompressible Navier–Stokes equations. The second term is simply rigid body rotation about the origin, which does not contribute to the chiral drift. Without loss of generality, we set $c = 0$. For the Taylor–Couette flow described by equation (1), the first coefficient is $b = -R_{\text{i}}^2 R_{\text{o}}^2 \omega_{\text{cell}} / (R_{\text{o}}^2 - R_{\text{i}}^2)$. The dynamics of planar particles in this circular flow field is governed by the following system of differential equations (see Supplementary Note 2 for derivation)

$$\begin{bmatrix} \dot{r}_o \\ r_o\dot{\theta}_o \\ \dot{\gamma} \end{bmatrix} = \sum_{\alpha,\beta} \frac{b}{r_\beta^2} \begin{bmatrix} \cos\phi & -\sin\phi & \cdot \\ \sin\phi & \cos\phi & \cdot \\ \cdot & \cdot & 1 \end{bmatrix} \begin{bmatrix} \tilde{\mathcal{R}}_{11}^{\prime\alpha\beta} & \tilde{\mathcal{R}}_{12}^{\prime\alpha\beta} \\ \tilde{\mathcal{R}}_{12}^{\prime\alpha\beta} & \tilde{\mathcal{R}}_{22}^{\prime\alpha\beta} \\ \tilde{\mathcal{C}}_{31}^{\prime\alpha\beta} & \tilde{\mathcal{C}}_{32}^{\prime\alpha\beta} \end{bmatrix} \begin{bmatrix} r_o\sin\phi - Y'_\beta \\ r_o\cos\phi + X'_\beta \end{bmatrix},$$

(13)

Here r_o and θ_o are polar coordinates specifying the position of the particle centre (Fig. 1b), γ and $\phi = \gamma - \theta_o$ are angles specifying the orientation of the particle (relative to the x axis and to the radial vector $\mathbf{r}_o$, respectively), $\tilde{\mathcal{R}}_{\alpha\beta} = \mathbf{K}^{-1} \cdot \mathcal{R}_{\alpha\beta}$ and $\tilde{\mathcal{C}}_{\alpha\beta} = \mathbf{\Omega}^{-1} \cdot \mathcal{C}_{\alpha\beta}$ are normalized resistance tensors, and r_β is the radial position of sphere β

$$r_\beta = \sqrt{r_o^2 + X_\beta'^2 + Y_\beta'^2 + 2r_o X'_\beta \cos\phi - 2r_o Y'_\beta \sin\phi}.$$

(14)

The position (X'_β, Y'_β) specifies the location of sphere β relative to the particle origin using a moving coordinate system, which is fixed to the particle and participates in its motion. Equation (13) governs the dynamics of the position (r_o, θ_o) and orientation (γ) of the particle in time. Because these equations depend only on the coordinates r_o and ϕ, we can combine the expressions for $\dot{\theta}_o$ and $\dot{\gamma}$ to obtain two differential equations for r_o and ϕ (cf. below).

Equation (13) are integrated numerically using a variable order Adams–Bashforth–Moulton solver with adaptive stepping and a relative error tolerance of 10^{-10} to avoid spurious drifting motions. For $b > 0$, particles rotate steadily in the clockwise direction ($\dot{\phi} < 0$) with an angular velocity that depends on their orientation and handedness (Supplementary Fig. 2b). The radial position of each particle oscillates in time with a period T equal to that of its rotation and defined by $\phi(t + T) = \phi(t) + 2\pi$. Importantly, each rotation contributes a small displacement in the radial direction, and the direction of this steady drifting motion depends on the chirality of the object (Supplementary Fig. 2c). Here, R particles move in the positive radial direction; S particles move in the negative radial direction; achiral particles exhibit no such drifting motion.

Because the oscillations in the radial direction are small, we can define a local drift velocity v_c by averaging the radial velocity over one period of rotation.

$$v_c(r_o) = \frac{1}{T} \int_0^T \dot{r}_o \, dt$$

(15)

Numerical evaluation of v_c reveals that the magnitude of v_c decreases monotonically with distance from the origin as $v_c \propto r_o^{-3}$ (Supplementary Fig. 2d). In the absence of other forces (for example, due to walls of a Taylor–Couette cell), S particles spiral into the origin, and R particles spiral outward indefinitely. In addition, the drift velocity increases quadratically with the size of the chiral particle as $v_c \propto L^2$, where L is a characteristic linear dimension of the particle (Supplementary Fig. 2e). Below, we develop an analytical approach to approximate v_c and confirm the scaling relations observed in the numerical simulations.

Scaling analysis. To obtain an analytical approximation for the chiral drift velocity, we first expand the governing equations in powers of the object size. Specifically, we modify the size-dependent particle quantities as $\mathbf{R}_\alpha \to \varepsilon\mathbf{R}_\alpha$ and $a \to \varepsilon a$ and expand the governing equation (13) in a power series in ε

$$\dot{\phi} = \frac{b}{r_o^2}\left(f_0(\phi) + \frac{\varepsilon}{r_o} f_1(\phi) + O(\varepsilon^2)\right),$$

(16)

$$\dot{r}_o = \frac{b}{r_o}\left(\frac{\varepsilon}{r_o} g_1(\phi) + \frac{\varepsilon^2}{r_o^2} g_2(\phi) + O(\varepsilon^3)\right).$$

(17)

Here, ε is a dummy value to be set to one at the end of the analysis; the functions $f_0(\phi)$, $f_1(\phi)$, $g_1(\phi)$ and $g_1(\phi)$ are given explicitly in Supplementary Note 3 and depend only on components of the tensors characterizing the hydrodynamic shape of the particle. Importantly, the approximation of equations (16) and (17) includes only those terms needed to reproduce the drifting motion of chiral particles (that is, no chiral drift is observed if the series is truncated with fewer terms, and the addition of more terms introduces no new qualitative behaviours). On the basis of this approximation, we now derive analytical estimates for the oscillation period and the chiral drift velocity.

At zeroth order in the object size (ε^0), the radial position of the object does not change in time, $r_o = \text{constant} + O(\varepsilon)$; however, the object rotates about its centre as described by

$$\dot{\phi}_0 = \frac{b f_0(\phi_0)}{r_o^2} = -\frac{b}{r_o^2}(1 + A\cos2\phi + B\sin2\phi),$$

(18)

where $\phi(t) = \phi_0(t) + O(\varepsilon)$ and the coefficients A and B are constants that depend on the shape of the particle (see Supplementary Note 3). Equation (18) implies that the angular velocity $\dot{\phi}_0$ oscillates with the particle's orientation ϕ reaching a maximum value when the major axis of the object is perpendicular to the streamlines and a minimum value when the major axis is parallel to the streamlines. Such periodic variations in the angular velocity are characteristic of the

motion of anisotropic particles in linear shear flows (for example, the Jeffrey orbits[23] of spheroidal particles). Here, the orientation of the particle's major axis is described by the angle ϕ_m with respect to the X' axis

$$\phi_m = \frac{1}{2}\text{atan2}(B, A),$$

(19)

where $\text{atan2}(y, x)$ is the four-quadrant inverse tangent function. Equation (18) can then be written as

$$\dot{\phi}_0 = -\frac{b}{r_o^2}(1 + C\cos[2(\phi - \phi_m)]),$$

(20)

where $C^2 = A^2 + B^2$. Integrating this equation over one complete rotation, we obtain the period of rotation

$$T = \pi r_o^2(\chi + \chi^{-1})/b + O(\varepsilon)$$

(21)

where χ is an effective aspect ratio of the particle defined as

$$\chi = \sqrt{\frac{1 - C}{1 + C}}.$$

(22)

This expression for the oscillation period exhibits the same dependence on the particle aspect ratio (χ) and the shear rate (b/r_o^2) as the Jeffrey orbits[23] for spheroidal particles in linear shear flows.

To approximate the chiral drift velocity and its scaling with particle size, we combine the approximate equations (16) and (17) and expand in powers of the particle size ε to obtain

$$\frac{1}{r_o}\frac{dr_o}{d\phi} = \frac{\varepsilon}{r_o}\left(\frac{g_1(\phi)}{f_0(\phi)}\right) + \frac{\varepsilon^2}{r_o^2}\left(\frac{f_0(\phi)g_2(\phi) - f_1(\phi)g_1(\phi)}{f_0^2(\phi)}\right) + O(\varepsilon^3).$$

(23)

Integrating over one period of oscillation ($0 \le \phi \le 2\pi$), the first term vanishes leaving

$$\Delta r_o = \frac{\varepsilon^2 K}{r_o} + O(\varepsilon^3),$$

(24)

where $\Delta r_o \ll r_o$ is the radial displacement after a single particle rotation and K is a chiral factor defined as

$$K = \int_0^{2\pi} \frac{f_0(\phi)g_2(\phi) - f_1(\phi)g_1(\phi)}{f_0^2(\phi)} d\phi,$$

(25)

which depends only on the shape of the particle. The chiral drift velocity v_c can now be approximated as the ratio of this displacement and the period of rotation.

$$v_c = \frac{\Delta r_o}{T} = \frac{bK\varepsilon^2}{\pi(\chi + \chi^{-1})r_o^3} + O(\varepsilon^3)$$

(26)

This is the central result of our analysis.

The chiral drift velocity scales with the size of the object as ε^2 and inversely with distance from the origin as r_o^{-3} in agreement with numerical predictions. Furthermore, the direction and magnitude of the drift velocity depends on the shape of the object as characterized by the chiral factor K. Importantly, the sign of K unambiguously determines the chirality of a 2D object; by arbitrary convention, we assign objects with $K < 0$ as S and $K > 0$ as R. Furthermore, it can be shown that for achiral particles with an additional plane of mirror symmetry, the chiral factor K is identically zero (see Supplementary Note 4).

Importantly, the chiral drift velocity derived in equation (26) requires the inclusion of hydrodynamic interactions between the spheres that make up the composite particle. In particular, the chiral factor K depends linearly on the sphere radius a as illustrated in Supplementary Fig. 3. This suggests that the chiral factor scales with the particle size L and the sphere size a as $K \sim aL$. Therefore, we can defined a dimensionless chiral factor: $\kappa = K/aL$, which is at most an order one quantity regardless of particle size; this is the chiral factor that is used in the main text. The dimensionless chiral factor κ depends only on the shape of the particle—for example, the largest drift velocities are achieved for asymmetric particles with large aspect ratios (Supplementary Fig. 4).

Effects of Brownian motion. To describe the effects of Brownian motion on the dynamics of planar particles moving in rotating flow fields, we start from the Langevin equation for translational and rotational motion,

$$\mathbf{m} \cdot \frac{d\mathcal{U}}{dt} = \mathcal{F}_H + \mathcal{F}_B,$$

(27)

where

$$\mathbf{m} = \begin{pmatrix} m\delta & \cdot \\ \cdot & \mathbf{I} \end{pmatrix}, \mathcal{U} = \begin{pmatrix} \mathbf{U} \\ \boldsymbol{\omega} \end{pmatrix}, \mathcal{F}_H = \begin{pmatrix} \mathbf{F}_H \\ \mathbf{T}_H \end{pmatrix}, \text{ and } \mathcal{F}_B = \begin{pmatrix} \mathbf{F}_B \\ \mathbf{T}_B \end{pmatrix}.$$

(28)

Here, $\mathbf{m}$ is a generalized mass/moment-of-inertia tensor (m is the mass, $\mathbf{I}$ is the moment of inertia), $\mathcal{U}$ is the particle translational/rotational velocity vector, $\mathcal{F}_H$ is the hydrodynamic force/torque vector, and $\mathcal{F}_B$ is the stochastic force that gives rise to Brownian motion[29]. At low Reynolds numbers, the hydrodynamic force/torque on the particle is well described by equation (9) or equivalently by equation (4) where

$$\mathcal{R}_{FU} = \begin{pmatrix} \mathbf{K} & \mathbf{C}^{\dagger} \\ \mathbf{C} & \mathbf{\Omega} \end{pmatrix}, \; \mathcal{C}_2 = \begin{pmatrix} ^2\mathbf{\Phi} \\ ^2\mathbf{\Gamma} \end{pmatrix}, \; \mathcal{C}_3 = \begin{pmatrix} ^3\mathbf{\Phi} \\ ^3\mathbf{\Gamma} \end{pmatrix} \text{and} \; \mathcal{C}_4 = \begin{pmatrix} ^4\mathbf{\Phi} \\ ^4\mathbf{\Gamma} \end{pmatrix}. \quad (29)$$

The stochastic force/torque $\mathcal{F}_B$ arises from thermal fluctuations and is characterized by

$$\langle \mathcal{F}_B \rangle = 0 \quad \text{and} \quad \langle \mathcal{F}_B(0)\mathcal{F}_B(t) \rangle = 2k_BT\mathcal{R}_{FU}\delta(t), \quad (30)$$

where k_BT is the thermal energy, the angle brackets denote an ensemble average, and $\delta(t)$ denotes a delta function. The magnitude of the Brownian forces is related to the viscous resistance to motion via the fluctuation–dissipation theorem.

The Langevin equation (27) can be integrated[29] in time to give the translational/rotational displacement of the particle in a small time Δt

$$\Delta\mathbf{x} = \mathbf{X}_B(\Delta t) + \int_t^{t+\Delta t} \mathcal{R}_{FU}^{-1} \cdot \left(\mathcal{C}_2 \cdot \mathbf{u}_o + \mathcal{C}_3 : (\nabla\mathbf{u})_o + \mathcal{C}_4 \vdots (\nabla\nabla\mathbf{u})_o + \; \dots \; \right) dt'. \quad (31)$$

Here, $\mathbf{X}_B(\Delta t)$ is a random displacement/rotation due to Brownian motion characterized by

$$\langle \mathbf{X}_B \rangle = 0 \quad \text{and} \quad \langle \mathbf{X}_B(0)\mathbf{X}_B(\Delta t) \rangle = 2k_BT\mathbf{R}_{FU}^{-1}\Delta t. \quad (32)$$

The second term of equation (31) represents the deterministic (non-Brownian) component of particle motion as detailed in the previous sections. Importantly, the time step Δt must be small compared with the time scale of deterministic particle motion (for example, much smaller than the period of the particle's Jeffery orbit, $\Delta t \ll T$) but also large compared with the Brownian relaxation time $\tau = m/6\pi\eta L$ (for example, $\tau \sim 10$ ps for an $L \sim 10$ nm particle in water). Furthermore, when integrating the deterministic contribution of equation (31), care must be taken to avoid numerical errors that can contribute to spurious drifting motions. We use a variable order Adams–Bashforth–Moulton solver with adaptive stepping and a relative error tolerance of 10^{-10} in all numerical simulations.

The data in Fig. 4 are generated by integrating the stochastic dynamics of a single particle through M rotations (that is, $0 \le \phi \le 2\pi M$) for different degrees of thermal agitation as characterized by the Péclet number, Pe. The M particle rotations correspond to a characteristic length $G = v_c TM$—the distance over which a particle drifts after M rotations. We used 'whole' particle rotations instead of arbitrarily specifying the length G to more accurately estimate the chiral drift velocity as the displacement due to chiral drift can be small compared with variations in the particle's radial position during a single particle rotation (especially for the relatively short time scales used in the numerical simulations). Such considerations are important even in the absence of Brownian motion.

References

1. Howard, D. W., Lightfoot, E. N. & Hirschfelder, J. O. Hydrodynamic resolution of optical isomers. *AIChE J.* **22**, 794–798 (1976).
2. Chen, P. L. & Zhang, Q. Y. Dynamical solutions for migration of chiral DNA-type objects in shear flows. *Phys. Rev. E* **84**, 056309 (2011).
3. Eichhorn, R. Enantioseparation in microfluidic channels. *Chem. Phys.* **375**, 568–577 (2010).
4. Marcos, H., Fu, H. C., Powers, T. R. & Stocker, R. Separation of microscale chiral objects by shear flow. *Phys. Rev. Lett.* **102**, 158103 (2009).
5. de Gennes, P. G. Mechanical selection of chiral crystals. *Europhys. Lett.* **46**, 827–831 (1999).
6. Kostur, M., Schindler, M., Talkner, P. & Hänggi, P. Chiral separation in microflows. *Phys. Rev. Lett.* **96**, 014502 (2006).
7. Makino, M. & Doi, M. Migration of twisted ribbon-like particles in simple shear flow. *Phys. Fluids* **17**, 103605 (2005).
8. Kim, Y. J. & Rae, W. J. Separation of screw-sensed particles in a homogeneous shear field. *Int. J. Multiph. Flow* **17**, 717–744 (1991).
9. Tencer, M. & Bielski, R. Mechanical resolution of chiral objects in achiral media: where is the size limit? *Chirality* **23**, 144–147 (2011).
10. Watari, N. & Larson, R. G. Shear-induced chiral migration of particles with anisotropic rigidity. *Phys. Rev. Lett.* **102**, 246001 (2009).
11. Maier, N. M., Franco, P. & Lindner, W. Separation of enantiomers: needs, challenges, perspectives. *J. Chromatogr. A* **906**, 3–33 (2001).
12. Pirkle, W. H. & Pochapsky, T. C. Considerations of chiral recognition relevant to the liquid-chromatographic separation of enantiomers. *Chem. Rev.* **89**, 347–362 (1989).
13. Chen, P. L. & Chao, C. H. Lift forces of screws in shear flows. *Phys. Fluids* **19**, 017108 (2007).
14. Ribó, J. M., Crusats, J., Sagues, F., Claret, J. & Rubires, R. Chiral sign induction by vortices during the formation of mesophases in stirred solutions. *Science* **292**, 2063–2066 (2001).
15. Rubires, R., Farrera, J. A. & Ribó, J. M. Stirring effects on the spontaneous formation of chirality in the homoassociation of diprotonated meso-tetraphenylsulfonato porphyrins. *Chem. Eur. J.* **7**, 436–446 (2001).
16. Aristov, M., Eichhorn, R. & Bechinger, C. Separation of chiral colloidal particles in a helical flow field. *Soft Matter* **9**, 2525–2530 (2013).
17. Grzybowski, B. A. & Whitesides, G. M. Dynamic aggregation of chiral spinners. *Science* **296**, 718–721 (2002).
18. Makino, M., Arai, L. & Doi, M. Shear migration of chiral particle in parallel-disk. *J. Phys. Soc. Jpn* **77**, 064404 (2008).
19. Mead, C. A. & Moscowitz, A. Some comments on the possibility of achieving asymmetric-synthesis from achiral reactants in a rotating vessel. *J. Am. Chem. Soc.* **102**, 7301–7302 (1980).
20. Amabilino, D. B. Supramolecular assembly: nanofibre whirlpools. *Nat. Mater.* **6**, 924–925 (2007).
21. Wolffs, M. *et al.* Macroscopic origin of circular dichroism effects by alignment of self-assembled fibers in solution. *Angew. Chem. Int. Ed.* **46**, 8203–8205 (2007).
22. Crusats, J., El-Hachemi, Z. & Ribó, J. M. Hydrodynamic effects on chiral induction. *Chem. Soc. Rev.* **39**, 569–577 (2010).
23. Jeffery, G. B. The motion of ellipsoidal particles in a viscous fluid. *Proc. R. Soc. Lond. Ser. A* **102**, 161–179 (1922).
24. Ho, B. P. & Leal, L. G. Inertial migration of rigid spheres in 2-dimensional unidirectional flows. *J. Fluid Mech.* **65**, 365–400 (1974).
25. Brenner, H. The stokes resistance of an arbitrary particle. *Chem. Eng. Sci.* **18**, 1–25 (1963).
26. Brenner, H. The stokes resistance of an arbitrary particle—II: an extension. *Chem. Eng. Sci.* **19**, 599–629 (1964).
27. Brenner, H. The stokes resistance of an arbitrary particle—III: Shear fields. *Chem. Eng. Sci.* **19**, 631–651 (1964).
28. Brenner, H. The stokes resistance of an arbitrary particle—IV: Arbitrary fields of flow. *Chem. Eng. Sci.* **19**, 703–727 (1964).
29. Brady, J. F. & Bossis, G. Stokesian dynamics. *Annu. Rev. Fluid Mech.* **20**, 111–157 (1988).
30. Durlofsky, L., Brady, J. F. & Bossis, G. Dynamic simulation of hydrodynamically interacting particles. *J. Fluid Mech.* **180**, 21–49 (1987).

Acknowledgements

This work was supported by the Non-equilibrium Energy Research Center, which is an Energy Frontier Research Center funded by the U.S. Department of Energy, Office of Science, Office of Basic Energy Sciences under grant number DE-SC0000989. T.M.H. was funded by the Human Frontier Science Program. We acknowledge Professor Alexander Z. Patashinski and Professor Wesley R. Burghardt for helpful discussions.

Author contributions

T.M.H. performed the experiments. K.J.M.B., T.M.H., P.S.S. and S.H.D. developed the model and performed the simulations. B.A.G. conceived and supervised the project, and T.M.H., K.J.M.B. and B.A.G. wrote the paper.

Additional information

Mapping multidimensional electronic structure and ultrafast dynamics with single-element detection and compressive sensing

Austin P. Spencer[1], Boris Spokoyny[1], Supratim Ray[1], Fahad Sarvari[1] & Elad Harel[1]

Compressive sensing allows signals to be efficiently captured by exploiting their inherent sparsity. Here we implement sparse sampling to capture the electronic structure and ultrafast dynamics of molecular systems using phase-resolved 2D coherent spectroscopy. Until now, 2D spectroscopy has been hampered by its reliance on array detectors that operate in limited spectral regions. Combining spatial encoding of the nonlinear optical response and rapid signal modulation allows retrieval of state-resolved correlation maps in a photosynthetic protein and carbocyanine dye. We report complete Hadamard reconstruction of the signals and compression factors as high as 10, in good agreement with array-detected spectra. Single-point array reconstruction by spatial encoding (SPARSE) Spectroscopy reduces acquisition times by about an order of magnitude, with further speed improvements enabled by fast scanning of a digital micromirror device. We envision unprecedented applications for coherent spectroscopy using frequency combs and super-continua in diverse spectral regions.

[1] Department of Chemistry, Northwestern University, 2145 Sheridan Road, Evanston, Illinois 60208, USA. Correspondence and requests for materials should be addressed to E.H. (email: elharel@northwestern.edu).

The ability to measure quantum correlations in complex systems with high spectral and temporal resolution provides deep physical insights into a wide range of phenomena from intermolecular dynamics in liquids to the electronic and vibrational structures of condensed phase molecular systems[1]. One powerful approach to directly measure intra- and inter-molecular couplings far from equilibrium is two-dimensional photon echo spectroscopy (2D PES)[2-4]. In 2D PES, coherences encoded in multiple time intervals are correlated, providing insight into the quantum states of the system and their interactions with the surroundings. 2D PES has been used to examine the dynamics of energy transfer in photosynthetic proteins[5-7], ultrafast dynamics of solute–solvent species[8] and intraband relaxation in semiconductors[9,10]. However, the advantages of performing 2D PES over one-dimensional spectroscopies come at a cost: increased acquisition time, additional experimental complexity and limited sensitivity using regions of the spectrum outside the visible and infrared. Here we introduce a method to overcome these limitations by combining sparse sampling and spatiotemporal encoding of nonlinear optical signals.

To make this connection, we begin by noting that the signal measured in 2D PES is typically sparse (that is, compressible). In most cases, the phase of the nonlinear signal, which is needed to extract the oscillation frequencies of the system, is measured by spectral interferometry in which an external reference field is coherently mixed with the signal and spectrally dispersed onto a detector[11-14]. Heterodyne detection yields an interferogram that is sparse in one or more Fourier domains, and it is this sparsity that we exploit using compressive sensing (CS) methods.

One consequence of CS theory is effectively to relax the Nyquist–Shannon sampling theorem[15], which enables perfect reconstruction of continuous signals (as a function of space, time, frequency, and so on) from samples acquired at a rate that is greater than the occupied bandwidth of the signal (irrespective of its distribution) rather than its total bandwidth. In effect, the Nyquist–Shannon sampling theorem relates how to sufficiently sample a signal's frequencies while CS relates how to sufficiently sample a signal's information[16]. CS has been shown to greatly reduce the number of measurements needed for signal recovery for a wide range of applications such as magnetic resonance imaging[17], nonlinear optical imaging[18], multidimensional spectroscopy[19,20], holography[21] and super-resolution microscopy[22], to name a few. In an analogous way, CS has been used to reduce the number of costly quantum mechanical calculations needed in computational studies[23-25]. One of the most promising applications of CS is in image reconstruction thanks to the potential to (i) utilize single-element detectors in regions of the spectrum where cameras perform poorly and (ii) bypass uniform sampling requirements imposed by array detectors when such sampling is not optimal.

To apply imaging-based CS to coherent spectroscopy, we first need to physically map the 2D spectrum to an image. In 2D PES, the signal is generated by exciting the sample with a series of ultrafast laser pulses, inducing a time-dependent polarization in the sample. The signal is spectrally dispersed onto an array detector, enabling multi-channel acquisition by detecting all signal frequencies, ω_t, simultaneously. The experiment is repeated for a range of coherence times, τ, and population times, T. Fourier transformation along τ yields the 2D Fourier transform (2DFT) spectrum $S(\omega_\tau, \omega_t; T)$ with parametric dependence on the waiting time, T.

Recently, some of the authors introduced a significantly higher-throughput sampling scheme called GRAPES[14,26,27] (GRadient Assisted Photon Echo Spectroscopy) that breaks with the traditional τ-scan approach. By incorporating a 2D array detector and a spatiotemporal gradient, all the τ delays may be sampled in parallel. The signal, emitted from the narrow focal line spatially overlapped with the excitation pulses in the sample, is spectrally resolved along an orthogonal direction, resulting in a direct image of the signal, $I(\tau, \omega_t)$. Through this imaging arrangement, a 2DFT spectrum is acquired for each laser shot upon Fourier transformation along τ. Combining the spatially encoded signal generated in GRAPES with a programmable spatial mask and a single-element detector enables the single-point array reconstruction by spatial encoding (SPARSE) spectroscopy method described here. This application of CS to 2DFT spectroscopy contrasts with prior approaches[19,20] involving sparse sampling of pulse time delays whereby data acquisition time is reduced by sampling a limited window of time delays.

In this work, we present 2DFT spectra of a carbocyanine dye and a photosynthetic pigment-protein complex measured using SPARSE spectroscopy. The accuracy of SPARSE-detected 2DFT spectra is evaluated based on comparison with conventional camera-detected 2DFT spectra. Reconstruction of spatial spectral interferograms is demonstrated using both CS and Hadamard methods, illustrating the robustness of CS retrieval. We show that in some cases, CS reconstruction requires only one-tenth of the complete set of Hadamard-encoded measurements for accurate interferogram recovery.

Results

Hadamard encoding and CS reconstruction. To verify the experimental methodology of SPARSE spectroscopy, we first implemented a 2D programmable version of Hadamard spectroscopy[28,29]. The Hadamard transform matrix, H_n, is analogous to the discrete Fourier transform matrix, but contains only binary ($+1$ and -1) elements. Hadamard sampling can be expressed as the linear problem $H_n x = y$, where x is a length-n signal vector and y is a length-n measurement vector. The unknown signal x can be reconstructed by performing the inverse Hadamard transform on y. To implement H_n experimentally, we use a digital micromirror device (DMD)[30], which is a 2D array of electro-mechanical mirror elements whose surface normal angles can be controlled between two binary states: $+12°$ ('on' state, 1) and $-12°$ ('off' state 0; see Methods for details). In this way, multiplication of the masks (each constructed from a different row (H_i) of the Hadamard matrix H_n) with the unknown image (x) occurs optically by reflecting the image formed at the spectrometer exit off of a spatial mask imprinted on the DMD. The reflected portions (that is, pixels) of the image are summed to form the observation $y_i = \sum H_{ij} x_j$ by focusing the reflected light with a lens onto a small active area photomultiplier tube (PMT) detector (Fig. 1). An example spatial mask (without the random inversions described in Method section) is shown in Fig. 2a.

As with the Fourier transform, Hadamard reconstruction requires Nyquist sampling to recover the signal. However, if the signal is sparse under a suitable unitary transform, then according to CS, an n-element signal may be faithfully reconstructed from fewer than n measurements. CS algorithms (for example, convex optimization and basis pursuit) minimize the L1-norm of the recovered signal subject to the constraint $Ax = y$, where A is an $n \times m$ observation matrix and y is a length-m measurement vector with $m \le n$. In this work, A is a pseudo-randomly chosen subset of the rows of a Hadamard matrix H_n, although other forms are possible such as a random matrix of 0 and 1. It is important to note that x itself does not necessarily have to be sparse, but rather it should be sparse in a suitable basis representation. To solve the constrained L1-norm minimization problem we used l_1-MAGIC, a program that uses standard interior-point methods to solve convex optimization problems[31].

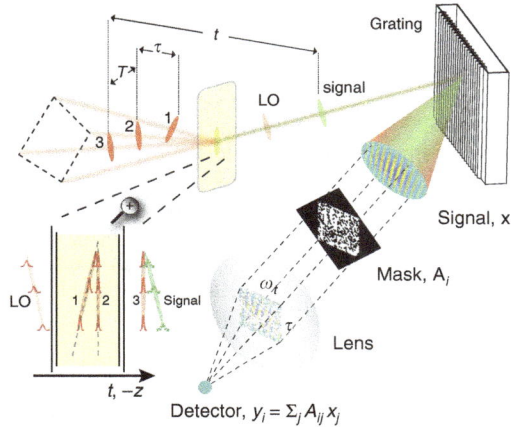

Figure 1 | A simplified schematic of single-shot three-pulse photon echo SPARSE spectroscopy. Three pulses (1, 2 and 3) generate a polarization in the sample which subsequently radiates a signal field that spatially interferes with a reference pulse (or local oscillator, LO) at the exit image plane of a spectrometer. The 2D signal-reference interferogram **x**, which spatially encodes the coherence time τ (inset shows pulse front tilts of each beam at the sample) and detection frequency ω_t, is spatially masked by a DMD (shown as transmissive instead of reflective for simplicity) and then focused by a lens onto a single-element detector. Each mask yields one intensity value on the detector, and by measuring the intensities for a sequence of different masks, the signal-reference interferogram can be retrieved through Hadamard or compressive sensing methods.

Detection schemes. Spatial spectral interferograms of a carbocyanine dye molecule, IR-144 and photosynthetic pigment-protein complex LH2 were acquired with a GRAPES apparatus using two distinct detection schemes: direct detection with a camera and SPARSE detection using a DMD spatial mask and PMT detector. In the direct case, a 2D array detector at the image plane of the spectrometer recorded 2D interferograms with one spatial dimension (along which τ is encoded) and one spectral dimension ($\lambda \propto 1/\omega_t$). In the second detection scheme, a DMD placed at the image plane selectively reflects light at each pixel towards ('on') or away ('off') from a PMT. A simplified schematic of this apparatus is depicted in Fig. 1. While the camera measures the light intensity incident on each pixel independently, the DMD–PMT detector measures the integrated intensity from all 'on' DMD pixels. Spatial spectral interferograms for IR-144 detected by both of these methods are shown in Fig. 2c. For SPARSE detection, two reconstruction methods are demonstrated, including reconstruction from a Nyquist-sampled set of Hadamard-encoded observations (Fig. 2c, middle panel) as well as CS reconstruction from a $10\times$ sub-Nyquist sampled (that is, undersampled) subset of the same Hadamard-encoded measurements (Fig. 2c, right panel). All three interferograms are nearly indistinguishable except for minor differences attributable to noise and reconstruction artifacts in the Hadamard and CS interferograms.

Since heterodyne detection by a time-delayed reference field yields a sinusoidal spectral interference pattern, the discrete cosine transform (DCT) was the natural choice for sparsifying transform. To test the validity of this transform, we explicitly compared the DCT of the CS and Hadamard reconstructed interferograms. To perform the 1D DCT, the interferogram is first flattened to a vector according to the order depicted in Supplementary Fig. 1. The DCT of the interferogram, shown in Fig. 2b, consists mostly of isolated groups of non-zero points. The

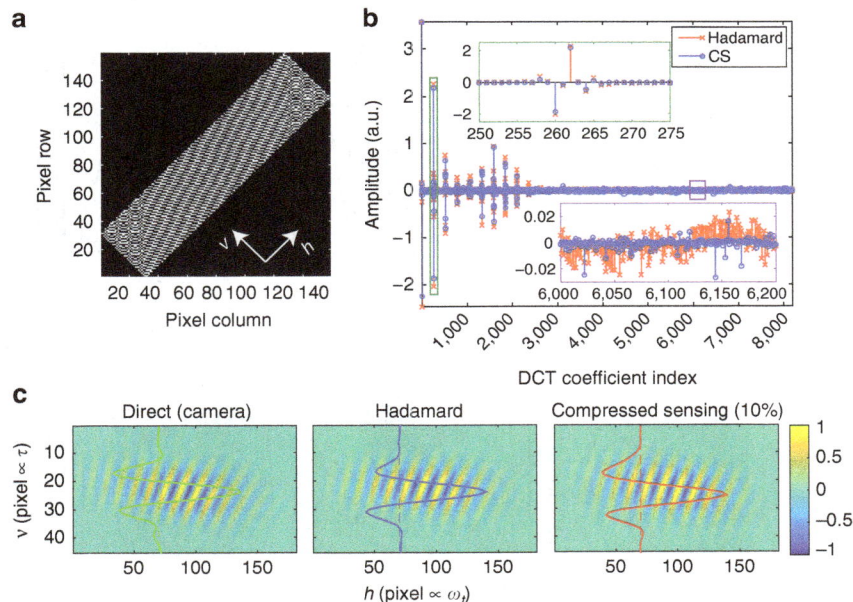

Figure 2 | Hadamard encoding and 1D DCT sparsifying transform for the reconstruction of spatial spectral interferograms. (**a**) A single representative Hadamard spatial mask (without random inversions) tiled into the 'active' region (Methods section) as it appears on the DMD. White arrows indicate horizontal (h) and vertical (v) lab coordinates. (**b**) Comparison of the 1D DCT of a Hadamard-retrieved flattened interferogram to that recovered by compressive sensing using convex optimization. Insets highlight two regions of the DCT interferogram: one at low spatial frequency (green, upper left) and one at high spatial frequency (purple, lower right). (**c**) Comparison of direct (camera), Hadamard and CS (10 × sub-Nyquist) detected interferograms after Fourier filtering, 45° rotation and cropping. The v axis is proportional to the spatially encoded τ dimension and the h axis is proportional to the detection frequencies, ω_t, in the 2DFT spectrum.

Hadamard measurement confirms that the DCT is effective at making the signal sparse and also demonstrates good agreement with the CS reconstruction. Notable differences appear in the high frequency DCT coefficients where the relatively constant, low amplitude features in the Hadamard reconstruction contrast with the higher, but less frequent spikes in the CS reconstruction. These deviations are expected due to the inability to exactly reconstruct uncorrelated noise from a sub-Nyquist set of measurements.

Comparison of 2DFT spectra. Absolute value 2DFT spectra of IR-144 and LH2 in Fig. 3 were constructed (Methods section) from spatial spectral interferograms collected using each of the three detection methods (camera, Hadamard and CS) described above. The Hadamard and CS (10% of Nyquist for IR-144 and 35% for LH2) reconstructions are faithful to the camera-detected spectra with respect to peak positions, although peak shapes and relative amplitudes do differ somewhat, especially in the case of LH2. The fraction of Hadamard-encoded spatial mask measurements used for CS reconstruction of LH2 and IR-144 interferograms was chosen such that variations between 2DFT spectra generated with different pseudo-randomly chosen measurement subsets were below the 10% level. The higher signal-to-noise ratio of IR-144 measurements enabled greater undersampling than for LH2. Statistical variations in CS reconstruction were explored at a range of sampling percentages to ensure that CS reconstructed 2DFT spectra are reproducible and that they converge to the Hadamard reconstruction as sampling approaches 100% (Supplementary Figs 2–15). Some differences between the SPARSE- and camera-acquired 2DFT spectra are expected based on the larger relative decrease in

responsivity at longer wavelengths of the PMT compared with the camera. This effect at least partially accounts for the differences in peak amplitudes and, perhaps to a lesser extent, peak shapes between 2DFT spectra acquired using a camera versus a PMT.

While IR-144 (Fig. 3a) has one spectrally broad transition centred ~ 800 nm (ref. 32), the spectrum of LH2 contains two absorption bands in the near-infrared, the B800 band at 800 nm $(12,500\,cm^{-1})$ and the B850 band at 850 nm $(11,765\,cm^{-1};$ ref. 33). Since the laser spectrum is sufficiently broad to excite both transitions, two diagonal peaks are expected at early waiting times $(T \approx 0)$. As the waiting time progresses, molecules initially excited to the higher energy B800 excited state relax to the B850 excited state, producing an off-diagonal cross peak in the spectrum. By $T = 1$ ps (Fig. 3b), much of the energy transfer has already taken place, leaving a diagonal peak and a cross peak at the B850 emission frequency and a weak diagonal peak at the B800 emission frequency. These features are readily observed in all three detection modalities.

Acquisition time comparison. Although a complete set of 8192 Hadamard-encoded spatial mask measurements is acquired in only 0.82 s, the overall speed of the experiment is limited by the need for signal averaging. Despite this, it takes < 3 min to acquire the Hadamard-encoded data needed to generate each 2DFT spectra in the middle column of Fig. 3. Taking into account the ability to undersample when using CS reconstruction, this acquisition time can be reduced by a factor ~ 3–10. Camera-detected 2DFT spectra, on the other hand, required only about 1 s of acquisition time due to the huge multi-channel advantage afforded by the pixel array sensor. In contrast to SPARSE and GRAPES spectroscopy, 2DFT spectroscopy

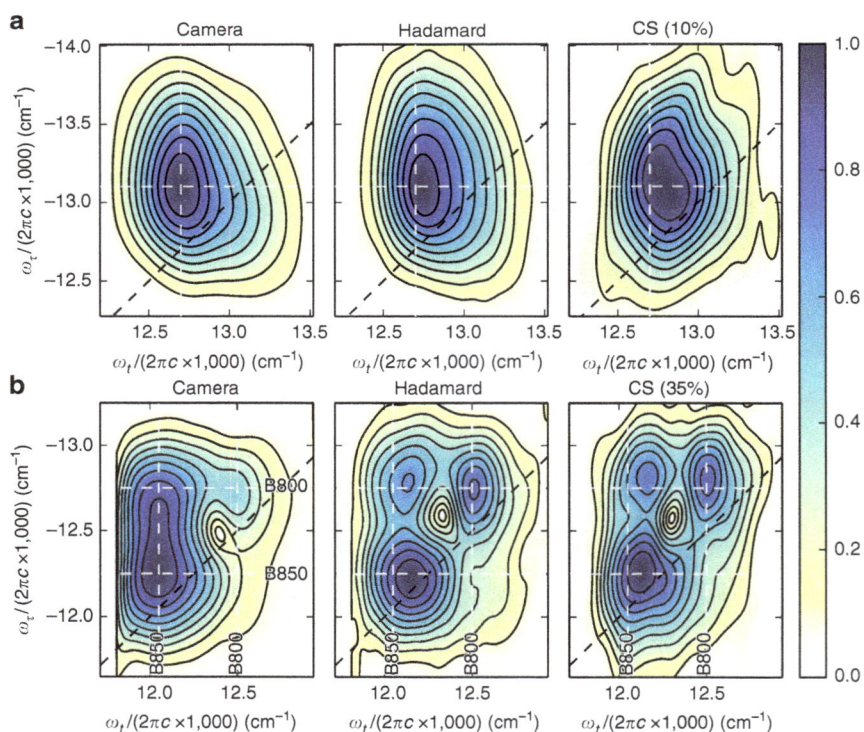

Figure 3 | Comparison of 2DFT spectra. Absolute value 2DFT spectra of (**a**) IR-144 cyanine dye ($T = 20$ fs) and (**b**) LH2 ($T = 1$ ps) detected directly by a camera versus SPARSE (DMD and PMT) detection using either the Hadamard transform (8,192 spatial masks) or compressed sensing with a subset of the Hadamard-encoded measurements (10% (819 spatial masks) for IR-144 and 35% (2,867 spatial masks) for LH2). Diagonal peaks arise from B800 and B850 bands corresponding to the excitation of ring subunits of bacteriochlorophyll pigments in the protein. The upper cross peak results from energy transfer from B800 to B850 in about 1 ps. Approximate locations of band centres are marked with dashed white lines for comparison between figure panels.

techniques that involve scanning pulse time delays can take up to 30 min for one scanned axis[3] or over an hour for two scanned axes[34]. While the most direct comparison for SPARSE spectroscopy would be to phase-sensitive optical detection of 2DFT spectra collected using a single-element detector and scanning both a pulse time delay (τ) and a spectral axis (ω_t), to our knowledge no such study has been reported. The increase in acquisition speed for SPARSE spectroscopy is afforded by the DMD's ability to perform fast 2D scanning of the GRAPES spatial spectral interferogram image instead of relying on slow scanning of pulse time delays.

Discussion

To conclude, we demonstrated the use of SPARSE spectroscopy for acquiring 2DFT spectra with sub-Nyquist sampling, reducing by at least an order of magnitude the number of measurements and, consequently, the data collection time. Array detectors impose uniform sampling even when it greatly oversamples the underlying information content of the image. By contrast, the adaptable nature of SPARSE spectroscopy enables, in principle, use of only the minimum number of measurements necessary to sample the information content of the signal given knowledge of the optimum sparsifying transform. While scientific-grade cameras can currently outperform SPARSE in the visible wavelength region thanks to their substantial multi-channel advantage (proportional to the number of pixels), sensitive cameras with high pixel densities are not available in many other spectral regions. Consequently, one of the most exciting applications of this technique is to expand coherent multidimensional spectroscopic methods to the THz and X-ray regimes where nonlinear spectroscopy is extremely challenging, or for use with frequency combs and supercontinuum sources for high-resolution metrology. Overcoming technical hurdles to detection in these spectral regions will be critically important for elucidating fundamental physical phenomena such as the nature of collective and coherent excitations in liquids and atom-specific electronic coherence in transition metal complexes.

Methods

Experiment. The 1,028 nm wavelength output of a 306 kHz repetition rate Yb:KGW laser system (PHAROS, Light Conversion) with self-contained oscillator and regenerative amplifier is used to pump a visible–near-infrared noncollinear optical parametric amplifier (ORPHEUS-N, Light Conversion), producing tunable pulses with ~35 fs duration. The noncollinear optical parametric amplifier output is spatially filtered through a 50-μm diameter pinhole before entering a four-arm interferometer. Four beams, arranged at the corners of a rhombus (or diamond), are focused into the sample cell by a 20-cm focal length cylindrical mirror (refer to Fig. 1 of ref. 14 for details on the beam geometry). The laser centre wavelength was tuned to 775 and 825 nm for measurements on IR-144 and LH2, respectively.

The signal and reference beams are imaged from the sample to the slit of a spectrograph (HR-320, Horiba) by a spherical lens. The spectrograph was modified to create a second output by adding a removable mirror between the second spherical mirror and the standard output. While the standard output houses a CMOS imaging sensor (Zyla 5.5 sCMOS 10-tap, Andor), the added second output is routed to the face of a DMD (DLP9500, Texas Instruments). Since the axes of rotation for the DMD micromirrors lie at a 45° angle relative to the axes of the 1,920 × 1,080 micromirror array, the DMD was rotated by 45° in the plane of its front face so that the reflected light would propagate in a plane parallel to the laser table. The output of the DMD is focused onto a PMT module (H12402, Hamamatsu). The PMT signal is amplified by a fast preamplifier (SR445A, Stanford Research Systems) and then integrated and digitized by a charge-integrating data acquisition (DAQ) system (IQSP518, Vertilon).

The DMD and DAQ are synchronized to the laser pulse train, enabling charge integration of individual laser pulses within a 50 ns window, as well as sub-100 ns control of DMD transition times. The DMD trigger signal, which initiates display of the next available spatial mask in its buffer, was generated by dividing the frequency of the laser clock signal by 32 using a microcontroller (ATmega328, Atmel), ensuring that DMD mask transitions are phase locked to the laser pulse train. In this configuration, each binary mask was held on the DMD for 32 consecutive laser shots, of which 28 laser shots were individually integrated and digitized by the DAQ with the four laser shots closest to the DMD transition being

discarded to ensure the measurements are not influenced by motion of the micromirrors. This results in a DMD frame rate of 9.6 kHz, close to its maximum rate of ~10 kHz. Integrated into the DMD control board is a field-programmable gate array, controller chip, driver and high-capacity on-board memory capable of storing ~15,000 binary spatial masks.

To perform image reconstruction using a DMD and single-element PMT detector, a set of binary masks was constructed from a 8192nd-order (($2^{13})^2 = 8192^2$ elements) Hadamard matrix whose columns had been randomly inverted. Such inversions make the mask appear spatially quasi-random, which has the benefit of reducing the influence of undesired diffractive contributions to the measured signal that would otherwise arise from variations in diffraction efficiency between different spatial masks. As opposed to full Hadamard multiplexing which requires balanced detection, an S-matrix–like implementation was used here wherein -1 elements of the Hadamard matrix are replaced by 0 such that only $+1$ elements are detected[29]. To provide adequate spatial resolution while also satisfying the memory limitations of the DMD (see Experimental Limitations and Optimization), the binary mask was limited to an 'active' region (or region of interest) on the DMD composed of 720 by 180 physical pixels, each containing a single micromirror. Groups of 4 by 4 physical pixels were binned (that is, treated as a single unit) to yield a 180 by 45 region containing 8,100 super-pixels. For each binary mask in the 8,192-frame sequence needed for image reconstruction, a row of the Hadamard matrix was tiled into the 8,100 super-pixel 'active' region of the DMD (Supplementary Fig. 1); the remainder of the DMD pixels were set to zero and unused Hadamard row elements were discarded. The 45° rotation of the DMD is compensated for by rotating the rectangular bounds of the active region of the spatial mask such that they lie diagonally in the DMD reference frame. This arrangement makes optimal use of the available 8,100 super-pixels since the signals of interest are typically elongated horizontally (in the lab reference frame) by the angular dispersion induced by the spectrograph grating. This rotation is apparent in the spatial mask shown in a since it is plotted in the DMD reference frame.

The sequence of $2^{13} = 8192$ unique DMD frames was repeatedly cycled through until 10–50 million laser shots had been measured, yielding 43–217 complete data sets that can be averaged and reconstructed into an image of the light intensity at the DMD face.

Data processing. The raw data, each element representing the integrated light intensity of a single laser shot with a given DMD mask, is first divided into complete sequences. Sequences are then averaged together to produce a single 8192-element sequence $\mathbf{y}$. To reconstruct the light intensity for each super pixel, the averaged sequence $\mathbf{y}$ is used to solve the equation $\mathbf{Ax} = \mathbf{y}$, where $\mathbf{A}$ is the randomly inverted Hadamard matrix used to construct the binary masks and $\mathbf{x}$ is a vector of super-pixel intensities. Finally, the image at the face of the DMD is obtained by tiling $\mathbf{x}$ into the active region of the DMD frame exactly as when done to produce the initial DMD masks. An example of an image reconstructed in this way is shown in Fig. 2c (middle panel).

To perform CS reconstruction, first a subset of the Hadamard-encoded measurements described above is pseudo-randomly chosen. The corresponding observation matrix for these measurements is transformed into a sparsifying basis using the 1D DCT along each row. This transformed observation matrix $\mathbf{A}$ is passed, along with its corresponding intensity measurements $\mathbf{y}$, to a convex optimization routine that minimizes the L1-norm of the solution $\mathbf{x}$ subject to the constraint $\mathbf{Ax} = \mathbf{y}$. The solution is inverse transformed using the 1D DCT to yield the pixel intensity values, which are subsequently tiled into the appropriate spatial arrangement as described above (Supplementary Fig. 1).

2DFT spectra are generated in the same way for both Hadamard and CS reconstructed interferograms. Interferograms first undergo a coordinate transform (involving a 45° rotation with cubic interpolation) from the DMD coordinate space to the camera coordinate space such that a pixel at a given wavelength and τ on the DMD maps to the equivalent wavelength and τ pixel position on the camera. This allows SPARSE-detected and camera-detected interferograms to be processed identically and to share the same pulse delay calibration information. Transformed interferograms are subsequently cropped to the extent of the active area of the DMD spatial mask. The τ axis is calibrated by measuring the spatial spectral interference between beams 1 and 2 using the spectrograph camera, as previously described[14]. The signal-beam 4 interferograms are (piecewise cubic) interpolated from a grid of equidistant wavelengths to a grid of equidistant frequencies. The interpolated interferograms are 2D Fourier transformed and filtered in the conjugate domain to select one of the two AC interference peaks. A 1D inverse Fourier transform along the t dimension (that is, the detection axis) then yields a distorted 2DFT spectrum which must be modified to correct for the crossing angle-induced wavefront tilt between the reference beam (beam 4) and beam 3, as previously reported[14]. Finally, the corrected 2DFT spectra are linearly interpolated to equalize the frequency grid spacing along both the ω_τ and ω_t dimensions. Camera-acquired interferograms are processed identically save for the initial coordinate transform. The camera-acquired interferograms are cropped to the extent of the DMD active area for comparison to SPARSE detection.

Experimental limitations and optimization. There are some properties of the experimental apparatus and methods presented here that are thought to limit the accuracy of 2DFT spectra collected using the implementation of SPARSE detection described here. First, collecting signal from a limited spatial area of the DMD causes the very low-lying wings of the signal to be lost (Fig. 2c). This limitation arises from the finite size of the DMD's on-board memory which, in turn, limits the number of unique DMD spatial masks that can be loaded at a given time. Since the number of pixels in a recovered interferogram image is related to the number of DMD spatial mask measurements, for a fixed number of measurements, there exists a tradeoff between an image's spatial resolution and its size. The spatial size of features in the interferogram image (for example, interference fringes, spectral lineshapes and transient dynamics) sets the minimum required spatial resolution. This spatial resolution requirement, combined with the limit in number of measurements, sets the maximum spatial extent of the region able to be sampled on the DMD. For Hadamard reconstruction, the on-board memory limits the number of recovered image pixel to $\sim 15,000$ since reconstruction of an n pixel image requires measurement of n unique DMD spatial masks. The 720 by 180 physical pixel region used corresponds to a 7.8 mm by 1.9 mm area, less than a quarter of the 16.6 mm by 4.6 mm area collected by the camera. This restriction could be alleviated either by increasing the on-board memory or by using CS reconstruction with $\leq 15,000$ DMD spatial mask chosen randomly from a larger Hadamard matrix (that is, $\mathbf{H}_n$ where $n > 15,000$).

Second, the small active area of the PMT (3 mm by 1 mm) requires careful alignment to ensure that all light reflected from the DMD is collected to avoid attenuating the edges of the interferogram. In addition, the small active area limits the maximum permissible energy per pulse incident on the detector to maintain detector linearity, which is critical for accurate interferogram reconstruction. To avoid detector saturation, likely due to space charge effects[35], lower than optimal reference beam pulse energies were used, limiting the heterodyne advantage of spectral interferometry.

Third, the measured signal is weighted by the spectral response of the PMT, which decreases at longer wavelengths. Although the same is true for the CMOS camera used here, the change in response is much larger for the PMT. Moving from 850 to 800 nm, the spectral response of the PMT increases by a factor of ~ 3 compared with a factor of ~ 1.5 increase for the camera. Overall, the PMT has a low quantum efficiency ($\sim 0.5\%$, compared with $\sim 28\%$ for the camera) which somewhat limits the signal-to-noise ratio of SPARSE detection. Signal-to-noise ratio could be improved by using a higher quantum efficiency detector such as an avalanche photodiode.

References

1. Mukamel, S. *Principles of nonlinear optical spectroscopy* (Oxford University Press, 1995).
2. Cho, M. H. Coherent two-dimensional optical spectroscopy. *Chem. Rev.* **108**, 1331–1418 (2008).
3. Hybl, J. D., Ferro, A. A. & Jonas, D. M. Two-dimensional Fourier transform electronic spectroscopy. *J. Chem. Phys.* **115**, 6606–6622 (2001).
4. Hamm, P., Lim, M., DeGrado, W. F. & Hochstrasser, R. M. The two-dimensional IR nonlinear spectroscopy of a cyclic penta-peptide in relation to its three-dimensional structure. *Proc. Natl Acad. Sci. USA* **96**, 2036–2041 (1999).
5. Harel, E., Long, P. D. & Engel, G. S. Single-shot ultrabroadband two-dimensional electronic spectroscopy of the light-harvesting complex LH2. *Opt. Lett.* **36**, 1665–1667 (2011).
6. Panitchayangkoon, G. *et al.* Long-lived quantum coherence in photosynthetic complexes at physiological temperature. *Proc. Natl Acad. Sci. USA* **107**, 12766–12770 (2010).
7. Collini, E., Wong, C. Y., Wilk, K. E., Curmi, P. M. G., Brumer, P. & Scholes, G. D. Coherently wired light-harvesting in photosynthetic marine algae at ambient temperature. *Nature* **463**, 644–647 (2010).
8. Zheng, J. & Fayer, M. D. Solute-solvent complex kinetics and thermodynamics probed by 2D-IR vibrational echo chemical exchange Spectroscopy. *J. Phys. Chem. B* **112**, 10221–10227 (2008).
9. Scholes, G. D. & Wong, C. Y. Biexcitonic fine structure of cdse nanocrystals probed by polarization-dependent two-dimensional photon echo spectroscopy. *J. Phys. Chem. A* **115**, 3797–3806 (2011).
10. Velizhanin, K. & Piryatinski, A. Probing Interband coulomb interactions in semiconductor nanostructures with 2d double-quantum coherence spectroscopy. *J. Phys. Chem. B* **115**, 5372–5382 (2011).
11. Lepetit, L., Cheriaux, G. & Joffre, M. Linear techniques of phase measurement by femtosecond spectral interferometry for applications in spectroscopy. *J. Opt. Soc. Am. B: Opt. Phys* **12**, 2467–2474 (1995).
12. Brixner, T., Mancal, T., Stiopkin, I. & Fleming, G. Phase-stabilized two-dimensional electronic spectroscopy. *J. Chem. Phys.* **121**, 4221–4236 (2004).
13. Shim, S. H., Strasfeld, D. B., Fulmer, E. C. & Zanni, M. T. Femtosecond pulse shaping directly in the mid-IR using acousto-optic modulation. *Opt. Lett.* **31**, 838–840 (2006).
14. Spencer, A. P., Spokoyny, B., Harel, E. & Enhanced-Resolution Single-Shot, 2DFT Spectroscopy by spatial spectral interferometry. *J. Phys. Chem. Lett.* 945–950 (2015).
15. Bracewell, R. N. *The Fourier Transform and Its Applications* 3rd edn. (McGraw Hill, 2000).
16. Donoho, D. L. Compressed sensing. *ITIT* **52**, 1289–1306 (2006).
17. Lustig, M., Donoho, D., Pauly, J. & Sparse, M. R. I. The application of compressed sensing for rapid MR imaging. *Magn. Reson. Med.* **58**, 1182–1195 (2007).
18. Cai, X. J., Hu, B., Sun, T., Kelly, K. F. & Baldelli, S. Sum frequency generation-compressive sensing microscope. *J. Chem. Phys.* **135** (2011).
19. Sanders, J. N. *et al.* Compressed sensing for multidimensional spectroscopy experiments. *J. Phys. Chem. Lett.* **3**, 2697–2702 (2012).
20. Dunbar, J. A., Osborne, D. G., Anna, J. M. & Kubarych, K. J. Accelerated 2D-IR using compressed sensing. *J. Phys. Chem. Lett.* **4**, 2489–2492 (2013).
21. Rivenson, Y., Stern, A. & Javidi, B. Overview of compressive sensing techniques applied in holography. *Appl. Opt.* **52**, A423–A432 (2013).
22. Zhu, L., Zhang, W., Elnatan, D. & Huang, B. Faster STORM using compressed sensing. *Nat. Methods* **9**, 721–723 (2012).
23. Andrade, X., Sanders, J. N. & Aspuru-Guzik, A. Application of compressed sensing to the simulation of atomic systems. *Proc. Natl Acad. Sci. USA* **109**, 13928–13933 (2012).
24. McClean, J. R. & Aspuru-Guzik, A. Compact wavefunctions from compressed imaginary time evolution. Preprint at http://arxiv.org/abs/1409.7358 (2014).
25. Sanders, J. N., Andrade, X. & Aspuru-Guzik, A. Compressed sensing for the fast computation of matrices: application to molecular vibrations. *ACS Cent. Sci.* **1**, 24–32 (2015).
26. Harel, E., Fidler, A. & Engel, G. Real-time mapping of electronic structure with single-shot two-dimensional electronic spectroscopy. *Proc. Natl Acad. Sci. USA* **107**, 16444–16447 (2010).
27. Spokoyny, B. & Harel, E. Mapping the Vibronic Structure of a Molecule by Few-Cycle Continuum Two-Dimensional Spectroscopy in a Single Pulse. *J. Phys. Chem. Lett.* **5**, 2808–2814 (2014).
28. Nelson, E. D. & Fredman, M. L. Hadamard Spectroscopy. *J. Opt. Soc. Am.* **60**, 1664 (1970).
29. Graff, D. K. Fourier and Hadamard: transforms in spectroscopy. *J. Chem. Educ.* **72**, 304 (1995).
30. Dudley, D., Duncan, W. & Slaughter, J. in SPIE Proceedings SPIE **4985**, 14–25 (MOEMS Display and Imaging Systems) (2003).
31. Candès, E. & Romberg, J. *l1-magic* v. 1.11California Institute of Technology, 2005).
32. Jonas, D. M. Two-dimensional femtosecond spectroscopy. *Annu. Rev. Phys. Chem* **54**, 425–463 (2003).
33. Cogdell, R., Gall, A. & Kohler, J. The architecture and function of the light-harvesting apparatus of purple bacteria: from single molecules to in vivo membranes. *Q. Rev. Biophys.* **39**, 227–324 (2006).
34. Tekavec, P. F., Lott, G. A. & Marcus, A. H. Fluorescence-detected two-dimensional electronic coherence spectroscopy by acousto-optic phase modulation. *J. Chem. Phys.* **127**, 214307 (2007).
35. HP., K. K. *Photomultiplier Tubes: Basics and Applications* 3rd edn (Hamamatsu Photonics K.K., 2006).

Acknowledgements

The work was supported by the Army Research Office (W911NF-13-1-0290), Air Force Office of Scientific Research (FA9550-14-1-0005) and the Packard Foundation (2013-39272) in part.

Author contributions

E.H. conceived the experiments. A.P.S. and B.S. conducted the experiments. S.R. developed the computer interface to the DMD. E.H., F.S. and B.S. developed CS implementation. A.P.S., B.S. and E.H. performed data analysis. E.H. and A.P.S. contributed to the preparation of manuscript.

Additional information

Pulse-density modulation control of chemical oscillation far from equilibrium in a droplet open-reactor system

Haruka Sugiura[1], Manami Ito[1], Tomoya Okuaki[1], Yoshihito Mori[2], Hiroyuki Kitahata[3] & Masahiro Takinoue[1,4]

The design, construction and control of artificial self-organized systems modelled on dynamical behaviours of living systems are important issues in biologically inspired engineering. Such systems are usually based on complex reaction dynamics far from equilibrium; therefore, the control of non-equilibrium conditions is required. Here we report a droplet open-reactor system, based on droplet fusion and fission, that achieves dynamical control over chemical fluxes into/out of the reactor for chemical reactions far from equilibrium. We mathematically reveal that the control mechanism is formulated as pulse-density modulation control of the fusion–fission timing. We produce the droplet open-reactor system using microfluidic technologies and then perform external control and autonomous feedback control over autocatalytic chemical oscillation reactions far from equilibrium. We believe that this system will be valuable for the dynamical control over self-organized phenomena far from equilibrium in chemical and biomedical studies.

[1] Department of Computational Intelligence and Systems Science, Tokyo Institute of Technology, 4259 Nagatsuta-cho, Midori-ku, Yokohama 226-8502, Japan. [2] Department of Chemistry, Faculty of Science, Ochanomizu University, 2-1-1 Ohtsuka, Bunkyo-ku, Tokyo 112-8610, Japan. [3] Department of Physics, Graduate School of Science, Chiba University, 1-33 Yayoi-cho, Inage-ku, Chiba 263-8522, Japan. [4] PRESTO, Japan Science and Technology Agency (JST), 4-1-8 Honcho Kawaguchi, Saitama 332-0012, Japan. Correspondence and requests for materials should be addressed to M.T. (email: takinoue.m.aa@m.titech.ac.jp).

L iving systems are achieved by complex chemical reaction dynamics far from equilibrium, such as gene expression networks, signalling networks, metabolic circuits and neural networks. The design, construction and control of artificial bio-inspired self-organized phenomena remain challenging in a wide range of science and engineering fields, such as the use of synthetic biology for understanding life[1], the fabrication of bio-inspired nano/microscale autonomous artificial systems[2] and the synthesis of dynamical microscale materials[3]. Recently, micrometre-sized reaction systems modelled on cellular systems have been actively studied, including chemical and biological reactions in microcompartments[4-19] and microfluidic devices [20-22], and artificial multicellular interactions[23-25].

Chemically open systems with well-controlled chemical fluxes into/out of the systems (that is, supply and dissipation of chemicals) are essential for complex chemical reactions far from equilibrium, to eliminate the increasing entropy in such systems[26]. The construction of microreactors with chemical fluxes is therefore necessary in microscale bio-inspired engineering. Open microreactors, such as semipermeable microcapsules[7,15], liposomes with nanopore proteins[9,25] and on-chip DNA compartment reactors[22], use passive substrate diffusion to generate chemical fluxes. In contrast, annular microchannels with valves and peristaltic mixers[20,21] produce chemical fluxes by the mechanical injection of solutions. Thus, the use of microreactors to generate chemical dynamics, including chemical oscillations based on constant chemical fluxes, has been successful. However, the time-variable chemical fluxes required for the external control and the feedback control depending on environment and inner reaction states, as in the case of living systems, have never been achieved in microscale systems. The development of useful and robust methods for the precise control of time-variable chemical fluxes is thus an important issue in microscale bio-inspired engineering.

In this paper, we report a microfluidic method that can control time-variable chemical fluxes into/out of a microreactor (Fig. 1). Our method is inspired by the universal molecular transportation systems in cells, which is based on vesicular fusion and fission observed in endo- and exo-cytotic processes, organellar vesicular transportation, and viral infection and budding via envelope. For example, a macrophage cell continually ingests and secretes solutions amounting to 25% of its volume each hour, while its own volume remains constant[27]. Similarly, our method achieves sustained chemical fluxes based on the repeated fusion and fission of microdroplets whose volumes remain constant. In this work, we use water-in-oil (W/O) microdroplets as microreactors. Droplet-based microfluidics enables rapid-response (for example, electrical and magnetic) manipulation of fluids without microfluidic mechanical components, such as valves. We show that the droplet open-reactor system is electrically controlled by the pulse-density modulation of fusion–fission timing, which enables precise control over time-variable chemical fluxes, including external control and autonomous feedback control. We believe that this system will facilitate innovations in chemical and biomedical studies in terms of the dynamical control of self-organized phenomena far from equilibrium.

Results

Mathematical analyses of the droplet open-reactor system. In general, the chemical reaction dynamics in a droplet open-reactor system (Fig. 1) is described as

$$\dot{u}_i = f_i(\mathbf{u}) + p(t; \mathbf{T}, \mathbf{w})k_i(c_i - u_i), \quad (1)$$

where t is time; u_i and c_i ($i = 1, 2, \cdots$) are the concentrations of chemicals U_i in a reactor droplet (reactor) and transporter droplets (transporters), respectively; k_i is the exchange rate of

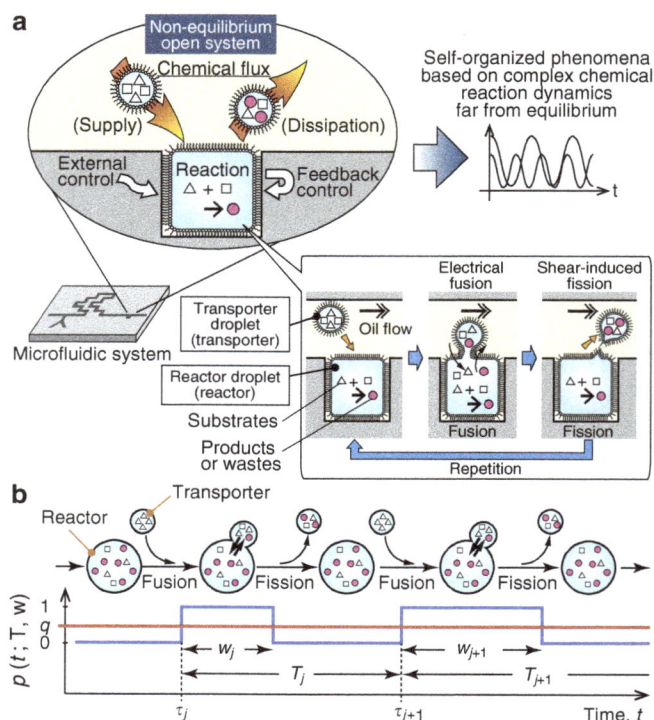

Figure 1 | Schematic diagram of chemical reactions far from equilibrium in a droplet open-reactor system controlled by pulse-density modulation. (**a**) In the droplet open-reactor system, the supply of substrates and the dissipation of products/wastes into/out of the reaction system are sustained, inducing self-organized phenomena based on complex chemical reaction dynamics far from equilibrium. The chemical reactions in the droplet open-reactor system are dynamically varied based on external control and feedback control. The droplet open-reactor system is based on the repeated fusion and fission of droplets. (**b**) A fusion–fission process and the pulse-density modulation concept. T_j and w_j are the interval and duration of j-th fusion–fission event, respectively. $p(t; \mathbf{T}, \mathbf{w})$ is a square pulse-train function used to express a fusion–fission process ($\mathbf{T} = \{T_j\}$; $\mathbf{w} = \{w_j\}$). q is the basal strength of the chemical fluxes. τ_j is the time at which the j-th fusion starts.

U_i caused by its diffusion during a fusion state; $f_i(\mathbf{u})$ ($\mathbf{u} = \{u_i\}$) expresses chemical reactions; $p(t; \mathbf{T}, \mathbf{w})$ expresses a fusion–fission process as a square pulse-train function with two discrete values, that is, 0 (non-fusion state) or 1 (fusion state) (Fig. 1b); $\mathbf{T} = \{T_j\}$ and $\mathbf{w} = \{w_j\}$ ($w_j < T_j$); and T_j and w_j are the interval and duration of j-th fusion–fission event, respectively.

First, a reaction dynamics in a droplet open-reactor system is investigated using a simple two-variable (u_1 and u_2) autocatalytic reaction:

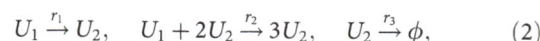

$$U_1 \xrightarrow{r_1} U_2, \quad U_1 + 2U_2 \xrightarrow{r_2} 3U_2, \quad U_2 \xrightarrow{r_3} \phi, \quad (2)$$

where r_1, r_2 and r_3 are reaction rates, and ϕ indicates the degradation of U_2. For this reaction system, we have $f_1(u_1, u_2) = -r_1u_1 - r_2u_1u_2^2$ and $f_2(u_1, u_2) = r_1u_1 + r_2u_1u_2^2 - r_3u_2$. This type of reaction is known to require well-controlled sustained chemical fluxes and is widely observed in systems ranging from physicochemical to cellular[26,28]. The time courses of u_2 for various fusion–fission periods are shown in Fig. 2a and Supplementary Fig. 1a ($T = 0.1$–10 min), where the fusion–fission events are assumed to be periodic (that is, $T_j = T$ and $w_j = w$ for all j). When T is relatively small ($T = 0.1$ and 1 min), the reaction exhibits its intrinsic behaviours: the limit cycle oscillation

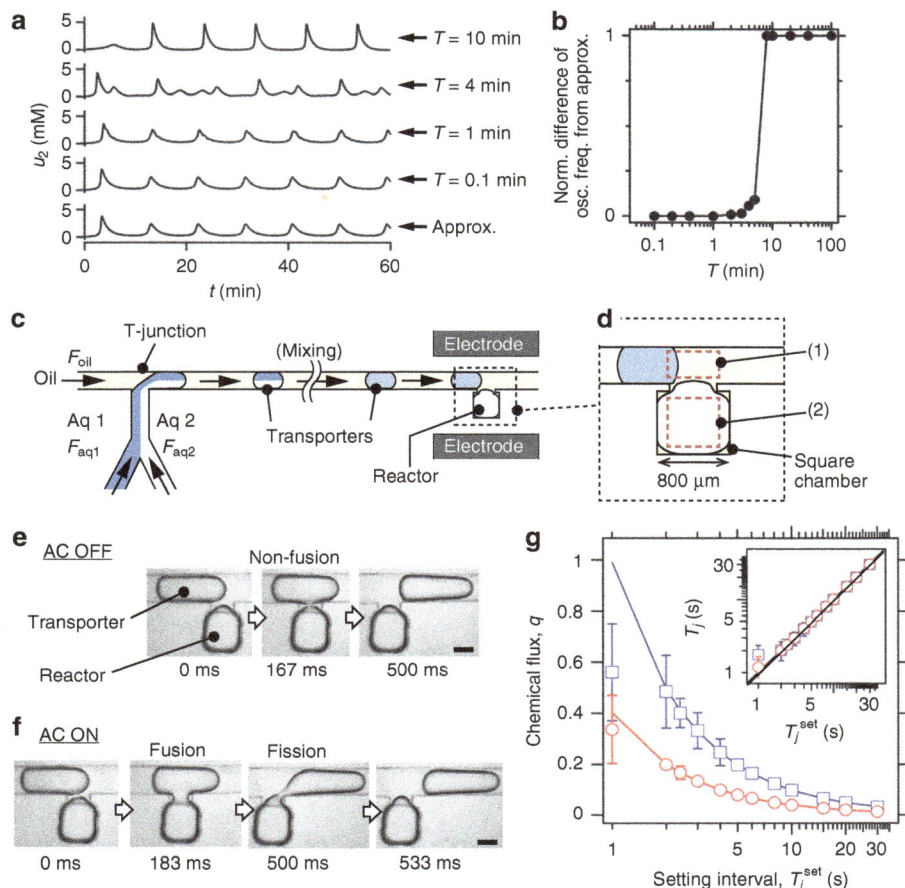

Figure 2 | Characterization of the droplet open-reactor system. (**a**) Results of numerical simulations of the autocatalytic reaction shown by equation (2) in the droplet open-reactor system. The simulations were performed using the general form indicated by equation (1) for $T = 0.1$–10 min, and using the approximate form described in equation (6) for 'approx.' (details are given in the Methods section). $T_j = T$ and $w_j = w$ for all j. $w/T = 0.25$ (fixed), which results in oscillation. (**b**) Normalized (Norm.) difference of the oscillation (osc.) frequency (freq.) from 'approx.' in **a**. The value 0 indicates no frequency shift, whereas 1 means that the oscillation frequency is equivalent to the frequency of the fusion $1/T$. Details are given in the Methods section. The solid line is provided as a guide for the eyes. (**c**) Design overview of the microfluidic system (details are given in Supplementary Fig. 2). Channel height, 500 μm. F_{oil}, flow rate of the oil phase; F_{aq1} and F_{aq2}, flow rate of the aqueous phases 1 and 2. (**d**) Enlarged view of the transporter and the reactor. The boxes outlined in red dashed lines indicate the areas in which (1) the detection of the transporter and (2) the fluorescence observation of the reactor were performed. (**e,f**) Bright-field microscope images of a fusion and fission event of a transporter and the reactor without and with a.c. voltage, respectively. Scale bars, 500 μm. (**g**) Control of q by T_j^{set}. Inset: control of T_j by T_j^{set}. The blue, red and black solid lines represent ideal values. Error bars: s.d. Sample size: 30 measurements. $F_{oil} = 30\,\mu l\,min^{-1}$ for **e** and **f**. $F_{oil} = 15\,\mu l\,min^{-1}$ (blue open square, $w = 0.99\,s$) and $30\,\mu l\,min^{-1}$ (red open circle, $w = 0.40\,s$) for **g**. Solutions in the transporter and reactor were water for **e** and **f**, and 0.1-mM Fl-Na for **g**.

(Fig. 2a) and the convergence to a steady state (Supplementary Fig. 1a). In contrast, when T is relatively large ($T \geq 4$ min), the reaction appears to be disturbed by the droplet fusion and fission. Thus, an appropriate range of fusion–fission periods is required to achieve chemical reactions in the droplet open-reactor system.

Next, we mathematically analyse the fusion–fission process. A single fusion–fission event is expressed by $H(t - \tau_j) - H(t - \tau_j - w_j)$, where τ_j is the time at which the j-th fusion starts (that is, $\tau_{j+1} - \tau_j = T_j$; Fig. 1b) and $H(t)$ is a step function: $H(t) = 0$ ($t < 0$) or 1 ($t \geq 0$). Thus,

$$p(t; \mathbf{T}, \mathbf{w}) = \sum_{j=-\infty}^{\infty} \left[H(t - \tau_j) - H(t - \tau_j - w_j) \right]. \quad (3)$$

In the simplest case, namely, in which the fusion–fission events are periodic ($T_j = T$ and $w_j = w$ for all j), we have

$$p(t; \mathbf{T}, \mathbf{w}) = \frac{w}{T} + \sum_{m=1}^{\infty} \frac{2}{\pi m} \sin\left(\frac{\pi m}{T} w\right) \cos\left(\frac{2\pi m}{T} t\right) \quad (4)$$

(details are given in Supplementary Note 1). Here τ_{mac} is defined as a characteristic time for the macroscopic dynamics of chemical reactions (for example, in the case of Fig. 2a, $\tau_{mac} \sim 10$ min). When $T \ll \tau_{mac}$, the second term in the right-hand side of the equation does not affect the reaction dynamics; thus,

$$p(t; \mathbf{T}, \mathbf{w}) \simeq w/T \equiv q, \quad (5)$$

where q is a dimensionless parameter expressing the ratio of fusion states in the fusion–fission process (Fig. 1b). Therefore, the reaction dynamics in the droplet open-reactor system can be described by the following approximate form:

$$\dot{u}_i = f_i(\mathbf{u}) + qk_i(c_i - u_i) \quad (6)$$

The reaction time course calculated using the approximate form (equation (6)) is shown in Fig. 2a (approx.) and is almost equivalent to the time courses of $T = 0.1$ and 1 min shown in Fig. 2a. Figure 2b shows the difference between the time courses calculated by equations (1) and (6). These results indicate that the approximate form described by equation (6) can predict the reaction dynamics when $T \ll \tau_{mac}$ (~ 10 min). Supplementary Fig. 1a (approx.) and 1b show the same result.

In summary, the reaction dynamics in the droplet open-reactor system is essentially regulated by q, which is the pulse density of the pulse-train function p. This type of control mechanism is called pulse-density modulation and is widely used in electrical devices and information–communication technologies because of its usefulness for parameter control.

Construction of a droplet open-reactor system. Figure 2c,d shows the design overview of a droplet-based microfluidic system constituting the droplet open-reactor system (design details are given in Supplementary Fig. 2). The reactor used for chemical reactions consisted of a W/O microdroplet that was fixed in a square chamber in a microchannel between a pair of electrodes. The transporters were W/O microdroplets flowing through the microchannel. The diameter of the reactor was ~ 800 µm, and its reaction volume was ~ 0.3 µl. The diameter of the transporters was ~ 500 µm, and the length along the flow was on the order of several hundred to one thousand micrometres, depending on the flow rate. The transporters were generated at the T-junction from two aqueous solutions (Supplementary Fig. 3a) and then delivered to the reactor after the solutions in the transporters were mixed through a zigzag channel[29] (Supplementary Figs 2a and 3b). Fusion of the reactor and transporters was controlled by applying an a.c. voltage between the electrodes[30–32]. The reactor remained in a non-fusion state in the absence of the a.c. voltage (Fig. 2e and Supplementary Movie 1) and fused with the transporters on the application of the a.c. voltage (Fig. 2f and Supplementary Movie 2). The fused droplets were immediately fissioned by the shear stress of the oil flow and returned to their non-fusion states[30]. The fusion–fission event was then repeated. Because the volume of the reactor was limited to the square chamber and the reactor was stabilized by its own surface tension, its reaction volume was kept almost constant.

In this microfluidic device, a droplet-fusion control programme was used to precisely control the fusion–fission interval T_j according to a set value T_j^{set} (Supplementary Fig. 4). The droplet-fusion control programme monitored the positions of the transporters and the fluorescence intensity of the reactor (Fig. 2d). On the basis of the monitoring information, the droplet-fusion control programme controlled the fusion and fission by switching the a.c. voltage on/off.

Using this system, we investigated the controllability of the fusion–fission process. The duration of the fusion state w was determined according to the oil flow rate F_{oil} (Supplementary Fig. 3c). Figure 2g shows the relationship between q and T_j^{set}, and its inset shows the relationship between T_j and T_j^{set}. When $T_j^{set} \geq 2$ s, $T_j = T_j^{set}$, and thus, T_j was successfully controlled. However, when $T_j^{set} < 2$ s, $T_j > T_j^{set}$ (inset) because the arrival interval of the transporters was comparable to T_j^{set}. As a result q was well controlled when $T_j^{set} \geq 2$ s. Supplementary Fig. 3d shows that the inner solution of the reactor (initially pure water) was completely exchanged with a fluorescein sodium (Fl–Na) solution in the transporters after several tens of fusion events with the transporters, indicating that ~ 1–2% of the chemicals were exchanged in each fusion–fission event. The inner solution of the reactor after a single fusion was homogenized within ~ 60 s, much faster than homogenization through the simple diffusion of molecules (~ 600 s). This faster mixing resulted from a rotating flow in the reactor induced by the oil flow in the microchannel (Supplementary Fig. 3e)[30].

Control over chemical reaction dynamics far from equilibrium. We investigated the controllability of chemical reaction dynamics

far from equilibrium using a droplet open-reactor system through bromate–sulfite–ferrocyanide (BSF) pH oscillation[33–35] (details are given in the Methods section, Supplementary Note 2 and Supplementary Table 1). This reaction consists of a time-delay negative feedback loop of the autocatalytic production of H^+ and the consumption of H^+ at low pH. This reaction requires BrO_3^-, $Fe(CN)_6^{4-}$ and SO_3^{2-} as substrates, and generates a limit cycle oscillation of pH only when appropriate chemical fluxes are maintained. In addition to the fact that the reaction mechanism has been well documented, the strict requirement for appropriate chemical fluxes is suitable for investigating the performance of an open reactor unlike other nonlinear chemical reactions, such as the Belousov–Zhabotinsky reaction[26], which exhibit relatively stable transient chemical oscillations even in chemically closed conditions.

In the experiments all the substrates were supplied to the reactor using transporters. The pH change in the reactor was observed via the fluorescence intensity change of a pH indicator (Fl–Na; Fig. 3a–c). The value of q was varied by changing T_j (w was fixed at 0.99 here and in further experiments). When q was high ($q = 0.33$), the reaction converged to a steady state at a higher pH (SSH; higher intensity) (Fig. 3a and Supplementary Movie 3). When q was low ($q = 0.05$), the reaction converged to a steady state at a lower pH (SSL; Fig. 3b and Supplementary Movie 4). Under the intermediate condition ($q = 0.17$), the intensity in the reactor exhibited pH oscillation (Fig. 3c and Supplementary Movie 5). Spatial heterogeneity was observed in these results, possibly because of the non-instantaneous mixing of the inner solution of the reactor (Supplementary Fig. 3e, Supplementary Note 3 and Supplementary Fig. 6)[36]. Figure 3d shows the time courses of the fluorescence intensity; bifurcation among SSH, pH oscillation and SSL was observed by changing q as a bifurcation parameter. The observed time courses are in semi-quantitative agreement with the numerical simulation results (Fig. 3e, details are given in Supplementary Note 2). Figure 3f shows a two-dimensional (2D) bifurcation diagram when q and $c_{SO_3^{2-}}$ were used as bifurcation parameters (all of the time-course data are shown in Supplementary Fig. 7). The result agrees with the 2D bifurcation diagram calculated using linear stability analysis (Fig. 3g, details are given in Supplementary Note 2). The bistable steady state shown in Fig. 3g was not experimentally observed because it was difficult to precisely control q within the bistable steady-state region (Fig. 2g, blue open squares; $T_j^{set} \leq 2$ s). In addition, we observed the similar reaction behaviours in arrayed multiple reactors (Supplementary Fig. 8). In summary, the results show that the chemical dynamics far from equilibrium was successfully controlled using the droplet open-reactor system.

Time-variable external control of chemical reaction dynamics. Here we extended equation (5) to the case in which T_j and w_j are variable for each fusion–fission event j:

$$p(t; \mathbf{T}, \mathbf{w}) \simeq w_j / T_j \simeq q(t), \qquad (7)$$

where $q(t)$ is a time-variable function (Fig. 4a). Figure 4b–e shows various waveforms of the chemical fluxes $q(t)$ that were produced through pulse-density modulation (details are given in the Methods section). When the pulse train of p had equally spaced intervals, q was constant (Fig. 4b). Figure 4c shows the resulting sinusoidal waveform. When T_j^{set} was randomly generated so that $q(t)$ followed a uniform distribution, q exhibited white noise (Fig. 4d). Figure 4e shows a saw-tooth wave with a 3-min period that demonstrates sharp switching of chemical fluxes. Similarly, different waveforms including square waves were produced (Supplementary Note 4 and Supplementary Fig. 9).

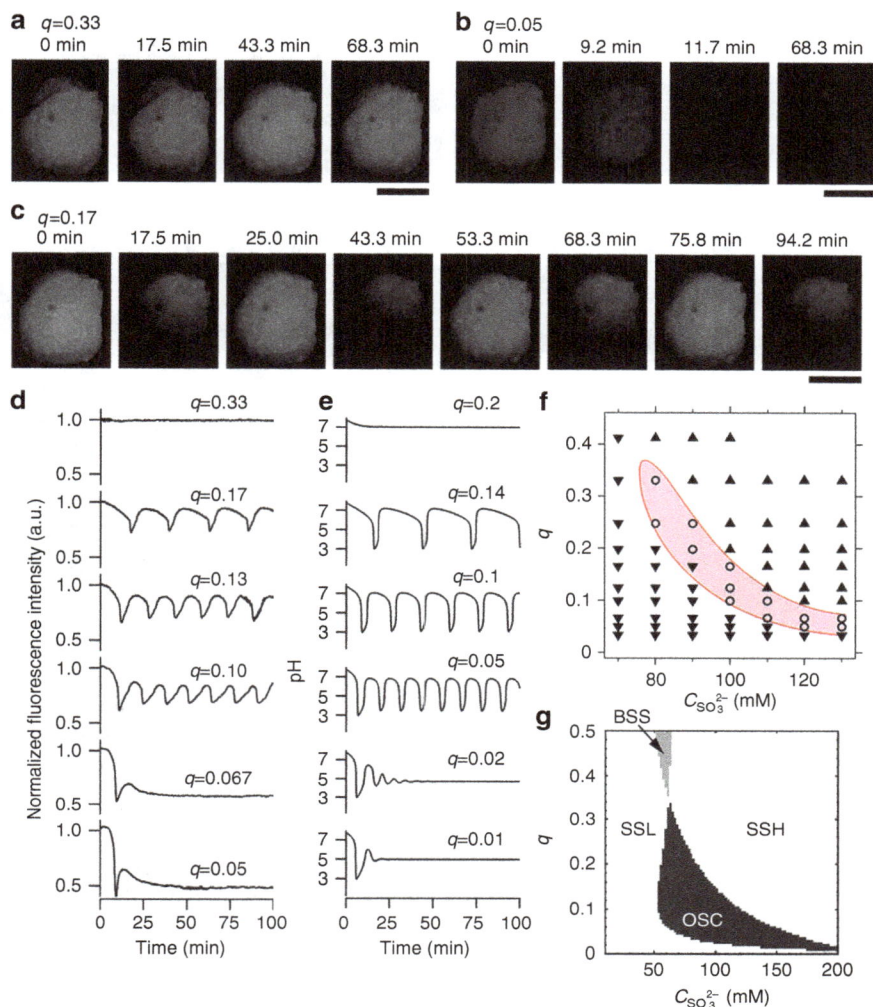

Figure 3 | Control of chemical oscillation dynamics using the droplet open-reactor system. (a–c) Fluorescence microscope images of the reactor. High intensity (white) indicates high pH, and vice versa (Supplementary Fig. 5). Scale bars, 500 μm. (a) Convergence to a steady state at a higher pH (SSH). (b) Convergence to a steady state at a lower pH (SSL). (c) Limit cycle oscillation of pH (OSC). (d) Time courses of fluorescence intensity in the reactor. (e) Numerical simulation of **d**. (f) 2D bifurcation diagram. Open circle: limit cycle oscillation; filled triangle: SSH; filled inverted triangle: SSL. (g) 2D bifurcation diagram calculated using linear stability analysis. White: monostable steady state of SSH or SSL; light grey: bistable steady state (BSS) of SSH and SSL; dark grey: OSC. $c_{SO_3^{2-}} = 100$ mM for **a–e**. The compositions of the solutions in the transporter and reactor are given in the Methods section.

We demonstrated external control of chemical reactions using time-variable chemical fluxes. First, Fig. 4f,g shows the BSF pH oscillation time courses and their Fourier power spectra obtained when the chemical fluxes included a sinusoidal external signal. When the periods of the added sinusoidal signal and the chemical oscillation were substantially different, additional periods appeared in the chemical oscillation ((2) and (3) in Fig. 4f,g; red arrows). In contrast, when they were similar, the period of the intrinsic chemical oscillation was entrained to that of the added sinusoidal signal ((4) in Fig. 4f,g; blue and red arrows). Next, we investigated the effect of the noise on the chemical reactions in SSL by changing the noise strength included in the chemical fluxes (Fig. 4h,i). When the noise was weak, noise-induced pulsed excitations randomly occurred ((2) in Fig. 4h,i). However, when the noise was relatively strong, the excitation timing was more coherent ((3) and (4) in Fig. 4h,i). We calculated then the degree of coherence, d_c (definition is given in the Methods section)[37,38]. d_c has a maximum at (3) in Fig. 4i, which suggests that noise was too strong and disturbed the pulsed excitations ((4) in Fig. 4i). This coherent phenomenon is called

coherence resonance[37,38]. Thus, based on these results, complex nonlinear chemical phenomena observed far from equilibrium can be quantitatively studied using the droplet open-reactor system.

Autonomous feedback control of chemical reaction dynamics. Finally, we investigated autonomous feedback control of chemical reaction dynamics far from equilibrium in the droplet open-reactor system, by extending the control method used for time-variable chemical fluxes. The feedback scheme is shown in Fig. 5a (the detailed algorithm is given in Supplementary Figs 10 and 11). Figure 5b shows the result of feedback control when the designated reaction state was set to 'oscillation with a period of 15 min'. The experiments started from SSH. The droplet-fusion control programme changed q every 40 min (Fig. 5b, upper graph) in response to the monitored reaction state. As a result, the chemical reaction system reached the designated state at ~200 min (Fig. 5b, lower graph). Figure 5c shows a long-term (>20 h) observation of the sustained pH limit cycle oscillation

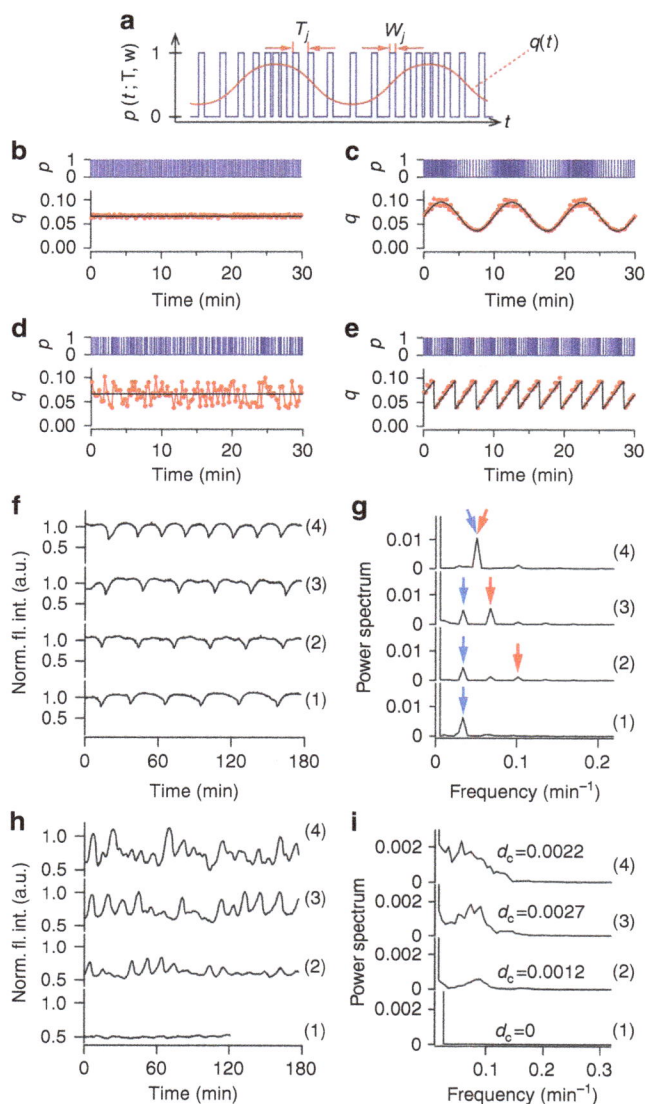

Figure 4 | Time-variable external control of chemical oscillation dynamics using the droplet open-reactor system. (**a**) Schematic diagram of the pulse-density modulation control of time-variable $q(t)$. When the pulse train of $p(t; \mathbf{T}, \mathbf{w})$ is denser, $q(t)$ is higher, and vice versa. (**b–e**) Generation of p and q. Blue lines: pulse trains of $p(t; \mathbf{T}, \mathbf{w})$; black lines: (**b,c,e**) theoretical curves of $q(t)$ and (**d**) theoretical average of $q(t)$; red dots and lines: $q(t)$ generated in experiments, calculated as $q(t = \tau_j) = w_j/T_j$. $w_j = 0.99$ s (fixed for all j). (**b**) Constant ($Z_C(t)$). (**c**) Sinusoidal wave ($Z_S(t)$, $A_q = 0.45$, $T_q = 10$ min). (**d**) White noise ($Z_N(t)$, $A_q = 0.45$). (**e**) Saw-tooth wave functions ($Z_{St}(t)$, $A_q = 0.45$, $T_q = 3$ min). $Z_C(t)$, $Z_S(t)$, $Z_N(t)$ and $Z_{St}(t)$ are described in detail in the Methods section. The baseline value of $q(t)$: $\bar{q} = 0.066$. (**f**) Entrainment of the chemical oscillation to the external sinusoidal signal. $\bar{q} = 0.062$. (1) $A_q = 0$ (without sinusoidal signal) and (2–4) $A_q = 0.45$. (2) $T_q = 10$ min, (3) $T_q = 15$ min and (4) $T_q = 20$ min. (**g**) Power spectra of **f**. (**h**) Noise-induced pulsed excitation when white noise was added to SSL. $\bar{q} = 0.03$. (1) $A_q = 0$ (without noise), (2) $A_q = 0.15$, (3) $A_q = 0.3$ and (4) $A_q = 0.45$. (**i**) Power spectra of **h**. $c_{SO_3^{2-}} = 100$ mM. The compositions of the solutions in the transporter and reactor are given in the Methods section. Norm. fl. int., normalized fluorescence intensity.

after applying the feedback control in Fig. 5b. This observation indicates that the droplet open-reactor system can stably 'incubate' the controlled dynamical chemical reactions far from equilibrium over the long term.

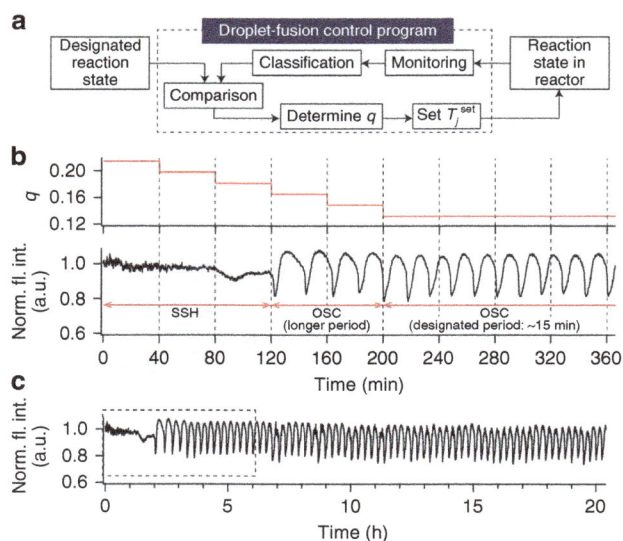

Figure 5 | Feedback control of BSF pH oscillation using the droplet open-reactor system. (**a**) Scheme of feedback control. The droplet-fusion control programme fluorescently monitors and classifies the current reaction state in the reactor (SSH, SSL and oscillation of pH). The droplet-fusion control programme compares the reaction state with a designated one; when they are different, the droplet-fusion control programme automatically changes q by changing T_j^{set} to obtain a reaction state closer to the designated one. (**b**) Time course of q (upper) and normalized fluorescence intensity (Norm. fl. int.; lower) during feedback control (designated period: 15 min). Initial condition $q = 0.215$ ($T_j = 4.62$ s). The change in q per step was 0.0165. (**c**) Long-term observation of oscillation after applying the feedback control in **b** (box indicated by dashed lines). The compositions of the solutions in the transporter and reactor are given in the Methods section.

Discussion

In this study, we developed a droplet open-reactor system that can finely and dynamically control chemical fluxes. Its control mechanism was mathematically formulated as pulse-density modulation control and was implemented using electrical control of the fusion and fission of droplets. Using the pulse-density modulation control, we produced various waveforms such as sinusoidal waves, saw-tooth waves and white noise. We first demonstrated control over a dynamical chemical reaction system far from equilibrium. The current system exhibited spatial heterogeneity of the chemical reaction (Fig. 3c and Supplementary Note 3) resulting from the non-instantaneous mixing of the inner solution because of the reactor size (several hundred micrometres in diameter). However, this issue will not occur in a smaller microreactor as we previously reported[30]. In addition, in some situations, the spatial heterogeneity may be utilized to investigate dynamically changing spatial patterns of chemical concentrations. In addition, the accurate chemical concentrations in the reactor could not be determined because of the volume measurement error ($\sim 9\%$) associated with the bright-field microscope images. However, the experimental error in the chemical concentration control is thought to be relatively small, as is suggested by the experiment illustrated in Fig. 5c, which shows the stable long-term limit cycle oscillation of pH. For highly quantitative analyses of reactions, especially in the case of using a smaller reactor droplet, a more accurate method for volume measurements, such as the use of a high-speed confocal microscope, will be required. Next, we performed external control of the reaction system, which has not been achieved previously at the microscale even though it is essential for the study of nonlinear chemical reactions and well studied in beaker-sized open reactors[39]. Finally, we demonstrated autonomous feedback control over the reaction system. In the feedback experiments, we used optical read-outs to monitor the reaction states, but alternative read-outs, such as electrical measurements, could be used in this system by introducing micropatterned electrodes. Combining multiple measurements to determine the reaction state will expand the abilities of this system.

To date, several microchannel-based open reactors have been reported[20–22]. Similar to these microfluidic open reactors, our droplet open-reactor system has the advantages of low sample consumption compared with conventional beaker-sized open reactors, facilitating microscale bio-inspired engineering. In the current setup, all chemicals flow into/out of the reactor, and selective chemical fluxes cannot be achieved, unlike on-chip DNA compartment reactors[22]. However, this capability will be improved by the immobilization of chemicals on solid surfaces, such as microbeads. In addition, the droplet open-reactor system has an advantage over previous open microreactors in terms of the controllability of chemical fluxes. The better controllability of this system is attributable to two-phase-flow microfluidics, in which the reaction solutions compartmentalized in droplets and the transporting fluids can be separately manipulated[40,41]. As a result, the addition/removal of chemicals can be achieved through pulse-density modulation control of the frequency of the digitalized droplet fusions that are electrically switched. Thus, manipulation of the whole solution in the channels and tubes is not required, and complicated microfluidic components, such as valves and mixers, are unnecessary. These characteristics allow open reactors to be combined with sophisticated control methods, such as the electrical[42–44] and optical[45] manipulation of fluids. In addition, because the on/off switching response was rapid compared with that of traditional beaker-sized open reactors and other microfluidic open reactors[20,21], sharp waveforms, such as saw-tooth and white noise waves, were also achieved. As a result, we successfully investigated the coherence resonance, which is usually difficult to identify because the intentionally produced noise must be properly controlled to prevent it from being buried in the intrinsic noise of the experimental system. This control was achieved by the precise and rapid control of chemical fluxes. However, based on the mathematical and experimental results, we identified a few limitations caused by the use of droplets as transporters. First, the controllability of q decreases at high values of q (Fig. 2g) because the fusion interval (T_j) cannot be less than the arrival time of the transporters. In addition, T_j values close to or exceeding the characteristic time of the macroscopic reaction dynamics cannot be used (Fig. 2b and Supplementary Fig. 1b) because the discreteness of the fusion–fission process affects the chemical reaction dynamics (that is, q that are too low cannot be used). In summary, there is an appropriate range of q necessary to control a chemical reaction dynamics in this system.

When this system is extended to coupled multiple reactors, it will be even more useful for the study of chemical reactions far from equilibrium. For example, multiple reactors linearly arrayed along a microchannel (for example, Supplementary Fig. 8a) could be diffusionally coupled with each other if semipermeable walls are constructed between the chambers. In addition, multiple reactors could contain different chemicals if the chemicals are injected through the top injection holes of the chambers (Supplementary Fig. 2) and then kept separate by fixation on solid surfaces, such as microbeads. Larger-scale coupled reactors arrayed in a 2D manner could be constructed and accessed through three-dimensional microchannels. This type of open-reactor system will facilitate studying the dynamical population behaviours of chemical reactions far from equilibrium[13].

We believe that the droplet open-reactor system can be applied to many complex reaction systems such as artificial DNA circuits and gene circuits[46–50], metabolic systems[51,52] and microchemostat-like reactors[53], in synthetic biology. In addition, the precise dynamical control of non-equilibrium chemical reactions will promote the rational design of oscillating enzymatic networks[54] and the production of complex and hierarchical materials, including biomineralization far from equilibrium[3]. In particular, our method has good compatibility with system control theory and computational intelligence because of the controllability achieved using pulse-density modulation. In the future, this method may therefore be applied to system control biology[55] based on the real-time monitoring and model-driven control of living cells or artificial cell-like systems[56–61] based on software–wetware hybrids.

Methods

Numerical simulations of the two-variable model. The parameter values used in the numerical simulation of equation (2) are as follows: $r_1 = 0.04$ min^{-1}, $r_2 = 1$ mM^{-2} min^{-1}, $r_3 = 1$ min^{-1}, $k_1 = k_2 = 0.4$ min^{-1}, $c_1 = 10$ mM and $c_2 = 0$ mM. The vertical axis in Fig. 2b indicates the degree of the oscillation-frequency shift, calculated using the maximum peak of the fast Fourier transformation spectra: $(f_{T, max} - f_{approx., max})/(f_{fusion} - f_{approx., max})$, where $f_{fusion} = 1/T$.

Fabrication of the microfluidic device. The fabrication details are given in Supplementary Note 5 and Supplementary Fig. 2. The microfluidic system was constructed using two poly(methyl methacrylate) plates (1 mm thickness) because poly(methyl methacrylate) microchannels are stable for long periods without liquid swelling and are amenable to repeated use. The microchannel was fabricated on the upper plate using a fine-milling machine (MDX-40A, Roland DG). The upper and lower plates were attached by thermal compression bonding. At the beginning of the experiments, a reactor was introduced into the square chamber through a top injection hole using a micropipette; the hole was then sealed with transparent cellophane tape. The oil phase and aqueous phases were then flowed using microsyringe pumps (LEGATO180, KD Scientific) and disposable syringes (10 ml, SS-10SZ, Terumo). The electrodes were connected to a function generator (WF1974, NF Corporation) through a voltage amplifier (M-2629B-2CH, MESS-TEK). The droplet-fusion control programme was developed using an image-processing module of the OpenCV library and Microsoft Visual C++ (Microsoft Corporation).

General experimental conditions. The oil phase consisted of mineral oil (Nacalai Tesque) containing 0.5% Span80 (Tokyo Chemical). Aqueous phases 1 and 2 were water for Fig. 2e,f. For Figs 2g and 4b–e, aqueous phase 1 was 0.2-mM Fl-Na (Sigma-Aldrich) and aqueous phase 2 was water. The aqueous phases used in the other cases are described in the Methods section reporting the BSF pH oscillation. $F_{aq1} = F_{aq2} = 10$ μl min^{-1} for all experiments. $F_{oil} = 15$ μl min^{-1} (that is, $w = 0.99$ s) for Figs 3–5. F_{oil} was varied for each experiment in Fig. 2. The a.c. voltage applied to effect droplet fusion was $V_{pp} = 300$ V (peak to peak; 1 kHz).

All the experiments were carried out using a fluorescence microscope (IX71 or IX81, Olympus). For the observations in Fig. 2e,f, a high-speed complementary metal-oxide-semiconductor camera (FASTCAM SA3 120K, Photron) was used. For the experiments in Figs 2g, 3–5, a digital camera (EOS 60D, Canon) was used.

The reproducibility of all experimental results was confirmed by performing at least two or three experiments.

Experiments and numerical simulations of BSF pH oscillation. The following is a simple reaction model of the BSF reaction that is called the Rabai–Kaminaga–Hanazaki model[34,35]:

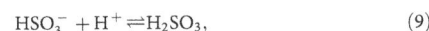

$$SO_3^{2-} + H^+ \rightleftharpoons HSO_3^-, \tag{8}$$

$$HSO_3^- + H^+ \rightleftharpoons H_2SO_3, \tag{9}$$

$$BrO_3^- + 3HSO_3^- \rightarrow 3SO_4^{2-} + Br^- + 3H^+, \tag{10}$$

$$BrO_3^- + 3H_2SO_3 \rightarrow 3SO_4^{2-} + Br^- + 6H^+, \tag{11}$$

$$BrO_3^- + 6Fe(CN)_6^{4-} + 6H^+ \rightarrow 6Fe(CN)_6^{3-} + Br^- + 3H_2O. \tag{12}$$

For Figs 3,4f,h and 5, aqueous phase 1 consisted of 150-mM KBrO$_3$ (Wako Pure Chemical) and 0.2-mM Fl-Na, whereas aqueous phase 2 was 30-mM K$_4$Fe(CN)$_6$ (Wako Pure Chemical), 15-mM H$_2$SO$_4$ (Nacalai Tesque) and Na$_2$SO$_3$ (Wako Pure Chemical). Thus, the final concentrations in the transporters were $c_{BrO_3^-} = 75$ mM, $c_{Fe(CN)_6^{4-}} = 15$ mM, $c_{H_2SO_4} = 7.5$ mM and 0.1-mM Fl-Na. $c_{SO_3^{2-}}$ was varied depending on the experiment being performed. The initial chemical concentrations in the reactor were the same as those in the transporters. All the experiments were carried out at $\sim 35\,°C$ using a thermo-plate (MATS-U52RA26, Tokai Hit) on the microscope.

We used the Rabai–Kaminaga–Hanazaki model for the numerical analyses in Fig. 3e,g. The numerical analyses were performed using Mathematica (Wolfram Research). The details are given in Supplementary Note 2.

Time-variable control over chemical fluxes by pulse-density modulation. To generate time-variable chemical fluxes, we used $T_j^{set} = w/q(t_j)$ and $q(t) = \bar{q}[1 + Z_q(t)]$, where $t_0 = 0$; $t_j = \sum_{l=0}^{j-1} T_l^{set}$; and $\bar{q}$ is the baseline value of $q(t)$. $Z_q(t)$ was varied according to designated functions as follows: $Z_C(t) = 0$ for constant functions; $Z_S(t) = A_q \sin(2\pi t/T_q)$ for sinusoidal wave functions; $Z_N(t) = A_q U(-1,1)$ for white noise functions, where $U(-1,1)$ indicates uniform random numbers between $[-1, 1]$; and $Z_{St}(t) = A_q[2R(t, T_q)/T_q - 1]$ for saw-tooth wave functions, where $R(t, T_q)$ gives the residue obtained when t is divided by T_q. Details are given in Supplementary Note 4.

Analysis of the noise-induced pulsed excitation. We produced the Fourier power spectra in Fig. 4i by five-point smoothing of the fast Fourier transformation results in Fig. 4h. From Fig. 4i, the degree of coherence[37,38] was calculated by $d_c \equiv h(\Delta f/f_p)^{-1}$, where f_p is the peak frequency, and h and Δf are the peak height and the peak width at half height, respectively.

References

1. Elowitz, M. & Lim, W. A. Build life to understand it. *Nature* **468**, 889–890 (2010).
2. Hagiya, M., Konagaya, A., Kobayashi, S., Saito, H. & Murata, S. Molecular robots with sensors and intelligence. *Acc. Chem. Res.* **47**, 1681–1690 (2014).
3. Noorduin, W. L., Grinthal, A., Mahadevan, L. & Aizenberg, J. Rationally designed complex, hierarchical microarchitectures. *Science* **340**, 832–837 (2013).
4. Oberholzer, T., Albrizio, M. & Luisi, P. L. Polymerase chain reaction in liposomes. *Chem. Biol.* **2**, 677–682 (1995).
5. Tsumoto, K., Nomura, S.-I. M., Nakatani, Y. & Yoshikawa, K. Giant liposome as a biochemical reactor: transcription of DNA and transportation by laser tweezers. *Langmuir* **17**, 7225–7228 (2001).
6. Yu, W. *et al.* Synthesis of functional protein in liposome. *J. Biosci. Bioeng.* **92**, 590–593 (2001).
7. Pautot, S., Frisken, B. J. & Weitz, D. A. Production of unilamellar vesicles using an inverted emulsion. *Langmuir* **19**, 2870–2879 (2003).
8. Nomura, S.-I. M. *et al.* Gene expression within cell-sized lipid vesicles. *Chembiochem* **4**, 1172–1175 (2003).
9. Noireaux, V. & Libchaber, A. A vesicle bioreactor as a step toward an artificial cell assembly. *Proc. Natl Acad. Sci. USA* **101**, 17669–17674 (2004).
10. Pietrini, A. V. & Luisi, P. L. Cell-free protein synthesis through solubilisate exchange in water/oil emulsion compartments. *Chembiochem* **5**, 1055–1062 (2004).
11. Chen, I. A., Salehi-Ashtiani, K. & Szostak, J. W. RNA catalysis in model protocell vesicles. *J. Am. Chem. Soc.* **127**, 13213–13219 (2005).
12. Kaneda, M. *et al.* Direct formation of proteo-liposomes by in vitro synthesis and cellular cytosolic delivery with connexin-expressing liposomes. *Biomaterials* **30**, 3971–3977 (2009).
13. Toiya, M., González-Ochoa, H. O., Vanag, V. K., Fraden, S. & Epstein, I. R. Synchronization of chemical micro-oscillators. *J. Phys. Chem. Lett.* **1**, 1241–1246 (2010).
14. Martino, C. *et al.* Protein expression, aggregation, and triggered release from polymersomes as artificial cell-like structures. *Angew. Chem. Int. Ed.* **51**, 6416–6420 (2012).
15. Nourian, Z., Roelofsen, W. & Danelon, C. Triggered gene expression in fed-vesicle microreactors with a multifunctional membrane. *Angew. Chem. Int. Ed.* **51**, 3114–3118 (2012).
16. Stano, P., D'Aguanno, E., Bolz, J., Fahr, A. & Luisi, P. L. A remarkable self-organization process as the origin of primitive functional cells. *Angew. Chem. Int. Ed.* **52**, 13397–13400 (2013).
17. Ichihashi, N. *et al.* Darwinian evolution in a translation-coupled RNA replication system within a cell-like compartment. *Nat. Commun.* **4**, 2494 (2013).
18. Hasatani, K. *et al.* High-throughput and long-term observation of compartmentalized biochemical oscillators. *Chem. Commun.* **49**, 8090–8092 (2013).
19. Weitz, M. *et al.* Diversity in the dynamical behaviour of a compartmentalized programmable biochemical oscillator. *Nat. Chem.* **6**, 295–302 (2014).
20. Galas, J.-C., Haghiri-Gosnet, A.-M. & Estévez-Torres, A. A nanoliter-scale open chemical reactor. *Lab Chip* **13**, 415–423 (2013).
21. Niederholtmeyer, H., Stepanova, V. & Maerkl, S. J. Implementation of cell-free biological networks at steady state. *Proc. Natl Acad. Sci. USA* **110**, 15985–15990 (2013).
22. Karzbrun, E., Tayar, A. M., Noireaux, V. & Bar-Ziv, R. H. Programmable on-chip DNA compartments as artificial cells. *Science* **345**, 829–832 (2014).
23. Villar, G., Heron, A. J. & Bayley, H. Formation of droplet networks that function in aqueous environments. *Nat. Nanotechnol.* **6**, 803–808 (2011).
24. Villar, G., Graham, A. D. & Bayley, H. A tissue-like printed material. *Science* **340**, 48–52 (2013).
25. Elani, Y., Law, R. V. & Ces, O. Vesicle-based artificial cells as chemical microreactors with spatially segregated reaction pathways. *Nat. Commun.* **5**, 5305 (2014).
26. Nicolis, G. & Prigogine, I. *Self-Organization in Nonequilibrium Systems: From Dissipative Structures to Order through Fluctuations* (Wiley, 1977).
27. Alberts, B. *et al. Molecular Biology of the Cell* 5th edn (Garland Science, 2007).
28. Strogatz, S. H. *Nonlinear Dynamics And Chaos: With Applications To Physics, Biology, Chemistry, And Engineering* (Westview, 2001).
29. Song, H., Tice, J. D. & Ismagilov, R. F. A microfluidic system for controlling reaction networks in time. *Angew. Chem. Int. Ed.* **42**, 768–772 (2003).
30. Takinoue, M., Onoe, H. & Takeuchi, S. Fusion and fission control of picoliter-sized microdroplets for changing the solution concentration of microreactors. *Small* **6**, 2374–2377 (2010).
31. Herminghaus, S. Dynamical instability of thin liquid films between conducting media. *Phys. Rev. Lett.* **83**, 2359–2361 (1999).
32. Priest, C., Herminghaus, S. & Seemann, R. Controlled electrocoalescence in microfluidics: targeting a single lamella. *Appl. Phys. Lett.* **89**, 134101 (2006).
33. Edblom, E., Luo, Y., Orbán, M., Kustin, K. & Epstein, I. R. Kinetics and mechanism of the oscillatory bromate-sulfite-ferrocyanide reaction. *J. Phys. Chem.* **93**, 2722–2727 (1989).
34. Rábai, G., Kaminaga, A. & Hanazaki, I. Mechanism of the oscillatory bromate oxidation of sulfite and ferrocyanide in a CSTR. *J. Phys. Chem.* **100**, 16441–16442 (1996).
35. Sato, N., Hasegawa, H. H., Kimura, R., Mori, Y. & Okazaki, N. Analysis of the bromate-sulfite-ferrocyanide pH oscillator using the particle filter: toward the automated modeling of complex chemical systems. *J. Phys. Chem. A* **114**, 10090–10096 (2010).
36. Epstein, I. R. The consequences of imperfect mixing in autocatalytic chemical and biological systems. *Nature* **374**, 321–327 (1995).
37. Gang, H., Ditzinger, T., Ning, C. Z. & Haken, H. Stochastic resonance without external periodic force. *Phys. Rev. Lett.* **71**, 807–810 (1993).
38. Miyakawa, K. & Isikawa, H. Experimental observation of coherence resonance in an excitable chemical reaction system. *Phys. Rev. E* **66**, 046204 (2002).
39. Dolník, M., Schreiber, I. & Marek, M. Experimental observations of periodic and chaotic regimes in a forced chemical oscillator. *Phys. Lett. A* **100**, 316–319 (1984).
40. Abate, A. R., Hung, T., Mary, P., Agresti, J. J. & Weitz, D. A. High-throughput injection with microfluidics using picoinjectors. *Proc. Natl Acad. Sci. USA* **107**, 19163–19166 (2010).
41. Nightingale, A. M., Phillips, T. W., Bannock, J. H. & de Mello, J. C. Controlled multistep synthesis in a three-phase droplet reactor. *Nat. Commun.* **5**, 3777 (2014).
42. Cho, S. K., Moon, H. & Kim, C.-J. Creating, transporting, cutting, and merging liquid droplets by electrowetting-based actuation for digital microfluidic circuits. *J. Microelectromech. Syst* **12**, 70–80 (2003).
43. Ahn, K. *et al.* Dielectrophoretic manipulation of drops for high-speed microfluidic sorting devices. *Appl. Phys. Lett.* **88**, 024104 (2006).
44. Link, D. R. *et al.* Electric control of droplets in microfluidic devices. *Angew. Chem. Int. Ed.* **45**, 2556–2560 (2006).
45. Baroud, C. N., de Saint Vincent, M. R. & Delville, J.-P. An optical toolbox for total control of droplet microfluidics. *Lab Chip* **7**, 1029–1033 (2007).
46. Elowitz, M. B. & Leibler, S. A synthetic oscillatory network of transcriptional regulators. *Nature* **403**, 335–338 (2000).
47. Montagne, K., Plasson, R., Sakai, Y., Fujii, T. & Rondelez, Y. Programming an in vitro DNA oscillator using a molecular networking strategy. *Mol. Syst. Biol.* **7**, 466 (2011).
48. Kim, J. & Winfree, E. Synthetic in vitro transcriptional oscillators. *Mol. Syst. Biol.* **7**, 465 (2011).
49. Takinoue, M., Kiga, D., Shohda, K.-I. & Suyama, A. RNA oscillator: limit cycle oscillations based on artificial biomolecular reactions. *New Generat. Comput.* **27**, 107–127 (2009).
50. Takinoue, M., Kiga, D., Shohda, K.-I. & Suyama, A. Experiments and simulation models of a basic computation element of an autonomous molecular computing system. *Phys. Rev. E* **78**, 041921 (2008).
51. Fung, E. *et al.* A synthetic gene-metabolic oscillator. *Nature* **435**, 118–122 (2005).
52. Danø, S., Sørensen, P. G. & Hynne, F. Sustained oscillations in living cells. *Nature* **402**, 320–322 (1999).
53. Danino, T., Mondragón-Palomino, O., Tsimring, L. & Hasty, J. A synchronized quorum of genetic clocks. *Nature* **463**, 326–330 (2010).
54. Semenov, S. N. *et al.* Rational design of functional and tunable oscillating enzymatic networks. *Nat. Chem.* **7**, 160–165 (2015).

55. Kitano, H. Innovative changes induced by system control theory (Japanese). *Exp. Med.* **33**, 100–106 (2015).
56. Szostak, J. W., Bartel, D. P. & Luisi, P. L. Synthesizing life. *Nature* **409**, 387–390 (2001).
57. Pohorille, A. & Deamer, D. Artificial cells: prospects for biotechnology. *Trends Biotechnol.* **20**, 123–128 (2002).
58. Noireaux, V., Bar-Ziv, R., Godefroy, J., Salman, H. & Libchaber, A. Toward an artificial cell based on gene expression in vesicles. *Phys. Biol.* **2**, P1–P8 (2005).
59. Luisi, P., Ferri, F. & Stano, P. Approaches to semi-synthetic minimal cells: a review. *Naturwissenschaften* **93**, 1–13 (2006).
60. Noireaux, V., Maeda, Y. T. & Libchaber, A. Development of an artificial cell, from self-organization to computation and self-reproduction. *Proc. Natl Acad. Sci. USA* **108**, 3473–3480 (2011).
61. Takinoue, M. & Takeuchi, S. Droplet microfluidics for the study of artificial cells. *Anal. Bioanal. Chem.* **400**, 1705–1716 (2011).

Acknowledgements

We thank Prof H. Onoe (Keio Univ.), Prof S. Takeuchi (Univ. of Tokyo), Prof K. Yoshikawa (Doshisha Univ.), Prof H. Noji (Univ. of Tokyo), Prof Y. Rondelez (Univ. of Tokyo), Dr André Estévez-Torres (CNRS) and Dr Masamune Morita (Tokyo Tech.) for fruitful discussions. This research was supported by PRESTO (Design and Control of Cellular Functions) of JST, and partly supported by a Grant-in-Aid for Challenging Exploratory Research (No. 26540150) and Scientific Research (B) (No. 26280097) from JSPS.

Author contributions

M.T. designed the project; H.S., M.T., M.I. and T.O. performed the experiments; M.T. and H.K. performed the theoretical analyses; M.T. wrote the manuscript.

Additional information

4

Native characterization of nucleic acid motif thermodynamics via non-covalent catalysis

Chunyan Wang[1], Jin H. Bae[1] & David Yu Zhang[1,2]

DNA hybridization thermodynamics is critical for accurate design of oligonucleotides for biotechnology and nanotechnology applications, but parameters currently in use are inaccurately extrapolated based on limited quantitative understanding of thermal behaviours. Here, we present a method to measure the $\Delta G°$ of DNA motifs at temperatures and buffer conditions of interest, with significantly better accuracy (6- to 14-fold lower s.e.) than prior methods. The equilibrium constant of a reaction with thermodynamics closely approximating that of a desired motif is numerically calculated from directly observed reactant and product equilibrium concentrations; a DNA catalyst is designed to accelerate equilibration. We measured the $\Delta G°$ of terminal fluorophores, single-nucleotide dangles and multinucleotide dangles, in temperatures ranging from 10 to 45 °C.

[1] Department of Bioengineering, Rice University, Houston, Texas 77030, USA. [2] Systems, Synthetic, and Physical Biology, Rice University, Houston, Texas 77030, USA. Correspondence and requests for materials should be addressed to D.Y.Z. (email: dyz1@rice.edu).

Nucleic acid biotechnology uses engineered oligonucleotides as therapeutic agents (for example, micro RNAs)[1-6] and diagnostics reagents (for example, PCR primers and next generation sequencing capture probes)[7-12]. Nucleic acid nanotechnology uses engineered oligonucleotides as building blocks for constructing precisely patterned structures[13-17] and complex spatio-temporal circuits[18-22]. In both fields, accurate rational design of oligonucleotides with intended thermodynamics properties is critical to achieving desired system function.

Errors in predicted $\Delta G°$ or melting temperature values result in undesirable non-specific interactions or unpredictable aggregation/assembly pathways. Currently, the best DNA design and folding software exhibit errors in predicted melting temperature of $\sim 1.5°C$ (ref. 23), corresponding to errors of roughly $3\,kcal\,mol^{-1}$. This indicates that there is significant inaccuracy even in common motifs such as DNA base stacks and dangles, an observation confirmed by chemical probing of nucleic acid folding structures[24].

Current thermodynamic parameters are extrapolated from experiments in very different temperatures and buffer conditions than typically used (for example, melt curves in $1\,M\,Na^+$; (refs 25,26). In addition, thermodynamics parameters for individual motifs need to be inferred through linear algebra decomposition or principal component analysis, because each melt curve provides the aggregate $\Delta H°$ and $\Delta S°$ of many motifs. Large numbers of experiments on different sequences statistically mitigate the latter source of unbiased error, but the former results in systematic errors that cannot be reduced via melt curve experiments.

Here we present a generalized method for measuring $\Delta G°$ values of individual DNA motifs in native conditions of interest. A reaction is designed with $\Delta G°$ equivalent to the $\Delta G°$ of the motif of interest, and the equilibrium concentrations of the reactant and product species are measured to numerically calculate the equilibrium constant K_{eq}. The innovation that enables this approach is a DNA catalyst system that accelerates forward and reverse rate constants by orders of magnitude, allowing rapid equilibration[27,28].

We believe our native catalysis method can be used to provide an updated database of nucleic acid thermodynamic parameters for more accurate prediction and design software. We have started this process by providing $\Delta G°$ values for (1) 80 terminal fluorophore parameters, (2) 160 single-nucleotide dangle parameters and (3) 140 multinucleotide dangle parameters. Through the course of our studies, we discovered that multinucleotide dangles destabilize nearby DNA duplexes, asymptoting at $+0.56\,kcal\,mol^{-1}$ above a single-nucleotide dangle for dangles 8 nt or longer.

Results

Nucleic acid motif thermodynamics.
The standard model of nucleic acid hybridization thermodynamics[25,26] is a local model, in which the structure of a nucleic acid molecule or complex is dissected into a number of non-overlapping motifs, such as base stacks, bulges and dangles. The standard free energy of formation of a molecule is calculated as the sum of its component motifs (Fig. 1a). In the past 30 years, the thermodynamics of many DNA and RNA motifs have been characterized[29-32] to inform bioinformatics software for nucleic acid structure prediction[33-36] and probe[37-39] or primer[40,41] or nanostructure[15,42] design. However, inaccuracies in the $\Delta G°$ of DNA motifs are compounded in the overall $\Delta G°$ of hybridization, and lead to errors in predicted hybridization affinities.

Traditionally, motif thermodynamics (standard free energy of formation $\Delta G°$) are characterized through melt curve analysis

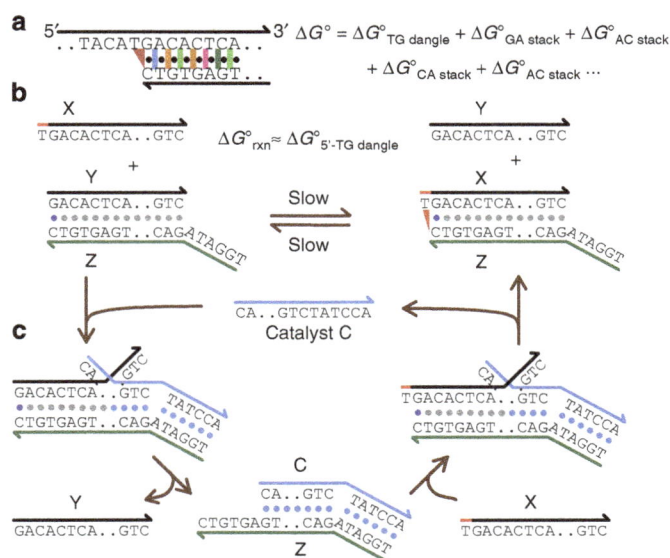

Figure 1 | DNA hybridization thermodynamics. (a) The standard model of nucleic acid hybridization decomposes any nucleic acid molecule into a number of non-overlapping motifs (for example, dangles and stacks), and predicts the standard free energy of formation of the molecule to be the sum of the $\Delta G°$ of the individual motifs. Traditionally, motif $\Delta G°$ values are characterized using melt curves, and then extrapolated to other temperatures and buffer conditions. (b) In principle, motif $\Delta G°$ can be measured in native isothermal conditions by observing the equilibrium concentrations of an appropriately designed $X + YZ \rightleftharpoons Y + XZ$ strand displacement reaction. For example, the shown reaction characterizes the $\Delta G°$ of the 5'-T dangle next to a closing G–C base pair; the partially double-stranded product differs from the partially double-stranded reactant by only the motif of interest. By designing the single-stranded reactants to possess negligible secondary structure, the reaction standard free energy $\Delta G°_{rxn}$ will closely approximate the dangle standard free energy $\Delta G°_{dangle}$. (c) A non-covalent strand displacement catalyst C can be designed to accelerate reaction equilibration to complete within minutes. The reverse reaction pathway is analogous to the forward reaction pathway shown here, and is also accelerated by the catalyst.

experiments[43,44]. These methods suffer from two major sources of inaccuracy: (1) yield and $\Delta G°$ measurement are inaccurate except near the melting temperature, and (2) linear algebra decomposition of motif thermodynamics reduces accuracy. Although recently melt curve analysis accuracy has been improved through the development of high-resolution melt (HRM; ref. 45), these limitations still exist. See Supplementary Note 1 for further discussion on melt curve inference of $\Delta G°$. Other methods, such as isothermal titration calorimetry[46], can be potentially used to provide higher resolution thermodynamic data, but are rarely used in practice because of the high nucleic acid concentration requirements necessitated by the small absolute values of $\Delta H°$.

Reaction construction.
Our method starts with the construction of a reaction whose reaction standard free energy $\Delta G°_{rxn}$ closely approximates the motif standard free energy $\Delta G°_{motif}$ (Fig. 1b). This is achieved by designing a duplex product bearing the motif of interest (XZ) and a duplex reactant differing from the duplex product by only the motif (YZ). To balance the reaction, a single-stranded reactant oligonucleotide (X), which is part of the duplex product but not the duplex reactant, is added to the left side of the reaction. Similarly, a single-stranded product oligonucleotide (Y) is added to the right side of the reaction.

The net reaction is $X + YZ \rightleftharpoons Y + XZ$, and

$$\Delta G_{rxn}^{\circ} = \Delta G_{XZ}^{\circ} + \Delta G_{Y}^{\circ} - \Delta G_{X}^{\circ} - \Delta G_{YZ}^{\circ} \quad (1)$$

$$= (\Delta G_{XZ}^{\circ} - \Delta G_{YZ}^{\circ}) + (\Delta G_{Y}^{\circ} - \Delta G_{X}^{\circ}) \quad (2)$$

The standard free energy of the motif (in Fig. 1b, a 5'-TG dangle) can be expressed as $\Delta G_{motif}^{\circ} = \Delta G_{XZ}^{\circ} - \Delta G_{YZ}^{\circ}$ and is equal to ΔG_{rxn}° when $\Delta G_{Y}^{\circ} = \Delta G_{X}^{\circ}$. One way to ensure the latter is approximately true is to design the sequences of X and Y to avoid any predicted secondary structure.

The value of ΔG_{rxn}° can be numerically computed from the observed equilibrium concentrations of the species:

$$\Delta G_{rxn}^{\circ} = -R\tau \ln(K_{eq}) \quad (3)$$

$$= -R\tau \ln \left(\frac{[Y][XZ]}{[X][YZ]} \right) \quad (4)$$

Here R is the universal gas constant (or Boltzmann constant), and τ denotes the temperature in Kelvin. Given the initial concentrations $[X]_0$, $[Y]_0$, $[XZ]_0$, and $[YZ]_0$, all four equilibrium concentrations can be computed from any one equilibrium concentration based on simple algebra. Thus, we can accurately determine ΔG_{motif}° using an accurate measure of the equilibrium concentration of any one of the four species.

Non-covalent catalysis. The above reaction, by itself, equilibrates slowly with a second order rate constant of roughly $1\,M^{-1}s^{-1}$ (refs 28,47,48), corresponding to a half-life of roughly 100 days at 100 nM concentrations. It is not practical to wait so long for the reaction to naturally reach equilibrium because nucleotide oxidation (for example, adenine to inosine) occurs on a similar time scale[49,50], and will bias the inferred thermodynamic results. To accelerate equilibration kinetics, we design a non-covalent strand displacement catalyst C, first introduced in ref. 27.

Figure 1c shows the catalysis reaction mechanism. C first binds to the single-stranded portion of Z in YZ, known as a toehold, and then displaces Y one nucleotide at a time through an unbiased random walk process known as branch migration. Because the elementary single-base replacement steps in branch migration are very fast ($10–100\,\mu s$ (ref. 51)), the entire displacement process occurs with a first order rate constant of about $1\,s^{-1}$ (ref. 28). Thus, the reaction is typically rate limited by the second order toehold binding process, and is similar to hybridization kinetics for sufficiently long toeholds[28]. When all nucleotides in C are base paired to their complementary nucleotides in Z, Y can spontaneously dissociate, leaving intermediate CZ. Subsequently, CZ can react with either X or Y in a similar manner as YZ initially reacted with C, releasing C to catalyse other reactions (Fig. 1c). Thus, catalyst C enables a rapid pathway for both the forward and the reverse reactions of $X + YZ \rightleftharpoons Y + XZ$ without changing the reaction's equilibrium constant (Supplementary Note 2).

Fluorophore characterization. In this manuscript, we rely exclusively on fluorescent non-denaturing polyacrylamide gel electrophoresis (PAGE) to quantitate the equilibrium concentrations. Fluorescent PAGE shows very low background and high linearity (Supplementary Figs 5–8, Supplementary Note 4), allowing us to map band intensity to species concentration via a simple scaling constant. It also allows us to verify that the catalytic mechanism is functioning as intended, and that no long-lived side products or intermediates are being formed (such as CXZ or CYZ).

We first study the effects of terminal fluorophores on stabilizing DNA duplexes, both because these interactions have been relatively poorly characterized[52,53] and because these are

necessary prerequisites for our studies of other motifs. We chose to focus our fluorophore studies on the carboxy-X-rhodamine (ROX) and Alexa532 fluorophores, because our past experience indicates that these are temperature robust and their per unit fluorescence does not change significantly over a period of months when properly stored.

Figure 2a shows the net reaction used to characterize the thermodynamic effects of a 5' ROX next to an adenine (A). We refer to the ROX-labelled species as the X species; species containing X (X and XZ) will appear as visible bands on a fluorescent PAGE, while all other bands will be invisible. Figure 2b shows the results from one fluorescent PAGE experiment characterizing 5' ROX-A at 25 °C in 12.5 mM Mg^{2+}. Gel band intensities were quantitated by software, and relative fluorescence units are listed below the gel; differences in pipetting volume and fluorophore quantum yield in ssDNA and dsDNA states resulted in different total fluorescences for different lanes; see Supplementary Fig. 14 and Supplementary Note 5 for discussion on quantitation and ΔG° calculation methodology.

Lanes 1 and 2 are controls showing the positions and intensities of the fluorescent XZ and X bands; species identities were verified with SYBR gold stained gels with a DNA ladder lane (Supplementary Fig. 19). In lane 3, X and Z are pre-mixed for 30 min and then Y is introduced and allowed to react with XZ for 3 h before PAGE. In lane 4, Y and Z are pre-mixed and X is subsequently introduced. The band distribution in lane 3 closely resembles that of lane 2, and the band distribution in lane 4 closely resembles lane 1, despite the fact that lanes 3 and 4 have the same composition of strands; this indicates that equilibrium is far from reached after 3 h of reaction. Lanes 5 and 6 are similar to lanes 3 and 4, except 0.1x of the catalyst C is introduced into each reaction after all other species are introduced, but before the 3 h incubation. The band distribution in lanes 5 and 6 are quantitatively similar to each other and intermediate between lanes 3 and 4, indicating that equilibrium has been achieved.

Lack of a higher molecular weight CXZ band indicates that CXZ and CYZ are short-lived and low in concentration at equilibrium. Under the assumption that CZ is likewise short-lived and insignificant in concentration (Supplementary Fig. 16), the equilibrium concentrations [Y] and [YZ] can be calculated given the initial concentrations $[X]_0$, $[Y]_0$, $[XZ]_0$ and $[YZ]_0$ from either [X] or [XZ]:

$$[Y] = [Y]_0 + ([X]_0 - [X]) \quad (5)$$

$$[YZ] = [YZ]_0 - ([X]_0 - [X]) \quad (6)$$

$$[Y] = [Y]_0 + ([XZ] - [XZ]_0) \quad (7)$$

$$[YZ] = [YZ]_0 - ([XZ] - [XZ]_0) \quad (8)$$

In a perfectly run reaction, the calculated equilibrium concentration [Y] should be the same regardless of whether it is calculated from [X] or [XZ], but in practice slight pipetting errors in the total amount of sample loaded per gel lane will lead to different values. A similar argument applies to [YZ] inference. We correct for these in our mathematical analysis and calculation of K_{eq} and ΔG° from gel band intensities (Supplementary Figs 16 and 21).

From the intensities of the gel bands in lanes 1, 2, 5 and 6, we calculate ΔG_{ROX-A}° to be $-0.07\,kcal\,mol^{-1}$ from lane 5 and $-0.31\,kcal\,mol^{-1}$ from lane 6. Five additional fluorescent PAGE experiments were run under the same conditions using separately prepared samples, to mitigate the effects of pipetting error in any one experiment, and the inferred ΔG° values are shown in Fig. 2c. These 12 independent ΔG° values are then combined into a single mean $\mu = \Delta G_{ROX-A}^{\circ} = -0.293\,kcal\,mol^{-1}$, corresponding to our best estimate of the true value of the parameter from our

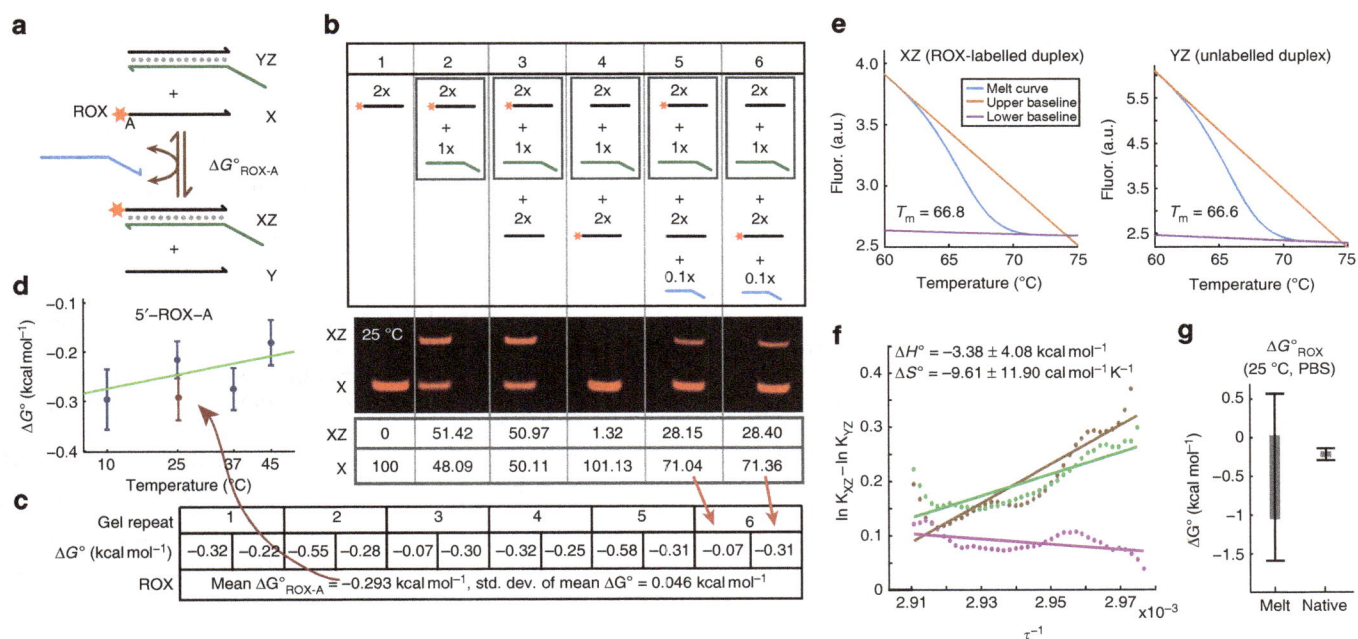

Figure 2 | Experimental results. (**a**) Reaction constructed to measure the ΔG° of a 5′ ROX fluorophore next to an adjacent DNA duplex with a terminal A nucleotide ($\Delta G^\circ_{\mathrm{ROX-A}}$). (**b**) Fluorescent PAGE demonstrating non-covalent catalysis. Relevant species were reacted at 25 °C in 12.5 mM Mg^{2+} buffer; 1x denotes ~300 nM. Lanes 3 and 4 show significant difference in band intensity distribution, indicating that equilibrium is not achieved after 3 h of reaction in the absence of the catalyst. In contrast, lanes 5 and 6 show that a sub-stoichiometric (0.1x) amount of catalyst drives the reaction to equilibrium. Listed numbers below the gel are quantitated gel band intensities (arbitrary fluorescence units). (**c**) Summary of five additional experimental repeats. From the 12 inferred ΔG° values, we obtain mean $\Delta G^\circ_{\mathrm{ROX-A}} = -0.293$ kcal mol^{-1}. (**d**) Experiments to measure $\Delta G^\circ_{\mathrm{ROX-A}}$ at 10, 25, 37 and 45 °C in PBS (blue), as well as in 12.5 mM Mg^{2+} (brown). The mean and mean s.d. of the ΔG° values are plotted against temperature; error bars show 1 s.d. The gels were always run at the same temperature as the hybridization reaction. The green line shows the best linear fit against the four sets of experiments in PBS, corresponding to $\Delta H^\circ = -0.81$ kcal mol^{-1} and $\Delta S^\circ = -1.89$ cal mol^{-1} K^{-1}. (**e**) Melt curve of blunt duplex and 5′ ROX-labelled duplex in PBS buffer, based on fluorescence from the Syto82 intercalating dye (Supplementary Figs 70–72 and Supplementary Note 7). Upper and lower baselines are fitted to infer hybridization yields and equilibrium constants at different temperatures. (**f**) Van't Hoff plot of $\ln(K_{eq})$ against τ^{-1} (Kelvin). (**g**) Melt curve analysis produced a ΔG° estimate with 14-fold higher s.d. than inferred using our native catalysis approach. Box shows 1 s.d. and error bars show 2 s.d.'s.

experiments. The s.d. of the mean is calculated assuming Gaussian distributed error in the individual ΔG° values, and

$$\sigma_\mu = \frac{1}{\sqrt{12}} \cdot \sqrt{\frac{\sum_i (\Delta G_i^\circ - \mu)^2}{12-1}} = 0.046 \, \mathrm{kcal \cdot mol^{-1}}.$$

In the course of calculating ΔG° values from gel band intensities, the initial concentrations $[X]_0$, $[Y]_0$, $[XZ]_0$ and $[YZ]_0$ are extremely important, and we found that conventional methods based on absorbance and extinction coefficient at 260 nm were insufficiently accurate. Instead, we use stoichiometry PAGE gels to determine the relative concentrations of X or Y to Z (Supplementary Fig. 9 and Supplementary Note 3). Our method systematically overestimates ΔG° values by ignoring the presence of the CZ, CXZ and CYZ intermediate species. However, because the amount of catalyst C is only 0.1x, the overestimation is bounded by 0.2 kcal mol^{-1} in an absolute worst case.

Temperature dependence. In addition to 25 °C in 12.5 mM Mg^{2+}, we also performed similar experiments (5–8 gels each) for the ROX-A fluorophore at temperatures of 10, 25, 37 and 45 °C in PBS (Sigma P5493, 0.01 M phosphate; 0.154 M NaCl; pH 7.4). The gels were always run at the same temperature as the hybridization reaction. Although the gel running buffer (TAE with no added salt) differs from the hybridization buffer, our control experiments (Supplementary Fig. 18) shows no significant difference in inferred ΔG° due to the transient change in buffer conditions before the start of the electrophoresis. The mean and mean s.d.'s of $\Delta G^\circ_{\mathrm{ROX-A}}$ in each of these conditions are plotted

against temperature in Fig. 2d. With the four separate series of experiments performed in PBS at different temperatures, we can use the relation $\Delta G^\circ = \Delta H^\circ - \tau \Delta S^\circ$ to infer ΔH° and ΔS° of the 5′-ROX-A using a linear fit.

Because each of the four mean ΔG° values at different temperatures has a different level of uncertainty (mean s.d.), it is inappropriate to use a standard least-squares fit. Instead, we obtain the best fit ΔH° and ΔS° values from a maximal likelihood standpoint, minimizing the sum of square z scores. The z score is calculated as:

$$z = \frac{\Delta G^\circ_{\mathrm{predicted}} - \Delta G^\circ_{\mathrm{observed}}}{\sigma_\mu} \tag{9}$$

where $\Delta G^\circ_{\mathrm{predicted}}$ is the value calculated as $\Delta H^\circ_{\mathrm{fit}} - \tau \Delta S^\circ_{\mathrm{fit}}$, $\Delta G^\circ_{\mathrm{observed}}$ is the mean experimental ΔG° at the relevant temperature, and σ_μ is the mean s.d. The green line in Fig. 2d shows the best fit $\Delta H^\circ = -0.81$ kcal mol^{-1} and $\Delta S^\circ = -1.89$ cal mol^{-1} K^{-1}. See Supplementary Fig. 17 for details on ΔH° and ΔS° fitting, including calculated of s.d.'s on best fit ΔH° and ΔS° values.

Our best fit linear regression to our measured ΔG° values often showed poor quality of fit, resulting in large s.d.'s in our inferred values of ΔH° and ΔS°. One possible explanation could be the temperature dependence of motif ΔH° and ΔS° values. Although the standard model of DNA hybridization[25] assumes temperature invariance of ΔH° and ΔS°, there is debate within the literature as to whether this is true for all hybridization motifs[54–58].

Comparison to melt curve analysis. To compare the quality of the thermodynamics parameters produced by our native catalysis against the standard-of-practice melt curve analysis approach, we also performed HRM analysis of the ROX-labelled XZ and unlabelled YZ molecules (Fig. 2e). Hybridization yields of the XZ and YZ molecules are inferred from the melt curves, and used to inform the numerical calculation of the equilibrium constants K_{XZ} and K_{YZ}, of the $X + Z \rightleftharpoons XZ$ and $Y + Z \rightleftharpoons YZ$ reactions, respectively. The difference in the logarithms of K_{XZ} and K_{YZ} corresponds to the equilibrium constant of the ROX-duplex motif, and is plotted against the inverse of the temperature in Kelvin in a van't Hoff plot. The best linear fits to the van't Hoff plot provides $\Delta H°$ and $\Delta S°$ values, from which $\Delta G°$ at a particular temperature can be calculated.

Triplicate runs of the melt curves of XZ and YZ gave relatively low quantitative agreement in the van't Hoff plot, due to the small changes in melting temperature and thermodynamics involved (Fig. 2f). The $\Delta G°$ value of the ROX-duplex interaction at 25 °C in PBS, based on melt curve parameters, is predicted with a mean of -0.51 kcal mol^{-1} with a mean s.d. of 0.54 kcal mol^{-1}. In contrast, our native catalysis method produces $\Delta G° = -0.22$ kcal mol^{-1}, with a mean s.d. of 0.038 kcal mol^{-1}. Although unbiased errors due to repeat experiments can be attenuated via statistically large sampling, the 14-fold higher s.d. of parameters produced by melt curves would necessitate 196-fold more experiments, on average, to generate the same mean s.d. as our native catalysis technique.

Single-nucleotide dangles. Although it is possible to construct a similar catalysed $X + YZ \rightleftharpoons XZ + Y$ reaction to directly measure dangle motif thermodynamics, it would be difficult to accurately quantitate the concentrations of any of the species, because X and Y would have very similar mobility, and likewise with XZ and YZ. Consequently, we chose to indirectly characterize the

dangle motifs by measuring the relative $\Delta G°$ of a dangle compared with a fluorophore (reaction 2 in Fig. 3a), and subtracting that value from the $\Delta G°$ of the fluorophore that we obtained earlier in Fig. 2 and related experiments. The resulting $\Delta G°_{rxn3} = \Delta G°_{rxn1} - \Delta G°_{rxn2}$ corresponds to the net reaction querying only the dangle, and closely approximates $\Delta G°_{dangle}$ that we ultimately wish to characterize. As a further benefit, because both $\Delta G°_{rxn1}$ and $\Delta G°_{rxn2}$ are overestimated by a similar amount due to the ignorance of the CZ, CYZ and CXZ species, $\Delta G°_{rxn3}$ becomes unbiased due to the cancellation of the bias terms. The disadvantage of such a 2-reaction approach is the increased s.d. on the $\Delta G°$ parameters from subtracting two distributions.

The left panel of Fig. 3b shows the summarized mean $\Delta G°_{rxn2}$ values at different temperatures in PBS based on a number of fluorescent PAGE experiments similar to that shown in Fig. 2b and the corresponding mean s.d.'s. The right panel shows the mean $\Delta G°_{rxn3}$ values obtained by subtracting the $\Delta G°_{rxn2}$ values in the left panel from the $\Delta G°_{rxn1}$ values shown in Fig. 2d. The mean s.d.'s of $\Delta G°_{rxn3}$ are calculated assuming that errors in $\Delta G°_{rxn1}$ and $\Delta G°_{rxn2}$ are independent, so that variance is additive:

$$\sigma_{\mu,rxn3} = \sqrt{\sigma_{\mu,rxn1}^2 + \sigma_{\mu,rxn2}^2} \qquad (10)$$

Figure 3c shows the mean and mean s.d. of $\Delta G°_{rxn3}$ for the same 5'-TG dangle assayed using the Alexa532 fluorophore, and Fig. 3d shows the overlay of the two sets of values. The thermodynamics of Alexa532 was likewise characterized for a variety of nearest neighbour bases and temperatures (Supplementary Table 4). The *consensus* $\Delta G°$ value that we attribute to the dangle at each temperature is calculated as the maximum likelihood value based

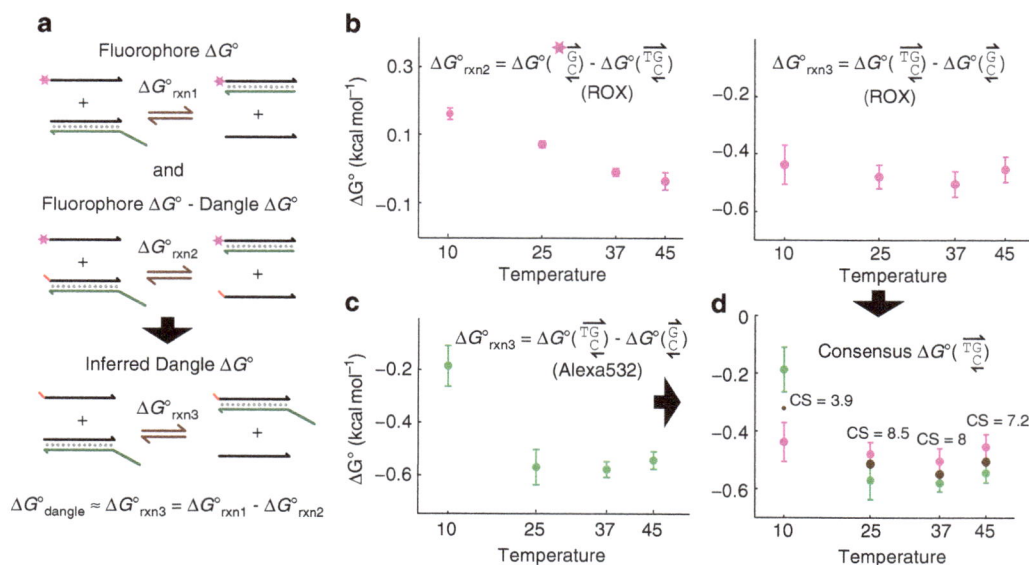

Figure 3 | Dangle $\Delta G°$ characterization. (a) To accurately measure dangle $\Delta G°$ using fluorescent PAGE, we design a second reaction with reaction standard free energy $\Delta G°_{rxn2}$ reflecting the effects of both the dangle of interest and the fluorophore. By numerically subtracting $\Delta G°_{rxn2}$ from $\Delta G°_{rxn1}$, we calculate the value of $\Delta G°_{rxn3}$ corresponding to the dangle. **(b)** Summary of mean and mean s.d. $\Delta G°_{rxn2}$ based on ROX, and the corresponding $\Delta G°_{rxn3}$ after subtracting from the relevant $\Delta G°_{rxn1}$ values from Fig. 2d. **(c)** $\Delta G°_{rxn3}$ values inferred using the Alexa532 fluorophore, rather than the ROX fluorophore; see Supplementary Table 4 for data on Alexa532-based measurements. **(d)** Consensus dangle $\Delta G°$ values are calculated as the maximal likelihood $\Delta G°$ value based on the independent readings from ROX and Alexa532 experiments. The CS represents 10 minus the sum of the squares of z score of the consensus $\Delta G°$ relative to the mean $\Delta G°$ values measured from ROX and Alexa532; a CS of eight indicates that the consensus $\Delta G°$ is within 1 s.d. of each mean $\Delta G°$. See Supplementary Table 6 for summary results of all 32 single-nucleotide dangles.

on minimizing the sum of square z scores. Mathematically,

$$z_{ROX} = \frac{\Delta G^\circ_{consensus} - \Delta G^\circ_{ROX}}{\sigma_{\mu, ROX}} \quad (11)$$

$$z_{A532} = \frac{\Delta G^\circ_{consensus} - \Delta G^\circ_{A532}}{\sigma_{\mu, A532}} \quad (12)$$

$$Error = z^2_{ROX} + z^2_{A532} \quad (13)$$

As a more intuitive metric of the agreement between the ROX and Alexa532, we define consensus score (CS) as:

$$CS = 10 - Error \quad (14)$$

The CS is a measure of how much two independently ordered sets of strands, using two distinct fluorophores, agree with each other in terms of inferred ΔG° parameters, out of a maximum score of 10. CS ≥ 8 indicates that the consensus ΔG° is within 1 s.d. of both the ROX- and the Alexa-based ΔG° values, and CS ≥ 0 indicates that the consensus ΔG° is within 2.24 s.d.'s.

Table 1 shows consensus ΔG° values of all 32 single-nucleotide dangles at 25 and 37 °C in PBS, and at 25 °C in 12.5 mM Mg^{2+}. Also shown is the corresponding s.d.'s and CS for each ΔG° parameter. Additional parameters are available for similar

experiments performed at 10 and 45 °C in PBS and best fit ΔH° and ΔS° values (Supplementary Table 7). Using measured ΔG° values in PBS at four different temperatures, we also were able to calculate best fit values of ΔH° and ΔS° for the dangle motifs explored. As with fluorophore ΔH° and ΔS° parameters, however, we caution the reader against too much reliance on these values, due to the poor quality of linear fits, and the possibility of temperature-dependent ΔH° and ΔS° values[54-58].

Error analysis. In the course of our experiments for the previous two subsections on characterizing fluorophore and single-nucleotide dangle ΔG° values, we performed at least five fluorescent PAGE gels for each fluorophore parameter (corresponding to 10 independent ΔG° measurements) and at least two fluorescent PAGE gels for each fluorophore-dangle parameter. These repeats were done to reduce the mean s.d. on ΔG° values.

Figure 4a plots as a histogram the distribution of the mean s.d.'s of ΔG° values for fluorophore interactions. $N = 80$ corresponds to the number of different ΔG° parameters measured: (2 fluorophores; 5' or 3'; 4 nearest neighbour nucleotides; 5 temperature/buffer conditions). These were collected from 429 separate fluorescent PAGE gels corresponding to 858 individual ΔG°

Table 1 | Consensus ΔG° of single-base dangles.

	ΔG°: 25 °C; PBS (kcal·mol^{-1})	ΔG°: 37 °C; PBS	ΔG°: 25 °C; 12.5 mM Mg^{2+}
AA/T	-0.549 ±0.045 (9.0)	-0.647 ±0.084 (7.1)	-0.595 ±0.080 (8.0)
AT/A	-0.483 ±0.038 (8.5)	-0.368 ±0.071 (4.7)	-0.511 ±0.045 (8.9)
AC/G	-0.835 ±0.042 (8.4)	-0.635 ±0.046 (9.6)	-0.777 ±0.097 (5.0)
AG/C	-0.678 ±0.107 (0.7)	-0.632 ±0.059 (7.0)	-0.604 ±0.112 (2.9)
TA/T	-0.566 ±0.111 (-2.5)	-0.432 ±0.086 (8.2)	-0.504 ±0.072 (9.4)
TT/A	-0.047 ±0.056 (8.5)	-0.155 ±0.056 (9.5)	-0.145 ±0.040 (10.0)
TC/G	-0.427 ±0.076 (4.2)	-0.364 ±0.064 (8.9)	-0.462 ±0.052 (10.0)
TG/C	-0.513 ±0.051 (8.5)	-0.548 ±0.038 (8.0)	-0.405 ±0.067 (1.8)
CA/T	-0.279 ±0.108 (2.6)	-0.224 ±0.123 (2.5)	-0.297 ±0.034 (10.0)
CT/A	-0.042 ±0.050 (9.8)	0.001 ±0.041 (9.7)	-0.105 ±0.033 (9.9)
CC/G	-0.519 ±0.038 (10.0)	-0.322 ±0.094 (5.6)	-0.443 ±0.068 (10.0)
CG/C	-0.603 ±0.097 (6.6)	-0.420 ±0.065 (9.8)	-0.467 ±0.033 (9.4)
GA/T	-0.389 ±0.042 (9.9)	-0.355 ±0.062 (6.4)	-0.468 ±0.058 (9.6)
GT/A	-0.121 ±0.045 (8.9)	-0.154 ±0.052 (9.0)	-0.148 ±0.084 (5.6)
GC/G	-0.576 ±0.049 (8.0)	-0.422 ±0.056 (9.5)	-0.601 ±0.093 (6.7)
GG/C	-0.703 ±0.033 (9.7)	-0.520 ±0.088 (-7.9)	-0.520 ±0.074 (-2.5)
AA/T	-0.144 ±0.045 (9.9)	-0.011 ±0.070 (9.4)	-0.328 ±0.127 (6.2)
TA/A	-0.324 ±0.077 (9.1)	-0.195 ±0.156 (4.4)	-0.267 ±0.113 (7.7)
CA/G	-0.588 ±0.063 (9.8)	-0.659 ±0.091 (7.0)	-0.622 ±0.097 (5.7)
GA/C	-0.131 ±0.058 (9.3)	-0.118 ±0.049 (10.0)	0.036 ±0.039 (9.9)
AT/T	0.139 ±0.135 (5.9)	0.183 ±0.046 (10.0)	0.008 ±0.237 (-7.1)
TT/A	0.091 ±0.072 (9.0)	0.223 ±0.168 (6.0)	0.024 ±0.146 (-2.0)
CT/G	-0.261 ±0.089 (4.1)	0.003 ±0.115 (8.0)	-0.239 ±0.117 (6.5)
GT/C	0.398 ±0.062 (9.2)	0.264 ±0.077 (8.0)	0.541 ±0.047 (9.3)
AC/T	0.310 ±0.101 (7.7)	0.485 ±0.060 (10.0)	0.117 ±0.060 (9.9)
TC/A	0.036 ±0.092 (10.0)	0.107 ±0.167 (3.8)	-0.069 ±0.068 (8.0)
CC/G	-0.155 ±0.073 (9.4)	-0.081 ±0.102 (8.5)	-0.100 ±0.219 (4.7)
GC/C	0.025 ±0.069 (7.8)	0.034 ±0.090 (9.2)	0.161 ±0.041 (9.9)
AG/T	0.093 ±0.036 (10.0)	0.203 ±0.063 (9.6)	0.095 ±0.164 (-3.3)
TG/A	-0.278 ±0.125 (0.5)	-0.066 ±0.121 (8.0)	-0.186 ±0.075 (7.9)
CG/G	-0.625 ±0.046 (9.9)	-0.544 ±0.142 (-3.0)	-0.503 ±0.125 (8.2)
GG/C	0.046 ±0.056 (10.0)	0.067 ±0.077 (8.0)	0.262 ±0.038 (7.9)

CS, consensus scores; ROX, carboxy-X-rhodamine.
Shown in parentheses are CS, a measure of internal agreement between our two independent sets of experiments based on the ROX and the Alexa532 fluorophores.
The last closing base pair of a duplex is shown in blue and the dangle nucleotide is shown in red.

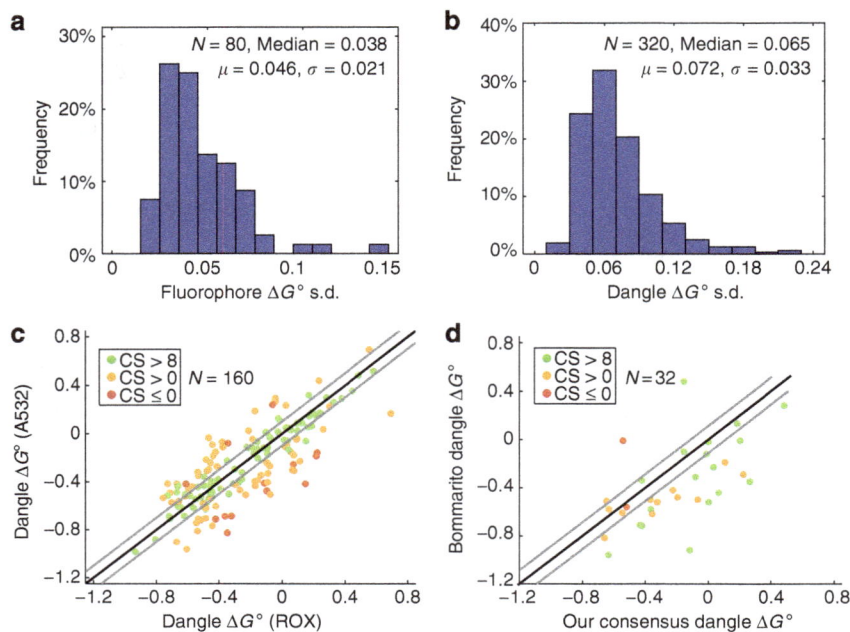

Figure 4 | Error analysis of native characterization methodology. (**a**) Distribution of mean s.d.'s of a fluorophore next to a duplex (Fig. 2). All 80 fluorophore $\Delta G°$ parameters we obtained have mean s.d. below 0.15 kcal mol^{-1}. (**b**) Distribution of mean s.d.'s of single-nucleotide dangles (Fig. 3). All 320 single-base dangle parameters have mean s.d. below 0.25 kcal mol^{-1}. Because dangle $\Delta G°$ values were obtained by subtracting two $\Delta G°$ values each with their own independent s.d.'s, the s.d.'s are larger than those in **a**. (**c**) Comparison of dangle $\Delta G°$ values obtained via ROX and via Alexa532. Black line shows equality and grey lines show ± 0.1 kcal mol^{-1} deviation. Of the 160 parameter pairs, 80 (50.0%) have CS > 8, 66 (41.25%) have CS between 0 and 8, and 14 (8.75%) have CS < 0. (**d**) Comparison of our consensus dangle $\Delta G°$ values with literature values by Bommarito *et al.*[59]

values measured. All 80 dangle parameters' mean s.d.'s were below 0.15 kcal mol^{-1}, with a mean of 0.046 kcal mol^{-1} and a s.d. (on the mean s.d.) of 0.021 kcal mol^{-1}.

Similarly, Fig. 4b plots as a histogram the distribution of the mean s.d.'s of the $\Delta G°$ values for $N = 320$ single-nucleotide dangle parameters: (2 fluorophores; 5' or 3'; 4 dangle nucleotides; 4 nearest neighbour nucleotides; 5 temperature/buffer conditions). These were collected from 723 separate fluorescent PAGE gels corresponding to 1446 individual $\Delta G°$ values measured. The mean and s.d. of the parameter s.d.'s here were 0.072 kcal mol^{-1} and 0.033 kcal mol^{-1}, respectively, and all parameter s.d.'s were below 0.25 kcal mol^{-1}. These parameter s.d. values are larger than those in Fig. 4a both because they include the variance from fluorophores and because of the relatively fewer independent $\Delta G°$ measurements per parameter (average 4.52 per dangle parameter, compared with average 10.73 per fluorophore parameter).

Figure 4c plots the dangle $\Delta G°$ parameters measured using the Alexa532 against the same parameters measured using ROX. The Alexa532- and ROX-based measurements of the dangle motif $\Delta G°$ represent independent experiments and assays, agreement between the final dangle $\Delta G°$ values (high CS) indicates high confidence in our methodology and consensus $\Delta G°$ value. Parameters with CS > 8 generally have no > 0.1 kcal mol^{-1} deviation between the ROX and Alexa532 measurements.

We observed 14 parameters (out of 160) with CS ≤ 0 (Supplementary Table S9-1), whereas a Gaussian distribution of errors would typically result in 4 out of 160 parameters (2.5%) possessing CS ≤ 0. Except one outlier at − 37.1, CS values ranged between − 8.7 and 10.0, with 91% of values being positive and 50% of values being 8 or higher. Parameters with low CS likely imply that one or more assumptions our methodologies required were incorrect for these parameters. For example, it is possible that the equilibrium concentrations of CZ are different for the

corresponding experiments using the two different fluorophores, resulting in bias in one or both $\Delta G°$ values. We believe that the consensus parameters with CS ≥ 8 are likely to be objectively correct, because while any of a number of problems could cause the mean $\Delta G°$ measured from Alexa532 to differ from that from ROX, it is unlikely for these values to be coincidentally similar to within 0.1 kcal mol^{-1} while being wrong.

Figure 4d plots the literature single-nucleotide dangle parameters by Bommarito *et al.*[59] against our consensus single-nucleotide dangle parameters at 37 °C. There is relatively little agreement between our values and Bommarito's values; only 9 out of 32 dangle parameters agreed to within 0.1 kcal mol^{-1}. The mean absolute error between our consensus values and Bommarito's values was 0.269 kcal mol^{-1}, and the root mean square error between them was 0.333 kcal mol^{-1}.

Furthermore, there is little correlation between our parameters' CS and their agreement with Bommarito's values. Our consensus 5' $\begin{pmatrix} GT \\ A \end{pmatrix}$ $\Delta G°$ has CS = 9.0, and is 0.601 kcal mol^{-1} below Bommarito's value, while our consensus 3' $\begin{pmatrix} GA \\ C \end{pmatrix}$ $\Delta G°$ has CS = 10.0, and is 0.802 kcal mol^{-1} above Bommarito's value. Because our method offers a more direct measure of dangle $\Delta G°$ parameters, compared to Bommarito's melt curve methods that had to deconvolute the thermodynamic contributions of many base stacks from that of dangles, we generally believe our parameters to be more accurate.

Another contribution to the discrepancy may be the difference in the 'core' sequences (for example, all nucleotides except the dangle and nearest neighbour, in our case) used for the two papers. The standard nearest neighbour model assumes that the identities of distal bases should have no effect on the motif thermodynamics, but both our own analyses (not shown) and experimental studies on coaxial dangles[60] suggest that this simplifying assumption may not be true.

a

First dangle nt forms stabilizing partial stack

Additional dangle nt adds destabilizing electrostatics (diminishing marginal effects)

d

Length (nt)	Approx. ΔG° (kcal mol^{-1})
1	ΔG°_1
2	ΔG°_1 +0.13
3	ΔG°_1 +0.30
4	ΔG°_1 +0.48
5	ΔG°_1 +0.51
6	ΔG°_1 +0.55
7	ΔG°_1 +0.53
9	ΔG°_1 +0.58
11	ΔG°_1 +0.56
13	ΔG°_1 +0.56
16	ΔG°_1 +0.56
21	ΔG°_1 +0.58

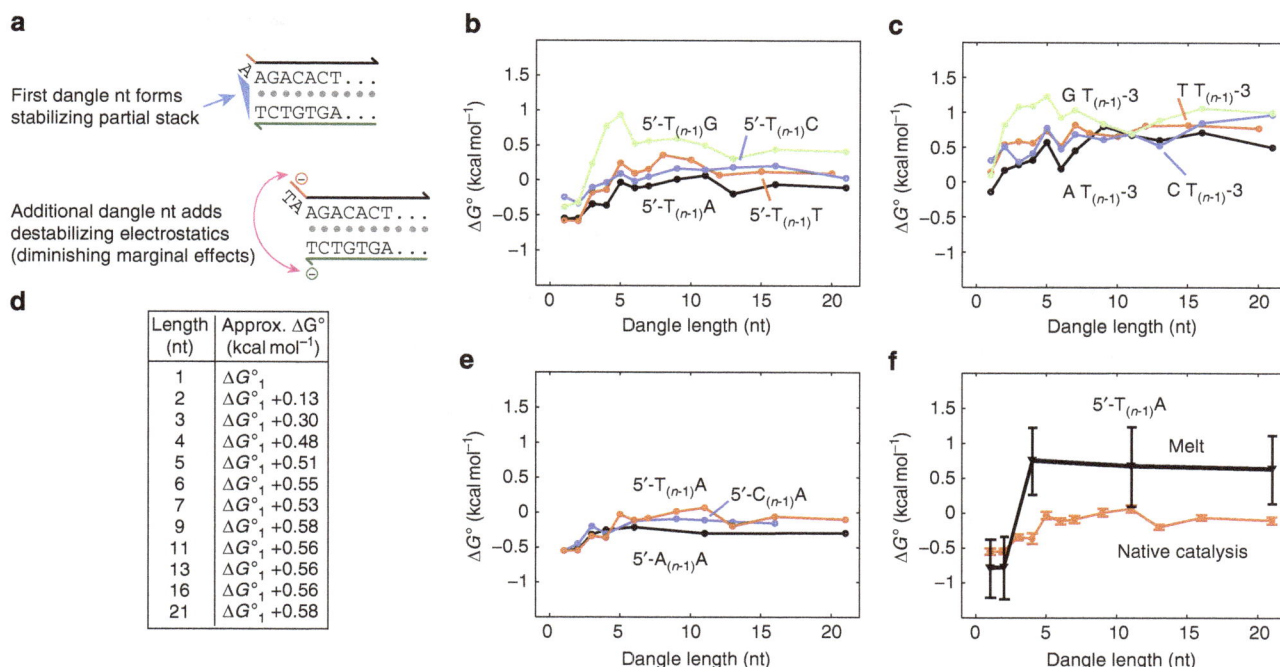

Figure 5 | Thermodynamics of multinucleotide dangles. (**a**) The standard model of DNA hybridization considers only the effects of the first dangle nucleotide, which is known to potentially form a partial stack with the last base pair of the adjoining helix. From a biophysics point of view, additional nucleotides of dangle likely contribute a destabilizing effect due to the electrostatic repulsion from the additional dangle bases and the nucleotides on the complementary strand. (**b**) Consensus ΔG° for 5′ multinucleotide dangles of different lengths at 25 °C in PBS. Each plotted data point represents the consensus value from experiments using ROX and using Alexa532, as in Fig. 3. (**c**) Consensus ΔG° for 3′ multinucleotide dangles of different lengths at 25 °C in PBS. (**d**) Approximate ΔG° of multinucleotide dangles based on length and ΔG° of the single-nucleotide dangle. (**e**) Consensus ΔG° for 5′ multinucleotide dangles of different lengths at 25 °C in PBS. The first dangle base was fixed to be A; the remainder were homopolymers of A, T, or C. (**f**) Comparison of multinucleotide dangle thermodynamics based on native catalysis versus melt curves. Identical buffers and oligonucleotides were used for the two sets of experiments. Inferred parameters are consistent with each other to within 2 s.d.'s, but melt curve-based parameters show 6- to 12-fold higher errors than the native catalysis method.

There is also a systematic bias in comparing the two sets of dangle parameters; on average, our consensus dangles parameters were 0.190 kcal mol^{-1} more positive. One possible contributing factor to the systematic bias is that Bommarito's experiments were performed in 1 M Na$^+$, whereas our experiments were in the physiological PBS (0.15 M Na$^+$) buffer. The standard model of DNA hybridization assumes that dangles ΔG° contributions are not affected by salinity, but we believe this may be incorrect.

Multinucleotide dangles. We next applied our native motif characterization method to the study of multinucleotide dangles. The standard model of DNA hybridization used by most commonly used software[33,36,40] considers only the effects of the first dangle nucleotide. Melt studies on multinucleotide dangles[61–64] reported inconsistent findings, with some reporting stabilizing effects of long dangles and others reporting destabilizing effects. As far as we are aware, there have not been any systematic studies on the effects of dangle length on ΔG°.

From a theoretical point of view, we believe that the first dangle nucleotide contributes two opposing effects on the stability of nearby duplexes (Fig. 5a). First, it can form a partial base stack with the terminal base pair, adding stability by increasing the interactions between the aromatic groups in the nucleoside bases[65]. Second, the negatively charged phosphodiester backbone is brought in close proximity to the negatively charged phosphodiester backbone of the complementary strand, reducing stability by increasing electrostatic repulsion. In buffers such as PBS or 12.5 mM Mg^{2+}, the distances involved (0.43 nm for

the dangle plus 2.36 nm for the B-DNA helix) are comparable to the Debye length (estimated to be roughly 2 nm (ref. 66)).

Except in the case where the dangle forms significant secondary structure, additional dangle nucleotides past the first do not contribute further base stacking, but continue to increase electrostatic repulsion, albeit with diminishing impact due to increasing distances involved. Consequently, we hypothesized that duplexes with longer dangles will be destabilized relative to single-nucleotide dangles. To verify our hypothesis, we measured $\Delta G^\circ_{\text{dangle}}$ for multinucleotide poly-T dangles with dangle length between 2 and 21. We chose to use poly-T, poly-A and poly-C multinucleotide dangles to ensure that there would be no secondary structure in the dangle. Because the first dangle nucleotide also contributes the partial base stack, we performed experiments for all four nucleotides as the first dangle nucleotide. We did not do any experiments on multinucleotide dangles with non-homogeneous sequence, because the secondary structure of dangles would have introduced additional complicating thermodynamics terms.

Figure 5b shows the consensus ΔG° values we measured for 5′ multinucleotide dangles, based on independent experiments with ROX and with Alexa532. For poly-T dangles of >8 nt, the value of ΔG° appears to asymptotically approach +0.56 kcal mol^{-1} above that of the single-nucleotide dangle ΔG°. Figure 5c shows the consensus ΔG° values we measured for 3′ multinucleotide dangles; all values follow similar trend versus dangle length, although all ΔG° values are more positive due to the weaker stabilizing effect of the first dangle nucleotide closest to the duplex. Figure 5d shows the averaged, smoothed value of multinucleotide dangle ΔG° compared with single-base dangle

ΔG°, and can be used as guidelines for estimating the ΔG° of long dangles; see Supplementary Tables 16–27 for summarized results. Figure 5e shows that other homopolymer dangles also exhibit a general asymptotically destabilizing effect as dangle length increases.

For comparison, Fig. 5f shows the ΔG° values of multinucleotide dangles inferred through HRM. The error bars on HRM-derived ΔG° values are 6- to 12-fold larger than those inferred based on native catalysis; simultaneously, the two different methods produce ΔG° values that are not inconsistent with each other (within 2 s.d.'s). Thus, native catalysis-based characterization represents a more precise method for measuring nucleic acid motif thermodynamics.

Discussion

In this manuscript, we applied a well-characterized dynamic DNA nanotechnology circuit, the non-covalent catalyst, to the problem of DNA motif thermodynamics characterization. One particular qualitative difference in our motif thermodynamics characterization method is the fact that ΔG° values of individual motifs are measured isothermally in native temperature and buffer conditions. This is especially desirable for DNA biotechnology and nanotechnology applications because DNA thermodynamics in buffers needed for enzymatic amplification and self-assembly are often extrapolated from experiments at other conditions.

Using our method, we characterized the ΔG° contributions of (1) the ROX and Alexa532 fluorophores next to duplexes, (2) single-nucleotide dangles, and (3) multinucleotide dangles. Our measured parameters possessed high precision due to multiple-independent repeats of each experiment; we typically achieved a mean s.d. of $< 0.1 \, \text{kcal mol}^{-1}$. For all dangle parameters, we used ROX- and Alexa532-labelled strands to independently produce two sets of mean ΔG° values for internal consistency verification.

Our measurements of fluorophore thermodynamics may possess systematic error of up to $0.2 \, \text{kcal mol}^{-1}$ due to the presence of intermediate CZ, but our measurements of dangles cancels out this bias through subtracting two ΔG° values, so dangle ΔG° values should have no systematic bias. There are four sources of unbiased errors in our experiments: pipetting errors, temperature errors, gel band quantitation errors and synthesis errors. All but the last should be independent from experiment to experiment, so repeat experiments will tend to reduce the errors in the mean values.

The single-nucleotide dangle parameters are the only parameters in our studies that have corresponding literature reported values, and there is significant disagreement between our consensus values and literature values[59]. Common sources of error between our approach and previous melt curve approaches include oligonucleotide synthesis impurities and concentration errors. Because DNA oligonucleotide synthesis have advanced significantly over the past 20 years, we believe that our results based on oligonucleotides synthesized by modern optimized techniques is more representative of true thermodynamics. To calibrate the effects of oligonucleotide synthesis errors, we performed the same experiments on a set of independently ordered strands; the ΔG° values measured using different synthesis preps agreed with each other to within $0.09 \, \text{kcal mol}^{-1}$, on average. In addition, we believe that literature reported extinction coefficients[67] are inaccurate at the 10% level, so that previous studies relying on absorbance at 260 nm for quantitating DNA concentration likely have a larger concentration error than our stoichiometric gel characterization protocol. Finally, our measurements directly interrogate the motif

of interest rather than rely on linear algebra decomposition, and we cross-confirmed our ΔG° across two independent sets of fluorophores. For all these reasons, we believe our values to be more accurate.

To the extent that there may be differences in the true values of the dangle ΔG°, such differences may reflect a limitation of nearest neighbour model itself. A more complex model considering distal base identities to holistically evaluate nucleic acid structure stability may be warranted, but would also require significantly more experiments on multiple core sequence instances of the DNA motifs.

From simple biophysics, we postulated that multinucleotide dangles should contribute an increasing destabilizing thermodynamic effect, towards an asymptotic limit. We confirmed this hypothesis experimentally, and discovered that the saturation length is roughly 8 nt, with ΔG° roughly $0.56 \, \text{kcal mol}^{-1}$ more positive than a single-nucleotide dangle. We suggest that dangle length should be part of the standard model for DNA hybridization and folding software, and that the table in Fig. 5d be used as an empirical formula for estimating multinucleotide dangle ΔG° values. However, we currently do not have a quantitative biophysical model that predicts our measured multinucleotide dangle thermodynamic parameters, and encourage interested parties to explore area of research, for example using molecular dynamics simulations.

Performing PAGE is more labour-intensive than running a melt curve, and this is a disadvantage of our overall workflow. However, our approach generated motif ΔG° parameters with s.d.'s 6- to 14-fold lower than those obtained from HRM experiments. Statistically, 36- to 196-fold more experimental repeats of HRM would be needed to generate parameters of the same precision. In addition, PAGE experiments could potentially be automated, and other readout mechanisms may be possible. Possible approaches for downstream readout include liquid chromatography and/or microfluidics approaches. Scaling the presented method will be vital in obtaining high accuracy ΔG° parameters for a variety of motifs in a variety of temperatures and buffer conditions. For example, many artificial nucleotides with desirable properties, such as 2'-O-methyl RNA[68], locked nucleic acids[69] and peptide nucleic acids[70,71], lack characterization beyond rudimentary rule-of-thumb melting temperature effects from base stacks. Establishment of an accurate and comprehensive nucleic acid motif thermodynamics database will help usher a new paradigm of knowledge-driven design of nucleic acid biotechnology and nanotechnology.

Methods

Reagents. All DNA oligonucleotides (oligos) were synthesized by and purchased from Integrated DNA Technologies (IDT; Coralville, IA). All oligos were HPLC purified by IDT, and quality controlled by capillary electrophoresis and electrospray ionization mass spectrometry. All oligos were received in solution form at roughly 100 μM concentration in 10 mM Tris-EDTA (pH 8.0) after HPLC purification. Secondary stocks were prepared by diluting DNA samples in 10 mM TE buffer with 12.5 mM $MgCl_2$ (pH 8.0) or PBS (10 mM, NaCl 154 mM pH 7.4). Oligo sequences are listed in Supplementary Tables 28–40. Single stranded 10-nt resolution DNA ladder was purchased from IDT. 10–330-bp double strand ladder was purchased from Thermo Scientific.

Chemicals and buffer used include: ammonium persulfate, magnesium chloride hexahydrate ($\geq$98%), Tween 20 (viscous liquid), glycerol ($\geq$99%), acrylamide and bis-acrylamide (19:1) ratio stock solution (m/v concentration of 40%), and TEMED (N,N,N',N'-tetramethylethylenediamine) ($\geq$99%). All chemicals were purchased from Sigma Aldrich unless otherwise noted. SYBR gold gel stain was supplied as 10,000x concentrated in DMSO from Life technologies. SYTO82 orange fluorescent nucleic acid stain was supplied as 5 mM solution in DMSO from Life technologies.

All the experiments were carried out in PBS (phosphate buffer saline) buffer (diluted from purchased 10x PBS) or 10 mM TE 12.5 mM $MgCl_2$ buffer which is prepared from 100x TE buffer (1.0 M Tris-HCl pH 8.0, 0.1 M EDTA) and solid magnesium chloride hexahydrate. All the reactions were performed in Peltier-temperature controlled thermocyclers. All samples were covered with aluminium foil during hybridization reaction to avoid potential photobleaching.

The gel apparatus was connected to a water bath to control temperature of inner chamber. The accuracy of temperature was calibrated with a thermocouple. Gel running buffer was always pre-incubated to the desired temperature before each experiment.

Hybridization and displacement reactions. Before analysis by gel electrophoresis, oligonucleotide molecules were allowed to undergo isothermal hybridization and/or strand displacement. All reactions occurred in 650 μl microcentrifuge tubes (Corning), in 50 μl total reaction volume; 1x concentration corresponds to 300 nM final concentration in all cases. Samples were incubated at the reported temperatures for either 20 h (1x PBS) or 3 h (TE-MgCl$_2$). A multi-thermocycler (Benchmark) was used for temperature control for reactions at 10 and 25 °C. A MasterCycler Personal (Eppendorf) with a heated lid (95 °C) was used for temperature control for reactions at 37 and 45 °C. All temperatures were verified to be accurate within 0.2 °C using a thermocouple sensor.

After completion of the hybridization and just prior to loading into gel, 15 μl of 30% glycerol (volume to volume) were added to each 50 μl sample to facilitate gel loading. We elected to use 30% glycerol rather than conventional gel loading buffer because the bromophenol blue or xylene cyanol FF chemicals appear to give significant signal in the ROX fluorescence channel; by using glycerol instead, we were able to minimize background signal.

Polyacrylamide gel preparation. Polyacrylamide gels (10%) were used in all experiments; 50 ml of acrylamide solution were prepared by mixing 32.0 ml ultrapure water (Millipore) with 5.0 ml 10x TAE buffer (Amresco Inc) and 12.5 ml (40% 19:1 bis:acrylamide) stock solution. A measure of 300 μl of 10% ammonium persulfate (m/v) and 28 μl TEMED were subsequently added to the acrylamide solution. Roughly 10 ml of the final solution was transferred into each 1.5 mm thick XCell SureLock gel cassette (Invitrogen) using 5 ml transfer pipettes. Ten-well gel combs (Invitrogen) were inserted into the cassettes to seal the gel from air and form the desired number of wells. Gels were allowed to polymerize for ∼40 min before samples were loaded.

Polyacrylamide gel electrophoresis. All PAGE experiments were performed using a Model 85–1010 gel box (Galileo Bioscience), with temperature regulated by an external water bath (Cleaver). For most experiments, 1x TAE was used as running buffer; 1x TAE and 12.5 mM MgCl$_2$ was used to verify that reduced salinity in running buffer did not significantly change inferred $\Delta G°$ values. Running buffer was prepared by dilution from 50x TAE buffer and solid magnesium chloride hexahydrate. The gel running temperature was always set to be the same as the temperature of reaction. The voltage and duration each 8 cm gel was run at varied based on temperature: 140 V for 90 min for 10 °C experiments, 140 V for 60 min for 25 °C experiments, 85 V for 65 min for 37 °C experiments, or 75 V for 65 min for 45 °C experiments. The voltages were set to minimize joule heating. In all cases, gel running time was controlled by the power supply (Major Science). Gels were run in parallel in sets of two on the same power supply.

Post-electrophoresis gel handling. At the end of electrophoresis experiments, the gel cassettes were cracked open using a gel cutter, and the gels were directly transferred to a Typhoon 9500 quantitative gel imager (GE Healthcare) for fluorescence imaging. For control experiments comparing observed bands to a DNA ladder, gels were further stained with 0.01% SYBR gold. The gel was transferred to staining solution (5 μl of SYBR gold into 50 ml of water), and stained for 30 min on a shaker. Then the gel was gently rinsed in deionized water for three times and transferred to scanner for imaging. Gels were scanned at 100 μm resolution with 450 V photomultiplier tuber voltage using appropriate laser and filter settings. Images were saved in both BMP and .gel format for subsequent software analysis.

Gel band quantitation and analysis. Gel images were analysed using the version 8.1 Image Quant TL 1D gel analysis software (GE Healthcare). Gel bands were automatically detected by the software, fluorescence background was subtracted and fluorescence intensity of bands were reported by normalizing the single-stranded F band to 100 arbitrary units (Supplementary Note 1). By default, lanes were defined to be full-width centred on each visible lane, and a 'rolling ball' background subtraction was used.

HRM experiments. SYTO 82 was used as reporter dye for HRM experiments in either 1x PBS or Tris-MgCl$_2$ buffer. 5p-Comp-T (300 nM) and the corresponding dangle strand (5′-A, 5′-T1A, 5′-T3A, 5′-T10A, 5′-T20A, 5′-A1A, 5′-A3A, 5′-A10A, or 5′-A20A) were mixed with 5 μM of SYTO 82, respectively. Each 10 μl sample was transferred to a quantitative PCR plate (Hard Shell PCR Plates 96-well thin-wall) for the melt experiment (CFX96 Real-Time System with C1000 Touch, Bio-Rad). The samples were heated to 95 °C, then cooled down to 30 °C, holding 10 s at each integer degree Celsius, with a ramp speed of 0.2 °C per sec. The melting process was measured by the fluorescence of SYTO 82. Melting temperatures determination and thermodynamic parameters calculation were described in Supplementary Note 11.

References

1. Alvarez-Erviti, L. *et al.* Delivery of siRNA to the mouse brain by systemic injection of targeted exosomes. *Nat. Biotechnol.* **29,** 341–345 (2011).
2. Lee, H. *et al.* Molecularly self-assembled nucleic acid nanoparticles for targeted in vivo siRNA delivery. *Nat. Nanotechnol.* **7,** 389–393 (2012).
3. Kasinski, A. L. & Slack, F. J. MicroRNAs en route to the clinic: progress in validating and targeting microRNAs for cancer therapy. *Nat. Rev. Cancer* **11,** 849–864 (2011).
4. Ibrahim, A. F. *et al.* MicroRNA replacement therapy for miR-145 and miR-33a is efficacious in a model of colon carcinoma. *Cancer Res.* **71,** 5214–5224 (2011).
5. Hildebrandt-Eriksen, E. S. *et al.* A locked nucleic acid oligonucleotide targeting microRNA 122 is well-tolerated in cynomolgus monkeys. *Nucleic Acid Ther.* **22,** 152–161 (2012).
6. Hoyer, J. & Neundorf, I. Peptide vectors for the nonviral delivery of nucleic acids. *Acc. Chem. Res.* **45,** 1048–1056 (2012).
7. Frampton, G. M. *et al.* Development and validation of a clinical cancer genomic profiling test based on massively parallel DNA sequencing. *Nat. Biotechnol.* **31,** 1023–1031 (2013).
8. Imperiale, T. F. *et al.* Multitarget stool DNA testing for colorectal-cancer screening. *New Engl. J. Med* **370,** 1287–1297 (2014).
9. Newton, C. R. *et al.* Analysis of any point mutation in DNA. The amplification refractory mutation system (ARMS). *Nucleic Acids Res.* **17,** 2503–2516 (1989).
10. Misale, S. *et al.* Emergence of KRAS mutations and acquired resistance to anti-EGFR therapy in colorectal cancer. *Nature* **486,** 532–536 (2012).
11. Morlan, J., Baker, J. & Sinicropi, D. Mutation Detection by Real-Time PCR: A Simple, Robust and Highly Selective Method. *PLoS ONE* **4,** e4584 (2009).
12. Tomita, N., Mori, Y., Kanda, H. & Notomi, T. Loop-mediated isothermal amplification (LAMP) of gene sequences and simple visual detection of products. *Nat. Protoc.* **3,** 877–882 (2008).
13. Rothemund, P. Folding DNA to create nanoscale shapes and patterns. *Nature* **440,** 297–302 (2006).
14. Seeman, N. C. Nanomaterials Based on DNA. *Annu. Rev. Biochem.* **79,** 65–87 (2010).
15. Afonin, K. *et al.* In vitro assembly of cubic RNA-based scaffolds designed in silico. *Nat. Nanotechnol.* **5,** 676–682 (2010).
16. Wei, B., Dai, M. & Yin, P. Complex shapes self-assembled from single-stranded DNA tiles. *Nature* **485,** 623–626 (2012).
17. Pinheiro, A. V., Han, D., Shih, W. & Yan, H. Challenges and opportunities for structural DNA nanotechnology. *Nat. Nanotechnol.* **6,** 763–772 (2011).
18. Stojanovic, M. N. *et al.* Deoxyribozyme-based ligase logic gates and their initial circuits. *J. Am. Chem. Soc.* **127,** 6914–6915 (2005).
19. Seelig, G., Soloveichik, D., Zhang, D. Y. & Winfree, E. Enzyme-free nucleic acid logic circuits. *Science* **314,** 1585–1588 (2006).
20. Qian, L. & Winfree, E. Scaling up digital circuit computation with DNA strand displacement cascades. *Science* **332,** 1196–1201 (2011).
21. Chirieleison, S. M., Allen, P. B., Simpson, Z. B., Ellington, A. D. & Chen, X. Pattern transformation with DNA circuits. *Nat. Chem.* **5,** 1000–1005 (2013).
22. Zhang, D. Y. & Seelig, G. Dynamic DNA nanotechnology using strand displacement reactions. *Nat. Chem.* **3,** 103–114 (2011).
23. Owczarzy, R., Moreira, B. G., You, Y., Behlke, M. A. & Walder, J. A. Predicting stability of DNA duplexes in solutions containing magnesium and monovalent cations. *Biochemistry* **47,** 5336–5353 (2008).
24. Deigan, K. E., Li, T. W., Mathews, D. H. & Weeks, K. M. Accurate SHAPE-directed RNA structure determination. *Proc. Natl Acad. Sci. USA* **106,** 97–102 (2009).
25. SantaLucia, Jr. J. & Hicks, D. The thermodynamics of DNA structural motifs. *Annu. Rev. Biophys. Biomol. Struct.* **33,** 415–440 (2004).
26. SantaLucia, J. A unified view of polymer, dumbbell, and oligonucleotide DNA nearest-neighbor thermodynamics. *Proc. Natl Acad. Sci. USA* **95,** 1460–1465 (1998).
27. Zhang, D. Y., Turberfield, A. J., Yurke, B. & Winfree, E. Engineering entropy-driven reactions and networks catalyzed by DNA. *Science* **318,** 1121–1125 (2007).
28. Zhang, D. Y. & Winfree, E. Control of DNA strand displacement kinetics using toehold exchange. *J. Am. Chem. Soc.* **131,** 17303–17314 (2009).
29. Miller, S., Jones, L. E., Giovannitti, K., Piper, D. & Serra, M. J. Thermodynamic analysis of 5′ and 3′ single- and 3′ double-nucleotide overhangs neighboring wobble terminal base pairs. *Nucleic acids Res.* **36,** 5652–5659 (2008).
30. Senior, M., Jones, R. A. & Breslauer, K. J. Influence of dangling thymidine residues on the stability and structure of two DNA duplexes. *Biochemistry* **27,** 3879–3885 (1988).
31. Matveeva, O. V. *et al.* Thermodynamic calculations and statistical correlations for oligo-probes design. *Nucleic acids Res.* **31,** 4211–4217 (2003).
32. O'Toole, A. S., Miller, S., Haines, N., Zink, M. C. & Serra, M. J. Comprehensive thermodynamic analysis of 3′ double-nucleotide overhangs neighboring Watson-Crick terminal base pairs. *Nucleic acids Res.* **34,** 3338–3344 (2006).
33. Zuker, M. Mfold web server for nucleic acid folding and hybridization prediction. *Nucleic Acids Res.* **31,** 3406–3415 (2003).

34. Markham, N. R. & Zuker, M. UNAFold: software for nucleic acid folding and hybridization. *Methods Mol. Biol.* **453**, 3–31 (2008).
35. Dirks, R. M., Bois, J. S., Schaeffer, J. M., Winfree, E. & Pierce, N. A. Thermodynamic analysis of interacting nucleic acid strands. *SIAM Rev.* **49**, 65–88 (2007).
36. Zadeh, J. N. *et al.* NUPACK: Analysis and design of nucleic acid systems. *J. Comput. Chem.* **32**, 170–173 (2011).
37. Zhang, D. Y., Chen, S. X. & Yin, P. Thermodynamic optimization of nucleic acid hybridization specificity. *Nat. Chem.* **4**, 208–214 (2012).
38. Wang, J. S. & Zhang, D. Y. Simulation-Guided DNA probe design for consistently ultraspecific hybridization. *Nat. Chem.* **7**, 545–553 (2015).
39. Wu, L. R. *et al.* Continuously tunable nucleic acid hybridization probes. *Nat. Methods.* **12**, 1191–1196 (2015).
40. OligoAnalyzer by Integrated DNA Technologies; available at http://www.idtdna.com/calc/analyzer.
41. Untergasser, A. *et al.* Primer3–new capabilities and interfaces. *Nucleic Acids Res.* **40**, e115 (2012).
42. Douglas, S. M. *et al.* Rapid prototyping of 3D DNA-origami shapes with caDNAno. *Nucleic Acids Res.* **37**, 5001–5006 (2009).
43. Breslauer, K. J., Frank, R., Blocker, H. & Marky, L. A. Predicting DNA duplex stability from the base sequence. *Proc. Natl Acad. Sci. USA* **83**, 3746–3750 (1986).
44. SantaLucia, J. & Turner, D. H. Measuring the thermodynamics of RNA secondary structure formation. *Biopolymers* **44**, 309–319 (1997).
45. Erali, M. & Wittwer, C. T. High resolution melting analysis for gene scanning. *Methods* **50**, 250–261 (2010).
46. Kamiya, M., Torigoe, H., Shindo, H. & Sarai, A. Temperature dependence and sequence specificity of DNA triplex formation: an analysis using isothermal titration calorimetry. *J. Am. Chem. Soc.* **118**, 4532–4538 (1996).
47. Reynaldo, L. P., Vologodskii, A. V., Neri, B. P. & Lyamichev, V. I. The kinetics of oligonucleotide replacements. *J. Mol. Biol.* **297**, 511–520 (2000).
48. Yurke, B. *et al.* DNA fuel for free-running nanomachines. *Phys. Rev. Lett.* **90**, 118102 (2003).
49. Lindahl, T. Instability and Decay of the Primary Structure of DNA. *Nature* **362**, 709–715 (1993).
50. Schroeder, G. K. & Wolfenden, R. Rates of spontaneous disintegration of DNA and the rate enhancements produced by DNA glycosylases and deaminases. *Biochemistry* **46**, 13638–13647 (2007).
51. Radding, C. M., Beattie, K. L., Holloman, W. K. & Wiegand, R. C. Uptake of homologous single-stranded fragments by superhelical DNA. *J. Mol. Biol.* **116**, 825–839 (1977).
52. Moreira, B. G. *et al.* Effects of fluorescent dyes, quenchers, and dangling ends on DNA duplex stability. *Biochem. Biophys. Res. Commun.* **327**, 473–484 (2005).
53. Marras, S. A. E., Kramer, F. R. & Tyagi, S. Efficiencies of fluorescence resonance energy transfer and contact-mediated quenching in oligonucleotide probes. *Nucleic Acids Res.* **30**, e122 (2002).
54. Rouzina, I. & Bloomfield, V. A. Heat capacity effects on the melting of DNA. 1. General aspects. *Biophys. J.* **77**, 3242–3251 (1999).
55. Wu, P., Nakano, S.-I. & Sugimoto, N. Temperature dependence of thermodynamic properties for DNA/DNA and RNA/DNA duplex formation. *Eur. J. Biochem.* **269**, 2821–2830 (2002).
56. Williams, M. C., Wenner, J. R., Rouzina, I. & Bloomfield, V. A. Entropy and heat capacity of dna melting from temperature dependence of single molecule stretching. *Biophys. J.* **80**, 1932–1939 (2001).
57. Chalikian, T. V., Volker, J., Plum, G. E. & Breslauer, K. J. A more unified picture for the thermodynamics of nucleic acid duplex melting: a characterization by calorimetric and volumetric techniques. *Proc. Natl Acad. Sci. USA* **96**, 7853–7858 (1999).
58. Petruska, J. & Goodman, M. F. Enthalpy and Entropy Compensation in DNA Melting Thermodynamics. *J. Biol. Chem.* **270**, 746–750 (1995).
59. Bommarito, S., Peyret, N. & SantaLucia, J. Thermodynamic parameters for DNA sequences with dangling ends. *Nucleic Acids Res.* **28**, 1929–1934 (2000).
60. Pyshnyi, D. V. & Ivanova, E. M. The influence of nearest neighbours on the efficiency of coaxial stacking at contiguous stacking hybridization of oligodeoxyribonucleotides. *Nucleosides Nucleotides Nucleic Acids* **23**, 1057–1064 (2004).
61. Doktycz, M. J., Paner, T. M., Amaratunga, M. & Benight, A. S. Thermodynamic stability of the 5' dangling-ended DNA hairpins formed from sequences 5'-(XY)2GGATAC(T)4GTATCC-3', where X, Y = A, T, G, C. *Biopolymers* **30**, 829–845 (1990).
62. Riccelli, P. V., Mandell, K. E. & Benight, A. S. Melting studies of dangling-ended DNA hairpins: effects of end length, loop sequence and biotinylation of loop bases. *Nucleic acids Res.* **30**, 4088–4093 (2002).
63. Liu, J. D., Zhao, L. & Xia, T. The dynamic structural basis of differential enhancement of conformational stability by 5'- and 3'-dangling ends in RNA. *Biochemistry* **47**, 5962–5975 (2008).
64. Ohmichi, T., Nakano, S.-I., Miyoshi, D. & Sugimoto, N. Long RNA dangling end has large energetic contribution to duplex stability. *J. Am. Chem. Soc.* **124**, 10367–10372 (2002).
65. Yakovchuk, P., Protozanova, E. & Frank-Kamenetskii, M. D. Base-stacking and base-pairing contributions into thermal stability of the DNA double helix. *Nucleic Acids Res.* **34**, 564–574 (2006).
66. Hsieh, C. C., Balducci, A. & Doyle, P. S. Ionic effects on the equilibrium dynamics of DNA confined in nanoslits. *Nano Lett.* **8**, 683–688 (2008).
67. Puglisi, J. D. & Tinoco, Jr. I. Absorbance melting curves of RNA. *Methods Enzymol.* **180**, 304–325 (1989).
68. Kierzek, E. *et al.* The influence of locked nucleic acid residues on the thermodynamic properties of 2'-O-methyl RNA/RNA heteroduplexes. *Nucleic Acids Res.* **33**, 5082–5093 (2005).
69. Kaur, H., Arora, A., Wengel, J. & Maiti, S. Thermodynamic, counterion, and hydration effects for the incorporation of locked nucleic acid nucleotides into DNA duplexes. *Biochemistry* **45**, 7347–7355 (2006).
70. Kushon, S. A. *et al.* Effect of secondary structure on the thermodynamics and kinetics of PNA hybridization to DNA hairpins. *J. Am. Chem. Soc.* **123**, 10805–10813 (2001).
71. Ratilainen, T., Holmen, A., Tuite, E., Nielsen, P. E. & Norden, B. Thermodynamics of sequence-specific binding of PNA to DNA. *Biochemistry* **39**, 7781–7791 (2000).

Acknowledgements
This work was funded by the Rice University startup fund to D.Y.Z., by the Welch Foundation Grant C-1862 to D.Y.Z., and by the Cancer Prevention Research Institute of Texas award RP140132 to D.Y.Z.

Author contributions
C.W. designed and conducted the experiments, analysed the data, and wrote the paper. J.H.B. analysed the data and wrote the paper. D.Y.Z. conceived the project, designed the experiments, analysed the data and wrote the paper.

Additional information

Competing financial interests: The authors declare no competing financial interests.

5

Electropolymerization on wireless electrodes towards conducting polymer microfibre networks

Yuki Koizumi[1], Naoki Shida[1], Masato Ohira[1], Hiroki Nishiyama[1], Ikuyoshi Tomita[1] & Shinsuke Inagi[1]

Conducting polymers can be easily obtained by electrochemical oxidation of aromatic monomers on an electrode surface as a film state. To prepare conducting polymer fibres by electropolymerization, templates such as porous membranes are necessary in the conventional methods. Here we report the electropolymerization of 3,4-ethylenedioxythiophene and its derivatives by alternating current (AC)-bipolar electrolysis. Poly(3,4-ethylenedioxythiophene) (PEDOT) derivatives were found to propagate as a fibre form from the ends of Au wires used as bipolar electrodes (BPEs) parallel to an external electric field, without the use of templates. The effects of applied frequency and of the solvent on the morphology, growth rate and degree of branching of these PEDOT fibres were investigated. In addition, a chain-growth model for the formation of conductive material networks was also demonstrated.

[1] Department of Electronic Chemistry, Interdisciplinary Graduate School of Science and Engineering, Tokyo Institute of Technology, 4259 Nagatsuta-cho, Midori-ku, Yokohama 226-8502, Japan. Correspondence and requests for materials should be addressed to S.I. (email: inagi@echem.titech.ac.jp).

Recently, there have been a number of interesting reports concerning bipolar electrochemistry, in which anodic and cathodic reactions take place simultaneously on both poles of a conductive material placed between a pair of driving electrodes[1–12]. The functioning of this special wireless electrode (or bipolar electrode (BPE)) is induced by an external electric field generated in a low concentration of a supporting electrolyte. Considering the potential energy diagram under the application of a direct current (DC) voltage (E) between the driving electrodes, the potential of BPEs is floating to an equilibrium value (E_{elec}) in a gradient of solution potential, consequently the anodic and cathodic overpotentials (ΔV_{BPE}) can drive redox reactions at each BPE in the same manner (Fig. 1)[3,5]. In one of the most significant studies, Kuhn *et al.* demonstrated the elegant surface modification of conductive particles[13–16], involving metal plating (cathodic reduction of metal ions) at one pole in conjunction with the electrochemical oxidation (electropolymerization) of pyrrole at the opposite pole to give Janus-type particles coated simultaneously with metal and a conducting polymer film (polypyrrole)[13].

Our approach in obtaining bifunctional particles, such as these, has been to employ alternating current (AC)-bipolar electrochemistry, through which a variety of glassy carbon (GC) particles modified with gold in a site selective manner have been synthesized[17]. The next challenge regarding this technology was to demonstrate the iterative electropolymerization of aromatic monomers on both poles of a BPE using this AC-bipolar system, and preliminary work using pyrrole as a monomer gave GC particles modified with polypyrrole films, as expected. However, to our surprise, the AC-bipolar electrolysis of 3,4-ethylenedioxythiophene (EDOT) resulted in the formation of polymer fibres at both poles of the GC-BPE, rather than film deposition. To the best of our knowledge, the propagation of conducting polymer microfibres from the very edges of wireless electrodes parallel to the external electric field and subsequent network formation have not previously been observed. So far, fibre/wire-like conducting polymers have been prepared by the conventional electropolymerization using templates such as porous membranes or seeds[18–21]. In another system of the electropolymerization of aromatic monomers with conventional

'wired' working electrodes using an AC voltage, conducting polymer film was deposited from two working electrodes along with the two-/three-dimensional formation of dendrites and connected each other[22,23]. However, it was difficult to control finely the size of dendrites, thus the method for the interconnection of electrodes was not satisfactory towards practical applications. In this context, this spontaneous propagation of conducting polymer microfibres from wireless electrodes is thus a new phenomenon and is worthy of further, detailed investigation.

Results

AC-bipolar electropolymerization. To assess the reaction sites on the BPE, gold (Au) wires ($\phi = 50\,\mu m$, length = 20 mm) placed 1 mm apart from one another were employed as BPEs and the reaction process was monitored using an optical microscope. The experimental configuration is summarized in Fig. 2a, consisting of a pair of platinum feeder electrodes (20×20 mm, distance: 60 mm) in 1 mM tetrabutylammonium perchlorate (Bu_4NClO_4)/acetonitrile (MeCN) containing 50 mM EDOT monomer and 5 mM benzoquinone (BQ) as a sacrificial reagent for reduction. During the functioning of the Au wires as BPEs, the electropolymerization of EDOT takes place at the anodic part of the Au wire, while the sacrificial reduction of BQ to hydroquinone simultaneously proceeds at the cathodic part of the wire (Supplementary Fig. 1). Based on the principles of bipolar electrochemistry (Fig. 1), the potential difference applied across a BPE (ΔV_{BPE}) can be estimated from the voltage between the driving electrodes (E) and the length of the wire, weighted by the cell factors (Supplementary Fig. 2 and Supplementary Equation 6)[17]. The redox reactions will proceed when ΔV_{BPE} is sufficiently above their potential difference value (ΔV_{min}, Fig. 2a).

On the application of AC voltage ($E = 30$ V, $\Delta V_{BPE} = 8.3$ V, 5 Hz, square wave alternating in polarity) between the driving electrodes, several polymer fibres were observed to propagate dendritically from the end of each Au wire. The resulting iterative AC-bipolar electropolymerization of EDOT generated poly(3,4-ethylenedioxythiophene) (PEDOT) fibres. The real-time observation of the gradual propagation of the PEDOT fibres revealed that the terminals of the growing fibres were activated for further electropolymerization (Supplementary Movie). After 90 s, the tip of one of the propagating fibres met the tip of a fibre growing from the other Au wire and the fibres connected to one another, bridging the 1 mm gap between the Au wires (Fig. 2b). Following this, the propagation of the fibres abruptly ceased, because the two Au wires were now connected by the conducting polymer fibres and hence behaved as a single BPE. Providing further evidence of this, the other ends of the Au wires, which were closer to the feeder electrodes, still worked as anodic and cathodic poles for the electropolymerization of EDOT, and further propagation of fibres was observed (Supplementary Fig. 3). During the bipolar electropolymerization experiment, PEDOT was not produced on the driving electrodes. Although the oxidation of EDOT should occur at the driving anode, the potential applied was probably too low (most of voltage applied was lost due to the potential drop in the solution) to conduct the effective electropolymerization.

When a smaller voltage ($E = 21$ V, $\Delta V_{BPE} = 5.8$ V) was applied, the propagation rate of the PEDOT fibre was apparently too slow to bridge the Au wires (Supplementary Fig. 4). However, under such moderate conditions, the degree of branching was also small. In contrast, the application of a larger voltage ($E = 39$ V, $\Delta V_{BPE} = 10.8$ V) resulted in similar propagation rate to that

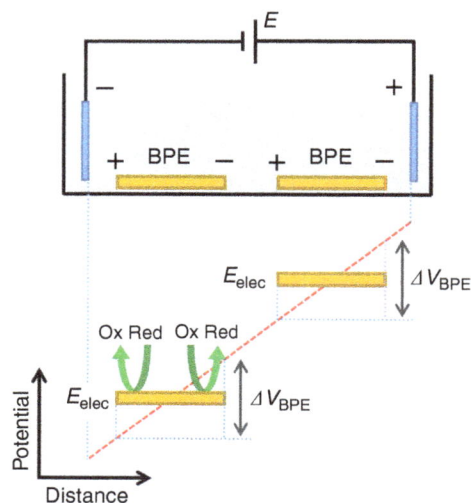

Figure 1 | Schematic illustration of the principle of bipolar electrochemistry. The potential difference at the electrode/solution interface varies across the length of BPEs according to the potential gradient applied to the solution; thus, the overpotentials (ΔV_{BPE}) can drive redox reactions at both sides.

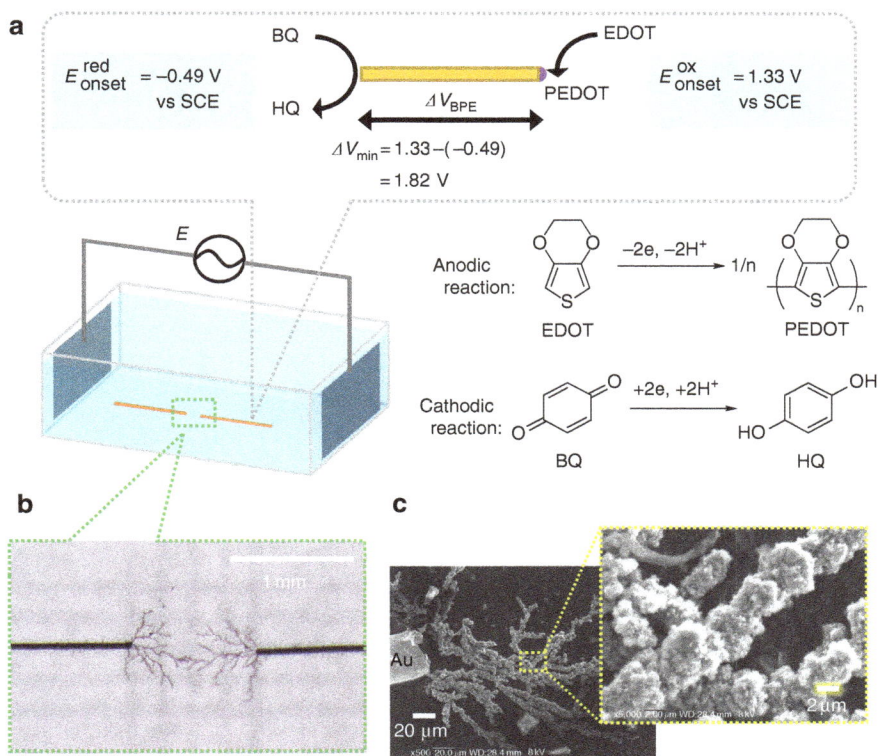

Figure 2 | AC-bipolar electropolymerization of EDOT. (a) Schematic representation of the electrochemical setup for AC-bipolar electrolysis including oxidative polymerization of EDOT and sacrificial reduction of BQ with Au wires ($\phi = 50\,\mu m$, 20 mm) as BPEs set in between Pt driving electrodes (20×20 mm, distance: 60 mm), **(b)** optical microscope image of PEDOT fibres bridging the 1 mm gap between Au wires ($\Delta V_{BPE} = 8.3$ V, 90 s) and **(c)** SEM images of the PEDOT fibres. HQ, hydroquinone; SCE, saturated calomel electrode.

with $\Delta V_{BPE} = 8.3$ V because the fibre formation process at the high ΔV_{BPE} application was diffusion-limited.

The PEDOT fibres and the connected Au wires were successfully transferred onto a carbon tape after carefully washing with MeCN and drying, and Fig. 2c shows scanning electron microscopy (SEM) images of the fibres. It was found that each fibre was composed of clusters connected in a linear fashion and had a diameter of ~ 3–$5\,\mu m$. Energy dispersive X-ray mapping of the same observation area indicated the presence of sulfur derived from the EDOT moiety and generated an image in good agreement with the SEM image of the fibres (Supplementary Fig. 5). Such fibre propagation with a uniform thickness is totally different from the case of the conventional AC-electropolymerization giving the irregular conducting polymer dendrites[22,23].

Fibre-propagation mechanism. Here we propose a possible propagation mechanism by which the PEDOT fibres are grown through AC-bipolar electropolymerization (Fig. 3). Initially, the external electric field generates Au-BPEs, at which oxidation of the EDOT monomer and reduction of BQ take place (Fig. 3a). The former results in polymerization of EDOT and, once the polymer has grown sufficiently, it becomes insoluble and deposits on one end of the Au wire. It is well known that the oxidation potential of PEDOT is less positive than that of EDOT monomer. During polymerization of EDOT under bipolar electrochemical conditions, the resulting polymer is typically doped and has cationic charges, and so can be electrophoresed under the influence of the external electric field. Consequently, the polymer is deposited not as a film but rather in an anisotropic morphology (Fig. 3b). Under a contrary electric field, Au-BPEs with opposite polarities are generated and similar electrode reactions take place

(Fig. 3c). Upon repeated cycling of the AC power supply, the edges of the PEDOT fibres, where the ΔV_{BPE} is the highest, grow in a manner parallel to the external electric field, serving as active sites for the electropolymerization because of their sufficiently high conductivity (Fig. 3d). Accordingly, the electrophoresis of the charged polymer species plays an important role in the fibre-propagation process. In a previous report on the copper wire formation between copper particles by the DC-bipolar electrolysis, the electrophoresis of copper ion in a low concentration of electrolyte was necessary to determine the propagation direction[24].

We next examined the AC-bipolar electropolymerization of pyrrole and thiophene as monomers; however, these did not give fibres but rather formed films covering the BPEs (Supplementary Fig. 6). This seems to be explained by the inactivation of the BPEs covered with the relatively low electric conductivity of these polymers compared to that of PEDOT. It should be noted that, under a DC voltage, there appeared no PEDOT fibre at the ends of the Au wires, while the migration of blue-coloured polymers parallel to the external electric field was observed. Alternatively, PEDOT films and clusters were continuously deposited on the anodic portion of the wire (Supplementary Fig. 7a). In the cases of pyrrole and thiophene under a DC voltage, the corresponding polymer films were formed but more slowly and thinly than that of PEDOT probably owing to the low conductivity of the polymers (Supplementary Fig. 7b,c). This supports the result of the unsuccessful fibre formation under AC-bipolar electropolymerization mentioned above.

When a higher frequency of 50 Hz for AC voltage was applied, the number of generated PEDOT fibres was increased and highly connected networks were obtained (Supplementary Fig. 8). The diameter of the resulting fibres was decreased to ~ 1–$2\,\mu m$

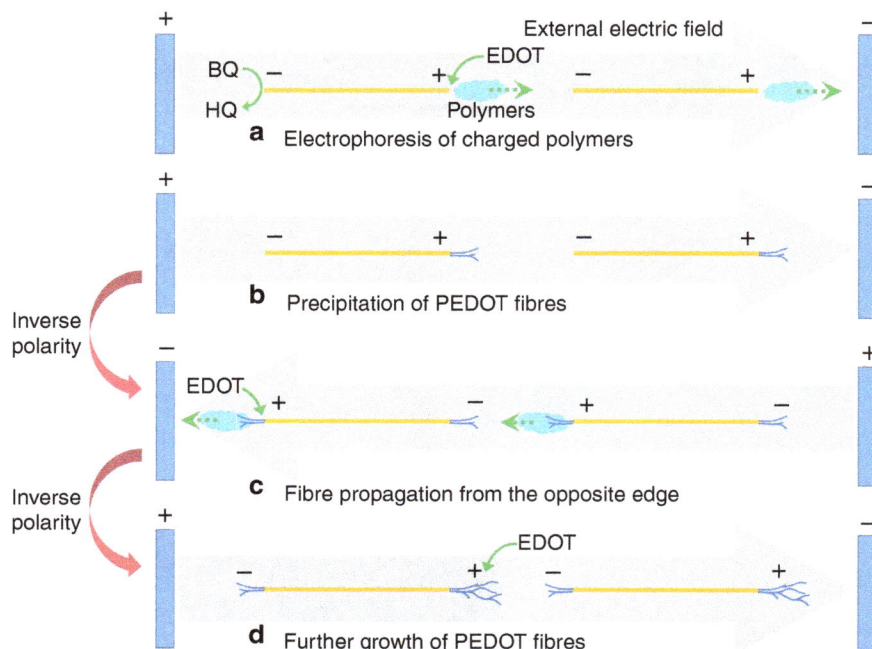

Figure 3 | Proposed PEDOT fibre-propagation mechanism. (**a**) Electrophoresis of charged polymers, (**b**) precipitation of PEDOT fibres, (**c**) propagation of fibres from the opposite end and (**d**) further growth of PEDOT fibres.

compared with that of the fibres prepared using a lower frequency of 5 Hz as evidenced by the SEM observation (Supplementary Fig. 9). Considering the propagation mechanism described in Fig. 3, the frequency is evidently an important factor determining fibre morphology. The main difference between the frequencies is in the diffusion length of the charged polymers during electrophoresis at each anodic moment. When applying a lower frequency, the diffusion length of the charged polymers is relatively long, and the local concentration of the active species for polymerization is lower. In contrast, a higher local concentration of the charged polymers is expected at higher frequencies. This model explains why the quantity of PEDOT fibres increased when applying a higher frequency. The difference in the diffusion length of the polymer also affects propagation rate of fibres. The lower frequency (5 Hz) accelerated it. In addition, the application of a much higher frequency (100 Hz) ended in failure of forming of any fibres under the conditions. Since it takes longer time to form electric double layers in low concentration of an electrolyte, such a higher frequency was not suitable for the AC-bipolar electropolymerization. The supporting salts were also found to affect the morphology and propagation rate of the fibres, as shown in Supplementary Fig. 9, with the complicated factors as follows. In general, counter ions, which compensate for the cationic state of PEDOT, play a crucial role in its polymerization rate and morphology[25,26]. Ionic conductivity should be related to the rate of the double layer formation. The solubility of the growing PEDOT seems to be another important factor in determining the size and the morphology of the fibres; the use of dichloromethane (CH_2Cl_2) in place of MeCN reduced the propagation rate such that a span of 540 s was required to bridge the Au wires (Supplementary Fig. 10).

Scope of EDOT monomers. To investigate the scope of the AC-bipolar electropolymerization, two additional EDOT derivatives were also assessed (Fig. 4). In the case of the EDOT-C1 monomer, rosary-like PEDOT fibres were obtained

with a smooth surface, while the polymerization of EDOT-C10 resulted in the formation of rod-like fibres. In both cases, the degree of branching of the fibres was decreased compared with the extent observed for EDOT. It was therefore possible to create versatile fibre structures using EDOT derivatives, although the relationship between the chemical structure of the monomer and the morphology of the resulting PEDOT fibres is still unclear at the present time.

Selective network formation. Towards the practical connection of an Au wire intersection with the PEDOT fibres, we prepared the setup under the application of electric fields with the different direction as shown in Supplementary Fig. 11. In the case of Supplementary Fig. 11a, only two Au wires put parallel to the electric field were active as BPEs and connected each other, while the other two Au wires were inactive to promote redox reactions because of their insufficient length crossing the electric field. On the other hand, the case of Supplementary Fig. 11b resulted in the successful interconnection of the Au wires in the different mode in accordance with the direction of the electric field. The conductive networks of the Au wires and the PEDOT fibres were achieved selectively.

From the foregoing findings, it seems possible to sequentially activate conductors with different lengths (that is, different ΔV_{BPE}) by the connection with PEDOT fibres once grown from an active BPE. Finally, we demonstrated the 'chain-growth model' for the formation of conducting networks using the bipolar electrochemical method. As shown in Fig. 5, three kinds of Au wires (W_1–W_3, with lengths of 20, 5 and 2 mm) were placed between the driving electrodes, 0.3 mm apart from one another. During the application of AC voltage ($E = 18$ V, 5 Hz) between the driving electrodes, the value of ΔV_{BPE} for these wires was 6.0, 1.5 and 0.6 V, respectively; thus, only W_1 was active as a BPE for the electropolymerization of EDOT. PEDOT, therefore, propagated from the end of W_1 and contacted the end of W_2 in a span of 90 s. W_2 then became active to initiate the propagation of PEDOT from its

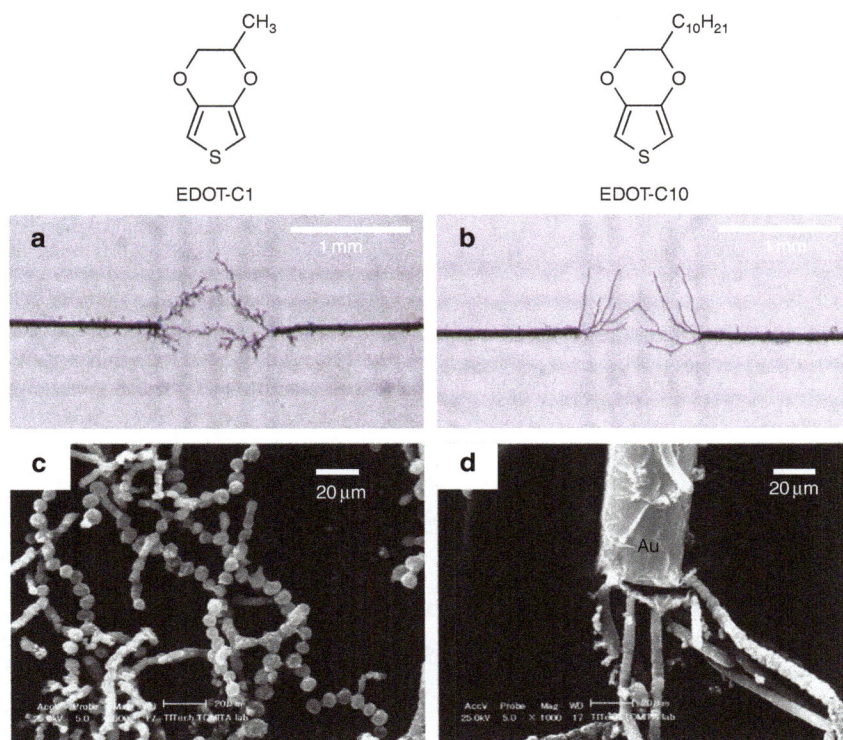

Figure 4 | Versatile fibre morphology of PEDOT derivatives. (a,b) Optical microscope and **(c,d)** SEM images of the polymer fibres obtained using EDOT-C1 **(a,c)** and EDOT-C10 **(b,d)** as monomers.

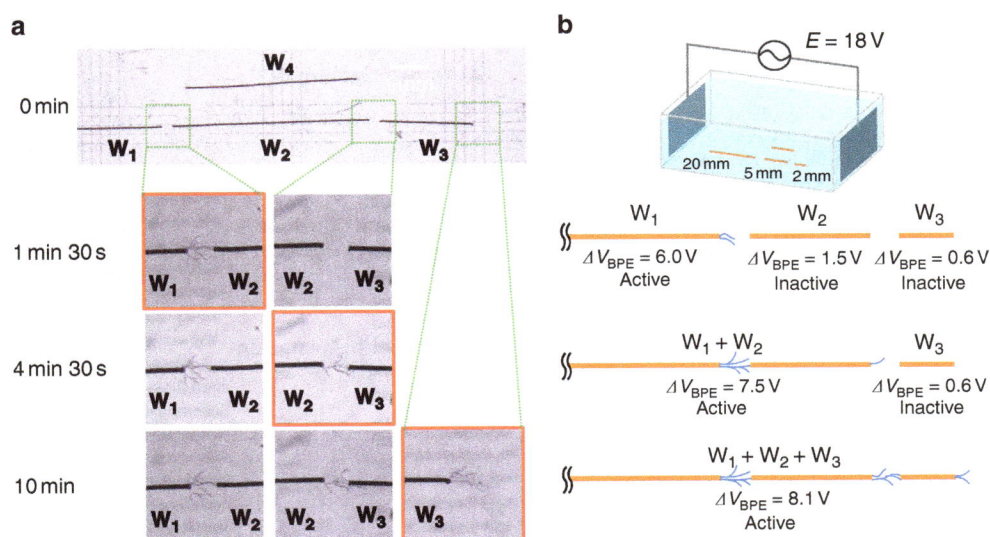

Figure 5 | Chain-growth model for the formation of conductive material networks. (a) Optical microscope images of Au wires (W_1: 20 mm, W_2: 5 mm, W_3: 2 mm and W_4: 4.5 mm) on the chain-growth model experiment by the AC-bipolar electropolymerization of EDOT. **(b)** Illustration showing the proposed mechanism of the chain-growth model.

right end owing to the sufficient ΔV_{BPE} value (7.5 V) of the combined BPE (W_1 and W_2), with the resulting fibre reaching W_3 after 270 s. Finally the right end of W_3 became an active site for further electropolymerization. Throughout the experiment, an additional wire (W_4) with a length of 4.5 mm acting as a reference remained intact since it received an electric field insufficient to drive the polymerization. This model demonstrated the one-directional propagation of a conducting network of metal wires and PEDOT fibres from an initiator by an electrochemical stimulus, taking advantages of the bipolar electrochemistry.

Discussion

We have successfully demonstrated the unusual electropolymerization behaviour of the EDOT monomer, leading to one-dimensional propagation from both ends of Au wires acting as BPEs without direct feeding of an electric potential under AC-bipolar electrochemical conditions. The electrophoresis of charged polymers evidently played an important role in obtaining various PEDOT microfibre structures. The morphology of the fibres, as well as the propagation rate and degree of branching were found to be dependent on the frequency, solvent and supporting electrolyte used. This spontaneous propagation of conducting polymer fibres in the absence of templates should be widely applicable to the creation of conductive material networks with a wireless process. The chain-growth model we demonstrated is a totally novel mode to activate a small conductor as a BPE under the application of a mild electric field.

Methods

Materials. All reagents and chemicals were obtained from commercial sources and used without further purification otherwise noted. EDOT-C10 was prepared according to a procedure defined in the literature[27]. Gold (Au) wires and platinum (Pt) plates were purchased from commercial sources.

Synthesis of EDOT-C1. EDOT-C1 was synthesized via transetherification similarly to the literature procedure[27] using 1,2-propanediol as a diol. Yield: 72%, yellowish oil. ^{1}H NMR (270 MHz, CDCl$_3$): δ 6.31 (s, 1H); 6.30 (s, 1H); 4.27 (m, 1H); 4.13 (dd, $J = 11.5$ Hz, 2.1 Hz, 1H); 3.82 (dd, $J = 11.5$ Hz, 8.5 Hz, 1H); and 1.34 (d, $J = 6.5$ Hz, 3H). ^{13}C NMR (68 MHz, CDCl$_3$): δ 142.16, 141.41, 99.30, 99.29, 70.00, 69.43 and 16.24. High resolution MS (EI): m/z [M$^+$] calculated for C$_7$H$_8$O$_2$S$_1$: 156.0245; found:156.0245.

A typical procedure for AC-bipolar electropolymerization. The bipolar electrolysis apparatus shown in Fig. 2a was employed, containing an electrolytic solution of MeCN (1 mM), EDOT (50 mM) and BQ (5 mM). An AC voltage ($E = 30$ V, 5 Hz) was applied between the driving electrodes for the desired time span at room temperature. Following electrolysis, the Au wires, now connected by PEDOT fibres, were carefully washed with solvent and dried.

Measurements. DC and AC power were supplied to the driving electrodes using an EC1000SA AC/DC power source (NF Corporation). Optical microscope observations were conducted with an Olympus SZX10 and a Keyence VHX-5000, and SEM observations were performed using a Shimadzu SS-550. Energy dispersive X-ray spectra were acquired with a Keyence Genesis XM2 and cyclic voltammetry measurements were carried out using an ALS 6005C Electrochemical Analyzer.

References

1. Loget, G., Zigah, D., Bouffier, L., Sojic, N. & Kuhn, A. Bipolar electrochemistry: from materials science to motion and beyond. *Acc. Chem. Res.* **46**, 2513–2523 (2013).
2. Fosdick, S. E., Knust, K. N., Scida, K. & Crooks, R. M. Bipolar electrochemistry. *Angew. Chem. Int. Ed.* **52**, 10438–10456 (2013).
3. Mavré, F. *et al.* Bipolar electrodes: a useful tool for concentration, separation, and detection of analytes in microelectrochemical systems. *Anal. Chem.* **82**, 8766–8774 (2010).
4. Inagi, S. & Fuchigami, T. Electrochemical post-functionalization of conducting polymers. *Macromol. Rapid Commun.* **35**, 854–867 (2014).
5. Chow, K.-F., Mavré, F., Crooks, J. A., Chang, B.-Y. & Crooks, R. M. A large-scale, wireless electrochemical bipolar electrode microarray. *J. Am. Chem. Soc.* **131**, 8364–8365 (2009).
6. Loget, G. & Kuhn, A. Propulsion of microobjects by dynamic bipolar self-regeneration. *J. Am. Chem. Soc.* **132**, 15918–15919 (2010).
7. Loget, G. & Kuhn, A. Electric field-induced chemical locomotion of conducting objects. *Nat. Commun.* **2**, 535 (2011).
8. Ulrich, C., Andersson, O., Nyholm, L. & Björefors, F. Formation of molecular gradients on bipolar electrodes. *Angew. Chem. Int. Ed.* **47**, 3034–3036 (2008).
9. Inagi, S., Ishiguro, Y., Atobe, M. & Fuchigami, T. Bipolar patterning of conducting polymers by electrochemical doping and reaction. *Angew. Chem. Int. Ed.* **49**, 10136–10139 (2010).
10. Ishiguro, Y., Inagi, S. & Fuchigami, T. Site-controlled application of electric potential on a conducting polymer "canvas". *J. Am. Chem. Soc.* **134**, 4034–4036 (2012).
11. Inagi, S., Nagai, H., Tomita, I. & Fuchigami, T. Parallel polymer reactions of a polyfluorene derivative by electrochemical oxidation and reduction. *Angew. Chem. Int. Ed.* **52**, 6616–6616 (2013).
12. Shida, N., Koizumi, Y., Nishiyama, H., Tomita, I. & Inagi, S. Electrochemically mediated atom transfer radical polymerization from a substrate surface manipulated by bipolar electrolysis: Fabrication of gradient and patterned polymer brushes. *Angew. Chem. Int. Ed.* **54**, 3922–3926 (2015).
13. Loget, G. *et al.* Versatile procedure for synthesis of Janus-type carbon tubes. *Chem. Mater.* **23**, 2595–2599 (2011).
14. Loget, G. & Kuhn, A. Bulk synthesis of Janus objects and asymmetric patchy particles. *J. Mater. Chem.* **22**, 15457–15474 (2012).
15. Loget, G., Roche, J. & Kuhn, A. True bulk synthesis of Janus objects by bipolar electrochemistry. *Adv. Mater.* **24**, 5111–5116 (2012).
16. Ongaro, M., Gambirasi, A., Favaro, M., Kuhn, A. & Ugo, P. Asymmetrical modification of carbon microfibers by bipolar electrochemistry in acetonitrile. *Electrochim. Acta* **116**, 421–428 (2014).
17. Koizumi, Y., Shida, N., Tomita, I. & Inagi, S. Bifunctional modification of conductive particles by iterative bipolar electrodeposition of metals. *Chem. Lett.* **43**, 1245–1247 (2014).
18. Martin, C. R. Nanomaterials: a membrane-based synthetic approach. *Science* **266**, 1961–1966 (1994).
19. Li, C., Bai, H. & Shi, G. Conducting polymer nanomaterials: electrosynthesis and applications. *Chem. Soc. Rev.* **38**, 2397–2409 (2009).
20. Zhang, X. & Manohar, S. K. Bulk synthesis of polypyrrole nanofibers by a seeding approach. *J. Am. Chem. Soc.* **126**, 12714–12715 (2004).
21. Shi, W., Liang, P., Ge, D., Wang, J. & Zhang, Q. Starch-assisted synthesis of polypyrrole nanowires by a simple electrochemical approach. *Chem. Commun.* 2414–2416 (2007).
22. Curtis, C. L., Ritchie, J. E. & Sailor, M. J. Fabrication of conducting polymer interconnects. *Science* **262**, 2014–2016 (1993).
23. Fujii, M., Arii, K. & Yoshino, K. Neuron-type polypyrrole device prepared by electrochemical polymerization method and its properties. *Synth. Met.* **71**, 2223–2224 (1995).
24. Bradley, J.-C. *et al.* Creating electrical contacts between metal particles using directed electrochemical growth. *Nature* **389**, 268–271 (1997).
25. Moradi, A., Emamgolizadeh, A., Omrani, A. & Rostami, A. A. Electropolymerization and characterization of 3,4-ethylenedioxy thiophene on glassy carbon electrode and study of ions transport of the polymer during redox process. *J. Appl. Polym. Sci.* **125**, 2407–2416 (2012).
26. Melato, A. I., Mendonca, M. H. & Abrantes, L. M. Effect of electropolymerisation conditions on the electrochemical, morphological and structural properties of PEDOTh films. *J. Solid State Electrochem.* **13**, 417–426 (2009).
27. Wolfs, M., Darmanin, T. & Guittard, F. Versatile superhydrophobic surfaces from a bioinspired approach. *Macromolecules* **44**, 9286–9294 (2011).

Acknowledgements

This study was financially supported by JSPS KAKENHI (Grant-in-Aid for Young Scientists (A), 26708013) and MEXT KAKENHI (Grant-in-Aid for Scientific Research on Innovative Areas 'New Polymeric Materials Based on Element-Blocks', 15H00724).

Author contributions

S.I. conceived and directed the project. Y.K., N.S. and M.O. designed and performed the experiments. S.I., H.N. and I.T. discussed results. S.I. and Y.K. wrote the manuscript.

Additional information

A simple and versatile design concept for fluorophore derivatives with intramolecular photostabilization

Jasper H. M. van der Velde[1], Jens Oelerich[2], Jingyi Huang[3], Jochem H. Smit[1], Atieh Aminian Jazi[1], Silvia Galiani[4], Kirill Kolmakov[5], Giorgos Guoridis[1], Christian Eggeling[4], Andreas Herrmann[3], Gerard Roelfes[2] & Thorben Cordes[1]

Intramolecular photostabilization via triple-state quenching was recently revived as a tool to impart synthetic organic fluorophores with 'self-healing' properties. To date, utilization of such fluorophore derivatives is rare due to their elaborate multi-step synthesis. Here we present a general strategy to covalently link a synthetic organic fluorophore simultaneously to a photostabilizer and biomolecular target via unnatural amino acids. The modular approach uses commercially available starting materials and simple chemical transformations. The resulting photostabilizer–dye conjugates are based on rhodamines, carbopyronines and cyanines with excellent photophysical properties, that is, high photostability and minimal signal fluctuations. Their versatile use is demonstrated by single-step labelling of DNA, antibodies and proteins, as well as applications in single-molecule and super-resolution fluorescence microscopy. We are convinced that the presented scaffolding strategy and the improved characteristics of the conjugates in applications will trigger the broader use of intramolecular photostabilization and help to emerge this approach as a new gold standard.

[1] Molecular Microscopy Research Group, Zernike Institute for Advanced Materials, University of Groningen, Nijenborgh 4, 9747 AG Groningen, The Netherlands. [2] Stratingh Institute for Chemistry, University of Groningen, Nijenborgh 4, 9747 AG Groningen, The Netherlands. [3] Department of Polymer Chemistry, Zernike Institute for Advanced Materials, University of Groningen, Nijenborgh 4, 9747 AG Groningen, The Netherlands. [4] MRC Human Immunology Unit, Weatherall Institute of Molecular Medicine, University of Oxford, Headley Way, Oxford OX3 9DS, UK. [5] Department NanoBiophotonics, Max-Planck-Institute of Molecular Medicine, Am Fassberg 1, 37077 Goettingen, Germany. Correspondence and requests for materials should be addressed to T.C. (email: t.m.cordes@rug.nl).

Organic fluorophores are a major driving force for the recent success of fluorescence-based methods, but they intrinsically suffer from transient excursions to dark states (blinking) and irreversible destruction (photobleaching)[1,2]. Both processes fundamentally limit their applicability and have, for a long time, hampered the development of advanced microscopy techniques with single-molecule sensitivity[2] or optical super-resolution <250 nm (refs 2–4). Lüttke and colleagues[5] introduced covalent binding of triplet-state quenchers and singlet-oxygen scavengers[6] to organic fluorophores as a strategy to reduce the above mentioned effects. Such photostabilizer–dye conjugates with intramolecular triplet-state quenching have 'self-healing'[7] or 'self-protecting'[8] properties, preventing photodamage without the use of solution additives. This non-invasive strategy has clear advantages compared with commonly used approaches, where micro- to millimolar concentrations of organic compounds are added to the buffer system[9–15].

Intramolecular photostabilization was recently revived independently by two groups[16–19], to reduce photobleaching and blinking even in demanding applications such as single-molecule fluorescence microscopy[16–20] or in vivo imaging[16]. It is also shown that efficient intramolecular photostabilization can be achieved without additives in the buffer system[18], minimizing potential influences on the biological system of interest. Applications of intramolecular photostabilization in biophysical or microscopy research are, however, still limited to proof-of-principle studies[5,6,16–19]. This is mainly due to the synthetic effort of photostabilizer–dye conjugates, which thus far requires a multi-step synthesis route[16,17]. These synthetic challenges represent a fundamental hurdle for researchers with limited organic chemistry experience to use this concept. Moreover, only a small number of bifunctional cyanine derivatives are currently available to synthesize photostabilizer–dye conjugates on specific biomolecular targets (DNA, RNA, proteins and antibodies)[16,17], strongly restricting the choice of fluorophore type (chemical structure, redox potential, water solubility and so on) and photophysical properties (colour, brightness, fluorescence lifetime and so on). Especially the available cyanine fluorophores suffer from limited brightness and signal-to-noise ratio (SNR) due to cis/trans isomerization[21], a fact that emphasizes the urgent need for a synthetic strategy to study and use other classes of organic fluorophores via intramolecular photostabilization[18–22].

Here we introduce a versatile and simple design concept to synthesize photostabilizer–dye conjugates on a specific biomolecular target using unnatural amino acids (UAAs)[23]. UAAs were chosen as a scaffold that links multiple chemical units, that is, the fluorophore and photostabilizer to a specific target. The presented conjugation strategy is based on well-known chemical reactions with (commercially available derivatives of) synthetic organic fluorophores, photostabilizers and UAAs, which can be bound as a single moiety to a biomolecular target. Depending on the UAA scaffold, the chemical nature of the functional groups can be N-hydroxysuccinimid esters (NHS), alkynes, azides or other bio-orthogonal reactive functionalities. Two different UAAs were used to bind different rhodamine, carbopyronine and cyanine fluorophores (Alexa555, RhodamineB, KK114, ATTO647N and Cy5) covalently to a photostabilizer on distinct biomolecular targets. The synthesized fluorophore derivatives comprise of either a reducing or oxidizing photostabilizer in the form of the antioxidant Trolox (TX) or a nitrophenyl group[10,12]. We characterized their photophysical properties with single-molecule fluorescence microscopy on the biomolecular target DNA and observed significant increases in photostability for all compounds including suppression of triplet-based blinking. Secondly, photostabilizer–dye conjugates were synthesized,

which allow labelling of biomolecules (DNA, antibodies and proteins) in a single step as with commercially available fluorophores. Finally, we demonstrate state-of-the-art applications of photostabilizer–dye conjugates in single-molecule Förster resonance energy transfer (smFRET) and super-resolution stimulated emission depletion (STED) microscopy, with significantly increased sensitivity and photostability of the conjugates compared to their non-stabilized parent fluorophores.

Results

Amino acids for intramolecular photostabilization. Figure 1 shows the central idea to use UAAs as a scaffold that links multiple chemical units, that is, fluorophore and photostabilizer onto a specific biomolecular target. The conjugation strategy is based on established chemical reactions (amide bond formation or click chemistry) using commercially available derivatives of synthetic organic fluorophores (F), photostabilizers (P) and UAAs.

As a proof-of-concept, we synthesized photostabilizer–dye conjugates of fluorophores that could thus far not be tested for intramolecular photostabilization due to limited scaffolding options. We focused on rhodamines and carbopyronines, that is, fluorophores from the ATTO and Alexa series, which are extremely popular for (life science) applications. (S)-Nitrophenyl-lalanine, NPA (Fig. 2), was used as a scaffold for the first generation of compounds, which consists of single-stranded DNA (ssDNA) as the biomolecular target, a commercially available organic fluorophore (Alexa555, ATTO647N and Cy5) and the p-nitrophenyl group of the known photostabilizer NPA[16,17,19] (Fig. 2, compounds 5, 6 and 7). The cyanine fluorophore Cy5 served as a 'positive control' experiment that permits to benchmark the effects of intramolecular photostabilization for different fluorophore classes with respect to published studies[16–20]. The used DNA and its specific base sequence was selected because of the detailed photophysical characterization of various fluorophores on this target[11,18,19,24–27].

The synthesis shown in Fig. 2 was conducted starting from commercially available fluorenylmethyloxycarbonyl (Fmoc)-protected NPA (1), which was converted into the corresponding NHS-ester derivative (2) to react with a 5′-aminoalkyl functionalized ssDNA, yielding 3. Subsequent Fmoc deprotection of 3 was followed by a reaction of the resulting primary amine (4) with the commercially available NHS-ester derivative of Alexa555, ATTO647N or Cy5, to yield 5, 6 and 7, respectively. All compounds comprise an organic fluorophore and a p-nitrophenyl group for intramolecular photostabilization linked to the biomolecular ssDNA target (Figs 1 and 2) and are abbreviated 'NPA fluorophore' throughout the text. The non-stabilized control compounds were obtained via a direct reaction of the NHS-ester-activated fluorophores with the respective amino-modified ssDNA-NH2. The DNA–photostabilizer–dye conjugates were isolated by high-performance liquid chromatography (HPLC; see Supplementary Fig. 1) and characterized by ultraviolet–visible absorption spectroscopy and mass

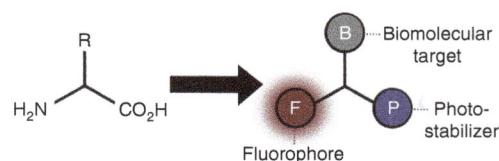

Figure 1 | Design concept for photostabilizer–dye conjugates. UAAs are used to combine an organic fluorophore covalently with a photostabilizer on a biomolecular target or linker structure.

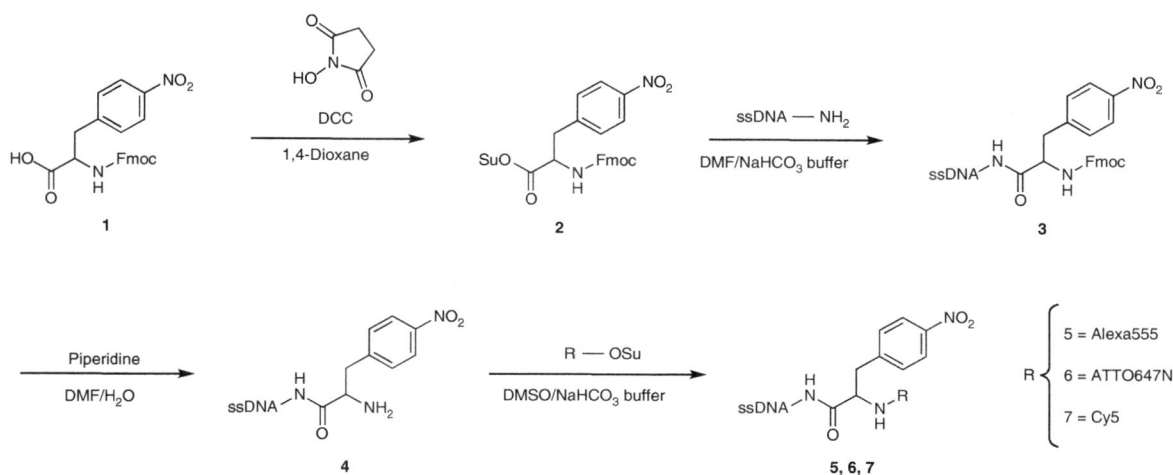

Figure 2 | Synthesis route towards photostabilizer–dye conjugates on nucleic acids. NPA-based fluorophores of Alexa555, ATTO647N and Cy5 on ssDNA. Fmoc, fluorenylmethyloxycarbonyl; DCC, N-N'-dicyclohexylcarbodiimide; Su, succinimide; DMF, dimethylformamide; NaHCO₃ buffer, 200 mM sodium hydrogen carbonate buffer pH 8.35; TBTA, tris[(1-benzyl-1H-1,2,3-triazol-4-yl)methyl]amine; TEAA buffer, 50 mM triethylammonium acetate buffer, pH 7.0.

spectrometry (Supplementary Figs 1 and 2). After purification, all compounds elute as single peaks in the HPLC chromatogram (Supplementary Fig. 1b,c) and display ultraviolet–visible absorption maxima characteristic for DNA (~ 260 nm) and the respective chromophore ($\sim 550/\sim 650$ nm) (Supplementary Fig. 2a). In addition, matrix-assisted laser desorption/ionization time-of-flight (MALDI-TOF) mass spectrometry reveals a significant mass increase of the DNA–dye conjugates compared with the non-modified DNA (Supplementary Fig. 2b). See the Methods section and the Supplementary Information for further details on chemical synthesis and characterization of functionalized oligonucleotides and reactive precursor molecules.

Next, single-molecule fluorescence microscopy was used to benchmark the potential of the scaffolding approach with respect to photophysical parameters. Confocal scanning microscopy[11,12,18,24] was used to investigate signal fluctuations and fluorescence lifetime, whereas total-internal-reflection fluorescence (TIRF) microscopy[11,18] was employed to obtain quantitative photophysical values such as bleaching lifetime, fluorescence count rate, total number of photons and SNR ratio. The different fluorophore derivatives were immobilized according to published procedures on a streptavidin-coated microscope coverslip by hybridization to form double-stranded DNA (dsDNA) comprising a 3'-terminal biotin unit (Supplementary Figs 3a–11a)[18,19]. All experiments described in this section were performed in the absence of oxygen, to minimize the convolution of triplet-state quenching by molecular oxygen and intramolecular photostabilization by NPA.

Using confocal microscopy, single immobilized fluorophores on DNA were identified as spots in the images and fluorescence time traces were recorded for each of the spots using single-photon counting detectors. The length of the fluorescence time traces gave an estimate of the photostability (the longer, the more stable), whereas fluctuations in the time traces indicated blinking due to transient population of dark states (such as the triplet state). The calculation of the autocorrelation function of the individual fluorescence time traces revealed the underlying time constants of the blinking kinetics (Fig. 3) and time-correlated single-photon counting was used to determine the fluorescence lifetime. The non-stabilized fluorophores, showed fast photobleaching and pronounced blinking on the corresponding dsDNA in deoxygenated PBS buffer (Fig. 3a,c,e and Supplementary Figs 3d, 5d and 7d), whereas the NPA derivatives showed a stable

fluorescence signal over extended periods of time (Fig. 3b,d,f and Supplementary Figs 4d, 6d and 7d).

Fluorescence time traces of Alexa555, a rhodamine-based fluorophore, were characterized by short observation times and pronounced blinking on the timescale of 22 ± 6 ms in deoxygenated PBS buffer (Fig. 3a and Supplementary Fig. 3). This is a typical behaviour for rhodamines and is caused by population of the triplet state[4]. The fluorescence lifetime of the sample was found to be 1.2 ± 0.3 ns (Supplementary Fig. 3). On conjugation of Alexa555 to the NPA scaffold (NPA–Alexa555, compound 5), the triplet-induced blinking diminished (negligible autocorrelation) and the overall fluorophore photostability and count rate increased to several seconds and 5–10 kHz, respectively, at ~ 0.3 kW cm^{-2} irradiance (Fig. 3b and Supplementary Fig. 4). Despite the overall increase in count rate (mainly due to abolishment of dark state transitions), the fluorescence lifetime of NPA–Alexa555 decreased to 0.9 ± 0.1 ns compared with the parent fluorophore, indicating the presence of singlet quenching by the NPA scaffold (Supplementary Fig. 4).

The photophysical behaviour of ATTO647N in deoxygenated PBS buffer was characterized by a mixture of short and long photobleaching times ranging from seconds to minutes with pronounced blinking on the millisecond timescale (Fig. 3c and Supplementary Fig. 5). The off-state lifetime associated with the blinking events was found to be 29 ± 5 ms and is attributed to the triplet state (Fig. 3c and Supplementary Fig. 5e)[4]. This and the observed fluorescence lifetime of 4.2 ± 0.4 ns are in agreement with literature[11,28]. Covalent binding of NPA to ATTO647N resulted in a homogeneous, bright, non-blinking and prolonged fluorescence emission (Fig. 3d). The fluorescence lifetime was reduced to 3.4 ± 0.2 ns for the majority of traces, again indicating dynamic singlet quenching by the NPA scaffold. Notably, $>50\%$ of the observed NPA–ATTO647N molecules did not photobleach within the 2 min observation period at excitation intensities of ≈ 0.66 kW cm^{-2} (see Fig. 3d and Supplementary Fig. 6). In addition, a small fraction ($<30\%$) of NPA–ATTO647N molecules showed a reduced lifetime of around 2.6 ns in combination with a lower count rate pointing to stronger singlet quenching (Supplementary Fig. 6). Direct switching of a single NPA–ATTO647N molecule between the two states was observed and will be a topic of future research[11].

Results for Cy5 (Fig. 3e,f) were obtained at higher excitation intensities (4 kW cm^{-2}), to allow for a direct comparison with

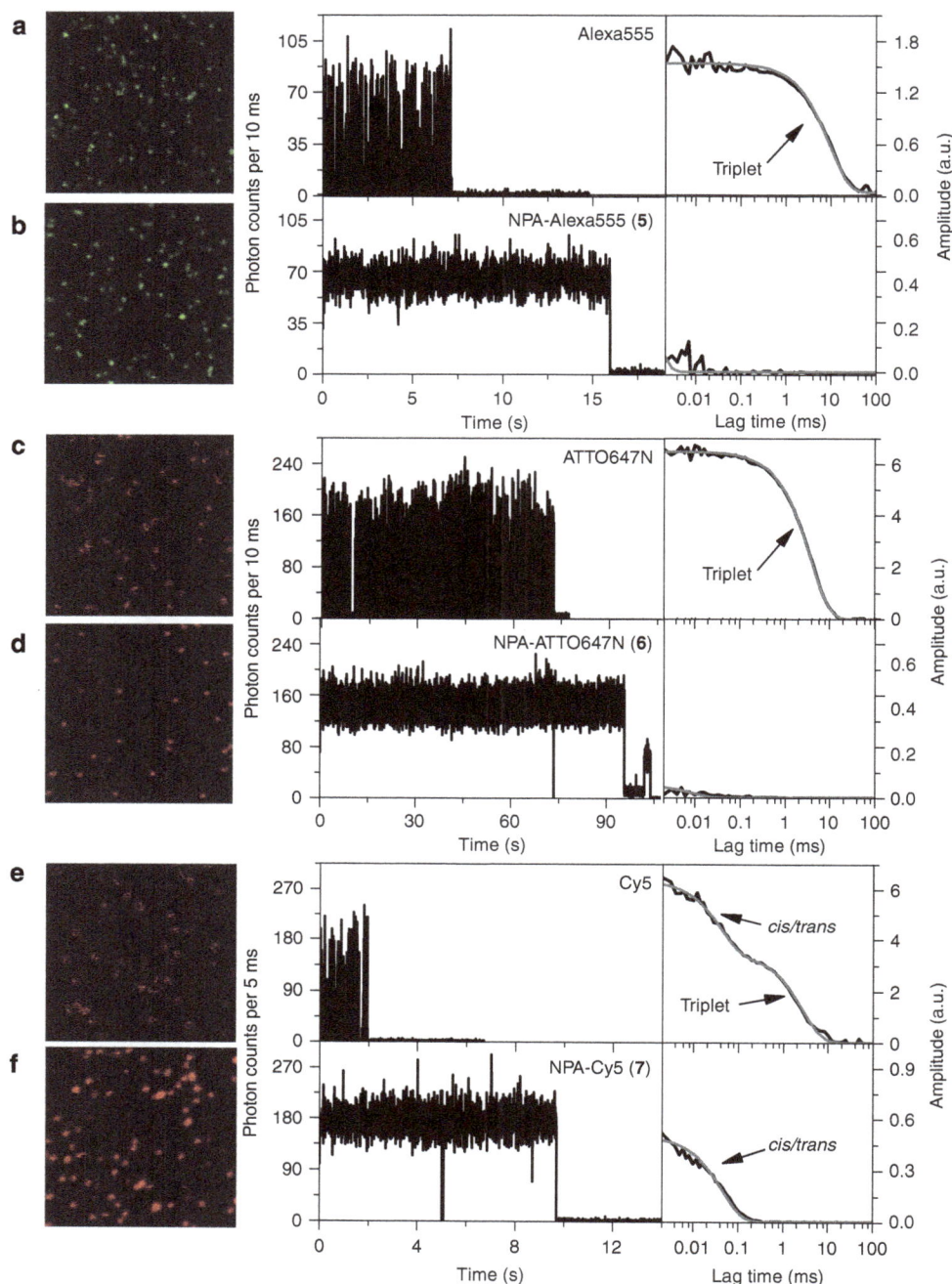

Figure 3 | Photophysical characterization of photostabilizer-dye conjugates with confocal microscopy. All data was recorded with a home-built confocal microscope with a sample in aqueous PBS buffer at pH 7.4 in the absence of oxygen. Each panel shows a representative overview image (10 × 10 μm, 50 nm pixel size, 2 ms per pixel) with spots from individual immobilized fluorophores (left), a fluorescence time trace (middle) and the corresponding autocorrelation decay in black, with according fits in grey (right). (**a,b**) Alexa555 and NPA-Alexa555 (image intensity scale from 3 to 60 counts, excitation intensity of $\approx 0.3\,kW\,cm^{-2}$ at 532 nm). (**c,d**) ATTO647N and NPA-ATTO647N (image intensity scale from 5 to 100 counts, excitation intensity of $\approx 0.66\,kW\,cm^{-2}$ at 640 nm). (**e,f**) Cy5 and NPA-Cy5 image intensity scale from 10 to 300 counts, excitation intensity of $4\,kW\,cm^{-2}$ at 640 nm. Further experimental details and data for each fluorophore can be found in the Methods section and Supplementary Figs 3–9.

previously published results using different approaches towards intramolecular photostabilization[29]. In agreement with earlier findings, the autocorrelation analysis of Cy5 fluorescent time traces[18] revealed the presence of two different photophysical processes that were attributed to triplet blinking (11 ± 4 ms) and *cis/trans* isomerization (54 ± 12 μs)[4,18,26,27]. The fluorescence lifetime of Cy5 was found to be 1.65 ± 0.15 ns (compare also Supplementary Fig. 7)[18,30]. NPA–Cy5 (**7**) under identical conditions revealed bright and prolonged fluorescence emission

with typical observation times of several seconds and, as revealed by the autocorrelation analysis, with negligible triplet-state blinking for ∼75% of the observed emitters (Fig. 3f). The remaining autocorrelation decay revealed the on–off transition due to *cis/trans* isomerization, with a lifetime in the order of ∼50–70 μs (see Supplementary Figs 7 and 8), which therefore still restricts the overall achievable count rate for the cyanine fluorophores. In addition, a mono-exponential fit does not fully describe the experimental autocorrelation decay (see the deviations between grey fit and black data in Fig. 3f), suggesting a remaining small triplet population with a lifetime $<100\,\mu s$. A smaller population of molecules ($<25\%$) showed intensity fluctuations on completely different timescales (Supplementary Fig. 8), which is consistent with earlier reports[18,19], and its origin remains to be explained mechanistically. The fluorescence lifetime of NPA–Cy5 was found to be $1.60 \pm 0.15\,ns$ and is similar to the parent fluorophore, indicating that any reduction in the signal brightness of NPA–Cy5 is due to static quenching, that is, transient formation of non-fluorescent complexes between photostabilizer and fluorophore rather than dynamic singlet quenching. These findings are consistent with a subtle blue shift in the Cy5-absorption spectrum when bound to NPA (Supplementary Fig. 2a).

To further quantify the improved performance of the photostabilizer–dye conjugates, we used single-molecule TIRF microscopy to determine various parameters with high statistics. Movies with $100\,ms$ integration time were recorded at laser excitation intensities of ≈ 50–$100\,W\,cm^{-2}$, that is, a significantly lower excitation intensity than in the confocal scanning microscopy experiments, yielding different count rates and SNR ratios. Figure 4a shows a typical example of a camera frame with single Cy5 fluorophores. The number of fluorescent molecules per video frame was determined and the decay in number of molecules over subsequent image frames was fitted with an exponential decay ($y(t) = C + A \times \exp(-t/\tau_{bleach})$), to obtain the photobleaching lifetime τ_{bleach} (Fig. 4a)[18]. Background-corrected single-molecule time traces were extracted and used to determine the fluorescence count rate in kHz (Fig. 4a, brightness), the SNR ratio (Fig. 4a) and the total number of detected photons before photobleaching (Fig. 4a, N_{total} = brightness $\times$ τ_{bleach}). The mean and s.d. of all values was derived from multiple ($n \geq 3$) independent experiments. We benchmarked the performance of the photostabilizer–dye conjugates against the antioxidant TX as a solution additive in the deoxygenated imaging buffer ($2\,mM$ after $20\,min$ ultraviolet treatment), which is a common standard for photostabilization in single-molecule experiments[10,12].

The data of the non-stabilized rhodamine Alexa555 revealed a brightness of $1.5 \pm 0.2\,kHz$, $N_{total} = 5.1 \pm 0.4 \times 10^4$ and an $SNR = 3.5 \pm 0.3$, whereas the photobleaching could be described by a mono-exponential decay with a time constant $\tau_{bleach} = 34 \pm 26\,s$ (Fig. 4b). An improvement for most of the photophysical parameters was found for both adding $2\,mM$ TX or Alexa555-NPA with up to a 20-fold increase in photostability, 25-fold increase in N_{total} and a 3-fold increase in SNR, whereas the brightness was found to be comparable.

Similar results were obtained for both the carbopyronine ATTO647N and the cyanine dye Cy5 (Fig. 4c and Supplementary Movie 1). The photobleaching time increased from $\tau_{bleach} = 138 \pm 93\,s$ (ATTO647N) to $212 \pm 52\,s$ (NPA–ATTO647N) and $298 \pm 45\,s$ (ATTO647N + $2\,mM$ TX), and from $\tau_{bleach} = 7.0 \pm 1.5\,s$ (Cy5) to $139 \pm 55\,s$ (NPA–Cy5) and $384 \pm 60\,s$ (Cy5 + $2\,mM$ TX). This results in reduced blinking and an increase in the total number of detected photons (N_{total}) and SNR from $N_{total} = 1.9 \pm 0.7 \times 10^5$ and $SNR = 1.6 \pm 0.2$ (ATTO647N) to $N_{total} = 8.6 \pm 0.4 \times 10^5$ and $SNR = 11.6 \pm 1.7$ (NPA–ATTO647N), and $N_{total} = 2.0 \pm 0.7 \times 10^6$ and $SNR = 17.4 \pm 2.4$ (ATTO647N + $2\,mM$ TX), as well as $N_{total} = 3.1 \pm 0.9 \times 10^4$ and $SNR = 3.0 \pm 0.7$ (Cy5) to $N_{total} = 6.4 \pm 1.6 \times 10^5$ and $SNR = 11.6 \pm 1.7$ (NPA–Cy5), and $N_{total} = 2.4 \pm 0.6 \times 10^6$ and $SNR = 9.5 \pm 1.2$ (ATTO647N + $2\,mM$ TX), that is, 20- to 50-fold increases.

The investigations in Figs 3 and 4 regarding the photophysical properties of rhodamines, carbopyronines and cyanines with intramolecular photostabilization revealed the following: (i) photobleaching and blinking could be efficiently removed using UAA scaffolding and (ii) all photophysical parameters were improved substantially compared with the parent fluorophore. (iii) Despite showing no decrease in brightness, rhodamines and carbopyronines showed singlet quenching, whereas cyanines show no reduction of the excited state lifetime. (iv) Solution-based healing using $2\,mM$ TX remained more efficient than intramolecular photostabilization.

Despite the slightly lower photostability of compounds with intramolecular photostabilization compared with solution additives such as TX, this approach has unique advantages that compensates for these shortcomings. The UAA approach is thus far the only possible method for photostabilization when organic fluorophores are used under live-cell conditions, in experiments where the addition of a diffusion-based photostabilizer is not tolerated due to its toxicity or when the photostabilizer has an unwanted influence on properties of the system of interest[31]. It is also the only viable option when diffusion-based photostabilization remains ineffective which could be caused by lack of collisions between the photostabilizer and the fluorophore[32,33].

UAAs as a general scaffold for photostabilizer–dye conjugates.

The scaffolding strategy presented in Fig. 2 is restricted by the availability of commercial UAAs with different (photostabilizing) residues and does not allow to use custom-made photostabilizers[19]. We hence set out to generalize the approach and to allow linkage of three arbitrary moieties. The amino acid propagylglycine (PG (**8**); Fig. 5) represents a more versatile scaffold that can link three chemical groups via NHS and click chemistry (here, a $[3+2]$-Huysgens cycloaddition between alkyne and azide derivatives). Hence, PG (**8**) provides the flexibility needed to combine any photostabilizer and fluorophore on a (bio)molecular target, assuming their availability as NHS- or click-reactive derivatives.

In the first synthetic step, racemic **8** was reacted with the NHS-ester derivative of TX (**9**). The resulting adduct **10** was converted into the corresponding NHS-ester derivative **11** and reacted with the $5'$-aminoalkyl functionalized oligonucleotide to yield **12**. Finally, the fluorophore Cy5 was bound using a copper-catalysed click reaction[34] yielding **13**, called TX–PG–Cy5 throughout the manuscript.

TX–PG–Cy5 (**13**) showed a very similar photophysical behaviour compared with NPA–Cy5, that is, removal of blinking and an increased photobleaching lifetime (Fig. 4d) with significant improvement factors (up to 11-fold). However, a higher heterogeneity was observed with $\sim 60\%$ of the molecules showing fluorescent traces with little intensity fluctuations (shorter bleaching lifetime; Supplementary Fig. 9). The remaining $\sim 40\%$ exhibited longer observation times with an increased amount of blinking events (Supplementary Fig. 9). A unifying fluorescence lifetime of $1.65 \pm 0.15\,ns$ was observed, which is equal to the parent Cy5. Several TX–PG–Cy5 molecules showed a typical behaviour for fluorophores with intramolecular photostabilization, where 'bleaching' or irreversible destruction of the stabilizer occurred before photobleaching of the fluorophore. Here the fluorophore drastically changed its emission pattern from stable to a blinking emission pattern that closely resembles the behaviour in the absence of photostabilizer (Supplementary Fig. 9d, bottom row, left panel: stable: 0–8.5 s; blinking: 8.5–10 s).

The PG-based fluorophore system showed to be a versatile scaffold where the photostabilizer and fluorophore can be varied independently of each other. This has the advantage that the fluorophore and photostabilizer can be matched, to obtain maximum photostabilization as was shown above for the fluorophore Cy5 and the photostabilizer TX.

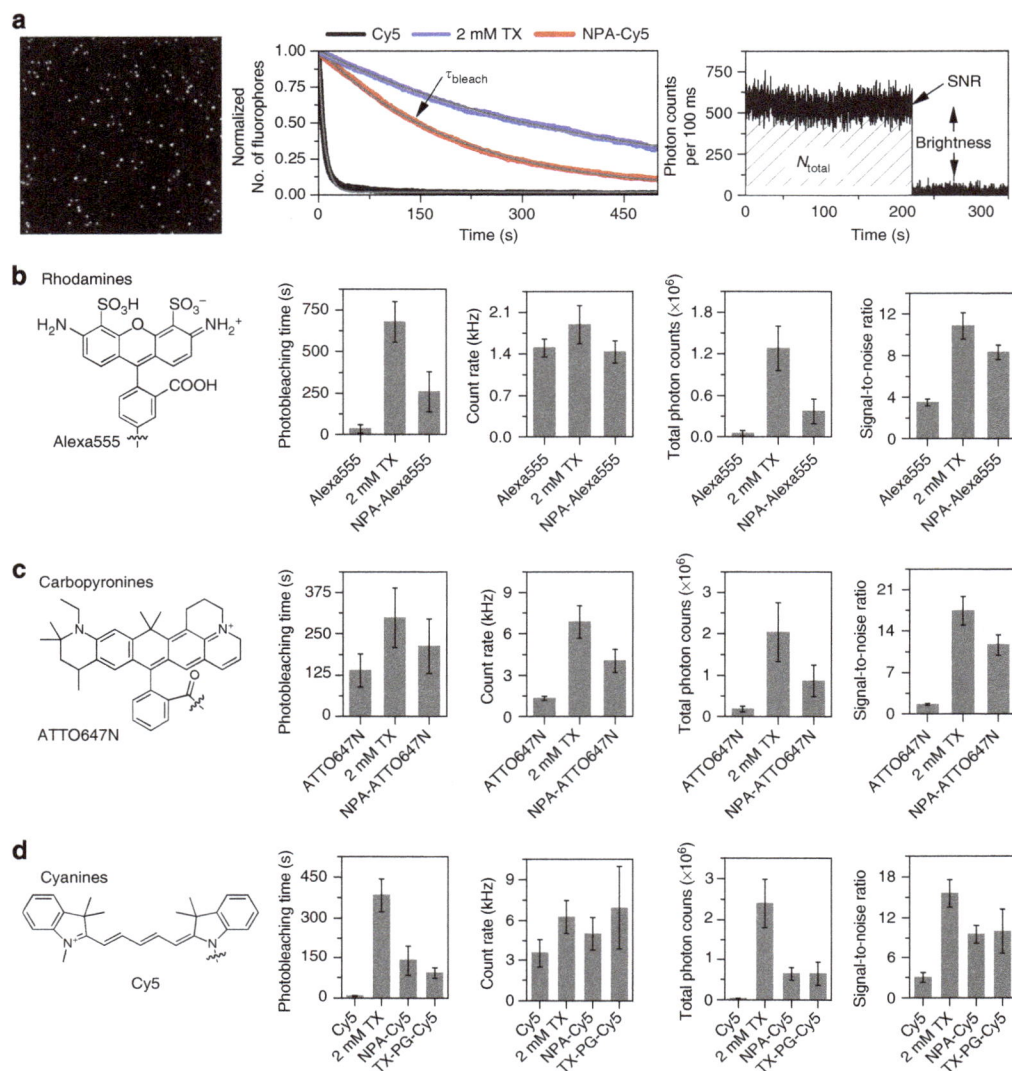

Figure 4 | Photophysical characterization of photostabilizer–dye conjugates and their parent fluorophores with single-molecule TIRF microscopy.
Data was recorded in aqueous PBS buffer at pH 7.4 in the absence of oxygen under continuous 640 nm excitation with ≈50 W cm^{-2} or 532 nm excitation with ≈20 W cm^{-2}. (**a**) Representative image frame (left panel) showing single fluorescent molecules (10 × 10 μm, exemplarily for Cy5). Subsequent images recorded over a period of 500 s showed an exponential decrease in the number of fluorescing molecules with a photobleaching lifetime τ_{bleach}, as shown for Cy5 (black), Cy5 with 2 mM TX (blue) and NPA–Cy5 (red) (middle panel). The curves shown were obtained by averaging over >5 TIRF movies. (**b,c,d**) Right panel: chemical structures (left panels) and respective photophysical parameters obtained from background-corrected fluorescence traces for Alexa555 (**b**, rhodamines), ATTO647N (**c**, carbopyronines) and Cy5 (**d**, cyanines). Values and error bars (s.d.) in bar graphs obtained from $N > 500$ molecules. For further details of the experimental techniques, data acquisition and analysis, see the Methods section.

Direct labelling of biomolecules. The synthesis and photophysical characterization of the photostabilizer–dye conjugates in the previous sections demonstrated the principle improvements in fluorescence emission properties by the UAAs scaffolding approach. Yet, our overall goal is to show the potential of the new fluorophores in biological research. For this, the most important step is labelling biomolecules such as proteins, nucleic acids or antibodies. As shown in 'UAAs as a general scaffold for photostabilizer–dye conjugates', this usually requires a complex (multi-step) synthesis process, demanding larger amounts of substances, that is, both fluorophore and biomolecular target. However, the amount of biomolecules is often limited. We hence reduced the chemical steps needed for labelling to a minimum. For this we altered the synthesis strategy of the photostabilizer–dye conjugates, introducing a functionalized group such as a NHS ester or a maleimide, which allow using the fluorophore derivative straightforward for labelling of a biomolecular target via

primary amines or cysteine residues (as is done conventionally). NPA was again used as a scaffold for this second generation of photostabilizer–dye conjugates.

Owing to its ease and cost-effective accessibility, we first used the NHS-ester derivative of the dye RhodamineB (**14**), which was subjected to NPA (**15**) under basic conditions to yield **16** (Fig. 6). Subsequent activation with NHS/N-N'-dicyclohexylcarbodiimide in dimethylformamide (DMF) gave the NHS-ester derivative of the NPA–RhodamineB conjugate (**17**). In a first step, we (as for the previous fluorophores) targeted this conjugate to ssDNA (**18**), which was done straightforward from **17** without further purification. The photophysical characteristics of the resulting compound were characterized when immobilized on glass in deoxygenated buffer.

Similar to the fluorophores studied before (see Fig. 3), RhodamineB showed strong blinking in deoxygenated PBS buffer with typical observation times until photobleaching on the

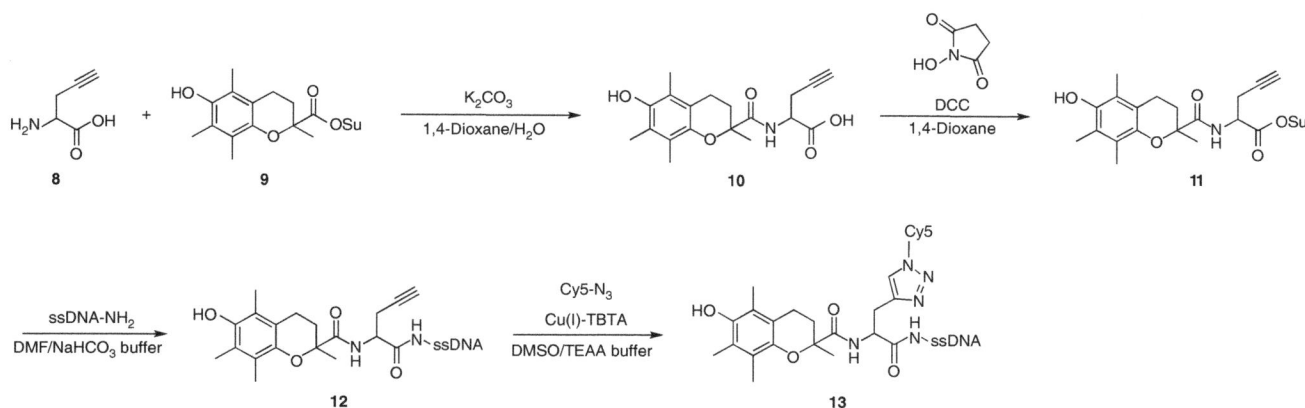

Figure 5 | Synthesis route to obtain TX–PG–Cy5 (13) on ssDNA. DCC, *N-N'*-dicyclohexylcarbodiimide; Su, succinimide; DMF, dimethylformamide; NaHCO₃ buffer, 200 mM sodium hydrogen carbonate buffer pH 8.35; TBTA, tris[(1-benzyl-1H-1,2,3-triazol-4-yl)methyl]amine; TEAA buffer, 50 mM triethylammonium acetate buffer, pH 7.0.

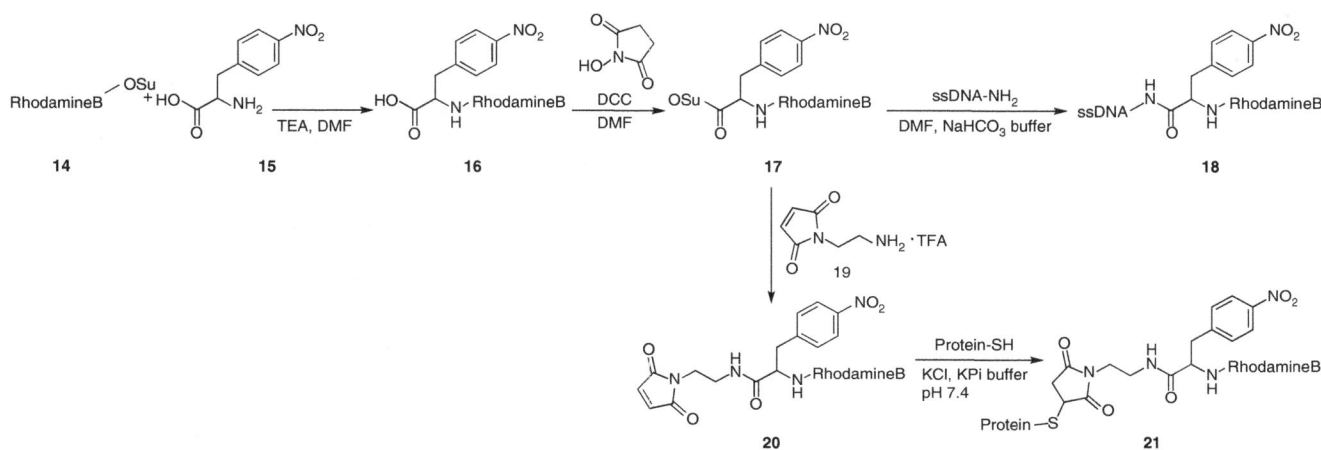

Figure 6 | Synthesis of reactive photostabilizer–dye conjugates of RhodamineB for direct labelling of primary amines and thiol residues. The strategy can be extended to other biochemical targets by a varying the linker of molecule 19.

timescale of 10–20 s (Supplementary Fig. 10). In contrast, RhodamineB displayed strongly heterogeneous blinking characteristics with very short or longer off times within one trace (Supplementary Fig. 10). Consequently, the autocorrelation function of this fluctuating signal could only be described by a bi-exponential decay with average off times peaking at 7 ± 4 ms and a significant fraction of values >20 ms, deviating from a normal distribution (Supplementary Fig. 10). This indicates the presence of multiple dark states or heterogeneous dye environments. Still the blinking off times were in the same range as determined before for the triplet state of the structurally related rhodamine fluorophores, for example, Alexa555 (22 ± 6 ms, Supplementary Fig. 3) or ATTO565 (6 ± 2 ms)[4] and are hence considered to be triplet related. Yet, the fluorescence lifetime of RhodamineB was found to be rather homogeneously distributed with an average of 3.1 ± 0.4 ns (Supplementary Fig. 10). Strikingly, NPA–RhodamineB on ssDNA (**18**) showed strongly reduced blinking, an increased photostability resulting in a stable and non-blinking emission pattern with observation times of up to minutes (Supplementary Fig. 11) and an increased brightness (Supplementary Fig. 12).

Next, we intended to use our NPA-based photostabilizer–dye conjugates for direct labelling of proteins. For this purpose, two different synthesis strategies were developed, accounting for the available quantity of the fluorophore. In the most straightforward case, that is, larger amounts of amine-reactive photostabilizer–dye conjugate are available (for example, >50 mg of compound **17**

was available), the NHS ester of the fluorophore (**17**) can be coupled directly with 2-maleimidoethylamine (**19**), to yield a maleimide derivative of, for example, NPA–RhodamineB (**20**) in Fig. 6. The second strategy is also feasible for small quantities of reactive fluorophore species (<10 mg) that could be due to high prices of commercially available precursors or complicated synthesis. ATTO647N is a good example for a fluorophore that is often used in demanding fluorescence applications but is not readily available in large amounts. For these cases we optimized the synthesis route as shown in Fig. 7, to yield a thiol-reactive derivative of ATTO647N containing the photostabilizer NPA (Fig. 7, compound **25**). As shown in 'Biomolecular FRET study with photostabilizer–dye conjugates', both maleimide derivatives can covalently bind to recombinant proteins via solvent-exposed cysteine residues (Fig. 6 compound **21** and Fig. 7 compound **26**).

Biomolecular FRET study with photostabilizer–dye conjugates. To show the benefits of intramolecular photostabilization in fluorescence applications with proteins, we studied the substrate-binding domain 2 (SBD2) of the *Lactococcus lactis* ABC transporter GlnPQ[35]. Using smFRET and alternating laser excitation (ALEX) spectroscopy, the conformational states of the protein were monitored (Fig. 8a, open unliganded and closed liganded). The structural rearrangements of SBD2 on ligand binding causes a change of ~0.9 nm regarding the distance between two selected amino acids in the protein[35]. As the FRET

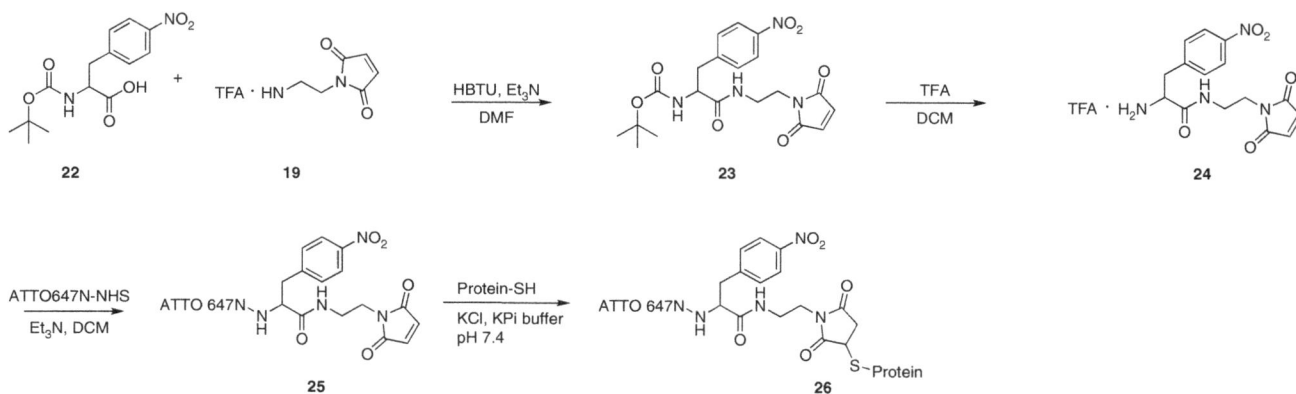

Figure 7 | Simplified synthesis of reactive photostabilizer-dye conjugates where only small quantities of fluorophore are available. The resulting NPA–ATTO647N conjugate can be used for direct labelling of thiol residues, for example, in proteins (compound **25**). The strategy can be extended to other biochemical targets by a variation of the linker molecule 19.

donor and acceptor are attached via maleimide chemistry at these positions in the protein (mutant of SBD2: T369C/S451C), the transfer efficiency E^* reports on the conformational state of the protein. As described previously[35], SBD2 was labelled stochastically using appropriate mixtures of donor and acceptor fluorophores (details see Methods). To understand the effects of intramolecular photostabilization in FRET-based assays, we used different fluorophore combinations: Cy3B or RhodamineB as donor fluorophores and ATTO647N as the acceptor. In experiments described below, either the donor (RhodamineB) or the acceptor (ATTO647N) was stabilized via covalent linkage to NPA (synthesis see Figs 6 and 7).

We used smFRET and ALEX spectroscopy, as described in refs 33,35,36, to investigate the photophysical properties of the fluorophores when bound covalently to the protein and the biomolecular function of SBD2. ALEX is a valuable tool for both purposes, as it allows to distinguish the desired protein molecules containing both fluorophores, thus monitoring protein conformation (donor–acceptor species; Fig. 8b, $0.9 < S < 0.4$), from those labelled with only one species not providing distance information (donor only, $S > 0.9$; acceptor only, $S < 0.4$). Although S relates to the relative fluorophore brightness and labelling stoichiometry, E^* indicates the FRET efficiency and thus distance between the donor and acceptor (with larger values indicating small donor–acceptor distances), which is the final read out of the protein conformation. In addition, ALEX reveals photophysical artefacts such as blinking of donor or acceptor in the form of bridges between the three different subpopulations[37]. For such measurements, the fluorescence emission of the donor $F(DD)$ under green excitation, that of the acceptor $F(DA)$ when excited via FRET from the donor and that of the acceptor via direct red excitation light $F(AA)$ was determined (see Methods and refs 33,35,36). In our experiments, individual biomolecules were studied for short time periods of a few milliseconds, while diffusing through the excitation volume of a confocal microscope. The challenge of such an experiment is to acquire intense fluorescent bursts during the short observation time under the high excitation intensities of $> 10\,\mathrm{kW\,cm^{-2}}$.

The combination of Cy3B and ATTO647N is known for excellent photophysical performance resulting in fast data acquisition and superior histogram quality in smFRET with little bleaching artefacts and narrow distributions (Fig. 8c). For apo-SBD2 we find mean E^* and S of 0.48 ± 0.08 and 0.59 ± 0.07, respectively, with these labels[35]. These values are in good agreement with our published work and the mean E^* correlates with the expected interprobe distance of $\sim 4.9\,\mathrm{nm}$ derived from the crystal structure.

Results of such quality are, however, only available when using 2 mM TX as a photostabilizer in solution, seen from comparison of Fig. 8c,d. Here we show data of apo-SBD2 (labelled with Cy3B/ATTO647N) in the presence and absence of TX. In agreement with Kong et al.[37], the high excitation intensities used in our experiments promote acceptor signal fluctuations, that is, blinking and/or bleaching. Cy3B and ATTO647N can hence be seen as a FRET couple where the acceptor photostability is limiting. This appears as a prominent bridge between the donor-only and donor-acceptor population (Fig. 8d), altering both E^*/S-values substantially. Closer inspection of the histogram reveals that a significant portion of the molecules show these unwanted photophysical effects. Under these conditions, neither mean E^* nor correction factors for accurate FRET determination are directly accessible. Besides the complete loss of information, the overall acquisition time has also increased in the absence of photostabilizer to obtain sufficient statistics from the relevant donor–acceptor species. It should be noted that such photophysical artefacts of the acceptor (Fig. 8d) are extremely problematic for data interpretation, as they suggest the existence of (non-biological) species in between the donor only and the actual FRET species (see $1D - E^*$ in Fig. 8d that can only be fitted by the sum of two Gaussians with $E^* = 0.19 \pm 0.07$ and 0.43 ± 0.1). Furthermore, it remains challenging to identify those when performing smFRET with green excitation in continuous-wave mode.

Strikingly, the bridge population can be removed by sole photostabilization of the acceptor fluorophore via scaffolding of ATTO647N to NPA. The ALEX data of apo-SBD2 with Cy3B/NPA–ATTO647N is shown in Fig. 8e. Here the bridge between the donor–acceptor and donor-only population is fully removed without the addition of stabilizer to the solution (Fig. 8d versus Fig. 8e). The resulting mean E^*/S values are changed compared to apo-SBD2 with Cy3B/ATTO647N (2 mM TX in solution, Fig. 8b) accounting for decreased acceptor brightness; mean E^* is now found at 0.34 ± 0.09 and mean S at 0.67 ± 0.08 (Fig. 8e). These differences are, however, expected considering the results from ATTO647N on DNA, where a decrease in the overall brightness is observed on conjugation to NPA (Fig. 4c). It should be mentioned that differences in the donor–acceptor population relative to donor and acceptor only comparing the samples Cy3B/ATTO647N and Cy3B/NPA–ATTO647N are not solely due to photophysics but also due to different labelling ratios of the protein.

Next, we used Cy3B/NPA–ATTO647N to study the biomolecular function of SBD2 (Fig. 8b). On addition of the ligand glutamine, the mean E^* changes from 0.36 ± 0.1 (open, interprobe

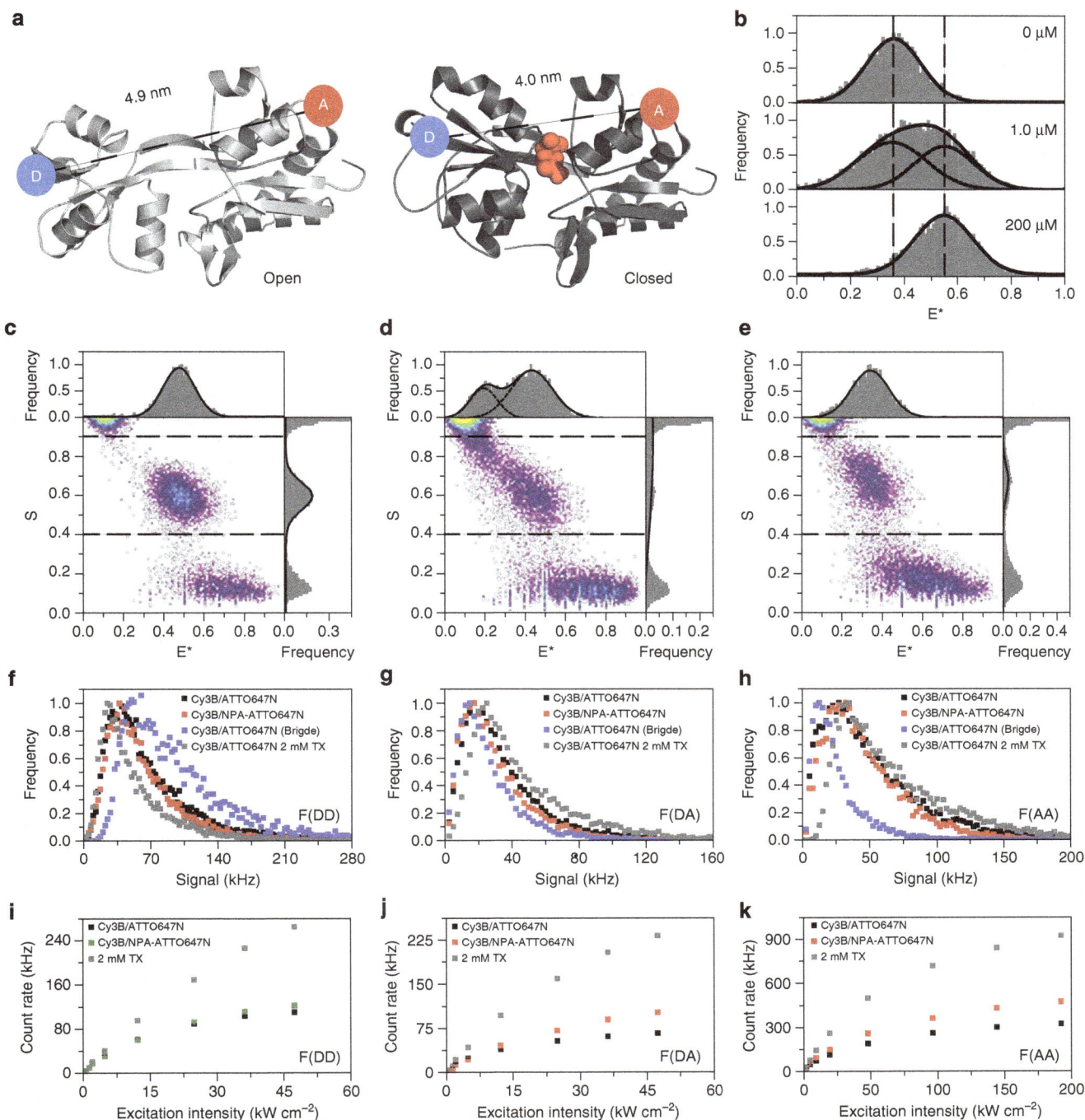

Figure 8 | Improving smFRET-ALEX measurements by using NPA-based fluorophore as FRET acceptor on the protein GlnPQ-SBD2. (**a**) Crystal structures of the SBD2 (T369C/S451C) open (left panel, PDB: 4KR5) and closed state (right panel, PDB:4KQP, after binding of the ligand glutamine shown in red) with label positions of donor (D) and acceptor (A). (**b**) Corresponding one-dimensional histograms of E^*-values for increasing amounts of ligand glutamine where an increased $E^* = 0.55 \pm 0.1$ (dashed line, closed conformation) becomes visible (as opposed to the open conformation, $E^* = 0.36 \pm 0.1$, dashed line). Excitation intensities of 30 kW cm^{-2} at 532 nm and 20 kW cm^{-2} at 640 nm; data evaluated with dual colour burst search and displayed with additional per-bin thresholds of $F(DD) + F(DA) + F(AA) > 75$ and minimal number of counts per bin in ALEX histogram of 3. (**c,d,e**) 2D histograms of joint pair values of S (labelling stoichiometry) and E^* (FRET efficiency, that is, interprobe distance) of Cy3B/ATTO647N in the presence (**c**) and absence (**d**) of 2 mM TX and Cy3B/NPA-ATTO647N (**e**) without ligand glutamine. Excitation intensities of 30 kW cm^{-2} at 532 nm and 20 kW cm^{-2} at 640 nm; data evaluated with all photon burst search and displayed with additional per-bin thresholds of $F(DD) + F(DA) + F(AA) > 100$ and minimal number of counts per bin in ALEX histogram of 3. (**f,g,h**) Histogram of fluorophore brightness values as determined from photon-counting histograms (PCHs) on single-molecule transits of labelled SBD2 diffusing through the observation volume, comparison of donor brightness $F(DD)$ (**f**), acceptor brightness when excited via FRET, $F(DA)$ (**g**) and acceptor brightness via direct red excitation $F(AA)$ (**h**). Excitation intensities of 30 kW cm^{-2} at 532 nm and 20 kW cm^{-2} at 640 nm. (**i,j,k**) Dependence of the mean count rate of $F(DD)$, $F(DA)$ and $F(AA)$ of the different samples for increasing excitation intensity, respectively.

distance of ~4.9 nm) to 0.55 ± 0.1 (closed, with a decreased interprobe distance of ~4.0 nm). A concentration of 200 µM saturates ligand binding and therefore results in a 100% population of the closed state (Fig. 8b). A ligand concentration of 1 µM, which is close to the K_d-value of the protein[38], consequently results in a mix of open and closed states (Fig. 8b).

As fluorophore brightness and the resulting photon budget ultimately determine the quality of the final histograms including the necessary measurement time, we quantitatively analysed Cy3B/ATTO647N and Cy3B/NPA–ATTO647N by means of photon-counting histograms. Figure 8f,g,h shows the three relevant photon streams used to determine E^* and S of Cy3B/ATTO647N (donor–acceptor: 0.9 > S > 0.4; bridge: 0.9 > S > 0.7) and Cy3B/NPA–ATTO647N (donor–acceptor: 0.9 > S > 0.4). $F(DD)$ shows that the strongest donor quenching and hence the most efficient FRET is found for the addition of TX to Cy3B/ATTO647N, whereas molecules in the bridge (Cy3B/ATTO647N, no TX) show inefficient donor quenching due to a non-functional acceptor (Fig. 8f). Cy3B/NPA–ATTO647N and healthy molecules from the Cy3B/ATTO647N population with no TX show a similar performance of the acceptor-based donor quenching. $F(DA)$ correlates directly with quenching in $F(DD)$ as seen in Fig. 8g. Again, the best performance is observed from Cy3B/ATTO647N in the presence of TX, while molecules in the bridge show the lowest counts. A striking difference between the bridge and all other conditions is seen in $F(AA)$, that is, direct excitation of the acceptor (Fig. 8h). The overall comparison suggest that NPA-based acceptor stabilization is sufficient to remove photophysical artefacts and hence make the smFRET useful for biomolecular studies. Further optimization of the data quality could, however, be obtained by additional donor stabilization of, for example, Cy3B in this case. This interpretation is further supported by excitation intensity-dependent count rates of Cy3B/ATTO647N and Cy3B/NPA–ATTO647N (Fig. 8i,j,k). NPA-based acceptor stabilization improves the saturation characteristics in all three channels but the addition of TX to the solution (resulting in stabilization of both donor and acceptor at the same time) remains superior in terms of achievable count rates.

To study the donor dependence in more detail, we repeated the above described experiments using RhodamineB/ATTO647N and NPA–RhodamineB/ATTO647N. Here, the photostability of the donor is the limiting factor for smFRET data quality. For RhodamineB/ATTO647N we find prominent donor blinking in the absence of photostabilizer (Supplementary Fig. 13). The bridge between the donor–acceptor and acceptor-only population can be removed by addition of TX in solution (Supplementary Fig. 13) or conjugation of RhodamineB to NPA (Supplementary Fig. 13). The overall magnitude of the observed effects is lower than for the case of acceptor bleaching/blinking in Fig. 8. Photon-counting histograms and the intensity dependence show a similar trend as before, that is, correlation between $F(DD)$ and $F(DA)$ with the wish to increase donor photostability as much as possible. Our data makes clear that NPA–RhodamineB can be used at significantly higher excitation intensities than RhodamineB, as triplet-state population was minimized and would allow for faster and better data acquisition (Supplementary Fig. 13).

Our results show that the simple addition of the NPA unit to the acceptor fluorophore is sufficient to gather reliable results from solution-based smFRET (Fig. 8). Although the addition of a stabilizer on the donor fluorophore can increase the overall available photon budget (which should always be maximized), the acceptor strictly requires a solution-based or covalently linked photostabilizer. We consequently suggest to stabilize the acceptor as a minimum and (if possible) also to stabilize the donor to

maximize available count rates and hence data quality. We finally note that all smFRET results of SBD2 with the different pairs of fluorophores shown here are in good agreement with biochemical data from isothermal calorimetry and with our previous results, supporting our hypothesis of an induced-fit-type mechanism in GlnPQ[35]. In addition, it confirms successful labelling of the protein with the custom-made NPA derivative, and that the results from the established FRET assay are indeed independent of the fluorophore pair that is used to monitor the protein conformation[35]. It is noteworthy that spectrally uncorrected apparent FRET is reported here accounting for absolute differences in mean E^*-values from varying Förster radius R_0 of the fluorophore pairs used or variations in the setup alignment[35].

Photostabilizer–dye conjugates in super-resolution microscopy. A common tool in cellular far-field fluorescence microscopy is the use of immunofluorescence, where the majority of cells are fixed and individual structures or proteins are tagged using specific primary antibodies and fluorophore-labelled secondary antibodies. In recent years, pioneering developments in far-field fluorescence microscopy, so-called super-resolution microscopes, have revolutionized cellular imaging, allowing the visualization of biological structures with nanometre resolution, that is, beyond the physical diffraction limit that thus far prevented to resolve structures with a precision better than ~250 nm[2]. Given the importance of photostability and brightness in such techniques, we investigated the potential of fluorophores with intramolecular photostabilization for immunofluorescence and specifically super-resolution imaging. For this, we have used the NHS ester of the dye KK114 (ref. 39) and its derivative NPA–KK114 (Supplementary Fig. 14) to directly tag secondary antibodies using standard procedures. These antibodies were used to specifically immunolabel nuclear pore complexes (NPCs) in fixed mammalian PtK2 cells. Figure 9a and Supplementary Fig. 15 depict that we could successfully apply NPA–KK114 to visualize the spatial distribution of these NPCs as shown in the representative confocal scanning images. However, it also becomes obvious that these complexes appear as rather large (>250 nm), blurred spots due to the diffraction limit. To increase resolution, we employed super-resolution STED microscopy[3,40], which in its most common application adds a STED laser to the confocal microscope that features a focal intensity distribution with a central zero, to allow imaging with subdiffraction spatial resolution. Figure 9b and Supplementary Fig. 15 clearly show the succesfull implementation of NPA–KK114 in STED microscopy. The NPCs were much better resolved and appeared as much smaller spots (85–90 nm, which is a reasonable value considering the use of primary and secondary antibodies). When compared with conventional KK114, the NPA–KK114 showed an increased photostability under STED conditions. Repeated scanning of the same area of the cell revealed that fading of fluorescence was reduced for NPA–KK114, while brightness was only subtly increased (Fig. 9d–f, Supplementary Fig. 15 and Supplementary Movie 2). The improvement in photobleaching resistance under STED conditions via an intramolecular mechanism highlights future directions towards improved dynamic STED imaging without the need of adding (potential toxic) chemical compounds[41].

Discussion
Our results show that effective intramolecular photostabilization and removal of (triplet induced) dark states can be achieved for NPA- and PG-based fluorophore derivatives with similar or better photostabilization effects as previously introduced photostabilizer–dye conjugates. This shows that the UAA scaffold

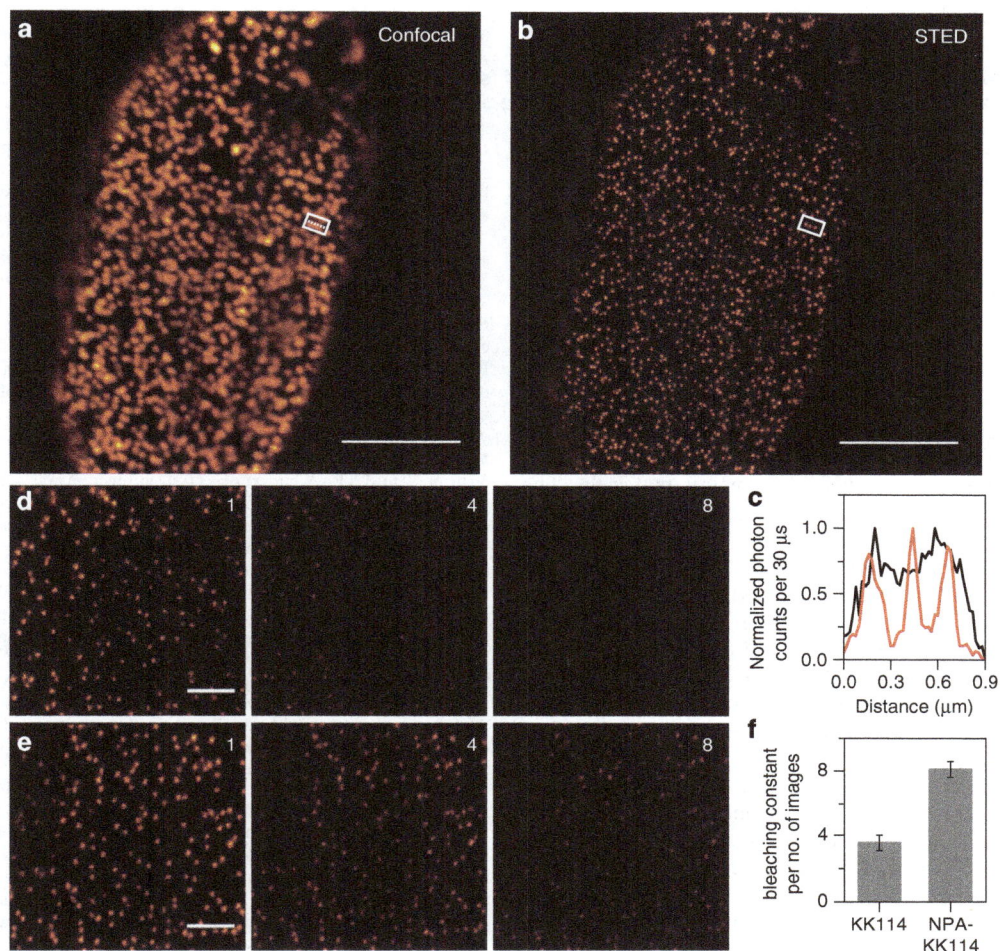

Figure 9 | Use of photostabilizer–dye conjugates in confocal and super-resolution STED microscopy of immunolabelled cells. Data from NPCs in fixed mammalian PtK2 cells immunolabelled with KK114 and NPA–KK114 Confocal (**a**) and STED (**b**) images of a representative cell stained with NPA–KK114 (scale bars, 5 μm). (**c**) Normalized intensity profile along the dashed line in the white boxes marked in the images of **a,b**, exemplifying how neighbouring NPCs can be much better resolved in the STED (red) compared with the confocal (black) recordings. (**d–f**) Repeated scanning of the same area of the cells in the STED mode indicates reduced photobleaching in the case of NPA–KK114 compared with KK114: images 1, 4 and 8 for KK114 (**d**) and NPA–KK114 (**e**) (scale bars, 1 μm), and (**f**) bleaching constant from an exponential fit in terms of number of images for KK114 and NPA–KK114 (total number of subsequent images recorded = 30, $n = 4$).

is a simple and useful tool to covalently link fluorophore, photostabilizer and biomolecular target in a modular manner. The presented approach has many advantages over existing strategies, as it is cost effective, requires low amounts of material, uses only well-known chemical transformations such as amide bond formation or click chemistry, requires standard purification procedures and has the potential to address various important biomolecular targets.

Moreover, our conjugation strategy opens the possibility to apply intramolecular photostabilization to fluorophore classes that could not be studied before. This possibility broadens our understanding of intramolecular photostabilization and might provide a basis for further improvement of the photostabilizer–dye conjugates by studying the effects of fluorophore properties such as charge, redox potential or absorption/emission wavelength. For certain types of fluorophores, for example, oxazines and perylenes, conjugation to NPA is expected to show effects different from those described here considering their interaction with antioxidants[4,24,25], DNA bases and amino acids such as tryptophane[42–44]. Intramolecular quenching of Alexa fluorophores observed with natural amino acids[45] even suggests that labelling of proteins at strategic locations, for example, in the

vicinity of aromatic amino acids, could be used for fluorophore stabilization without the use of UAAs, provided that a suitable combination of fluorophore and amino acids can be found.

Intramolecular photostabilization remains, however, less effective compared to the use of diffusion-based photostabilization (see Fig. 4) as reported before[18]. This shortcoming of intramolecular photostabilization is compensated by the fact that it is often the only viable option for applications such as live-cell imaging, for assays where the addition of photostabilizer remains ineffective or is not tolerated by the biological system. We are convinced that future work—also facilitated by the presented synthesis strategy—will allow solving this problem. We hypothesize that the properties of photostabilizer–dye conjugates can be optimized by parameters such as the linking geometry between fluorophores and photostabilizer. This is supported by published data of Cy5 derivatives, where we linked the aromatic nitro group to the fluorophore in a different way[19], compared with the compounds shown in this study (compound **7**). These architectures show a significantly higher photostability using basically the same photostabilizer moiety. We are in the process of varying the linker length of NPA-based fluorophore derivatives, to optimize the effects of intramolecular photostabilization and to become fully competitive with solution additives as shown in ref. 19.

We finally suggest that UAA scaffolding could be extended beyond photostabilization, to provide a general framework for the manipulation of fluorophore properties. Potentially interesting UAAs feature antioxidants, triplet sensitizers, photoswitchable molecules to induce blinking for localization-based super-resolution microscopy[2-4,46,47], alter water solubility[48] or the affinity to membranes and residues of natural amino acids such as thiols (see Lui *et al.*[23] for an overview of UAA residues). UAAs with 'clickable' functionalities could be used to link the fluorophore to a biomolecular target and an affinity tag (strep-/his-tag) to simplify purification of labelled protein species.

In summary, we introduced a versatile and simple design concept to synthesize photostabilizer–dye conjugates on a specific target using UAAs. The strategy is based on a straightforward and modular synthesis route using amide bond formation and click chemistry with commercially available starting materials. The UAAs NPA and PG were used as a scaffold to link rhodamine, carbopyronine and cyanine fluorophores covalently to a photo-stabilizer on a biomolecular target, such as DNA, antibodies and proteins in our case study. We are convinced, however, that other targets (RNA, affinity tags and so on) can also be labelled via similar means, while maintaining the positive effects of photo-stabilization on these targets. All studied compounds show a considerable increase in photostability and a suppression of triplet-based blinking compared to the non-stabilized parent fluorophore. With this, we are the first to test intramolecular photostabilization for various classes of organic fluorophores and show that intramolecular triplet-state quenching is a generally applicable method. The approach allows labelling of biomolecules in a single step with the same effort as for commercially available reactive fluorophore derivatives. Finally, photostabilizer–dye conjugates were used in two state-of-the-art applications, that is, the study of conformational changes in proteins via smFRET and super-resolution imaging using STED microscopy. We are convinced that the presented strategy will stimulate broader use of intramolecular photostabilization and help to emerge this strategy to the new gold standard for photostabilization.

Methods

Synthesis of photostabilizer-dye conjugates and precursors. All reagents were purchased from commercial suppliers and used without further purification, unless stated otherwise. Cy5 fluorophores were obtained from Lumiprobe (Germany), Alexa555-NHS was obtained from Life Technologies (USA) and ATTO647N-NHS from ATTOTEC (Germany). Synthetic oligomers (NH_2-C6-5′-TAA TAT TCG ATT CCT TAC ACT TAT ATT GCA TAG CTA TAC G-3′) were received in HPLC-purified quality from IBA or Eurofins (Germany). TX–NHS **9** was synthesized following a published procedure[49]. A Varian 400 (400 and 100 MHz) and Varian 500 MHz were used to record ^{1}H-NMR and ^{13}C-NMR spectra. Chemical shifts (δ) are denoted in p.p.m. using residual solvent peaks as internal standard ($\delta_H = 7.26$ and $\delta_C = 77.0$ for CDCl$_3$; $\delta_H = 3.31$, 4.78 and $\delta_C = 49.15$ for CD$_3$OD). High-resolution mass spectra were recorded on an Orbitrap XL (Thermo Fisher Scientific; ESI pos. or neg. mode). Liquid chromatography–mass spectra were recorded on Waters Acquity H-class UPLC equipped with TUV detector and a Xevo G2 TOF mass detector from Waters Chromatography BV. Flash chromatography was performed using a Grace Reveleris Flash System (40 μm silica column).

Purification of functionalized oligonucleotides. The functionalized oligonucleotides were purified on an Illustra NAP 5 column loaded with Sephadex G-25 DNA Grade material obtained from GE Healthcare. Illustra NAP 5 columns were equilibrated with 3 × 5 ml of eluent before use (50 mM triethylammonium acetate (TEAA) buffer, pH 7.0, or water for Fmoc-protected oligonucleotide **3** and **12**). After the reaction, the oligonucleotide sample was diluted with eluent to a final volume of 0.5 ml and added to the column. After the sample was loaded, 1 mL of eluent was added and the purified oligonucleotide was collected in one portion. The oligonucleotide solution was lyophilized (Christ Alpha 2–4 LD plus freeze dyer) directly after collection.

Isolation and characterization of functionalized oligonucleotides. Reversed-phase HPLC (rp-HPLC) analysis and preparative purifications (isolation) was performed on a Shimadzu LC-10AD VP machine equipped with Waters Xterra MS C18 column (3.0 × 150 mm, particle size 3.5 μm) and Waters Xterra MS C18 prep

column (7.8 × 150 mm, particle size 10 μm) using a gradient of acetonitrile/TEAA buffer (50 mM, pH 7.0). Gradient 1: 05/95, 0–10 min, to 65/35 at 60 min, to 75/25 at 65 min, to 05/95 at 75 min for 15 min. Gradient 2: 05/95, 0–10 min, to 35/65 at 60 min, to 65/35 at 65 min, to 05/95 at 70 min for 20 min. Flow 0.5 ml min^{-1} analytical run or 1.0 ml min^{-1} preparative run. The DNA was isolated by collecting the major peak of interest (see Supplementary Fig. 1). The resulting compounds were characterized by ultraviolet–visible absorption spectroscopy (Supplementary Fig. 2a) and MALDI-TOF mass spectrometry (Supplementary Fig. 2b). Spectra were recorded on an ABI Voyager DE-PRO MALDI-TOF (delayed extraction reflector) Biospectrometry Workstation mass spectrometer.

Synthesis of ssDNA-NPA-Alexa555/ATTO647N/Cy5. Step 1: the lyophilized ssDNA NH_2 was resuspended in MilliQ water and adjusted to 80 μM in 0.2 M NaHCO$_3$ buffer, pH 8.35. To 100 μl of this solution, the same volume of a 20 mg ml^{-1} solution of **2** in DMF was added and the mixture was vortexed thoroughly. If necessary, additional DMF was added in 10 μl portions, to obtain a clear solution. After the reaction at room temperature overnight, **3** was purified on Illustra NAP 5 column (*vide supra*, MilliQ water was used as eluent for purification on Illustra NAP 5 columns, to prevent partial deprotection of the Fmoc group), and isolated by preparative rp-HPLC (Supplementary Fig. 2, gradient 1, *vide supra*), to yield ssDNA NPA–Fmoc **3**. Step 2: Fmoc deprotection of **3** was performed as follows. The HPLC-purified and lyophilized oligonucleotide was resuspended in 50 μl of 50 mM TEAA buffer, pH 7.0. Deprotection was achieved by addition of 40 μl DMF and 10 μl piperidine. The mixture was vortexed and allowed to react for 2 h at room temperature. The deprotected oligonucleotide was purified on Illustra NAP 5 column (*vide supra*) and lyophilized to yield **4**. Step 3: coupling with fluorophore NHS was achieved as follows. Cy5–NHS as obtained from Lumiprobe, Alexa555–NHS was obtained from Life Technologies and ATTO647N–NHS from ATTOTEC; all samples were evenly distributed into ~300 nmol portions inside a glove box. Each portion was sealed with Parafilm and kept in the dark at −18 °C. Lyophilized **4** (~2 nmol) was resuspended in 50 μl of 0.2 M NaHCO$_3$ buffer, pH 8.35. To this solution, one portion of fluorophore NHS dissolved in 10 μl of dimethyl sulfoxide (DMSO) was added and the mixture was vortexed thoroughly. After incubation overnight the oligonucleotide was purified on Illustra NAP 5 column (*vide supra*) and isolated by preparative rp-HPLC (gradient 1, *vide supra*) to yield **5**, **6** and **7** (see Fig. 2 and Supplementary Fig. 1).

Synthesis of ssDNA TX–PG–Cy5 (13). Step 1: the lyophilized ssDNA was resuspended in MilliQ water and adjusted to 80 μM in 0.2 M NaHCO$_3$ buffer, pH 8.35. To 100 μl of this solution, the same volume of a 20 mg ml^{-1} solution of **11** in DMF was added and the mixture was vortexed thoroughly. If necessary, additional DMF was added in 10 μl portions, to obtain a clear solution. After the reaction at room temperature overnight, the oligonucleotide was purified on Illustra NAP 5 column and isolated by preparative rp-HPLC (Supplementary Fig. 1, gradient 2, *vide supra*) to yield **12**. Step 2: coupling with Cy5-N_3 was achieved following a modified manufacturer protocol (Lumiprobe). After every addition step the mixture was vortexed briefly. The HPLC-purified and -lyophilized oligonucleotide **12** (~2 nmol) was resuspended in 30 μl of 2 M TEAA buffer, pH 7.0 and added into a 0.5 ml tube. To this solution, 8 μl MilliQ water and 7 μl of DMSO were added. A 15-nmol portion of Cy5-N_3 was dissolved in 40 μl of DMSO and added. To this mixture, 10 μl of 5 mM stock solution of ascorbic acid in MilliQ water was added. The solution was degassed by a stream of N_2 for 30 s and finally 5 μl of a 10-mM stock solution of Cu(II)-tris[(1-benzyl-1H-1,2,3-triazol-4-yl)methyl]amine in 55% DMSO/MilliQ water was added. The vial was flushed with N_2, closed and sealed with Parafilm. After incubation overnight the oligonucleotide was purified on Illustra NAP 5 column (*vide supra*) and isolated by preparative rp-HPLC (*vide supra*) to yield **13** (Supplementary Fig. 1).

Synthesis of ssDNA Alexa555/ATTO647N. Step 1: Alexa555–NHS was obtained from Life Technologies and ATTO647N–NHS from ATTOTEC; all were evenly distributed into ~300 nmol portions inside a glove box. Each portion was sealed with Parafilm and kept in the dark at −18 °C. ssDNA NH_2 (~2 nmol) was resuspended into 50 μl of 0.2 M NaHCO$_3$ buffer, pH 8.35. To this solution, one portion of Alexa555–NHS/ATTO647N–NHS dissolved in 10 μl of DMSO was added and the mixture was vortexed thoroughly. In case of RhodamineB, an NHS crude mixture in 100 μl of DMF was added to ssDNA NH_2 and the mixture was vortexed thoroughly. After incubation overnight the oligonucleotide was purified with an Illustra NAP 5 column (*vide supra*) and isolated by preparative rp-HPLC (gradient 1, *vide supra*).

Synthesis of ssDNA NPA–RhodamineB. Step 1: the lyophilized ssDNA-NH_2 was resuspended in MilliQ water and adjusted to 20 μM in 0.2 M NaHCO$_3$ buffer, pH 8.35. To this solution, RhoB–NPA–NHS (**20**) in 100 μl of DMF was added and the mixture was vortexed thoroughly. After incubation overnight the oligonucleotide was purified on Illustra NAP 5 column (*vide supra*) and isolated by preparative rp-HPLC (gradient 1, *vide supra*) to yield **18** (see Fig. 6). The yield was found to be 25% for coupling NPA–RhodamineB–NHS to ssDNA.

Synthesis of ssDNA RhodamineB. Step 1: the lyophilized ssDNA-NH_2 was resuspended in MilliQ water and adjusted to 20 μM in 0.2 M NaHCO$_3$ buffer, pH 8.35. To this solution, RhodamineB–NHS in 100 μl of DMF was added and the mixture was vortexed thoroughly. After incubation overnight the oligonucleotide was purified on Illustra NAP 5 column (*vide supra*) and isolated by preparative rp-HPLC (gradient 1, *vide supra*) (see Supplementary Fig. 1).

Labelling of GlnPQ–SBD2. SBD2 cysteine-containing derivative was obtained as described in ref. 35 and stored at −20 °C in 100-μl aliquots of 20–40 mg ml^{-1} in 50 mM KPi, pH 7.4, 50 mM KCl and 50% glycerol plus 1 mM dithiothreitol (DTT).

Stochastic labelling with maleimide derivatives of donor and acceptor fluorophores was carried out on ~5 nmol of protein; SBD_2 derivatives were labelled with RhodamineB–/NPA–RhodamineB–maleimde (donor) and KK114–maleimide (acceptor) in a ratio of protein:donor:acceptor of 1:6:8. Briefly, purified proteins were treated with DTT (10 mM, 30 min), to fully reduce oxidized cysteines. After dilution of the protein sample to a DTT concentration of 1 mM, the reduced protein was bound to a Ni^{2+}-Sepharose resin (GE Healthcare) and washed with ten column volumes of 50 mM KPi, pH 7.4, 150 mM KCl and 5% glycerol (buffer B). Simultaneously, the applied fluorophore stocks (50 nmol (s)) dissolved in 5 µl of anhydrous DMSO were added at appropriate amounts to buffer B and immediately applied to the protein bound to the Ni^{2+}-Sepharose resin (keeping the final DMSO concentration below 1%). The resin was incubated overnight and kept at 4 °C (under mild agitation). After labelling, unreacted dye was removed by sequential washing with ten column volumes of buffer B, and this was followed by 100 column volumes of 50 mM KPi, pH 7.4, 1 M KCl and 50% glycerol. The protein was eluted in 1 ml of 50 mM KPi, pH 7.4, 150 mM KCl, 5% glycerol and 500 mM imidazole, and was analyzed on a Superdex-200 column (GE Healthcare) equilibrated with 50 mM KPi, pH 7.4, and 200 mM KCl.

Microscopy and sample preparation. *Sample preparation and surface immobilization of oligonucleotides.* Immobilization and investigation of single fluorophores was achieved using a dsDNA scaffold comprising two 40-mer oligonucleotides, that is, ssDNA fluorophore and ssDNA biotin. Sequences of both oligomers were adapted from refs 12,24,25. As a non-stabilized control we used ssDNA fluorophore (Cy5-C6-5′-TAA TAT TCG ATT CCT TAC ACT TAT ATT GCA TAG CTA TAC G-3′; as received from Eurofins and IBA). Single immobilized fluorophore molecules were studied in Lab-Tek eight-well-chambered cover slides (Nunc/VWR, The Netherlands) with a volume of 750 µl as described in ref. 11. After cleaning with 0.1% hydrofluoric acid (HF) and washing with PBS buffer (one PBS tablet was dissolved in deionized water yielding a 10 mM phosphate buffer with 2.7 mM potassium chloride and 137 mM sodium chloride at pH 7.4; Sigma Aldrich, The Netherlands), each chamber was incubated with a mixture of 5 mg per 800 ml BSA and 1 mg per 800 ml BSA/biotin (Sigma Aldrich) at 4 °C in PBS buffer overnight. After rinsing with PBS buffer, a 0.2-mg ml^{-1} solution of streptavidin was incubated for 10 min and subsequently rinsed with PBS buffer.

The immobilization of dsDNA was achieved via a biotin–streptavidin interaction using pre-annealed dsDNA with the aim to observe single emitters for prolonged time periods and to guaranty free rotation of fluorophores. For this, 5–50 µl of a 1 µM solution of ssDNA fluorophore, ssDNA-NPA-fluorophore or ssDNA-TX-PG-Cy5 was mixed with the complementary ssDNA-biotin at the same concentration (Biotin-5′-CGT ATA GCT ATG CAA TAT AAG TGT AAG GAA TCG AAT ATT A-3′, used as received from IBA). The respective mixtures of two oligomers were heated to 98 °C for 4 min and cooled down to 4 °C with a rate of 1 °C min^{-1} in annealing buffer (500 mM sodium chloride, 20 mM TRIS-HCL and 1 mM EDTA at pH 8). The treated LabTek coverslides were incubated with a 50–100 pM solution of pre-annealed dsDNA for 1–2 min, leading to a typical surface coverage of fluorophore-labelled dsDNA as shown in Figs 3 and 4 and Supplementary Figs 3–11.

All single-molecule experiments were carried out at room temperature (22 ± 1 °C). Oxygen was removed from the buffer system using an oxygen-scavenging system (PBS, pH 7.4, containing 10% (wt/vol) glucose and 10% (vol/vol) glycerine, 50 µg ml^{-1} glucose oxidase, 100–200 µg ml^{-1} catalase and 0.1 mM Tris(2-carboxyethyl)phosphine hydrochloride (TCEP)). As shown before, such low concentrations of the reducer TCEP has no noticeable effect on the photophysics of organic fluorophores[25,50] and hence do not convolute with effects from intramolecular stabilization. Glucose oxidase catalase[11,29] was used instead of a combination of protocatechuic acid and protocatechuate-3,4-dioxygenase[29], to avoid convolution of inter- and intramolecular photostabilization with protocatechuic acid[51]. When adding TX to the imaging buffer (PBS with 2 mM TX), the latter was irradiated with ultraviolet light for 30 min before adding 10% (wt/vol) glucose and 10% (vol/vol) glycerine, 50 µg mL^{-1} glucose oxidase, 100–200 µg mL^{-1} catalase and 0.1 mM TCEP.

Confocal scanning microscopy and data analysis. A custom-built confocal microscope, described in ref. 18, was used to study fluorescence properties of organic fluorophores on the level of single molecules. Briefly, excitation was achieved with a spectrally filtered laser beam of a pulsed supercontinuum source (SuperK Extreme, NKT Photonics, Denmark) with an acousto-optical tunable filter (AOTFnc-VIS, EQ Photonics, Germany), leading to ≈2 nm broad excitation pulses centred at 640 nm or 532 nm. The spatially filtered beam was coupled into an oil-immersion objective (× 60, numerical aperture (NA) 1.35, UPLSAPO 60XO mounted on an IX71 microscope body, both from Olympus, Germany) by a dichroic beam splitter (zt532/642rpc, AHF Analysentechnik, Tuebingen, Germany). Surface scanning was performed using a XYZ-piezo stage with 100 × 100 × 20 µm range (P-517-3CD with E-725.3CDA, Physik Instrumente, Germany). Fluorescence was collected by the same objective, focused onto a 50-µm pinhole and detected by two avalanche photodiodes (τ-spad, <50 dark counts per second, Picoquant, Germany) with appropriate spectral filtering (green: HC582/75, red: ET700/75 AHF, both from Analysentechnik). The detector signal was registered using a Hydra Harp 400 ps event timer and a module for time-correlated single-photon counting (both from Picoquant). The data were evaluated using custom-made LabVIEW software[24,25]. Blinking kinetics were extracted from fluorescent time traces in the form of ON and OFF times according to established procedures[24]. Fluorescence lifetimes were determined using time-correlated single-photon counting as described before[18].

TIRF microscopy including data analysis. Widefield fluorescence imaging was conducted on an inverted microscope (Olympus IX-71 with UPlanSApo × 100, NA 1.49, Olympus, Germany) in objective type TIRF) configuration as described before[18]. Images were collected with a back-illuminated electron multiplying charge-coupled device camera (512 × 512 pixel, C9100-13 (Hammamatsu, Japan), in combination with either ET585/50 or ET700/75 (AHF Analysentechnik)). Excitation from a diode laser (Sapphire and Cube (Coherent, Germany), filtered either by ET535/70 or ZET640/10 (Chroma, USA)) was at 532 nm and 640 nm with ≈50 W cm^{-2} at the sample location. To quantitatively characterize photostability, we imaged areas with the size of ≈25 × 35 µm containing >100 molecules. A movie was typically recorded for 300–600 s with an integration time of 100 ms. Fluorescent time traces were extracted by identifying pixels, which showed at least 2–3 s.d. above background noise (s.d. of all pixels over all frames of the movie) and summing the intensity in a 3 × 3 pixel area. Neighbouring peaks closer than 5 pixels were not taken into account (see typical examples of fluorophore density and fluorescent time traces in Supplementary Figs 3–11). The number of fluorescent spots in each frame image was determined using an absolute threshold criterion. The number of fluorescent emitters per image were than plotted over time and fitted to a mono-exponential decay $y(t) = C + A \times e^{-c \times t}$ (with $c = 1/\tau_{bleach}$ and τ_{bleach} being the characteristic bleaching time constant). For some samples with a more complicated behaviour, a double exponential decay of similar form was used and τ_{bleach} was calculated according to $\tau_{bleach} = A_1 \times \tau_1 + A_2 \times \tau_2$ with amplitude normalization to 1. Bleaching times and associated s.d. were derived from multiple repeats of the same experiment on different days, where each compound was tested in ≥5 movies. The SNR ratio was determined using fluorescent time traces, by dividing the s.d. of the signal before photobleaching with the average fluorescence intensity during that period. The total number of detected photons before photobleaching was calculated by multiplying the obtained count rate and photobleaching lifetime.

Single-molecule FRET and ALEX spectroscopy. For data acquisition we used the same home-built confocal microscope as described above. Fluorescence photons arriving at the two detection channels (donor detection channel D or acceptor detection channel A) were assigned to either donor- or acceptor-based excitation, based on their photon arrival time as described previously[33,35,36]. Three relevant photon streams were extracted from the data corresponding to donor-based donor emission $F(DD)$, donor-based acceptor emission $F(DA)$ and acceptor-based acceptor emission $F(AA)$. During diffusion, stoichiometry S and apparent FRET efficiencies E^* were calculated for each fluorescent burst above a certain threshold yielding a two-dimensional (2D) histogram. Uncorrected FRET efficiency E^* monitors the proximity between the two fluorophores and is calculated according to $E^* = F(DA)/(F(DD) + F(DA))$. Stoichiometry S is defined as the ratio between the overall fluorescence intensity during the green excitation period over the total fluorescence intensity during both green and red periods, and describes the ratio of donor-to-acceptor fluorophores in the sample: $S = (F(DA) + F(DD)/(F(DD) + F(DA) + F(AA))$. Using published procedures to identify bursts corresponding to single molecules, we obtained bursts characterized by three parameters (M, T and L). A fluorescent signal is considered a burst provided it meets the following criteria: a total of L photons are collected, having M neighbouring photons within a time interval of T microseconds. We applied different burst search algorithms on the data sets: the sum of all three detection channels (all photons) using parameters $M = 15$, $T = 500$ µs and $L = 50$ or a dual-colour burst search with $M = 12$, $T = 500$ µs and $L = 30$; additional per-bin thresholding removed spurious changes in fluorescence intensity and selected for intense single-molecule bursts; precise treatment of each data set is given in each figure caption. Binning of the detected bursts into a 2D E^*/S histogram yielded subpopulations that can be separated according to their S-values. Photon-counting histograms were obtained using similar thresholds as for the 2D E^*/S histograms.

ALEX experiments were carried out at room temperature using 25–50 pM of double-labelled protein in imaging buffer: 50 mM KPi, pH 7.4, 150 mM KCl, pH 7.4, containing 10% (wt/vol) glucose and 10% (vol/vol) glycerine, 50 µg ml^{-1} glucose oxidase, 100–200 µg ml^{-1} catalase and 0.1 mM TCEP. For experiments with TX added to the imaging buffer, 2 mM TX, 50 mM KPi, pH 7.4, and 150 mM KCl were irradiated with ultraviolet light for 30 min before adding 10% (wt/vol) glucose and 10% (vol/vol) glycerine, 50 µg ml^{-1} glucose oxidase, 100–200 µg ml^{-1} catalase and 0.1 mM TCEP. Titration experiments were done via adding increasing concentrations of ligand to the imaging buffer (0, 1.0 and 200 µM of glutamine).

Count rate versus excitation intensity. The count rate per molecule dependence on the excitation intensity was recorded using 20–100 nM of labelled SBD2 protein in imaging buffer (10 mM KPi, 2.7 mM KCl, 137 mM NaCl, pH 7.4, 10% w/v glucose, 50 µg ml^{-1} glucose oxidase, 100–200 µg ml^{-1} catalase, and 0.1 mM TCEP, both in the presence and the absence of 2 mM aged TX. The count rate per molecule was acquired via excitation from the 543 and 633 nm laser on a LSM 710 confocal laser scanning microscope (Carl Zeiss, Jena, Germany) through a C-apochromat × 401.20 w Korr M27 water-immersion objective (NA = 1.2). Excitation light from both lasers (at different intensities) was coupled into the objective by a dichroic beam splitter (MBS 458/543/633). Fluorescence was collected through the same objective with appropriate spectral filtering (NFT 635 VIS and LP 580).

STED and confocal microscopy. STED microscopy and the corresponding confocal microscopy was performed on an Abberior Instruments Resolft microscope (Abberior Instruments, Goettingen, Germany) to which a Ti:Sa STED laser was added (MaiTai, Newport-Spectra Physics; as outlined in detail in ref. 52). Excitation was performed at 640 nm (LDH-D-C-640P laser diode, Picoquant) and STED at 780 nm, at 80 MHz. Pulsing of the excitation laser was triggered by the Ti:Sa laser using a photodiode (APS-100-01), an amplifier (CON-TTL, both from Becker & Hickl, Berlin, Germany) along with a ps-delay unit (either Abberior Instruments or LAS-015617, MPD/Laser2000 UK). A donut-shaped intensity distribution of the STED laser focus was realized by the incorporation of a vortex phase plate (VPP-1a, RPC Photonics, Rochester, NY) into the collimated STED beam. Laser focusing and fluorescence collection was done by an oil-immersion microscope objective (UPlanSapo × 100/1.4 oil, Olympus, Japan). The collected fluorescence was detected by an avalanche photodiode (SPCM-AQRH-13, Excelitas Technologies) with a 650–690 nm bandpass filter (670/40, AHF Analysentechnik).

Immunolabelling. Gm5756T human fibroblast were fixed with 3% paraformaldehyde for 10 min, rinsed several times in PBS, permeabilized in 0.1% Triton X-100 for 10 min, rinsed again several times in PBS and blocked in 2% BSA/5% FCS in PBS for 1 h. Samples were incubated with primary mouse antibodies, to stain nuclear pore diluted 1:500 in blocking buffer for 1 h at room temperature. Coverslips were washed five times in 1% BSA in PBS and then incubated for 30 min with goat anti-mouse antibodies (Abberior GmbH, Gottingen, Germany) diluted 1:250 in blocking buffer. Samples were washed five times in 1% BSA in PBS and mounted in the buffer system using an oxygen-scavenging system as used before in the single-molecule experiments.

References

1. Weiss, S. Fluorescence spectroscopy of single biomolecules. *Science* **283**, 1676–1683 (1999).
2. Ha, T. & Tinnefeld, P. Photophysics of fluorescent probes for single-molecule biophysics and super-resolution imaging. *Annu. Rev. Phys. Chem.* **63**, 595–617 (2012).
3. Hell, S. W. Far-field optical nanoscopy. *Science* **316**, 1153–1158 (2007).
4. Vogelsang, J. *et al.* Make them blink: probes for super-resolution microscopy. *Chemphyschem.* **11**, 2475–2490 (2010).
5. Liphardt, B., Liphardt, B. & Luttke, W. Laser dyes with intramolecular triplet quenching. *Opt. Commun.* **38**, 207–210 (1981).
6. Liphardt, B., Liphardt, B. & Luttke, W. Laser dyes III: concepts to increase the photostability of laser dyes. *Opt. Commun.* **48**, 129–133 (1983).
7. Tinnefeld, P. & Cordes, T. 'Self-healing' dyes: intramolecular stabilization of organic fluorophores. *Nat. Methods* **9**, 426–427 (2012).
8. Blanchard, S. C. Reply to "Self-healing' dyes: intramolecular stabilization of organic fluorophores'. *Nat. Methods* **9**, 427–428 (2012).
9. Dave, R., Terry, D. S., Munro, J. B. & Blanchard, S. C. Mitigating unwanted photophysical processes for improved single-molecule fluorescence imaging. *Biophys. J.* **96**, 2371–2381 (2009).
10. Rasnik, I., McKinney, S. A. & Ha, T. Nonblinking and long-lasting single-molecule fluorescence imaging. *Nat. Methods* **3**, 891–893 (2006).
11. Vogelsang, J. *et al.* A reducing and oxidizing system minimizes photobleaching and blinking of fluorescent dyes. *Angew. Chem. Int. Ed. Engl.* **47**, 5465–5469 (2008).
12. Cordes, T., Vogelsang, J. & Tinnefeld, P. On the mechanism of Trolox as antiblinking and antibleaching reagent. *J. Am. Chem. Soc.* **131**, 5018–5019 (2009).
13. Widengren, J., Chmyrov, A., Eggeling, C., Löfdahl, P.-A. & Seidel, C. a. M. Strategies to improve photostabilities in ultrasensitive fluorescence spectroscopy. *J. Phys. Chem. A* **111**, 429–440 (2007).
14. Lemke, E. A. *et al.* Microfluidic device for single-molecule experiments with enhanced photostability. *J. Am. Chem. Soc.* **131**, 13610–13612 (2009).
15. Cordes, T., Maiser, A., Steinhauer, C., Schermelleh, L. & Tinnefeld, P. Mechanisms and advancement of antifading agents for fluorescence microscopy and single-molecule spectroscopy. *Phys. Chem. Chem. Phys.* **13**, 6699–6709 (2011).
16. Altman, R. B. *et al.* Cyanine fluorophore derivatives with enhanced photostability. *Nat. Methods* **9**, 68–71 (2012).
17. Altman, R. B. *et al.* Enhanced photostability of cyanine fluorophores across the visible spectrum. *Nat. Methods* **9**, 428–429 (2012).
18. Van der Velde, J. H. M. *et al.* Mechanism of intramolecular photostabilization in self-healing cyanine fluorophores. *Chemphyschem.* **14**, 4084–4093 (2013).
19. Van der Velde, J. H. M. *et al.* The power of two: covalent coupling of photostabilizers for fluorescence applications. *J. Phys. Chem. Lett.* **5**, 3792–3798 (2014).
20. Zheng, Q. *et al.* Ultra-stable organic fluorophores for single-molecule research. *Chem. Soc. Rev.* **43**, 1044–1056 (2014).
21. Levitus, M. & Ranjit, S. Cyanine dyes in biophysical research: the photophysics of polymethine fluorescent dyes in biomolecular environments. *Q. Rev. Biophys.* **44**, 123–151 (2011).
22. Juette, M. F. *et al.* The bright future of single-molecule fluorescence imaging. *Curr. Opin. Chem. Biol.* **20**, 103–111 (2014).
23. Liu, C. C. & Schultz, P. G. Adding new chemistries to the genetic code. *Annu. Rev. Biochem.* **79**, 413–444 (2010).
24. Vogelsang, J., Cordes, T., Forthmann, C., Steinhauer, C. & Tinnefeld, P. Controlling the fluorescence of ordinary oxazine dyes for single-molecule switching and superresolution microscopy. *Proc. Natl Acad. Sci. USA* **106**, 8107–8112 (2009).
25. Vogelsang, J., Cordes, T. & Tinnefeld, P. Single-molecule photophysics of oxazines on DNA and its application in a FRET switch. *Photochem. Photobiol. Sci.* **8**, 486–496 (2009).
26. Stein, I. H. *et al.* Linking single-molecule blinking to chromophore structure and redox potentials. *Chemphyschem.* **13**, 931–937 (2012).
27. Steinhauer, C., Forthmann, C., Vogelsang, J. & Tinnefeld, P. Superresolution microscopy on the basis of engineered dark states. *J. Am. Chem. Soc.* **130**, 16840–16841 (2008).
28. Kasper, R., Heilemann, M., Tinnefeld, P. & Sauer, M. in *Biophotonics 2007 Optics in Life Science* 6633_71 (OSA Publishing, 2007).
29. Aitken, C. E., Marshall, R. A. & Puglisi, J. D. An oxygen scavenging system for improvement of dye stability in single-molecule fluorescence experiments. *Biophys. J.* **94**, 1826–1835 (2008).
30. Cordes, T. *et al.* Single-molecule redox blinking of perylene diimide derivatives in water. *J. Am. Chem. Soc.* **132**, 2404–2409 (2010).
31. Alejo, J. L., Blanchard, S. C. & Andersen, O. S. Small-molecule photostabilizing agents are modifiers of lipid bilayer properties. *Biophys. J.* **104**, 2410–2418 (2013).
32. Mickler, M., Hessling, M., Ratzke, C., Buchner, J. & Hugel, T. The large conformational changes of Hsp90 are only weakly coupled to ATP hydrolysis. *Nat. Struct. Mol. Biol.* **16**, 281–286 (2009).
33. Cordes, T. *et al.* Sensing DNA opening in transcription using quenchable Förster resonance energy transfer. *Biochemistry* **49**, 9171–9180 (2010).
34. Hein, J. E. & Fokin, V. V. Copper-catalyzed azide-alkyne cycloaddition (CuAAC) and beyond: new reactivity of copper(I) acetylides. *Chem. Soc. Rev.* **39**, 1302–1315 (2010).
35. Gouridis, G. *et al.* Conformational dynamics in substrate-binding domains influences transport in the ABC importer GlnPQ. *Nat. Struct. Mol. Biol.* **22**, 57–64 (2014).
36. Kapanidis, A. N. *et al.* Fluorescence-aided molecule sorting: analysis of structure and interactions by alternating-laser excitation of single molecules. *Proc. Natl Acad. Sci. USA* **101**, 8936–8941 (2004).
37. Kong, X., Nir, E., Hamadani, K. & Weiss, S. Photobleaching pathways in single-molecule FRET experiments. *J. Am. Chem. Soc.* **129**, 4643–4654 (2007).
38. Fulyani, F. *et al.* Functional diversity of tandem substrate-binding domains in ABC transporters from pathogenic bacteria. *Structure* **21**, 1879–1888 (2013).
39. Kolmakov, K. *et al.* Red-emitting rhodamine dyes for fluorescence microscopy and nanoscopy. *Chemistry* **16**, 158–166 (2010).
40. Hell, S. W. & Wichmann, J. Breaking the diffraction resolution limit by stimulated emission: stimulated-emission-depletion fluorescence microscopy. *Opt. Lett.* **19**, 780 (1994).
41. Kasper, R. *et al.* Single-molecule STED microscopy with photostable organic fluorophores. *Small* **6**, 1379–1384 (2010).
42. Marmé, N., Knemeyer, J.-P., Sauer, M. & Wolfrum, J. Inter- and intramolecular fluorescence quenching of organic dyes by tryptophan. *Bioconjug. Chem.* **14**, 1133–1139 (2003).
43. Doose, S., Neuweiler, H. & Sauer, M. A close look at fluorescence quenching of organic dyes by tryptophan. *Chemphyschem.* **6**, 2277–2285 (2005).
44. Doose, S., Neuweiler, H. & Sauer, M. Fluorescence quenching by photoinduced electron transfer: a reporter for conformational dynamics of macromolecules. *Chemphyschem.* **10**, 1389–1398 (2009).
45. Chen, H., Ahsan, S. S., Santiago-Berrios, M. B., Abruña, H. D. & Webb, W. W. Mechanisms of quenching of Alexa fluorophores by natural amino acids. *J. Am. Chem. Soc.* **132**, 7244–7245 (2010).
46. Schwering, M. *et al.* Far-field nanoscopy with reversible chemical reactions. *Angew. Chem. Int. Ed. Engl.* **50**, 2940–2945 (2011).
47. Nanguneri, S., Flottmann, B., Herrmannsdörfer, F., Thomas, K. & Heilemann, M. Single-molecule super-resolution imaging by tryptophan-quenching-induced photoswitching of phalloidin-fluorophore conjugates. *Microsc. Res. Tech.* **00**, 1–7 (2014).
48. Yang, S. K., Shi, X., Park, S., Ha, T. & Zimmerman, S. C. A dendritic single-molecule fluorescent probe that is monovalent, photostable and minimally blinking. *Nat. Chem.* **5**, 692–697 doi:10.1038/nchem.1706 (2013)).
49. Koufaki, M. *et al.* Synthesis of chroman analogues of lipoic acid and evaluation of their activity against reperfusion arrhythmias. *Bioorg. Med. Chem.* **12**, 4835–4841 (2004).
50. Vaughan, J. C., Dempsey, G. T., Sun, E. & Zhuang, X. Phosphine quenching of cyanine dyes as a versatile tool for fluorescence microscopy. *J. Am. Chem. Soc.* **135**, 1197–1200 (2013).

51. Le Gall, A. *et al.* Improved photon yield from a green dye with a reducing and oxidizing system. *Chemphyschem.* **12**, 1657–1660 (2011).

52. Clausen, M. P. *et al.* Pathways to optical STED microscopy. *NanoBioImaging* **1**, 1–12 (2013).

Acknowledgements

This work was financed by the Zernike Institute for Advanced Materials, the Centre for Synthetic Biology (University of Groningen, start-up grant to T.C.). J.H.M.v.d.V. acknowledges Ubbo-Emmius funding (University of Groningen). J.O. and G.R. thank the NRSC-C and the European Commission (ERC Starting Grant Number 280010) for financial support. A.H. thanks the European Commission (ERC Starting Grant and STREP project MICREAGENTS) and the Netherlands Organization for Scientific Research (NWO-Vici) for funding. C.E. and S.G. acknowledge funding by the MRC, BBSRC and ESPRC. G.G. acknowledges support by an EMBO (long-term fellowship ALF 47-2012) and is currently supported by a NWO-VENI grant (grant number 722.012.012). We thank L.J. Ugen, E.M. Warszawik and E. Ploetz for discussions. We thank S.W. Hell and V. Belov for the generous gift of KK114 carboxylic acid and C. Wurm for help with labelling antibodies. Finally, we are grateful to A.M. van Oijen for enduring and generous support, friendly advice and thoughtful comments on the manuscript.

Author contributions

T.C. conceived the study. J.H.M.v.d.V., J.O., A.H., G.R. and T.C. designed experiments and synthesis routes. J.H.M.v.d.V., J.H. and A.A.J. performed single-molecule imaging and ALEX experiments. J.O., J.H., J.H.S., K.K., A.H. and G.R. provided new reagents including chemical characterization. S.G. and C.E. performed STED microscopy. G.G. provided proteins for FRET studies. J.H.M.v.d.V. and T.C. analysed data and all authors contributed to writing the paper.

Additional information

Design of crystal-like aperiodic solids with selective disorder–phonon coupling

Alistair R. Overy[1,2], Andrew B. Cairns[1,3], Matthew J. Cliffe[1,4], Arkadiy Simonov[1], Matthew G. Tucker[2,5] & Andrew L. Goodwin[1]

Functional materials design normally focuses on structurally ordered systems because disorder is considered detrimental to many functional properties. Here we challenge this paradigm by showing that particular types of strongly correlated disorder can give rise to useful characteristics that are inaccessible to ordered states. A judicious combination of low-symmetry building unit and high-symmetry topological template leads to aperiodic 'procrystalline' solids that harbour this type of disorder. We identify key classes of procrystalline states together with their characteristic diffraction behaviour, and establish mappings onto known and target materials. The strongly correlated disorder found in these systems is associated with specific sets of modulation periodicities distributed throughout the Brillouin zone. Lattice dynamical calculations reveal selective disorder-driven phonon broadening that resembles the poorly understood 'waterfall' effect observed in relaxor ferroelectrics. This property of procrystalline solids suggests a mechanism by which strongly correlated topological disorder might allow independently optimized thermal and electronic transport behaviour, such as required for high-performance thermoelectrics.

[1] Department of Chemistry, University of Oxford, Inorganic Chemistry Laboratory, South Parks Road, Oxford OX1 3QR, UK. [2] Diamond Light Source, Chilton, Oxfordshire, OX11 0DE, UK. [3] European Synchrotron Radiation Facility, 71 avenue des Martyrs, 38043 Grenoble, France. [4] Department of Chemistry, University of Cambridge, Lensfield Road, Cambridge CB2 1EW, UK. [5] ISIS Facility, Rutherford Appleton Laboratory, Harwell Oxford, Didcot, Oxfordshire OX11 0QX, UK. Correspondence and requests for materials should be addressed to A.L.G. (email: andrew.goodwin@chem.ox.ac.uk).

The relationship between building block geometry and bulk material structure is one of the cornerstones of structural science. By way of example, the solid phases of elemental Xe (ref. 1), C_{60} (ref. 2) and human adenovirus[3] are all structurally related, not by virtue of any particular chemical similarity but because each of these phases reflects the same solution to the problem of packing weakly interacting spherical objects in three-dimensional (3D) space. The reticular approach to understanding zeolite and metal–organic framework topologies is related from a conceptual viewpoint, because it links the geometries of molecule-like components (coordination polyhedra, molecular linkers) to the 3D architectures formed by their assemblies[4,5]. The importance of structure in determining the physical properties of solids is what then gives sense to the approach of developing new types of functional materials through informed design of constituent building blocks. It is precisely this type of 'ground-up' approach that has recently been exploited in the rational design of, for example, solid oxide fuel-cell cathodes and room-temperature multiferroic candidates[6,7].

To a large extent, the rich structural information accessible using crystallographic techniques has focussed effort on the design of crystalline materials. There are also obvious functional advantages to the long-range periodicity characteristic of crystals, because it governs useful correlated properties—including the lattice dynamics, electronic states, and charge, orbital and magnetic order. Moreover, crystal symmetry is central to mechanical properties such as piezoelectricity and ferroelasticity, and is clearly pivotal in determining phase transition behaviour. While certain building block geometries allow—or can even force[8,9]—non-crystalline assemblies, the link to function is usually much less clear in these cases. Indeed the received wisdom is that disorder is something to be avoided, despite increasingly strong empirical evidence that links disordered states to advanced functionalities[10]. So the development of disorder–property relationships, and the eventual control over these properties through suitable building block design have emerged as key challenges in the field.

Here we develop an approach of intentionally designing functional disordered materials by focussing on systems in which structural disorder is extremely strongly correlated. The type of disorder we consider is similar to that found in ice, and our paper begins by developing a generalization of ice-like states to arbitrary materials geometries. We proceed to establish a link between the geometry of structural building blocks and the propensity for specific types of strongly correlated disorder in the resulting material assembly. This is the design element of our approach. Having made this connection, we suggest a number of physical realizations of these correlated disordered states. Our paper concludes by demonstrating how the correlated disorder deliberately engineered within one representative affects its lattice dynamics in a highly specific manner. This is the functional element of our approach because the effect we observe suggests, for example, a fundamentally new way of optimizing thermoelectric response.

Results

Generalization of ice rules. Our starting point is the simple toy model of square ice, in which water molecules are arranged on a square lattice and oriented so as to satisfy sensible hydrogen-bonding rules (Fig. 1a). As there is no unique way of satisfying these rules the system is disordered, even if the molecule orientations are far from random. As in the real-world examples of cubic ice (I_c) and its nano-confined variants[11], this system is characterized by a degenerate manifold of structural ground states[12]. An idea we will come to develop is that this propensity

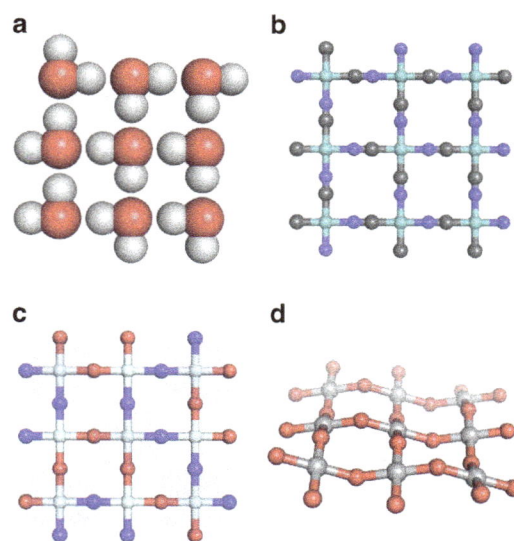

Figure 1 | Correlated disorder in square ice analogues. (**a**) A configurational fragment of square ice. The water molecules are arranged on a square lattice and are oriented so as to satisfy local hydrogen-bonding rules: each molecule accepts two hydrogen bonds and donates two hydrogen bonds. There is no unique solution to satisfying these local constraints, and so the square ice state is configurationally disordered. (**b**) A square-planar transition-metal cyanide configuration that maps onto the ice-like state in **a**. Here each metal cation (coloured green) is coordinated by two nitrogen atoms (blue) and two carbon atoms (grey) such that each N atom is opposite to a C atom. (**c**) Transition metal oxynitrides adopt a related structure[14], in which square-grid layers consist of metal cations coordinated by two nitrogen atoms (blue) and two oxygen atoms (red). The topological equivalence to the square ice configuration can be seen by alternately considering O–M–O and N–M–N orientations for neighbouring metal centres (shaded regions)[26]. (**d**) Displacive modulation of a square MO_2 lattice by in-plane $[MO_4]$ rotations gives configurations that again map onto the square ice state. The correspondence relates displacements above and below the plane with, respectively, O and N atoms of the oxynitride configuration shown in **c**.

for disorder is encoded in the combination of the symmetry of the water molecule (that is, the structural building block) and the lattice on which the water molecules are arranged (here enforced by the directionality of the chemical interactions between building blocks). Any system that shares these geometric features will be characterized by the same configurational degeneracy. So, for example, replacing O–H...O linkages by the M–C–N–M motif found in transition-metal cyanides gives a mapping that—in 3D—relates head-to-tail cyanide disorder in $Cd(CN)_2$ to water molecule orientations in cubic ice (Fig. 1b)[13]. The question of O/N ordering within square grid layers of transition-metal oxynitrides presents a related problem, which maps onto the square ice model following geometry inversion from one site to the next (Fig. 1c)[14]. These examples involve compositional or orientational modulations of the square lattice, but the same ideas are well-known to translate to a variety of modulation types, many of which are key to material function: for example; displacive, electronic, charge density, spin density, orbital and spin orientation (Fig. 1d). For ice-like disorder on the diamond lattice, these mappings are well-established in the literature: hence the 'Coulomb phases'[15] of charge[16], orbital[17] and spin[18] ices.

In seeking to generalise ice-like states, we take our lead from the reticular chemistry approach for generating network structures[19,20]. The key idea here is that lattice topologies can be considered in terms of the assembly of nodes and linkers (Fig. 2a): the geometry of the node determines the possible topologies of the corresponding lattice. In this context, the various square-ice systems of Fig. 1 can be considered perturbations of the square lattice in which two adjacent linkers are distinguished among the four that meet at each square node. If we identify three linkers rather than two, then we generate a distinct family of disordered configurations that—from a reticular chemistry perspective— might be considered to arise from the linking of T-shaped building blocks (Fig. 2b). As for the square ices, there are many possible realizations of this same state: one mapping is to Anderson's resonance valence bond (RVB) description of singlet pair formation in cuprate superconductors[21]; another is to so-called 'domino' tilings of the plane[22].

There are in total just six cases to be considered for perturbations of the square lattice. Two of these are trivial (distinguishing either four or zero linkers); two are related to one another (distinguishing one linker being the same as distinguishing three); and the two cases that remain distinguish different pairings of the four linkers, as shown in Fig. 2c,d. We have

already met the first of these cases in the guise of square ice (Fig. 1c); the second case—in which the linkers distinguished are opposite one another—is ordered and results in symmetry breaking of the underlying square lattice. So there is a nontrivial relationship between perturbations of the node symmetry and the resulting configurational degeneracy.

Ice-like configurations are not confined to perturbations of the square lattice. Equivalent states for the hexagonal, triangular, diamond, cubic and pyrochlore nets are enumerated in Fig. 2e–s and Supplementary Figs 1–37. The extent of disorder can be deduced from the corresponding diffraction patterns, which contain structured diffuse scattering in cases where there is strong correlated disorder (Fig. 2; ref. 10). What emerges is that a substantive fraction of these systems admit large configurational entropies, with a complex relationship between node geometry and extent of correlated disorder. So as to provide insight into this relationship, we generalise Pauling's approximation for the configurational entropy of ice[12,14]:

$$S_{\text{config}} \simeq R \ln\left(\frac{n}{2^{d/2}}\right). \quad (1)$$

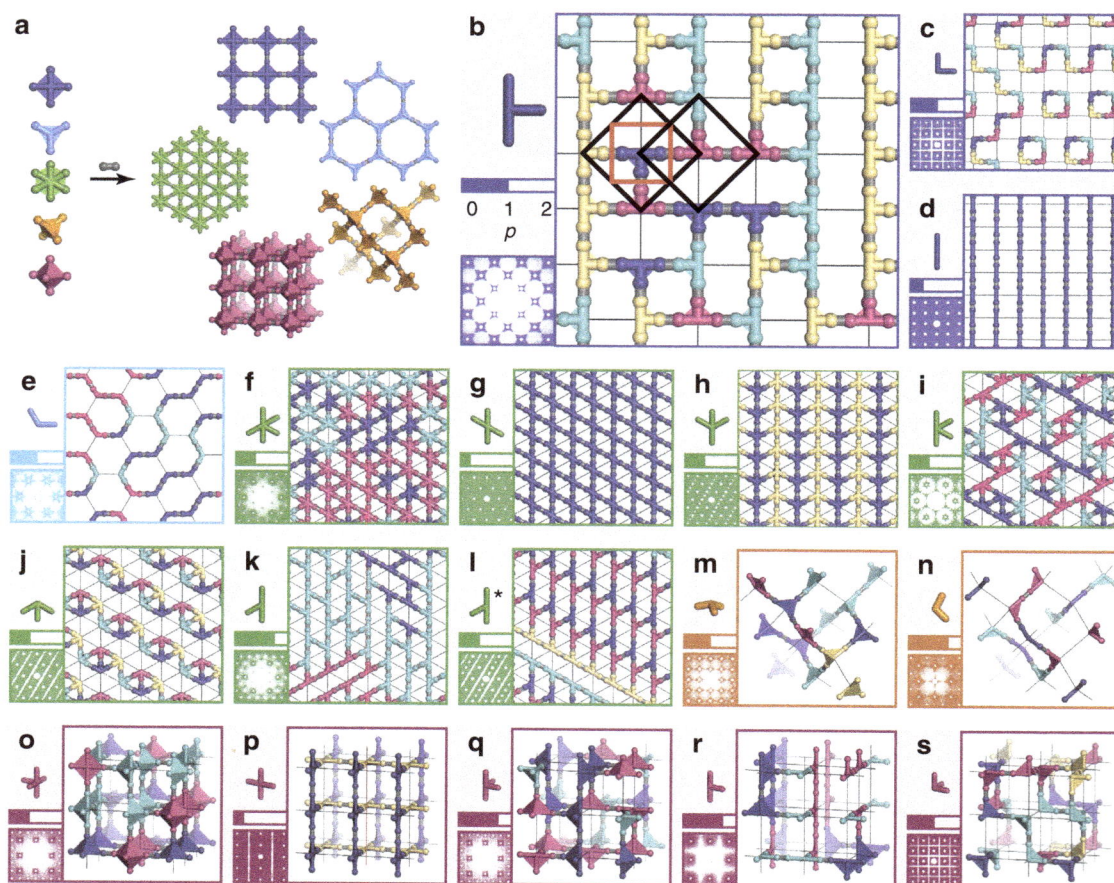

Figure 2 | Reticular design approach to generating procrystalline networks. (a) High-symmetry building blocks connect to form familiar two- and three-dimensional networks: square (indigo), hexagonal (blue), triangular (green), diamondoid (orange) and cubic (red). (b–s) Distinguishing different possible subsets of linkers for these high-symmetry lattices gives a variety of ordered and disordered states, which are grouped here according to the parent lattice. For each panel, the perturbed node geometry is shown in the top-left corner, followed immediately below by a representation of the Pauling number p, a qualitative indicator of the propensity for disorder. A representative network configuration is shown as the main image, with nodes coloured according to their orientation. One suitable projection of the corresponding X-ray diffraction pattern is given in the bottom-left corner (see Supplementary Figs 1–37 for further details). For the configuration shown in b two overlapping neighbourhoods are outlined in black. The Dirichlet–Voronoi cell of the neighbourhood lattice is shown in red; the neighbourhood itself is generated by augmenting this cell to include connected latticed points (see Supplementary Figs 38 and 39 and Supplementary Note 1 for further details). The asterisk in l indicates that a single enantiomer of the node geometry is used (cf. k); this node is chiral when constrained to lie in two dimensions. Further discussion, including extension to the pyrochlore lattice, is given in Supplementary Tables 2 and 3 and Supplementary Note 2.

Here d represents the underlying node connectivity and n corresponds to the number of symmetry-equivalent node perturbations of a given type. The value of n can often be determined by inspection, but it is given more rigorously by the ratio of the orders of the point groups of the parent and perturbed node geometries; for example, $n = |D_{4h}|/|C_{2v}| = 4$ for square ice. We call $p = n/2^{d/2}$ the Pauling number, with the significance that maximising p maximises the propensity for disorder. In this way one expects low-symmetry perturbations of high-symmetry lattices to lead to states of the greatest configurational entropy, a qualitative relationship that is borne out in practice (Fig. 2). From a materials design perspective, what we are saying is that building block geometry and the arrangement of the interactions between building blocks can together encode for specific types of correlated structural disorder.

The procrystalline state. Common to many of the configurations of Fig. 2 is the absence of translational periodicity characteristic of the crystalline state. For a given configuration, every node experiences the same local environment—and hence it is not meaningful to think of these structures as defective in the vernacular sense. Yet there is no unit cell and space group symmetry that properly describes the topological connectivity. We proceed to argue that these systems should not be considered as crystals, but form a separate class of aperiodic solid with its own characteristics. We will use the term 'procrystalline' to describe this state and to emphasise that conventional crystals might be seen as a special case of the definitions that follow.

The procrystalline state is a dense, overlapping packing of identical fundamental structural units (we use the term 'neighbourhoods'), which are positioned periodically but orientationally permuted as permitted by the point symmetry of the neighbourhood geometry. For magnetic systems, these permutations may involve time reversal operations as realized in, for example, the Ising spin ices[23]. In simple cases the neighbourhood corresponds to the Dirichlet–Voronoi cell of the underlying lattice augmented to include neighbouring, correlated lattice points (Fig. 2b; Supplementary Note 1 and Supplementary Figs 38 and 39). Whereas crystals correspond to the special case in which the neighbourhood orientations are themselves periodic, the more general procrystalline state allows for discrete orientational disorder (cf. the continuous orientational degrees of freedom in, for example, plastic crystals and superionics). Any such disorder will always be correlated since neighbourhoods overlap.

Because their underlying neighbourhood lattice is periodic, all procrystals admit a Bragg diffraction pattern and have a well-defined reciprocal lattice. This diffraction pattern can be analysed using conventional crystallographic approaches but doing so yields a structural model in which neighbourhoods are averaged over their different possible orientations and all information regarding orientational correlation is lost; for example, the states represented in Fig 2r,s share identical Bragg diffraction patterns in spite of their distinct local symmetries. Like crystals, procrystalline phases are characterized by macroscopic point symmetry that can be as high as that of the neighbourhood lattice. Yet, unlike crystals, they can support a complete absence of any point or translational symmetry at the microscopic level. It is the existence of a periodic 3D reciprocal lattice that distinguishes procrystals from incommensurate and quasicrystalline phases, and which also guarantees a well-defined Brillouin zone and Bloch-wave-like description of phonon and electronic states.

Physical realizations of procrystallinity. The structures of a number of well- and lesser-known materials can be thought of in precisely these terms. This is true by construction for any phase

with ice-like disorder; in addition to the various systems described above, the family of ferroelectric phases related to KH_2PO_4 is an obvious additional example[24]. Similarly well-established are the statistical mechanical models of RVB[25] and loop[26] states, which correspond to procrystalline lattices with, respectively, one and two linkers distinguished for each node. Hence, physical realizations of either class also fall under our definition (for example, TaS_2 (ref. 27) and $SrTaO_2N$ (ref. 28)). A less obvious example is the assembly of p-terphenyl-3,5,3', 5'-tetracarboxylic acid molecules on pyrolytic graphite to form a hydrogen-bonded network related to the procrystalline lattice illustrated in Fig. 2e (Fig. 3a,b)[29]. This arrangement maps onto the so-called 'rhombus' or 'lozenge' tiling, which in turn corresponds at once to both the Ising triangular antiferromagnet and the RVB description of π-bonding in graphene[30-32]. These equivalences are straightforwardly seen in reciprocal space: Fourier transform of the scanning tunnelling microscopy image of Fig. 3b reveals the same distribution of diffuse scattering and 'pinch point' features expected from our simple geometric model (Fig. 3c,d). A further example is the pattern of correlated Nb off-centre displacements found in the high-temperature cubic phase of $KNbO_3$ (ref. 33). Here the mapping is to the procrystalline lattice illustrated in Fig. 2s, as reflected (again) in the diffuse scattering distribution observed in single-crystal diffraction measurements (Fig. 3e–g).

We expect the link between characteristic diffuse scattering patterns and particular procrystalline states to aid in diagnosing and understanding a range of problems of correlated disorder[10,34]. For powder diffraction measurements, often the only signature of this diffuse scattering is the presence of hkl-dependent anisotropic peak shape broadening. This is the case, for example, in the scattering patterns of $Pd(CN)_2$ and $Pt(CN)_2$; a procrystalline structural model based on connected square-planar $[M(C/N)_4]$ units provides the first convincing description of their diffraction behaviour (Fig. 3h,i)[35]. In other cases, procrystalline states (if not necessarily recognized as such) have been inferred from the combination of a disordered average structure and clear signatures of local distortions that can persist only if suitably correlated. Examples include Jahn Teller distortions in the high-temperature orbital-disordered phase of $LaMnO_3$ (refs 36,37) and the high-pressure amorphous phase of ZrW_2O_8 (ref. 38). So there is good evidence that a variety of procrystalline phases do exist, even if their structures are difficult to interpret using established crystallographic approaches.

Disorder–phonon coupling. But what of the link between correlated structural disorder and function? In principle, the existence of a well-defined Brillouin zone allows coupling between the structural modulations that characterize the procrystalline state and other physical properties that depend on periodicity—for example, the lattice dynamics and electronic band structure[39]. We tested for coupling of this type using as our example a two-dimensional oxynitride lattice (Fig. 1c). The idea was to set up a simple harmonic lattice dynamical model in which we assigned different equilibrium values and stiffnesses to M–O and M–N bond lengths, and/or to O–M–O, N–M–O and N–M–N bond angles, and then to determine the extent to which correlated compositional order might affect the phonon spectrum. Conventional lattice dynamical calculations are designed for periodic (crystalline) structures, and so we made use of a supercell lattice dynamical approach in order to treat compositional disorder explicitly. We first benchmarked our calculations by determining the 'mean-field' phonon dispersion expected for an average of the different force constant values; we found perfect agreement between our supercell lattice dynamical calculations and those obtained using a conventional implementation of the general utility lattice programme (GULP)[40] (Fig. 4a, Supplementary Fig. 40 and Supplementary Methods 1). For random distributions of equal numbers of O and N atoms, the basic phonon structure was similar to the mean-field case, with a slight broadening of phonon frequencies throughout the Brillouin Zone (Fig. 4b). This result is in agreement with *ab initio* molecular dynamics studies of the related problem of the lattice dynamics of configurational glasses[41]. By contrast, for the procrystalline arrangement there was a dramatic dispersion in energy of the acoustic branches that we observed most noticeably for wave-vectors near the zone corner (Fig. 4c,d). The qualitative similarity to the 'waterfall' phonons observed in thermoelectrics and relaxor ferroelectrics is striking, and suggests a plausible origin for the phenomenon in those systems[42,43].

So this is our key result: strongly correlated structural disorder allows selective control over physical properties that depend on periodicity. The implication for systems where thermal and electronic conductivities are mediated by, respectively, phonons and electronic states localised in different regions of the Brillouin zone is that disorder–phonon coupling offers a means of selectively reducing thermal conductivity (inversely proportional to phonon bandwidth) without affecting charge transport behaviour. This is an attractive design strategy for developing next-generation thermoelectrics, and one that contrasts with the use of 'rattlers' which are indiscriminate in their **k**-space coupling[44].

Discussion

Our reticular chemistry methodology suggests a number of synthetic routes for realising new classes of functional procrystalline solids. Metal-organic frameworks are an obvious platform, given they offer the requisite control over building unit geometry and their energetics tend to be dominated by local interactions[45,46]. While there is reduced scope for coupling between structural disorder and electronic behaviour in these systems, porosity percolation will certainly be affected by disorder and may in turn govern sorption, mechanical and ion storage properties[47,48]. In more conventional inorganics, local symmetry lowering can be achieved by covalency effects (as in mixed-anion perovskites) or by first- or second-order Jahn Teller distortion (as in the chalcogenide thermoelectrics and perovskite ferroelectrics). Moreover, because our analysis is essentially geometric in nature, there is clear scope to extend these concepts to magnetic or electronic states, or indeed to the macroscopic scale. The recent demonstration that disordered metamaterials can show strong structural coupling to light scattering processes is an example relevant to the generation of modern photonics[49]. Thinking beyond the ground-state properties of procrystals, we anticipate the existence of novel collective and hidden degrees of freedom that promise a rich physics of their own; for example, topological excitations[50] and/or 'hidden order' transitions between distinct local symmetries[51].

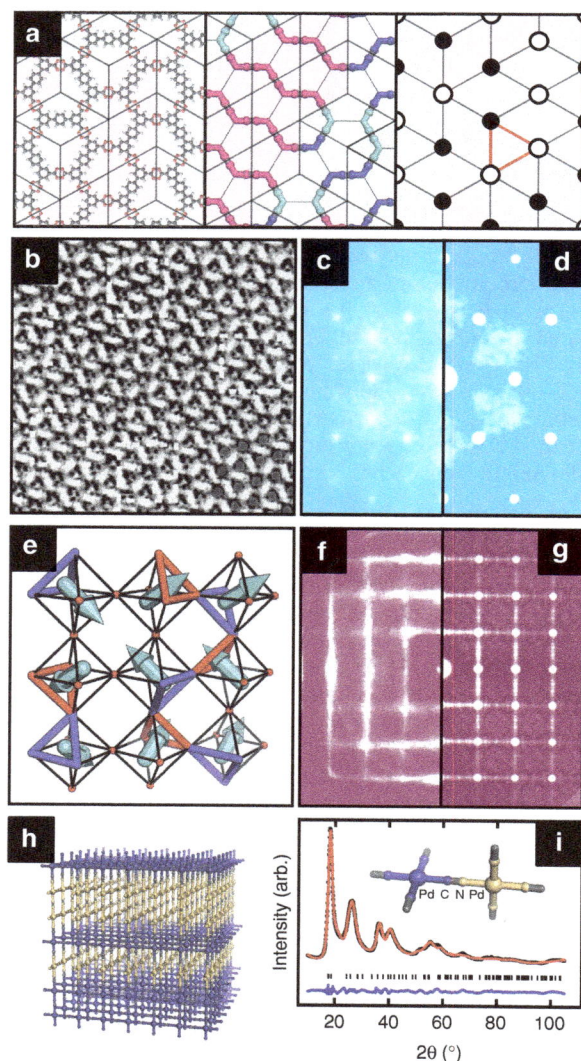

Figure 3 | Representative physical realizations of procrystalline states. (**a**) Molecules of *p*-terphenyl-3,5,3',5'-tetracarboxylic acid (TPTC) self assemble on pyrolytic graphite to form a hexagonal procrystalline network related to that represented in Fig. 2e (ref. 29). One possible arrangement is shown here (left), together with our topological abstraction (centre); molecule orientations in the former map onto the 'missing' linkers of the latter. Shown in background is the configurationally-equivalent rhombus tiling, which further relates to the triangular Ising antiferromagnet (right): opposite vertices of each rhombus are decorated with pairs of 'spin-up' (white circles) and 'spin-down' (black circles) states such that each triplet of neighbouring spins (red triangle) contains at least one state of each kind. (**b**) A scanning tunnelling microscopy image of the corresponding experimental state. Adapted from ref. 29. Reprinted with permission from AAAS. (**c**) The Fourier transform of the image in **b** showing regions of structured diffuse scattering. (**d**) The strikingly similar scattering pattern calculated for the hexagonal procrystalline network of Fig. 2e. (**e**) In the high-temperature cubic phase of $KNbO_3$, Nb^{5+} ions displace towards one face of their octahedral coordination environment such that neighbouring Nb centres displace in the same sense relative to the vector joining the pair. Octahedral faces alternately near to (red) and away from (blue) the displaced Nb centres gives a procrystalline net related to that shown in Fig. 2s. (**f**) Structured diffuse scattering observed in single-crystal X-ray diffraction measurements of $KNbO_3$. Adapted from ref. 33. Reprinted with permission of the International Union of Crystallography. (**g**) Diffuse scattering calculated for the cubic procrystalline network of Fig. 2s. (**h**) A procrystalline model for the structure of $Pd(CN)_2$ and $Pt(CN)_2$ based on the state represented in Fig. 2p: the two possible orientations of square-planar $M(C/N)_4$ nodes are shown in indigo and gold. (**i**) Comparison of the Rietveld fit for this structural model of $Pd(CN)_2$ (red lines) and the experimental X-ray powder diffraction data of ref. 35 (black points, $\lambda = 1.54$ Å); tick marks indicate the positions of parent Bragg reflections and the difference (data–fit) is shown in blue (see Supplementary Fig. 42, Supplementary Tables 4 and 5, Supplementary Note 3 and Supplementary Method 2 for further discussion).

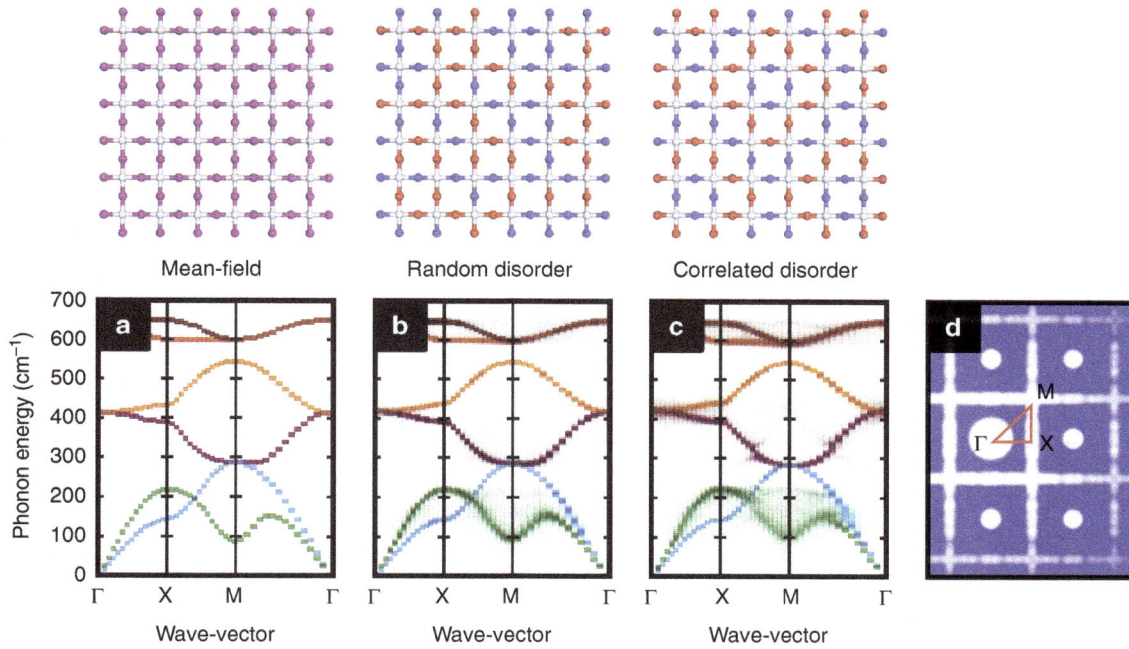

Figure 4 | Selective disorder–phonon coupling in a procrystalline network. (**a**) Mean-field phonon dispersion curves for a two-dimensional oxynitride lattice, determined using the SCLD approach described in the text. (**b**) The phonon dispersion determined using SCLD for configurations in which equal numbers of O and N atoms are distributed randomly across the O/N sites of the same oxynitride lattice. There is a slight broadening of phonon frequencies relative to (**a**). (**c**) A third set of phonon dispersion curves, again determined using SCLD but for configurations in which O and N atoms have been distributed according to the correlated disorder (procrystalline) model of Fig. 2c. There is now substantive 'waterfall'-like phonon broadening around the M point. (**d**) The calculated single crystal scattering pattern for the procrystalline model of Fig. 2c, demonstrating that this particular state is characterized by modulations that—like the phonon broadening—are localised around the M point of the BZ. BZ, Brillouin zone; SCLD, supercell lattice dynamics.

Methods

Procrystal lattice generation. Representative procrystalline configurations were generated using a suite of custom Monte Carlo codes. For each given lattice type, a supercell of the corresponding crystallographic cell was generated and periodic boundary conditions were applied [Supplementary Table 1]. Linkers were randomly assigned one of two initial states $e = \pm 1$. A fictitious configurational energy of typical form:

$$E = \sum_{i \in \text{nodes}} \left| \left(\sum_{j \in \text{linkers}} e_{ij} \right) - \bar{e} \right|^2 \quad (2)$$

was calculated, where e_{ij} represents the state of the jth linker of the ith node and the expectation value $\bar{e}$ is a function of the particular node geometry of interest. By way of example, $\bar{e} = 0$ for ice-like states on the diamond lattice, since a local '2-in-2-out' configuration is described by two linkers with $e = +1$ and two with $e = -1$. The form of equation (2) is more complex for some combinations of lattice and node geometry, but the relationship that $E = 0$ if and only if every node adopts the same local geometry was maintained throughout our study. Monte Carlo minimisation proceeded via the usual Metropolis algorithm[52], with moves $e_{ij} \rightarrow -e_{ij}$. Because we are interested in defect-free procrystalline states, and because the absolute value of the energy term in equation (2) is not physically meaningful, we terminated our Monte Carlo minimisation not at equilibrium but only when $E = 0$. We note that in physical realizations of procrystalline states higher-order correlations may bias towards specific subsets of the $E = 0$ configurational space explored in this first-order Monte Carlo approach. All relevant code is available by request.

Diffuse scattering calculations. Physical realizations of the various procrystalline networks generated as described above were produced by placing Nb atoms at every node position and O atoms at those linker sites for which $e_{ij} = +1$. Linker sites for which $e_{ij} = -1$ were left vacant. Powder and single crystal X-ray diffraction patterns were generated from individual atomic configurations using the programs CrystalDiffract and SingleCrystal, respectively. Note that the diffuse scattering evident in our diffraction patterns contains contributions only from the linker sites, and its intensity is proportional to the difference in scattering power for $e_{ij} = \pm 1$ occupancies. Consequently, all simple substitutional realizations of a given procrystalline state share the same diffuse scattering pattern, up to a constant factor.

Lattice dynamical calculations. Phonon calculations made use of the GULP program[40]. Supercells of a fictitious two-dimensional NbON square lattice containing $30 \times 30 \times 1$ unit cells (space group symmetry Pm; relative atom coordinates Nb$(\frac{1}{2}, \frac{1}{2}, 0)$, O/N1$(\frac{1}{2}, 0, 0)$, O/N2$(0, \frac{1}{2}, 0)$) were constructed in three ways. First, a 'mean-field' configuration was generated in which all O/N sites were treated as hybrid atoms. Second, random assignment of O/N sites to equal numbers of O and N atoms gave a 'random disorder' configuration. And, third, a suitable procrystalline configuration of the type illustrated in Fig. 1c was used to assign O and N atoms such that each Nb centre was coordinated by exactly two O and two N atoms in a *cis* arrangement; we refer to this as the 'correlated disorder' configuration. In all cases the same set of simple harmonic potentials was used to calculate the lattice enthalpy

$$E_{\text{latt}} = \frac{1}{2}k_1 \sum_{ij} \left(r_{ij} - r_e \right)^2 + \frac{1}{2}k_2 \sum_{ijk} \left(\theta_{ijk} - \theta_e \right)^2 + k_3 \sum_{ijk} \left(1 + \cos \theta_{ijk} \right)^2. \quad (3)$$

Here $k_1 = 10\,\text{eV Å}^{-2}$ is the stiffness of the Nb–O/N bond, $k_2 = 2\,\text{eV rad}^{-2}$ is the stiffness of the O/N–Nb–O/N angle, and $k_3 = 0.1\,\text{eV}$ is the stiffness of the Nb–O/N–Nb angle. This potential model was chosen because it gave a phonon spectrum with physically sensible mode frequencies, good separation between branches, and realistic low-energy mode features (e.g., low-energy tilts at the zone corner). Ultimately, we found that the difference between the phonon dispersion for random and correlated disorder configurations was sensitive to variation in any or all of the k_i, r_e, θ_e and atomic masses for O and N atoms. For the phonon dispersion curves illustrated in Fig. 4 we used the very simplest case in which only the value of θ_e was distinguished: specifically, we used values of 75°, 90° and 105° for N–Nb–N, N–Nb–O and O–Nb–O angles in each of the random and correlated disorder configurations and a common value of $\theta_e = 90°$ for the mean-field configuration. In all cases, the same equilibrium bond length $r_e = 1\,\text{Å}$ was used, corresponding to half the Nb...Nb separation in the configurations. Likewise all O and N atoms were assigned the same effective mass of 15 a.m.u.

Atomic coordinates were relaxed and phonon calculations performed at the Γ point ($\mathbf{k}_{\text{supercell}} = (0, 0, 0)$) of the supercell which corresponds to the $\mathbf{k}$-vector grid of $\frac{1}{30}$ in the reduced mean-field unit cell. The normal modes were calculated as eigenvalues of the dynamical matrix. Despite nominal Pm symmetry, not all of the phonons with atomic displacements along z could be fully separated from in-plane displacements (most probably due to numerical errors). Thus the eigenvalues were calculated using the 'scipy' python library from the two-dimensional component of dynamical matrix which calculated by the GULP. The contribution of each eigenvector $\mathbf{e}_i$ to all possible $\mathbf{k}$-points was calculated by projecting the eigenvectors at each wave-vector $\mathbf{k}$ in the following way:

$$e_i|_{\mathbf{k}} = \sum_{j\alpha} \left| \sum_{\ell} e_i(j\ell, \alpha) \exp[i\mathbf{k} \cdot \mathbf{r}(\ell)] \right|^2, \tag{4}$$

where j indexes the atoms in each unit cell ℓ of the supercell, and $\alpha \in \{x, y\}$. For the mean-field case the projection gave phonon dispersion curves indistinguishable from those obtained using a conventional single-cell GULP calculation (Supplementary Fig. 40 and Supplementary Methods 2). In the case of the random and correlated disorder configurations, the phonon dispersion curves shown in Fig. 4 represent an average over the results obtained for five independent configurations. All relevant code is available by request.

Optical Fourier transform. The scanning tunnelling microscopy image shown in Fig. 3b was Fourier transformed using the Java applet 'Diffraction and Fourier transform'[53]. The input data were converted to grayscale, inverted, and optimized for brightness and contrast; the output data were corrected for sample orientation and symmetrised (plane symmetry *p6m*). The actual image used and its raw transform are shown in Supplementary Fig. 41.

References

1. Natta, G. & Nasini, A. G. The crystal structure of xenon. *Nature* **125**, 457 (1930).
2. David, W. I. F. *et al.* Crystal structure and bonding of ordered C$_{60}$. *Nature* **353**, 147–149 (1991).
3. Reddy, V. S., Natchiar, K., Stewart, P. L. & Nemerow, G. R. Crystal structure of human adenovirus at 3.5 Å resolution. *Science* **329**, 1071–1075 (2010).
4. Meier, W. M. *Molecular Sieves* 10–27 (Soc. of Chem. and Ind., 1968).
5. Eddaoudi, M. *et al.* Modular chemistry: secondary building units as a basis for the design of highly porous and robust metal–organic carboxylate frameworks. *Acc. Chem. Res.* **34**, 319–330 (2001).
6. Dyer, M. S. *et al.* Computationally assisted identification of functional inorganic materials. *Science* **340**, 847–852 (2013).
7. Pitcher, M. J. *et al.* Tilt engineering of spontaneous polarization and magnetization above 300 K in a bulk layered perovskite. *Science* **347**, 420–424 (2015).
8. Steinhardt, P. J. & Jeong, H.-C. A simpler approach to Penrose tiling with implications for quasicrystal formation. *Nature* **382**, 431–433 (1996).
9. Damasceno, P. F., Engel, M. & Glotzer, S. C. Predictive self-assembly of polyhedra into complex structures. *Science* **337**, 453–457 (2012).
10. Keen, D. A. & Goodwin, A. L. The crystallography of correlated disorder. *Nature* **521**, 303–309 (2015).
11. Algara-Siller, G. *et al.* Square ice in graphene nanocapillaries. *Nature* **519**, 443–445 (2015).
12. Pauling, L. The structure and entropy of ice and of other crystals with some randomness of atomic arrangement. *J. Am. Chem. Soc.* **57**, 2680–2684 (1935).
13. Fairbank, V. E., Thompson, A. L., Cooper, R. I. & Goodwin, A. L. Charge-ice dynamics in the negative thermal expansion material Cd(CN)$_2$. *Phys. Rev. B* **86**, 104113 (2012).
14. Camp, P. J., Fuertes, A. & Attfield, J. P. Sub-extensive entropies and open order in perovskite oxynitrides. *J. Am. Chem. Soc.* **134**, 6762–6766 (2012).
15. Henley, C. L. The 'Coulomb phase' in frustrated systems. *Ann. Rev. Cond. Matt. Phys* **1**, 179–210 (2010).
16. Shoemaker, D. P. *et al.* Atomic displacements in the charge ice pyrochlore Bi$_2$Ti$_2$O$_6$O′ studied by neutron total scattering. *Phys. Rev. B* **81**, 144113 (2010).
17. Chern, G.-W. & Wu, C. Orbital ice: an exact Coulomb phase on the diamond lattice. *Phys. Rev. E* **84**, 061127 (2011).
18. Bramwell, S. T. & Gingras, M. J. P. Spin ice state in frustrated magnetic pyrochlore materials. *Science* **294**, 1495–1501 (2001).
19. Wells, A. F. *Structural Inorganic Chemistry* (Clarendon Press, 1984) 5th edn.
20. Yaghi, O. M. *et al.* Reticular synthesis and the design of new materials. *Nature* **423**, 705–714 (2003).
21. Anderson, P. W. The resonating valence bond state in La$_2$CuO$_4$. *Science* **235**, 1196–1198 (1987).
22. Kasteleyn, P. W. The statistics of dimers on a lattice. I. The number of dimer arrangements on a quadratic lattice. *Physica* **27**, 1209–1225 (1961).
23. Bramwell, S. T. & Harris, M. J. Frustration in Ising-type spin models on the pyrochlore lattice. *J. Phys. Cond. Matt.* **10**, L215–L220 (1998).
24. Slater, J. C. Theory of the transition in KH$_2$PO$_4$. *J. Chem. Phys.* **9**, 16–33 (1941).
25. Anderson, P. W. Resonating valence bonds: a new kind of insulator? *Mater. Res. Bull.* **8**, 153–160 (1973).
26. Kondev, J. Liouville field theory of fluctuating loops. *Phys. Rev. Lett.* **78**, 4320–4323 (1997).
27. Tosatti, E. & Fazekas, P. On the nature of the low-temperature phase of 1T-TaS$_2$. *J. Phys. Coll.* **37**, 165–168 (1976).
28. Günther, E., Hagenmayer, R. & Jansen, M. Strukturuntersuchungen an den Oxidnitriden SrTaO$_2$N, CaTaO$_2$N and LaTaON$_2$ mittels Neutronen- und Röntgenbeugung. *Z. Anorg. Allg. Chem.* **626**, 1519–1525 (2000).
29. Blunt, M. O. *et al.* Random tiling and topological defects in a two-dimensional molecular network. *Science* **322**, 1077–1081 (2008).
30. Wannier, G. H. Antiferromagnetism. The triangular Ising net. *Phys. Rev.* **79**, 357–364 (1950).
31. Fisher, M. E. Statistical mechanics of dimers on a plane lattice. *Phys. Rev.* **124**, 1664–1672 (1961).
32. Kasteleyn, P. W. Dimer statistics and phase transitions. *J. Math. Phys.* **4**, 287–293 (1963).
33. Comès, R., Lambert, M. & Guinier, A. Désordre linéaire dans les cristaux (cas du silicium, du quartz, et des pérovskites ferroélectriques). *Acta Cryst. A* **26**, 244–254 (1970).
34. Welberry, T. R. & Weber, T. One hundred years of diffuse scattering. *Cryst. Rev.* **22**, 2–78 (2016).
35. Hibble, S. J. *et al.* Structures of Pd(CN)$_2$ and Pt(CN)$_2$: intrinsically nanocrystalline materials? *Inorg. Chem.* **50**, 104–113 (2011).
36. Božin, E. S. *et al.* Local structural aspects of the orthorhombic to pseudo-cubic phase transformation in La$_{1-x}$Ca$_x$MnO$_3$. *Physica B* **385-386**, 110–112 (2006).
37. Ahmed, M. R. & Gehring, G. A. Volume collapse in LaMnO$_3$ studied using an anisotropic Potts model. *Phys. Rev. B* **79**, 174106 (2009).
38. Keen, D. A. *et al.* Structural description of pressure-induced amorphization in ZrW$_2$O$_8$. *Phys. Rev. Lett.* **98**, 225501 (2007).
39. Valla, T. *et al.* Anisotropic electron—phonon coupling and dynamical nesting on the graphene sheets in superconducting CaC$_6$ using angle-resolved photoemission spectroscopy. *Phys. Rev. Lett.* **102**, 107007 (2009).
40. Gale, J. D. GULP: a computer program for the symmetry-adapted simulation of solids. *J. Chem. Soc., Faraday Trans.* **93**, 629–637 (1997).
41. Fang, H., Dove, M. T., Rimmer, L. H. N. & Misquitta, A. J. Simulation study of pressure and temperature dependence of the negative thermal expansion in Zn(CN)$_2$. *Phys. Rev. B* **88**, 104306 (2013).
42. Gehring, P. M., Park, S.-E. & Shirane, G. Soft phonon anomalies in the relaxor ferroelectric Pb(Zn$_{1/3}$Nb$_{2/3}$)$_{0.92}$Ti$_{0.08}$O$_3$. *Phys. Rev. Lett.* **84**, 5216–5219 (2000).
43. Delaire, O. *et al.* Giant anharmonic phonon scattering in PbTe. *Nat. Mater.* **10**, 614–619 (2011).
44. Christensen, M. *et al.* Avoided crossing of rattler modes in thermoelectric materials. *Nat. Mater.* **7**, 811–815 (2008).
45. Cairns, A. B. & Goodwin, A. L. Structural disorder in molecular framework materials. *Chem. Soc. Rev.* **42**, 4881–4893 (2013).
46. Cliffe, M. J. *et al.* Correlated defect nanoregions in a metal–organic framework. *Nat. Commun.* **5**, 4176 (2014).
47. Wessells, C. D., Huggins, R. A. & Cui, Y. Copper hexacyanoferrate battery electrodes with long cycle life and high power. *Nat. Commun.* **2**, 550 (2011).
48. Cliffe, M. J., Hill, J. A., Murray, C. A., Coudert, F.-X. & Goodwin, A. L. Defect-dependent colossal negative thermal expansion in UiO-66(Hf) metal–organic framework. *Phys. Chem. Chem. Phys.* **17**, 11586–11592 (2015).
49. Riboli, F. *et al.* Engineering of light confinement in strongly scattering disordered media. *Nat. Mater.* **13**, 720–725 (2014).
50. Bak, P. Phenomenological theory of icosahedral incommensurate ('quasiperiodic') order in Mn–Al alloys. *Phys. Rev. Lett.* **54**, 1517–1519 (1985).
51. Toudic, B. *et al.* Hidden degrees of freedom in aperiodic materials. *Science* **319**, 69–71 (2008).
52. Metropolis, N., Rosenbluth, A. W., Rosenbluth, M. N., Teller, A. H. & Teller, E. Equation of state calculations by fast computing machines. *J. Chem. Phys.* **21**, 1087–1092 (1953).
53. Schoeni, N. & Chapuis, G. Fourier Transform, http://escher.epfl.ch.fft/ (2006).

Acknowledgements

A.R.O., A.B.C., M.J.C., A.S. and A.L.G. gratefully acknowledge financial support from the EPSRC (Grant EP/G004528/2), the ERC (Grant 279705), from the Diamond Light Source to A.R.O., and from the Leverhulme Trust and Swiss National Science Foundation to A.S. This project has received funding from the European Union's Horizon 2020 research and innovation programme under the Marie Skłodowska-Curie grant agreement No. 641887 (Project Acronym: DEFNET-ETN). We are grateful to U. Waghmare (Bangalore), D.A. Keen (ISIS), W.J.K. Fletcher (Oxford) and J.A.M. Paddison (Georgia Tech.) for relevant discussions.

Author contributions

All authors contributed to the development of the conceptual basis for describing the procrystalline state. A.R.O. carried out the Monte Carlo simulations and single-crystal diffuse scattering calculations. A.R.O. and A.S. performed the lattice dynamical calculations. A.B.C. carried out the powder diffraction refinements. A.R.O., M.G.T. and A.L.G. conceived the study. A.L.G. wrote the paper with input from all authors.

Additional information

Competing financial interests: The authors declare no competing financial interests.

8

Ionic polarization-induced current–voltage hysteresis in $CH_3NH_3PbX_3$ perovskite solar cells

Simone Meloni[1,2], Thomas Moehl[3], Wolfgang Tress[3,4], Marius Franckevičius[3,5], Michael Saliba[4], Yong Hui Lee[4], Peng Gao[4], Mohammad Khaja Nazeeruddin[4], Shaik Mohammed Zakeeruddin[3], Ursula Rothlisberger[1,2] & Michael Graetzel[3]

$CH_3NH_3PbX_3$ (MAPbX$_3$) perovskites have attracted considerable attention as absorber materials for solar light harvesting, reaching solar to power conversion efficiencies above 20%. In spite of the rapid evolution of the efficiencies, the understanding of basic properties of these semiconductors is still ongoing. One phenomenon with so far unclear origin is the so-called hysteresis in the current–voltage characteristics of these solar cells. Here we investigate the origin of this phenomenon with a combined experimental and computational approach. Experimentally the activation energy for the hysteretic process is determined and compared with the computational results. First-principles simulations show that the timescale for MA$^+$ rotation excludes a MA-related ferroelectric effect as possible origin for the observed hysteresis. On the other hand, the computationally determined activation energies for halide ion (vacancy) migration are in excellent agreement with the experimentally determined values, suggesting that the migration of this species causes the observed hysteretic behaviour of these solar cells.

[1] Laboratoire de Chimie et Biochimie Computationnelles, ISIC, FSB-BCH, École Polytechnique Fédérale de Lausanne (EPFL), Lausanne CH-1015, Switzerland. [2] National Competence Center of Research (NCCR) MARVEL—Materials' Revolution: Computational Design and Discovery of Novel Materials, Lausanne CH-1015, Switzerland. [3] Laboratory of Photonics and Interfaces, ISIC, Swiss Federal Institute of Technology (EPFL), Lausanne CH-1015, Switzerland. [4] Group for Molecular Engineering of Functional Materials, ISIC-Valais, Swiss Federal Institute of Technology (EPFL), Lausanne CH-1015, Switzerland. [5] Center for Physical Sciences and Technology, Savanorių Avenue 231, Vilnius LT-02300, Lithuania. Correspondence and requests for materials should be addressed to T.M. (email: Thomas.moehl@epfl.ch) or to U.R. (email: Ursula.roethlisberger@epfl.ch).

The perovskite MAPbI3 (methylammonium lead triiodide) and its halide analogues have emerged as one of the new and very interesting absorber materials for highly efficient solar cells[1–5]. Because of the ease of processing and wide range of applications (solar cells, photodetectors[6–10] and lasing[11–14]) they have attracted intense attention in the research community. Currently, the increase of efficiency of the solar cell devices is one of the main focuses[15,16], but the understanding of the basic operation principles of devices containing this material is also slowly evolving[17–19]. One of the main open questions is the origin of the observed hysteresis of the current-voltage (JV) curve of MAPbI3-based solar cells, first reported by Dualeh et al.[18], and investigated in more detail by Snaith et al.[20], Tress et al.[17], O Regan et al.[21] and other authors[22–24]. This effect complicates the determination of the 'real' solar-to-electrical power conversion efficiency of such devices and can make 'bad cells look good' as presented in a recent publication by Christians et al.[25] Moreover, hypotheses were put forward suggesting a fundamental link between hysteresis and the limited long-term stability of halide perovskites[17]. Several ideas for the peculiar phenomenon underlying hysteresis have been proposed, the main advocated views involving either a ferroelectric effect[26–30] or ionic (vacancy) movement[17,31] inside the perovskite as probable cause.

Possible ferroelectric effects can be caused by the orientation of the organic (dipolar) cations, namely MA, or induced by deformation of the inorganic framework. In both cases the crystal acquires a net dipole moment. Under the effect of the external potential the MA molecules align with the external electric field, and the dipole moment of aligned MA molecules produces a balancing field lowering the effective field acting on the charge carriers. If the characteristic timescale necessary for the alignment is of the same order as the scanning time of the external potential, the effective potential acting on the charge carriers depends on the 'history' of the experiment, resulting in the hysteresis. Although several theoretical[27,30,32,33] and experimental[26,34–36] studies have been undertaken in this direction, a recent investigation of Fan et al.[35] showed that no room temperature ferroelectric behaviour could be measured.

An alternative hypothesis is that the hysteresis is due to the movement of ionic species[17,24,31]. It is well known from the literature that inorganic perovskites, like CsPbBr3 or CsPbCl3, are excellent halide conductors[37,38]. Moveable ions (or their vacancies) will induce a retarded reaction towards the change of electronic charge distribution of such a device under operation that could explain the so-called slow time component. Under the influence of an external biasing field or the 'built-in' potential of the device, the migration of ionic vacancies may result in a net

charge accumulation in certain regions of the MAPbX3 or at its contacts. The change of the concentration profile of the vacancies produces a balancing internal counterfield acting on the electronic charge carriers. Again, if the characteristic time of polarization of the sample is associated to the migration of ionic species is of the same order of magnitude as the potential scanning time, this phenomenon will result in a hysteresis of the JV curve.

Previous experimental studies[39,40] have estimated the phenomenon at the origin of the observed hysteresis to take place on a timescale of microseconds to seconds. The relatively long timescale of this phenomenon suggests that it is a thermally activated process characterized by an activation energy sizably higher than the thermal energy available to the system. Here we have used a combined experimental and theoretical approach to determine this activation energy, and to identify the causative process. Experimentally we have determined the activation energy of the hysteretic process from the temperature-dependent measurement of the JV curve for MAPbI3 and MAPbBr3. By density functional based simulations we have determined the activation energy of different ion (vacancy) migrations in the crystal lattice as well as the characteristic rotational time of the MA ions. This combined approach allowed us to establish the general nature of the phenomenon and its activation energy. Our results support the hypothesis that hysteresis is due to halide ion (vacancy) migration induced polarization of the perovskite layer and exclude a ferroelectric effect due to the alignment of the MA ions.

Results

Experimental determination of the activation energy. To determine the activation energy, E_a, of the process underlying the hysteretic behaviour, we have chosen a simple approach consisting in analysing the effect of the temperature on the JV curve of perovskite solar cell devices under illumination. A typical plot of different JV curves under illumination is presented in Fig. 1. The efficiencies of the devices with iodide were in the range of 10–14% PCE. Bromide devices had lower efficiencies, mainly because of the low J_{sc} (about 3–5 mA cm^{-2}), resulting in a power conversion efficiency (PCE) of 2–4%.

The general protocol for most of the measurements was (if no other conditions are indicated): (i) 50 s waiting at reverse potential (-0.5 V); (ii) scanning the JV curve with 50 mV s^{-1} to 1.1 V forward bias (in the following denoted as 'forward scan'); (iii) 50 s waiting at 1.1 V forward bias; and (iv) scanning the JV curve with 50 mV s^{-1} back to -0.5 V reverse bias ('reverse scan').

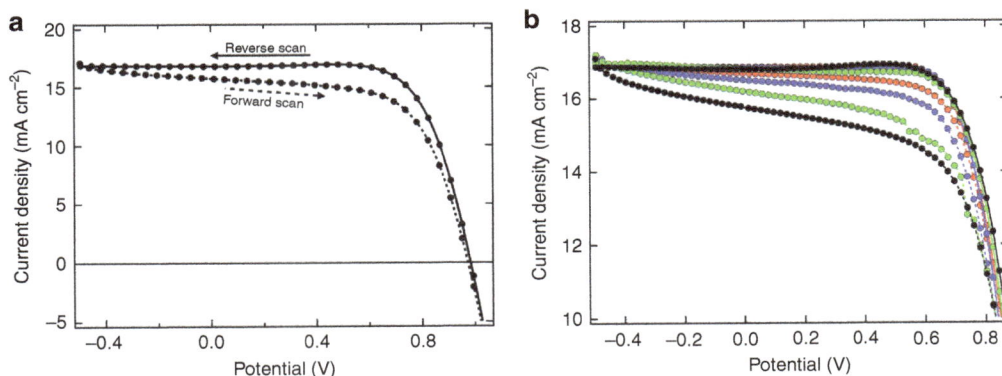

Figure 1 | JV curves of an iodide-based device. (a) At 1 sun at $-15°$ C **(b)** at different temperatures (for a better comparison all curves are scaled to reach the same J_{sc} for the reverse scans). The voltage scan rate is 50 mV s^{-1} between -0.5 to 1.1 V. Between each forward and backward scan the potential was kept constant for 50 s to let the device equilibrate (please see text for further explanation). Red (20 °C); blue (5 °C); green (-5 °C); and black (-15 °C).

The measurements have been performed at several different temperatures (normally between 20 and −20 °C). Figure 1b shows the dependence of the JV curve on the temperature. The measurements were started at 20 °C. After each measurement cycle, the JV curve of the device was re-measured at 20 °C to ensure that the device did not degrade during the measurements. Several features can be observed in the JV curve in Fig. 1b. The reverse scans show minor dependence on the temperature exhibiting generally a similar shape. Furthermore, one can observe a 'bump' immediately before reaching a plateau of the current. This bump appeared for most of the devices though shape and change with temperature was not further investigated (please, see also text below regarding the phenomenon). The forward scans show a stronger dependence on the temperature. The slope of the overall current increases just as if the shunt resistance decreases with temperature. In other words, at room temperature the hysteresis between forward and reverse scan is small. Upon decreasing the temperature, the hysteresis between forward and reverse scan increases. To extract the E_a, we measured the hysteresis as the difference of the current at a given voltage along the backward and forward scan, $\Delta I = J_B(V) - J_F(V)$, and studied its dependence on the temperature.

The simultaneous presence of several types of charge carriers (electrons, holes and ionic defects) in the system, the possibility that polarization affects the carrier dynamics in the perovksite layer or at the interface with the contacts, altering absorption, transport and recombination properties, makes it impossible to derive an analytical expression of the dependence of ΔI as a function of the temperature. A numerical solution, on the other hand, requires the determination of the dependence of the generation rate, diffusion coefficient, recombination rate and surface recombination velocity on temperature and the polarization of the perovskite layer. Moreover, in view of simulating hysteretic behaviour as a function of the temperature and sweeping rate, the solution of the time-dependent transport-reaction problem is required, and not the simpler steady-state solution of the time-independent problem. All this renders also the numerical solution option for interpreting experimental results out of reach at the moment. Thus, similarly to other authors (see, for example, Eames et al.[24] and Yang et al.[41]), we have used an empirical relation between ΔI and T:

$$\frac{1}{\Delta I} = A \times e^{-\frac{E_a}{k_B T}} + C \quad (1)$$

with A as prefactor, k_B as Boltzmann's constant, T as temperature and C as a constant. The reason for using the inverse of the difference between the backward and forward photocurrent is

that this current difference is reduced when the process generating the hysteresis relaxes more quickly to the stationary condition during the voltage scanning. In other words, ΔI is expected to have an inverse proportionality to the ionic current ($\Delta I \sim 1/\Delta I_{ionic}$), which depends on the corresponding diffusion coefficient, typically having an Arrhenius-like dependence on the temperature. Intuitive arguments supporting the empirical relation above between ΔI and E_a will be further elaborated in the Discussion section. In the following, we show that this relation is, indeed, obeyed by the experimental data. In addition, we will use results of atomistic simulations to show that the experimentally determined activation energy is, indeed, associated to a microscopic diffusion process.

Figure 2a shows ΔI as a function of the potential for an iodide-based device at different temperatures. Generally ΔI increases at lower temperature. One can also observe the strong increase of ΔI in the high forward bias region originating from the already mentioned 'bump' often observed in the reverse scan of perovskite solar cells. In Fig. 2b $\ln(1/\Delta I)$ is plotted against $1/T$ for selected potentials in the range of −0.2 to 0.4 V. By fitting these data with the Arrhenius-like equation, E_a is determined from the slope of $\ln(1/\Delta I)$ vs $1/T$ (similar plots for the bromide-based devices are shown in Supplementary Fig. 1).

The activation energy determined at high forward (> 0.4 V) and reverse (< −0.2 V) bias range showed the highest deviation from the E_a values at potentials in the medium range. This has two-fold reasons. One is the already mentioned bump in the reverse scan, as also observed by Tress et al. The origin of this bump is, presumably, the initially low field in the device for the reverse scan and, therefore, the low driving force for any kind of process which then sets in towards lower forward voltages. Therefore only the E_a values determined in the bias range −200 to 400 mV have been used for the calculation of the averaged E_a of Table 1. The hysteresis effect of the devices with bromide is generally less pronounced, which also leads to a higher error in the estimation of the activation energy from experimental data (vide infra). As a result, some Br-based devices did not seem to show the expected dependence on the temperature, for example, showing very-low or negative activation energies.

The current density may also have an impact on the hysteretic behaviour. To compare the hysteresis at similar J_{sc} in the two halide systems we have measured the iodide-based devices also at low light intensities (0.1–0.2 sun, see Supplementary Fig. 2). We did not find any significant differences due to different light intensities in the E_a's shown in Fig. 3. In this figure, one can also clearly observe the lower activation energy for the devices with bromide. The different activation energies for the different

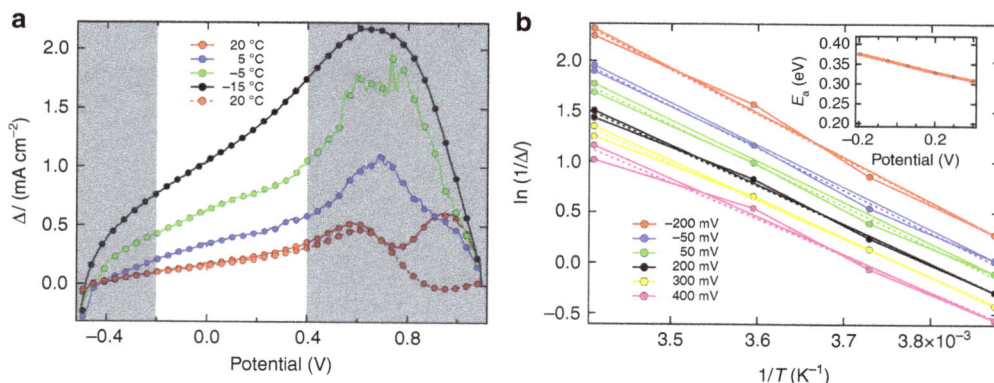

Figure 2 | ΔI and potential dependent activation energies. (a) The current difference between forward and reverse voltage scan and the potential window (0.2–0.4 V) used for the fitting by the Arrhenius-type equation. **(b)** Plot of $\ln(1/\Delta I)$ vs the inverse of the temperature for selected values of the potential (inset shows the determined activation energy independence of the potential).

Table 1 | Averaged activation energies.

Perovskite	Light intensity	E_a (meV)
MAPbI$_3$	High (~1 sun)	314 (±48)
	Low (0.1-0.2 sun)	341 (±42)
	All iodide devices	333 (±47)
MAPbBr$_3$	~1sun	168 (±43)

Determined average activation energies for iodide and bromide-based MAPbX$_3$ devices. Values averaged between −200 and 400 mV.

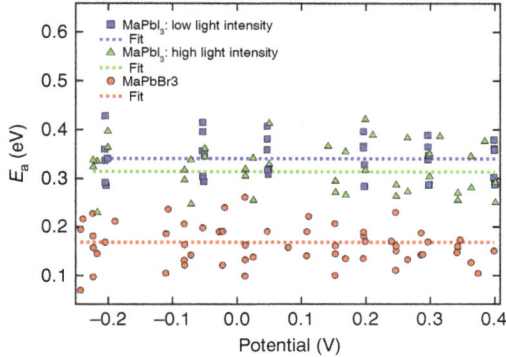

Figure 3 | Potential dependent activation energies of different samples. Collection of the activation energies at different potential for the different samples measured.

devices with iodide (high and low light intensities) and with bromide are given in the Supplementary Fig. 3.

ΔI might also depend on the scan velocity and on the scan bounds of the voltage interval next to the already mentioned light intensity. It is clear that hysteresis can only be observed if the scan rates are performed within times similar to the characteristic timescale of the underlying phenomenon. In fact, it was shown by Tress et al. that if the scan velocities are too fast ($\geq 100{,}000\,mV\,s^{-1}$) or too slow ($< 10\,mV\,s^{-1}$), no hysteresis is observed[17]. Therefore, we have tested different measurement conditions—different scan rates (50, 100 and 200 mV s^{-1}, see Supplementary Fig. 4a) and scan bounds (-0.5, -0.2 and 0 V, Supplementary Fig. 4b)—but could not observe any significant changes in E_a.

Noticing that the major difference between the forward and backward scan of the JV curves at different temperatures is the slope of the JV curve along the forward scan, we also computed the activation energy associated to this resistance-like term at 0 V. The barriers obtained from ΔI and the slope (in principle equal to an Arrhenius-type relation of $\ln(1/R)$ vs $1/T$) are consistent (see Supplementary Figs 1d, 2d and 5a). This justifies the approach of taking either ΔI or the slope to measure the activation energy of the slow process as hysteresis in this voltage range is mainly governed by a rate limited process that does not strongly depend on the actual voltage applied during the scan. Thus, the determined E_a directly reflects the activation energy of this slow process, which is reacting retarded to the change of the applied voltage (as will be explained in more detail in the discussion section and in the gedankenexperiments in the last part of the SI).

Extracting the activation energies for the different devices by the procedure(s) described yields a clear trend (see Supplementary Figs 3 and 6). The iodide-based devices show activation energies of in average 333 ± 47 meV, and the bromide-based devices of about 168 ± 43 meV. When using the

determination over the slope of the forward scan (at 0 V), the tendency is similar with E_a being 275 ± 19 meV for the iodide-based perovskite devices and 176 ± 43 meV for the bromide-based perovskite devices (see Supplementary Fig. 5b). As mentioned above, the relative error on the estimation of E_a is larger for the bromide devices but the tendency is clear. The E_a dependence on the halide rules out that E_a describes a temperature-activated transport in the contacting materials, which are the same for all devices. In addition, changed transport properties of the contacting materials should take effect as series resistance under high forward bias reducing the fill factor. The results reported above clearly indicate that the nature of the halide significantly affects the hysteresis via the barrier of the associated thermally activated process, which is lower for bromide than for iodide. This suggests that the process underlying hysteresis involves movements of halide ions (or their vacancies).

Simulations of the ferroelectric effect. Present experiments show that hysteresis is due to a 'thermally activated' process, with an associated barrier in the range of ~0.1–0.4 eV, depending on the type of perovskite used in the device and the conditions of the experiment. To identify what is the microscopic process causing hysteresis, in particular, whether it is due to ferroelectricity or ionic polarization, we performed two types of simulations. First, we performed ~30 ps long first-principles (on the basis of density functional theory) molecular dynamics simulations (MD) at various temperatures ($T = 100$, 200, 300 and 400 K) starting from the tetragonal MAPbI$_3$ crystal phase. The computational sample consisted of a $2 \times 2 \times 2$ supercell of the simple tetragonal analogue of the experimental body-centered tetragonal crystal[42], containing 32 stoichiometric units (384 atoms).

The system equilibrated at different temperatures is able to assume different crystalline phases with a trend consistent with experimental results[42,43]. At 100 and 200 K, the atoms arrange in an orthorhombic-like phase with non-negligible values of all the three tilting angles. At 300 K, two of the three tilting angles are reduced and the structure becomes tetragonal-like. Finally at 400 K, all the three tilting angles are approximately 0° and the structure is cubic-like. More details on the temperature-dependent simulations are also provided in the Supplementary Notes 1 (Supplementary Figs 7–12).

Albeit the relatively short simulation times of 30 ps, the computational results suggest that a ferroelectric origin of hysteresis in unlikely. Hysteresis induced by ferroelectricity might be due to either a break of symmetry of the PbI$_3$ lattice, perhaps induced or enhanced by the lack of inversion symmetry in the crystal due to the MA cation, or by the alignment of the polar MA molecules[27,44]. A break of symmetry in the PbI$_3$ lattice should result in a histogram of the Pb–I bond distances (< 3.2 Å) with multiple maxima. The computed distribution ($g_{PbI}(r)$, Supplementary Fig. 7) shows no evidence of this feature at any of the temperatures investigated, suggesting that there is no break of symmetry in the PbI$_3$ framework whatever the crystal phase of the sample is.

Furthermore, we investigated the possibility that ferroelectricity originates from a persistent preferential alignment of MA ions. Under the effect of the external bias plus built-in potential MA molecules might align with the overall electric field and produce a counterfield. To determine the characteristic rotational reorientation time, the time correlation function of the C–N unit vector, $\langle \boldsymbol{d}(t) \cdot \boldsymbol{d}(0) \rangle$ ($\boldsymbol{d} = \boldsymbol{r}_N - \boldsymbol{r}_C / |\boldsymbol{r}_N - \boldsymbol{r}_C|$), is computed, and it is fitted with a double exponential decay, $\langle \boldsymbol{d}(t)\boldsymbol{d}(0) \rangle = a_1 e^{-t/\tau_1} + a_2 e^{-t/\tau_2}$. The time for MA molecules to loose memory of their initial orientation, τ_1, is on the picoseconds timescale at 200–400 K (see Fig. 4). At 100 K the reorientation time is much longer, probably because of the fact that complete rotation of MA ions is hindered at this temperature. These results are consistent with previous first-principles calculation using a different, more qualitative, approach to determine the characteristic reorientation time[39,45], recent classical MD simulations[46] and experimental data[47-50]. The consistency with classical MD results on bigger samples suggest that no relevant finite size effects affect the estimated reorientation times. From the dependence of τ_1 on the temperature in the range 200–400 K, we estimated an activation energy for MA reorientation of 0.042 eV (inset of Fig. 4), consistent with experimental results[50].

If the dynamics of the MA molecules is uncorrelated, that is, if they rotate independently from each other, the characteristic orientational correlation time determined in the simulations is also the time the sample takes to polarize under the action of a bias. This leads to the conclusion that polarization due to dipole alignment takes place on the picosecond timescale. This is too short a time to account for the JV hysteresis, that is associated to a process with a characteristic time in the milliseconds-to-seconds range[39,51]. To estimate the amount of correlation between MA molecules, we computed the spatial correlation function, $\langle d_i \cdot d_j \rangle$ with d_i and d_j unit C–N vectors of molecules i and j (Supplementary Fig. 12). The spatial correlation function is 1 if two MA ions have the same orientation during the simulation, -1 if they have opposite orientation and 0 if the orientation of one is independent of that of the others. MD simulations results show that at room temperature the correlation between MA ions is small. This, indeed, is consistent with the fact that all the MA ions undergo a quick decorrelation, as shown by the time autocorrelation curves of individual molecules, $\langle d_i(t) \cdot d_i(0) \rangle$ (see Supplementary Fig. 10). Thus, we expect that the alignment of the sample takes place on the timescale of the rotational reorientation time of a single MA ion, that is, on the picosecond timescale estimated above. Further information about these simulations are available in the SI (Supplementary Figs 9, 10 and 11).

Recent Monte Carlo simulations[32] on a lattice model of MAPbI$_3$ have estimated the alignment under the action of an electric field to take place in 10^4 spin steps. Making an accurate estimation of the reorientation time from Monte Carlo steps is not simple, as there is no one-to-one correspondence between a Monte Carlo step per MA molecule and the time MA ions would take to cover the corresponding reorientation. However, an approximate upper limit estimation of the polarization time of the sample can be obtained by assuming that a global Monte Carlo step, that is, one step for each molecule in the sample, costs a time corresponding to the rotational reorientation time, τ_1. This would yield an alignment time of ~50 ns (10^4 steps × ~5 ps), still too short for MA-reorientation-related ferroelectricity to be responsible for hysteresis.

Simulations of the ionic migration. To probe the second hypothesis, that is, that a bias-induced stepwise migration of ions might result in a change of their local concentration and then in a balancing internal counterfield, we performed MD simulations on systems containing a single MA$^+$, Pb^{2+} or I$^-$ vacancy, respectively. The reason for focusing on vacancies is that previous experiments have shown that ionic transport in Br and Cl perovskites is most likely assisted by this type of defect.

During the 10 ps of first-principles MD, no spontaneous ionic vacancy migration was observed. This is not surprising as it is known that ionic/vacancy migration in MAPbX$_3$-related materials is slower than the timescale of our simulations[37,52,53]. The longer timescale of ionic migration processes makes this second phenomenon a more plausible candidate as source of the observed JV hysteresis. To assess this, we performed string simulations[54] aimed at computing the vacancy-driven ionic migration path and the associated energy barrier, E_a, for MAPbI$_3$ and MAPbBr$_3$. In the case of MAPbBr$_3$, for which the stable phase at room temperature is cubic, we considered both tetragonal and cubic structures. This allows to distinguish between the effect of halide substitution and phase change on the migration barriers. In the case of cubic MAPbBr$_3$ we observed no qualitative changes with respect to the tetragonal case, and the activations energies typically are between the values estimated for the corresponding tetragonal system.

Moreover, to estimate the possible effect on E_a arising from the arbitrary choice of the initial orientation of MA, with its high orientational mobility, we also investigate two systems containing the spherically isotropic Cs$^+$ monovalent cation, namely CsPbI$_3$ and CsPbBr$_3$. For these systems we considered the tetragonal phase. Because of the crystal symmetry, MA$^+$/Cs$^+$ and Pb^{2+} ionic migration can take place either along or perpendicularly to the tetragonal axis, and the barriers of these two processes can differ. Here we consider both cases: axial (a) and equatorial (e) migration (see Fig. 5). It is also important to remark that tetragonal perovskite crystals contain two non-equivalent I/Br sites. The first, denominated axial in the following, is the one in which the halide ion forms the Pb–X–Pb triplex oriented along the tetragonal axis (see Fig. 5). The second, the equatorial site, is the one in which the halide forms Pb–X–Pb laying on the plane orthogonal to the tetragonal axis. Halide migration can take place from an equatorial to another equatorial site, e2e, or from an equatorial to an axial site, e2a (or vice versa, a2e). Stroboscopic images of the migration processes mentioned above for the case of MAPbI$_3$ are shown in Fig. 6a–f.

The migration barriers for all the processes and systems mentioned above are reported in Table 2. The comparison of the migration paths of the X$^-$, Pb^{2+} and A$^+$ ions easily explains the difference in the migration energy of the various species. Figure 6a,b shows that a very-small distortion of the crystal structure accompanies halide migration, while migration of A$^+$ (Fig. 6c,d) and Pb^{2+} (Fig. 6e,f) requires a significant even though local rearrangement of the crystal structure.

The migration of halide ions essentially affects only the PbX$_6$ units involved in the event, with negligible distortions of the rest of the lattice. The migration barrier does not change significantly replacing MA$^+$ with Cs$^+$. In particular, the trend of the migration barrier with the chemical nature of the halide is

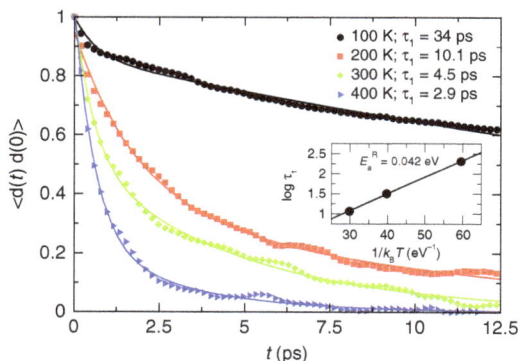

Figure 4 | Time-dependent autocorrelation function of the unit C–N vectors. The autocorrelation function has been computed averaging over the microcanonical ensemble sampled by constant number of particles, volume and energy (NVE) first-principles MD simulations performed at four different energies corresponding to average temperatures of 100, 200, 300 and 400 K. In the inset log τ_1 vs $1/_{k_B T}$ is shown, together with the linear fitting from which the rotational reorientation activation energy is obtained.

69

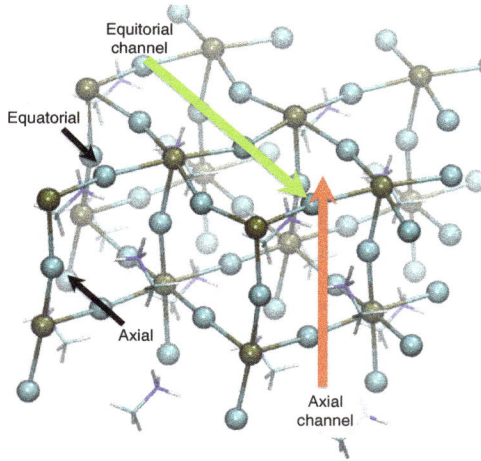

Figure 5 | Propagation channels. Periodic crystalline (defect free) MAPbI₃ sample. Cyan and brown spheres represent I and Pb ions, respectively. MA is shown as sticks. Periodic boundary conditions are applied in all the crystallographic directions. Black arrows point to equatorial and axial I⁻ ions. Red and green arrows indicate the axial and equatorial channels for MA⁺ and Pb²⁺ ion migration.

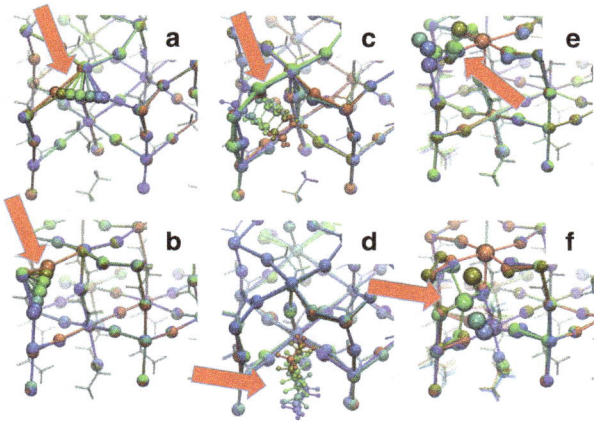

Figure 6 | Stroboscopic images of the ionic migration paths in MAPbI₃. (a–f) Paths for the corresponding migration events in MAPbBr₃ are analogous. The colours of the frames go from blue (initial states) to green (intermediate states) to red (final states). (a,b) Migration path of I⁻ along the equatorial-to-equatorial and equatorial-to-axial channels, respectively. (c,d) Migration paths of MA⁺ along the equatorial and axial channels, respectively. (e,f) Migration paths of Pb²⁺ along the axial and equatorial channels. Both paths are rather complicated. The details are described in the main text.

confirmed. This suggests that our results for MAPbX₃ samples are not significantly biased by the arbitrary choice of the initial orientation of the MA ions. The axial vacancy is significantly less stable than the equatorial one ($\Delta E \sim 0.1$–0.2 eV), and a vacancy in the axial state is expected to migrate towards an equatorial one. Thus, a complete migration event along the axial channel, bringing the vacancy from an equatorial site to another, requires two steps: an e2a migration followed by an a2e one. Thus, this channel requires the crossing of two barriers. In the case of MA-perovskites, the first barrier, e2a (450–460 meV), is sizably higher than the single barrier of the equatorial channel, e2e (200–280 meV), and the one-step e2e migration channel will give the major contribution to the halide transport in perovskites. It is, then, polarization activation energy of this channel determining the halide transport in perovskites.

The migration of the monovalent cation requires the opening of the PbX₃ framework separating the initial and final A⁺ sites. The shape of this framework is different in the axial and equatorial directions. Thus, the activation energy and the migration path, including the orientation of the MA ions along it, depend on the channel, whether equatorial or axial. Given the high orientation disorder of MA⁺, it is expected that under experimental conditions this ion migrates following paths characterized by different orientations of the C-N bond and, possibly, different migration barriers. We expect that the migration barrier of Cs⁺ represents a lower bound for that of MA⁺, consistent with the fact that Cs⁺ is smaller than MA⁺ (as suggested by the larger size of the lattice of MA-perovskites with respect to Cs-perovskites computed in simulations).

Also the migration of Pb²⁺ requires a significant distortion of the crystal structure. This is necessary to let the Pb²⁺ ion leave its original site and enter into the new one. This explains the high barrier for the migration of Pb²⁺.

Walsh et al.[55] have shown that the defects in MAPbI₃ are formed according to the Schottky mechanism, in which the amount of iodide and cation defect concentrations have the same order of magnitude. In particular, according to Walsh et al. the most probable point defect formation reaction is

$$nil \rightarrow V'_{MA} + V°_I + MAI \qquad (2)$$

with an associated formation enthalpy of 0.08 eV. These results, taken together with the calculated activation barriers for defect migration, show that not only $V°_I$ is the most mobile defect, but it is also present in similar concentrations with the other defects considered in the present work, and that, indeed, the ionic mobility is the limiting step in the vacancy-driven polarization of the perovskite sample. Thus, considering the migration barriers of all the ionic species shown in Table 2, at room temperature the timescale of the migration of the MA⁺ and Pb²⁺ ions is $\sim e^{(\Delta E^*_{A/Pb} - \Delta E^*_X)/k_B T} \geq e^{10}$ larger than the one for X⁻ ions. This is consistent with the experiments by Yang et al.[41] with solid-state electrochemical cells of MAPbI₃, in which it has been shown that the active mobile ion is iodide.

Indeed, the I⁻ and Br⁻ computational migration barriers along the e2e channel are in very good agreement with the experimental activation energies. In addition, the effect of halide substitution follows the experimental trend discussed in the previous section, with a lower barrier for halide migration in Br than in I perovskites. The good match between experimental and computational results, both in the absolute value of the migration barriers and in the trend with halogen substitution, together with experimental results reported previously[17,39], strongly support the hypothesis that hysteresis is due to the polarization of the perovskite layer associated to halide-vacancy migration. As proposed in refs 17,39, halide ions/vacancies migrate in the same direction as the corresponding charge carriers. Since ions are not extracted at the contacts, they accumulate at the interface of the electrodes producing a balancing potential that reduces the efficiency of collection of charge carriers. In extreme cases, and in the absence of the compact-TiO₂ hole-blocking layer, the internal balancing potential can significantly reduce the V_{oc} (ref. 17).

Discussion

Before discussing experimental and computational results, it is worth giving the intuitive arguments at the basis of the empirical relation between ΔI and T (Eq. (1)). A first observation is that polarization of the sample due to ionic movements, or to the alignment of MA molecules, produces a counterfield opposing to

Table 2 | Activation barriers for ionic migration.

	E_a (meV)		
	$A^+ = Cs^+$		$A^+ = MA^+$
APbI3			
Vacancy I⁻			
e2e	360		**280**
e2a	290 (170)		450 (130)
Vacancy A⁺			
e	590		1,120 (880)
a	1,160		700 (600)
Vacancy Pb²⁺			
e	810		1,390
a	990		1,780

	Tetragonal		Cubic
APbBr3			
Vacancy Br⁻			
e2e	270	**200**	220
e2a	290 (160)	460 (90)	
Vacancy A⁺			
e	700	1,130 (950)	1,010
a	1,200	800 (700)	
Vacancy Pb²⁺			
e	940	1,350	1,620
a	1,220	1,800	

In the tetragonal phases, there are different possible migration channels. For the e2a case of halide migration, where the initial and final states have a different energy, the barriers of the inverse process, a2e, are reported in parentheses. The e2e halide migration channel (reported in bold in the table) is expected to be the most efficient ionic transport path, and it is the corresponding activation energy that must be compared with the experimental values (168 meV for MAPbBr₃ and 333 meV for MAPbI₃). The initial and final states of the MA⁺ migration are non-equivalent, due to non-equivalent MA orientation in the initial and final states. Also in this case we report in parentheses the activation energy of the reverse process.

the external bias plus built-in voltage. This results in a reduction of the measured current. A possible explanation is that reducing the overall field for extracting charge carriers increases electron–hole recombination. A second observation is that in this work we focus on a temperature range, in which hysteresis shrinks with the temperature (see below). In this case, the faster is the relaxation to the stationary condition from the preset state along the voltage scanning, the lower is the hysteresis. Thus, we assume that hysteresis is in inverse proportion to the relaxation rate. The relaxation rate is related to the diffusion coefficient, in the case of ionic polarization, or the reorientation time, in the case of ferroelectric polarization. In both cases, the dependence on the temperature is via an Arrhenius-like relation characterized by an activation energy. These arguments are summarized in Eq. (1). The values determined for the activation energy in our experiments are in close agreement with the values reported by several other authors using different experimental approaches[24,37,38,41,56], supporting the validity of such an empirical $\Delta I - E_a$ relation.

The comparison between experimental and theoretical results suggests that hysteresis is due to ionic movements rather than ferroelectricity. Here we propose a possible mechanism explaining the experimental observations based on this hypothesis. The effect of ionic displacement is the polarization of the perovskite layer at the contacts, which eventually changes the characteristic properties of the device. A similar effect has been exploited in the field of piezophototronics[57,58]. It has been shown that the accumulation of cations or anions at the interface of, for example, p–n junctions can significantly alter the band bending and can also affect the conduction and valence band edge positions of the involved semiconductors (or work functions of the contacting materials). In perovskite-based solar cells, mobile ions can

similarly accumulate at the interface with the contacting materials under the action of the electric field generated by the space charge region at the contacts. The effect is minimized close to V_{oc}, where the built-in potential is approximately balanced by the external bias and a minimum force is acting on the ions. Thus we can assume that at V_{oc} ions are distributed almost uniformly in the perovskite layer. In contrast, when the external bias is zero (and the internal field is high), ions accumulate at the contact(s). The switching from the polarized to unpolarized state is not instantaneous, and this results into the hysteresis.

However, considering a broader temperature range, the dependence of the hysteresis with temperature is more complex: In the temperature regime below ~ 180 K, an increase in the hysteresis can be observed as shown by Zhang et al.[59] Since the perovskite is in its ferroelectric crystal phase, with the J_{sc} being very low and the overall solar cell efficiency going under 0.1% the reasons for the hysteretic behaviour are more complex and its interpretation goes beyond the scope of this manuscript. At 180 K, nearly no hysteresis is visible in the JV curve. At this temperature the ions do not move significantly on the timescale of the scanning plus dwell time (~ 60–80 s, depending on the scanning rate). In absence of polarization, hysteresis cannot be observed as the state of the system at the given voltage is independent of scan direction. Increasing the temperature increases the mobility of ions and, thus, the polarization during the dwell time at the starting point of the voltage scanning. During the voltage scan, which lasts for maximum ~ 30 s in the case of the slowest scan rate (50 mV s⁻¹), the system cannot reach the stationary condition at each voltage and hysteresis increases with T (in the temperature regime with 180 K $< T < \sim 240$ K). There exists a temperature, $T > \sim 240$ K, at which the system is able to approach a stable polarization during the dwell time. Thus, any further increase of the temperature does not significantly increase the initial polarization before the scan. However, mobility keeps increasing with the temperature and the system can more closely approach the voltage-dependent stationary ionic distribution during the scan. This results in a reduction of the hysteresis. This behaviour is shown in Supplementary Figs 1a and 2a, which show a slight decrease of the measured current with increasing temperature in the backward scan, and a complementary increase in the forward scan. Summarizing, a complex dependence of the hysteresis on temperature is observed, and this is due to the interplay of the effect of the increased mobility of the ions on the polarization during the dwell time at the starting points of the scanning, and the relaxation during the scanning. The effect of the temperature between 80 and 360 K has been investigated in more detail by Zhang et al.[59]

The outlined model leading to the observed hysteresis can explain the characteristics of perovskites solar cells measured in this and other recent works. Tress et al.[17] have observed a shift in the forward and backward JV curves in the dark by changing the pre-conditioning potentials. This device was made without a TiO₂ blocking layer. In this case, the charge polarization can act as a term balancing the built-in potential and/or changing the work function of the fluorine-doped tin oxide (FTO) due to the different ionic distribution at this interface, and, thus, shifting the dark current curves. Almora et al.[60] have reported JV curves in the dark presenting a capacitive loop. The (retarded) variation of ionic polarization during the voltage scans can result into a change in the extension or charging of the space charge zone, which is equivalent to a varying capacitance along the backward and forward scan. Our experiments and calculations, as well as the above mentioned experiments, interpreted in light of the model described here, suggest that the JV characteristics of perovskite solar cells under illumination are strongly influenced through a modulation of the current by ionic polarization. We have made two gedankenexperiments that are presented in the Supplementary Notes 2 highlighting in more detail the thoughts we have presented in this paragraph (see also Supplementary Fig. 13).

The possibility that the width of the space charge region, edge positions of the semiconductor bands (which also includes the contacting semiconductors, for example, the TiO_2) or the work function of the conductive substrate (for example, FTO as shown by Tress et al.) are altered by changes of the ionic environment, leads to a complicated interplay of the different materials. We want to stress that the qualitative model outlined above and in the SI requires additional investigation to identify the detailed effect of ion accumulation on band bending and band edge position, and ultimately on charge separation and collection.

In summary, here we have shown that the hysteresis observed in JV curves of different $MAPbX_3$ perovskite solar cells is due to a thermally activated process and we have determined the associated activation energy. The variation in the processing parameters for the preparation of the perovskite layer, as well as different measurement conditions (JV-scan rate and value of the potential at which hysteresis is measured) did not significantly affect the determined values for the activation energy, demonstrating the independence of the origin of the hysteretic effect on the processing and measurement conditions. Consistently, the activation energy for the bromide- and iodide-based devices showed an average of 168 ± 43 and 333 ± 47 meV, respectively.

We paralleled the experimental investigation with first-principles MD simulations to determine the characteristic time for (re-)orienting MA molecules. We could show that this characteristic time, in the picosecond range, is too short for the process being associated with JV hysteresis. Furthermore, we determined the activation energy of the migration of vacancies of the various ionic species forming the perovskite and show that the lowest activation energy for vacancy migration is the one for halides. Its values matches well with the experimentally determined activation barriers involved in hysteresis. The dependence of this barrier on the type of halide computed in the simulations is also in agreement with the experimental trend. Present experimental and computational results strongly support the hypothesis put forward in ref. 17 that hysteresis is due to polarization of ionic charges in the perovskite layer under the influence of the built-in and applied potential. The mobility of the other possible ionic species (MA^+ and Pb^{2+}) than the halides is much lower, and we do not expect them to give any significant contribution in the polarization of devices in experiments with a scanning rate in the range of 10 to 10,000 mV s^{-1}.

Methods

General methodological information. Refer to the Supplementary Information for experimental section, figures on E_a for different scan velocity and scan bounds, different preparation methods and light intensities. Additional computational details and results, and further discussion of the outlined model is available free of charge via the Internet at http://pubs.acs.org.

References

1. Kojima, A., Teshima, K., Shirai, Y. & Miyasaka, T. Organometal halide perovskites as visible-light sensitizers for photovoltaic cells. *J. Am. Chem. Soc.* **131**, 6050 (2009).
2. Im, J. H., Lee, C. R., Lee, J. W., Park, S. W. & Park, N. G. 6.5% Efficient perovskite quantum-dot-sensitized solar cell. *Nanoscale* **3**, 4088–4093 (2011).
3. Kim, H. S. et al. Lead iodide perovskite sensitized all-solid-state submicron thin film mesoscopic solar cell with efficiency exceeding 9%. *Sci. Rep.* **2**, 591 (2012).
4. Burschka, J. et al. Sequential deposition as a route to high-performance perovskite-sensitized solar cells. *Nature* **499**, 316 (2013).
5. Lee, M. M., Teuscher, J., Miyasaka, T., Murakami, T. N. & Snaith, H. J. Efficient hybrid solar cells based on meso-superstructured organometal halide perovskites. *Science* **338**, 643–647 (2012).
6. Moehl, T. et al. Strong photocurrent amplification in perovskite solar cells with a porous TiO_2 blocking layer under reverse bias. *J. Phys. Chem. Lett.* **5**, 3931–3936 (2014).
7. Lee, Y. et al. High-performance perovskite-graphene hybrid photodetector. *Adv. Mater.* **27**, 41–46 (2015).
8. Hu, X. et al. High-performance flexible broadband photodetector based on organolead halide perovskite. *Adv. Funct. Mater.* **24**, 7373–7380 (2014).
9. Dou, L. T. et al. Solution-processed hybrid perovskite photodetectors with high detectivity. *Nat. Commun.* **5**, 5404 (2014).
10. Xia, H. R., Li, J., Sun, W. T. & Peng, L. M. Organohalide lead perovskite based photodetectors with much enhanced performance. *Chem. Commun.* **50**, 13695–13697 (2014).
11. Deschler, F. et al. High photoluminescence efficiency and optically pumped lasing in solution-processed mixed halide perovskite semiconductors. *J. Phys. Chem. Lett.* **5**, 1421–1426 (2014).
12. Dhanker, R. et al. Random lasing in organo-lead halide perovskite microcrystal networks. *Appl. Phys. Lett.* **105**, 151112 (2014).
13. Sutherland, B. R., Hoogland, S., Adachi, M. M., Wong, C. T. O. & Sargent, E. H. Conformal organohalide perovskites enable lasing on spherical resonators. *ACS Nano* **8**, 10947–10952 (2014).
14. Xing, G. C. et al. Low-temperature solution-processed wavelength-tunable perovskites for lasing. *Nat. Mater.* **13**, 476–480 (2014).
15. Jeon, N. J. et al. Compositional engineering of perovskite materials for high-performance solar cells. *Nature* **517**, 476–480 (2015).
16. Jeon, N. J. et al. Solvent engineering for high-performance inorganic–organic hybrid perovskite solar cells. *Nat. Mater.* **13**, 897–903 (2014).
17. Tress, W. et al. Understanding the rate-dependent J-V hysteresis, slow time component, and aging in CH3NH3PbI3 perovskite solar cells: the role of a compensated electric field. *Energy Environ. Sci.* **8**, 995–1004 (2015).
18. Dualeh, A. et al. Impedance spectroscopic analysis of lead iodide perovskite-sensitized solid-state solar cells. *ACS Nano* **8**, 362–373 (2014).
19. Gonzalez-Pedro, V. et al. General working principles of CH3NH3PbX3 perovskite solar cells. *Nano Lett.* **14**, 888–893 (2014).
20. Snaith, H. J. et al. Anomalous hysteresis in Perovskite solar cells. *J. Phys. Chem. Lett.* **5**, 1511–1515 (2014).
21. O'Regan, B. C. et al. Opto-electronic studies of methylammonium lead iodide perovskite solar cells with mesoporous TiO_2; separation of electronic and chemical charge storage, understanding two recombination lifetimes, and the evolution of band offsets during JV hysteresis. *J. Am. Chem. Soc.* **137**, 5087–5099 (2015).
22. Azpiroz, J. M., Mosconi, E., Bisquert, J. & De Angelis, F. Defect migration in methylammonium lead iodide and its role in perovskite solar cell operation. *Energy Environ. Sci.* **8**, 2118–2127 (2015).
23. Haruyama, J., Sodeyama, K., Han, L. & Tateyama, Y. First-principles study of ion diffusion in perovskite solar cell sensitizers. *J. Am. Chem. Soc.* **137**, 10048–10051 (2015).
24. Eames, C. et al. Ionic transport in hybrid lead iodide perovskite solar cells. *Nat. Commun.* **6**, 7497 (2015).
25. Christians, J. A., Manser, J. S. & Kamat, P. V. Best practices in perovskite solar cell efficiency measurements. avoiding the error of making bad cells look good. *J. Phys. Chem. Lett.* **6**, 852–857 (2015).
26. Wei, J. et al. Hysteresis analysis based on the ferroelectric effect in hybrid perovskite solar cells. *J. Phys. Chem. Lett.* **5**, 3937–3945 (2014).
27. Frost, J. M. et al. Atomistic Origins of high-performance in hybrid halide perovskite solar cells. *Nano Lett.* **14**, 2584–2590 (2014).
28. Yuan, Y. B., Xiao, Z. G., Yang, B. & Huang, J. S. Arising applications of ferroelectric materials in photovoltaic devices. *J. Mater. Chem. A* **2**, 6027–6041 (2014).
29. Juarez-Perez, E. J. et al. Photoinduced giant dielectric constant in lead halide perovskite solar cells. *J. Phys. Chem. Lett.* **5**, 2390–2394 (2014).
30. Stroppa, A. et al. Tunable ferroelectric polarization and its interplay with spin-orbit coupling in tin iodide perovskites. *Nat. Commun.* **5**, 5900 (2014).
31. Xiao, Z. et al. Giant switchable photovoltaic effect in organometal trihalide perovskite devices. *Nat. Mater.* **14**, 193–198 (2015).
32. Frost, J. M., Butler, K. T. & Walsh, A. Molecular ferroelectric contributions to anomalous hysteresis in hybrid perovskite solar cells. *Appl. Mater.* **2**, 081506 (2014).
33. Liu, S. et al. Ferroelectric domain wall induced band gap reduction and charge separation in organometal halide perovskites. *J. Phys. Chem. Lett.* **6**, 693–699 (2015).
34. Chen, H. W., Sakai, N., Ikegami, M. & Miyasaka, T. Emergence of hysteresis and transient ferroelectric response in organo-lead halide perovskite solar cells. *J. Phys. Chem. Lett.* **6**, 164–169 (2015).
35. Fan, Z. et al. Ferroelectricity of CH3NH3PbI3 Perovskite. *J. Phys. Chem. Lett.* **6**, 1155–1161 (2015).

36. Kutes, Y. *et al.* Direct observation of ferroelectric domains in solution-processed CH$_3$NH$_3$PbI$_3$ perovskite thin films. *J. Phys. Chem. Lett.* **5**, 3335–3339 (2014).

37. Mizusaki, J., Arai, K. & Fueki, K. Ionic conduction of the perovskite-type halides. *Solid State Ionics* **11**, 203–211 (1983).

38. Kuku, T. A. Ionic transport and galvanic cell discharge characteristics of CuPbI3 thin films. *Thin Solid Films* **325**, 246–250 (1998).

39. Mosconi, E., Quarti, C., Ivanovska, T., Ruani, G. & De Angelis, F. Structural and electronic properties of organo-halide lead perovskites: a combined IR-spectroscopy and ab initio molecular dynamics investigation. *Phys. Chem. Chem. Phys.* **16**, 16137–16144 (2014).

40. Zhao, Y. *et al.* Anomalously large interface charge in polarity-switchable photovoltaic devices: an indication of mobile ions in organic-inorganic halide perovskites. *Energy Environ Sci.* **8**, 1256–1260 (2015).

41. Yang, T.-Y., Gregori, G., Pellet, N., Grätzel, M. & Maier, J. The significance of ion conduction in a hybrid organic–inorganic lead-iodide-based perovskite photosensitizer. *Angew Chem. Int. Ed. Engl.* **54**, 7905–7910 (2015).

42. Stoumpos, C. C., Malliakas, C. D. & Kanatzidis, M. G. Semiconducting tin and lead iodide perovskites with organic cations: phase transitions, high mobilities, and near-infrared photoluminescent properties. *Inorg. Chem.* **52**, 9019–9038 (2013).

43. Trots, D. M. & Myagkota, S. V. High-temperature structural evolution of caesium and rubidium triiodoplumbates. *J. Phys. Chem. Solids* **69**, 2520–2526 (2008).

44. Frost, J. M., Butler, K. T. & Walsh, A. Molecular ferroelectric contributions to anomalous hysteresis in hybrid perovskite solar cells. *Appl. Mater.* **2**, 081506 (2014).

45. Carignano, M. A., Kachmar, A. & Hutter, J. Thermal effects on CH$_3$NH$_3$PbI$_3$ perovskite from *ab initio* molecular dynamics simulations. *J. Phys. Chem. C* **119**, 8991–8997 (2015).

46. Mattoni, A., Filippetti, A., Saba, M. I. & Delugas, P. Methylammonium rotational dynamics in lead halide perovskite by classical molecular dynamics: the role of temperature. *J. Phys. Chem. C* **119**, 17421–17428 (2015).

47. Poglitsch, A. & Weber, D. Dynamic disorder in methylammoniumtrihalog-enoplumbates(Ii) observed by millimeter-wave spectroscopy. *J. Chem. Phys.* **87**, 6373–6378 (1987).

48. Jamnik, J. & Maier, J. Treatment of the impedance of mixed conductors—equivalent circuit model and explicit approximate solutions. *J. Electrochem. Soc.* **146**, 4183–4188 (1999).

49. Leguy, A. M. A. *et al.* The dynamics of methylammonium ions in hybrid organic-inorganic perovskite solar cells. *Nat. Commun.* **6**, 7124 (2015).

50. Chen, T. *et al.* Rotational dynamics and its relation to the photovoltaic effect of CH$_3$NH$_3$PbI$_3$ perovskite. Preprint at http://arXiv:150602205 (2015).

51. Zhao, Y. *et al.* Anomalously large interface charge in polarity-switchable photovoltaic devices: an indication of mobile ions in organic-inorganic halide perovskites. *Energy Environ. Sci.* **8**, 1049 (2015).

52. Hoshino, H., Yokose, S. & Shimoji, M. Ionic conductivity of lead bromide crystals. *J. Solid State Chem.* **7**, 1–6 (1973).

53. Schoonman, J. The ionic conductivity of pure and doped lead bromide single crystals. *J. Solid State Chem.* **4**, 466–474 (1972).

54. E, W., Ren, W. & Vanden-Eijnden, E. Simplified and improved string method for computing the minimum energy paths in barrier-crossing events. *J. Chem. Phys.* **126**, 164103 (2007).

55. Walsh, A., Scanlon, D. O., Chen, S., Gong, X. G. & Wei, S.-H. Self-regulation mechanism for charged point defects in hybrid halide perovskites. *Energy Environ. Sci.* **8**, 995–1004 (2015).

56. Hoke, E. T. *et al.* Reversible photo-induced trap formation in mixed-halide hybrid perovskites for photovoltaics. *Chem. Sci.* **6**, 613–617 (2015).

57. Pan, C. *et al.* Enhanced Cu2S/CdS coaxial nanowire solar cells by piezo-phototronic effect. *Nano Lett.* **12**, 3302–3307 (2012).

58. Wang, Z. L. in *Microtechnology and MEMS* (Springer, 2012).

59. Zhang, H. *et al.* Photovoltaic behaviour of lead methylammonium triiodide perovskite solar cells down to 80 K. *J. Mater. Chem. A* **3**, 11762–11767 (2015).

60. Almora, O. *et al.* Capacitive dark currents, hysteresis, and electrode polarization in lead halide perovskite solar cells. *J. Phys. Chem. Lett.* **6**, 1645–1652 (2015).

Acknowledgements

We greatly acknowledge Guido Rothenberger, Alessandro Mattoni and Carlo Massimo Casciola for valuable hints and discussions during the preparation of the manuscript. M.G. acknowledges a Sciex fellowship under Project Code 13.194. Financial support from SNF-NanoTera (SYNERGY), Swiss Federal Office of Energy (SYNERGY) and CCEM-CH in the 9th call proposal 906: CONNECT PV is gratefully acknowledged. M.G. and T.M. acknowledge the European Research Council (ERC) for an Advanced Research Grant (ARG no. 247404) funded under the 'Mesolight' project. M.K.N. thank the Swiss Federal Office for Energy (Energy center special funds), and the European Union Seventh Framework Program under grant agreement numbers 604032 of the MESO, 308997 of the NANOMATCELL. U.R. acknowledges funding from the Swiss National Science Foundation via individual grant No. 200020-146645, the NCCRs MUST and MARVEL, and support from the Swiss National Computing Center (CSCS) and the CADMOS project for computing resources. We acknowledge PRACE for awarding us access to computer resources at Supermuc at LRZ, Germany.

Author contributions

S.M., U.R., T.M. and M.G. conceived, designed and led the study. S.M. carried out the computational calculations. M.F., Y.L. and M.S. prepared devices. T.M. carried out the measurements and did the data analysis with the help of W.T. and P.G. synthesized the methyl ammonium halides. S.M., T.M. and W.T. wrote the manuscript with the help of U.R. and M.G., S.M.Z. and M.K.N. contributed to the discussions.

Additional information

High-contrast and fast electrochromic switching enabled by plasmonics

Ting Xu[1,2,3,*], Erich C. Walter[2,3,*], Amit Agrawal[2,3,*], Christopher Bohn[2], Jeyavel Velmurugan[2,3], Wenqi Zhu[2,3], Henri J. Lezec[2] & A. Alec Talin[2,4]

With vibrant colours and simple, room-temperature processing methods, electrochromic polymers have attracted attention as active materials for flexible, low-power-consuming devices. However, slow switching speeds in devices realized to date, as well as the complexity of having to combine several distinct polymers to achieve a full-colour gamut, have limited electrochromic materials to niche applications. Here we achieve fast, high-contrast electrochromic switching by significantly enhancing the interaction of light—propagating as deep-subwavelength-confined surface plasmon polaritons through arrays of metallic nanoslits, with an electrochromic polymer—present as an ultra-thin coating on the slit sidewalls. The switchable configuration retains the short temporal charge-diffusion characteristics of thin electrochromic films, while maintaining the high optical contrast associated with thicker electrochromic coatings. We further demonstrate that by controlling the pitch of the nanoslit arrays, it is possible to achieve a full-colour response with high contrast and fast switching speeds, while relying on just one electrochromic polymer.

[1] National Laboratory of Solid State Microstructures, College of Engineering and Applied Sciences and Collaborative Innovation Center of Advanced Microstructures, Nanjing University, 22 Hankou Road, Nanjing 210093, China. [2] Center for Nanoscale Science and Technology, National Institute of Standards and Technology, Gaithersburg, Maryland 20899, USA. [3] Maryland Nanocenter, University of Maryland, College Park, Maryland 20742, USA. [4] Sandia National Laboratories, Livermore, California 94551, USA. * These authors contributed equally to this work. Correspondence and requests for materials should be addressed to T.X. (email: xuting@nju.edu.cn) or to H.J.L. (email: henri.lezec@nist.gov) or to A.A.T. (email: aatalin@sandia.gov).

Advances in flat-panel display technology over the last-decade have enabled a new generation of portable electronic equipment. Ultra-thin, flexible and low-power-consuming displays drive the rapid growth in electronic readers, reconfigurable signage and a host of other consumer electronics. Current mainstream commercial devices depend on electrophoretic displays (EPDs)[1], liquid crystal displays[2] and organic light-emitting diodes[3]. EPDs, collectively known as electronic ink, provide very sharp images, bistability and excellent outdoor readability. However, EPDs have limited potential for displaying full-colour images, and their response time and refresh rate are too slow for sophisticated interactive applications and video. Although liquid crystal displays and organic light-emitting diodes do enjoy full-colour scheme and shorter response time, they are not bistable, and their high-cost and complex manufacture remain a challenge. Other emerging display technologies based on photonic crystals[4-6] and quantum dots[7-9] still require significant development.

Discovered in the late 1960s, electrochromic materials, including various transition metal oxides and conducting polymers, show a reversible colour change on electrochemical reduction or oxidation by application of a small voltage[10]. Electrochromic materials are particularly attractive for flexible display applications because of their vibrant colours, low-cost and relatively simple processing requirements[11-14]. Nevertheless, long switching times, typically on the order of seconds, have limited electrochromic materials to niche applications[15]. In general, the switching time τ of electrochromic devices is limited by ionic diffusion in the electrochromic material ($\tau \propto L^2/D$, where L is the film thickness and D is the diffusivity). Switching times on the order of tens of milliseconds have been demonstrated for certain multilayer-thin electrochromic polymer films, although at a cost of substantially reduced optical contrast, in general $<30\%$ (ref. 16). In addition to speed, a typical multi-colour electrochromic display requires at least three separate electrochromic materials to provide the colours necessary to make an additive, red–green–blue, or a subtractive, cyan–magenta–yellow, colour palette. These three separate layers can require up to six layers of transparent conductors to operate, raising production complexity, cost and further limiting the switching contrast.

With recent advancements in nanofabrication and optical characterization techniques, surface plasmon polariton (SPP)-based nanoscale photonic and optoelectronic devices have generated considerable interest[17]. SPPs are photon-induced collective charge oscillations that are able to sustain the propagation of optical frequency electromagnetic waves at an interface between a dielectric medium and a metal. The tight spatial confinement and high local field intensity associated with SPPs have enabled operation of nano-optical devices beyond the optical diffraction limit[18]. More recently, plasmonic nanostructures, such as metallic nanohole arrays and nanoparticles, have been used to explore electro-optic switching[19] and electrochemical tunability of localized surface plasmon resonances[20-22].

Here, using plasmonic nanoslit arrays, we demonstrate high contrast, fast monochromatic and full-colour electrochromic switching using two different electrochromic polymers, polyaniline (PANI) and poly(2,2-dimethyl-3,4 propylenedioxythiophene) (PolyProDOT-Me$_2$). Unlike transition-metal-oxide electrochromic materials, which are usually sputter coated, or inorganic polymers such as Prussian blue, which often form nanocrystals and are therefore difficult to deposit uniformly over high-aspect ratio features[23], both PANI and PolyProDOT-Me$_2$ polymers can be electrodeposited as conformal, extremely thin coatings on metal structures with well-controlled thicknesses[24-25], favouring scatter-free propagation of SPPs with maximum interaction with the electrochromic films. As a result, the plasmonic electrochromic switchable configurations retain the advantages of both fast switching speed and high optical contrast.

Results

Plasmon-enhanced monochromatic electrochromic switching.
The working electrode designed for monochromatic operation incorporates an Au film patterned with a nanoslit array and conformally coated with a thin layer of PANI ('Au-nanoslit', Fig. 1a). The Au-nanoslit electrode is immersed in an electrolyte solution, along with a Pt counter electrode and reference electrode. A voltage applied to the working electrode causes electrons (from the metal) and ions (from the electrolyte) to either flow in (reduction) or out (oxidation) of the polymer, thus changing its state of charge and, concurrently, its optical absorption characteristics[26]. Light normally incident on the Au-nanoslit array couples to SPPs travelling both as surface waves along the illuminated Au surface and as guided modes in the nanoslits, with field maxima occurring at the Au-polymer interfaces. The electrochromic material's effective optical thickness for transmission is thus close to that of the slit depth, which can far exceed the physical thickness of the electrochromic layer. Therefore, strong optical absorption can be realized using a thin electrochromic polymer layer, while simultaneously achieving a fast switching time due to correspondingly small charge-propagation distance (which is normal to the film surface and orthogonal to the direction of light propagation). This combination of efficient optical modulation and switching speed cannot be achieved with a more conventional planar, unpatterned configuration provided by an equally thin electrochromic polymer film coated on a semi-transparent thin Au film illuminated at normal incidence ('Reference', Fig. 1b). Achieving an effective electrochromic optical thickness and switching contrast comparable to that of the plasmonic nanoslit

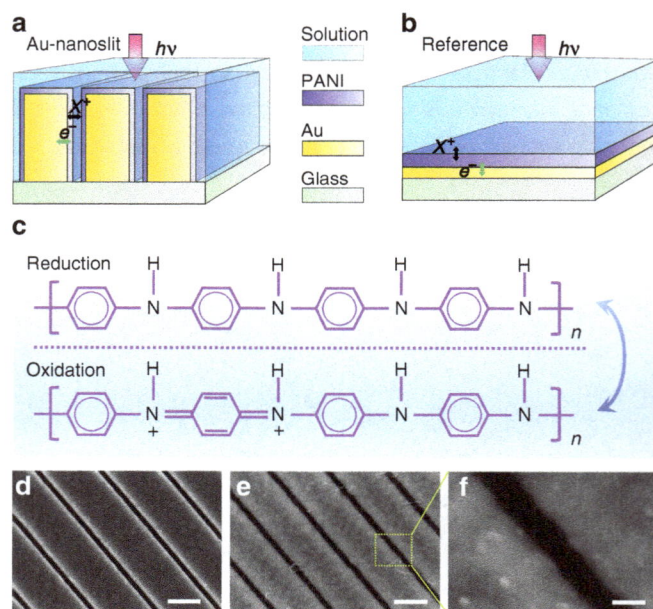

Figure 1 | Plasmonic electrochromic electrodes. Schematic diagram of a plasmonic electrochromic electrode incorporating (**a**) Au-nanoslit array and (**b**) reference planar electrochromic electrode. The pitch of the Au-nanoslit array is 500 nm. The depth and width of the slit is 60 and 250 nm, respectively. (**c**) Chemical structures of PANI in the reduced and oxidized form. SEM images of the fabricated Au-nanoslit electrode (**d**) before and (**e**) after deposition of a PANI to a thickness $d \approx 15$ nm. (**f**) Magnified SEM image from **e**. Scale bars, 300 nm (**d,e**). Scale bar, 100 nm (**f**).

structure implies the use of a much thicker polymer layer, leading to longer charge diffusion distances and slower switching speeds. The chemical structures of PANI in the oxidized and reduced form are shown in Fig. 1c.

Patterned electrodes corresponding to the Au-nanoslit geometry are fabricated by sputter deposition of a 250-nm-thick Au film onto a 25 mm × 25 mm borosilicate glass substrate pre-coated with a 5-nm-thick Ti adhesion layer, followed by focused ion beam (FIB) milling of a nanoslit array (nominal slit width $w = 60$ nm; pitch $P = 500$ nm) over a 10 μm × 10 μm area. Scanning electron microscope (SEM) images of the resulting array, before and after deposition of PANI by potentio-dynamic cycling to a thickness $d \approx 15$ nm, are shown in Fig. 1d–f. Other Au-nanoslit arrays are coated with similarly prepared PANI films of thicknesses d ranging from ≈ 5 nm to ≈ 25 nm. Reference planar electrodes are also fabricated by depositing PANI films under identical conditions to those of Au-nanoslit electrodes on glass substrates coated with 25-nm-thick Au films. A custom-built photoelectrochemical cell is used to switch the PANI films between the reduced form ('ON' state, applied voltage $V_{ON} = -0.2$ V versus Ag/AgCl) and oxidized form ('OFF' state, applied voltage $V_{OFF} = 0.3$ V versus Ag/AgCl). Within these potential limits, the PANI films are not further oxidized to the emeraldine base or the pernigraniline states, to avoid polymer degradation over thousands of switching cycles. The transmission spectra of PANI film with different applied voltages are shown in Supplementary Fig. 1. A HeNe laser working at $\lambda = 632.8$ nm in transverse-magnetic (TM) polarization (electric field perpendicular to the slit length) is used to illuminate both Au-nanoslit

and reference planar electrodes at normal incidence, while the polymer is cycled between its clear ('ON') and absorbing ('OFF') states. The transmitted light is collected using an inverted optical microscope and its intensity is measured using a Si-photodiode connected to the output ports of the microscope.

In Fig. 2a, we show the transmitted light intensity (I) versus time (t) at $\lambda = 632.8$ nm for the Au-nanoslit and reference planar electrodes, where an abrupt step transition in the applied voltage from -0.2 V versus Ag/AgCl to 0.3 V versus Ag/AgCl is applied at time $t = 15$ s. Both electrodes are coated with a PANI film of thickness $d \approx 25$ nm. As expected, the Au-nanoslit electrode exhibits significantly higher optical modulation amplitude than the reference planar one. The absolute optical transmission through the Au-nanoslit electrode at wavelength of 632.8 nm in the reduced state is ≈ 10 %. Figure 2b plots the optical switching contrast, defined as $\gamma = (I_{ON} - I_{OFF})/I_{ON}$, as a function of the PANI film thickness d for various Au-nanoslit and reference planar electrodes, where I_{ON} and I_{OFF} refer to, respectively, the transmitted intensity in the reduced and oxidized form of the PANI films. In the case of the Au-nanoslit geometry, γ monotonically increases as a function of film thickness, linearly at smaller thicknesses and with a noticeable roll off beyond $d = 15$ nm, owing to the onset of negligible transmission in the OFF state as a result of near-full absorption by the polymer fully filling the slits. These trends are substantiated by finite-difference-time-domain simulations replicating the light transmission through PANI-coated Au-nanoslit and reference electrodes, using the published refractive index values of as-deposited PANI films measured by *in-situ* ellipsometry[27]. Simulations for a given

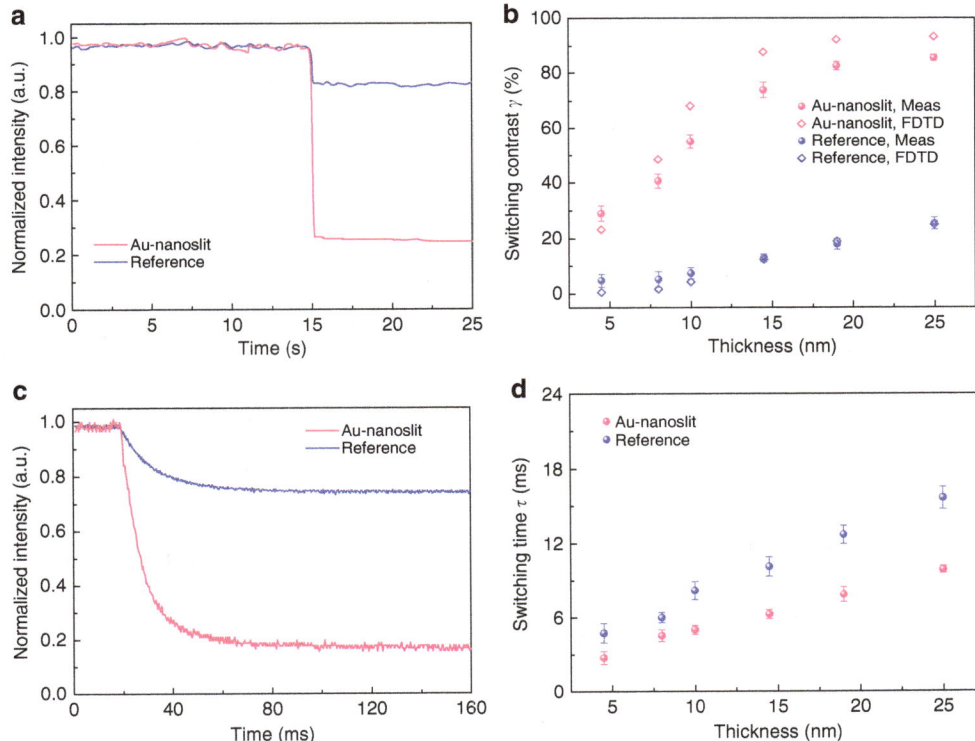

Figure 2 | **Experimental results for Au-nanoslit and reference planar electrodes. (a)** Transmitted light intensity as a function of time for, respectively, Au-nanoslit and reference planar electrodes coated with 25 nm-thick PANI layer, given a step transition in applied voltage at $t = 15$ s, from -0.2 V (clear) to 0.3 V (absorbing) state. **(b)** Experimentally measured and numerically simulated switching contrast γ as a function of PANI thickness for Au-nanoslit and reference planar electrodes. The refractive indices of PANI film in clear and oxidized forms at different thickness used in simulations are taken from ref. 21 and fitted by Boltzmann and exponential functions. Error bars, s.d. for repeated experimental measurements (four in total). **(c)** Transmitted light intensity (measured using photodiode) as a function of time for Au-nanoslit and reference planar electrodes, each coated with 25-nm-thick PANI films. **(d)** Switching time τ for Au-nanoslit and reference planar electrodes as a function of PANI film thickness. Error bars, s.d. for repeated measurements (four in total).

transmission state and under identical illumination conditions (Supplementary Fig. 2) indicate that the light intensity is significantly enhanced within the PANI films coating the slit sidewalls of the Au-nanoslit structure, relative to the case of the PANI film coating the planar surface of the reference structure, resulting in a higher optical modulation efficiency per unit interaction length.

Figure 2c displays the temporal switching characteristics of the Au-nanoslit and unpatterned reference electrodes, each coated with PANI films of nominal thickness 25 nm. The Au-nanoslit electrode exhibits a faster switching time ($\tau \approx 9$ ms) compared with that of the reference electrode ($\tau \approx 14$ ms). The switching time τ is calculated by fitting a decaying-exponential function to the transmitted intensity in each case using the expression $I(t) = A + \gamma \exp(-t/\tau)$, where $I(0) = A + \gamma = 1$ is the normalized intensity as measured on the photodiode at ON state, A is the steady-state normalized intensity at OFF state ($t \gg \tau$) and γ is the switching contrast. Figure 2d summarizes the measured switching time, τ, for various Au-nanoslit and reference electrodes plotted as a function of the PANI film thickness. For any given PANI film thickness, the switching times for the Au-nanoslit electrodes are unexpectedly lower than those of the corresponding reference ones. The lower values of τ for the Au-nanoslit electrodes possibly result from polymer deposition on the slit sidewalls to values that are lower than the nominal value assumed for deposition on a planar surface.

The ratio of optical contrast to the switching time, γ/τ, plotted in Supplementary Fig. 3, defines a useful figure-of-merit (FOM) for electrochromic switching. The experimental FOM values for the Au-nanoslit electrode are approximately one order of magnitude higher than those of the corresponding unpatterned reference electrode over the explored PANI film thickness range $d \approx 5$ nm to $d \approx 25$ nm, achieving a maximum for $d \approx 15$ nm. Conversely, the corresponding FOM for the reference planar electrode increases monotonically over the same thickness range. Supplementary Fig. 4a displays an SEM image of a large-area Au-nanoslit electrode composed of a 10×10 matrix of individual nanoslit arrays (each $10\,\mu m \times 10\,\mu m$ in area), coated with a 15-nm-thick PANI layer. Transmission of red light ($\lambda = 632.8$ nm) through the electrodes in the ON state (applied voltage $V_{ON} = -0.2$ V versus Ag/AgCl) and OFF state (applied voltage $V_{OFF} = 0.3$ V versus Ag/AgCl) is illustrated in Supplementary Fig. 4b,c, respectively. Real-time switching of red light transmitted through the Au-nanoslit electrode as the applied voltage is repeatedly stepped between V_{ON} and V_{OFF} is demonstrated in Supplementary Movie 1. These results summarize the high optical contrast and fast switching speed associated with the plasmonic Au-nanoslit electrode designed for monochromatic operation in the red.

Plasmon-enhanced full-colour electrochromic switching. In addition to the monochromatic switching, we fabricate another set of plasmonic electrodes using a hybrid geometry including a nitride planar waveguide, a slit-patterned Al metal substrate and an electrochromic polymer coating, to demonstrate full-colour and fast switching capability across the entire visible range. It has been shown that nanoslit arrays fabricated with deep-sub-wavelength slit widths on opaque metal films can function as spectral filters[28-33] and their peak spectral transmission can be tuned by altering the period of the array. Efficient visible-frequency operation of the Au-nanoslit device described in the previous section is limited to the red spectral range, because SPP propagation losses along Au surfaces are higher at shorter wavelengths[34]. Furthermore, the high cost of Au makes it impractical for use in consumer display applications. To overcome these challenges, we use a modified plasmonic device geometry leveraging Al coated with a thin layer of PolyProDOT-Me$_2$, to achieve a full-colour electrochromic optical response. We use Al because it is a low-cost, earth-abundant metal that supports SPPs with low optical losses in the ultraviolet and visible regions of the spectrum. In addition, previous work on plasmonic devices fabricated using Al also confirms that this metal is a good candidate for SPP-based colour filters[29-31]. Furthermore, we use PolyProDOT-Me$_2$ because it is a high-colouration-efficiency electrochromic polymer that exhibits broadband optical absorption[35]. The peak absorbance of the PolyProDOT-Me$_2$ lies in the centre of visible spectrum ($\lambda_{peak} \approx 570$ nm), making it suitable for fast switching applications across the entire visible range. Finally, the electrodeposition and electrochemical cycling conditions for PolyProDOT-Me$_2$ are completely compatible with Al.

The schematic diagram of the Al-nanoslit electrode is shown in Fig. 3a. In contrast to the monochromatic Au-nanoslit structure described earlier, a Si$_3$N$_4$ waveguide is added as a buffer layer underneath the Al nanoslit array to further narrow the filtered spectral linewidth and increase its colour purity[32]. Al-nanoslit electrodes of various slit-array periods P are prepared via physical vapour deposition of a 170-nm-thick Si$_3$N$_4$ waveguide layer on a $25\,mm \times 25\,mm$ borosilicate glass substrate, followed by sputter deposition of a 250-nm-thick Al film. Nanoslit arrays with nominal slit width ≈ 70 nm wide are then patterned through the Al using FIB milling to form multiple slit arrays each covering areas of $10\,\mu m \times 10\,\mu m$, with P ranging from 240 to 390 nm in steps of 30 nm. The thin native oxide that readily forms on the Al surface can block charge flow necessary for electrodeposition (as well as for subsequent electrochemical cycling); to inhibit formation of such an insulating layer, a 4-nm-thick conformal coating of Pt is deposited using atomic layer deposition, before polymer electrodeposition. Finally, a 15-nm-thick layer of PolyProDOT-Me$_2$ is electrodeposited onto the surface, conformally coating the nanoslit arrays. The chemical structures of PolyProDOT-Me$_2$ in the oxidized and reduced form are shown in Fig. 3b.

In Fig. 3c,d, we show the experimentally measured optical transmission spectra of the fabricated Al-nanoslit electrodes, along with corresponding optical micrographs, for both transmitting ON (applied voltage $V_{ON} = 0.2$ V versus Ag wire) and absorbing OFF (applied voltage $V_{OFF} = -0.6$ V versus Ag wire) states of the polymer. In contrast to the uniformly dark colours exhibited in the OFF state, the Al-nanoslit electrodes in the ON state show, as a function of period, an assortment of vivid colours covering the entire visible spectrum. The experimentally measured absolute transmission at filtered wavelengths in the ON state ranges from 13 to 18%. The switching contrast of each Al-nanoslit electrode averaged over the entire visible spectrum,

$$\gamma = \frac{\int (I_{ON}(\lambda) - I_{OFF}(\lambda))d\lambda}{\int I_{ON}(\lambda)d\lambda},$$ ranges from 73 to 90%. The experimentally

measured switching time for the Al-nanoslit electrode is $\tau \approx 80$ ms. We expect that optimization of the photoelectrochemical cell, such as decreasing the non-patterned electrode area (thus, decreasing the cell capacitance) and decreasing electrolyte resistance will further decrease the switching time.

Discussion

The switching speed of the plasmonic electrochromic electrodes demonstrated here, on the order of tens of milliseconds, is comparable to pixel switching speeds in commercial displays and is compatible for use in sophisticated applications requiring dynamic switching. Although the On-state (reduced state) absolute transmission efficiency of the electrodes are lower than those of commercial liquid crystal pixels, we expect that it can be further

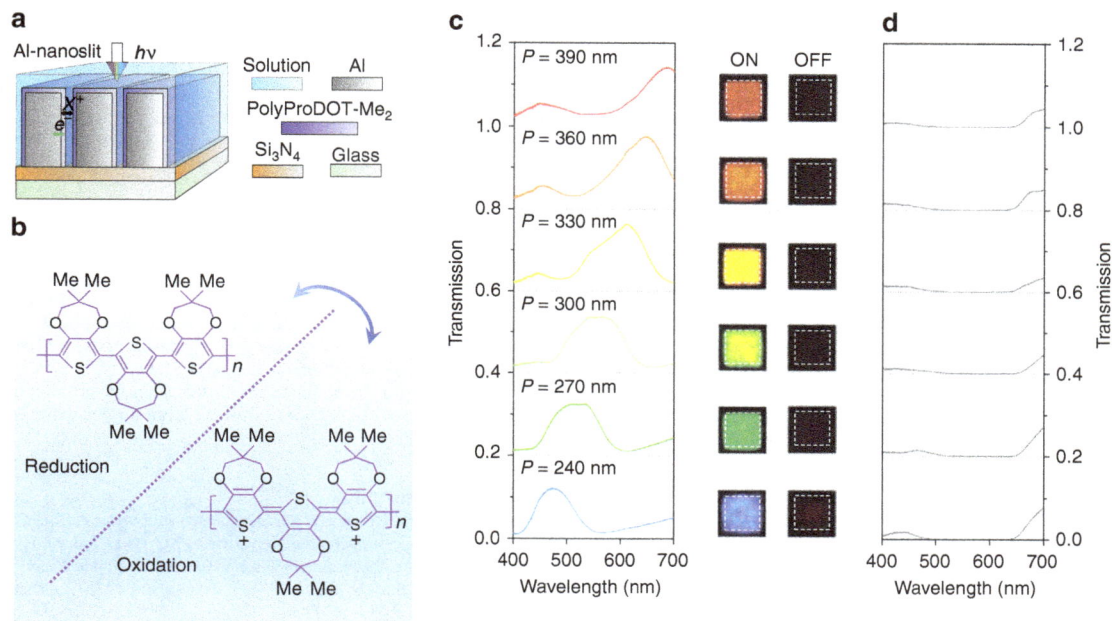

Figure 3 | Full-colour plasmonic electrochromic electrodes. (a) Schematic diagram of a plasmonic electrochromic electrode incorporating Al-nanoslit array. The pitch of six Al-nanoslit arrays ranges from 240 to 390 nm, as a step of 30 nm. The thickness of Al layer and Si_3N_4 waveguide layer is 250 and 170 nm, respectively. **(b)** Chemical structures of PolyProDOT-Me$_2$ in the oxidized and reduced form. **(c,d)** Optical transmission spectra of PolyProDOT-Me$_2$-coated Al-nanoslit structures with respective values of slit period $P = 240$, 270, 300, 330, 360 and 390 nm, along with corresponding optical micrographs of device areas imaged in transmission. Transmission spectra and micrographs for **(c)** ON and **(d)** OFF states of the polymer are displayed, respectively.

improved by optimizing plasmonic nanostructures or using more efficient polymers. The plasmonic electrochromic electrodes implemented here, each only tens of micrometres in lateral dimension, provide a potential pathway for achieving switchable pixels that are approximately one to two orders of magnitude smaller than those used in current high-definition displays[36]. In particular, the Al-based plasmonic electrochromic electrodes display high contrast and rapid modulation over the full-colour gamut using only one electrochromic polymer. Circumventing the need for multiple electrochromic polymers avoids potential material incompatibilities and considerably simplifies the fabrication process while reducing manufacturing costs.

In conclusion, Au and Al metallic nanoslit arrays conformally coated with electrochromic polymers are shown to enhance the optical performance of the polymer materials, yielding the electrodes combining the fast charge transport properties of an ultra-thin film and the high optical absorption properties of a thick film. Based on these concepts, we use two ordinary electrochromic polymers, PANI and PolyProDOT-Me$_2$, integrated with periodic metallic nanoslit arrays to demonstrate both monochromatic and full-colour fast switching with high optical contrast. Furthermore, the simple and elegant geometry of the proposed configurations can be extended to large areas for mass production on a flexible substrate through techniques such as roll-to-roll nanoimprint lithography[37–38] or nanotransfer printing[39]. Finally, the contrast and speed enhancements observed for the conformal coating of thin electrochromic material can be translated to any optically sensitive material with thickness-dependent properties such as the charge diffusion length, with applications ranging from catalysis to photovoltaics.

Methods

Plasmonic electrode preparation. Au and Al films are prepared by direct-current sputtering onto pre-cleaned ultra-flat glass slides coated with a 5-nm-thick Ti adhesion layer. The deposition rate for Au and Al is $R_{Au} \approx 0.36$ nm s^{-1} and $R_{Al} \approx 0.25$ nm s^{-1}, respectively. For full-colour electrochromic switches, a 170-nm-

thick Si_3N_4 layer is deposited by sputtering on glass substrate before the deposition of Al film. The deposition rate for Si_3N_4 is $R_{Si_3N_4} \approx 0.15$ nm s^{-1}. All the depositions are performed at room temperature. Subwavelength slits are prepared by FIB milling using a dual-beam (FIB/SEM) system (Ga$^+$ ions, 40 pA beam current, 30 keV beam energy).

Polymer synthesis and characterization. PANI is synthesized electrochemically from a 2-M HNO$_3$ solution containing 15 mM aniline. The films are deposited using potentio-dynamic cycling from -0.2 to 1.05 V versus Ag/AgCl, at a cycling rate of 30 mV s^{-1}. A Pt mesh is used as a counter electrode and a Ag/AgCl is used as a reference electrode. Film thickness is controlled by varying the number of cycles (Supplementary Fig. 5). PolyProDOT-Me$_2$ is synthesized inside of an Ar glovebox with an Al film from a 10-mM solution of monomer in 0.1 M tetrabutylammonium perchlorate/acetonitrile at a constant potential of 1.3 V versus Ag wire. All films are washed with monomer-free electrolyte solution. Following deposition, polymer-coated electrodes are characterized using cyclic voltammetry in solution consisting of 0.1 mol l^{-1} HNO$_3$ and 1 mol l^{-1} NaNO$_3$ for PANI and 0.1 mol l^{-1} LiClO$_4$ in a mixture (2:1) of dimethyl carbonate and ethylene carbonate for PolyProDOT-Me$_2$. Atomic force microscopy (AFM) is used to confirm polymer thicknesses on planar portions of the substrates and scanning electron microscopy is used to image the surface morphology.

Spectroelectrochemical measurements. All spectroelectrochemical measurements are conducted in custom-built cells consisting of Pt-coated glass electrodes serving as counter electrodes and a micro Ag/AgCl electrode (PANI) or Ag wire (PolyProDOT-Me$_2$) used as reference electrodes. All measurements are taken with the same electrolyte solutions used for cyclic voltammetry and in the same cell geometry. In optical transmission experiments, samples are irradiated with a HeNe laser and tungsten halogen bulb, and observed using both upright and inverted optical microscopes. The transmitted light is collected with an amplified photodiode or a spectrophotometer.

References

1. Comiskey, B., Albert, J. D., Yoshizawa, H. & Jacobson, J. An electrophoretic ink for all-printed reflective electronic displays. *Nature* **394**, 253–255 (1998).
2. Kawamoto, H. The history of liquid-crystal-displays. *Proc. IEEE* **90**, 460–500 (2002).
3. Tang, C. W. & Vanslyke, S. A. Organic electroluminescent diodes. *Appl. Phys. Lett.* **51**, 913–915 (1987).
4. Fudouzi, H. & Xia, Y. Photonic papers and inks: color writing with colorless materials. *Adv. Mater.* **15**, 892–896 (2003).

5. Arsenault, A. C., Puzzo, D. P., Manners, I. & Ozin, G. A. Photonic-crystal full-color displays. *Nat. Photon.* **1**, 468–472 (2007).
6. Kim, H. *et al.* Structural color printing using a magnetically tunable and lithographically fixable photonic crystal. *Nat. Photon.* **3**, 534–540 (2009).
7. Kim, L. *et al.* Contact printing of quantum dot light-emitting devices. *Nano Lett.* **8**, 4513–4517 (2008).
8. Cho, K. S. *et al.* High-performance crosslinked collideal quantum-dot light emitting diodes. *Nat. Photon.* **3**, 341–345 (2009).
9. Kim, T. H. *et al.* Full-color quantum dot displays fabricated by transfer printing. *Nat. Photon.* **5**, 176–182 (2011).
10. Deb, S. K. & Chopoorian, J. A. Optical properties and color-center formation in thin films of molybdenum trioxide. *J. Appl. Phys.* **37**, 4818–4825 (1966).
11. Rosseinsky, D. R. & Mortimer, R. J. Electrochromic systems and the prospects for devices. *Adv. Mater.* **13**, 783–793 (2001).
12. Somani, P. R. & Radhakrishnan, S. Electrochromic materials and devices: present and future. *Mater. Chem. Phys.* **77**, 117–133 (2002).
13. Argun, A. A. *et al.* Multicolored electrochromism in polymers: structures and devices. *Chem. Mater.* **16**, 4401–4412 (2004).
14. Mortimer, R. J., Dyer, A. L. & Reynolds, J. R. Electrochromic organic and polymeric materials for display applications. *Displays* **27**, 2–18 (2006).
15. Graham-Rowe, D. Electronic paper rewrites the rulebook for display. *Nat. Photon.* **1**, 248–251 (2007).
16. Jain, V., Yochum, J. M., Montazami, R. & Heflin, J. R. Millisecond switching in solid state electrochromic polymer devices fabricated from ionic self-assembled multilayers. *Appl. Phys. Lett.* **92**, 033304 (2008).
17. Zayatsa, A. V., Smolyaninovb, I. I. & Maradudinc, A. A. Nano-optics of surface plasmon polaritons. *Phys. Rep.* **408**, 131–314 (2005).
18. Ozbay, E. Plasmonics: merging photonics and electronics at nanoscale dimensions. *Science* **311**, 189–193 (2006).
19. Dintinger, J., Klein, S. & Ebbesen, T. W. Molecule-surface plasmon interactions in hole arrays: enhanced absorption, refractive index changes, and all-optical switching. *Adv. Mater.* **18**, 1267–1270 (2006).
20. Leroux, Y. *et al.* Active plasmonic devices with anisotropic optical response: a step toward active polarizer. *Nano Lett.* **9**, 2144–2148 (2009).
21. Stockhausen, V. *et al.* Giant plasmon resonance shift using Poly(3,4-ethylenedioxythiophene) electrochemical switching. *J. Am. Chem. Soc.* **132**, 10224–10226 (2010).
22. König, T. A. F. *et al.* Electrically tunable plasmonic behavior of nanocube–polymer nanomaterials induced by a redox-active electrochromic polymer. *ACS Nano* **8**, 6182–6192 (2014).
23. Agrawal, A. *et al.* An integrated electrochromic nanoplasmonic optical switch. *Nano Lett.* **11**, 2774–2778 (2011).
24. Guiseppi-Elie, A. *et al.* Growth of electropolymerized polyaniline thin films. *Chem. Mater.* **5**, 1474–1480 (1993).
25. Welsh, D. M., Kumar, A., Meijer, E. W. & Reynolds, J. R. Enhanced contrast ratios and rapid switching on electrochromics based on poly(3,4-propylenedioxythiophene) derivatives. *Adv. Mater.* **11**, 1379–1382 (1999).
26. Carpi, F. & Rossi, D. D. Colours from electroactive polymers: electrochromic, electroluminescent and laser devices based on organic materials. *Opt. Laser Technol.* **38**, 292–305 (2006).
27. Cruz, C. & Ticianelli, E. A. Electrochemical and ellipsometric studies of polyaniline films grown under cycling conditions. *J. Electroanal. Chem.* **428**, 185–192 (1997).
28. Lux, E., Genet, C., Skauli, T. & Ebbesen, T. W. Plasmonic photon sorters for spectral and polarimetric imaging. *Nat. Photon.* **2**, 161–164 (2008).
29. Xu, T., Wu, Y. K., Luo, X. G. & Guo, L. J. Plasmonic nanoresonators for high-resolution color filtering and spectral imaging. *Nat. Commun.* **1**, 59 (2010).
30. Chen, Q. & Cumming, D. S. R. High transmission and low color cross-talk plasmonic color filters using triangular-lattice hole arrays in aluminum films. *Opt. Express* **18**, 14056–14062 (2010).
31. Xu, T. *et al.* Structural colors: from plasmonic to carbon nanostructures. *Small* **7**, 3128–3136 (2011).
32. Kaplan, A. F., Xu, T. & Guo, L. J. High efficiency resonance-based spectrum filters with tunable transmission bandwidth fabricated using nanoimprint lithography. *Appl. Phys. Lett.* **99**, 143111 (2011).
33. Yokogawa, S., Burgos, S. P. & Atwater, H. A. Plasmonic color filters for CMOS image sensor applications. *Nano Lett.* **12**, 4349–4354 (2012).
34. Maier, S. A. *Plasmonics: Fundamentals and Applications* (Springer, 2007).
35. Invernale, M. A. *et al.* Variable-color poly(3,4-propylenedioxythiophene) electrochromics from precursor polymers. *Polymer* **51**, 378–382 (2010).
36. Lien, A., Cai, C., John, R. A., Galligan, J. E. & Wilson, J. 16.3″QSXGA high resolution wide viewing angle TFT-LCDs based on ridge and fringe-field structures. *J. Displays* **22**, 9–14 (2001).
37. Ahn, S. H. & Guo, L. J. High-speed roll-to-roll nanoimprint lithography on flexible plastic substrates. *Adv. Mater.* **20**, 2044–2049 (2008).
38. Ahn, S. H. & Guo, L. J. Large-area roll-to-roll and roll-to-plate nanoimprint lithography: a step toward high-throughput application of continuous nanoimprinting. *ACS Nano* **3**, 2304–2310 (2009).
39. Chen, Q., Martin, C. & Cumming, D.R.S. Transfer printing of nanoplasmonic devices onto flexible polymer substrates from a rigid stamp. *Plasmonics* **7**, 755–761 (2012).

Acknowledgements

T.X., E.C. W., A.A., J.V and W.Z. acknowledge support under the Cooperative Research Agreement between the University of Maryland and the National Institute of Standards and Technology, Center for Nanoscale Science and Technology, Award 70NANB10H193, through the University of Maryland. A.A.T. was supported by the Nanostructures for Electrical Energy Storage (NEES), an Energy Frontier Research Center funded by the U.S. Department of Energy, Office of Science, Basic Energy Sciences under Award number DESC0001160. Sandia is a multi-programme laboratory operated by Sandia Corporation, a Lockheed Martin Company, for the U.S. DOE National Nuclear Security Administration under Contract DE-AC04-94AL85000. T.X. acknowledges support from the Thousand Talents Program for Young Professionals, Collaborative Innovations Center of Advanced Microstructures and the Fundamental Research Funds for the Central Universities.

Author contribution

T.X., E.W. and A.A. designed, fabricated and characterized the devices. C.B, J.V and W.Z. helped with synthesis and deposition of the polymers. T.X., H.L and A.T. directed the project. All authors discussed the results and contributed to the manuscript.

Additional information

Competing financial interests: The authors declare no competing financial interests.

Transition state theory demonstrated at the micron scale with out-of-equilibrium transport in a confined environment

Christian L. Vestergaard[1,†], Morten Bo Mikkelsen[1], Walter Reisner[1,†], Anders Kristensen[1] & Henrik Flyvbjerg[1]

Transition state theory (TST) provides a simple interpretation of many thermally activated processes. It applies successfully on timescales and length scales that differ several orders of magnitude: to chemical reactions, breaking of chemical bonds, unfolding of proteins and RNA structures and polymers crossing entropic barriers. Here we apply TST to out-of-equilibrium transport through confined environments: the thermally activated translocation of single DNA molecules over an entropic barrier helped by an external force field. Reaction pathways are effectively one dimensional and so long that they are observable in a microscope. Reaction rates are so slow that transitions are recorded on video. We find sharp transition states that are independent of the applied force, similar to chemical bond rupture, as well as transition states that change location on the reaction pathway with the strength of the applied force. The states of equilibrium and transition are separated by micrometres as compared with angstroms/nanometres for chemical bonds.

[1]Department of Micro- and Nanotechnology, Technical University of Denmark, DK-2800 Kgs. Lyngby, Denmark. † Present addresses: Aix Marseille Université, Université de Toulon, CNRS, CPT, UMR 7332, 13288 Marseille, France (C.L.V.); Department of Physics, McGill University, Montreal, Quebec, Canada H3A 2T8 (W.R.). Correspondence and requests for materials should be addressed to C.L.V. (email: cvestergaard@gmail.com).

Transition state theory (TST), with its scenario of a reaction pathway through a free-energy landscape (Fig. 1), provides concepts for understanding how thermally activated processes take place. Its development can be traced back to the second half of the nineteenth century[1], notably to 1889 when Arrhenius proposed his famous empirical relation between the reaction rate r of an irreversible chemical reaction and temperature T:

$$r \propto e^{-\frac{\Delta \mathcal{F}^{\ddagger}}{k_B T}}. \qquad (1)$$

Here $\Delta \mathcal{F}^{\ddagger}$ is the height of the free-energy barrier separating the initial state (reactants) and the end state (product), and k_B is the Boltzmann constant. Theoretical efforts to describe such reactions led to the development of TST in the second half of the 1930s, notably by Eyring, Polanyi, Evans and Wigner[1-3]. TST for elementary chemical reactions assumes, as Wigner summarized it, statistical mechanics, classical motion of atomic nuclei, adiabatically changing electronic states and what has become known as TST's fundamental assumption, fundamental dynamical assumption or no-recrossing assumption. When the reaction process is described by a single reaction coordinate x, as in Fig. 1, the no-recrossing assumption states that if x crosses the point of maximal free energy—the 'transition state' $x^{\ddagger}$—from left to right, it does not recross from right to left. This is plausible if motion in x is inertial, as in chemical reactions between colliding gas molecules.

To investigate the validity of TST, Kramers introduced in 1940 a model, which has become known as 'Kramers' problem'[4,5]. This model relaxes TST's no-recrossing assumption. It considers the reaction to be described by a fictive particle undergoing Brownian motion with more or less friction in the free-energy landscape along the reaction coordinate (Fig. 1). The case of large friction does not model colliding gas molecules, but reactants diffusing in a liquid. It differs from Wigner's TST by having x diffuse across $x^{\ddagger}$ with multiple recrossing expected from its trajectory of Brownian motion. In many cases Kramers' formalism allows calculation of the proportionality factor between the reaction rate in equation (1) and the Boltzmann factor, but it does not change the exponential dependence on the barrier height in equation (1).

Despite their simplicity, TST and Kramers' model are surprisingly successful at predicting chemical reaction rates, and they are unrivalled at providing conceptual insight into how such reactions occur. Though devised to describe chemical reactions (Fig. 1a), where reaction pathways are measured in fractions of angstroms and reaction times in femtoseconds[6], their formalism has been extended to processes taking place at timescales and length scales that are orders of magnitude longer. At the nanometre scale the formalism has been applied to rupture of chemical bonds[7-11] (Fig. 1b), protein (un)folding[12-14] and RNA unzipping[15] under both constant and time-dependent loads[16-18]; at the micron scale it has been applied to polymers crossing entropic barriers[19-22]. Han and Craighead notably showed that

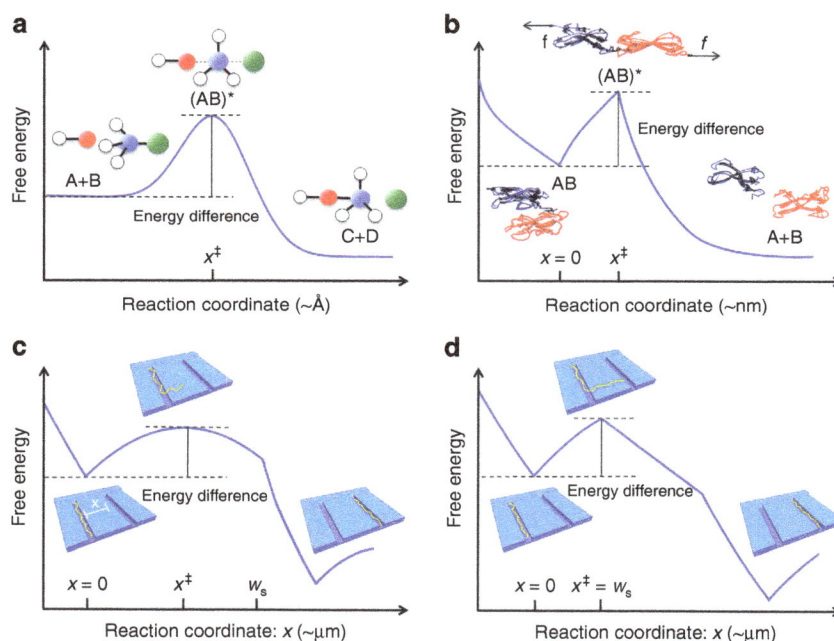

Figure 1 | The scenario of TST. A free-energy landscape is traversed by a reaction pathway that is parameterized by a reaction coordinate;[1,2] typical length scales of reaction pathways are given in parentheses. Insets portray physical situations corresponding to (quasi) equilibrium and transition states (at $x^{\dagger}$). (**a,c**) Transition states that will change location on the reaction pathway with the strength of an applied force, exemplified by (**a**) a chemical reaction (with transition state (AB)*), and in the present study (**c**) a polymer crossing an entropic barrier in the form of nanoslit separating two nanogrooves, where the transition state lies inside the nanoslit. (**b,d**) Sharp transition states that are independent of an applied force, similar to the situation in chemical bond rupture; here exemplified by (**b**) the separation of two binding proteins under an external force (with transition state (AB)*), and in the present study (**d**) a polymer crossing through a nanoslit where the transition state is located at the end of the nanoslit. In **c,d** the reaction coordinate x parameterizes the continuous shifting of DNA in a transition from the upstream to the downstream nanogroove. Specifically, it measures the extension of the leading end of the DNA into the nanoslit, until this end of the DNA enters the next nanogroove, which happens at $x = w_s$. For $0 < x < w_s$, x is approximately proportional to the contour length ℓ of the DNA that has left the upstream nanogroove. After the leading end of the DNA has entered the downstream nanogroove, we let x denote a fixed fraction of the contour length ℓ of the DNA that has left the upstream nanogroove, the same fraction as x denoted for $0 < x < w_s$. Note that we need not know the value of this fraction, and its existence can be an approximation. The qualitative picture described here still captures the essence of Fig. 2's experimental observations of transitions. The trailing end of the DNA leaves the upstream nanogroove when ℓ equals the full contour length of the DNA molecule, L_{DNA}. We denote the value of x at that point by $x = x_{DNA}$. After this point, we let x denote x_{DNA} plus the distance that the DNA molecule's trailing end has moved into the nanoslit. After the DNA has completely entered the next nanogroove, the landscape repeats itself as from $x = 0$ (see also Fig. 3).

TST describes the mean waiting time before translocation of a randomly coiled DNA molecule from one micro-groove to another through a nanoslit, driven by an external electric field. The transition state occurs where the external force that squeezes the coil into the slit balances entropic recoil forces[19,20].

We have here replaced the microgrooves of Han and Craighead, in which the DNA assumes a bulk coiled conformation, with nanogrooves that force the molecule to extend linearly and transversely to the axis of propagation (Figs 1c,d and 2). This ensures that the molecule can escape via clearly defined excursions of its end points into the nanoslit (Fig. 2c), leading to a well-defined single reaction coordinate, the DNA strand's extension into the nanoslit (Fig. 2d). The corresponding free-energy landscape for the DNA is (a) one dimensional; (b) tuneable; (c) so large that we can see the DNA moving through it, from one quasi-equilibrium state, over a free-energy barrier, into another quasi-equilibrium state, and so on repeatedly (Fig. 2e); (d) periodic, so each escape is an independent repetition of the same process, which allows us to accumulate good statistics; and (e) so simple that we find a closed formula for transition rates.

We study the translocation of DNA strands between nanogrooves driven by an external flow (Figs 1c,d and 2). For sufficiently weak flow, Brownian motion dominates the dynamics of the DNA to such a degree that its translocation between neighbouring nanogrooves is diffusive[23]. As one increases the flow, and hence its force on the DNA, the system transitions smoothly from the 'diffusion-dominated regime' to a 'force-dominated regime' in which translocation is irreversible and described by TST; see below. In the force-dominated regime, Brownian motion still plays a pivotal role by providing the fluctuations that let the DNA cross the entropic barrier separating neighbouring grooves. However, after the DNA has crossed the transition state, translocation is effectively deterministic and dictated by the external force field of the imposed flow.

We show theoretically, using TST, and verify experimentally by measuring the waiting time in each nanogroove that in the force-dominated regime two distinct sub-regimes exist for the transition of the DNA molecule from one groove to the next: (i) for large separation between grooves and high flow speeds, the transition state lies inside the nanoslit (Fig. 1c). Its location is determined by the balance between entropic and drag forces[19,20,22] and thus changes with the applied flow speed. (ii) As we decrease the external field (the flow), the location of the transition state moves in the field's direction (downstream), until it reaches the end of the nanoslit. Below the critical field strength at which this happens, the transition state does not move further downstream. It remains fixed at the width of the nanoslit (Fig. 1d). In this previously unobserved low-force (yet force dominated) regime, the transition state is independent of the external field. Both the initial and transition states are here 'sharp'—that is, the derivative of the free energy with respect to the reaction coordinate is not continuous at these points; it changes value abruptly. This is why the initial and transition states do not move along the reaction coordinate when we alter the external field. The dynamics in the low-force regime is consequently described by the Bell–Evans[7–9] model for chemical bond breaking under external load.

Results

Transition state theory for DNA translocation via a nanoslit.

We consider a DNA strand trapped in a nanogroove. Thermal fluctuations will now and then move one of its ends into the nanoslit (Fig. 2c,d). Let x denote the position of this end inside the nanoslit, measured in the direction parallel to the flow

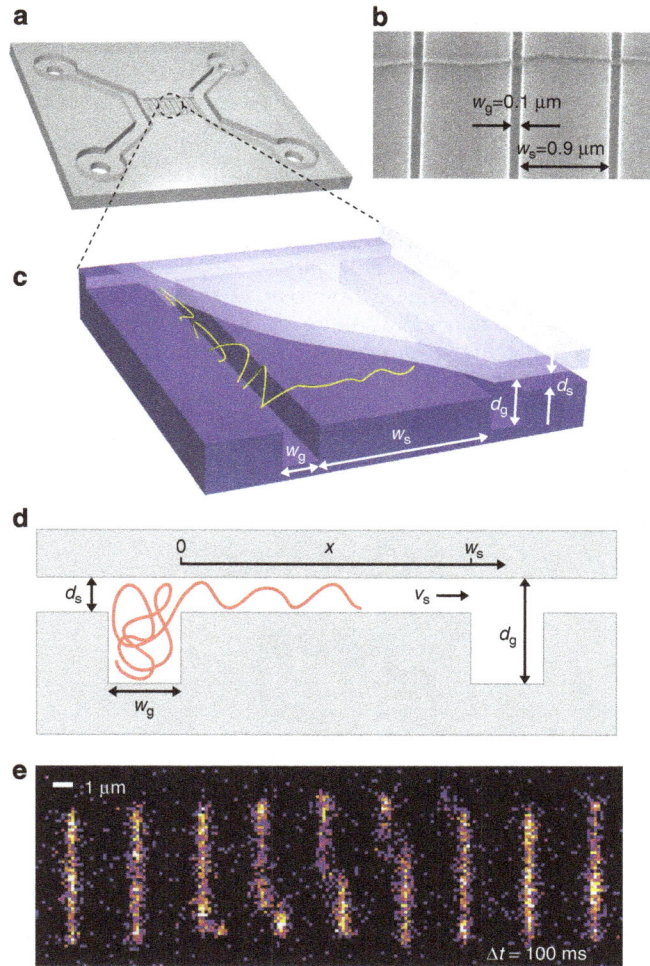

Figure 2 | Experimental set-up. (a) Schematic drawing of the microfluidic device containing the nanogroove array. A hydrodynamic flow (from left to right) is induced in the chip by imposing a pressure difference ΔP over the chip. Fluorescently labelled λ-DNA molecules (48.5 kb, $L_{DNA} = 21\,\mu m$) were introduced into the nanogroove array by the flow. **(b)** Electron micrograph of a section of a nanogroove array. **(c)** Schematic representation of a DNA strand trapped in a nanogroove and attempting to cross the nanoslit separating two grooves. The nanogroove geometry extends the DNA molecule transversally to the flow direction. Consequently, escape into the nanoslit is initiated by an end of the DNA. This vastly simplifies the dynamics compared with other entropic trapping geometries where polymers tend to form herniations inside the nanoslit. **(d)** Same as **c**, but showing a cross-section perpendicular to the nanogrooves. The extent of the DNA molecule's end inside the slit in the direction of the flow is called x. The hydrodynamic drag force on the DNA is proportional to x and v_s, where v_s is the mean speed of the buffer flow inside the nanoslit. The relevant dimensions of the nanogroove array are the height of the nanoslit, $d_s \approx 50\,nm$; the total height of a nanogroove plus the nanoslit, $d_g \approx 150\,nm$; the width of a nanogroove, $w_g \approx 100\,nm$; and the width of a nanoslit separating two grooves, $w_s = 0.4, 0.9, 1.9, 3.9\,\mu m$. **(e)** Montage of fluorescence images of a DNA molecule performing a sidewinder transition from one groove to the next[21]. The timelapse between consecutive images is 0.1 s. The fluorescence intensity is indicated with false colours. Uneven fluorescence of DNA in nanochannels is due to thermal fluctuations in the density of DNA in channels, where it coils a little as indicated in **c**. The lower fluorescence of DNA where it connects two channels in frames 3–6 (counting left to right) is due to the DNA being stretched in the slit, as indicated in **c**. Frame 7 shows the very last part of the transition between channels/barrier crossing. **a,b** and **e** are adapted from ref. 21.

(Fig. 2d). This x is our reaction coordinate. Thus, $x = 0$ denotes the equilibrium state of a DNA strand trapped in a nanogroove. Similarly, $x^\ddagger$ denotes the transition state for crossing into the next groove downstream. $x = 0$ and $x^\ddagger$ correspond to local minima and maxima, respectively, of the free-energy landscape experienced by a DNA strand moving through the chip (Fig. 3a).

The drag force f pulling at the DNA in the slit is proportional to the length ℓ of this DNA, $f = \gamma v_s \ell$, where v_s is the mean flow speed and γ is the effective drag coefficient of the DNA inside the slit (Fig. 2d). Since the flow in the chip is laminar, v_s is proportional to the pressure drop over the microchip, ΔP, which we control experimentally. Assuming that ℓ is approximately proportional to the DNA's extension parallel to the flow, x, we find that the drag force on the part of the DNA strand inside the nanoslit is proportional to $x\Delta P$. Thus, the decrease in free energy associated with the hydrodynamic drag force on the DNA is proportional to $x^2\Delta P$.

Note that the part of the DNA that rests in the nanogroove also experiences a drag force. It is proportional to the flow speed

inside the nanogroove, $v_g \approx v_s/(1 + d_g/d_s) = v_s/3$. This force, however, does not contribute to the free-energy difference along the reaction coordinate, since the part of the DNA inside the nanogrove does not move downstream with the drag force it experiences in the groove.

The decrease in entropy caused by the introduction of the DNA into a nanoslit, where it is more confined than in a nanogroove, gives rise to an entropic recoil force that tends to pull the DNA back out of the slit. This decrease in entropy is proportional to x (ref. 24).

Introduction of an end of the DNA strand into a nanoslit thus changes its free energy by the amount $\Delta \mathcal{F} = k_B T(bx - ax^2\Delta P/2)$ compared with its equilibrium state, where the whole strand resides in the nanogroove, $x = 0$ (Fig. 3b)[19,20,24,25]. Here a and b are constants of proportionality that depend, respectively, on the mean drag coefficient on the DNA inside the nanoslit and the increase in entropy per unit length of DNA introduced into the nanoslit.

TST then predicts that the waiting times of a DNA strand in a groove are exponentially distributed (Fig. 4 and Supplementary Fig. 1) with a mean value that is given by $\tau = \tau_0 \exp[bx^\ddagger - a(x^\ddagger)^2\Delta P/2]$. Two regimes exist, separated by a critical pressure difference ΔP_{crit} (Fig. 3c): (i) a high-force regime, characterized by $\Delta P > \Delta P_{crit}$, with the transition state inside the nanoslit at $x^\ddagger = b/(a\Delta P)$, and the dynamics of barrier crossing independent of w_s; and (ii) a low-force regime, characterized by $\Delta P < \Delta P_{crit} = b/(aw_s)$, with the transition state given by $x^\ddagger = w_s$, where w_s is the width of the nanoslit.

The mean trapping time is thus given by

$$\tau = \begin{cases} \tau_0 \exp(bw_s - aw_s^2\Delta P/2) & \text{for } \Delta P \leq \Delta P_{crit}, \\ \tau_0 \exp[b^2/(2a\Delta P)] & \text{for } \Delta P \geq \Delta P_{crit}, \end{cases} \quad (2)$$

where the prefactor τ_0 is related to the effective timescale of the motion along the reaction coordinate[1-4]. Equation (2) shows that for $\Delta P < \Delta P_{crit}$, the trapping time is described by the Bell–Evans model and $\log(\tau/\tau_0)$ is a first-degree polynomial in ΔP (Fig. 5a,c,e). For $\Delta P > \Delta P_{crit}$, equation (2) shows that $\log(\tau/\tau_0)$ is proportional to $1/\Delta P$ (Fig. 5b,d,f), as also observed in refs 19,20. At $\Delta P = \Delta P_{crit}$ we have a continuous transition between the two distinct regimes (Fig. 5a,b). The values found for the parameters of equation (2) (Table 1) are connected with microscopic physical

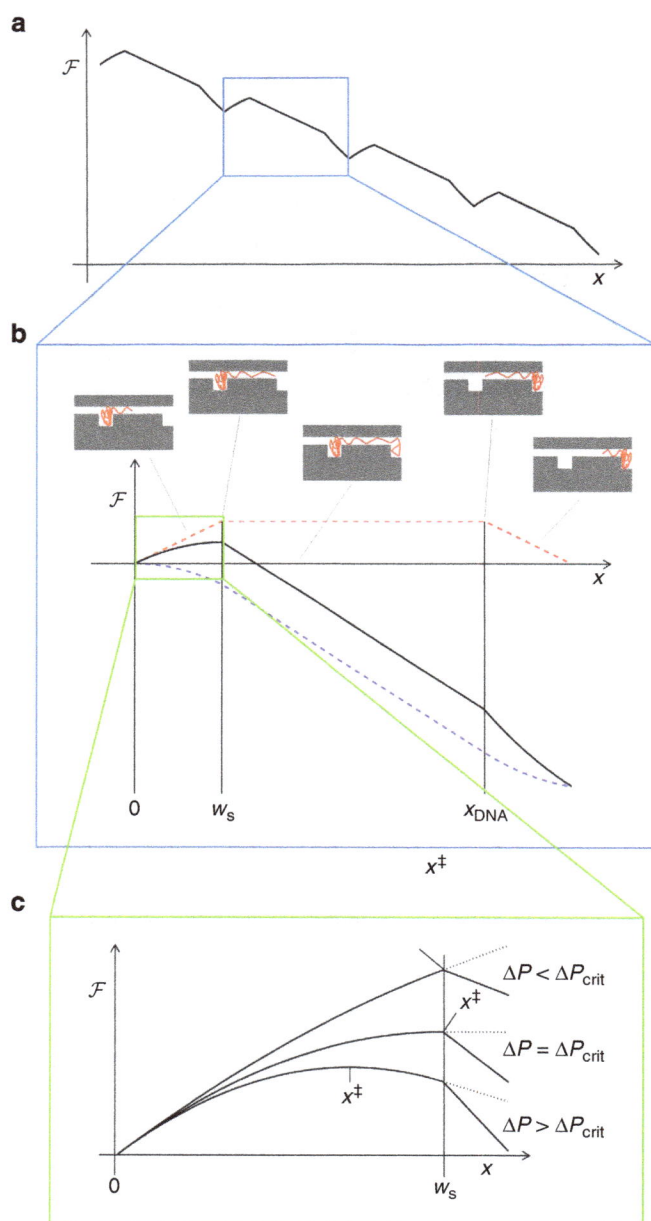

Figure 3 | Energy landscape experienced by a DNA molecule in the nanogroove array. (a) Free energy $\mathcal{F}(x)$ experienced by the molecule when driven by the force field from the buffer flow through a series of entropic traps. **(b)** Free energy $\mathcal{F}(x)$ experienced by a the molecule during a single transition between adjacent grooves. Insets show schematic drawings of the physical situation corresponding to five values of x. Here x is the x coordinate inside the nanoslit of the leading end of the DNA strand until that end descends into the next nanogroove. From that point and until the next state of quasi-equilibrium has been reached, further increase in x describes the length of DNA that has entered the next nanogroove. Thus, $x = 0$ corresponds to the equilibrium state in which the whole strand resides in the left nanogroove. At $x = w_s$ the leading end enters the next nanogroove. At $x = x_{DNA}$, the trailing end of the DNA leaves the upstream nanogroove. At $x = x_{DNA} + w_s$, the DNA is again in quasi-equilibrium in the next groove, and the energy landscape repeats itself downstream from there as shown in **a**. Red dashed line: contribution to the free energy due to loss of entropy; blue dashed line: loss of potential energy of the DNA due to higher hydrodynamic drag on the part of the strand inside the nanoslit; black full line: total free-energy difference $\mathcal{F}(x)$. **(c)** Zoom on the energy landscape for $0 < x \leq w_s$. The transition state at $x^\ddagger \leq w_s$ is the point with maximal free energy. For $\Delta P \leq \Delta P_{crit}$, $x^\ddagger = w_s$; for $\Delta P > \Delta P_{crit}$, $x^\ddagger < w_s$. (Note that we have assumed that $x_{DNA} \geq w_s$, which is always true in the present study. For DNA too short to span the width of the slit, $x_{DNA} < w_s$, and TST predicts $x^\ddagger \leq x_{DNA}$.)

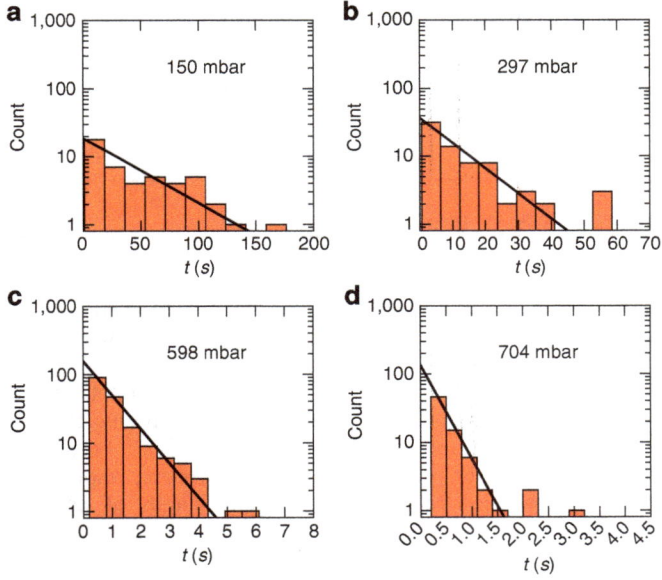

Figure 4 | Example distributions of waiting times in a groove. Measured waiting times in nanogrooves separated by nanoslits of width $w_s = 0.4\,\mu m$ (chip 1 below) for different values of the pressure difference imposed over the chip, spanning most of the parameter range explored in experiments here. Histograms shown agree well with single exponential fits (solid lines—obtained from maximum likelihood estimation, see Methods). Apparent 'outliers' arise from finite statistics in the tails and should be there, as their numbers agree with the expected numbers given by the areas under the tails of the theoretical distributions. Numbers of measured transition events are (**a**) 47, (**b**) 72, (**c**) 180 and (**d**) 73.

quantities. The value for τ_0 suggest that the timescale of relaxation of the DNA inside the slit is of the order of milliseconds, while a and b are determined by the DNA's effective drag coefficient and persistence length, the degree of stretching and the effective confinement energy of the DNA inside the slit. The latter four quantities cannot be found from our values for a and b alone. But by using that the drag coefficient and persistence length for DNA under similar conditions were found to be $\gamma \approx 1-2\,fN\,s\,\mu m^{-2}$ and $\ell_p \approx 40\,nm$, respectively, we can give rough estimates of the degree of stretching and effective confinement energy (see the section 'Microscopic interpretation of parameter values' in Methods). We find that the degree of stretching of the DNA inside the slit is 30–50%, and the effective confinement energy of the DNA inside the slit is 0.4–$0.6k_BT$ per persistence length ($\ell_p \approx 40\,nm$) of DNA introduced into the slit.

Finally, by renormalizing τ and ΔP as $u = (\tau/\tau_0)^{1/(bw_s)}$ and $\xi = \Delta P/\Delta P_{crit}$, we find that all data fall on the same curve (Fig. 5g) given by

$$u = \begin{cases} \exp(1-\xi/2) & \text{for } \xi \leq 1, \\ \exp[1/(2\xi)] & \text{for } \xi \geq 1. \end{cases} \quad (3)$$

Why TST works here. We made several simplifying assumptions to derive equations (2) and (3). These assumptions hold for our experiments for the following reasons.

(i) We assume that the DNA strand is in a state of thermal quasi-equilibrium when trapped in a nanogroove with its ends occasionally, randomly entering the nanoslit—more specifically, each point of the free-energy landscape inside the trap should be visited with a probability given by its Boltzmann factor. The validity of this assumption depends on the timescale of relaxation

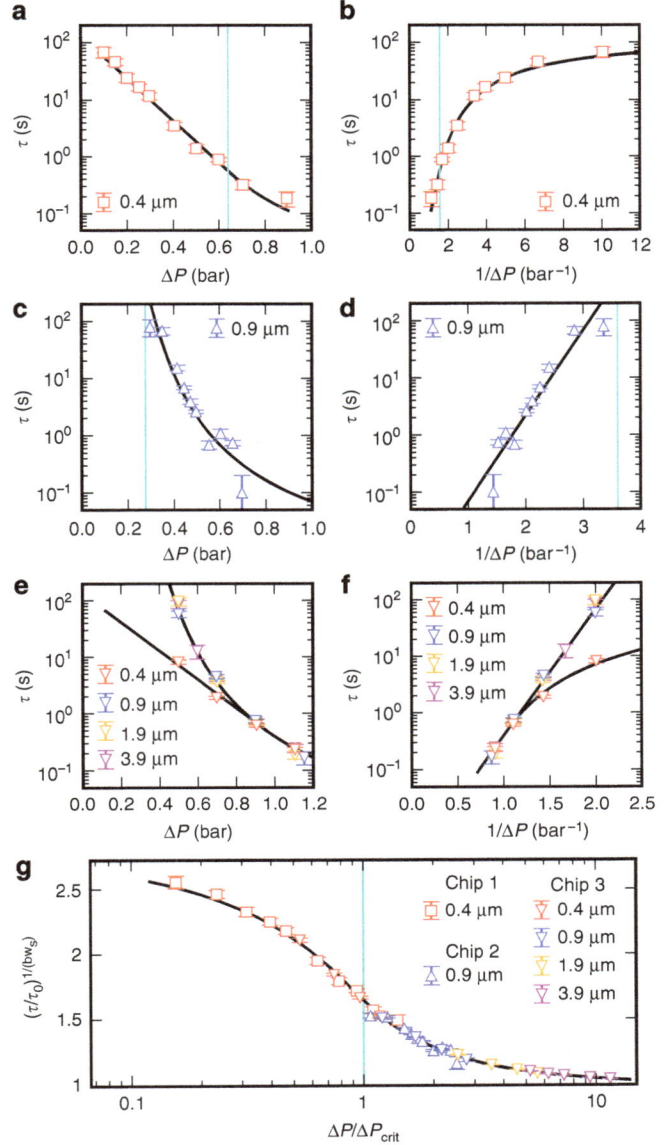

Figure 5 | Average lifetimes of quasi-stationary states in various force fields. Experimentally measured average residence times τ of DNA strands in a nanogroove as function of the pressure drop ΔP over the microfluidic chip. Different symbols correspond to different chips, different colors correspond to different nanoslit widths, see legends; chip 3 has different nanoslit widths at different places and hence does several distinct experiments all on the same chip. Data were collected for three different chips: (**a,b**) chip 1 (642 transition events), (**c,d**) chip 2 (1,604 events) and (**e,f**) chip 3 (2,873 events). (**a,c,e**) τ as function of ΔP—data follow a straight line for $\Delta P < \Delta P_{crit}$ and a hyperbolic curve for $\Delta P > \Delta P_{crit}$; (**b,d,f**) τ as function of $1/\Delta P$—data follow a straight line for $1/\Delta P < 1/\Delta P_{crit}$ and a hyperbolic curve for $1/\Delta P > 1/\Delta P_{crit}$. Symbols: experimental data, mean ± s.e.m. (examples of full distributions are shown in Fig. 4 and Supplementary Fig. 1); black lines: single fit of the theory (equation (2)) to all data in **a**–**f** (see Methods section); blue vertical lines: ΔP_{crit} for **a,c,e**, $1/\Delta P_{crit}$ for **b,d,f**. (**g**) All data from the three microfluidic chips fall on a single curve given by equation (3), independent of chip geometry when expressed in dimensionless variables $\Delta P/\Delta P_{crit}$ and $(\tau/\tau_0)^{1/(bw_s)}$.

of the DNA strand in a trap, τ_{relax}, being much shorter than the average time to escape from the trap.

For λ-DNA in a nanogroove with cross-section $\approx 100 \times 150\,nm^2$ (similar to here), $\tau_{relax} \sim 1\,s$ (ref. 26). Inside the slit, the relaxation

Table 1 | Estimated parameter values (top) and estimated values of related microscopic quantities (bottom).

τ_0 (ms)	a_1 ($\mu m^{-2}\,mbar^{-1}$)	a_2 ($\mu m^{-2}\,mbar^{-1}$)	a_3 ($\mu m^{-2}\,mbar^{-1}$)	b (μm^{-1})
2.2 (68% CI (0.6, 7.6))	0.11 ± 0.03	0.11 ± 0.01	0.07 ± 0.01	27 ± 4
	$\langle x/\ell \rangle$	$\langle x/\ell \rangle$	$\langle x/\ell \rangle$	$\delta\mathcal{F}_{conf}$ ($k_B T/\ell_p$)
	0.4–0.5	0.3–0.5	0.4–0.6	0.4–0.6

CI, confidence interval.
Microscopic quantities were estimated using $\gamma = 1$–$2\,fN\,s\,\mu m^{-2}$ and $\ell_p = 40$ nm as described in the section 'Microscopic interpretation of parameter values' in Methods.

time is much smaller than this due to additional confinement[26], the tension on the strand due to drag and the short length of DNA that is inside the slit (the relaxation timescales as $\sim 1/L$)[26]. Confinement reduces the relaxation time by a factor ~ 5 (ref. 26). Less than one-tenth of the DNA is in the slit before the transition state is traversed, as the longest distance separating the initial and transition states is 0.8 μm, further reducing τ_{relax} in the slit by a factor ~ 10.

These two effects alone then reduce the relaxation time to ~ 10 ms, while drag reduces it further, in agreement with the fitted value of τ_0 being of the order of milliseconds. This is fast enough compared with typical waiting times and experimental resolution that we may consider the DNA to be in quasi-equilibrium in the energy landscape before crossing the barrier; the exponential distribution of recorded waiting times to escape confirms this (Fig. 4 and Supplementary Fig. 1), while any 'inertial' effects making a second transition more probable immediately after a transition has occurred, say, due to incomplete relaxation in the trap, would result in an excess of counts for low waiting times.

(ii) We do not need the no-recrossing assumption, since Kramers' problem with large friction covers our case. It is approximately valid, however, if the free-energy landscape is steep and drops far, starting just past the transition state, and this condition is satisfied in our experiment.

Consider the case of lowest force ($w_s = 0.4$ μm and $\Delta P = 100$ mbar), where the free-energy landscape is the least steep. The height of the free-energy barrier here is $\Delta\mathcal{F} \approx 10 k_B T$, and the slope of the energy landscape to the right of the transition state is $\approx -5 k_B T$ per μm. Since recrossing happens with a probability that is proportional to the Boltzmann factor, x needs not be much larger than $x^{\ddagger}$ for recrossings to become highly unlikely, and translocation is hence effectively irreversible in our experiments.

Our data confirm this understanding. Recrossings over the transition state leads to 'dynamical corrections' of τ_0: it is expected to scale with the width of the free-energy barrier[4]. For $\Delta P \gg \Delta P_{crit}$, the barrier is essentially a parabolic potential, and one finds $\tau_0 \sim \Delta P^{-1/2}$, while for $\Delta P \ll \Delta P_{crit}$, the landscape is essentially rectilinear around the barrier, and one finds $\tau_0 \sim \Delta P^{-1}$ (ref. 4). For our experiments, this dependence on ΔP is so weak compared with the exponential factor in equation (2) that Fig. 5 shows agreement between data and theory without these dynamical corrections.

Further assumptions used here are the following: the shape of the free-energy landscape of the DNA in the nanoslit was derived assuming (iii) that nonlinear effects of hydrodynamic self-screening of the DNA in the slit is negligible; (iv) that the degree of stretching of the DNA in the slit does not depend on the amount of DNA in the slit; (v) that the increase in confinement energy caused by introduction of an end of the DNA into the slit scales linearly with the amount of DNA contour introduced; and (vi) that escape over the barrier through formation of a hernia, that is, a hairpin-like protrusion, of DNA in the slit happens so rarely that it can be ignored.

The effects of screening (iii) and uneven stretching (iv) tend to cancel each other, while both are diminished by the high degree of stretching of the DNA inside the nanoslit.

Assumption (v) that the free-energy cost per unit length of escaping contour is constant can be justified by noting that the free energy of confinement $\delta\mathcal{F}_{conf}^{(slit)}$ and $\delta\mathcal{F}_{conf}^{(groove)}$ both scale linearly with contour present in the slit and groove, respectively. Thus, the cost in free energy per unit length of contour, $\left(\delta\mathcal{F}_{conf}^{(slit)} - \delta\mathcal{F}_{conf}^{(groove)} \right)/\Delta\ell$, of driving contour length $\Delta\ell$ from the groove to the slit is constant. This linear scaling is fundamental and will hold regardless of the specific confinement regime as long as the size of the slit- or groove-confined polymer is much larger than the size of a 'statistical blob' in the slit or groove (which is true for the λ-DNA used here). To see this, simply note that for a confined polymer of contour length L, with $k_B T$ stored per blob, $\mathcal{F} = k_B T (L/L_{blob}) \sim L$. For a semiflexible chain, this linear scaling of confinement free energy with contour has been explicitly demonstrated in ref. 27 (for a slit) and ref. 28 (for a nanochannel, which approximates a nanogroove geometry well in as much as both force DNA into linearly extended configurations).

Finally, we assumed (vi) that escape over the entropic barrier through formation of a hernia inside the slit is an event sufficiently rare to be ignored. This escape event would lead to a non-exponential distribution of waiting times since the timescale of escape via this mechanism differs from that of end-induced escape. The reasons we do not observe escape via hernias here are threefold. First, as a hernia may form anywhere along a DNA strand, the rate of escape via hernias scales linearly with the length of the DNA[19,21,29,30]. Since λ-DNA is relatively short, this suppresses escape via herniation. Second, the free energy of a herniation inside the slit is more than twice that of an end (for an ideal flexible chain, it is exactly twice as high)[29,30]. So the probability of finding a hernia (as opposed to its multiplicity) is the square of the probability of finding a given end extending equally far into the slit. The latter being small, the former is very small. Third, since the DNA in the groove is stretched and the timescale of herniation is much faster than DNA relaxation in the groove (compare $\tau_0 \sim 1$ ms with $\tau_{relax} \sim 1$ s), introduction of a hernia into the nanoslit must stretch the DNA in the groove close to the hernia, thus increasing the free-energy barrier against escape via herniation even more. In contrast, introduction of an end into the nanoslit does not decrease the entropy per unit length of the DNA remaining in the groove.

Discussion

The use of an external pressure gradient to control and understand translocation of molecules by nanofluidic flows is poorly represented in the literature. So are simple models of such processes and their experimental verification.

Here we have shown that TST describes translocation of DNA driven by a hydrodynamic flow through a nano-confined environment that forms a series of entropic traps. We observed two distinct regimes: (i) a high-force regime in which the free-energy barrier is parabolic around the transition state. The transition state consequently moves along the reaction coordinate when the external force is altered; and (ii) a low-force regime in which the transition state is sharp and thus does not move when the force is altered. Observation of this low-force regime was made possible by reducing the barrier width considerably compared with earlier experiments[19,20,22]. A simple order-of-magnitude calculation shows that one would have to wait on the order of 100 years to see a single translocation in the geometries used in refs 19,20,22 (see the section 'Size matters' in Methods).

The applicability of TST to DNA translocation over entropic barriers relies on two conditions on the energy landscape describing the barriers: (i) the barrier separating two traps must be sufficiently high for quasi-equilibrium to exist before translocation; and (ii) the barrier must be steep enough beyond the transition state for recrossing not to occur, effectively.

If (i) is not satisfied, for example, for very high force, the motion becomes (partially) ballistic[21], leading to a non-exponential distribution of waiting times with an excess of short waits. If (ii) is not satisfied, for example, for very low force, Brownian motion dominates over drift, so the escape process is no longer irreversible[23] and hence ill defined as an 'escape.'

Microscopic 'bottom-up' models for barrier crossing of ideal Rouse polymers (polymers without bending rigidity and excluded volume effects)[29–33] yield an expression (equation (25) of ref. 30) for the rate of barrier crossing that is similar to our simple TST result (equation (1)). In these models the polymer crosses the barrier by stretching out in a 'kink' configuration, if it is long enough, since this lowers its free energy. An experimental demonstration of such stretching is provided in ref. 34, which shows DNA stretching where it crosses a potential barrier created by a conservative thermophoretic force field.

Note, however, that the physical mechanism responsible for the stretching of DNA in our nanoslit is entirely different. In refs 29–33, and in essence also in ref. 34, each monomer experiences the same potential energy barrier. Stretching lowers the energy barrier for the whole polymer by placing fewer monomers on top of the barrier, while stretching also costs a decrease in entropy, since some degrees of freedom are suppressed. The polymer stretches across the barrier when this decreases its potential energy by more than the ensuing cost in entropic free energy. However, this stretching is just a side effect of having the potential energy barrier; it is an adaption of the polymer to it. The barrier remains a potential energy barrier, while it affects the configurational entropy of the polymer.

In our experiments, on the other hand, DNA entering the nanoslit is less free to move thermally than it is in the nanogroove. The ensuing cost in entropic free energy constitutes the whole free-energy barrier. Thus, this barrier is entropic in nature. (Note that Figs 4 and 5 in ref. 30 resemble ours, but are artefacts resulting from displaying the DNA in an extra, non-existing dimension.) The final results are similar, because it is only the total free energy (containing both the energetic and entropic contributions) that matters. It should be interesting to extend the models of refs 29–33 to a non-smooth energy landscape to see if these microscopic models can predict the crossover between the high- and low-force regimes observed here.

The methods developed here might be useful also in the study of translocation of other biomolecules with more complex topologies (RNA, proteins, circular DNA or branched polymers) and in biological phenomena such as chromatin translocation in the cell nucleus and nuclear export. In particular, the quadratic shape of the energy landscape seen here is predicted also for more complex polymers, by scaling arguments[25,35]. It would be interesting to investigate experimentally whether in

consequence the statistics of barrier crossing for such more complex polymers also is described by the simple formulas derived here.

The study presented here demonstrates the wide applicability of TST and the Bell–Evans model, in particular to out-of-equilibrium transport in confined environments. It is, to the best of our knowledge, the first time that the Bell–Evans model for barrier crossing under external load has been demonstrated on the micron scale and for polymers crossing entropic barriers. The fact that the process can be monitored with video microscopy should appeal to anyone who teaches or has been taught TST.

In general, our study may serve as a reminder that TST applies where thermal activation is possible, irrespective of length scales. Its rates are dominated by the factor given in equation (1) above, so the Boltzmann energy sets the scale, while length scales are irrelevant in a first approximation. This physics insight can be used deliberately in engineering, in microfluidic handling of polymers, particles or cells. Or, if ignored, it might cause problems.

This fact that TST applies on any length scale where thermal activation is possible, is a small demonstration of the universality of many physical theories. TST is more universal than that, however: The Brownian motion at its core needs not be thermal in origin[36], so TST can describe other random processes as well.

Methods

Device fabrication and experimental set-up. The nanofluidic devices were fabricated from fused silica wafers (JINSOL)[37]. Electron beam lithography in zep520A resist was used to define the nanogrooves, and photolithography in AZ5214E resist was used to define the nanoslit and the inlet channels. The structures were transferred to the silica through CF_4:CHF_3 reactive ion etching, and the channels were closed by fusion bonding of a 157-μm-thick fused silica coverslip to the wafer surface. Experiments were performed using λ-phage DNA (48.5 kb, New England Biolabs) stained with the fluorescent dye YOYO-1 (Molecular Probes) at a ratio of 1 dye molecule per 5 base pairs. A buffer solution of $0.5 \times$ TBE (0.445 M Tris-base, 0.445 M boric acid and 10 mM EDTA) with 3% beta-mercaptoethanol was used. The buffer was driven through the nanogroove channel by applying air pressure, controlled to a precision of 1 mbar, to the inlets of the device. The DNA molecules were observed using an epi-fluorescence microscope (Nikon Eclipse TE2000-U) with $\times 60$ and $\times 100$ oil immersion 1.4 NA objectives. Movies of the DNA were recorded at up to 10 fps with an electron multiplying charge-coupled device camera (Cascade II, Photometrics). Identification of DNA molecules in the sidewinder state was performed using two distinct criteria[21]: (i) the DNA rested at least two frames in a groove between two transitions; and (ii) during a transition, both the DNA contour that connects two grooves and the DNA inside each groove were stretched, along the flow direction and the nanogroove, respectively. Waiting times were defined as the durations between the time at which the DNA found its equilibrium configuration inside a nanogroove after a transition and the time at which the leading end of the DNA had crossed over into the next nanogroove. In total, 5,119 sidewinder transitions were observed (chip 1: 642; chip 2: 1,604; and chip 3: 2,873).

Length and persistence length of YOYO-1-stained DNA. Intercalating YOYO dye affects both the length L and the persistence length ℓ_p of DNA. The dye increases L by approximately a factor 1.3 for a concentration of 1 YOYO-1 molecule per 5 base pairs[38]. For λ-DNA this leads to $L \approx 21$ μm. There is controversy in the literature as to its effect on ℓ_p, however, with some studies reporting an increase[39,40] in ℓ_p and others[38,41,42] reporting a decrease in ℓ_p. We here use the most recent results of ref. 38, which, for a dye concentration of 1 YOYO-1 molecule per 5 base pairs, gives $\ell_p \approx 40$ nm.

Note that using the results of refs 39,40, which give $\ell_p = 65$ nm, does not change the conclusions presented in the sections 'Microscopic interpretation of parameter values' and 'Size matters' below. The following argument shows this.

For $\ell_p \approx 65$ nm, we find from our estimated value of b an effective confinement energy per persistence length inside the nanoslit of $\delta\mathcal{F}_{conf} \approx 0.6 - 1.0 k_B T$, which should be compared with $\delta\mathcal{F}_{conf} \approx 0.33 k_B T$ from equations (6) and (7). The change of entropy due to the introduction of one persistence length of DNA into the nanoslit in the microarray of refs 19,20 is found to be $\delta\mathcal{F}_{conf} \approx 0.57 k_B T$ and $\delta\mathcal{F}_{conf} \approx 0.36 k_B T$ for ref. 22. For $\ell_p \approx 65$ nm, the confinement energy in the geometries of refs 19,20,22 would thus be roughly twice as high as for $\ell_p = 40$ nm, and the argument used below would then give that the expected waiting time for a single DNA strand to cross the barrier at critical field strength is $\sim 10^{10}$ years or more.

Calculating the mean of exponentially distributed data. As described in ref. 21, the DNA may at random switch from the 'sidewinder state', in which it is trapped for a time t_i in thermal equilibrium in a nanogroove, before it escapes to the next nanogroove, to a 'tumbleweed state', in which the DNA strand moves through the array without getting trapped in the grooves. Thus, there are no waiting times in the tumbleweed state; motion is continuous. The interaction with a groove may, however, slow the speed of the DNA in the tumbleweed state. This may falsely be detected as a waiting time by our movie-analysis software. As a filter against such false positives, we discard all waiting times that are shorter than twice the time Δt between frames in the movie. This means, however, that we also reject some true positives: DNA strands that are in the sidewinder state, but escape sooner than $2\Delta t$. Thus, the average $\bar{t} = \sum_i t_i/n$ of the measured dwell times is a biased estimate of $\tau = \langle t_i \rangle$. Instead we use the unbiased (maximum likelihood) estimator $\hat{\tau} = \bar{t} - 2\Delta t$.

Parameter estimation. Equation (2) is our theory for the observed waiting times. We fit the parameters $\theta = (a_1, a_2, a_3, b, \tau_0)$ of this theory to data $(\hat{\tau}_n \pm \sigma_n, w_{s,n}, \Delta P_n)_{n=1}^N$ using weighted least squares with weights $1/\hat{\sigma}_n^2$. Here $\hat{\sigma}_n$ is the empirically estimated s.e.m. of $\hat{\tau}_n$. We fit simultaneously to the data from all three microchips. In this fit, the parameters (a_1, a_2, a_3) are allowed individual values for each chip, since they depend on the hydraulic resistance, which differs between chips. The parameters b and τ_0 depend only on the DNA, solvent and temperature. They should not differ between chips, so we fit values shared by all chips. Fitting to data presented in Fig. 5, we obtain the estimates given in Table 1.

The variances of errors on fitted parameter values were estimated as

$$\mathrm{var}(\hat{\theta}_i) = \hat{\chi}^2_{\mathrm{norm}} \left(\hat{J}\hat{\Sigma}^{-1}\hat{J}^{\top}\right)^{-1}_{ii}, \tag{4}$$

where $\hat{\chi}^2_{\mathrm{norm}} = \hat{\chi}^2/(N-5)$ is $\hat{\chi}^2$ per degree of freedom, $\hat{\Sigma}$ is the covariance matrix of $\hat{\tau}_n$ with entries given by $\hat{\Sigma}_{mn} = \hat{\sigma}_n^2 \delta_{m,n}$ and $\hat{J}$ is the Jacobian of the vector $\tau(\hat{\theta}) = (\tau_n(\hat{\theta}))_{n=1}^N$, with entries given by $\hat{J}_{in} = \partial \tau_n(\hat{\theta})/\partial \theta_i$.

Microscopic interpretation of parameter values. The parameter values returned by the fit described above can be interpreted at the microscopic level. Thus, the value obtained for τ_0 means that the effective timescale of DNA motion in the free-energy potential inside the slit is of the order of milliseconds.

The values found for a is connected with the degree of stretching of the DNA inside the nanoslit. From our definition of a and x, we have $\gamma v_s \ell^2/(2k_B T) \approx a\Delta P x^2/2$. The mean degree of stretching of the DNA inside the slit is then given by

$$\langle x/\ell \rangle \approx \sqrt{\gamma v_s/(a\Delta P k_B T)}. \tag{5}$$

Here $v_s \approx \Delta P/(d_s w_{ch} R_{hyd})$, where $w_{ch} = 50\,\mu m$ is the chip width perpendicular to the flow, and $R_{hyd} \approx 12\eta N w_s/(d_s^3 w_{ch})$ is the hydraulic resistance over the chip[43], where N is the number of slits in the chip and η is the dynamic viscosity of water. Since equation (5) depends on the value of γ, our estimate for a does not directly give us the degree of stretching of the DNA inside the nanoslit. However, the drag coefficient for a flow parallel to the DNA backbone has been found previously for similar conditions to be $\gamma_{||} \approx 1$ fN s μm^{-1} (J.N. Pedersen, personal communication). For DNA segments aligned perpendicular to the flow, $\gamma_{\perp} = 2\gamma_{||}$, so we expect the effective drag coefficient on DNA inside the slit to lie between these two values.

We use this range of values for γ in the following and that $\eta = 10^{-3}$ Pa · s, $k_B T = 4.1$ fN μm and $d_s = 0.05\,\mu m$. For chip 1 we have $N = 900$, $w_s = 0.4$ and $a = 0.11$, yielding $\langle x/\ell \rangle \approx 0.4 - 0.5$. For chip 2 we have $N = 450$, $w_s = 0.9$ and $a = 0.11$, yielding $\langle x/\ell \rangle \approx 0.3 - 0.5$. For chip 3 we have $N = 375$, average slit width $\overline{w_s} = 1.1$ and $a = 0.07$, yielding $\langle x/\ell \rangle \approx 0.4 - 0.6$.

Finally, we may use the above estimate of the degree of stretching inside the nanoslit and the value of b to give a rough estimate of the effective confinement energy of the DNA inside the slit compared with the groove. Since b gives the effective confinement energy in the slit per μm along the reaction coordinate, we may find the confinement energy per persistence length ℓ_p as $\delta\mathcal{F}_{\mathrm{conf}} \approx b\langle x/\ell \rangle \ell_p \approx 0.4 - 0.6 k_B T$ by using $\ell \approx 40$ nm and the values for $\langle x/\ell \rangle$ found above.

We may compare this result with the expected difference in confinement energies between the slit and groove, $\delta\mathcal{F}_{\mathrm{conf}} = \delta\mathcal{F}^{(\mathrm{slit})}_{\mathrm{conf}} - \delta\mathcal{F}^{(\mathrm{groove})}_{\mathrm{conf}}$. The DNA in the groove and in the slit is in a crossover regime between the De Gennes and Odjik regimes. Here interpolation formulas for the confinement energy were recently determined from a combination of high precision simulations and experiments. The free energy of confinement per persistence length of DNA in a channel (which approximates the nanogroove geometry) is[28]

$$\delta\mathcal{F}^{(\mathrm{groove})}_{\mathrm{conf}} = \frac{k_B T(\pi^2/3)(\ell_p/d_s)^2}{\left[5.147(\ell_p/d_s)^2 + 3.343\ell_p/d_s + 1\right]^{2/3}}, \tag{6}$$

while the confinement energy in the nanoslit is[27,44]

$$\delta\mathcal{F}^{(\mathrm{slit})}_{\mathrm{conf}} = \frac{k_B T(\pi^2/3)(\ell_p/d_s)^2}{\left[5.146(\ell_p/d_s)^2 + 1.984\ell_p/d_s + 1\right]^{2/3}}. \tag{7}$$

The expected confinement energy per ℓ_p is thus $\delta\mathcal{F}_{\mathrm{conf}} \approx 0.3 k_B T$, which is somewhat smaller than the energy estimated from b. This difference may be explained by the flow stretching the DNA inside the slit, such that b reflects not only the confinement energy due the walls of the slit but also contains a term from the stretching of the DNA due to the flow. Table 1 collects our estimates for microscopic quantities.

Size matters. In the present experiment, the barrier width is significantly smaller than in earlier similar experiments[19,20,22]. This makes all the difference for our ability to observe the low-force regime ($\Delta P < \Delta P_{\mathrm{crit}}$).

In refs 19,20, the potential wells of the microarray with the smallest dimensions were separated by $w_s \approx 2\,\mu m$ of nanoslit, while (ref. 22) studied DNA translocation through nanopores of 6 μm or more. This made observation of the low-force regime impossible according to the following order-of-magnitude estimate of the time—100 years or more—that one would have to wait for a single DNA molecule to traverse the nanoslit/nanopore in these geometries. Our argument hinges on the fact that the mean trapping time of a DNA strand at critical pressure difference ΔP_{crit} depends exponentially on bw_s, that is, $\tau_{\mathrm{crit}} = \tau_0 \exp(bw_s/2)$. Thus, a linear increase in b or w_s leads to an exponential increase in the waiting time.

The entropic traps (microgrooves) of refs 19,20 had dimensions $d_g \approx 1\,\mu m$ by $w_g \approx 2\,\mu m$. Since the radius of gyration, $R_g \approx 0.5\,\mu m$, of YOYO-stained λ-DNA is smaller than both d_g and w_g, the DNA was essentially in bulk conformation there. The increase in free energy due to confinement caused by the introduction of a unit length ℓ of DNA from bulk into the nanoslit was thus simply equal to its confinement free energy in the slit, given by equation (7).

For a height of $d_s \approx 0.09\,\mu m$ of the nanoslits of refs 19,20, this gives $\delta\mathcal{F}_{\mathrm{conf}} \approx 0.32 k_B T$, which is comparable to the confinement energy in our set-up of $\delta\mathcal{F}_{\mathrm{conf}} \approx 0.28 k_B T$ found from equations (6) and (7). So while the DNA strand is more confined in the nanoslit in our set-up than in the one of refs 19,20 ($d_s = 50$ nm in our case versus $d_s \approx 90$ nm in refs 19,20), the preconfinement in the nanogroove lowers the relative confinement energy here such that the entropic recoil force is similar in magnitude to, or lower than, the one in refs 19,20.

In the present experiment, we have $\tau_{\mathrm{crit}} \approx 0.5$ s from $\tau_0 = 0.002$ s and $b = 27\,\mu m^{-1}$, found by fitting equation (2) to data (see 'Parameter estimation' above). We assume that τ_0 does not change much for different geometries, and since $b \propto \delta\mathcal{F}_{\mathrm{conf}}$, we may assume that b was roughly the same in the set-up of refs 19,20 as here. Thus, since their narrowest nanoslit measured 2 μm, the expected mean waiting time at critical field strength would in their geometry be $\tau_{\mathrm{crit}} \sim 100$ years. Similarly, for the least constricted geometry of ref. 22, we have $\delta\mathcal{F}_{\mathrm{conf}} = 0.17 k_B T$ (approximating their nanopore as a nanochannel), that is, around half of the confinement energy of the present study. However, since their pore measured 10 μm, we find here $\tau_{\mathrm{crit}} \sim 10^{20}$ years.

References

1. Laidler, K. J. & King, M. C. The development of transition-state theory. *J. Phys. Chem.* **87,** 2657–2664 (1983).
2. Laidler, K. J. A lifetime of transition-state theory. *Chem. Intell.* **4,** 39–47 (1998).
3. Garrett, B. C. Perspective on 'The transition state method". *Theor. Chem. Acc.* **103,** 200–204 (2000).
4. Kramers, H. Brownian motion in a field of force and the diffusion model of chemical reactions. *Physica* **7,** 284–304 (1940).
5. Mel, V. The Kramers problem: fifty years of development. *Phys. Rep.* **209,** 1–71 (1991).
6. Zewail, A. H. Laser femtochemistry. *Science* **242,** 1645–1653 (1988).
7. Zhurkov, S. Kinetic concept of strength of solids. *Int. J. Fract. Mech.* **1,** 311–323 (1965).
8. Bell, G. I. Models for the specific adhesion of cells to cells. *Science* **200,** 618–627 (1978).
9. Evans, E. Probing the relation between force-lifetime-and chemistry in single molecular bonds. *Annu. Rev. Biophys. Biomol. Struct.* **30,** 105–128 (2001).
10. Freund, L. Characterizing the resistance generated by a molecular bond as it is forcibly separated. *Proc. Natl Acad. Sci. USA* **106,** 8818–8823 (2009).
11. Tian, Y. & Boulatov, R. Comparison of the predictive performance of the Bell-Evans, Taylor-expansion and statistical-mechanics models of mechanochemistry. *Chem. Commun.* **49,** 4187–4189 (2013).
12. Schuler, B. & Clarke, J. Biophysics: rough passage across a barrier. *Nature* **502,** 632–633 (2013).
13. Chung, H. S. & Eaton, W. A. Single-molecule fluorescence probes dynamics of barrier crossing. *Nature* **502,** 685–688 (2013).
14. Woodside, M. T. & Block, S. M. Folding in single riboswitch aptamers. *Science* **180,** 2006–2009 (2008).
15. Dudko, O. K., Mathé, J., Szabo, A., Meller, A. & Hummer, G. Extracting kinetics from single-molecule force spectroscopy: nanopore unzipping of DNA hairpins. *Biophys J.* **92,** 4188–4195 (2007).
16. Dudko, O., Hummer, G. & Szabo, A. Intrinsic rates and activation free energies from single-molecule pulling experiments. *Phys. Rev. Lett.* **96,** 1–4 (2006).
17. Dudko, O. K., Hummer, G. & Szabo, A. Theory, analysis, and interpretation of single-molecule force spectroscopy experiments. *Proc. Natl Acad. Sci. USA* **105,** 15755–15760 (2008).

18. Dudko, O. K. Single-molecule mechanics: new insights from the escape-over-a-barrier problem. *Proc. Natl Acad. Sci. USA* **106**, 8795–8796 (2009).

19. Han, J., Turner, S. & Craighead, H. Entropic trapping and escape of long DNA molecules at submicron size constriction. *Phys. Rev. Lett.* **83**, 1688–1691 (1999).

20. Han, J., Turner, S. & Craighead, H. Erratum: entropic trapping and escape of long DNA molecules at submicron size constriction. *Phys. Rev. Lett.* **86**, 9007 (2001).

21. Mikkelsen, M. B. *et al.* Pressure-driven DNA in nanogroove arrays: complex dynamics leads to length-and topology-dependent separation. *Nano Lett.* **11**, 1598–1602 (2011).

22. Auger, T. *et al.* Zero-mode waveguide detection of flow-driven DNA translocation through nanopores. *Phys. Rev. Lett.* **113**, 028302 (2014).

23. Hoogerheide, D. P., Albertorio, F. & Golovchenko, J. A. Escape of DNA from a weakly biased thin nanopore: experimental evidence for a universal diffusive behavior. *Phys. Rev. Lett.* **111**, 248301 (2013).

24. De Gennes, P.-G. *Scaling Concepts in Polymer Physics* (Cornell Univ. Press, 1979).

25. Sakaue, T., Raphaël, E., de Gennes, P.-G. & Brochard-Wyart, F. Flow injection of branched polymers inside nanopores. *Europhys. Lett.* **72**, 83 (2005).

26. Reisner, W. *et al.* Statistics and dynamics of single DNA molecules confined in nanochannels. *Phys. Rev. Lett.* **94**, 196101 (2005).

27. Chen, J. & Sullivan, D. Free energy of a wormlike polymer chain confined in a slit: crossover between two scaling regimes. *Macromolecules* **39**, 7769–7773 (2006).

28. Tree, D. R., Wang, Y. & Dorfman, K. D. Extension of DNA in a nanochannel as a rod-to-coil transition. *Phys. Rev. Lett.* **110**, 208103 (2013).

29. Sebastian, K. L. Kink motion in the barrier crossing of a chain molecule. *Phys. Rev. E* **61**, 3245 (2000).

30. Sebastian, K. L. & Paul, A. K. R. Kramers problem for a polymer in a double well. *Phys. Rev. E* **62**, 927 (2000).

31. Lee, S. K. & Sung, W. Coil-to-stretch transition, kink formation, and efficient barrier crossing of a flexible chain. *Phys. Rev. E* **63**, 021115 (2001).

32. Sebastian, K. L. & Debnath, A. Polymer in a double well: dynamics of translocation of short chains over a barrier. *J. Phys. Condens. Matter* **18**, S283 (2006).

33. Debnath, A., Paul, A. K. R. & Sebastian, K. L. Barrier crossing in one and three dimensions by a long chain. *J. Stat. Mech. Theor. Exp.* **11**, P11024 (2010).

34. Pedersen, J. N. *et al.* Thermophoretic forces on DNA measured with a single-molecule spring balance. *Phys. Rev. Lett.* **113**, 268301 (2014).

35. De Gennes, P. G. Flexible polymers in nanopores. *Adv. Polym. Sci.* **138**, 92 (1999).

36. D'Anna, G. *et al.* Observing brownian motion in vibration-fluidized granular matter. *Nature* **424**, 909–912 (2003).

37. Reisner, W. *et al.* Nanoconfinement-enhanced conformational response of single DNA molecules to changes in ionic environment. *Phys. Rev. Lett.* **99**, 058302 (2007).

38. Maaloum, M., Muller, P. & Harlepp, S. DNA-intercalator interactions: structural and physical analysis using atomic force microscopy in solution. *Soft Matter* **9**, 11233–11240 (2013).

39. Quake, S. R., Babcock, H. & Chu, S. The dynamics of partially extended single molecules of DNA. *Nature* **388**, 151–154 (1997).

40. Makita, N., Ullner, M. & Yoshikawa, K. Conformational change of giant DNA with added salt as revealed by single molecular observation. *Macromolecules* **39**, 6200–6206 (2006).

41. Sischka, A. *et al.* Molecular mechanisms and kinetics between DNA and DNA binding ligands. *Biophys J.* **88**, 404–411 (2005).

42. Murade, C. U., Subramaniam, V., Otto, C. & Bennink, M. L. Force spectroscopy and fluorescence microscopy of dsDNA-YOYO-1 complexes: implications for the structure of dsDNA in the overstretching region. *Nucleic Acids Res.* **38**, 3423–3431 (2010).

43. Bruus, H. *Theoretical Microfluidics* (Oxford Univ. Press, 2007).

44. Klotz, A. R. *et al.* Measuring the confinement free energy and effective width of single polymer chains via single-molecule tetris. *Macromolecules* **48**, 5028–5033 (2015).

Acknowledgements

The work was partly funded by the Danish Council for Strategic Research Grant 10-092322 (PolyNano), by the Danish Council for Technology and Innovation through the Innovation Consortium OCTOPUS, by the Danish Research Council for Technology and Production, FTP Grant 274-05-0375 and by the European Commission-funded project READNA (Contract HEALTH-F4-2008-201418).

Author contributions

C.L.V., M.B.M., W.R., A.K. and H.F. designed the study; M.B.M. performed the experiments; C.L.V. and M.B.M. analysed the experimental data; C.L.V. derived theoretical results; C.L.V., M.B.M. and W.R. produced figures; C.L.V. and H.F. wrote the paper.

Additional information

Sub-10 nm rutile titanium dioxide nanoparticles for efficient visible-light-driven photocatalytic hydrogen production

Landong Li[1,2,*], Junqing Yan[1,2,*], Tuo Wang[1,3], Zhi-Jian Zhao[1,3], Jian Zhang[4], Jinlong Gong[1,3] & Naijia Guan[1,2]

Titanium dioxide is a promising photocatalyst for water splitting, but it suffers from low visible light activity due to its wide band gap. Doping can narrow the band gap of titanium dioxide; however, new charge-carrier recombination centres may be introduced. Here we report the design of sub-10 nm rutile titanium dioxide nanoparticles, with an increased amount of surface/sub-surface defects to overcome the negative effects from bulk defects. Abundant defects can not only shift the top of the valence band of rutile titanium dioxide upwards for band-gap narrowing but also promote charge-carrier separation. The role of titanium(III) is to enhance, rather than initiate, the visible-light-driven water splitting. The sub-10 nm rutile nanoparticles exhibit the state-of-the-art activity among titanium dioxide-based semiconductors for visible-light-driven water splitting and the concept of ultra-small nano-particles with abundant defects may be extended to the design of other robust semiconductor photocatalysts.

[1] Collaborative Innovation Center of Chemical Science and Engineering, Tianjin 300072, China. [2] Key Laboratory of Advanced Energy Materials Chemistry of Ministry of Education, College of Chemistry, Nankai University, Tianjin 300071, China. [3] Key Laboratory for Green Chemical Technology of Ministry of Education, School of Chemical Engineering and Technology, Tianjin University, 92 Weijin Road, Nankai District, Tianjin 300072, China. [4] Department of New Energy Technology, Ningbo Institute of Materials Technology and Engineering, Chinese Academy of Sciences, Tianjin, Ningbo 315201, China. * These authors contributed equally to this work. Correspondence and requests for materials should be addressed to J.G. (email: jlgong@tju.edu.cn).

Since the discovery of water photolysis on a TiO$_2$ photo-anode in the 1970s[1], semiconductor photocatalysis has attracted significant attention due to its promising applications in environment remediation and solar energy conversion in the past decades[2-5]. TiO$_2$ is the initial semiconductor photocatalyst investigated and it is still regarded as a benchmark photocatalyst under ultravoilet irradiation due to its intrinsic high activity. However, TiO$_2$ is a type of wide band-gap semiconductor and it only adsorbs ultravoilet light, which greatly limits its practical applications[6]. Accordingly, persistent efforts have been made to narrow the band gap of TiO$_2$ to extend its working spectrum to the visible light region, that is, so-called band-gap engineering.

Doping with metal or non-metal elements is known as a feasible means to tune the electronic structure of TiO$_2$ and to introduce new states into the TiO$_2$ band gap for visible light response[7-17]. In general, doping with non-metal elements, for example, C, N and S, can build acceptor states above the valence band from the p states of non-metal ions[16], and doping with metal elements, for example, Fe and Cr, can build donor states below the conduction band[2]. For doped TiO$_2$, the lattice defects induced by the dopants will unavoidably introduce new charge-carrier trapping and recombination centres, which might correspondingly show a degrading effect on the photocatalytic activity[7]. Moreover, doped TiO$_2$ materials are usually not adequate catalysts for photocatalytic hydrogen production from water splitting, although their superior activity in photocatalytic oxidation reactions have been well documented[6,18]. Self-doping with Ti^{3+} was further developed for narrowing the band of TiO$_2$ without the introduction of unwanted carrier recombination centres from dopants[19-21], which consequently exhibited good stability and considerable activity for photocatalytic hydrogen production under visible light[20,21]. Recently, hydrogenation of crystalline TiO$_2$ was disclosed as a new approach to enhance the visible as well as infrared light absorption of pristine TiO$_2$ and subsequently triggered great interest[22-28]. The surface disorder of hydrogenated black TiO$_2$, instead of bulk Ti^{3+}, was proposed to be responsible for the extended light absorption[23,24]. The hydrogenated TiO$_2$ exhibited remarkable activity in the photocatalytic hydrogen production from water splitting under full solar irradiation. However, considering that a sharp decline in the hydrogen production rate by two orders of magnitude (for example, from 10 to 0.1 mmol h^{-1} g^{-1})[22,25] was observed if ultraviolet light was filtered out ($\lambda > 400$ nm), the original hydrogen production activity should more probably come from the enhanced charge separation and transportation under ultraviolet irradiation rather than the extended visible light absorption[29].

The possibility of extending the working spectrum of TiO$_2$ to the visible light region is generally acknowledged based on both experimental observations and theoretical calculations[7-32]. The common point of different strategies lies in disrupting the integrity of ordered lattice structure of pristine TiO$_2$ (rutile or anatase) and building new states within the band gap for photoexcitation with lower energy. However, despite recent achievements, a simple strategy to TiO$_2$-based semiconductors for efficient visible-light-driven photocatalytic hydrogen production is still challenging. Indeed, the successful photocatalyst system should fulfill all the requirements simultaneously: (i) narrowed band gap for visible light response; (ii) delicately designed band edge positions to realize photocatalytic redox reaction; and (iii) high efficiency for charge-carrier separation to promote photocatalytic activity.

Here we present a direct hydrolysis route to sub-10 nm rutile TiO$_2$ nanoparticles for efficient photocatalytic hydrogen production under visible light irradiation. The simple strategy leads to state-of-the-art photocatalytic activity among TiO$_2$-based semiconductors, and the simplified rutile TiO$_2$ semiconductor system can provide information on the essence of defect-induced visible light photocatalytic activity.

Results

Preparation and characterization of rutile TiO$_2$ samples.
Anatase and rutile TiO$_2$, both with tetragonal structure, are commonly used in photocatalytic reactions. Rutile TiO$_2$ has a band gap $ca.$ 0.2 eV lower than that of anatase (3.0 versus 3.2 eV), and this could be crucial to the band-gap narrowing its working spectrum to the visible light region. Bulk rutile TiO$_2$ can be obtained via the high temperature calcination of anatase TiO$_2$ at temperatures higher than 773 K, while rutile TiO$_2$ nanostructures can be prepared via a hydrothermal route[33,34] or a direct hydrolysis route[35-37]. In the present work, the direct hydrolysis of TiCl$_4$, free of any additives, is employed to prepare rutile TiO$_2$ nanoparticles.

Experimentally, the hydrolysis of TiCl$_4$ solution (TiCl$_4$/H$_2$O = 1:3 v/v) can produce rutile TiO$_2$ nanoparticles below 10 nm. We need to emphasize that the rapid hydrolysis and evaporation not only ensure the formation of rutile TiO$_2$ nanoparticles instead of nanorods in the presence of concentrated chloride ions but also create abundant defects in the as-obtained rutile TiO$_2$ nanoparticles (vide infra).

As shown in Fig. 1a, the sample prepared via hydrolysis route followed by calcination at 473 K, T-1, and the reference sample prepared via hydrothermal route followed by calcination at 473 K, T-3, give the typical diffraction patterns of rutile TiO$_2$ (JCPDF#21-1276), indicating that pure rutile phase is obtained. The crystallinity of T-1 is relatively low and the crystallite size is calculated to be $ca.$ 9 nm according to the Scherrer equation from the broadening of rutile (110) reflection. Calcination at 673 K can, on the one hand, greatly enhance the crystallinity of rutile TiO$_2$ but result in a significant increase in the crystallite size of T-2 (Table 1). Raman spectra of the samples with relative high crystallinity, that is, T-2 and T-3, show three Raman-active modes of multi-proton process (230 cm^{-1}), E_g (440 cm^{-1}) and A_{1g} (610 cm^{-1}), corresponding to the tetragonal space group of $P4_2/mnm$[38]. For T-1, an obvious blue shift and broadening of the E_g band (~ 25 cm^{-1}, Fig. 1b) is clearly observed due to the photon-confinement effects induced by defects[23,39].

X-ray photoelectron spectroscopy (XPS) analysis is performed to study the surface and sub-surface chemical states (in the depth up to 3 nm) of rutile TiO$_2$ samples. In Ti $2p$ spectra (Fig. 1c), two peaks at binding energy of 458.6 and 464.1 eV, assignable to $2p_{3/2}$ and $2p_{1/2}$, respectively, of Ti^{4+} in TiO$_2$ (ref. 40), are observed for all rutile TiO$_2$ samples, indicating the identical chemical state of Ti atoms in these samples. No obvious Ti^{3+} signals could be observed in Ti $2p$ spectra. In O 1-s spectra (Fig. 1d), a well-formed peak at 529.6 eV and two shoulders at 531.7 and 533.1 eV are observed for samples prepared by hydrolysis. The binding energy value of 529.6 eV is attributed to the lattice oxygen in TiO$_2$ (ref. 23), while the binding energy values of 531.7 and 533.1 eV are attributed to bridging hydroxyls and physisorbed water, respectively[41,42]. Hydrothermal synthesized sample T-3 exhibits a week peak shape at 531.7 eV corresponding to bridging hydroxyl groups, revealing its fine crystal structure with few surface defects. The presence of bridging surface hydroxyls in rutile TiO$_2$ samples is further supported by Fourier transform infrared spectroscopy (FTIR) with the observation of strong broad band located at 3,425 cm^{-1} (Supplementary Fig. 1). It is interesting to note that the percentages of hydroxyls in total oxygen species for T-1 is slightly higher than that of T-2, both much higher than that of

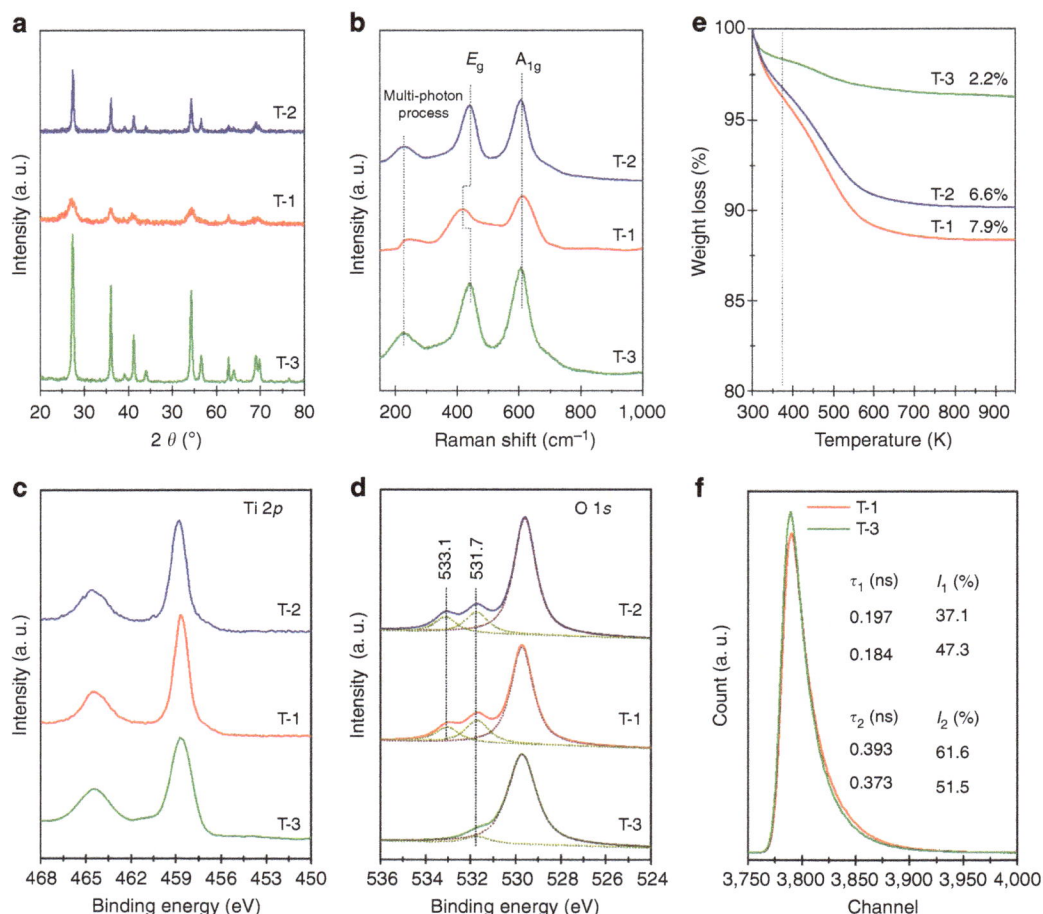

Figure 1 | Spectroscopy charatcerization of rutile TiO₂ samples. (**a**) XRD patterns of rutile TiO₂ samples. (**b**) Raman spectra of rutile TiO₂ samples. (**c,d**) Ti 2p and O 1s XPS of rutile TiO₂ samples. (**e**) Thermogravimetry analysis of rutile TiO₂ samples. (**f**) Positron annihilation lifetime spectra of rutile TiO₂ samples with corresponding positron lifetime and relative intensity in the inset.

Table 1 | Physico-chemical properties of TiO₂ samples under study.

Sample	Preparation strategy	Crystalline phase	Crystallite size (nm)		BET (m²g⁻¹)	OH groups (%)		Color
T-1	Hydrolysis at 373 K Calcined at 473 K	Rutile	9*	8†	101	15.7‡	18.8§	Light grey
T-2	Hydrolysis at 373 K Calcined at 673 K	Rutile	32*	27†	86	14.3‡	15.9§	White
T-3	Hydrothermal at 453 K Calcined at 473 K	Rutile	23*	16 × 70†	55	4.9‡	5.2§	White
T-4	Hydrothermal at 373 K Calcined at 473 K	Anatase	6*	7†	208	/	/	White

TEM, transmission electron microscopy; TG, thermogravimetry; XP, X-ray photoelectron; XRD, X-ray diffraction.
*Calculated by Scherrer equation from XRD patterns.
†TEM observations.
‡Calculated from O 1-s XP spectra.
§Calculated from TG analysis in the temperature range of 373–973 K.

hydrothermal-synthesized T-3 (Table 1). This is further confirmed by the thermogravimetry analysis in Fig. 1e, where weight losses of 7.9% and 2.2% in the temperature range of 373–973 K are observed for T-1 and T-3, respectively.

Positron annihilation is a well-established technique to study the defects in semiconductor materials and the lifetime of the positron is able to give information on the nature of various defects down to the p.p.m. level[43]. We have employed this technique to investigate the defects in rutile TiO₂ samples

prepared via direct hydrolysis and hydrothermal synthesis, and the results are shown in Fig. 1f. Two different fitted curves are clearly observed and the corresponding fitting parameters, lifetime, τ_1, τ_2 and τ_3, with relative intensity, I_1, I_2 and I_3, are summarized in the inset. The longest lifetime component τ_3 is attributed to the annihilation of orthopositronium atoms formed in very large voids presented in the material[44]. The lifetime components τ_1 and τ_2 have been demonstrated to indicate the existence of point defects in the bulk phase and the defects located on the surface of the samples, respectively[45,46]. The lifetime τ_1 and τ_2 of T-1 is comparable with that of T-3, which indicates the similar nature of surface and bulk defects in both samples, as the nature of defects can influence the local electron density and subsequently influence the positron lifetime[47]. The value of I_2/I_1, reflecting the intensity ratio of surface to bulk defects, is calculated to be 1.66 for T-1, obviously higher than the 1.09 of T-3, indicating the rich of surface defects in sub-10 nm rutile TiO_2 nanoparticles prepared via hydrolysis.

The morphology and structure of rutile TiO_2 samples are analysed by transmission electron microscopy (TEM), as shown in Fig. 2a. T-1 appears as aggregates of nanoparticles with an average diameter of $ca.$ 8 nm. Calcination at 673 K results in a distinct increase in the crystallite size of rutile TiO_2 from 8 to 27 nm, basically in agreement with X-ray diffraction results. The reference sample T-3 appears as nanorods with an average diameter of 16 nm and a length of 70 nm. For rutile TiO_2, (110) facet is the most stable facet with lowest formation energy[48] and it is observed as the dominant facet for all rutile TiO_2 samples, as illustrated by the lattice fringes with a spacing of 0.325 nm in the high-resolution TEM images (Fig. 2b). In addition, disordered structure induced by defects, can be observed for rutile TiO_2

samples prepared via direct hydrolysis (T-3). Structural analysis based on high-resolution TEM image is further performed to provide visualized information on the existence of defects in rutile TiO_2 samples under study (Fig. 2c). For T-3, clearly resolved and well-defined lattice fringes are observed and the distance between the adjacent lattice planes is equivalent to standard spacing, revealing the high crystallinity and the formation of fine crystals. For T-1, different types of defects, including intrinsic bulk defects (different distances between adjacent lattice planes) and surface hydroxyls (higher intensity trace, as marked with circle in Fig. 2c), can be clearly observed. Calcination at 673 K results in an increase in the crystallinity and the corresponding decrease in the defect degree. However, defects can still be observed, especially at the edges of nanoparticles (Fig. 2c), which is consistent with XPS results. Electron energy loss spectra (EELS) is further performed on T-1 and T-3, to give a qualitative interpretation of electronic states, as shown in Fig. 2d. For Ti $L_{2,3}$ edge, the Ti $3d$ character splits into two groups: the threefold t_{2g} and the twofold e_g orbitals, owing to the octahedral coordination with O atom forming s-type and π-type bonds[49]. The $t_{2g} - e_g$ splitting in Ti $L_{2,3}$ edge of T-1 and T-3 is quite similar, ruling out the existence of Ti with different electronic states, that is, Ti^{3+}[23]. A noticeable intensity decrease of the L_2 peak of T-1 should be originated from the existence of defects[50,51].

Band gap states of as-prepared rutile TiO_2 samples. The optical properties of as-prepared rutile TiO_2 samples are investigated by diffuse reflectance ultraviolet–visible spectroscopy and the results are shown in Fig. 3a. T-3 shows a band-edge absorption around 410 nm, typical for rutile TiO_2 with a band-gap energy of $ca.$ 3.0 eV. In contrast, T-1 shows obvious visible light absorption up to 600 nm, which should be originated from the existence of abundant defects. Calcination at 673 K will eliminate some of the defects (as discussed in the previous section) and therefore significantly reduce the absorption in visible light region. The plots of transformed Kubelka–Munk function versus the energy of light (Fig. 3b) give band-gap energies of 2.74, 2.84 and 2.98 eV for T-1, T-2 and T-3, respectively. Based on its light-absorption feature, the sub-10 nm rutile TiO_2 nanoparticles, T-1, should be an active photocatalyst under visible light irradiation.

To further address the relative band structure of rutile TiO_2 samples under study, the flat-band potential is measured using the electrochemical method in 0.5 M $NaSO_4$ solution (pH 6.8) and the Mott–Schottky plots are shown in Fig. 3c. The Mott–Schottky plots of all the rutile TiO_2 samples show a positive slope, which is typical for n-type semiconductors. The conduction band bottom of all rutile TiO_2 samples under study is quite similar ($ca.$ − 0.10 eV versus normal hydrogen electrode) according to the Mott–Schottky plots. Moreover, T-1 shows a relatively smaller slope than T-3, indicating a faster charge transfer and a higher donor density[29,52]. The carrier densities (N_d) of the samples can be calculated from the slopes of Mott–Schottky plots according to equation (1):[29]

$$N_d = 2/(e_0\varepsilon\varepsilon_0)\left[d(1/C^2)/dV\right]^{-1} \tag{1}$$

where e_0 is the electron charge (1.6×10^{-19} C), ε is the dielectric constant of rutile TiO_2, ε_0 is the permittivity of vacuum (8.86×10^{-12} F m^{-1}) and V is the applied bias at the electrode. The carrier densities are calculated to be 1.10×10^{18}, 0.95×10^{18} and 0.45×10^{18} cm^{-3} for T-1, T-2 and T-3, respectively. As the defect sites always act as the electron donors, a higher carrier density implies the existence of a larger amount of defects.

The valence band positions of rutile TiO_2 samples are determined by linear extrapolation of the leading edges of valence

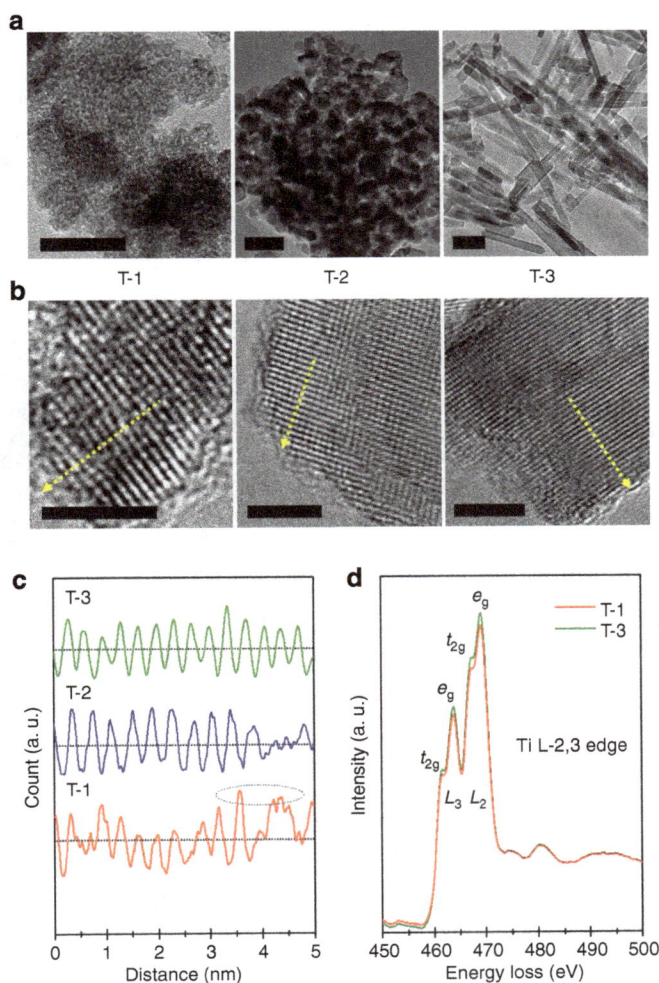

Figure 2 | Morphology and structure of rutile TiO_2 samples. (a) Overview TEM images of rutile TiO_2 samples. Scale bar, 50 nm. (b) High-resolution TEM (HRTEM) images of rutile TiO_2 samples. Scale bar, 4 nm. (c) Structural analysis based on HRTEM images, analysis region and orientation marked in the HRTEM images. (d) Ti $L_{2,3}$ edge EELS of rutile TiO_2 samples.

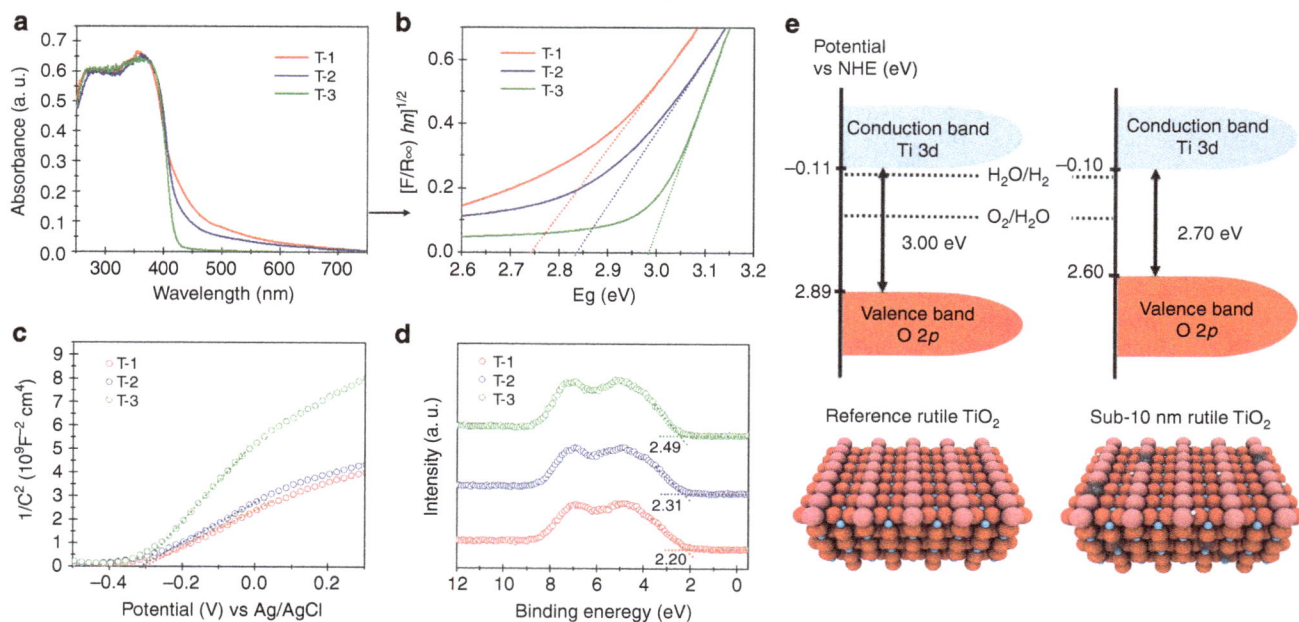

Figure 3 | Band-gap states of rutile TiO₂ samples. (a,b) Diffuse reflectance ultraviolet–visible spectra of rutile TiO_2 samples. (c) Mott–Schottky plots of rutile TiO_2 samples. (d) VB XPS of rutile TiO_2 samples. (e) Band energy diagram of reference (T-3) and sub-10 nm (T-1) rutile TiO_2.

band (VB) XPS spectra to the base lines. As shown in Fig. 3d, apparent valence band values of 2.20, 2.31 and 2.49 eV are observed for T-1, T-2 and T-3, respectively. The scatterings of data relative to the fittings are proved to be <0.05 eV by repeated experiments. After calibration with reference Fermi level, actual valence band values of 2.60, 2.71 and 2.89 eV versus normal hydrogen electrode can be obtained for T-1, T-2 and T-3, respectively. Compared with T-3, T-1 shows ca. 0.3 eV upward shift in the valence band top, which is the origin of its visible absorption.

To further clarify the influence of defects on the band structure of rutile, hybrid density functional theory (DFT) calculations were done with (2×2) rutile (110) slab models. The shifts of valence band top are summarized in Supplementary Table 1. The existence of surface defects, both bridging hydroxyls and bridging oxygen vacancy on rutile (110), indeed induces upward shift of valence band top. Moreover, higher hydroxyl coverage results in stronger shift, which agrees with the experimental observation that the valence band top of T-1, with more bridging hydroxyl, is higher in T-3. However, a qualitative comparison is not easy in this study due to the simplified model used in DFT, that is, (1) extended surface in calculation versus nanoparticle in experiment and (2) single-type defects in calculation versus coexisted multi-type defects in experiment.

Based on the results from Mott–Schottky plots (Fig. 3c) and VB XPS analysis (Fig. 3d), the band gaps are calculated to be 2.70, 2.82 and 3.00 eV for T-1, T-2 and T-3, respectively. These values are in good accordance with the band-gap values directly measured by ultraviolet–visible spectroscopy (Supplementary Table 2). The detailed band diagrams of reference and sub-10 nm rutile TiO_2 are depicted in Fig. 3e. Compared with reference T-3, sub-10 nm T-1 exhibits similar conduction band bottom but distinct higher valence band top. Therefore, a band-gap narrowing of ca. 0.3 eV could be achieved, which indicates that T-1 is a promising visible-light photoactive semiconductor. Moreover, the band-edge position reveals that the T-1 is suitable for photocatalytic hydrogen production, as its conduction band bottom is more negative than the reduction potential of H_2O/H_2 and its valence band top is more positive than the oxidation potential of O_2/H_2O.

Photocatalytic hydrogen production from water splitting. The photocatalytic activity of rutile TiO_2 samples (with 1 wt.% Pt as co-catalyst, Supplementary Fig. 2) was measured using the amount of hydrogen production from water splitting under different irradiations, and the time course of hydrogen evolution is shown in Fig. 4. Under ultraviolet light (Fig. 4a), T-1 exhibits the highest mass-specific activity (24.7 mmol h⁻¹ g⁻¹), followed by T-3 (15.4 mmol h⁻¹ g⁻¹) and then T-2 (9.2 mmol h⁻¹ g⁻¹). It is surprising to note that T-1 with more defects exhibits distinctly higher photocatalytic activity than T-2 with less defects, considering that defects are generally regarded as the charge-carrier trapping and recombination centres. Under visible light (400 nm $< \lambda <$ 780 nm, Fig. 4b), T-1 exhibits the highest mass-specific activity (932 µmol h⁻¹ g⁻¹), followed by T-2 (372 µmol h⁻¹ g⁻¹) and then T-3 (117 µmol h⁻¹ g⁻¹), consistent with their ability to absorb the visible light. Indeed, the hydrogen evolution rate achieved on T-1 is much greater, that is, more than five times higher, than other TiO_2-based photocatalysts ever reported. A direct comparison between TiO_2-based materials for visible-light-driven hydrogen production is summarized in Supplementary Table 3. Under the irradiation of full-spectrum simulator, that is, sunlight air mass T-1 also exhibits a remarkable photocatalytic activity with a hydrogen evolution rate of 1,954 µmol h⁻¹ g⁻¹, 4.2 and 5.6 times higher than that of T-2 and T-3, respectively (Fig. 4c). This value is also more than three times higher than reference TiO_2 P25. For a more objective comparison between the photocatalytic efficiencies for hydrogen production under visible light, the apparent quantum yield (QY) was calculated. The highest QY of 3.52% is obtained on T-1 sample under monochromatic light at $\lambda = 405$ nm, followed by 1.40% of T-2 and 0.36% of T-3, consistent with the trend in photocatalytic activity under visible light (Fig. 4b). Moreover, high QY of 1.74% can be obtained on T-1 sample under monochromatic light at $\lambda = 420$ nm. To our knowledge, these are the highest QYs ever reported for stable oxide semiconductors under comparable conditions[18,53], although they are still lower than other systems, for example, CdSe nanocrystals capped with dihydrolipoic acid[54] and Ni-decorated CdS nanorods[55].

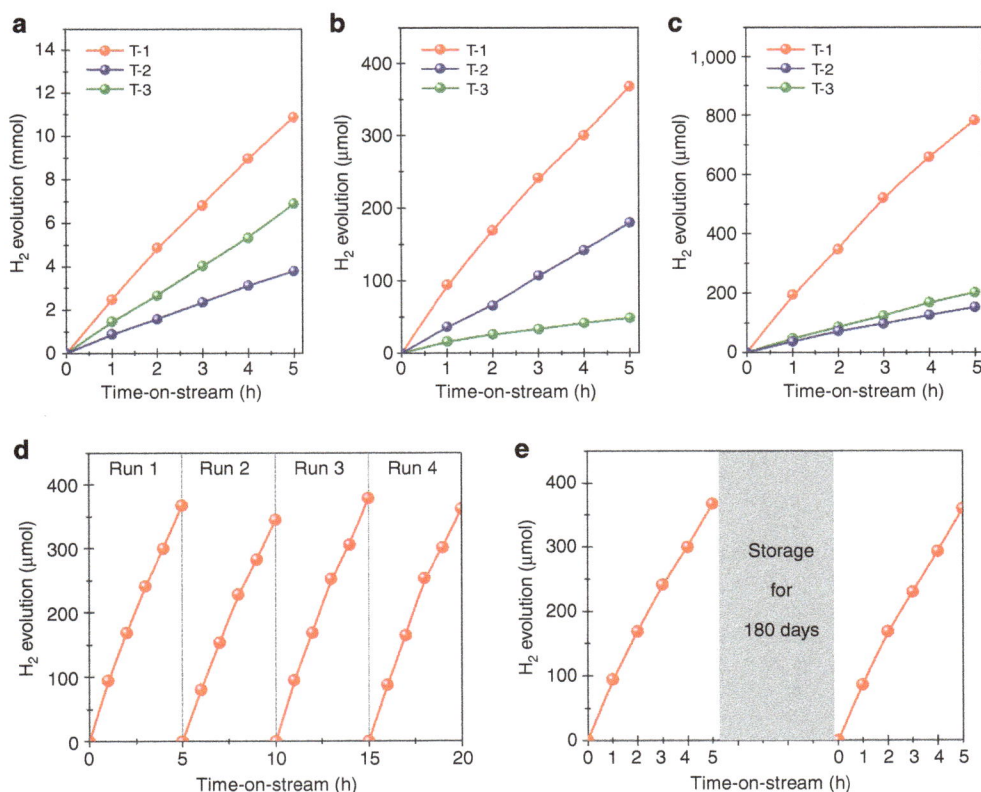

Figure 4 | Photocatalytic H₂ production over platinized rutile TiO₂. (**a**) Under ultraviolet light, 320 nm $< \lambda <$ 400 nm. (**b**) Under visible light, 400 nm $< \lambda <$ 780 nm. (**c**) Under sunlight air mass 1.5 irradiation. (**d**) Cycling tests of T-1 under visible light. (**e**) Photcatalytic activity of T-1 after long-term storage, under visible light. Pt (1 wt%) is *in situ* photo-deposited on the surface of samples as the co-catalyst and 10 ml methanol is used as sacrificial reagent for water splitting.

In addition to its remarkable photocatalytic activity, T-1 exhibits very good stability as a photocatalyst. As shown in Fig. 4d, no noticeable decrease in the activity for photocatalytic hydrogen production can be observed in the cycling tests. Moreover, no obvious activity loss can be observed for T-1 even after conventional storage (sealed in glass sample bottle with finger tight) for as long as 180 days (Fig. 4e). Therefore, the remarkable activity and stability make T-1 a promising semiconductor photocatalyst for hydrogen production from water splitting under visible light or solar light.

Functionalities of defects in rutile TiO₂ nanoparticles. For sub-10 nm rutile nanoparticles T-1, the presence of abundant defects is clearly revealed by Raman, FTIR, XPS, photoluminescence and positron annihilation lifetime spectroscopy. The absence of reduced Ti states, for example, Ti^{3+}, is confirmed by Ti 2p XPS (Fig. 1c) and Ti $L_{2,3}$-edge EELS (Fig. 2d). Therefore, the existence of oxygen vacancies or Ti interstitials, which are considered as major defects in TiO₂ contributing to band-gap states, could be expressly excluded. This is also supported by the fact that our rutile TiO₂ sample is exposed to ambient conditions with water and oxygen that would easily heal the oxygen vacancies and reduced Ti sites. Meanwhile, the presence of bridging hydroxyls is confirmed by O 1-s XPS (Fig. 1d) and FTIR analysis (Supplementary Fig. 1). As the bridging hydroxyls are known to come from the dissociative adsorption of water on oxygen vacancies[56,57], the detectable hydroxyls should indicate the ever-existing oxygen vacancies, together with Ti^{3+}. That is, oxygen vacancies could be created during the synthesis of sub-10 nm rutile TiO₂ nanoparticles via fast hydrolysis, but they immediately

react with water to form hydroxyls ($O_b - vac + O_b + H_2O \rightarrow 2OH_b$, where $O_b - vac$ means bridging oxygen vacancy, O_b means bridging oxygen and OH_b means bridging hydroxyl). In this context, hydroxyls should be the major defects in sub-10 nm rutile TiO₂ nanoparticles, also under ambient conditions, and the concentration of hydroxyls represents the defect degree in samples. The results from positron annihilation (Fig. 1f) and photoluminescence spectroscopy (Supplementary Fig. 3) further indicate the presence of similar types of surface defects, that is, hydroxyls, of rutile TiO₂ prepared via both hydrolysis and hydrothermal routes. Thus, the only difference in rutile TiO₂ samples under study lies in their defect degrees, which directly influence the amount of upward shift in the valence band top (Supplementary Table 1), and are responsible for the band-gap narrowing. The upward shift in the valence band top due to the existence of abundant defects in T-1 should be ascribed to the band bending[30], most probably associated with charge imbalance induced by the defect energy levels[58]. Calcination at elevated temperature will eliminate some of the defects in T-1 and, therefore, reduce the extent of upward shift in the valence band top of T-2 (Fig. 3d).

On the other hand, the defects in TiO₂ may greatly influence the efficiency of charge-carrier separation and the corresponding photocatalytic activity. The bulk defects will introduce charge-carrier trapping and recombination sites, and show degradation effects on the photocatalytic activity. In contrast, surface and sub-surface defects may contribute to charge-carrier separation, as the charges at surface or shallow traps become available for photocatalytic reaction. For most TiO₂-based semiconductor materials reported, for example, doped TiO₂, high defect degrees play a negative role on the photocatalytic activity, because the number of bulk defects is generally much higher than that of surface and sub-surface defects.

As discussed above, more surface and sub-surface defects are desired for band-gap narrowing, while less bulk defects are wanted to promote the charge carrier separation. In general, the degrading effects from bulk defects will negate the increased visible light absorption from band-gap narrowing induced by surface defects and low visible-light-driven photocatalytic activity will be observed. In this study, we present a new concept to solve the problem by reducing the size of TiO_2 nanoparticles. With decreasing particle size, the ratio of surface to bulk increases, indicating the increased ratio of surface/sub-surface to bulk defects (Fig. 1f). Band-gap broadening due to quantum size effect, expressed as equation (2), could be neglected with TiO_2 particle size of $> 3\,nm^{59,60}$.

$$\Delta E = \frac{h^2\pi^2}{2R^2}\left|\frac{1}{m_e} + \frac{1}{m_h}\right| - \frac{1.786e^2}{\varepsilon R} - 0.248E_{Ry}^* \qquad (2)$$

For T-1 with an average diameter of *ca.* 8 nm employed in this study (Fig. 2a), the positive effects from surface/sub-surface defects is strong enough to overcome the negative effects from bulk defects due to the relative high percentage of surface/sub-surface. This is well confirmed by its remarkable photocatalytic activity in hydrogen production under ultraviolet light (Fig. 4a).

Taking the functionalities of defects on both band-gap narrowing and charge-carrier separation into consideration, we successfully develop the sub-10 nm rutile TiO_2 nanoparticles for efficient visible-light-driven photocatalytic hydrogen production. The visible light response of sub-10 nm rutile TiO_2 nanoparticles is first created via the introduction of abundant defects and then promoted by the specific defect distribution in nanoparticles and the use of co-catalyst Pt as well. If rutile TiO_2 with smaller particles size can be synthesized, higher visible-light-driven photocatalytic activity could be expected.

Role of Ti^{3+} in photocatalytic hydrogen production. The role of Ti^{3+} in the visible-light photocatalytic activity of TiO_2-based materials has been intensively debated in the literature[17,20–24,61]. The major issue is that the appearance of Ti^{3+} is always accompanied by defects and, therefore, will inevitably compound the problem. For our as-prepared rutile TiO_2 samples, the existence of lower oxidation states of Ti, that is, Ti^{3+}, can be excluded based on XPS analysis. However, the presence of Ti^{3+} can be clearly observed for all rutile samples on ultraviolet irradiation, as confirmed by the distinct electron spin resonance (ESR) signal at $g_1 = 1.972$ and $g_2 = 1.949$ (Fig. 5a)[20,62]. We then measured the quasi *in situ* ultraviolet–visible spectra of T-1 and T-3 for further information. As shown in Fig. 5b, the presence of

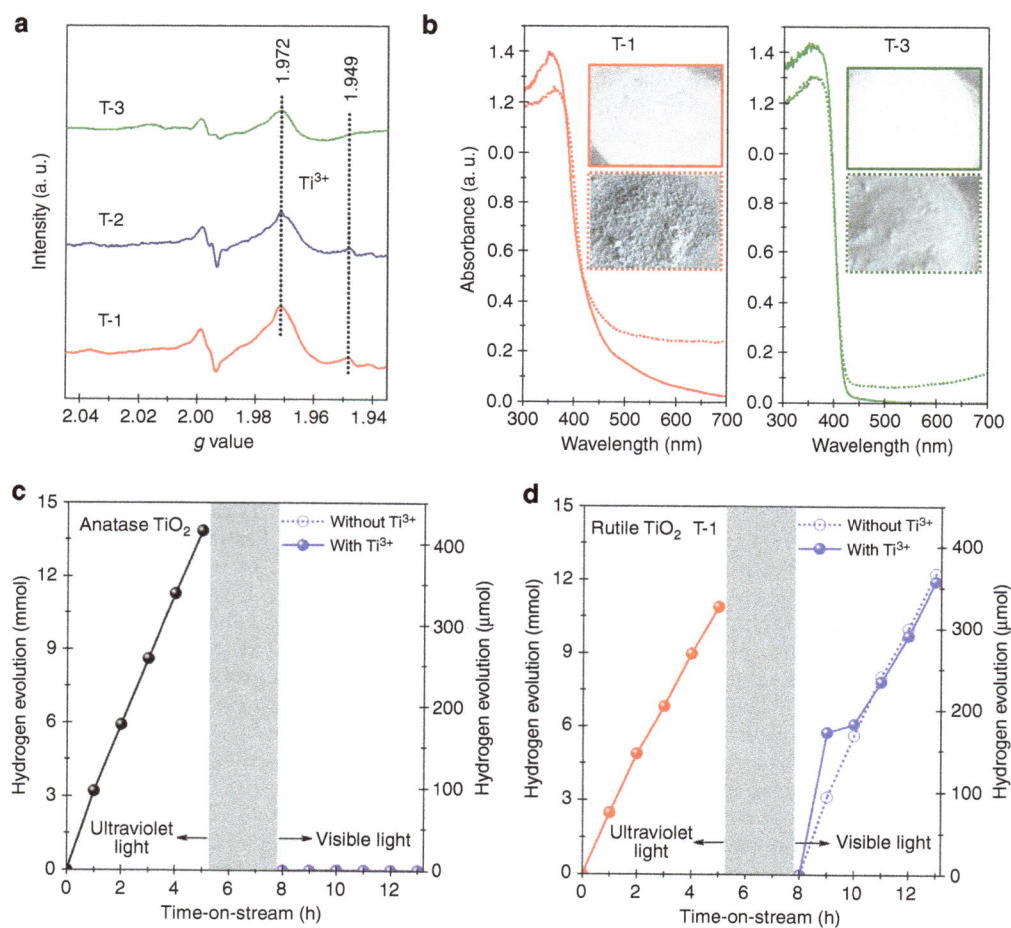

Figure 5 | Existence and role of Ti^{3+} in photocatalytic hydrogen production. (a) ESR spectra of rutile TiO_2 samples under ultraviolet irradiation. (b) Quasi *in situ* ultraviolet–visible spectra of platinized rutile TiO_2 during photocatalytic hydrogen production from water splitting under ultraviolet irradiation; solid line: ultraviolet–visible spectrum before irradiation; dashed line: ultraviolet–visible spectrum after ultraviolet irradiation for 5 min; inset: photographs of samples before and after 5 min reaction). (c,d) Step responses of photocatalytic activity of platinized anatase and rutile TiO_2 to the change of irradiation light from ultraviolet light ($400\,nm > \lambda > 320\,nm$) to visible light ($780\,nm > \lambda > 400\,nm$).

Ti^{3+} can be verified by the additional absorption in the visible light region for both T-1 and T-3 after 5 min time-on-stream ultraviolet irradiation to create a sufficiently detectable amount of Ti^{3+} (ref. 63). Judging from the intensity of absorption intensity, we propose that more Ti^{3+} species are presented in sub-10 nm rutile TiO_2 nanoparticles T-1 than reference T-3 on ultraviolet irradiation, which is clearly confirmed by the different intensities of Ti^{3+} signals in ESR spectra (Fig. 5a). That is, the presence of defects in rutile TiO_2 should facilitate Ti^{3+} formation during photocatalytic reaction under ultraviolet irradiation.

To obtain a direct evidence on the role of Ti^{3+} in the photocatalytic hydrogen production, step-response photocatalytic experiments are performed. Under ultraviolet light, anatase TiO_2 nanoparticles (Supplementary Fig. 4) are highly active in photocatalytic water splitting with a hydrogen evolution rate of $31.2 \, mmol \, h^{-1} \, g^{-1}$ (Fig. 5c), slightly higher than the $24.7 \, mmol \, h^{-1} \, g^{-1}$ from sub-10 nm rutile TiO_2 nanoparticles (Fig. 5d). During photocatalytic reaction under ultraviolet irradiation, the presence of Ti^{3+} is quite obvious with blue colouration (inset of Fig. 5b and Supplementary Fig. 5)[63,64]. After a 5-h photocatalytic reaction under ultraviolet radiation, the reaction system was either evacuated to preserve Ti^{3+} or exposed to oxygen and then evacuated to eliminate Ti^{3+}. In both cases, the anatase TiO_2 does not exhibit any detectable activity in subsequent photocatalytic hydrogen production tests under visible light (Fig. 5c). Therefore, it is rational to propose that the existence of Ti^{3+} itself will not initiate the visible light photocatalytic activity for hydrogen production. This is consistent with the fact that the band-gap state created by Ti^{3+} $3d^1$ state is about 1.0 eV below the conduction band of TiO_2 (ref. 32), which is obviously not suitable for H_2O reduction to hydrogen (Fig. 2e). Sub-10 nm rutile TiO_2, that is, T-1, is treated in the same manner and it exhibits remarkable photocatalytic activity under visible light after both treatments (Fig. 5d), which is ascribed to the surface defects, that is, hydroxyls, enabling the upward shift of valence band top of rutile TiO_2. We also observe that T-1 with a significant Ti^{3+} concentration exhibits much higher initial visible-light photocatalytic activity than that without Ti^{3+} (time-on-stream of 1 h, hydrogen evolution rate from 967 to $1,763 \, \mu mol \, h^{-1} \, g^{-1}$). The photocatalytic activity becomes almost the same after a time-on-stream of 2 h. The presence of Ti^{3+} can enhance the initial visible-light photocatalytic activity most probably due to the promoted methanol adsorption and/or the enhanced light absorbance and scattering[65], and the promotional effects gradually fade away with the elimination of Ti^{3+}. Consequently, we conclude that the presence of Ti^{3+} cannot initiate visible-light photocatalytic activity in hydrogen production, but Ti^{3+} is able to enhance the visible-light photocatalytic activity due to the promoted methanol adsorption.

Discussion

In summary, we have successfully developed a simple and scalable hydrolysis route to sub-10 nm rutile TiO_2 nanoparticles with diameters below 10 nm. We achieve a state-of-the-art hydrogen evolution rate of $932 \, \mu mol \, h^{-1} \, g^{-1}$ under visible light ($> 400 \, nm$) and $1,954 \, \mu mol \, h^{-1} \, g^{-1}$ under simulated solar light with platinized sub-10 nm rutile TiO_2 nanoparticles (1 wt.% Pt) among TiO_2-based semiconductors. Spectroscopic characterization results clearly confirm the existence of abundant surface defects, that is, hydroxyls, in the as-prepared rutile TiO_2 nanoparticles, which is responsible for the band-gap narrowing. The sub-10 nm particle size increases the percentage of surface/sub-surface defects compared with bulk defects, which results in enhanced charge-carrier separation. Moreover, we have revealed that the existence of defects in rutile TiO_2 is in favour of Ti^{3+}

formation during the photocatalytic reaction under ultraviolet irradiation. The presence of Ti^{3+} cannot initiate the visible-light photocatalytic activity, but it is able to enhance the visible-light-driven water splitting most likely due to the promoted reagent adsorption. The results presented here provide new insights into the functionalities of defects in visible-light photocatalytic activity. The concept of ultra-small nanoparticles should be useful for the future design of robust semiconductor photocatalysts with tunable band gap.

Methods

Preparation of rutile TiO₂. All of the chemical reagents of analytical grade were purchased from Alfa Aesar Chemical Co. and used as received without further purification. In a typical hydrolysis synthesis of rutile TiO_2, 10 ml titanium tetrachloride ($TiCl_4$) was dropwise added into 30 ml ice water under stirring to prepare a transparent $TiCl_4$ aqueous solution. After further stirring for 30 min, the $TiCl_4$ aqueous solution was rapidly heated to 373 K (within 5 min) to remove the water and hydrogen chloride (Note: Rapid heating is crucial to obtain rutile TiO_2 nanoparticles). The obtained white solid was thoroughly washed with deionized water, followed by drying in air at 353 K for 24 h and calcination in a muffle furnace at temperatures of 473 or 673 K for 2 h. The final products are denoted as T-1 and T-2, respectively. For reference, rutile TiO_2 was also prepared via a hydrothermal route. Typically, 10 ml of 1 M $TiCl_4$ aqueous solution was added to 50 ml water and the resulting solution was directly transferred into a 75-ml Teflon-lined autoclave for static crystallization at 453 K for 24 h. The resulting precipitates after crystallization were separated from the liquid phase by centrifugation, thoroughly washed with water, dried at 353 K for 24 h, subjected to calcination at 473 K for 2 h and denoted as T-3. For the synthesis of anatase TiO_2, $TiCl_4$ was dropwise added into ice water under stirring to prepare a $TiCl_4$ aqueous solution with concentration of $1 \, mol \, l^{-1}$. Next, 30 ml of $TiCl_4$ aqueous solution was mixed with 30 ml of KOH solution ($1 \, mol \, l^{-1}$) and the resulting solution was transferred into a 75-ml Teflon-lined autoclave for static crystallization at 373 K for 24 h. The resulting precipitates after crystallization were separated from the liquid phase by centrifugation, thoroughly washed with water, dried at 353 K for 24 h, subjected to calcination at 473 K for 2 h and denoted as T-4.

Characterization techniques. The specific surface areas of TiO_2 samples were determined through N_2 adsorption/desorption isotherms at 77 K collected on a Quantachrome iQ-MP gas adsorption analyser. The X-ray diffraction patterns of TiO_2 samples were recorded on a Bruker D8 ADVANCE powder diffractometer using Cu-Kα radiation ($\lambda = 0.1542 \, nm$) at a scanning rate of 4° per min in the region of $2\theta = 20–80°$. Raman analysis was carried out on a Renishaw InVia Raman spectrometer and the spectra were obtained with the green line of an Ar-ion laser (514.53 nm) in micro-Raman configuration. Diffuse reflectance ultraviolet–visible spectra of TiO_2 samples (ca. 20 mg diluted in ca. 80 mg BaSO₄) were recorded in the air against BaSO₄ in the region of 200–700 nm on a Varian Cary 300 ultraviolet–visible spectrophotometer. For the so-called quasi in situ ultraviolet–visible spectroscopy, 50 mg platinized TiO_2 (1 wt.% Pt) was mixed with 10 ml methanol aqueous solution (10%) in a home-made quartz reaction chamber, evacuated and then sealed for analysis. The diffuse reflectance ultraviolet–visible spectra against BaSO₄ were recorded before and after ultraviolet light irradiation for 10 min on the Varian Cary 300 ultraviolet–visible spectrophotometer. The concentration of surface hydroxyls in rutile TiO_2 was analysed by thermogravimetry on a Setram Setsys 16/18 thermogravmetric analyser. In a typical measurement, 0.1 g of rutile TiO_2 sample was heated in an Al_2O_3 crucible from 300 to 973 K with a constant heating rate of $10 \, K \, min^{-1}$ and under flowing Ar of $20 \, ml \, min^{-1}$. TEM images were taken on a FEI Tecnai G2 F20 electron microscope at an acceleration voltage of 200 kV and a Tecnai G2 F30 electron microscope at an acceleration voltage of 300 kV. A few drops of alcohol suspension containing the sample were placed on a carbon-coated copper grid, followed by evaporation at ambient temperature. EELS spectra of samples were collected using a GIF Tridiem 863 analyser. The lattice structural analysis was done using the Digital Micrograph software. XPS spectra of samples were recorded on a Kratos Axis Ultra delay line detector spectrometer with a monochromated Al-Kα X-ray source (hv = 1486.6 eV), hybrid (magnetic/electrostatic) optics and a multi-channel plate and delay line detector. All spectra were recorded using an aperture slot of $300 \times 700 \, \mu m$. Survey spectra were recorded with a pass energy of 160 eV and high-resolution spectra with a pass energy of 40 eV. Accurate binding energies ($\pm 0.1 \, eV$) were determined with respect to the position of the adventitious C 1s peak at 284.8 eV. VB XPS of samples were measured on PHI Quantera XPS Scanning Microprobe spectrometer using Al-Kα X-ray source (hv = 1486.6 eV). The energy scales are aligned by using the Fermi level of the XPS instrument (4.10 eV versus absolute vacuum value). Mott–Schottky plots were obtained using a three-electrode cell electrochemical workstation (IVIUM CompactStat). The saturated Ag/AgCl and platinum foil ($2 \times 2 \, cm^2$) were used as the reference electrode and the counter electrode, respectively. The sample of 1 mg TiO_2 was dispersed in 1 ml anhydrous ethanol and then evenly grinded to a slurry. The slurry was spread onto indium tin oxide glass and the exposed area was kept at $0.25 \, cm^2$. The prepared indium tin oxide/samples were dried overnight under ambient conditions and then used as the working electrode. The measurements were carried out at a fixed frequency of 1 kHz in 0.5 M Na_2SO_4 solution in the dark. Positron annihilation experiments were performed on a fast–slow coincidence ORTEC system with a time resolution of 187 ps full width at half maximum. The sample powder was pressed into a disk (diameter: 10.0 mm, thickness: 1.0 mm). A 5×10^5-Bq source of ^{22}Na was sandwiched between two identical sample disks. Measured spectra were analysed by the computer programme LT9.0, with source correction to evaluate the lifetime component τ_i and corresponding intensity I_i, using equation (3):

$$N(t) = \sum_{i=1}^{k+1} \frac{I_i}{\tau_i} \exp\left(-\frac{t}{\tau_i}\right) \qquad (3)$$

ESR was carried out on a JEOL JES-FA200 cw-EPR spectrometer (X-band) with a microwave power of 1.0 mW and a modulation frequency of 100 kHz. Mn-Marker was used as an internal standard for the measurement of the magnetic field. In a typical experiment, sample of 150 mg was placed in a quartz ESR tube and evacuated at room temperature for 2 h. After cooling down to 110 K, the tube was analysed under ultraviolet irradiation.

Photocatalytic hydrogen production from water splitting. Photocatalytic water splitting with methanol as sacrificial agent was performed in a top-irradiation-type Pyrex reaction cell connected to a closed gas circulation and evacuation system under the irradiation of Xe lamp (PLS-SXE, wavelength: 300–2,500 nm) with different optical reflector or filter (UVREF: 320–400 nm, ca. 83 mW cm^{-2}; UVCUT400: 400–780 nm, ca. 80 mW cm^{-2}) or Air Mass 1.5 (ca. 100 mW cm^{-2}). In a typical experiment, catalyst sample of 100 mg was suspended in ca. 100 ml 10% methanol aqueous solution (10 ml methanol and 90 ml H_2O) in the reaction cell. After being evacuated for 30 min, the reactor cell was irradiated by the Xe lamp at 200 W, at a constant temperature of 313 K under stirring. The gaseous products were analysed by an on-line gas chromatograph (Varian CP-3800) with a thermal conductivity detector. For all TiO_2 samples, 1 wt.% Pt was in situ photo-deposited from 1 M H_2PtCl_6 aqueous solution as co-catalyst, resulting a pH value of 6.5 in the reaction system. The apparent QY was measured using the same experimental setup for the photocatalytic hydrogen generation, but with additional band-pass filters to obtain monochromatic light at $\lambda = 405$ nm or $\lambda = 420$ nm (together with UVCUT400). The power density was measured to be ca. 2.0 ($\lambda = 405$ nm) or 1.5 mW cm^{-2} ($\lambda = 420$ nm) using a calibrated photodiode and the QY was calculated by equation (4):

$$QY = \frac{\text{Number of reacted electrons}}{\text{Number of incident photons}} * 100\% = \frac{\text{Number of evolved } H_2 \text{ molecules} * 2}{\text{Number of incident photons}} * 100\%$$
$$(4)$$

DFT calculations. DFT calculations were performed with the plane-wave-based Vienna ab initio simulation package VASP[66] at hybrid level in form of HSE06 (ref. 67). The (2×2) rutile $TiO_2(100)$ was represented by periodic slab models of four tri-TiO_2 layer thick. The valence wave functions were expanded in a plane-wave basis with a cutoff energy of 400 eV, while the interaction between the atomic cores and the electrons was described by the projector augmented wave method[68]. A Monkhorst–Pack mesh[69] of $2 \times 1 \times 1$ was used to sample the Brillouin zone during the geometry optimization (to 5×10^{-4} eV pm^{-1}), while the mesh was increased to $4 \times 2 \times 1$ in density of state analysis. The bands of different defect slabs were aligned to the same vacuum level.

References

1. Fujishima, A. & Honda, K. Electrochemical photolysis of water at a semiconductor electrode. *Nature* **238**, 37–38 (2008).
2. Hoffmann, M. et al. Environmental applications of semiconductor photocatalysis. *Chem. Rev.* **95**, 69–96 (1995).
3. Thompson, L. & Yates, J. Surface science studies of the photoactivation of TiO_2-new photochemical processes. *Chem. Rev.* **106**, 4428–4453 (2006).
4. Ravelli, D. et al. Photocatalysis. A multi-faceted concept for green chemistry. *Chem. Soc. Rev.* **39**, 1999–2011 (2009).
5. Hernandez-Alonso, M. et al. Development of alternative photocatalysts to TiO_2: challenges and opportunities. *Energy Environ. Sci.* **2**, 1231–1257 (2009).
6. Chen, X. & Mao, S. S. Titanium dioxide nanomaterials: synthesis, properties, modifications and application. *Chem. Rev.* **107**, 2891–2959 (2007).
7. Asahi, R. et al. Visible-light photocatalysis in nitrogen-doped titanium oxides. *Science* **293**, 269–271 (2001).
8. Khan, S. U., Al-Shahry, M. & Ingler, W. B. Efficient photochemical water splitting by a chemically modified n-TiO_2. *Science* **297**, 2243–2245 (2002).
9. Sakthivel, S. & Kisch, H. Daylight photocatalysis by carbon-modified titanium dioxide. *Angew. Chem. Int. Ed.* **42**, 4908–4911 (2003).
10. Park, J. H., Kim, S. & Bard, A. J. Novel carbon-doped TiO_2 nanotube arrays with high aspect ratios for efficient solar water splitting. *Nano Lett.* **6**, 24–28 (2006).
11. Hoang, S. et al. Visible light driven photoelectrochemical water oxidation on nitrogen-modified TiO_2 nanowires. *Nano Lett.* **12**, 26–32 (2002).

12. Zhao, W. et al. Efficient degradation of toxic organic pollutants with $Ni_2O_3/TiO_{2-x}B_x$ under visible irradiation. *J. Am. Chem. Soc.* **126**, 4782–4783 (2004).
13. Umebayashi, T. et al. Band gap narrowing of titanium dioxide by sulfur doping. *Appl. Phys. Lett.* **81**, 454–456 (2002).
14. Borgarello, E. et al. Visible light induced water cleavage in colloidal solutions of chromium-doped titanium dioxide particles. *J. Am. Chem. Soc.* **104**, 2996–3002 (1982).
15. Liu, M. et al. Energy-level matching of Fe(III) ions grafted at surface and doped in bulk for efficient visible-light photocatalysts. *J. Am. Chem. Soc.* **135**, 10064–10072 (2013).
16. Chen, X. & Burda, C. The electronic origin of the visible-light absorption properties of C-, N- and S-doped TiO_2 nanomaterials. *J. Am. Chem. Soc.* **130**, 5018–5019 (2008).
17. Hoang, S. et al. Enhancing visible light photo-oxidation of water with TiO_2 nanowire arrays via cotreatment with H_2 and NH_3: synergistic effects between Ti^{3+} and N. *J. Am. Chem. Soc.* **134**, 3659–3662 (2012).
18. Chen, X. et al. Semiconductor-based photocatalytic hydrogen generation. *Chem. Rev.* **110**, 6503–6570 (2010).
19. Justicia, I. et al. Designed self-doped titanium oxide thin films for efficient visible-light photocatalysis. *Adv. Mater.* **14**, 1399–1402 (2002).
20. Zuo, F. et al. Self-doped Ti^{3+} enhanced photocatalyst for hydrogen production under visible light. *J. Am. Chem. Soc.* **132**, 11856–11857 (2010).
21. Zuo, F. et al. Active facets on titanium (III)-doped TiO_2: an effective strategy to improve the visible-light photocatalytic activity. *Angew. Chem. Int. Ed.* **124**, 6223–6226 (2012).
22. Chen, X. et al. Increasing solar absorption for photocatalysis with black hydrogenated titanium dioxide nanocrystals. *Science* **331**, 746–750 (2011).
23. Naldoni, A. et al. Effect of nature and location of defects on bandgap narrowing in black TiO_2 nanoparticles. *J. Am. Chem. Soc.* **134**, 7600–7603 (2012).
24. Chen, X. et al. Properties of disorder-engineered black titanium dioxide nanoparticles through hydrogenation. *Sci. Rep.* **3**, 1510 (2013).
25. Zheng, Z. et al. Hydrogenated titania: synergy of surface modification and morphology improvement for enhanced photocatalytic activity. *Chem. Commun.* **48**, 5733–5735 (2012).
26. Hu, Y. Highly efficient photocatalyst-hydrogenated black TiO_2 for solar splitting of water to hydrogen. *Angew. Chem. Int. Ed.* **51**, 2–5 (2012).
27. Wang, Z. et al. Visible-light photocatalytic, solar thermal and photoelectrochemical properties of aluminium-reduced black titania. *Energy Environ. Sci.* **6**, 3007–3014 (2013).
28. Yang, C. et al. Core-shell nanostructured "Black" rutile titania as excellent catalyst for hydrogen production enhanced by sulfur doping. *J. Am. Chem. Soc.* **135**, 17831–17838 (2013).
29. Wang, G. et al. Hydrogen-treated TiO_2 nanowire arrays for photoelectrochemical water splitting. *Nano Lett.* **11**, 3026–3033 (2011).
30. Tao, J., Luttrell, T. & Batzill, M. A Two-dimensional phase of TiO_2 with a reduced bandgap. *Nat. Chem.* **3**, 296–300 (2011).
31. Ariga, H. et al. Surface-mediated visible-light photo-oxidation on pure TiO_2 (001). *J. Am. Chem. Soc.* **131**, 14670–14672 (2009).
32. Di Valentin, C. & Pacchioni, G. Electronic structure of defect states in hydroxylated and reduced rutile TiO_2 (110) surfaces. *Phys. Rev. Lett.* **97**, 166803 (2006).
33. Hosono, E. et al. Growth of submicrometer-scale rectangular parallelepiped rutile TiO_2 films in aqueous $TiCl_3$ solutions under hydrothermal conditions. *J. Am. Chem. Soc.* **126**, 7790–7791 (2004).
34. Aruna, S. T., Tirosh, S. & Zaban, A. Nanosize rutile titania particle synthesis via a hydrothermal method without mineralizers. *J. Mater. Chem.* **10**, 2388–2391 (2000).
35. Yurdakal, S. et al. Nanostructured rutile TiO_2 for selective photocatalytic oxidation of aromatic alcohols to aldehydes in water. *J. Am. Chem. Soc.* **130**, 1568–1569 (2008).
36. Li, Y., Fan, Y. & Chen, Y. A novel method for preparation of nanocrystalline rutile TiO_2 powders by liquid hydrolysis of $TiCl_4$. *J. Mater. Chem.* **12**, 1387–1390 (2002).
37. Wang, W. et al. Synthesis of rutile (α-TiO_2) nanocrystals with controlled size and shape by low-temperature hydrolysis: effects of solvent composition. *J. Phys. Chem. B* **108**, 14789–14792 (2004).
38. Ohsaka, T., Izumi, F. & Fujiki, Y. Raman spectrum of anatase, TiO_2. *J. Raman Spectr.* **7**, 321–324 (1978).
39. Gupta, S. K. et al. Titanium dioxide synthesized using titanium chloride: size effect study using Raman spectroscopy and photoluminescence. *J. Raman Spectr.* **41**, 350–355 (2010).
40. Scanlon, D. O. et al. Band alignment of rutile and anatase TiO_2. *Nat. Mater.* **12**, 798–801 (2013).
41. Sham, T. K. & Lazarus, M. S. X-ray photoelectron spectroscopy (XPS) studies of clean and hydrated TiO_2 (rutile) surfaces. *Chem. Phys. Lett.* **68**, 426–432 (1979).
42. Ketteler, G. et al. The nature of water nucleation sites on TiO_2 (110) surfaces revealed by ambient pressure X-ray photoelectron spectroscopy. *J. Phys. Chem. C* **111**, 8278–8282 (2007).

43. Krause-Rehberg, R. & Leipner, H. S. *Positron Annihilation in Semiconductors* (Springer, 1999).
44. Dutta, S. *et al.* Defect dynamics in annealed ZnO by positron annihilation spectroscopy. *J. Appl. Phys.* **98**, 053513 (2005).
45. Jiang, X. *et al.* Characterization of oxygen vacancy associates within hydrogenated TiO₂: a positron annihilation study. *J. Phys. Chem. C* **116**, 22619–22624 (2012).
46. Kong, M. *et al.* Tuning the relative concentration ratio of bulk defects to surface defects in TiO₂ nanocrystals leads to high photocatalytic efficiency. *J. Am. Chem. Soc.* **133**, 16414–16417 (2012).
47. Liu, X. *et al.* Oxygen vacancy clusters promoting reducibility and activity of ceria nanorods. *J. Am. Chem. Soc.* **131**, 3140–3141 (2009).
48. Henrich, V. E. & Cox, A. F. *The Surface Science of Metal Oxides* (Cambridge University Press, 1993).
49. Yaoshiya, M., Tanaka, I., Kaneko, K. & Adachi, H. First principles calculation of chemical shifts in ELNES/NEXAFS of titanium oxides. *J. Phys. Condens. Mater.* **11**, 3217–3228 (1999).
50. Wang, C. M. *et al.* Crystal and electronic structure of lithiated nanosized rutile TiO₂ by electron diffraction and electron energy-loss spectroscopy. *Appl. Phys. Lett.* **94**, 233116 (2009).
51. Diebold, U. The surface science of titanium dioxide. *Surf. Sci. Rep.* **48**, 53–229 (2003).
52. Su, F. *et al.* Dendritic Au/TiO₂ nanorod arrays for visible-light driven photoelectrochemical water splitting. *Nanoscale* **5**, 9001–9009 (2013).
53. Osterloh, F. E. Inorganic materials as catalysts for photochemical splitting of water. *Chem. Mater.* **20**, 35–54 (2008).
54. Han, Z. *et al.* Robust photogeneration of H₂ in water using semiconductor nanocrystals and a nickel catalyst. *Science* **238**, 1321–1324 (2012).
55. Simon, T. *et al.* Redox shuttle mechanism enhances photocatalytic H₂ generation on Ni-decorated CdS nanorods. *Nat. Mater.* **13**, 1013–1018 (2014).
56. Bikondoa, O. *et al.* Direct visualization of defect-mediated dissociation of water on TiO₂ (110). *Nat. Mater.* **5**, 189–192 (2006).
57. Pang, C. L., Lindsay, R. & Thornton, G. Chemical reactions on rutile TiO₂(110). *Chem. Soc. Rev.* **37**, 2328–2353 (2008).
58. Holt, D. B. & Yacobi, B. G. *Extended Defects in Semiconductors* (Cambridge University Press, 2007).
59. Brus, L. Quantum crystallites and nonlinear optics. *Appl. Phys. A* **53**, 465–474 (1991).
60. Kormann, C., Bahnemann, D. W. & Hoffmann, M. R. Preparation and characterization of quantum-size titanium dioxide. *J. Phys. Chem.* **92**, 5196–5201 (1988).
61. Xing, M. *et al.* Self-doped Ti³⁺-enhanced TiO₂ nanoparticles with a high-performance photocatalysis. *J. Catal.* **297**, 236–243 (2013).
62. Fenoglio, I., Greco, G., Livraghi, S. & Fubini, B. Non-UV-induced radical reactions at the surface of TiO₂ nanoparticles that may trigger toxic responses. *Chem. Eur. J.* **15**, 4614–4621 (2009).
63. Ma, J. *et al.* Photohole trapping induced platinum cluster nucleation on the surface of TiO₂ nanoparticles. *J. Phys. Chem. C* **118**, 1111–1117 (2014).
64. Yan, J. *et al.* Understanding the effect of surface/bulk defects on the photocatalytic activity of TiO₂: anatase versus rutile. *Phys. Chem. Chem. Phys.* **15**, 10978–10988 (2013).
65. Pang, C. L., Lindsay, R. & Thornton, G. Structure of clean and adsorbate-covered single-crystal rutile TiO₂ surfaces. *Chem. Rev.* **113**, 3887–3948 (2013).
66. Kresse, G. & Furthmüller, J. Efficient iterative schemes for *ab initio* total-energy calculations using a plane-wave basis set. *Phys. Rev. B* **54**, 169–186 (1996).
67. Krukau, A. V., Vydrov, O. A., Izmaylov, A. F. & Scuseria, G. E. Influence of the exchange screening parameter on the performance of screened hybrid functionals. *J. Chem. Phys.* **125**, 224106 (2006).
68. Blöchl, P. E. Projector augmented-wave method. *Phys. Rev. B* **50**, 17953–17979 (1994).
69. Monkhorst, H. J. & Pack, J. D. Special points for Brillouin-zone integrations. *Phys. Rev. B* **13**, 5188–5192 (1976).

Acknowledgements

We acknowledge the National Science Foundation of China (21006068, 21222604 and U1463205), Specialized Research Fund for the Doctoral Program of Higher Education (20120032110024), the Scientific Research Foundation for the Returned Overseas Chinese Scholars (MoE), the Ministry of Education of China (IRT-13R30 and IRT-13022) and the Program of Introducing Talents of Discipline to Universities (B06006 and B12015) for financial support.

Author contributions

J.G., L.L. and J.Y. conceived and designed the experiments, analysed the results and wrote the manuscript. J.G., L.L., J.Y. and T.W. performed the experiments and analysed the results. Z.Z. and J.G. carried out DFT calculations and analysed results. All authors contributed to the discussions of the results in this manuscript.

Additional information

Competing financial interests: The authors declare no competing financial interests.

Visualizing the orientational dependence of an intermolecular potential

Adam Sweetman[1], Mohammad A. Rashid[1], Samuel P. Jarvis[1], Janette L. Dunn[1], Philipp Rahe[1] & Philip Moriarty[1]

Scanning probe microscopy can now be used to map the properties of single molecules with intramolecular precision by functionalization of the apex of the scanning probe tip with a single atom or molecule. Here we report on the mapping of the three-dimensional potential between fullerene (C_{60}) molecules in different relative orientations, with sub-Angstrom resolution, using dynamic force microscopy (DFM). We introduce a visualization method which is capable of directly imaging the variation in equilibrium binding energy of different molecular orientations. We model the interaction using both a simple approach based around analytical Lennard–Jones potentials, and with dispersion-force-corrected density functional theory (DFT), and show that the positional variation in the binding energy between the molecules is dominated by the onset of repulsive interactions. Our modelling suggests that variations in the dispersion interaction are masked by repulsive interactions even at displacements significantly larger than the equilibrium intermolecular separation.

[1] School of Physics and Astronomy, University of Nottingham, Nottingham NG7 2RD, UK. Correspondence and requests for materials should be addressed to A.S. (email: adam.sweetman@nottingham.ac.uk).

The nature of intermolecular interactions underpins a vast array of physical and chemical phenomena, and is a scientific theme that straddles the disciplines of physics, chemistry and biology. Particular impetus has been given to the study of intermolecular forces at the single-molecule level due to the stunning advances in ultrahigh resolution scanning probe imaging pioneered by Gross et al.[1]. Three-dimensional (3D) force maps were acquired over planar organic molecules that bore a striking resemblance to the classic textbook 'ball-and-stick' models. These advances were first realized via the controllable functionalization of the scanning probe tip with a single pre-selected atom or molecule, which provides a unique level of control with which to investigate the atomic and molecular scale properties of matter, and also helps to eliminate the most troublesome aspect of scanning probe experiments, that is, the uncertainty surrounding the tip structure.

Although this tip functionalization strategy is now commonly applied to single CO molecules to allow intramolecular imaging[1,2], the technique has application well beyond imaging, and similar protocols have also been used to study the electronic[3,4] and mechanical[5] properties of single molecules trapped in the tip-sample junction, and to quantitatively measure intermolecular interactions[6-8]. There has also been considerable interest centred around the possibility of using this technique to directly visualize intermolecular interactions[9], although considerable debate surrounds the interpretation of these results[2,10-12].

In this paper, we discuss the results of a series of experiments—and their interpretation on the basis of both simple analytical potentials and DFT—that map the orientational dependence of the 3D potential between two-complex molecules. By measuring the full 3D potential we are able to apply a novel visualization method that directly shows the variation in the equilibrium binding energy for the molecular system for different relative orientations of the molecules. We also discuss the feasibility of detecting the variation in dispersion forces due to molecular rotation via DFM.

Results

Experimental results. Figure 1 shows representative constant height Δf images, taken from a 3D grid, acquired at decreasing tip-sample separation over three surface-adsorbed C_{60} molecules in different orientations, using a C_{60}-terminated tip.

At larger separations a featureless circular attractive interaction is observed (Fig. 1b), but on closer approach intricate intramolecular features are resolved (Fig. 1c,d), followed by their intense 'sharpening' (Fig. 1e,f). This evolution in contrast is similar to the onset of sub-molecular features during imaging of planar molecules with flexible tips[1,12]. However, because in this experiment both molecules have a complex structure, the intramolecular features in these images cannot be easily assigned to the molecular structure of the surface molecule as is the case for images taken with simple (that is, atomic point-like) tip terminations.

Converting the acquired Δf grid into a map of potential allows us to create similar constant height images of the tip-sample force and potential (Fig. 2a,b and Supplementary Fig. 1). Although constant height slices of force and energy provide the closest visual analogue to how the data are collected, these images necessarily conflate the value of the tip-sample energy and the topographic height of the molecule at a given position. Consequently, topographically higher features dominate the constant height image due to their being effectively shifted in z, even if the range of energies at these positions is identical to other locations over the molecule.

Representative single $U(z)$ curves may be extracted (Fig. 2c) and allow a selection of the energy minimum values at different positions to be observed, but this is an indirect, and not necessarily intuitive, method of analysing the variation in intermolecular potential across the molecules.

In Fig. 2d, we instead show an image constructed by searching each vertical column in the 3D data set (that is, each $U(z)$ curve) for the value of the potential energy minimum, and then projecting this minimum value over the xy plane of the grid, which we hereafter refer to as a 'U_{min}' image. This provides an

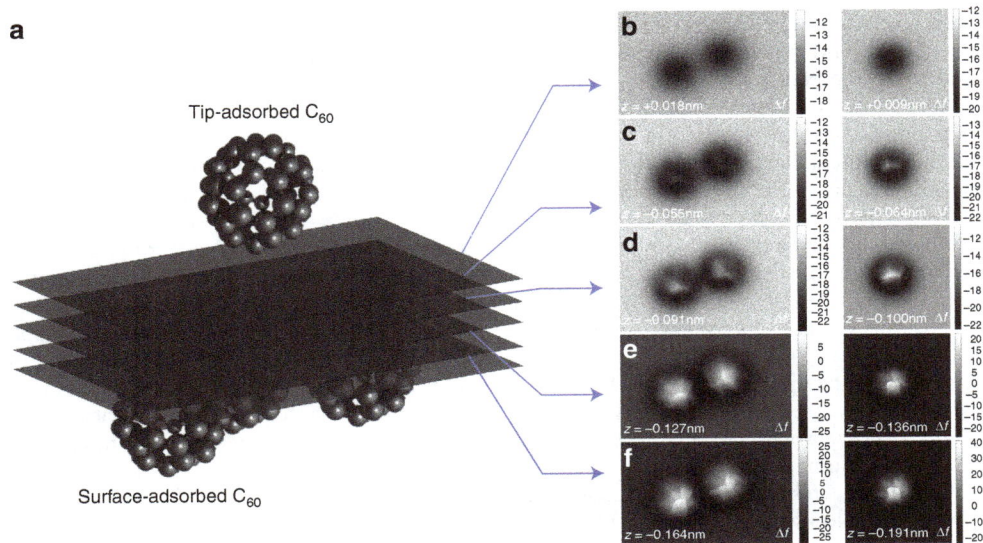

Figure 1 | Experimental data acquisition protocol. (a) cartoon showing the method of data acquisition for 3D potential mapping—a single C_{60} molecule is attached to the tip of the scanning probe microscope and brought close to a group of surface-adsorbed molecules. Constant height scans are acquired at decreasing tip-sample separation, with active drift compensation between each scan, and the variation in the frequency shift Δf measured. **(b-f)** representative Δf images (in Hz) at decreasing tip-sample height. Tip-sample heights shown for each image are given relative to the Δf set point used for atom-tracking over the molecule. The slightly different z heights for the two data sets result from the slightly different tracking heights used in each case. Image sizes: $3.5 \times 2.2 \, nm^2$ and $2.5 \times 2.5 \, nm^2$.

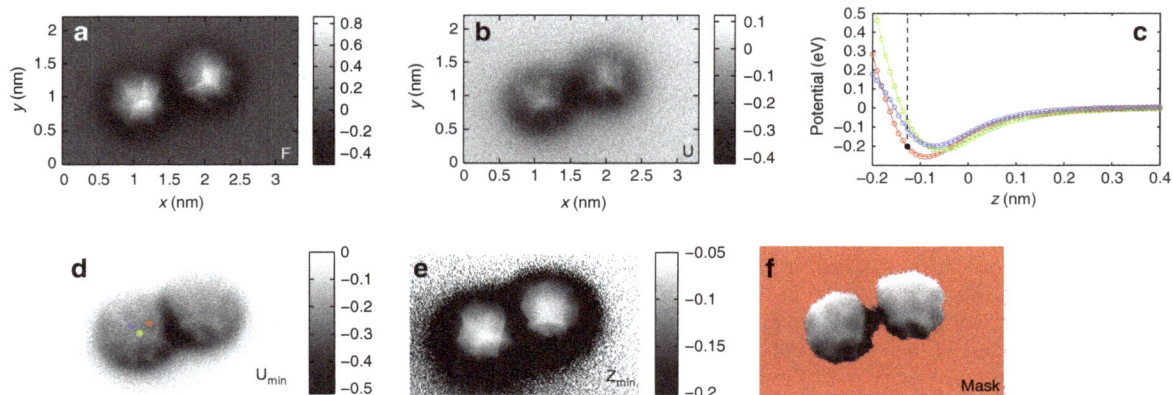

Figure 2 | Experimental measurement of variation in potential between C_{60} molecules in different orientations. Constant height images of (**a**) force (in nN) and (**b**) energy (in eV). (**c**) Representative $U(z)$ curves taken at different positions across the left hand C_{60} molecule, dotted line shows the height of the force and energy slices shown in **a,b**. (**d**) Image showing the variation in the value of the energy at the minimum in the $U(z)$ curve (in eV) at each position in the grid. The positions of the curves shown in **c** are marked. (**e**) As for **d** but showing instead the z height at which the minimum occurs, note that the black regions indicate parts of the grid where the minimum is found at the lowest tip-sample separation (that is, no turnaround detected). (**f**) Variation in energy minimum masked using the minimum in z position. Red shading indicates locations where the minimum in the intermolecular potential is not present in the $U(z)$ curve.

immediate and intuitive way of visualizing the strength of the equilibrium interaction as the relative position of the tip- and sample-adsorbed molecules is varied. We note that, because of the near-unique high-rotational symmetry of the C_{60} molecules, displacements in the xy plane should be equivalent to changes in rotational orientation.

We highlight here that some care must be taken in the interpretation of these images, as the value of the minimum in the potential energy curve only has a directly interpretable physical meaning when the actual minimum of the potential is present in a given $U(z)$ curve (that is, the turn-around in the $U(z)$ curve is present in the data set). If the minimum is not reached, then the closest point of approach will usually be identified as the minimum value. We therefore also map the height of the potential energy minimum in terms of z, which yields a complementary map of z_{min}. By masking the U_{min} map with the z_{min} map we can exclude those curves which do not contain the $U(z)$ turn-around, and visualize only the region of the image, which can be interpreted directly as representing the intermolecular interaction minimum (Fig. 2f). Application of this visualization technique also reveals a gradient in the value of the minimum in the potential across the molecule, most likely related to an asymmetric mounting of the molecule on the tip. Since this gradient directly affects the spread in the energy values we therefore only discuss the variation observed in the region located over the centre of the molecule, where the variation due to the gradient is small compared to the variation produced by the changes in molecular orientation (Supplementary Figs 16–18).

The same technique may also be applied to the 3D force field and raw 3D Δf measurements (see Supplementary Fig. 3). Although these maps do not have such a direct physical interpretation as for the minimum in the potential, they still provide an extremely powerful technique for visualizing the relative interaction over the molecule. Interestingly, we note a strong qualitative similarity in the appearance of these images and recent data acquired using a profile-corrected constant height technique by Moreno et al.[13]. We also note that Mohn et al.[14] recovered a pssuedo-topographic Δf image from a 3D data set, and experimentally it has been shown how to operate in the $\Delta f = 0$ regime[15], which might, in principle, produce similar imaging if applied to intermolecular measurements. Critically,

however, none of these earlier works directly measured and visualised the physical quantity of interest here: the variation in the value of the minimum in the intermolecular potential.

Our data demonstrate that as the relative orientations of the tip and surface C_{60} molecules are varied the potential minimum between the two molecules varies of the order 60 meV. A key question is therefore—what is the origin of this variation? A common approach to evaluating the C_{60}–C_{60} intermolecular interaction is to model molecular energy variation using the Girifalco potential[7,16], but this simplified model assumes a uniform spherical interaction, and does not give any information about sub-molecular variation in the potential. In particular, given the extended 3D nature of the molecule, it is not immediately clear how the attractive and repulsive components of the intermolecular potential contribute to the variation in the magnitude, and position, of the energy minimum. Following recent studies investigating the variation in dispersion force as a function of molecular size[6], there is also an open question as to whether the difference in the dispersion interaction can be observed for changes in the orientation of extended molecules. C_{60}, with its near-spherical symmetry represents a particularly important test bed for this hypothesis.

Computational results. To interpret our results, we modelled our experimental system with two different approaches (as described in the Methods section). First, we used a simple Lennard–Jones (L-J) potential for two C_{60} molecules, coupled with a modified version of the flexible tip model introduced by Hapala et al.[12], to simulate the C_{60} interaction. The simple nature of the model means that it is computationally inexpensive and thus can be exploited to generate high-resolution 3D data sets of comparable data density to those we obtain experimentally (Fig. 3d–g). Second, to test the validity of our empirical model, we compare the results of the L-J calculations to simulations of the same C_{60}–C_{60} interaction performed using the *ab initio* CP2K DFT code. The significant computational cost of the *ab initio* simulations precludes the calculation of a full 3D grid as for the L-J simulations, and we therefore instead compare 2D xz slices taken across the centre of the molecule–molecule interaction (Fig. 3b,c). In this comparison, we modelled a prototypical high symmetry orientation (hexagon face on hexagon face, hereafter referred to as

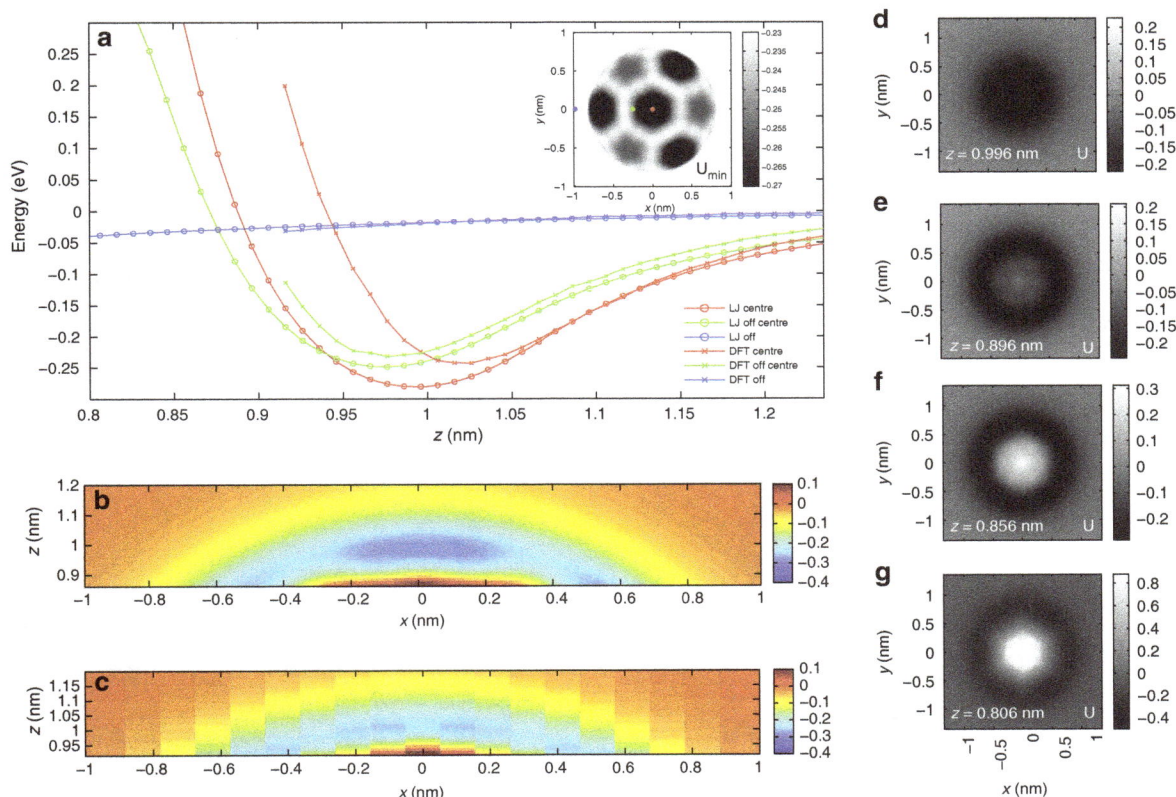

Figure 3 | Modelling the C_{60}–C_{60} interaction. (a) Comparison of potentials calculated by L–J modelling and DFT. z heights are defined as the vertical separation of the two molecules measured from the molecular centres. Curves are shown for the same initial xy coordinates in both simulations. Inset: Complete xy U_{min} (in eV) image for L–J simulation. **(b,c)** xz plot of calculated energies for L–J and DFT simulations, respectively. **(d–g)** Constant height xy energy images (in eV) from L–J simulations, relative heights labelled on each image.

Hex–Hex) for the two molecules using both simulations methods. In general, we find good agreement between the two techniques, noting in particular that the potential gradients in both the attractive and repulsive branches of the potential curve are very similar. We do, however, observe some quantitative differences between the empirical L–J simulations and *ab initio* DFT. Specifically, for the L–J parameters chosen ($\epsilon_a = 2.5$ meV and $r_a = 1.966$ Å (ref. 16)), the maximum depth and width of the well is slightly larger that for the DFT simulations, as is the variation in the range of U_{min} (ΔU_{min} for L–J xz plot ~ 50 meV compared with ~ 20 meV for the DFT xz plot over the same range (Supplementary Figs 20–22)). We note, however, that variation in the positions of the minimum is almost identical (ΔZ_{min} for L–J xz plot ~ 0.09 nm, compared with $\Delta Z_{min} \sim 0.11$ nm for the DFT xz plot). Furthermore, it is clear that tuning the choice of L–J parameters based on the DFT results could improve the quantitative agreement between the two simulation methods, but here we prefer to use those L–J parameters derived from previous experimental work and which are also consistent with our earlier publications, rather than arbitrarily adjusting the L–J parameters. These results imply that while the L–J model is a simplification of the complex intermolecular interaction, it nonetheless appears to be sufficient to model much of the essential physics underpinning the variation in intermolecular potential.

Although we stress that the high-symmetry Hex–Hex configuration used in the simulations is not the configuration of the C_{60} molecules in the experimental data set shown in Fig. 2, we nonetheless observe a number of qualitatively similar features in both the simulations and experiment. In particular, the simulations reproduce the 'sharpening' of the features observed in

the constant-height experimental images, in line with the sharpening reported for CO-terminated tips. In addition, the appearance of the simulated U_{min} image is qualitatively similar to that acquired in the experiment, which reveals the complex variation in potential minimum as the molecular positions are varied. Interestingly, the L–J simulations overestimate the depth of the potential relative to the DFT calculations, but better reproduce the variation in U_{min} observed experimentally, with a variation of ~ 50 to 60 meV in the U_{min} image depending on molecular orientation (Supplementary Figs 16–18). We also note that simulations performed with other tip-sample molecular configurations, such as those found for C_{60} adsorption on the Si(111)-7×7 substrate, produce much more complex patterns in the constant height, and U_{min}, images (see Supplementary Figs 7–13), qualitatively similar to those observed experimentally.

Discussion

Because of the simple additive, and analytical, nature of the L–J model, it is possible to decompose the interaction into its attractive and repulsive components, and ascertain if we might in principle be able to observe rotational variation in the dispersion interaction between the C_{60} molecules. To assess the relative influence of the repulsive and attractive elements of the potential on the value of the potential minimum we investigated the change in the potential for several orientations of the tip and sample C_{60} (Fig. 4a). We then plot the modulus (i.e. the absolute value) of the differences in the total energies, and the separate energies from the r^6 and r^{12} terms, between these orientations and the high symmetry 'Hex–Hex' configuration (Fig. 4b), and then extract the differences in these ΔU_{r^6} and $\Delta U_{r^{12}}$ terms (Fig. 4c). Specifically,

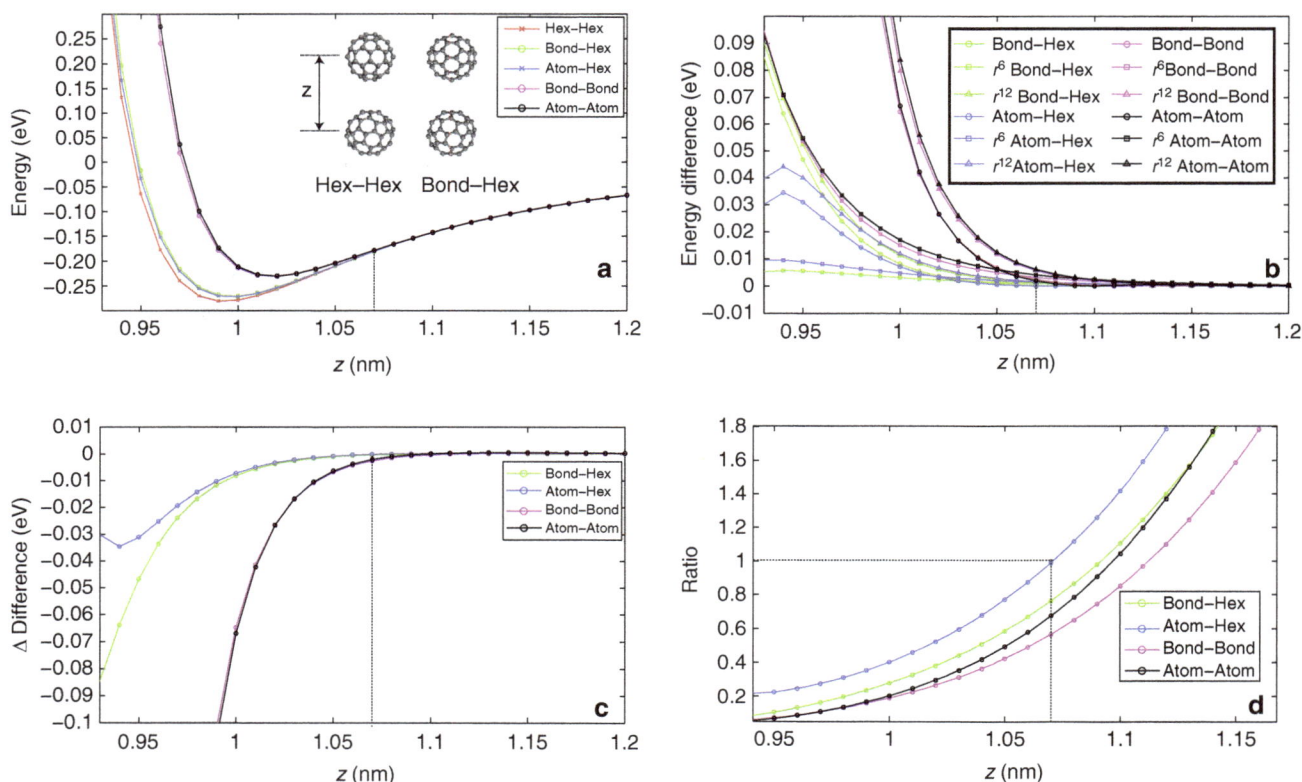

Figure 4 | Investigating the origin of the variation in potential energy minimum. (a) Variation in potential energy between two C_{60} molecules due to variation in rotational orientation, the terms in the legend refer to the facing part of the molecule. Inset, representative ball-and-stick models showing two of the simulated orientations. **(b)** Modulus of the difference in U_{total}, U_{r^6}, and $U_{r^{12}}$, between stated orientation and the Hex–Hex orientation. **(c)** Difference of variation in U_{r^6} and $U_{r^{12}}$ terms (that is, $|\Delta U_{r^6}| - |\Delta U_{r^{12}}|$). Negative values indicate the $U_{r^{12}}$ term dominates the difference in energy between the two configurations. **(d)** Ratio of variation in U_{r^6} and $U_{r^{12}}$ terms. Values above 1 indicate the U_{r^6} term has a larger contribution. Dotted lines indicate the separation at which the U_{r^6} term first becomes dominant.

we define $|\Delta U_{r^6}| - |\Delta U_{r^{12}}|$ as the modulus of the difference of the ΔU_{r^6} term (between the stated orientation and the Hex–Hex orientation), minus the modulus of the difference in the $\Delta U_{r^{12}}$ terms (between the stated orientation and the Hex–Hex orientation).

This quantity gives the relative influence of the two terms in defining the difference in the energy curves between the two orientations. If the difference in r^6 terms at a given separation is greater (that is, dispersion forces vary significantly between different orientations), then this quantity will be positive. If the difference in r^{12} terms is large (that is, Pauli forces vary significantly between different orientations), then it will be negative. If both quantities contribute equally to the difference in energy, then the term will be approximately zero. Surprisingly, we observe that the difference is negative and, consequently, the differences observed in the total energies, even in the part of the well where the potential gradient is positive, are dominated by repulsive interactions. Here we wish to make it explicitly clear that, for intermolecular separations greater than the equilibrium value (that is, before the energy 'turn-around'), the magnitude of the r^6 term is indeed larger in all cases, and dominates the r^{12} term, but that contribution of the r^6 term is very similar for all the orientations.

It must be noted, however, that the interplay between the two terms is somewhat subtle. If we examine the ratio of the differences (Fig. 4d), then it is clear that the r^6 term does begin to dominate the difference in the energies at around 1.06–1.1 nm separation. However, by reference to (Fig. 4c) it becomes clear that at this separation the difference in the potentials is < 5 meV, that is, below our experimental sensitivity. Therefore, our

modelling suggests that at the point at which the potential curves for different orientations become experimentally distinguishable, the difference between them is dominated by repulsive, rather than dispersive, interactions. As such it seems likely that although the magnitude of the variation in energy due to the variation in dispersion interaction under rotation of the molecule might in principle be within the noise limit of current DFM techniques, its direct measurement will always be hindered by the intrinsic convolution of the variation in energy due to repulsive forces, even at intermolecular separations significantly greater than the equilibrium value, where the gradient in the potential is positive.

We have presented 3D mapping of the variation in intermolecular interaction under changes in rotational orientation of a complex molecule with sub-Angstrom resolution via the functionalization of a scanning probe tip. Using a novel visualization method we can directly observe the variation in value and position of the minimum of the potential energy as the orientation of the molecules is varied. By comparison of our results to both simple analytical and *ab initio* simulations, we are able to show that the variation in binding energy across the molecule is dominated by the onset of repulsive interactions between the front-most parts of the molecules. Surprisingly, we also find that variation in the net attractive part of the potential due to rotation of the molecules is still dominated by the repulsive forces, and the majority of the molecule only adds a uniform background to the potential. We anticipate that similar experimental techniques to those described here could be utilized to intuitively visualise the reactivity across complex interatomic and intermolecular potentials, including molecules with polarized or hydrogen bonding end groups.

Methods

Experimental methods. Clean Si(111)-7 × 7 surfaces were prepared by flash annealing a silicon wafer to 1,200 °C, rapid cooling to 900 °C and then slow cooling to room temperature. A low coverage of C_{60} was prepared by depositing the molecules from a home made tantalum pocket deposition source onto the room temperature substrate. Post-deposition, the sample was transferred into the scan head of an Omicron Nanotechnology LT DFM operating in UHV at cryogenic temperatures, and left to cool to 5 K before imaging.

Commercial qPlus sensors (Omicron Nanotechnology GmbH) with electrochemically etched tungsten wire tips were introduced into the scan head without any further preparation. The sensors were first prepared on clean Si(111)-7 × 7 surfaces by standard STM techniques until good STM/DFM resolution was achieved. Single C_{60} molecules were transferred to the tip by close approach to surface-adsorbed molecules, and the functionalization of the tip was checked by inverse imaging of the tip adsorbed molecule on the surface adatoms (Supplementary Figs 2,4 and 5)[7]. In all experiments an oscillation amplitude (A_0) of 110 pm was used, and the tip-sample bias was set to 0 V. Three-dimensional Δf volumes over the molecules were collected via the 'slice' method[17] and site specific (short-range) Δf values were extracted using the 'on–off' method[18,19] then converted to potentials using the Sader–Jarvis algorithm[20]. Due to the long acquisition times required, residual thermal drift and piezoelectric creep were corrected using a custom atom-tracking and scripting setup[21,22]. Further details on the experimental setup, data processing steps and additional experimental data sets may be found in the Supplementary Methods.

Flexible tip model and simulated spectroscopy procedure. To simulate DFM images, we adapted the method proposed by Hapala *et al.*[12] to model the interaction between a sample and a CO-functionalized DFM tip. In our simulation the functionalized tip is assumed to consist of a tip base (outermost atom of the tip) and a probe. The probe is the flexible end of the model tip, and is allowed to move around the tip base. In our simulation, the probe is a C_{60} molecule consisting of 60 carbon atoms acting as a single effective probe particle attached to the tip base (Supplementary Fig. 6). Each atom in the probe experiences three forces; (i) a Lennard–Jones (L–J) force due to the tip base, (ii) a sum of all pairwise L–J forces due to interactions with atoms in the sample and (iii) a lateral harmonic force from the tip base. The net force on the probe is calculated by summing up all the forces experienced by each atom on the probe. The L–J interactions between atoms α and β are written as

$$\mathbf{F}_{\alpha\beta}(\mathbf{R}) = 12\epsilon_{\alpha\beta}\mathbf{R}\left(\frac{r_{\alpha\beta}^{12}}{r^{14}} - \frac{r_{\alpha\beta}^{6}}{r^{8}}\right) \quad (1)$$

$$U_{\alpha\beta}(r) = \epsilon_{\alpha\beta}\left(\frac{r_{\alpha\beta}^{12}}{r^{12}} - \frac{2r_{\alpha\beta}^{6}}{r^{6}}\right), \quad (2)$$

where $r = |\mathbf{R}|$ is the distance between atoms α and β, $\epsilon_{\alpha\beta} = \sqrt{\epsilon_\alpha \epsilon_\beta}$ is the pair-binding energy and $r_{\alpha\beta} = r_\alpha + r_\beta$ is the equilibrium separation of the two atoms with ϵ_α and r_α being the atomic parameters. In our calculations the L–J parameters for the carbon atoms were set to $\epsilon_\alpha = 2.5$ meV and $r_\alpha = 1.966$ Å to ensure consistency with the work of Girifalco *et al.*[16] and our own earlier work[7]. For the tip base a value of $r_\alpha = 5.0$ Å was chosen, in order to take into account the larger size of the C_{60} molecule. The probe lateral stiffness and apex L–J parameter were set to $k_{xy} = 0.5$ N m^{-1} and $\epsilon_\alpha = 1,000$ meV, respectively (Supplementary Figs 14–15). We acquired the simulation data by scanning the sample laterally with a step of $\Delta x, \Delta y = 0.1$ Å. At each lateral position we placed the tip base at an initial separation $z_0 = 22$ Å from the surface molecule and approached the sample (in our simulations another C_{60} molecule) in steps of $\Delta z = 0.1$ Å until $z = 17.5$ Å allowing the probe position to be relaxed at each step due to the combined force of the sample and tip base. Note, however, that for ease of comparison to the DFT simulations all molecular separations discussed in the paper are given relative to the initial vertical core-core separation of the probe C_{60} from the surface C_{60}.

Density functional theory. DFT calculations were performed using the same initial high symmetry geometry (as described in main text) as the L–J simulations using the open source CP2K/Quickstep code[23,24] utilising a hybrid Gaussian and plane-wave method[25]. Goedecker, Teter and Hutter pseudopotentials[26] and the Perdew Burke Ernzerhof generalized gradient approximation method[27] were used with a 300 Ry plane-wave energy cutoff. To account for dispersion interactions we employed the Grimme DFT-D3 method[28], which well reproduced the C_{60}–C_{60} pair potential (Supplementary Fig. 19) A double-zeta Gaussian basis set plus polarization (DZVP-MOLOPT)[29] was used with a force convergence criterion for geometry relaxation of 0.05 eV Å^{-1}. Geometry relaxation was carried out by allowing all atoms to relax other than the hexagonal faces of each molecule furthest apart from one another (to simulate attachment to the surface/tip).

References

1. Gross, L., Mohn, F., Moll, N., Liljeroth, P. & Meyer, G. The chemical structure of a molecule resolved by atomic force microscopy. *Science* **325**, 1110–1114 (2009).
2. Jarvis, S. P. Resolving intra-and inter-molecular structure with non-contact atomic force microscopy. *Int. J. Mol. Sci.* **16**, 19936 (2015).
3. Schull, G., Frederiksen, T., Arnau, A., Sanchez-Portal, D. & Berndt, R. Atomic-scale engineering of electrodes for single-molecule contacts. *Nat. Chem.* **6**, 23–27 (2010).
4. Schull, G., Frederiksen, T., Brandbyge, M. & Berndt, R. Passing current through touching molecules. *Phys. Rev. Lett.* **103**, 11–14 (2009; 0910.1281).
5. Hauptmann, N. *et al.* Force and conductance during contact formation to a C_{60} molecule. *New. J. Phys.* **14**, 073032 (2012).
6. Wagner, C. *et al.* Non-additivity of molecule-surface van der Waals potentials from force measurements. *Nat. Commun.* **5**, 5568 (2014).
7. Chiutu, C. *et al.* Precise orientation of a single C_{60} molecule on the tip of a scanning probe microscope. *Phys. Rev. Lett.* **108**, 268302 (2012).
8. Sun, Z., Boneschanscher, M. P., Swart, I., Vanmaekelbergh, D. & Liljeroth, P. Quantitative atomic force microscopy with carbon monoxide terminated tips. *Phys. Rev. Lett.* **106**, 46104 (2011).
9. Zhang, J. *et al.* Real-space identification of intermolecular bonding with atomic force microscopy. *Science* **342**, 611–614 (2013).
10. Sweetman, A. M. *et al.* Mapping the force field of a hydrogen-bonded assembly. *Nat. Commun.* **5**, 3931 (2014).
11. Hämäläinen, S. K. *et al.* Intermolecular contrast in atomic force microscopy images without intermolecular bonds. *Phys. Rev. Lett.* **113**, 186102 (2014).
12. Hapala, P. *et al.* Mechanism of high-resolution STM/AFM imaging with functionalized tips. *Phys. Rev. B* **90**, 085421 (2014).
13. Moreno, C., Stetsovych, O., Shimizu, T. K. & Custance, O. Imaging three-dimensional surface objects with submolecular resolution by atomic force microscopy. *Nano Lett.* **15**, 2257 (2015).
14. Mohn, F., Schuler, B., Gross, L. & Meyer, G. Different tips for high-resolution atomic force microscopy and scanning tunneling microscopy of single molecules. *Appl. Phys. Lett.* **102**, 073109 (2013).
15. Rode, S., Schreiber, M., Kühnle, A. & Rahe, P. Frequency-modulated atomic force microscopy operation by imaging at the frequency shift minimum: The dip-df mode. *Rev. Sci. Instrum.* **85**, 043707 (2014).
16. Girifalco, L. a. Interaction potential for carbon (C_{60}) molecules. *J. Phys. Chem.* **95**, 5370–5371 (1991).
17. Neu, M. *et al.* Image correction for atomic force microscopy images with functionalized tips. *Phys. Rev. B* **89**, 205407 (2014).
18. Lantz, M. A. *et al.* Quantitative measurement of short-range chemical bonding forces. *Science* **291**, 2580–2583 (2001).
19. Sweetman, A. & Stannard, A. Uncertainties in forces extracted from non-contact atomic force microscopy measurements by fitting of long-range background forces. *Beilstein J. Nanotechnol.* **5**, 386–393 (2014).
20. Sader, J. E. & Jarvis, S. P. Accurate formulas for interaction force and energy in frequency modulation force spectroscopy. *Appl. Phys. Lett.* **84**, 1801 (2004).
21. Abe, M. *et al.* Room-temperature reproducible spatial force spectroscopy using atom-tracking technique. *Appl. Phys. Lett.* **87**, 173503 (2005).
22. Rahe, P. *et al.* Flexible drift-compensation system for precise 3D force mapping in severe drift environments. *Rev. Sci. Instrum.* **82**, 063704 (2011).
23. Hutter, J, Iannuzzi, M., Schiffmann, F. & VandeVondele, J. CP2K: atomistic simulations of condensed matter systems. *WIREs Comput. Mol. Sci.* **4**, 1525 (2013).
24. VandeVondele, J. *et al.* Quickstep: Fast and accurate density functional calculations using a mixed gaussian and plane waves approach. *Comput. Phys. Commun.* **167**, 103–128 (2005).
25. Lippert, G., Jurg, H. & Parinello, M. A hybrid gaussian and plane wave density functional scheme. *Mol. Phys.* **92**, 477–488 (1997).
26. Goedecker, S., Teter, M. & Hutter, J. Separable dual-space gaussian pseudopotentials. *Phys. Rev. B* **54**, 1703–1710 (1996).
27. Perdew, J. P., Burke, K. & Ernzerhof, M. Generalized gradient approximation made simple. *Phys. Rev. Lett.* **77**, 3865–3868 (1996).
28. Grimme, S., Antony, J., Ehrlich, S. & Krieg, H. A consistent and accurate ab initio parametrization of density functional dispersion correction (dft-d) for the 94 elements h-pu. *J. Chem. Phys.* **132**, 154104 (2010).
29. VandeVondele, J. & Hutter, J. Gaussian basis sets for accurate calculations on molecular systems in gas and condensed phases. *J. Chem. Phys.* **127**, 114105 (2007).

Acknowledgements
We gratefully acknowledge valuable discussions with P. Hapla and S. Hämäläinen regarding implementation of the L–J model, and important comments on the manuscript by J. Leaf. A.S. gratefully acknowledges the support of the Leverhulme Trust via fellowship ECF-2013-525. S.P.J. thanks the Engineering and Physical Sciences Research Council (EPSRC) for the award of fellowship EP/J500483/1, and The Lever-hulme Trust for the award of fellowship ECF-2015-005, respectively. P.J.M. thanks EPSRC and the Leverhulme Trust, respectively, for Grants No. EP/G007837/1 and F00/114 BI. M.R.A. acknowledges funding from the University of Nottingham via the Vice-Chancellor's Scholarship for Research. P.R. received funding from the People Programme (Marie Curie Actions) of the European Union's Seventh Framework Programme (FP7/2007–2013) under REA grant agreement no. [628439]. We also acknowledge the support of the University of Nottingham High Performance Computing Facility (in particular, Dr Colin Bannister).

Author contributions

A.S. and P.M. conceived and designed the experiments. A.S. carried out the experiments and analysed the experimental data. A.S., M.A.R., J.L.D and S.P.J. designed and ran the L–J calculations. S.P.J. performed and analysed the DFT calculations. P.R. designed the atom tracking equipment and scripting. A.S. and P.M. wrote the paper. All authors read and commented on the final manuscript.

Additional information

Three-dimensional controlled growth of monodisperse sub-50 nm heterogeneous nanocrystals

Deming Liu[1], Xiaoxue Xu[1,2], Yi Du[3], Xian Qin[4], Yuhai Zhang[5], Chenshuo Ma[1], Shihui Wen[1,2], Wei Ren[1,2], Ewa M. Goldys[1], James A. Piper[1], Shixue Dou[3], Xiaogang Liu[4,5] & Dayong Jin[1,2]

The ultimate frontier in nanomaterials engineering is to realize their composition control with atomic scale precision to enable fabrication of nanoparticles with desirable size, shape and surface properties. Such control becomes even more useful when growing hybrid nanocrystals designed to integrate multiple functionalities. Here we report achieving such degree of control in a family of rare-earth-doped nanomaterials. We experimentally verify the co-existence and different roles of oleate anions (OA$^-$) and molecules (OAH) in the crystal formation. We identify that the control over the ratio of OA$^-$ to OAH can be used to directionally inhibit, promote or etch the crystallographic facets of the nanoparticles. This control enables selective grafting of shells with complex morphologies grown over nanocrystal cores, thus allowing the fabrication of a diverse library of monodisperse sub-50 nm nanoparticles. With such programmable additive and subtractive engineering a variety of three-dimensional shapes can be implemented using a bottom-up scalable approach.

[1] Laboratory of Advanced Cytometry, ARC Centre of Excellence for Nanoscale BioPhotonics, Department of Physics and Astronomy, Macquarie University, Sydney, New South Wales 2109, Australia. [2] Faculty of Science, Institute for Biomedical Materials and Devices, University of Technology Sydney, New South Wales 2007, Australia. [3] Institute for Superconducting and Electronic Materials, Innovation Campus, University of Wollongong, New South Wales 2522, Australia. [4] Institute of Materials Research and Engineering, 3 Research Link, Singapore 117602, Singapore. [5] Department of Chemistry, National University of Singapore, 3 Science Drive 3, Singapore 117543, Singapore. Correspondence and requests for materials should be addressed to X.L. (email: chmlx@nus.edu.sg) or to D.J. (email: dayong.jin@uts.edu.au).

Nanocrystal engineering, design and fabrication of nano-crystals with desirable size, shape[1-6], surface properties[7] and composition[8,9] is attracting growing interest due to its essential role in fundamental research and commercial relevance. Rare-earth-doped upconversion nanocrystals have recently emerged as the new generation of functional nanomaterials, because they exhibit exceptional optical, magnetic and chemical properties underpinning their diverse applications. In particular, alkaline rare-earth fluoride (AREF$_4$) nanocrystals[10-12], including hexagonal-phase β-NaYF$_4$, β-NaGdF$_4$, β-NaNdF$_4$ or β-NaLuF$_4$ are used in full-colour displays[12,13], photovoltaics[14], security inks[15], forensic science[16], autofluorescence-free biomolecular sensing[17-19], multimodal *in vivo* bio-imaging (fluorescence, magnetic resonance imaging, X-ray, SPECT and so on.)[20] and theranostics[17,21-23]. A trial-and-error approach is frequently used to produce nanoparticles with spherical, rod-like or other shapes[24-26] by varying dopant concentrations and/or constituent materials[27], reaction time and temperature[28-31]. This random sampling of vast, multidimensional parameter space, needs to be done rationally, with proper understanding of the underpinning growth mechanisms.

Here we find that oleate anions (OA$^-$), the dissociated form of oleic acid molecules (OAH), have variable, dynamic roles in mediating the growth of AREF$_4$ nanocrystals. This allows us to introduce a molecular approach to tailoring the shape and composition of AREF$_4$ nanocrystals. This new method is based on a selective epitaxial core–shell growth process in the presence of oleic acid, commonly used as a surfactant during the synthesis of β-AREF$_4$ nanocrystals[32]. Drawing inspiration from the recently discovered co-existence of oleic acid molecules (OAH) and their dissociated form, oleic acid ions (OA$^-$) in the binary systems of PbS[33] and PbSe nanocrystals[34], we hypothesize that the change in the ratio of OA$^-$ to OAH could influence the interaction of these ligands with the particle surface and hence the resulting morphology. Our computational modelling (Fig. 1, Supplementary Figs 1–6, Supplementary Notes 1 and 2

and Supplementary Table 1) and experimental results (Figs 2–4, Supplementary Figs 7–35, Supplementary Tables 2 and 3 and Supplementary Notes 3–18) demonstrate that the preferential affinity of OAH and OA$^-$ to different crystalline facets dictates the formation of nanocrystals of different shape. Importantly, we demonstrate that the precise control over the shell thickness and the particle shape can be achieved by deliberately switching the passivation, additive and subtractive roles of these surfactants.

Results

Computational modelling. To quantify the surface coordination chemistry between β-NaYF$_4$ surface and OAH and OA$^-$ ligands, we performed first-principles calculations based on density functional theory using CASTEP (CAmbridge Serial Total Energy Package)[35]. As shown in Fig. 1b and Supplementary Fig. 1, we treated the (001) and (100) planes of the β-NaYF$_4$ nanocrystals terminated with specific atomic arrangement as the most stable facets according to the calculated surface energies. Considering that the oxygen moiety in the ligands has a strong binding affinity to Y^{3+} ions at the particle surface[36], we modelled the interactions between the OAH and OA$^-$ molecules and the Y^{3+} ions under a number of conditions, such as different adsorption configurations (Supplementary Figs 2 and 3 and Supplementary Note 1), ligand chain length and ligand coverage (Supplementary Figs 4 and 5 and Supplementary Note 2). The key conclusion from these simulations is that OA$^-$ preferentially binds to RE^{3+} ions exposed on the (100) facet of the hexagonal fluoride nanocrystal, with a much higher binding energy (-35.4 eV) than on the (001) facet (-21.8 eV). It should be noted that the OAH molecule binds with a higher probability to the (001) facet than the (100) facet and has relatively small binding energies of -9.4 eV and -4.6 eV, respectively, on each of these facets (Supplementary Table 1). Our charge analysis (Supplementary Fig. 6) further indicates that such selective binding is attributed to the difference in the atomic arrangements of these two facets (Fig. 1b), giving rise to different charge transfer paths between the ligands and the surface ions.

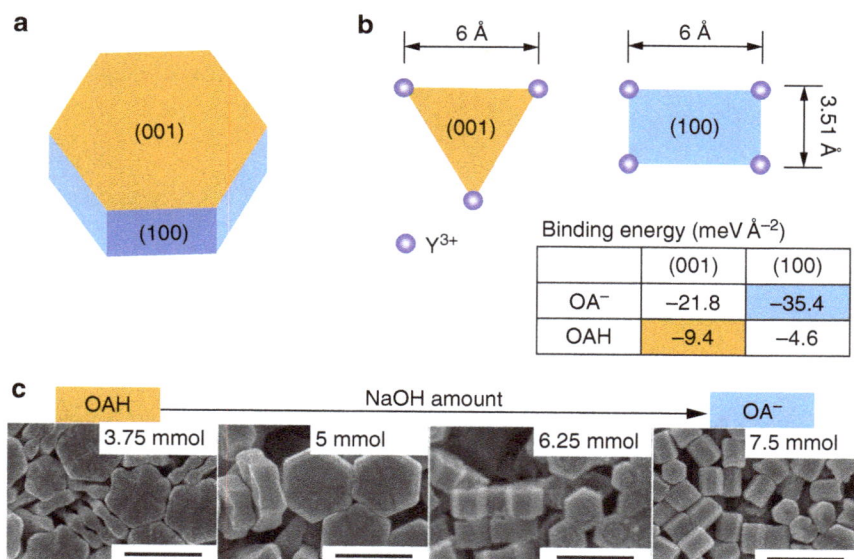

Binding energy (meV Å^{-2})		
	(001)	(100)
OA$^-$	−21.8	−35.4
OAH	−9.4	−4.6

Figure 1 | Preferred molecular bonding models of OA$^-$ and OAH. (a) The schematic shape of a β-NaYF$_4$ nanocrystal chosen as the core for directional epitaxial growth in this work. The hexagonal cylinder consists of the (001) facets at the ends and identical (100) and (010) facets around the cylinder sides. **(b)** The Y^{3+} arrangements and binding energies (see insert table) of OAH and OA$^-$ on the most stable (001) and (100) facets. The Y^{3+} atoms form equilateral triangles with a length of 6 Å in the relaxed (001) surface, while rectangles are observed in the (100) surface with a shorter length of 3.51 or 3.69 Å; **(c)** SEM characterization of submicron-sized nanocrystals synthesized using the hydrothermal route (detailed synthesis is included in the method; scale bar, 500 nm).

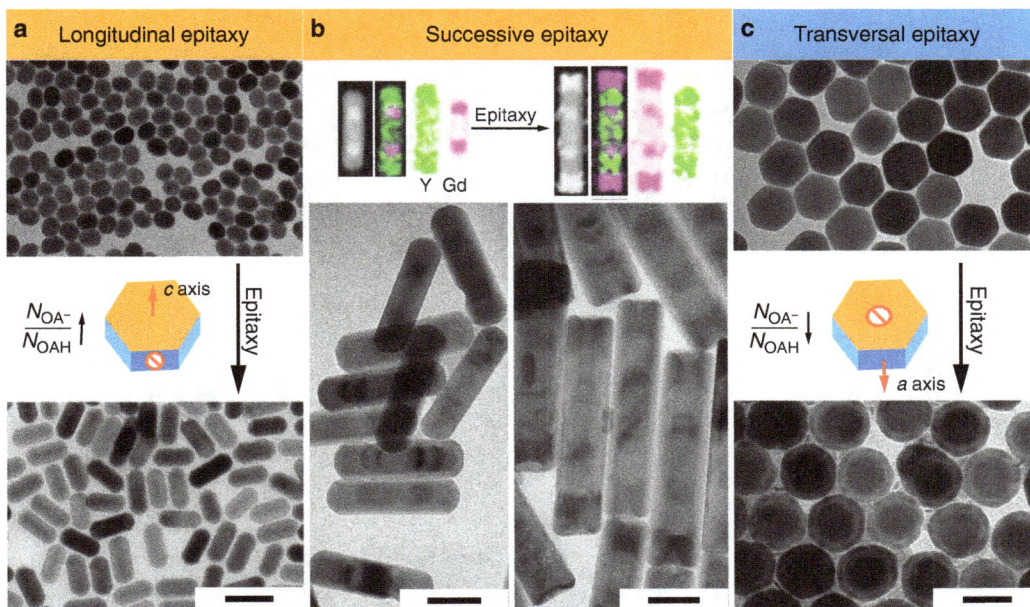

Figure 2 | Physical characterization of epitaxial growth of NaReF$_4$ NCs. (**a**) NaYF$_4$ core and homogenous NaYF$_4$ NCs after epitaxial growth of NaYF$_4$ in longitudinal direction with 0.5 mmol NaOH and 9.5 mmol OA at 310 °C for 1 h; (**b**) five-section and seven-section 'bamboo-shaped' NaYF$_4$/NaGdF$_4$ NRs formed by successive heterogeneous growth of periodical shells of NaGdF$_4$–NaYF$_4$ and NaGdF$_4$–NaYF$_4$–NaGdF$_4$ onto NaYF$_4$ core in the longitudinal direction, with 0.5 mmol NaOH and 0.4 mmol KOH and 9.5 mmol OA at 310 °C. Upper part of the panel shows elemental mapping of Y and Gd; (**c**) NaYF$_4$ core and heterogeneous NaYF$_4$/NaGdF$_4$ NCs after epitaxial growth of NaGdF$_4$ in the transversal direction with 0.15 mmol NaOH and 19 mmol OA at 290 °C for 3 h; the dimensions of individual nanocrystal were analysed statistically and included in the Supplementary Figs 10, 20–24. Scale bar, 50 nm.

Controlled epitaxial growth direction. The binding preferences of OAH and OA$^-$ molecules to different facets were first used to induce longitudinal epitaxial growth. We demonstrated (Fig. 1c) that sub-micrometre-sized NaYF$_4$ crystals of different aspect ratios could be prepared by tuning the concentration ratio of OA$^-$ to OAH in the hydrothermal synthesis system. As shown in Supplementary Fig. 7, higher concentrations of OA$^-$ encourage epitaxial growth along a longitudinal direction. A similar effect was observed in the synthesis of sub-50 nm NaYF$_4$ nanoparticles prepared by a co-precipitation method. Figure 2a,b show that high concentration of NaOH leads to longitudinally grown nanoparticles because of a large concentration of passivating OA$^-$ ions on the (100) facets (Supplementary Figs 8–10). The zeta potential of $+20$ mV for NaYF$_4$ nanocrystals after the removal of ligands (Supplementary Fig. 11) shows that the RE^{3+} cations are more abundant on the crystal surfaces than the F$^-$ ions. We further systemically studied other possible factors that could influence the epitaxial shell growth (experimental details in Supplementary Methods), including the reaction temperature (Supplementary Fig. 12 and Supplementary Note 3), the oleic acid concentration (Supplementary Fig. 13 and Supplementary Note 4), the F$^-$ ion concentration (Supplementary Fig. 14 and Supplementary Note 5) and the Na$^+$ concentration (Supplementary Fig. 15 and Supplementary Note 6). From these results, we confirm that the ratio of OA$^-$/OAH is a key factor that determines the epitaxial shell growth direction. However other parameters also have an effect on the growth speed or can change the OA$^-$/OAH ratio that indirectly affects the direction of growth. To rule out the effect of OH$^-$ on longitudinal growth, we added sodium oleate as the sodium source instead of hydroxide and identical results were obtained (Supplementary Fig. 16 and Supplementary Note 7). Supplementary Figures 17 and 18 further confirm that high ratio of OA$^-$/OAH directs longitudinal deposition of heterogeneous shells (NaGdF$_4$) on the end surfaces of NaYF$_4$ core. Interestingly, subtractive growth (dissolution) is observed from their side (100) surfaces. This results in concurrent decrease of the core width from 26 to 18 nm, thus producing dumbbell-shaped nanocrystals (Supplementary Note 8).

Moreover, we found that the addition of KOH further accelerates longitudinal growth rate (Supplementary Fig. 19 and Supplementary Note 9) due to a higher dissociation constant of KOH than NaOH, which increases the dissociation of OAH producing more OA$^-$. With the aid of KOH, heterogeneous 'bamboo-shaped' nanorods (NRs) with sharp edges were formed in a stepwise manner with a length of up to 173 nm (Fig. 2b, Supplementary Fig. 21 and Supplementary Note 10). The interesting one-dimension architecture of 'bamboo-shaped' NRs suggests that integrated multiple functionalities can be built. Thus our new platform enables rational design and facile synthesis of multiple sections of rare-earth-doped heterogeneous materials and investigation of their interactions and functions within a single integrated rod. We were also able to induce transversal epitaxial growth by increasing the amount of OAH and reducing the amount of NaOH. At a reaction temperature of 290 °C, the transversal growth was observed and NaGdF$_4$ rings of 7-nm-thick formed around the NaYF$_4$ cores formed without a measurable change in the longitudinal direction (Fig. 2c, Supplementary Figs 23 and 24 and Supplementary Note 11). Notably, the dissolution of the (100) facets of the cores took place as well, and the width of the core was, again, reduced from 49 to 30 nm at both ends. The observed dissolution always occurred on the (100) facets in both cases of longitudinal and transversal growth. This is consistent with the strong chelating character of OA$^-$ on the (100) facet, and with the fact that NaYF$_4$ is dissolved faster than NaGdF$_4$ because NaYF$_4$ is comparably less energetically stable[12]. To shed more light on this issue, we provided more evidence in the Supplementary Fig. 25 and Supplementary Note 12 to show that the dissolution of core is caused by the thermal stability difference

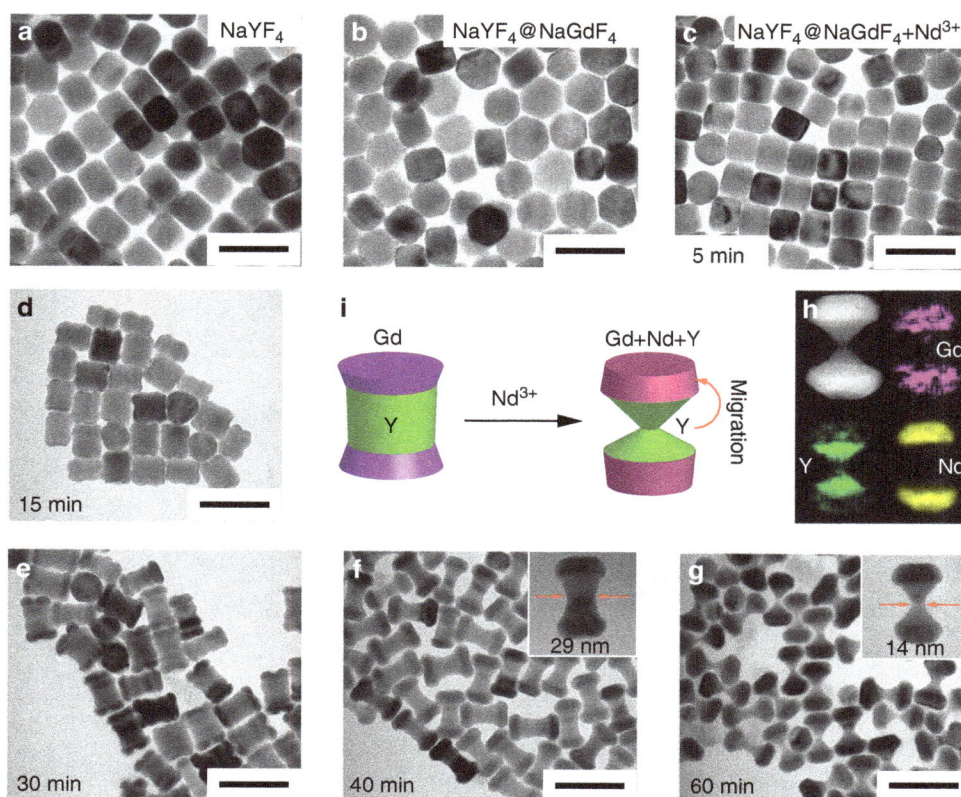

Figure 3 | Evolution of morphology and composition in migration growth. (**a,b**) NaGdF$_4$ growth along the longitudinal direction onto the ends of the NaYF$_4$ core; (**c**) transmission electron microscope image of the sample stopped 5 min after reacting with NaGdF$_4$/NaYF$_4$ nanocrystals in the presence of Na$^+$, K$^+$, Nd^{3+}, OA$^-$ and in the absence of F$^-$ at 310 °C, dissolution occurs first; (**d–g**) real-time monitoring of the epitaxial growth of NaNdF$_4$ along the longitudinal direction onto NaYF$_4$–NaGdF$_4$ nanocrystals, involving the dissolution of NaYF$_4$ and NaGdF$_4$ from the transversal surfaces of the crystal and their subsequent re-growth onto the NaNdF$_4$ nanocrystals in the presence of Na$^+$, K$^+$, Nd^{3+}, OA$^-$ and absence of F$^-$ ions at 310 °C. (**h**) HAADF-STEM image with elemental mapping results of the samples stopped after 60 min of reaction to confirm the distributions of Y^{3+}, Gd^{3+}, Nd^{3+} ions within a single NaYF$_4$/NaGdF$_4$/NaNdF$_4$ nanocrystal. (**i**) schematic processes of dissolution of NaYF$_4$/NaGdF$_4$ and the sequent epitaxial growth of NaNdF$_4$ in the longitudinal direction and the migration growth of F$^-$, Y^{3+} and Gd^{3+} ions (scale bar, 100 nm).

between core and shell materials in presence of OA$^-$ which leads to higher binding strength on the side surfaces. By comparing growth of NaTbF$_4$ as shell or NaYbF$_4$ as shell on a NaYF$_4$ core (Supplementary Fig. 25), we demonstrate that the dissolution of the core requires the shell materials to have higher thermal stability than the core material. Larger difference of thermal stability between core and shell result in a higher dissolution rate.

Controlled migration growth. By combining the approaches of longitudinal and transversal growth and selective dissolution with consideration of lattice mismatch (Supplementary Tables 2 and 3), we synthesized a variety of three-dimensional (3D) hybrid nanostructures (Supplementary Figs 26–34). Figure 3 shows a typical example of real-time evolution of morphology and composition of the NaYF$_4$/NaGdF$_4$/NaNdF$_4$ NCs, including the dissolution process of the NaYF$_4$/NaGdF$_4$ nanocrystals and subsequent longitudinal growth of NaNdF$_4$. The dissolution of NaYF$_4$/NaGdF$_4$ is initiated by the OA$^-$ adsorbed on the surface of the nanocrystals. The concomitant depletion of dissolved F$^-$ ions used for longitudinal growth of NaNdF$_4$ in the presence of high concentration of OA$^-$ facilitates the dissolution of NaYF$_4$/NaGdF$_4$ nanocrystals and this, in turn, promotes longitudinal growth of NaNdF$_4$. Following the dissolution of the Y^{3+} and Gd^{3+} ions from the surface of NaYF$_4$–NaGdF$_4$ nanocrystals, these ions then participate in the epitaxial growth of NaNdF$_4$ nanocrystals, as evidenced by the elemental mapping (Fig. 3h). Moreover, our real-time sampling transmission electron

microscope data further confirmed the underpinning mechanism (Fig. 3a–g, Supplementary Figs 26–28). The size of nanocrystal core decreased significantly in the first 5 min, indicating that the dissolution rate of the nanocrystals is faster than their growth rate. After 15 min, new material started to form at the top and at the bottom ends of the core with simultaneous decrease of the nanocrystal core width. This observation rules out 'surface mobility' ('atom diffusion') as the possible driving force behind the formation of the final shell, otherwise it is expected that the dissolution of NaYF$_4$ and growth of NaNdF$_4$ would occur at the same time. The only mechanism which explains the shape of this nanocrystal is that the absence of F$^-$ source in the reaction solution at its beginning prevents growth of NaNdY$_4$ until the concentration of released F$^-$ source exceeds a certain threshold.

Our control experiments (Supplementary Fig. 29 and Supplementary Note 15) further support the mechanism of OA$^-$-induced dissolution in which a firm bonding of the surfactant OA$^-$ to the surface RE^{3+} cations is the main factor responsible for the removal of the surface crystalline layers (experimental details in Supplementary Methods). As shown in Supplementary Fig. 29, we applied transversal growth approach to first grow a layer of NaGdF$_4$ on the side surfaces of NaYF$_4$ core. We see that smaller mismatch of NaGdF$_4$ versus NaNdF$_4$ compared with the NaYF$_4$ versus NaNdF$_4$ fails to direct the transversal migration growth of the NaNdF$_4$ on the side surfaces of NaGdF$_4$. Instead, dissolution occurs in the first 10 min of the reaction (Supplementary Fig. 29a,b) and both dissolution from

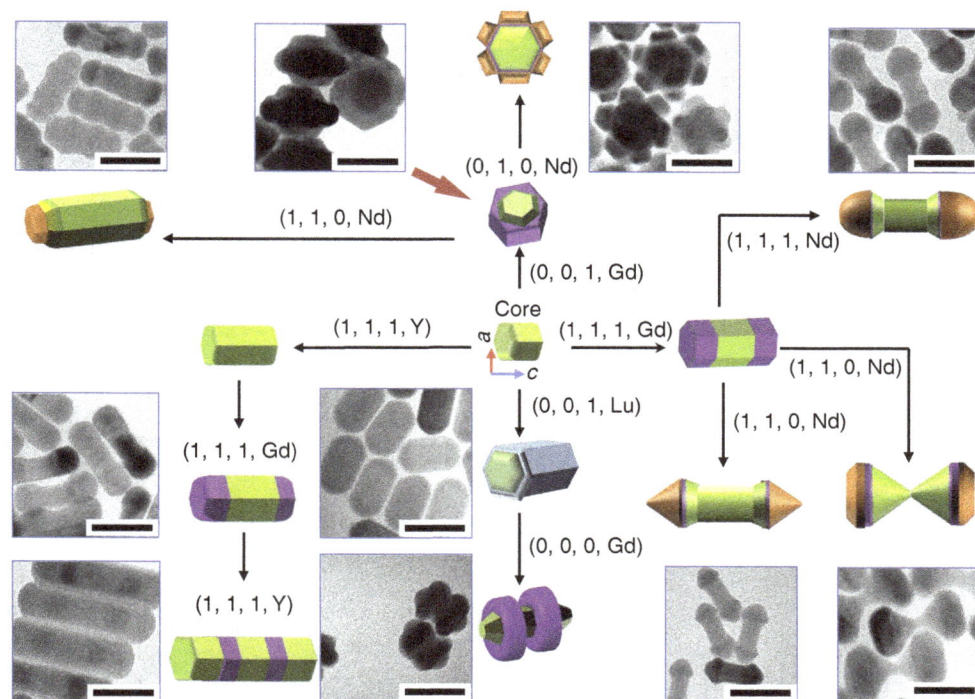

Figure 4 | Programmable routes for fabricating 3D nano-architectures. The four digital condition codes (R, T, F and RE) represent different reaction conditions where: $R = 0$, represents a low ratio of OA^-/OAH; $R = 1$, represents a high ratio of OA^-/OAH; $T = 0$, where the temperature is at 290 °C; $T = 1$, where the temperature is at 310 °C; $F = 0$, which designates the absence of an F^- ion source; $F = 1$, indicates the presence of an F^- ion source; RE = Y, with a rare earth Y^{3+} ion source; RE = Gd, with Gd^{3+} ion source; RE = Lu, with a Lu^{3+} source; RE = Nd, with Nd^{3+} source. By combining these different growth processes into a synthesis procedure, a variety of complex $NaREF_4$ nanostructures are fabricated as shown in the transmission electron microscope images, including hourglass shaped $NaYF_4/NaGdF_4/NaNdF_4$ nanocrystals, $NaYF_4/NaGdF_4/NaNdF_4$ nano-flowers, $NaYF_4/NaLuF_4$ co-axial nano-cylinders, $NaYF_4/NaLuF_4/NaGdF_4$ nanoscale spins with double rings, and $NaYF_4/NaGdF_4/NaNdF_4$ nano-dumbbells with smooth or sharp ends (scale bar, 50 nm).

the side surfaces and epitaxial growth of $NaNdF_4$ on the end surfaces of $NaGdF_4/NaYF_4$ cores result in a thinner and longer nanocrystal.

Guided by the principle that the ratio of OA^-/OAH controls the direction of epitaxial shell growth, we further demonstrated (as shown in Supplementary Fig. 30 and Supplementary Note 16) that a low ratio of OA^-/OAH at a lower temperature directs the migration growth along transverse direction. This enables the formation of heterogeneous $NaYF_4/NaGdF_4/NaNdF_4$ nanocrystals in the shape of a flower, although in this case the dissolution process on the side surfaces of nanocrystals is much less efficient because there are too few OA^- ligands bound to RE^{3+} cations on the (100) facet. Two additional experiments (Supplementary Note 17) demonstrate that well established parameters, such as reagent concentration, can be further applied to fine-tune our programmable protocols for other types of heterogeneous nanocrystals. During the formation of hourglass-shaped nanocrystals, the decrease in the amount of Nd^{3+} source is found to hinder the migration growth process and yield sharper tips (Supplementary Fig. 31), whereas a supply of additional F^- ions in the reaction increases the diameter of dumbbell ends with round tips (Supplementary Fig. 32). Such level of fine tuning to grow progressively sharper tips may suggest future rational methods, for example to optimize tip-sensitive physical and biochemical properties of NRs.

Figure 4 shows an array of heterogeneous $NaREF_4$ nanostructures synthesized by carrying out specific sequences of longitudinal, transversal growth, selective dissolution and directional migration growth of epitaxial shells in the presence of various OA^-/OAH ratios. To the best of our knowledge, these sub-50 nm nanoparticles are the smallest 3D objects

prepared by a bottom–up additive and subtractive process. To illustrate the application of this novel method we designed and synthesized multifunctional $NaYF_4/NaLuF_4/NaGdF_4$ heterogeneous nanocrystals with two $NaGdF_4$ rings on a $NaLuF_4/NaYF_4$ NRs (Supplementary Figs 33 and 34 and Supplementary Note 18). The hexagonal-phase $NaYF_4$ nanocrystal is an efficient luminescence upconversion material[37]. The addition of $NaLuF_4$ enables X-ray computed tomography[38], whereas using $NaGdF_4$ enables magnetic resonance imaging[39]. To the best of our knowledge, this work presents the first controlled fabrication of sub-50 nm 3D shaped heterogeneous nanocrystals logically programmed by the combinational approaches of OA^--assisted longitudinal growth, transversal growth and selective crystalline facet dissolution with consideration of crystallographic mismatch rates.

Discussion

The nanoscale engineering capability presented in this work enables quantitative studies which are virtually impossible by conventional approaches. We anticipate that optical properties of these nanostructures can be designed to precisely promote or inhibit inter-particle energy transfer. Similarly, magnetic properties may be optimized to enhance magnetic resonance imaging by correlating the morphology with the surface distribution of magnetic signals. In addition, such hybrid nanomaterials may be used as a platform for transporting biologically important molecules across cell membranes. Furthermore, access to a new library of precisely controlled shapes of nanoparticles provide a novel approach for the targeted delivery in nanomedicine where optimized morphologies of these nanoscale molecular carriers will yield greater efficiencies. This process could be further facilitated by harnessing the anisotropic properties of different types of nanoparticles that permit diverse surface functionalizations and multi-modal bio-conjugations. The concept presented in this work may further advance our current capabilities of nanoscale programmable and reproducible engineering of new classes of heterogeneous materials in scalable quantities. Our findings may lead to a new class of multi-functional nanomaterials and provide the groundwork for developing previously unforeseen applications of nanoparticles with complex programmable shapes and surface properties.

Methods

Hydrothermal synthesis of NaYF$_4$ crystal. The β-NaYF$_4$ disks were synthesized via a slightly modified hydrothermal reaction. In a typical experiment, NaOH (3.75 mmol) was first dissolved into 1.5 mL of double distilled water, followed by the addition of OA (7.5 mmol) and ethanol (2.5 mL) while undergoing vigorous stirring. Thereafter, an aqueous solution of NaF (0.5 M; 2 mL) was added to form a turbid mixture. Subsequently, a 1.2 mL aqueous solution of YCl$_3$ (Yb^{3+}/Tm^{3+} = 10/0.5 mol%; 0.2 M) was added and the solution was stirred for 20 min. The resulting mixture was then transferred into a 14 mL Teflon-lined autoclave and heated to 220 °C and the temperature maintained for 12 h. After cooling down to room temperature, the reaction product was isolated by centrifugation and washed with ethanol. In this work, different amounts of NaOH were added to adjust the ratio of OA$^-$/OAH by its reaction with OAH to form OA$^-$.

NaYF$_4$ nanocrystal cores. In a typical procedure, 4 ml of methanol solution of YCl$_3$ (2.0 mmol) was magnetically mixed with OA (38 mmol) and ODE (93 mmol) in a 100-ml three-neck round-bottom flask. The mixture was then degassed under the Ar flow and then heated to 150 °C for 30 min to form a clear solution, before cooling to room temperature. 15 ml of methanol solution containing NH$_4$F (8 mmol) and NaOH (5 mmol) was added to the solution of YCl$_3$ in OA and ODE and stirred for 60 min. The mixture solution was slowly heated to 110 °C and kept at 110 °C for 30 min to completely remove methanol and any residual water. The mixture solution was then quickly heated to the reaction temperature of 300 °C and aged for 1 h. After the solution was left to cool down to room temperature, ethanol was added to precipitate the nanocrystals. The product was washed with cyclohexane, ethanol and methanol for at least 4 times, before the final NaYF$_4$ nanocrystals were re-dispersed in 10 ml cyclohexane in preparation for their further use.

Longitudinal growth of NaYF$_4$ NRs. YCl$_3$ (0.2 mmol) in 1 ml methanol solution was magnetically mixed with OA (9.5 mmol) and ODE (25 mmol) in a 50-ml three-neck round-bottom flask. The mixture was degassed under Ar flow and heated to 150 °C for 30 min to form a clear solution, and then cooled to room temperature. Methanol solution (5 ml) containing NH$_4$F (0.8 mmol) and NaOH (0.5 mmol) was added and stirred for 60 min. The solution was slowly heated to 110 °C and kept at 110 °C for 30 min to completely remove methanol and residual water. The solution was then injected with 0.2 mmol NaYF$_4$ of nanocrystals in cyclohexane and the mixture kept at 110 °C for another 10 min to evaporate the cyclohexane. Then, the reaction mixture was quickly heated to 310 °C and aged for 1 h.

NaGdF$_4$/NaYF$_4$ nano-dumbbells. GdCl$_3$ (0.2 mmol) in 1 ml methanol solution was magnetically mixed with OA (9.5 mmol) and ODE (25 mmol) in a 50-ml three-neck round-bottom flask. The mixture was degassed under an Ar flow and heated to 150 °C for 30 min to form a clear solution, and then cooled to room temperature. Methanol solution (4 ml) containing NH$_4$F (0.8 mmol) and NaOH (0.5 mmol) was added to the OA and ODE solution and stirred for 60 min. The solution is slowly heated to 110 °C and kept at 110 °C for 30 min to remove methanol and the remaining water completely. Then, 0.2 mmol of NaYF$_4$ core nanocrystals in cyclohexane was injected into the reaction solution. After holding the reaction temperature at 110 °C for further 10 min to evaporate all cyclohexane, the mixture was quickly heated to 310 °C and aged for 1 h.

NaGdF$_4$/NaYF$_4$ NRs by adding KOH. GdCl$_3$ (0.2 mmol) in 1 ml of methanol solution was magnetically mixed with OA (9.5 mmol) and ODE (25 mmol) in a 50-ml three-neck round-bottom flask. The mixture was degassed under Ar flow and heated to 150 °C for 30 min to form a clear solution, before cooling to room temperature. Methanol solution (5 ml) containing NH$_4$F (0.8 mmol), KOH (0.4 mmol) and NaOH (0.5 mmol) was added into the OA and ODE solution and stirred for 60 min. The solution was slowly heated to 110 °C and kept at 110 °C for 30 min to remove the methanol and water completely. The reaction mix was then injected with 0.2 mmol of NaYF$_4$ core nanocrystals in cyclohexane, into the reaction solution. After holding the reaction mix at 110 °C for further 10 min to evaporate all cyclohexane, the mixture was heated rapidly to 310 °C before aging for 1 h at this temperature.

NaYF$_4$/NaGdF$_4$/NaYF$_4$ NCs in a bamboo-like shape. 0.2 mmol of YCl$_3$ in 1 ml of methanol solution was magnetically mixed with OA (9.5 mmol) and ODE (25 mmol) in a 50-ml three-neck round-bottom flask. The mixture was degassed under Ar flow and heated to 150 °C for 30 min to form a clear solution, and then cooled to room temperature. Methanol solution (5 ml) containing NH$_4$F (0.8 mmol), KOH (0.4 mmol) and NaOH (0.5 mmol) was added into the OA and ODE solution and stirred for 60 minutes. The solution was slowly heated to 110 °C and kept at 110 °C for 30 min to remove the methanol and water completely. The reaction solution was then injected with 0.2 mmol of NaYF$_4$/NaGdF$_4$ NRs in cyclohexane solution. After the reaction at 110 °C for a further 10 min to evaporate all the cyclohexane, the reaction mixture was quickly heated to 310 °C and held at this temperature for 1 h.

NaYF$_4$/NaGdF$_4$/NaYF$_4$/NaGdF$_4$ NCs in a bamboo-like shape. The same procedure for synthesizing NaYF$_4$/NaGdF$_4$/NaYF$_4$ NCs in bamboo-like shape was repeated, and then followed by the injection of 0.2 mmol of the five-section NaYF$_4$/NaGdF$_4$/NaYF$_4$ nano-bamboos which acted as the core, all in cyclohexane solution, into the reaction solution. After holding at 110 °C for a further 10 min to evaporate all cyclohexane, the reaction mixture was quickly heated to 310 °C and held again for 1 h.

NaYF$_4$/NaGdF$_4$/NaNdF$_4$ NCs in an hourglass shape. NdCl$_3$ (0.4 mmol) in 2 ml of methanol solution was magnetically mixed with OA (9.5 mmol) and ODE (25 mmol) in a 50-ml three-neck round-bottom flask. The mixture was degassed under Ar flow and heated to 150 °C for 30 min to form a clear solution, and then cooled to room temperature. Methanol solution (5 ml) containing KOH (0.8 mmol) and NaOH (0.8 mmol) was added and stirred for 60 min. The solution was slowly heated to 110 °C and kept at 110 °C for 30 min to completely remove the methanol and some of the water. It was then injected with 0.1 mmol 50 nm × 60 nm NaYF$_4$/NaGdF$_4$ nano-prisms particles, in a solution of cyclohexane. After having been kept at 110 °C for another 10 min to evaporate all cyclohexane, the reaction mixture was quickly heated to 310 °C. Samples (500 ul) of the reaction solution were collected each time with a syringe at 5, 15, 30, 40, 50 and 60 min after the start of the reaction.

Transversal growth of NaGdF$_4$ shell onto NaYF$_4$ core. GdCl$_3$ (0.1 mmol) in 1 ml methanol solution was magnetically mixed with OA (19.0 mmol) and ODE (18.7 mmol) in a 50-ml three-neck round-bottom flask. The mixture was degassed under Ar flow and heated to 150 °C for 30 min to form a clear solution, and then cooled to room temperature. Methanol solution (3 ml) containing NH$_4$F (0.4 mmol) and NaOH (0.15 mmol) was added into the OA and ODE solution and stirred for 60 min. The solution was slowly heated to 110 °C and kept at 110 °C for 30 min to remove completely the methanol and water. Then 0.1 mmol of the NaYF$_4$ cores in cyclohexane solvent were injected into the reaction mix. After being kept at 110 °C for further 10 min to evaporate all cyclohexane, the reaction mixture was quickly heated up to 290 °C and held at that temperature for 3 h.

Synthesis of NaYF$_4$/NaGdF$_4$/NaNdF$_4$ NCs in flower shape. NdCl$_3$ (0.1 mmol) in 1 ml of methanol solution was magnetically mixed with OA (19 mmol) and ODE (18.7 mmol) in a 50-ml three-neck round-bottom flask. The mixture was degassed under Ar flow and heated to 150 °C for 30 min to form a clear solution, and then cooled to room temperature. Methanol solution (5 ml) containing NaOH (0.6 mmol) was added and stirred for 60 min. The solution was slowly heated to 110 °C and kept at 110 °C for 30 min to completely remove the methanol and some of the water. Then, the reaction mix was injected with 0.1 mmol of 50 nm NaYF$_4$/NaGdF$_4$ nano-prisms prisms (NaGdF$_4$ growing on the lateral faces of NaYF$_4$ nanocrystal), suspended in a cyclohexane solution. After holding at 110 °C for another 10 min to evaporate all cyclohexane, the reaction mixture was quickly heated to 300 °C. samples (500 ul) of the reaction solution were collected each time with a syringe after 10, 25 and 45 min of the reaction time.

Synthesis of NaYF$_4$/NaGdF$_4$/NaNdF$_4$ sharp-end dumbbell. NdCl$_3$ (0.1 mmol) in 1 ml of methanol solution was magnetically mixed with OA (9.5 mmol) and ODE (25 mmol) in a 50-ml three-neck round-bottom flask. The mixture was degassed under Ar flow and heated to 160 °C for 30 min to form a clear solution, and then cooled to room temperature. Methanol solution (5 ml) containing KOH (0.2 mmol) and NaOH (0.2 mmol) was added and stirred for 60 min. Note: in this reaction no NH$_4$F was added to the solution. The solution was slowly heated to 110 °C and kept at 110 °C for 30 min to remove the methanol and the water completely. It was then injected with 0.1 mmol of NaYF$_4$/NaGdF$_4$ NR particle in suspended in cyclohexane solvent into the reaction solution. After holding at 110 °C for a further 10 min to evaporate all cyclohexane, the reaction mixture was quickly heated to 310 °C and held at this temperature for a further 30 min.

Synthesis of NaYF$_4$/NaGdF$_4$/NaNdF$_4$ round-end dumbbell. NdCl$_3$ (0.1 mmol) in 1 ml of methanol solution was magnetically mixed with OA (9.5 mmol) and

ODE (25 mmol) in a 50-ml three-neck round-bottom flask. The mixture was degassed under Ar flow and heated to 160 °C for 30 min to form a clear solution, and then cooled to room temperature. Methanol solution (5 ml) containing NH$_4$F (0.3 mmol), KOH (0.2 mmol) and NaOH (0.2 mmol) was added and the mixture was stirred for 60 min. The solution was slowly heated to 110 °C and kept at 110 °C for 30 min to remove the methanol and the water completely. Then, it was injected with 0.1 mmol of NaYF$_4$/NaGdF$_4$ NRs suspended in cyclohexane into the reaction solution. After being held at 110 °C for further 10 min to evaporate all cyclohexane, the reaction mixture was quickly heated to 310 °C and held for 30 min at this temperature.

Synthesis of pure α-NaGdF$_4$ NCs. Methanol solution (2 ml) of GdCl$_3$ (1.0 mmol) was magnetically mixed with OA (19 mmol) and ODE (47 mmol) in a 100-ml three-neck round-bottom flask. The mixture was degassed under Ar flow and heated to 150 °C for 30 minutes to form a clear solution, and then cooled to room temperature. Methanol solution (10 ml) containing NH$_4$F (4 mmol) and NaOH (2.5 mmol) was added and stirred for 60 min. Then, the solution was slowly heated to 110 °C and kept at 110 °C for 30 min to remove the methanol and water completely. After that, the reaction mixture was quickly heated to 240 °C and aged for 45 min.

Synthesis of NaLuF$_4$/NaYF$_4$ NRs. LuCl$_3$ (0.1 mmol) in 1 ml method solution was magnetically mixed with OA (19 mmol) and ODE (25 mmol) in a 50-ml three-neck round-bottom flask. The mixture was degassed under Ar flow and heated to 150 °C for 30 min to form a clear solution, and then cooled to room temperature. Methanol solution (2 ml) containing NaOH (0.15 mmol) and 0.4 mmol NH$_4$F was added and stirred for 60 min. The solution was slowly heated to 110 °C and kept at 110 °C for 30 min to completely remove the methanol and some of the water. It was then injected with 0.4 mmol of NaYF$_4$ seed particles in a cyclohexane solution. After holding the reaction mix at 110 °C for a further 10 min to evaporate cyclohexane, the reaction mixture was quickly heated to 290 °C and held at that temperature for a further 1 h.

Synthesis of NaLuF$_4$/NaYF$_4$ NRs with NaGdF$_4$ double-ring. GdCl$_3$ (0.1 mmol) in 1 ml methanol solution was magnetically mixed with OA (19.0 mmol) and ODE (18.7 mmol) in a 50-ml three-neck round-bottom flask. The mixture was degassed under Ar flow and heated to 150 °C for 30 min to form a clear solution, and then cooled to room temperature. Methanol solution (2 ml) containing NaOH (at 0.15 mmol) was added and stirred for 60 min. The solution was slowly heated to 110 °C and kept at 110 °C for 30 min to completely remove the methanol and some of the water. It was then injected with 0.1 mmol of the NaYF$_4$/NaLuF$_4$ seed particles, in a cyclohexane solution, into the reaction solution. After having been held the reaction mix at 110 °C for another 10 min to evaporate cyclohexane, the reaction mixture was quickly heated to 300 °C. It was then, injected with 0.02 mmol of α-NaGdF$_4$ nanocrystals into the reaction system. This was done every 10 min for 5 times at 300 °C. The reaction mix was held at this temperature for another 10 min after the last injection.

References

1. Jia, G. H. et al. Couples of colloidal semiconductor nanorods formed by self-limited assembly. Nat. Mater. 13, 302–308 (2014).
2. Jones, M. R., Seeman, N. C. & Mirkin, C. A. Programmable materials and the nature of the DNA bond. Science 347, 1260901 (2015).
3. Ye, X. et al. Seeded growth of metal-doped plasmonic oxide heterodimer nanocrystals and their chemical transformation. J. Am. Chem. Soc. 136, 5106–5115 (2014).
4. Yu, H. et al. Dumbbell-like bifunctional Au-Fe3O4 nanoparticles. Nano Lett. 5, 379–382 (2005).
5. Jun, Y. W., Choi, J. S. & Cheon, J. Shape control of semiconductor and metal oxide nanocrystals through nonhydrolytic colloidal routes. Angew. Chem. Int. Ed. 45, 3414–3439 (2006).
6. Hill, J. P. et al. Self-assembled hexa-peri-hexabenzocoronene graphitic nanotube. Science 304, 1481–1483 (2004).
7. Yin, Y. & Alivisatos, A. P. Colloidal nanocrystal synthesis and the organic-inorganic interface. Nature 437, 664–670 (2005).
8. Zeng, H. & Sun, S. H. Syntheses, properties and potential applications of multicomponent magnetic nanoparticles. Adv. Funct. Mater. 18, 391–400 (2008).
9. Qian, H. F., Zhu, Y. & Jin, R. C. Atomically precise gold nanocrystal molecules with surface plasmon resonance. Proc. Natl Acad. Sci. USA 109, 696–700 (2012).
10. Zheng, W. et al. Lanthanide-doped upconversion nano-bioprobes: electronic structures, optical properties, and biodetection. Chem. Soc. Rev. 44, 1379–1415 (2015).
11. Wang, X., Zhuang, J., Peng, Q. & Li, Y. D. A general strategy for nanocrystal synthesis. Nature 437, 121–124 (2005).
12. Wang, F. et al. Simultaneous phase and size control of upconversion nanocrystals through lanthanide doping. Nature 463, 1061–1065 (2010).
13. Deng, R. et al. Temporal full-colour tuning through non-steady-state upconversion. Nat. Nanotechnol. 10, 237–242 (2015).
14. Ramasamy, P. & Kim, J. Combined plasmonic and upconversion rear reflectors for efficient dye-sensitized solar cells. Chem. Commun. 50, 879–881 (2014).
15. Lu, Y. Q. et al. Tunable lifetime multiplexing using luminescent nanocrystals. Nat. Photon. 8, 33–37 (2014).
16. Wang, J. et al. Near-infrared-light-mediated imaging of latent fingerprints based on molecular recognition. Angew. Chem. 126, 1642–1646 (2014).
17. Huang, P. et al. Lanthanide-doped LiLuF$_4$ upconversion nanoprobes for the detection of disease biomarkers. Angew. Chem. Int. Ed. 53, 1252–1257 (2014).
18. Zhao, J. et al. Single-nanocrystal sensitivity achieved by enhanced upconversion luminescence. Nat. Nanotechnol. 8, 729–734 (2013).
19. Gargas, D. J. et al. Engineering bright sub-10-nm upconverting nanocrystals for single-molecule imaging. Nat. Nanotechnol. 9, 300–305 (2014).
20. Liu, Q. et al. F-18-labeled magnetic-upconversion nanophosphors via rare-earth cation-assisted ligand assembly. ACS Nano 5, 3146–3157 (2011).
21. Idris, N. M. et al. In vivo photodynamic therapy using upconversion nanoparticles as remote-controlled nanotransducers. Nat. Med. 18, 1580–1585 (2012).
22. Wang, C., Cheng, L. & Liu, Z. Upconversion nanoparticles for photodynamic therapy and other cancer therapeutics. Theranostics 3, 317–330 (2013).
23. Kang, X. J. et al. Lanthanide-doped hollow nanomaterials as theranostic agents. Wiley Interdiscip. Rev. Nanomed. Nanobiotechnol. 6, 80–101 (2014).
24. Mai, H. X., Zhang, Y. W., Sun, L. D. & Yan, C. H. Highly efficient multicolor up-conversion emissions and their mechanisms of monodisperse NaYF$_4$: Yb,Er core and core/shell-structured nanocrystals. J. Phys. Chem. C 111, 13721–13729 (2007).
25. Shen, J. et al. Tunable near infrared to ultraviolet upconversion luminescence enhancement in (alpha-NaYF4:Yb,Tm)/CaF2 core/shell nanoparticles for in situ real-time recorded biocompatible photoactivation. Small 9, 3213–3217 (2013).
26. Wen, H. et al. Upconverting near-infrared light through energy management in core-shell-shell nanoparticles. Angew. Chem. Int. Ed. 52, 13419–13423 (2013).
27. Johnson, N. J. & van Veggel, F. C. Lanthanide-based hetero-epitaxial core-shell nanostructures: the compressive vs. tensile strain asymmetry. ACS Nano 8, 10517–10527 (2014).
28. Zhang, C. & Lee, J. Y. Prevalence of anisotropic shell growth in rare earth core-shell upconversion nanocrystals. ACS Nano 7, 4393–4402 (2013).
29. Johnson, N. J. J. & van Veggel, F. C. J. M. Sodium lanthanide fluoride core-shell nanocrystals: a general perspective on epitaxial shell growth. Nano Res. 6, 547–561 (2013).
30. Xia, Y., Xiong, Y., Lim, B. & Skrabalak, S. E. Shape-controlled synthesis of metal nanocrystals: simple chemistry meets complex physics? Angew. Chem. Int. Ed. 48, 60–103 (2009).
31. Zhuang, Z., Peng, Q. & Li, Y. Controlled synthesis of semiconductor nanostructures in the liquid phase. Chem. Soc. Rev. 40, 5492–5513 (2011).
32. Wang, F., Deng, R. R. & Liu, X. G. Preparation of core-shell NaGdF$_4$ nanoparticles doped with luminescent lanthanide ions to be used as upconversion-based probes. Nat. Protoc. 9, 1634–1644 (2014).
33. Zherebetskyy, D. et al. Hydroxylation of the surface of PbS nanocrystals passivated with oleic acid. Science 344, 1380–1384 (2014).
34. Bealing, C. R., Baumgardner, W. J., Choi, J. J., Hanrath, T. & Hennig, R. G. Predicting nanocrystal shape through consideration of surface-ligand interactions. ACS Nano 6, 2118–2127 (2012).
35. Clark, S. J. et al. First principles methods using CASTEP. Z. Kristallogr. 220, 567–570 (2005).
36. Sui, Y. Q., Tao, K., Tian, Q. & Sun, K. Interaction between Y^{3+} and oleate ions for the cubic-to-hexagonal phase transformation of NaYF4 nanocrystals. J. Phys. Chem. C 116, 1732–1739 (2012).
37. Yi, G. et al. Synthesis, characterization, and biological application of size-controlled nanocrystalline NaYF4: Yb, Er infrared-to-visible up-conversion phosphors. Nano Lett. 4, 2191–2196 (2004).
38. Xia, A. et al. Gd^{3+} complex-modified NaLuF$_4$-based upconversion nanophosphors for trimodality imaging of NIR-to-NIR upconversion luminescence, X-Ray computed tomography and magnetic resonance. Biomaterials 33, 5394–5405 (2012).
39. Hou, Y. et al. NaGdF$_4$ nanoparticle-based molecular probes for magnetic resonance imaging of intraperitoneal tumor xenografts in vivo. ACS Nano 7, 330–338 (2012).

Acknowledgements

We thank D. Birch in Microscope Unit at Macquarie University and D.R.G. Mitchell in Electron Microscopy Centre at the University of Wollongong for their help and valuable discussion on transmission electron microscope characterizations. This project is primarily supported by the Australian Research Council (ARC) Future Fellowship Scheme (FT 130100517; D.J.), ARC Centre of Excellence Scheme through Centre for Nanoscale BioPhotonics, Macquarie University Research Fellowship Scheme (X.X.), ARC LIEF grant (LE120100104, Y.D.), ARC Discovery Project (DP140102581, Y.D.), China

Scholarship Council CSC scholarships (no.201206170136, D.L., C.M.) and the Agency for Science, Technology and Research (A*STAR; grant no. 1231AFG028, X.L.).

Author contributions

D.J. and X.L. conceived the project and supervised the research; X.X., D.L., D.J. and X.L. designed the experiments; D.L., C.M., Y.Z. and S.W. conducted synthesis; X.Q. and X.X. conducted crystallography analysis and computational modelling; D.L., Y.D., S.D., W.R. and X.X. conducted characterizations and analysis; D.L., X.X., X.Q. and Y.Z. prepared figures and supplementary information sections; D.J., D.L., X.X., E.M.G., X.Qin. and X. Liu wrote the manuscript. All authors contributed to data analysis, discussions and manuscript preparation.

Additional information

Competing financial interests: The authors declare no competing financial interests.

14

Electrochemically driven mechanical energy harvesting

Sangtae Kim[1], Soon Ju Choi[2], Kejie Zhao[3], Hui Yang[4], Giorgia Gobbi[3,5], Sulin Zhang[4] & Ju Li[1,3]

Efficient mechanical energy harvesters enable various wearable devices and auxiliary energy supply. Here we report a novel class of mechanical energy harvesters via stress–voltage coupling in electrochemically alloyed electrodes. The device consists of two identical Li-alloyed Si as electrodes, separated by electrolyte-soaked polymer membranes. Bending-induced asymmetric stresses generate chemical potential difference, driving lithium ion flux from the compressed to the tensed electrode to generate electrical current. Removing the bending reverses ion flux and electrical current. Our thermodynamic analysis reveals that the ideal energy-harvesting efficiency of this device is dictated by the Poisson's ratio of the electrodes. For the thin-film-based energy harvester used in this study, the device has achieved a generating capacity of 15%. The device demonstrates a practical use of stress-composition–voltage coupling in electrochemically active alloys to harvest low-grade mechanical energies from various low-frequency motions, such as everyday human activities.

[1] Department of Materials Science and Engineering, Massachusetts Institute of Technology, Cambridge, Massachusetts 02139, USA. [2] Department of Mechanical Engineering, Massachusetts Institute of Technology, Cambridge, Massachusetts 02139, USA. [3] Department of Nuclear Science and Engineering, Massachusetts Institute of Technology, Cambridge, Massachusetts 02139, USA. [4] Department of Engineering Science and Mechanics, Pennsylvania State University, University Park, Pennsylvania 16802, USA. [5] Politecnico di Milano, Department of Mechanical Engineering, Milan, 20156, Italy. Correspondence and requests for materials should be addressed to J.L. (email: liju@mit.edu).

Efficient energy-harvesting devices, which convert energies otherwise wasted to electricity, help decentralize power generation and reduce the distance of electricity transmission. During the last decade, enormous efforts have been dedicated to the development of a variety of energy harvesters, capable of harvesting energy of various forms[1-5]. In mechanical energy harvesting alone, several types of energy generators have been demonstrated, such as piezoelectric[6], electrokinetic[7-9] or triboelectric generators[10], and enabled a wide range of applications[11-16]. Advances in processing techniques such as virus-directed designs[17] or block copolymer self-assembly[18] have also been reported. However, these energy generators are most efficient for vibrational energy harvesting at a relatively high frequency (~ 20–$100\,Hz$), and inherently limited in the low-frequency regime (0.5–5 Hz) where everyday human activities such as walking take place.

Herein we report a new type of mechanical energy harvester operative in the low-frequency regime. The device uses the stress-composition coupling in electrochemically active materials, such as partially Li-alloyed Si or Ge (refs 19–21). The coupling between mechanical stress and lithiation thermodynamics and kinetics[22] has been widely recognized in high-capacity anodes of lithium ion (Li$^+$) batteries, but was usually regarded as an adverse effect[19,23,24]. Here we demonstrate that mechanical bending induces different stress states in two identical partially Li-alloyed Si electrodes, which drives Li$^+$ migration and generates electricity. The prototype generator demonstrates power density of $0.48\,\mu W\,cm^{-2}$ at 0.3 Hz. Our thermodynamic and mechanics analyses lay a theoretical foundation for the device design and optimization.

Results

Working principle and device design. Figure 1 illustrates the working principle of the energy harvester, consisting of two partially Li-alloyed electrodes sandwiching an electrolyte. In the initial stress-free condition, the two electrodes are isopotential (point A in Fig. 1a and I in Fig. 1b). Bending the device generates net tension in one electrode and compression in the other (points B and C in Fig. 1a and II in Fig. 1b). The asymmetric stress creates a chemical potential difference that drives Li$^+$ ion migration from the compressed to the tensed electrode through the electrolyte (see Supplementary Note 1 for the analysis). At the same time, to maintain charge neutrality, electrons flow in the outer circuit, also from the compressive to the tensile sides, generating electrical power. The Li$^+$ migration continues until the potential difference vanishes (points B' and C' in Fig. 1a and III in Fig. 1b), establishing new equilibrium states on the two electrodes with different Li concentration. When the external stresses are removed by unbending the device, the chemical potential shifts on the electrodes (from point B' to point B'' and from point C' to point C'' in Fig. 1a and III–IV in Fig. 1b). The difference in lithium concentration between the electrodes drives Li$^+$ ion migration in the opposite direction (from point B'' and C'' to A in Fig. 1a and IV–I in Fig. 1b), thus discharging the device. The device goes back to its original equilibrium state and may go through this cycle multiple times provided that it operates in the viscoelastic regime without any irreversible damage. The electrical energy generated is equivalent to the potential difference multiplied by the amount of Li$^+$ migrated, as illustrated as red-colored area in Fig. 1a.

Guided by this vision, we developed a prototype generator, consisting of two identical electrodes sandwiching a separator soaked with electrolyte. We used amorphous Li$_x$Si ($x \sim 3.1$) thin film as the electrodes for its mechanical flexibility and reasonable lithiation and delithiation rates[25]. Ethylene carbonate mixed with ethyl methyl carbonate, LiPF$_6$ and micro-porous polypropylene monolayer[26] were used as the electrolyte, lithium salt and separator,

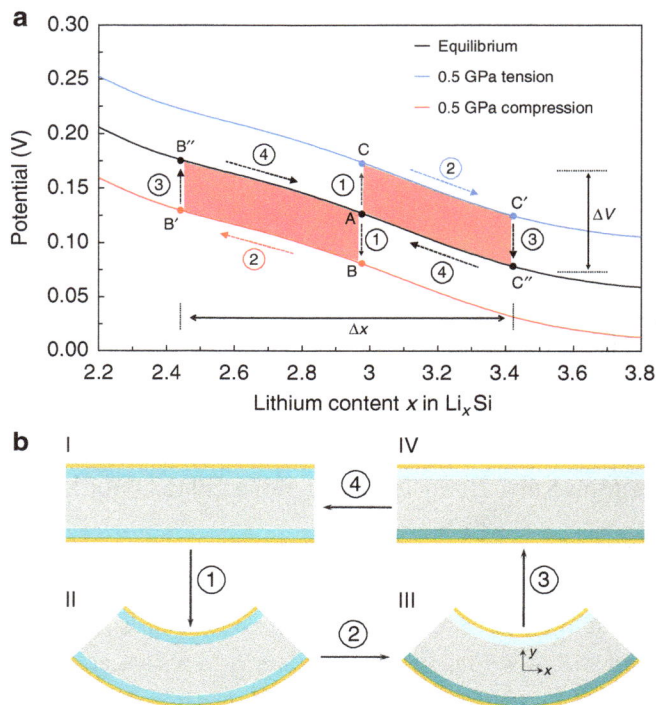

Figure 1 | The working principle behind the mechanical energy harvesting device. (a) Thermodynamic perspective on bending-unbending cycle. ① On introducing different stress states by bending, a chemical potential difference develops between two electrodes. ② When the electrodes are connected by an external circuit, new equilibrium under the stress states are established by Li$^+$ migration. ③ Once the stresses are removed, the lithiation states return back to the original equilibrium state. ④ The area covered by this cycle in red measures the energy output obtained. **(b)** Schematics of the cross-section of the device in operation.

respectively. We selected polyimide as the substrate to which the electrodes attach and Ag current collectors for their strong adhesion and stretchability[27]. The polyimide substrates were encapsulated by castable rubber, such as polydimethylsilane (PDMS) or polyurethane. Figure 2 shows the schematics of the device and the atomistic view of the active region. Each Li$_x$Si electrode is 249 nm thick, about two orders of magnitude thinner than the separator layer (25 μm). The thin-film configuration of the device allows large-curvature bending. On bending, the top, compressed electrode becomes an anode, while the bottom, tensed electrode a cathode. The device functions as an energy harvester, and because stress-driven Li$^+$ ion diffusion conforms to the Onsager linear-response behaviour, it is expected to exhibit a decent efficiency even with miniscule loads (that is, no threshold behaviour), converting mechanical energy input into electrical energy output.

Mechanics analysis of the device. Mechanics analysis provides insight into the energy conversion efficiency. Figure 1b illustrates the bending geometry. Bending the thin-film device generates compressive and tensile strains of equal magnitude ($\varepsilon_{xx} = \pm h/R$) on the top and bottom electrodes, respectively, where R is the radius of curvature and h the half thickness of the thin-film device. The stress state is obtained under the assumption of elastic deformation. Assuming a plane-stress condition along the y direction (see Fig. 1b for the coordinates), the stress on the bottom electrode can be written as follows,

$$\sigma_{xx} = \frac{E}{(1-v^2)}\frac{h}{R}, \quad \sigma_{zz} = \frac{vE}{(1-v^2)}\frac{h}{R}, \quad \sigma_{yy} = 0, \quad (1)$$

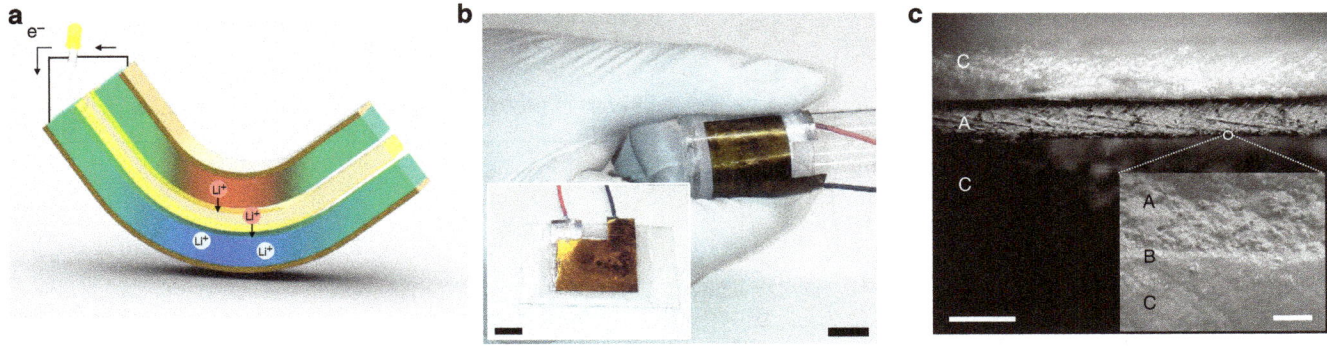

Figure 2 | A prototype of the mechanical energy harvester. (a) Schematic view of the device design. Compressed region is illustrated in red while the tensile region is illustrated in blue. Lithium ions migrating from the compressed plate to the tensile plate are shown with arrows. The electrolyte soaked separator is drawn in yellow. (b) An image of the actual device with a bending unit. Both scale bars indicate 1 cm. (c) Cross sectional image of the device showing polypropylene electrolyte layer (A in the figure), Li_xSi electrode on Ag current collector (B in the figure) and polyimide adhesion layer (C in the figure). The scale bars on the left and right indicate 40 and 2 μm, respectively.

where E is the Young's modulus and v the Poisson's ratio. The chemical potential difference between the two electrodes is only related to the difference in the hydrostatic stress $\sigma_{hydro} \equiv \mathrm{Tr}(\sigma)/3$

$$\Delta\mu = \Omega_{Li}\left(\frac{E}{1-v}\right)\frac{2h}{3R} \qquad (2)$$

Where $\Omega_{Li} = 14.95$ Å^3 is the estimated partial molar volume of Li in Li_xSi (refs 28,29). This gives a pressure sensitivity of 93 mV per GPa, close to a recent experimental measurement of 110 mV per GPa. (ref. 24).

It is noted that the deviatoric part of the stress tensor $\sigma_{deviatoric} \equiv \sigma - \sigma_{hydro}I$ does not couple to the chemical potential, where I is the identity tensor. Instead, the deviatoric stress $\sigma_{deviatoric}$ induces shear deformation which is volume conservative, thus does not generate electrical energy. The total strain-energy (U_{strain}) can be decoupled into the hydrostatic part and the deviatoric part, as:

$$U_{strain} = \frac{E}{2(1-v^2)}\left(\frac{h}{R}\right)^2 = U_{hydro} + U_{deviatoric}, \qquad (3)$$

with

$$U_{hydro} = \frac{\sigma_{hydro}^2}{2B}, \quad B = \frac{E}{3(1-2v)},$$

where B is the bulk modulus. Only the hydrostatic component U_{hydro} can be used for electricity generation, which also varies on lithium insertion and extraction. Assuming all of U_{hydro} can be used for electricity generation, while all of $U_{deviatoric}$ is wasted, an ideal efficiency can be expressed as

$$\eta \equiv \frac{U_{electrical}}{U_{strain}} = \frac{U_{hydro}}{U_{strain}} = \frac{(1-2v)(1+v)}{3(1-v)}. \qquad (4)$$

For $v = 0.25$ (ref. 30), an idealized efficiency of 27.8% is obtained. Interestingly, the energy conversion efficiency is independent of the Young's modulus, and maximizes at $v = 0$. The total amount of lithium that is expected to transport across the electrolyte to completely relax σ_{hydro} is

$$\Delta N_{Li} = \frac{V_{one\ side}}{\Omega_{Li}}\frac{\mathrm{Tr}(\sigma)}{3B} = \frac{V_{one\ side}}{\Omega_{Li}}\left(\frac{1-2v}{1-v}\right)\left(\frac{h}{R}\right), \qquad (5)$$

where $V_{one\ side}$ is the volume of one Li_xSi electrode affected by the radius of curvature and $V_{one\ side}/\Omega_{Li}$ the amount of Li present in the affected volume determined by previous reports[28]. It is noted that a thicker electrode increases the amount of migrating lithium and hence the capacity, however, it may also increase the

probability for developing structural inhomogeneities such as cracks during lithiation or bending.

Electrical energy output of the device. Just like photovoltaic energy harvesters, one could characterize a mechanical energy harvester by open-circuit voltage or short-circuit current. Figure 3a shows the open-circuit voltage, obtained by bending the device in the same direction for 30 s and resting the device for another 30 s with bending force released. The applied radius of curvature was ~1 cm, corresponding to a maximum tensile stress of 0.018 GPa generated in the electrodes. Since the yield stress of amorphous Li_xSi is ~1 GPa, the material deforms within its elastic regime[31]. We note that there exists a background potential (nonzero rest potential), which might be due to the side reactions such as a solid electrolyte interface layer formation or any unintentional inhomogeneity in composition between the two electrodes. The open-circuit voltage increases on bending and recovers its resting potential once the bending force is released. This trend is consistent with the relationship between the chemical potential difference and the applied radius of curvature ($\Delta\mu \sim R^{-1}$), shown in equation (2). We used experimentally measured value of Young's modulus (25 GPa) for lithiated amorphous silicon thin film, interpolated with rule of mixture for the specific composition we used[32]. Table 1 shows the predicted hydrostatic stress, voltage and measured voltage values at six different radii of curvature. The measured open-circuit voltages agree well with the predicted values.

Figure 4 shows the measured short-circuit current as the device is bent with a radius of curvature of 0.2 cm and then relaxed by releasing the bending forces. The bending and relaxation periods were 10 s each. Bending induces a sharp rise in the current, suggesting the stress-driven Li^+ migration inside. When holding the bending at a fixed radius of curvature, the current signal quickly reaches a maximum beyond which it decays gradually. This decay is due to the cancellation between the externally applied bending stress and Li insertion/extraction-induced stress in the electrodes. Specifically, externally applied bending creates tensile stress on one electrode and compressive on the other, generating a chemical potential difference that drives Li diffusion from the compressive to the tensile sides against the Li concentration gradient. At the same time, Li extraction from the compressed electrode and insertion into the tensed electrode attenuate both the compressive and tensile stresses, corresponding to a reduced driving force for Li^+ diffusion and to the gradual current decay. The current signal vanishes when the chemical potential between the two electrodes vanishes. The full

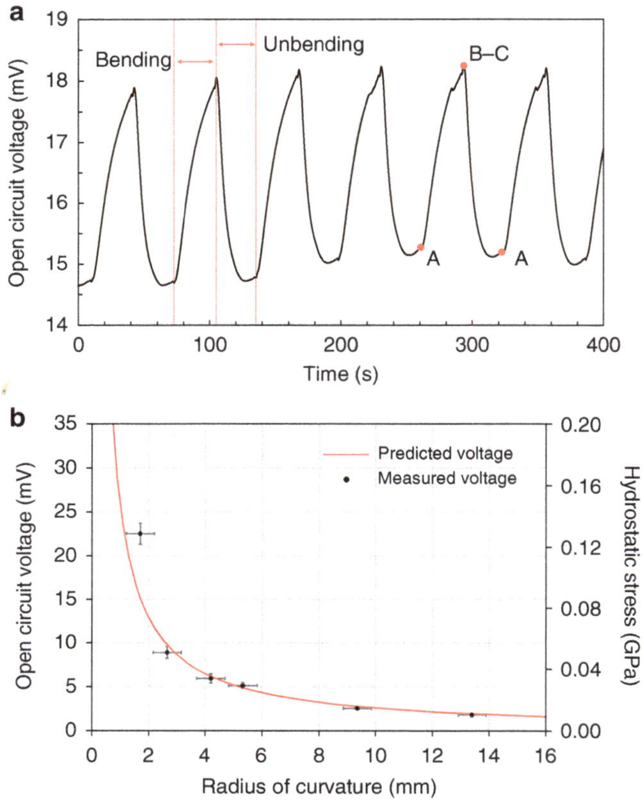

a

b

Figure 3 | Open circuit voltages measured during bending tests. (a) The open-circuit voltage measured from simple bending of the device. The measured values show clear voltage peaks during bending and releasing the device, each with 30 s interval. Each alphabetical points correspond to the bending geometry illustrated in Fig. 1b. **(b)** The predicted open-circuit voltage and hydrostatic stress according to the radii of curvature, operated in the elastic regime. s.e. resulted from at least five measurements for each radius of curvature is included. The measured voltage values agree well with the predicted values.

Table 1 | The predicted and measured voltage according to the radius of curvature values.

Radius of curvature (mm)	Hydrostatic stress (GPa)	$V_{predicted}$ (mV)	$V_{measured}$ (mV)
13.4	0.010	1.9	1.8
9.4	0.015	2.8	2.5
5.3	0.026	4.9	5.1
4.2	0.033	6.2	5.9
2.7	0.051	9.6	8.9
1.7	0.081	15.2	22.5

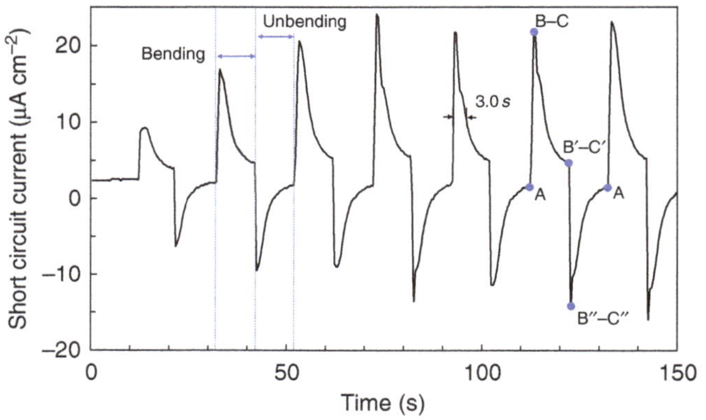

Figure 4 | Short circuit current density during bending tests. Bending was maintained at 2.0 mm radius of curvature with 10 s intervals. The positive peaks correspond to the current during bending and negative peaks to the current during unbending.

width at half maximum (time for peak current value to drop to half) was 3.0 s on average, which is two orders of magnitude greater than a typical piezoelectric device in similar geometry, promising the applicability of the device in the low-frequency regime[33].

On releasing the device from external bending force, Li insertion/extraction-generated stress difference along with the Li concentration gradient in the electrodes serves as the driving force for the backward Li$^+$ migration, corresponding to a sharp current increase in the opposite direction. As Li$^+$ migration continues, the stress and concentration difference between the two electrodes drop, so as the driving force for Li$^+$ diffusion, leading to a reduced current signal. The current signal vanishes when the two electrodes recover to their original, isopotential state. We also performed measurements by alternating the bending directions and the results were consistent with the trends described above, as illustrated in Supplementary Note 2.

The amount of Li$^+$ ions that migrate during each bending cycle is equivalent to the area under a current peak. In Fig. 4, the area under a peak is $\sim 73\,\mu C$. The volume of the affected area during the experiment is estimated to be $1.56 \times 10^{-5}\,cm^3$. (The bending geometry during experiment is shown in Supplementary Note 3.) According to the analysis above, the expected Li$^+$ ion migration is equivalent to $487\,\mu C$. The reversible lithium migration of $73\,\mu C$ is $\sim 15\%$ of the theoretically predicted amount. The reduced generating capacity per cycle may be caused

by 'self-discharging', that is, the electrons diffuse out of the bent region to the nearby flat regions within the same Li$_x$Si electrode without going to the external circuit, which actually causes the flat regions to curve. Such self-discharging behaviour is predicted to occur whenever there are unequal bending curvatures in the lateral direction (gradient in the bending curvature) and is expected to be a significant cause in low experimentally measured efficiency of the device. (The experimentally measured efficiency is estimated in Supplementary Note 4.) This internal electrochemical dissipation by self-discharging can be greatly reduced by applying a uniform radius of curvature throughout the device.

Device durability during repeated bending test. Figure 5 shows the data from two kinds of repetitive bending fatigue tests. In repeated open-circuit voltage tests, we observe that not only the peak height but also the background voltage are reduced over cycling, despite bent at moderate radius of curvature (10 mm). This is due to the damage accumulation in the electrodes. In open-circuit voltage measurements, lithium ions cannot migrate between the electrodes to relax the stresses developed in the electrodes. As a result, damages are accumulating in the electrodes in the form of pore formation, or even fracture. Since the films can no longer sustain the elastic stress, the peak heights decrease. In addition, it is expected that the electrode under repetitive tension exhibits different damage accumulation from that under repetitive compression. This difference results in the potential difference between the two electrodes and changes the background potential. In repeated short-circuit current tests, in contrast, we observe no major degradation in the peak height. The data in Fig. 5b shows reliable current generation during 1,500 repeated bending cycles. The background current is slowly moving towards zero during cycling in a stable manner. Data from repeated bending under different radii of curvature is available in Supplementary Note 5. These results indicate that the electrode materials maintain their homogeneity during cyclic loading of 1,500 cycles. The reliable performance of the device during short-circuit fatigue test is enabled as the migrating lithium ions act as effective stress reliever in the electrode, much the same way as in Nabarro–Herring creep[34]. It might thus be possible to adapt our active mechanical–passive electrochemical fatigue tests as a diagnostic tool to characterize damage creation, repair and accumulation inside electrode materials, as a complementary technique to the electrochemically driven battery cycling testing paradigm that monitors voltage-induced mechanical strain (such as film curvature change).

Since our generator should operate in the elastic regime to minimize damage accumulation, the yield surface of electrode material sets the limit on the voltage output. The atomic volume

of lithium is known to be approximately constant in a wide range of lithium alloys independent of lithium content[28], and thus, an alloy with the greatest yield strength promises the highest voltage output. It is also noteworthy that our device does not cause composition change nearly as large as that in battery electrode reactions, therefore many candidate alloys that do not cycle well as battery electrodes may still be excellent candidates for our generator device. Alloys of larger ions such as sodium or potassium may also be used to provide higher voltage output, as a higher atomic volume promises higher voltage.

Discussions

Since the electricity generation is driven by Li^+ ion diffusion, the effect of bending rate and frequency can be understood based on the characteristic time scales of Li^+ ion diffusion. While there are time scales for mechanical equilibrium and diffusion equilibrium, the mechanical equilibrium in the elastic range is established very fast. Thus, potential bending rate effect comes from the competition between applied strain rate and diffusion rate. Using a typical value of $D_{Li} = 10^{-10}\,cm^2\,s^{-1}$ and letting 250 nm be the characteristic diffusion distance, we estimate that the characteristic diffusion time is about 6 s. As long as strain rate exceeds Li^+ ion diffusion rate characterized by the width of the short-circuit current signal, we observe identical current curves with similar full width at half maximum. If strain rate is slower, a current signal with extended width and reduced peak current is expected. In either case, the total energy output or repeatability is not expected to change owing to the bending rate. Similarly, slow bending frequency would allow sufficient time for Li^+ ions to migrate between the electrodes and does not affect the output energy. These are demonstrated experimentally and are illustrated in Supplementary Note 6. If bent at a frequency so high that Li^+ ions do not have sufficient time to migrate between electrodes, the electrodes are not able to relieve stress by lithium insertion/extraction and will eventually fail by fatigue. This is in part equivalent to low temperature condition in Nabarro–Herring creep. This frequency, however, far exceeds the human activity timescale. The device is therefore not well suited for vibrational energy harvesting at high frequency ($\gg 100\,Hz$).

The energy output generated from the device in general exhibits a greater amount of current and less voltage when compared with the ceramic piezoelectric generators. Table 2 shows the comparison of peak power and energy output operated at 0.3 Hz frequency for known piezoelectric generators and our device. The piezoelectric generators cited in Table 2 have similar geometry to our device and the difference in energy output comes mainly from the material properties. The comparison is based on the area of the thin-film generators without considering film thickness, as the microstructure of the materials as well as the effect of thickness on performance differ significantly among the reported generators. In Table 2, PMN-PT (ref. 11) refers to $(1-x)Pb(Mg_{1/3}Nb_{2/3})O_3–xPbTiO_3$ and is a single crystalline thin film, while $BaTiO_3$ (ref. 33) is a nanoparticle composite. KNLN

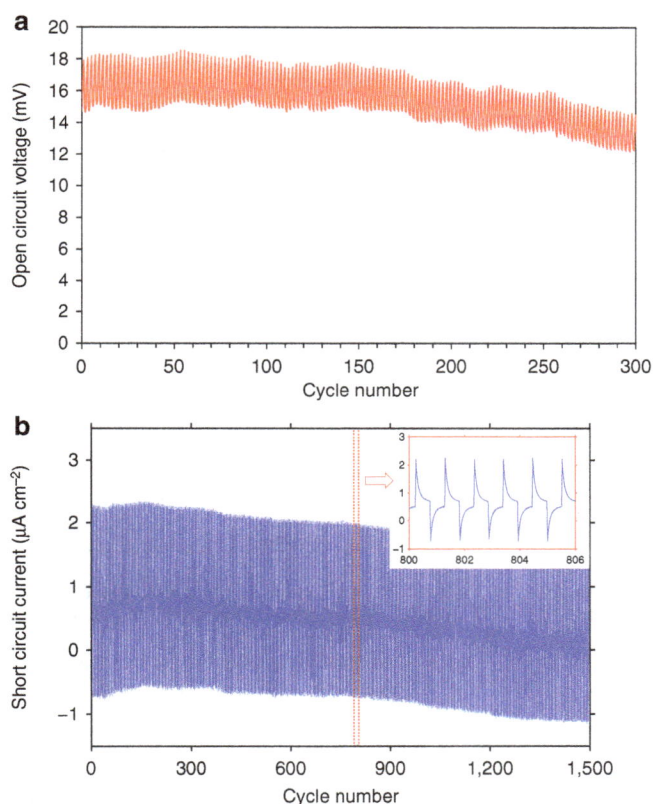

Figure 5 | Electricity generated during repeated bending tests. (a) The open-circuit voltage at 10 mm radius of curvature and **(b)** short-circuit current collected during repeated bending tests at 4.0 mm radius of curvature. The nested figure shows the zoomed-in view of the 800th–807th bending cycle.

Table 2 | Comparison of our energy harvester to piezoelectric generators in similar form factor.

	Active material thickness (µm)	Active area (cm²)	V_{oc} (V)	I_{sc} (µA)	Peak power (µW cm⁻²)	Energy output at 0.3 Hz (µJ cm⁻² s⁻¹)
Our device	0.25 × 2	0.63	0.023	14.5	0.53	0.476
PMN-PT	8.4	2.89	8.2	145	411	12.3
BaTiO₃	250	12	3.2	0.35	0.093	0.00278
ZnO nanowire[15]	0.05	1	20	6	120	3.60
KNLN	250	9	12	1.2	1.60	0.0480

(ref. 35) refers to $072(K_{0.480}Na_{0.535})NbO_3–0.28LiNbO_3$ and is a nanoparticle composite. Peak power is considered to be the direct product of short-circuit current and open-circuit voltage. The energy output at 0.3 Hz is calculated, assuming that the average peak width (full width at half maximum) of piezoelectric generators is 100 ms and that the peak width of our device is 3 s. As shown in Table 2, the peak power for our device is in general less than those of piezoelectric generators. Nevertheless, when operated under low frequency such as 0.3 Hz, the amount of energy generated per second is comparable to the best non-lead containing piezoelectric generators in the same form factor; our device outperforms those made of $BaTiO_3$ or KNLN by at least one order of magnitude. The PMN-PT-based generators and ZnO (ref. 15) nanowire generators outperform our device; nevertheless, these devices are elaborately optimized. PMN-PT generators consist of 20-µm-thick single crystalline thin film, grown and processed from an ingot of the material and sapphire substrate. The ZnO nanowire generators were optimized by coating the surface with layer-by-layer self-assembled polymer layer. Considering that both generators have gone through a significant amount of optimization, our device, with further optimization in material selection and architecture design, holds promise as an efficient energy harvester at low frequencies.

We have developed a novel type of mechanical energy harvester based on the fact that lithium ions have finite size $(14.95\ \text{Å}^3)$, and thus would migrate across an electrolyte membrane under a pressure difference–akin to reverse osmosis in seawater desalination. But because lithium ions carry $+e$ charge, it would drive electron flow in the outer circuit, same as how a typical battery would work. The device achieves long current pulse duration, which has not been achieved by other types of mechanical energy-harvesting devices. Owing to this characteristic, the device exhibits higher average energy output than most other piezoelectric generators when operating at low frequencies. This work also opens avenues for optimizing electrochemical devices coupled to mechanical stress for sensing and actuation, as well as the development of active mechanical–passive electrochemical tests as an alternative diagnostic protocol to study damage creation and accumulation in electrochemomechanically active solids.

Methods

Flexible substrate choice and electrode deposition. We chose Kapton Polyimide (PI) film as the flexible substrate. PI films provide flexibility, chemical compatibility with lithium ion electrolytes and strong adhesion between the film and electrode materials[27]. The device fabricated on several other flexible substrates resulted in electrode film cracking, little electrical conductivity or delamination during electrochemical lithiation step. On PI substrate, current collector consisting of 15 nm Cr and 100 nm Ag is deposited using electron beam evaporator at 1 and $2\ \text{Å s}^{-1}$ rate, respectively. On Ag thin film, amorphous Si electrode of 75 nm (Kurt Lesker, p-type) is deposited using e-beam evaporator. Thick silicon film (75 nm) was selected carefully to avoid potential structural inhomogeneities such as pores and cracks according to previous studies[25]. A 3 nm Ti adhesion layer was used between Ag and Si. A small region of Ag was left free of silicon for electrical connection.

Electrochemical lithiation. Lithium was inserted into silicon by electrochemical lithiation. An electrochemical cell consisting of the Si on PI, electrolyte (ethylene carbonate:ethyl methyl carbonate = 1:1 with $LiPF_6$, Novolyte) soaked separator (Celgard 2500) and thick Li foil (Alfa Aesar, 1.5 mm thick) was lithiated using Gamry Reference 3000 potentiostat at 0.5 C, up to 80% of the theoretical capacity ($2,880\ \text{mAh g}^{-1}$ out of $3,600\ \text{mAh g}^{-1}$). This corresponds to the film composition of approximately $Li_{3.1}Si$. The low cutoff voltage was set to 0.05 V to avoid nucleation of crystalline phase[36] and ensure that the film remains as homogeneous as possible as an amorphous Li–Si alloy. Previous studies have reported that amorphous Si thin films lithiated at C/10 remain smooth to nanometre scale[37]. Control experiments performed with films lithiated at C/2 and C/10 produced identical result within the scope of error and all lithiation was performed at C/2 in this study.

Device assembly and testing. The electrode prepared above was mounted onto thin PDMS film (Sylgard 184, Dow Corning Chemical) for easy handling and encapsulation. PDMS films were prepared by mixing the elastomer with hardener at 10:1 mass ratio and curing at 65 °C for 2 h. Paraffin wax was dissolved onto PDMS as illustrated in previous report[38] to avoid gas and vapour permeability in PDMS. When extra inhibition to gas permeation was required, for example, in repetitive bending test, the entire device was encapsulated in mylar bags often used to manufacture pouch type secondary batteries. The electrode on PI was bonded to cured PDMS by using a thin layer of uncured PDMS as glue. The electrode was then cut in half to comprise the bottom and top electrodes of the energy-harvesting device. The two electrodes were placed on top of each other, separated by a layer of electrolyte-soaked separator (Celgard 2500) and were sealed using uncured PDMS on the sides.

To eliminate any possible difference in electrochemical potential or composition between the top and bottom electrode, the two electrodes were left connected in short-circuit via external wire for at least 2 h.

Bending tests were performed either by finger tapping or by servo motor (HS-7966HB, Hitec). A custom repeatable bending station was constructed using the servo motor, a 32 pitch gear (Servocity), a pinion and Arduino controller. The strain rate was ∼ 1% per second.

References

1. Xu, S. *et al.* Self-powered nanowire devices. *Nat. Nanotechnol.* **5**, 366–373 (2010).
2. Snyder, G. J. & Toberer, E. S. Complex thermoelectric materials. *Nat. Mater.* **7**, 105–114 (2008).
3. Priya, S. & Inman, D. *Energy Harvesting Technologies* (Springer US, 2009).
4. Beeby, S. P., Tudor, M. J. & White, N. M. Energy harvesting vibration sources for microsystems applications. *Meas. Sci. Technol.* **17**, R175–R195 (2006).
5. Ravindran, S. K. T., Huesgen, T., Kroener, M. & Woias, P. A self-sustaining micro thermomechanic-pyroelectric generator. *Appl. Phys. Lett.* **99**, 104102 (2011).
6. Anton, S. R. & Sodano, H. A. A review of power harvesting using piezoelectric materials (2003–2006). *Smart Mater. Struct.* **16**, R1–R21 (2007).
7. Lu, M.-C., Satyanarayana, S., Karnik, R., Majumdar, A. & Wang, C.-C. A mechanical-electrokinetic battery using a nano-porous membrane. *J. Micromechanics Microengineering* **16**, 667–675 (2006).
8. Kim, D.-K., Duan, C., Chen, Y.-F. & Majumdar, A. Power generation from concentration gradient by reverse electrodialysis in ion-selective nanochannels. *Microfluid. Nanofluidics* **9**, 1215–1224 (2010).
9. Moon, J. K., Jeong, J., Lee, D. & Pak, H. K. Electrical power generation by mechanically modulating electrical double layers. *Nat. Commun.* **4**, 1487 (2013).
10. Fan, F.-R., Tian, Z.-Q. & Lin Wang, Z. Flexible triboelectric generator. *Nano Energy* **1**, 328–334 (2012).
11. Hwang, G.-T. *et al.* Self-powered cardiac pacemaker enabled by flexible single crystalline PMN-PT piezoelectric energy harvester. *Adv. Mater.* **26**, 4880–4887 (2014).
12. Park, K.-I. *et al.* Highly-efficient, flexible piezoelectric PZT thin film nanogenerator on plastic substrates. *Adv. Mater.* **26**, 2514–2520 (2014).
13. Wang, S. *et al.* Maximum surface charge density for triboelectric nanogenerators achieved by ionized-air injection: methodology and theoretical understanding. *Adv. Mater.* **26**, 6720–6728 (2014).
14. Fan, F. R. *et al.* Highly transparent and flexible triboelectric nanogenerators: performance improvements and fundamental mechanisms. *J. Mater. Chem. A* **2**, 13219 (2014).
15. Hu, Y., Lin, L., Zhang, Y. & Wang, Z. L. Replacing a Battery by a nanogenerator with 20 V Output. *Adv. Mater.* **24**, 110–114 (2012).
16. Hill, D., Agarwal, A. & Tong, N. *Assessment of Piezoelectric Materials for Roadway Energy Harvesting. California Energy Commission.* Report No: CEC-500-2013-007 (DNV KEMA Energy & Sustainability, Oakland, California, USA, 2014).
17. Jeong, C. K. *et al.* Virus-directed design of a flexible $BaTiO_3$ nanogenerator. *ACS Nano* **7**, 11016–11025 (2013).
18. Jeong, C. K. *et al.* Topographically-designed triboelectric nanogenerator via block copolymer self-assembly. *Nano Lett.* **14**, 7031–7038 (2014).
19. Sheldon, B. W., Soni, S. K., Xiao, X. & Qi, Y. Stress contributions to solution thermodynamics in Li-Si alloys. *Electrochem. Solid State Lett.* **15**, A9–A11 (2011).
20. Gu, M. *et al.* Bending-induced symmetry breaking of lithiation in germanium nanowires. *Nano Lett.* **14**, 4622–4627 (2014).
21. Yang, H., Liang, W., Guo, X., Wang, C.-M. & Zhang, S. Strong kinetics-stress coupling in lithiation of Si and Ge anodes. *Extreme Mech. Lett.* **2**, 1–6 (2015).
22. He, K. *et al.* Transitions from near-surface to interior redox upon lithiation in conversion electrode materials. *Nano Lett.* **15**, 1437–1444 (2015).

23. Zhao, K. *et al.* Lithium-assisted plastic deformation of silicon electrodes in lithium-ion batteries: a first-principles theoretical study. *Nano Lett.* **11**, 2962–2967 (2011).

24. Sethuraman, V. A., Srinivasan, V., Bower, A. F. & Guduru, P. R. In situ measurements of stress-potential coupling in lithiated silicon. *J. Electrochem. Soc.* **157**, A1253 (2010).

25. Ohara, S., Suzuki, J., Sekine, K. & Takamura, T. A thin film silicon anode for Li-ion batteries having a very large specific capacity and long cycle life. *J. Power Sources* **136**, 303–306 (2004).

26. Cannarella, J. *et al.* Mechanical properties of a battery separator under compression and tension. *J. Electrochem. Soc.* **161**, F3117–F3122 (2014).

27. Sim, G.-D. *et al.* Improving the stretchability of as-deposited Ag coatings on poly-ethylene-terephthalate substrates through use of an acrylic primer. *J. Appl. Phys.* **109**, 073511 (2011).

28. Obrovac, M. N., Christensen, L., Le, D. B. & Dahn, J. R. Alloy design for lithium-ion battery anodes. *J. Electrochem. Soc.* **154**, A849 (2007).

29. Zhao, K. *et al.* Reactive flow in silicon electrodes assisted by the insertion of lithium. *Nano Lett.* **12**, 4397–4403 (2012).

30. Shenoy, V. B., Johari, P. & Qi, Y. Elastic softening of amorphous and crystalline Li–Si Phases with increasing Li concentration: A first-principles study. *J. Power Sources* **195**, 6825–6830 (2010).

31. Soni, S. K., Sheldon, B. W., Xiao, X., Bower, A. F. & Verbrugge, M. W. Diffusion mediated lithiation stresses in Si thin film electrodes. *J. Electrochem. Soc.* **159**, A1520–A1527 (2012).

32. Hertzberg, B., Benson, J. & Yushin, G. Ex-situ depth-sensing indentation measurements of electrochemically produced Si–Li alloy films. *Electrochem. Commun.* **13**, 818–821 (2011).

33. Park, K.-I. *et al.* Flexible nanocomposite generator made of BaTiO3 nanoparticles and graphitic carbons. *Adv. Mater.* **24**, 2999–3004 (2012).

34. Nabarro, F. R. N. Steady-state diffusional creep. *Philos. Mag.* **16**, 231–237 (1967).

35. Jeong, C. K., Park, K.-I., Ryu, J., Hwang, G.-T. & Lee, K. J. Large-area and flexible lead-free nanocomposite generator using alkaline niobate particles and metal nanorod filler. *Adv. Funct. Mater.* **24**, 2620–2629 (2014).

36. Obrovac, M. N. & Christensen, L. Structural changes in silicon anodes during lithium insertion/extraction. *Electrochem. Solid State Lett.* **7**, A93 (2004).

37. Soni, S. K., Sheldon, B. W., Xiao, X. & Tokranov, A. Thickness effects on the lithiation of amorphous silicon thin films. *Scr. Mater.* **64**, 307–310 (2011).

38. Ren, K., Zhao, Y., Su, J., Ryan, D. & Wu, H. Convenient method for modifying poly(dimethylsiloxane) to be airtight and resistive against absorption of small molecules. *Anal. Chem.* **82**, 5965–5971 (2010).

Acknowledgements

J.L. acknowledges the support by NSF CBET-1240696. S.Z. acknowledges the support by NSF CMMI-0900692. S.K. acknowledges financial support from Samsung Scholarship Foundation and S.J.C. from Kwanjeong Educational Foundation. The device fabrication was performed in part at the Center for Nanoscale Systems at Harvard University, a member of the National Nanotechnology Infrastructure Network, which is supported by the National Science Foundation under NSF award no. ECS-0335765. S.K. and S.J.C. appreciate partial support from MIT MADMEC contest and helpful discussions from Michael Tarkanian, Soonwon Choi and Akihiro Kushima, and Vincent Chevrier for the potential-composition data on amorphous Li$_x$Si.

Author contributions

J.L. conceived the project. S.K., S.J.C. and J.L. designed the device. S.J.C. and S.K. fabricated samples and performed electrochemical tests. All authors analysed the data and contributed to the discussion. S.K., S.Z. and J.L. wrote the manuscript.

Additional information

15

Optimal metal domain size for photocatalysis with hybrid semiconductor-metal nanorods

Yuval Ben-Shahar[1], Francesco Scotognella[2], Ilka Kriegel[2], Luca Moretti[2], Giulio Cerullo[2], Eran Rabani[3,4] & Uri Banin[1]

Semiconductor-metal hybrid nanostructures offer a highly controllable platform for light-induced charge separation, with direct relevance for their implementation in photocatalysis. Advances in the synthesis allow for control over the size, shape and morphology, providing tunability of the optical and electronic properties. A critical determining factor of the photocatalytic cycle is the metal domain characteristics and in particular its size, a subject that lacks deep understanding. Here, using a well-defined model system of cadmium sulfide-gold nanorods, we address the effect of the gold tip size on the photocatalytic function, including the charge transfer dynamics and hydrogen production efficiency. A combination of transient absorption, hydrogen evolution kinetics and theoretical modelling reveal a non-monotonic behaviour with size of the gold tip, leading to an optimal metal domain size for the most efficient photocatalysis. We show that this results from the size-dependent interplay of the metal domain charging, the relative band-alignments, and the resulting kinetics.

[1] The Institute of Chemistry and Center for Nanoscience and Nanotechnology, The Hebrew University of Jerusalem, Edmond Safra Campus Givat-Ram, Jerusalem 91904, Israel. [2] Dipartimento di Fisica, IFN-CNR, Politecnico di Milano, 20133 Milan, Italy. [3] Department of Chemistry, University of California and Lawrence Berkeley National Laboratory, Berkeley, California 94720-1460, USA. [4] The Sackler Institute for Computational Molecular and Materials Science, Tel Aviv University, Tel Aviv 69978, Israel. Correspondence and requests for materials should be addressed to G.C. (email: Giulio.Cerullo@ polimi.it) or to E.R. (email: Eran.Rabani@Berkeley.edu) or to U.B. (email: Uri.Banin@mail.huji.ac.il).

The synergistic optical and chemical properties of semiconductor-metal hybrid nanoparticles lead to light-induced charge separation[1,2], opening the path for their function as photocatalysts in redox reactions, including hydrogen generation by water splitting[3–7], generation of radicals and photodegradation of organic contaminants[8]. The complexity of the photocatalytic cycle, in particular for the fuel generating water splitting reaction, requires a reductionist approach to address separately the effect of size, shape and composition of each component towards the rational design of an efficient photocatalytic system.

The role of the exciton dynamics, related to the semiconductor component, has been investigated in particular in prototypical TiO_2-based systems and in colloidal semiconductor-metal hybrid nanorods (NRs). These studies pointed out to the potential of tailoring the semiconductor component in the hybrid nanoparticles, permitting fine tuning of the light absorption and electron–hole dissociation as preliminary steps to charge separation and catalysis[9–11]. The metal co-catalyst component has a particularly important role in the photocatalytic cycle. The charge separation from the semiconductor is directly affected by the metal domain characteristics and the actual catalysis occurs on its surface[12,13]. Control over the size of the metal domain and its optimization is an essential parameter for the rational photocatalytic system design. The size dependence of catalysis on bare Au islands deposited on titania was investigated revealing sharp optimal catalytic performance for CO oxidation at island thickness of ~ 2 atomic layers corresponding to ~ 3 nm in diameter[14], and in the seminal work of Goodman and co-workers, this was attributed to a metal to non-metal transition[15]. In relation to photocatalytic activity, early studies discussed the effect of size on Fermi-level equilibration related to charging of the metal domain following irradiation of the system[16,17]. The effect of the metal co-catalyst size, particularly on the hydrogen generation, was addressed in the context of TiO_2-Au and CdS-Pt with multiple metal domains[18,19]. While the former observed no size effect for the Au domain varying between 3 and 12 nm, the latter has reported optimal hydrogen generation in extremely small Pt clusters (~ 50 atoms). However, a detailed mechanistic description, particularly in view of these diverse behaviours, is missing and is important for further development of such hybrid nanoparticles in the context of photocatalysis.

In this work we address the effect of the metal co-catalyst size on the photocatalysis process, using hybrid semiconductor-metal nanoparticles with a single catalytic domain as a model system. By a combination of ultrafast transient absorption measurements, hydrogen generation yield study and theoretical modelling, we observe and explain a non-monotonic metal domain size dependence of the hydrogen generation efficiency.

Results

Size-controlled Au tip growth on CdS NRs.
CdS NRs of 31.6 nm length and 3.9 nm diameter were synthesized using modification of a previously reported procedure employing seeded growth[20] (Supplementary Fig. 1). Site-selective Au deposition on a single-rod apex, with high control of the metal tip size, was achieved by combining two different synthetic approaches consecutively (Fig. 1a). First, a dark reaction was used to obtain site-selective small metal islands growth on the apexes of the CdS NRs (refs 21,22). As can be seen in the transmission electron microscopy (TEM) image in Fig. 1c, a single small metal-tip of ~ 1.5–1.8 nm diameter grows on one apex with narrow size distribution ($\sim 7\%$). This selective growth takes place due to the favoured surface reactivity of the $(00\bar{1})$ facet of the CdS rod, that encourages the heterogeneous nucleation of gold. Furthermore,

the use of long alkyl chain amines such as octadecylamine, instead of shorter ones, minimizes Au nucleation on the less reactive facets such as the sides of the rod. Owing to the length of the alkyl chain a phase transition to a static phase occurs at lower temperatures thus blocking the Au precursor's access to the NR surfaces[23]. These small Au domains serve as seeds for the second step, using light-induced Au growth under inert atmosphere and at low temperature, 2–4 °C (Supplementary Methods). This approach allows for size control by changing the irradiation times and the Au^{3+}/NRs ratio. Figure 1d–f show TEM images of CdS-Au hybrid nanoparticles with different Au tip sizes and with narrow size distribution (Fig. 1g). Figure 1b presents the absorption spectra of the bare CdS NRs and CdS-Au hybrid nanoparticles with different Au tip sizes. All spectra exhibit a similar sharp rise at 460 nm related to the onset of the CdS NRs absorption. Several absorption features are seen to the blue of the onset, related to the band gap and to higher excited optical transitions of the CdS NRs, which signify the good size monodispersity of the sample. A plasmon peak develops at 540 nm, correlated with the increase of the metal tip size. Phase transfer to aqueous solution was performed with polyethyleneimine which was recently reported as high performance surface coating for photocatalytic applications and provides good colloidal stability[11].

Transient absorption measurements of charge separation. The effect of the different Au domain sizes on charge carrier dynamics was studied using broadband ultrafast transient absorption spectroscopy. In Fig. 2a we show the time-resolved differential transmission ($\Delta T/T$) spectra for the CdS NRs and the series of Au tip sizes. Following 450 nm optical excitation and formation of excited electron–hole pairs, the $\Delta T/T$ spectra at early delay times reveal a pronounced bleach around 450 nm in both CdS NRs and CdS-Au hybrid NRs attributed to depletion of the first excitonic transition in the CdS rods due to electron state filling[24]. In addition, for the larger Au tips, a broad bleach feature develops ~ 540 nm, which corresponds to the plasmonic feature of the Au tips[25]. The decay of the plasmonic feature, by electron–phonon scattering, showed no size dependence and the measured lifetime for all Au domain sizes was 1.5–1.7 ps (Supplementary Figs 6–8) consistent with prior reports for colloidal Au nanoparticles[25,26].

Comparison between the normalized transient absorption dynamics for the bleach recovery in the spectral region of the CdS band gap exciton is presented in Fig. 2b for the different Au tip sizes. In the bare CdS NRs (upper panel, Fig. 2b), a fast decay component with a small amplitude is observed, and is related to residual cooling of the electrons to the CdS conduction band edge. This is followed by a long decay component, corresponding to exciton recovery to the ground state. These bleach recovery kinetics were fitted to a bi-exponential function and resulted in a time constant for electron–hole recombination of 3.4 ns (Supplementary Table 3). At the presence of a metal tip a clear additional timescale is introduced, slower than the rapid cooling and faster than the electron–hole recombination. The amplitude of the intermediate decay process increases with increasing metal domain size. We assign this additional timescale to the electron transfer from the CdS to the metal tip. The measured kinetic decays were fitted to a tri-exponential function (Supplementary Note 3). The charge transfer rates extracted from this fit were found to be 16 ps for 6.2 nm, 29 ps for 4.8 nm, 103 ps for 3.0 nm and 770 ps for 1.6 nm Au tips sizes, with amplitudes decreasing from 40 to 15% for smaller tip sizes.

Photocatalytic hydrogen production efficiency measurements. The photocatalytic activity of the same hybrid nanoparticles was

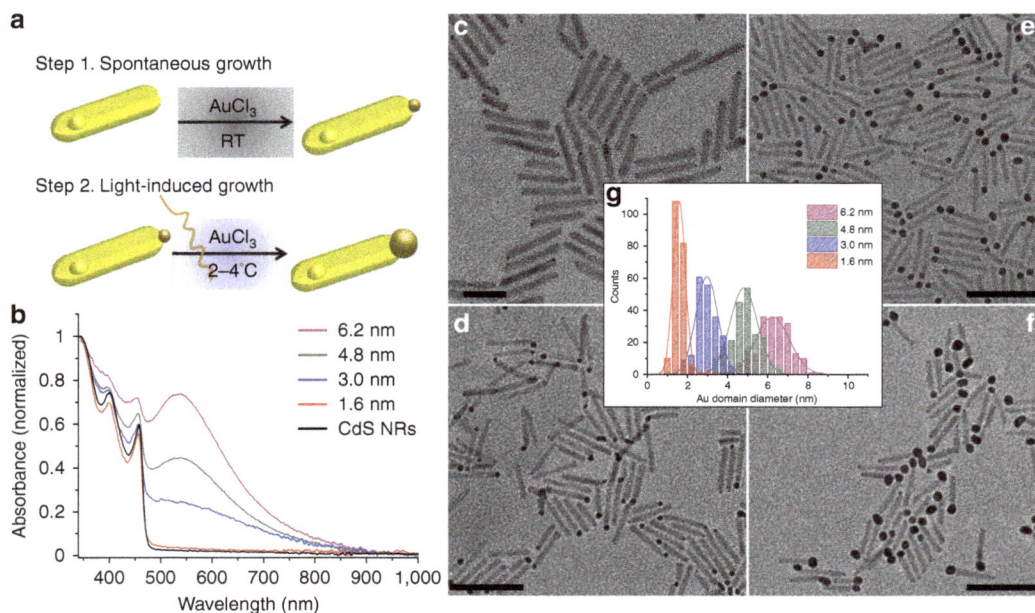

Figure 1 | Growth of size-controlled semiconductor-metal nanohybrids. (**a**) A scheme of the two-step metal growth deposition. (**b**) UV-vis absorbance spectra of CdS-Au hybrid nanoparticles showing the development of the plasmonic feature ~540 nm as the Au tip size increases. TEM images of CdS-Au hybrid nanoparticles with 1.5 ± 0.2 nm Au tip size after 1 h dark synthesis (**c**) and light-induced synthesis for 30 min with various CdS:Au molar ratio leading to Au tip size of of 3.0 ± 0.5 nm (**d**) 4.8 ± 0.7 nm (**e**), and 6.2 ± 0.8 nm (**f**). Scale bars; **c**, 20 nm; **d–f**, 50 nm. (**g**) size distribution histogram of the Au metal tip diameters.

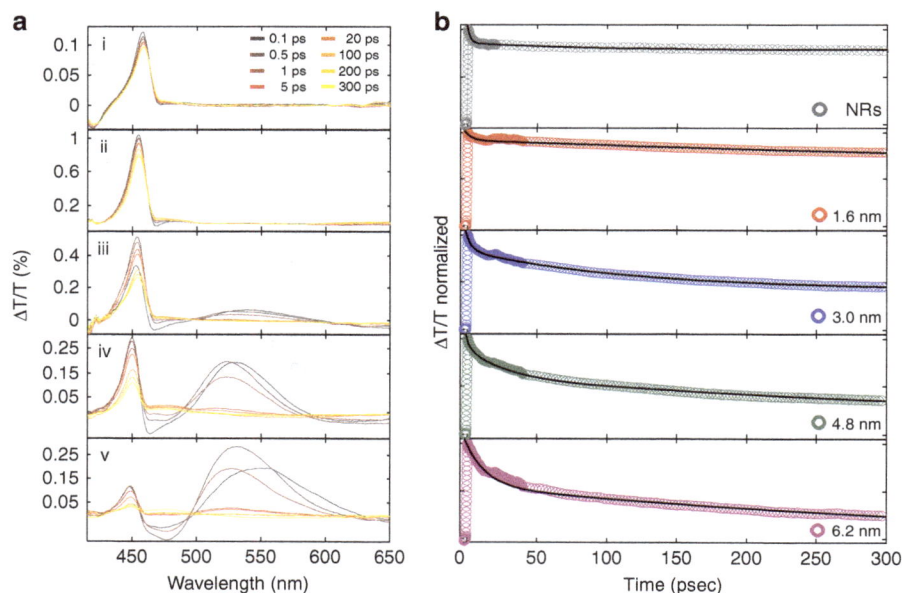

Figure 2 | Ultrafast spectroscopy of nanohybrids. (**a**) Transient absorption spectra of CdS NRs (i) and CdS-Au hybrid nanoparticles for different Au metal tip sizes including 1.6 nm (ii), 3.0 nm (iii), 4.8 nm (iv) and 6.2 nm (v) at 450 nm excitation. (**b**) Corresponding normalized transient absorption dynamics of the bleach recovery at 450 nm, attributed to the first excitonic transition of the CdS NR component for CdS NRs and CdS-Au hybrid nanoparticles with different Au metal tip sizes.

measured via the light-induced reduction of water at 405 nm to produce hydrogen in the presence of Na_2S-$9H_2O$ and Na_2SO_3, acting as sacrificial hole scavengers as depicted schematically in Fig. 3a (inset)[27]. The amount of hydrogen gas produced was determined using a gas chromatograph equipped with a thermal conductivity detector. Figure 3a, blue curve, shows the hydrogen generation rate versus Au tip size. A weak dependence, observed for the two smallest sizes, is followed by a marked decrease of the rate in the larger tips, overall by a factor of nearly 10 comparing the maximum and the minimum rates.

These results represent the actual hydrogen evolution rates, but they do not account for photons that are absorbed directly by the intraband transitions of the metal tip and do not contribute to the generation of hydrogen due to their rapid relaxation. The total absorption of such CdS-Au hybrid nanoparticles can be considered in first approximation as the superposition of the contributions of the exciton and the plasmon[28]. The red curve in Fig. 3a corrects for this, by normalizing the rates to the semiconductor component absorption. The actual concentration of each sample was obtained by inductively coupled plasma mass

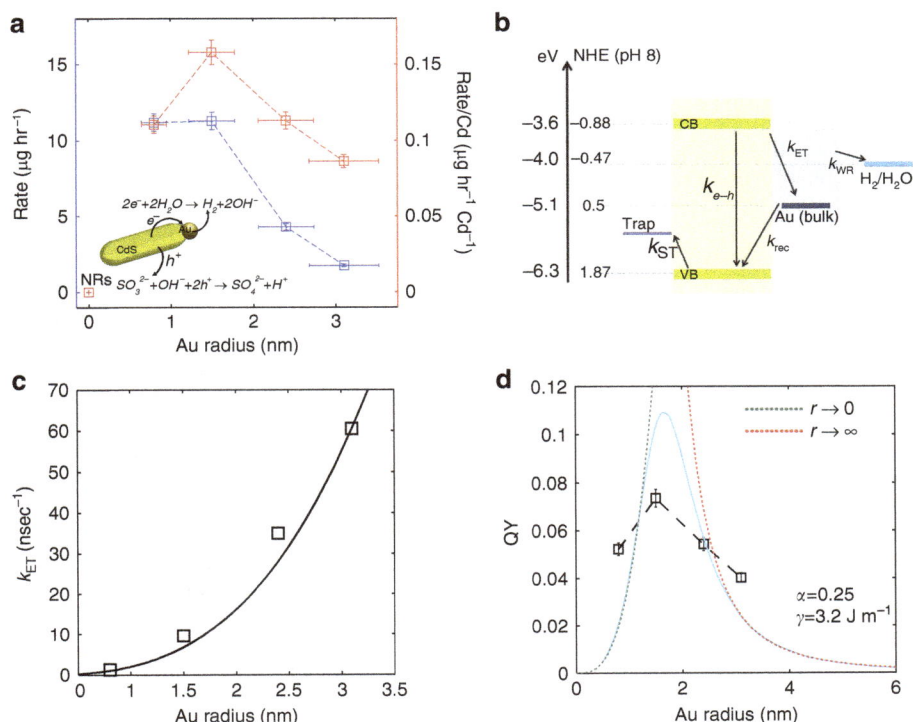

Figure 3 | Understanding size-dependent hydrogen production yield. (**a**) Hydrogen production rate (blue) and Cd normalized rate (red) curves as a function of Au size domain in the hybrid nanoparticles. Negligible rates are measured for the CdS NRs. (**b**) Energy band alignment diagram with relevant photocatalytic processes kinetic routes. (**c**) Measured semiconductor-metal electron transfer (k_{ET}) rates (squares) and fitting modified Fermi golden rule model for this process (solid line). (**d**) Measured QY (black squares connected by dashed line) along with the non-monotonic kinetic model behaviour (blue solid line). Green and red dotted lines present limiting behaviours of the model for zero and infinite metal domain sizes, respectively. Error bars in **a** and **d** indicate the Au tip size distribution and the uncertainty in the hydrogen production rate.

spectrometry (ICP-MS) analysis to determine the Cd content. Therefore the red curve is normalized to the overall Cd content, which is proportional to the contribution of the semiconductor to the nanohybrid absorption. This normalization is therefore expressing more cleanly the essential metal domain size effect in hydrogen reduction. It reveals a non-monotonic dependence in which an intermediate Au tip size provides the optimal hydrogen evolution rate for the photocatalytic water reduction and the highest value for the hydrogen evolution quantum yield (QY).

Additional aspect of hydrogen generation rate normalization can be considered to isolate the Au metal domain effect on the efficiency of this photocatalytic process. This may be considered by normalizing the hydrogen production rate to the loading of the co-catalyst component[18]. X-ray photoelectron spectroscopy measurements on the hybrid nanoparticles (HNPs) with different Au metal domain sizes show a dominant doublet at the Au 4f binding energy (BE) consistent with the zero-valent Au peak (84.0 eV), which designates the metallic nature of these domains. Intensities of the Au $4f_{7/2}$ and $4f_{5/2}$ peaks show a pronounced increase in the intensity correlated to the larger metal domain sizes (Supplementary Fig. 4). Quantification of the atomic concentration percent of each element in the HNPs structure presented in Supplementary Table 2 allows the calculation of the Cd:Au ratio in each HNPs sample. The nearly 10-fold increase of this relative atomic concentration of the Au between the different metal domain sizes indicates the significantly higher loading of the Au metal on the CdS semiconductor NR structure with increased Au tip size, while the minor change in the Cd or S atomic concentration implies lack of substantial change in the rod dimensions. Consequently, normalization of the hydrogen production rate by the Cd:Au ratio is considered as

normalization to the Au metal loading. The results of such correction for the measured hydrogen generation rates are fully consistent with the CdS absorption normalization obtained by ICP-MS discussed above, and reveal a similar non-monotonic behaviour as function of Au metal domain size (Supplementary Fig. 5 and Note 2).

Discussion

To rationalize the size dependence of the non-monotonic photocatalytic behaviour along with the charge transfer dynamics we propose a minimalistic model consisting of several kinetic steps, as sketched in Fig. 3b and Supplementary Fig. 9. The photo-excited electron in the semiconductor can relax by electron transfer to the co-catalyst metal domain (with a rate given by k_{ET}) or by electron–hole recombination (with rate k_{e-h}). The latter is rather slow[1] compared with the other processes and is assumed to be constant independent of the Au tip size ($k_{e-h} = \frac{1}{5}$ n sec^{-1}). The electron on the metal tip can either promote water reduction reaction (k_{WR}) or undergo back recombination with the hole that is left behind on the semiconductor domain (k_{rec}), which is modelled by the hole transfer from the semiconductor to the metal tip. An additional important process is the trapping of the hole (k_{ST}) with a rate that is comparable to that of electron transfer into the metal tip[29]. Once the hole is trapped it blocks the electron transfer channel by the electron–hole Coulomb interaction, which also leads to a localization of the electron[11,30].

The efficiency of hydrogen production can be obtained in a closed form and is given by the $t \to \infty$ of the solution to the master equation (Supplementary Note 4):

$$QY = \frac{k_{ET}k_{WR}}{(k_{WR} + k_{rec})(k_{ET} + k_{ST} + k_{e-h})} \qquad (1)$$

to make contact with the measured hydrogen production efficiency the different rates appearing in equation (1) need to be determined. In fact, some of these rates can be determined independently by combining the transient absorption measurements with a suitable theoretical model. The semiconductor-metal electron transfer (k_{ET}) and the back recombination (k_{rec}) rates can be described by Auger processes with values given by Fermi's golden rule:

$$k_{ET} = \frac{4}{3\hbar} r^3 |t_e|^2 \left(\frac{2m_e^*}{\hbar^2}\right)^{\frac{3}{2}} \sqrt{\varepsilon_c + \phi(r) + \varepsilon_F(r)} \qquad (2)$$

and

$$k_{rec} = \frac{4}{3\hbar} r^3 |t_h|^2 \left(\frac{2m_h^*}{\hbar^2}\right)^{\frac{3}{2}} \sqrt{\varepsilon_v + \phi(r) + \varepsilon_F(r)} \qquad (3)$$

where r is the metal domain radius, $\hbar$ is the Planck's constant divided by 2π, m_e^* and m_h^* are the effective mass of the electron and the hole in the metal, t_e and t_h are the electron and hole tunnelling matrix elements (assumed to be independent of r), ε_c, ε_v, $\phi(r)$ and $\varepsilon_F(r)$ are the semiconductor conduction and valance band energies, the metal work-function and the metal Fermi-level measured from the bottom of the energy band. Note that both rates depend on the metal tip radius r through the steep dependence of the density of states on the tip volume (r^3), and weakly through the dependence of the Fermi energy[31] ($\varepsilon_F(r)$) and work function[32,33] on r

$$\phi(r) = \phi_{bulk} - \frac{2\gamma v_M}{zFr} \qquad (4)$$

where γ is the surface tension, v_M is the molar volume, z is number of transferred electrons, F is Faraday's constant.

The water reduction reaction on the metal co-catalyst is described by the cathodic rate in the Butler–Volmer model[34] for redox reactions. The electron transfer process is given by a Marcus-like expression[35]:

$$k_{WR} = k_{WR}^0 \exp\left(-\frac{\alpha eF}{RT}(\varepsilon_W - \phi(r))\right) \qquad (5)$$

where k_{WR} is the electron reduction rate of an absorbed water molecule, k_{WR}^0 is the standard rate constant, R is the gas constant, T is the temperature, α is the electron transfer coefficient which determines the symmetry of the transition state (for example, $\alpha = \frac{1}{2}$ corresponds to a transition state with equal contributions from the reactants and products), and ε_W is the water reduction potential. The anodic rate for the hydrogen oxidation (back reaction) can be neglected because the hydrogen concentration is small compared with the proton concentration at the experimental pH level. In contrast to the electron transfer and recombination rates given by equations (2) and (3), k_{WR} depends exponentially on size of the metal domain through the dependence of $\phi(r)$.

We now have a working model to analyze the governing factors in the metal domain size effect on electron transfer and photocatalytic efficiency, and compare them with the experimental results. First, in Fig. 3c we determine k_{ET} by fitting the measured electron transfer rates obtained from the transient absorption to equation (2). The overall size dependence of k_{ET} is in good agreement with the theoretical r^3 prediction. The only fitting parameter used is the tunnelling matrix element, $t_e = 5.6 \times 10^{-5}$ eV. This seemingly small value is consistent with the results of Kamat on semiconductor-metal oxide hybrid nanoparticles[36]. The remaining parameters were taken from the literature and are provided in Supplementary Note 4.

Next, we calculate the efficiency of hydrogen production and compare the prediction to the experimental values, as shown in Fig. 3d. To capture the essential non-monotonic behaviour observed experimentally, we studied the dependence of the QY on the various parameters as shown in detail in the Supplementary Figs 10–12. In particular, the dependence on the hole tunnelling matrix element, t_h, was studied (Supplementary Fig. 10) and a reasonable description of the measured efficiencies is obtained for $t_h = 1.1 \times 10^{-6}$ eV, up to 50-folds smaller than t_e to yield meaningful results. This low value corresponds to an electron residence time on the Au tip sufficiently long to allow for effective water reduction. Indeed, the back recombination rate under these parameters for the different metal sizes is in the range of sub-µsec to several µsec for large to small Au domain sizes respectively, consistent with experimental results for Pt tips on CdS rods[1].

Two other parameters are needed for the qualitative description of the experimental results by the model. For the metal surface tension, γ, several reported values for gold nanoparticles were tested (Supplementary Fig. 11)[32,37,38]. Reasonable surface tension values (γ) are between 2.5 and 4 J m^{-1}. Higher surface tension values result in high work function values for the metal domain above the semiconductor conduction band minimum leading to vanishing electron injection rates. Furthermore, in larger tip sizes the efficiency is overestimated due to large overpotential relative to the standard water reduction potential at the experimental pH. Finally, for the electron transfer coefficient, α, the fits yield values around $\alpha = 0.25$ (Supplementary Fig. 12), in the range of values reported in the literature for similar systems[34,39]. This implies that the transition state for the water reduction is closer to the reactants.

A particular kinetic model solution is presented in Fig. 3d (solid blue curve) and manifests the non-monotonic dependence of the QY on the Au tip size consistent with the normalized experimental QY (squares connected by dashed line). The non-monotonic dependence arises from the opposing behaviour of the QY in the limits of $r \to 0$ and $r \to \infty$. For the former, we find that QY $\propto r^3$ increases with the tip radius, as the rate determining step is electron injection into the metal tip. For the latter, QY $\propto \exp\left(-\frac{2\gamma ev_M \alpha}{zRT}\frac{1}{r}\right)$ decreases exponentially with the tip radius where the rate determining step is the reduction of water as the back hole transfer competes with the reduction. Hence, an intermediate Au tip size provides an optimum balancing between the charge separation rate and efficiency, and the back recombination competing with the water reduction.

In conclusion, an optimal metal domain size is found for photocatalytic water reduction using hybrid semiconductor-metal NRs. The optimal value is explained in terms of competing processes where for small tips the hydrogen evolution QY is mainly determined by the rate of electron injection to the metal tip, whereas for large tips it is determined mainly by the water reduction on the metal surface. These two limits show opposite dependence on the metal domain size, leading to an optimal value as explained by a minimalist and general kinetic model with parameters fitted to reproduce qualitatively the experimental results, in particular after normalizing out the metal domain absorption effects. Thus, the behaviour is general and not limited to the metal type or the reduction reaction, and can be used for rational design of photocatalysts based on hybrid semiconductor-metal nanostructures.

Methods

Materials and syntheses. All chemical precursors used for this study were purchased from Sigma Aldrich, PCI Synthesis, and Merck. See Supplementary Methods for more details.

Synthesis and phase transfer of CdS NRs and CdS-Au hybrid nanoparticles are described in details in Supplementary Figs 1,2 and Supplementary Methods.

Nanoparticle characterization. TEM and high resolution scanning transmission electron microscope characterization was performed using a Tecnai T12 G2 Spirit and Tecnai F20 G2, respectively. All size statistics are done with 'Scion image' programme on 200 particles. Absorption was measured with a JASCO V-570 UV-vis-near IR spectrophotometer. Extinction coefficient values of the nanoparticles were calculated using a previously reported method[40].

Inductively coupled plasma mass spectrometry measurements. Following hydrogen generation kinetic measurements a 100 ml of HNPs solution was etched overnight in 1 ml of 69% nitric acid. Following sonication, 100 µl of the HNPs solution was mixed with 3.35 ml of three distilled water and analysed by ICP-MS (c × 7500, Agilent) for Cd. The quantity of Cd in each solution was calculated using external calibration with standard Cd solutions (Supplementary Table 1).

Hydrogen evolution rate and efficiency measurements. To determine and measure the evolved hydrogen gas from the photocatalytic reaction using the HNP model systems, the following set-up is used. The photocatalysts were dispersed in three distilled water solution (2 ml; optical density, OD ~ 1 at 405 nm). The photocatalyst solution was placed in a quartz cuvette and hole scavengers, Na_2S-$9H_2O$ and Na_2SO_3, (typically 0.05 and 0.07 M, respectively), were added to the water. The solution is purged with argon for 20 min and stirred. The hybrid nanoparticles were then illuminated with 40 mW 405 nm laser. Aliquots of the reaction vessel head space were taken using a gas tight syringe at different time intervals and detected and quantified using Varian gas chromatograph (model 6820) equipped with a molecular sieve (5 Å) packed column and a thermal conductivity detector. The resulting chromatograms and hydrogen concentration are obtained by the comparison to a calibration curve of known hydrogen amounts (Supplementary Fig. 3 and Note 1).

Ultrafast transient absorption measurements. The laser system employed for ultrafast transient absorption was based on a Ti-Sapphire chirped pulse amplified source, with maximum output energy of ~ 800 µJ, 1 kHz repetition rate, central wavelength of 800 nm and pulse duration of about 150 fs. Excitation pulses at 400 nm were obtained by doubling the fundamental frequency in a β-barium borate crystal while other pump photons at different wavelength were generated by non-collinear optical parametric amplification in β-barium borate, with pulse duration ~ 100 fs. Pump pulses were focused to 175 µm diameter spot. Probing was achieved in the visible range by using white light generated in a thin sapphire plate, and in the UV-visible range by using a thin Calcium Fluoride plate. Chirp-free transient transmission spectra were collected by using a fast optical multichannel analyser with dechirping algorithm. The measured quantity is the normalized transmission change, ΔT/T.

References

1. Wu, K. F., Zhu, H. M., Liu, Z., Rodriguez-Cordoba, W. & Lian, T. Q. Ultrafast charge separation and long-lived charge separated state in photocatalytic CdS-Pt nanorod heterostructures. *J. Am. Chem. Soc.* **134**, 10337–10340 (2012).
2. Kamat, P. V. Manipulation of charge transfer across semiconductor interface. a criterion that cannot be ignored in photocatalyst design. *J. Phys. Chem. Lett.* **3**, 663–672 (2012).
3. Kudo, A. & Miseki, Y. Heterogeneous photocatalyst materials for water splitting. *Chem. Soc. Rev.* **38**, 253–278 (2009).
4. Maeda, K. & Domen, K. Photocatalytic water splitting: recent progress and future challenges. *J. Phys. Chem. Lett.* **1**, 2655–2661 (2010).
5. Wilker, M. B., Schnitzenbaumer, K. J. & Dukovic, G. Recent progress in photocatalysis mediated by colloidal II–VI nanocrystals. *Isr. J. Chem.* **52**, 1002–1015 (2012).
6. Han, Z. J., Qiu, F., Eisenberg, R., Holland, P. L. & Krauss, T. D. Robust photogeneration of H-2 in water using semiconductor nanocrystals and a nickel catalyst. *Science* **338**, 1321–1324 (2012).
7. Banin, U., Ben-Shahar, Y. & Vinokurov, K. Hybrid semiconductor-metal nanoparticles: from architecture to function. *Chem. Mater.* **26**, 97–110 (2014).
8. Mills, A., Davies, R. H. & Worsley, D. Water-purification by semiconductor photocatalysis. *Chem. Soc. Rev.* **22**, 417–425 (1993).
9. Amirav, L. & Alivisatos, A. P. Photocatalytic hydrogen production with tunable nanorod heterostructures. *J. Phys. Chem. Lett.* **1**, 1051–1054 (2010).
10. O'Connor, T. et al. The effect of the charge-separating interface on exciton dynamics in photocatalytic colloidal heteronanocrystals. *ACS Nano* **6**, 8156–8165 (2012).
11. Ben-Shahar, Y. et al. Effect of surface coating on the photocatalytic function of hybrid CdS-Au nanorods. *Small* **11**, 462–471 (2015).
12. Yu, P. et al. Photoinduced ultrafast charge separation in plexcitonic CdSe/Au and CdSe/Pt nanorods. *J. Phys. Chem. Lett.* **4**, 3596–3601 (2013).
13. Khon, E. et al. Improving the catalytic activity of semiconductor nanocrystals through selective domain etching. *Nano Lett.* **13**, 2016–2023 (2013).
14. Bamwenda, G. R., Tsubota, S., Nakamura, T. & Haruta, M. The influence of the preparation methods on the catalytic activity of platinum and gold supported on TiO(2) for CO oxidation. *Catal. Lett.* **44**, 83–87 (1997).
15. Valden, M., Lai, X. & Goodman, D. W. Onset of catalytic activity of gold clusters on titania with the appearance of nonmetallic properties. *Science* **281**, 1647–1650 (1998).
16. Wood, A., Giersig, M. & Mulvaney, P. Fermi level equilibration in quantum dot-metal nanojunctions. *J. Phys. Chem. B* **105**, 8810–8815 (2001).
17. Subramanian, V., Wolf, E. E. & Kamat, P. V. Catalysis with TiO2/gold nanocomposites. Effect of metal particle size on the Fermi level equilibration. *J. Am. Chem. Soc.* **126**, 4943–4950 (2004).
18. Murdoch, M. et al. The effect of gold loading and particle size on photocatalytic hydrogen production from ethanol over Au/TiO₂ nanoparticles. *Nat. Chem.* **3**, 489–492 (2011).
19. Schweinberger, F. F. et al. Cluster size effects in the photocatalytic hydrogen evolution reaction. *J. Am. Chem. Soc.* **135**, 13262–13265 (2013).
20. Carbone, L. et al. Synthesis and micrometer-scale assembly of colloidal CdSe/CdS nanorods prepared by a seeded growth approach. *Nano Lett.* **7**, 2942–2950 (2007).
21. Mokari, T., Sztrum, C. G., Salant, A., Rabani, E. & Banin, U. Formation of asymmetric one-sided metal-tipped semiconductor nanocrystal dots and rods. *Nat. Mater.* **4**, 855–863 (2005).
22. Mokari, T., Rothenberg, E., Popov, I., Costi, R. & Banin, U. Selective growth of metal tips onto semiconductor quantum rods and tetrapods. *Science* **304**, 1787–1790 (2004).
23. Menagen, G., Macdonald, J. E., Shemesh, Y., Popov, I. & Banin, U. Au growth on semiconductor nanorods: photoinduced versus thermal growth mechanisms. *J. Am. Chem. Soc.* **131**, 17406–17411 (2009).
24. Wu, K. F., Rodriguez-Cordoba, W. E., Yang, Y. & Lian, T. Q. Plasmon-induced hot electron transfer from the Au Tip to CdS Rod in CdS-Au nanoheterostructures. *Nano Lett.* **13**, 5255–5263 (2013).
25. Link, S. & El-Sayed, M. A. Spectral properties and relaxation dynamics of surface plasmon electronic oscillations in gold and silver nanodots and nanorods. *J. Phys. Chem. B* **103**, 8410–8426 (1999).
26. Ahmadi, T. S., Logunov, S. L. & El-Sayed, M. A. Picosecond dynamics of colloidal gold nanoparticles. *J. Phys. Chem.* **100**, 8053–8056 (1996).
27. Bao, N. Z., Shen, L. M., Takata, T. & Domen, K. Self-templated synthesis of nanoporous CdS nanostructures for highly efficient photocatalytic hydrogen production under visible. *Chem. Mater.* **20**, 110–117 (2008).
28. Shaviv, E. et al. Absorption properties of metal-semiconductor hybrid nanoparticles. *ACS Nano* **5**, 4712–4719 (2011).
29. Wu, K. F. et al. Hole removal rate limits photodriven H-2 generation efficiency in CdS-Pt and CdSe/CdS-Pt semiconductor nanorod-metal tip heterostructures. *J. Am. Chem. Soc.* **136**, 7708–7716 (2014).
30. Berr, M. J. et al. Delayed photoelectron transfer in Pt-decorated CdS nanorods under hydrogen generation conditions. *Small* **8**, 291–297 (2012).
31. Jakob, M., Levanon, H. & Kamat, P. V. Charge distribution between UV-irradiated TiO(2) and gold nanoparticles: determination of shift in the Fermi level. *Nano Lett.* **3**, 353–358 (2003).
32. Plieth, W. J. Electrochemical properties of small clusters of metal atoms and their role in surface enhanced raman-scattering. *J. Phys. Chem.* **86**, 3166–3170 (1982).
33. Redmond, P. L., Hallock, A. J. & Brus, L. E. Electrochemical Ostwald ripening of colloidal Ag particles on conductive substrates. *Nano Lett.* **5**, 131–135 (2005).
34. Zhao, J., Holmes, M. A. & Osterloh, F. E. Quantum confinement controls photocatalysis: a free energy analysis for photocatalytic proton reduction at CdSe nanocrystals. *ACS Nano* **7**, 4316–4325 (2013).
35. Marcus, R. A. Chemical and electrochemical electron-transfer theory. *Annu. Rev. Phys. Chem.* **15**, 155–196 (1964).
36. Tvrdy, K., Frantsuzov, P. A. & Kamat, P. V. Photoinduced electron transfer from semiconductor quantum dots to metal oxide nanoparticles. *Proc. Natl Acad. Sci. USA* **108**, 29–34 (2011).
37. Jiang, Q., Liang, L. H. & Zhao, D. S. Lattice contraction and surface stress of fcc nanocrystals. *J. Phys. Chem. B* **105**, 6275–6277 (2001).
38. Nanda, K. K., Maisels, A. & Kruis, F. E. Surface tension and sintering of free gold nanoparticles. *J. Phys. Chem. C* **112**, 13488–13491 (2008).
39. Meng, Y., Aldous, L., Belding, S. R. & Compton, R. G. The hydrogen evolution reaction in a room temperature ionic liquid: mechanism and electrocatalyst trends. *Phys. Chem. Chem. Phys.* **14**, 5222–5228 (2012).
40. Shaviv, E., Salant, A. & Banin, U. Size dependence of molar absorption coefficients of CdSe semiconductor quantum rods. *Chemphyschem.* **10**, 1028–1031 (2009).

Acknowledgements

We thank Dr Vitaly Gutkin from the Unit for Nanocharacterization at the Hebrew University for assistance in the X-ray photoelectron spectroscopy measurements. The research leading to these results has received funding from The Israel Science Foundation (grant no. 1560/13) and the Ministry of Science, Technology and Space, Israel & the Directorate General for Political and Security Affairs of the Ministry of Foreign Affairs, Italy. U.B. thanks the Alfred & Erica Larisch memorial chair. G.C. acknowledges support by the EC under Graphene Flagship (contract no. CNECT-ICT-604391). Y.B.S. acknowledges support by the Ministry of Science, Technology and Space, Israel & the Camber Scholarship.

Author contributions

Y.B.S. and U.B. conceived and designed the experiments, Y.B.S. preformed the experiments, F.S., I.K., L.M. and G.C. contributed to transient absorption measurements and interpretation, Y.B.S., U.B. and E.R. provided and interpreted the kinetic model, Y.B.S., U.B., E.R. and G.C. co-wrote the paper.

Additional information

Competing financial interests: The authors declare no competing financial interests.

16

The hydrogen-bond network of water supports propagating optical phonon-like modes

Daniel C. Elton[1,2] & Marivi Fernández-Serra[1,2]

The local structure of liquid water as a function of temperature is a source of intense research. This structure is intimately linked to the dynamics of water molecules, which can be measured using Raman and infrared spectroscopies. The assignment of spectral peaks depends on whether they are collective modes or single-molecule motions. Vibrational modes in liquids are usually considered to be associated to the motions of single molecules or small clusters. Using molecular dynamics simulations, here we find dispersive optical phonon-like modes in the librational and OH-stretching bands. We argue that on subpicosecond time scales these modes propagate through water's hydrogen-bond network over distances of up to 2 nm. In the long wavelength limit these optical modes exhibit longitudinal–transverse splitting, indicating the presence of coherent long-range dipole–dipole interactions, as in ice. Our results indicate the dynamics of liquid water have more similarities to ice than previously thought.

[1] Department of Physics and Astronomy, Stony Brook University, Stony Brook, New York 11794-3800, USA. [2] Institute for Advanced Computational Science, Stony Brook University, Stony Brook, New York 11794-3800, USA. Correspondence and requests for materials should be addressed to D.C.E. (email: daniel.elton@stonybrook.edu) or to M.F.-S. (email: maria.fernandez-serra@stonybrook.edu).

The local structure and dynamics of liquid water as a function of temperature remains a source of intense research and lively debate[1-6]. A thus far unrecognized discrepancy exsits between the peak assignments reported in Raman spectra with those reported in dielectric/infrared spectra. Although early experimentalists fit the Raman librational band with two peaks[7], it is better fit with three (Supplementary Table 1)[8-12]. Previously these three peaks were assigned to the three librational motions of the water molecule—twisting ($\approx 435\,cm^{-1}$), rocking ($\approx 600\,cm^{-1}$) and wagging ($\approx 770\,cm^{-1}$)[8,9,11]. However, when comparing these assignments with infrared and dielectric spectra, one runs into a serious discrepancy. One expects to find the two higher frequency modes to be present, since only the rocking and wagging librations are infrared active. The twisting libration, consisting of a rotation of the hydrogen atoms around the C2 axis, is not infrared active since it does not affect the dipole moment of the molecule. Instead, infrared spectra show two peaks at 380 and 665 cm^{-1} (ref. 13), and similarly dielectric spectra show peaks at 420 and 620 cm^{-1} (ref. 14), in disagreement with this assignment.

The assignment of longitudinal optical phonon modes to Raman spectra can be made by looking at the longitudinal dielectric susceptibility. This method has been used previously to assign longitudinal phonon modes to the Raman spectra of ice Ih[15-17], cubic ice[18] and vitreous GeO$_2$ and SiO$_2$ (ref. 19). It has previously been shown that the librational peak in the longitudinal dielectric susceptibility of water is dispersive[20], and Bopp and Kornyshev noted that the dispersion relation has the appearance of an optical phonon mode[21]. The longitudinal mode in the dielectric susceptibility is equivalent to the dispersive mode discovered by Ricci et al.[22] in the spectrum of hydrogen-density fluctuations[22].

Comparison of peak positions in longitudinal and transverse dielectric susceptibilities often reveals longitudinal–transverse (LO–TO) splitting. LO–TO splitting indicates the presence of long-range dipole–dipole interactions in the system. One way to understand LO–TO splitting is through the Lyddane–Sachs–Teller (LST) relation[23]:

$$\frac{\omega_{LO}^2}{\omega_{TO}^2} = \frac{\varepsilon(0)}{\varepsilon_\infty} \quad (1)$$

Although this relation was originally derived for a cubic ionic crystal it was later shown to have very general applicability[24,25], and has been applied to disordered and glassy solids[17,26,27]. To apply this equation to water we must use a generalized LST relation, which takes into account all of the optically active modes in the system and the effects of dampening[24]. The generalized LST relation is[24]:

$$\prod_i \frac{\omega_{LDi}}{\omega_{TDi}} \prod_j \frac{|\bar{\omega}_{Lj}|^2}{\omega_{Tj}^2} = \frac{\varepsilon(0)}{\varepsilon_\infty} \quad (2)$$

Here the index i runs over the Debye peaks in the system and the index j runs over the number of damped harmonic oscillator peaks. The longitudinal frequencies of the damped harmonic oscillators must be considered as complex numbers ($\bar{\omega}_{Li} = \omega_{Li} + i\gamma_i$), where γ_i is the dampening factor.

As shown by Barker, the generalized LST equation can be understood purely from a macroscopic point of view[24], so by itself it yields little insight into microscopic dynamics. LO–TO splitting can be understood from a microscopic standpoint via the equation[28,29]:

$$\omega_{Lk}^2 - \omega_{Tk}^2 = \frac{4\pi C}{3v}\left(\frac{\partial \mu}{\partial Q_k}\right)^2 \quad (3)$$

Here v is the volume per unit cell, Q_k is the normal coordinate of

mode k, and C is a prefactor which depends on the type of lattice and the boundary conditions of the region being considered (for an infinite cubic lattice, $C=1$). Equation 3 shows that LO–TO splitting is intimately related to crystal structure, and it has been used to evaluate the quasi-symmetry of room temperature ionic liquids[30].

In this work we show how the dielectric susceptibility can be used to probe water's local structure and dynamics. Our work solves the aforementioned peak assignment discrepancy. We find that the lowest frequency librational Raman peak ($\approx 435\,cm^{-1}$) is a transverse optical phonon-like mode while the highest frequency peak ($\approx 770\,cm^{-1}$) is a longitudinal optical phonon-like mode. This explains why the highest frequency Raman mode does not appear in infrared or dielectric experiments, since such experiments only report the transverse response. We show that the transverse counterpart also exhibits dispersion. We argue that these dispersive modes are due to optical phonons that travel along the H-bond network of water. Our results indicate that not only does water exhibit LO–TO splitting, but also that its dependence with temperature is anomalous. We suggest that this measurement provides an alternative probe to evaluate structural changes in liquid water as a function of temperature.

Results

LO–TO splitting from experimental data. As in our previous work[31] we compared results from a rigid (TIP4P/ε) model, a flexible model (TIP4P/2005f), and a flexible and polarizable model (TTM3F) in all of our analyses.

We wish to study the k dependence of the dielectric susceptibility, where $k = 2\pi/\lambda$. $k-$ dependence cannot be probed directly by experiment, but in the limit of infinite wavelength ($k \to 0$) the longitudinal and transverse dielectric susceptibilities can be obtained from the dielectric function via the following relations[32,33]:

$$\chi_L(k \to 0, \omega) = 1 - \frac{1}{\varepsilon(\omega)} \quad (4)$$

$$\chi_T(k \to 0, \omega) = \varepsilon(\omega) - 1 \quad (5)$$

Note that the transverse susceptibility is what one normally calls susceptibility. The dielectric function can be obtained from the index of refraction $n(\omega)$ and extinction coefficient $k(\omega)$ as:

$$\begin{aligned}\varepsilon'(\omega) &= n^2(\omega) - k^2(\omega) \\ \varepsilon''(\omega) &= 2n(\omega)k(\omega)\end{aligned} \quad (6)$$

These equations allow us to use previously published experimental data[34,35] to calculate the imaginary part of the longitudinal response. We find significant LO–TO splitting in the librational and stretching bands (Fig. 1).

Polarization correlation functions. The normalized longitudinal and transverse polarization correlation functions are defined as:

$$\Phi_{L/T}(k,t) \equiv \frac{\langle P_{L/T}(k,t) \cdot P_{L/T}^*(k,0)\rangle}{\langle P_{L/T}(k,0) \cdot P_{L/T}^*(k,0)\rangle} \quad (7)$$

The correlation functions found for TIP4P/ε at small small k are shown in Fig. 2. Since TIP4P/ε is a rigid model, only librational motions are present. The addition of flexibility and polarizability add additional high-frequency oscillations to the picture (Supplementary Fig. 1). In the small wavenumber regime ($k < 1.75$ Å) there is a damped oscillation, which corresponds to the collective librational phonon-like mode. This damped oscillation is superimposed on an underlying exponential relaxation in both the transverse and longitudinal cases. In the

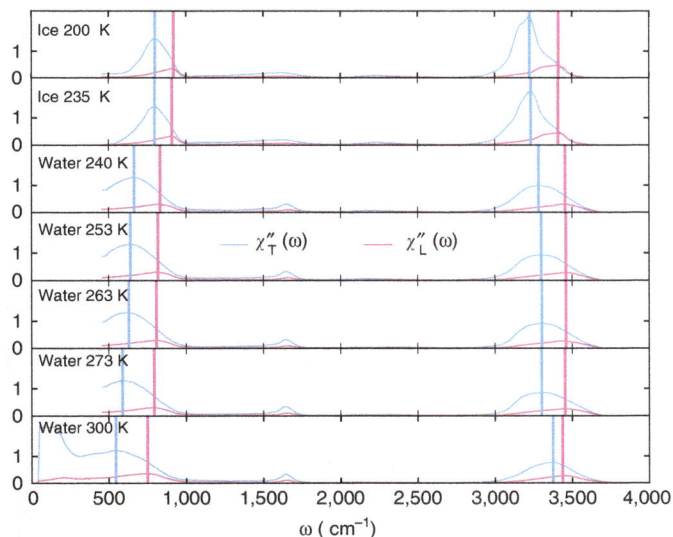

Figure 1 | Dielectric susceptibilities of ice and water. Computed from index of refraction data using equations 4 and 6. data from 210 to 280 K comes from aerosol droplets[34] while the data at 300 comes from bulk liquid[35].

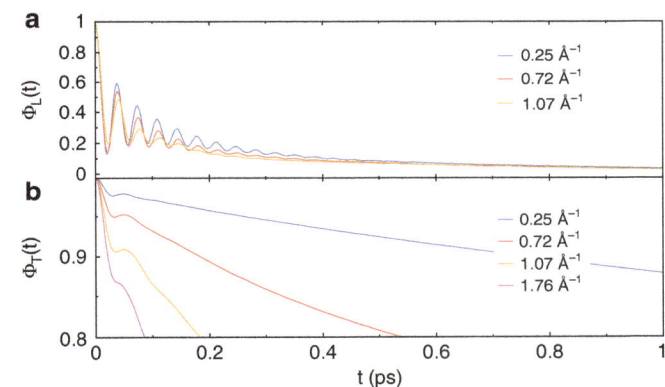

Figure 2 | Polarization correlation functions. Longitudinal (**a**) and transverse (**b**) polarization correlation functions (see equation 7) for TIP4P/ε, a rigid model. The oscillations at small k come from the collective librational mode, which is much more pronounced in the longitudinal case.

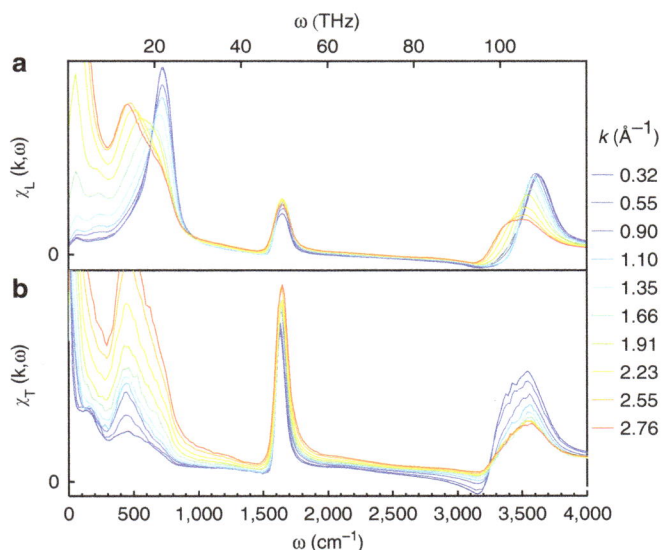

Figure 3 | Imaginary part of longitudinal and transverse susceptibility. (**a**) Transverse and (**b**) longitudinal susceptibility from simulations with TTM3F at 300 K. In the longitudinal spectra both the librational ($\sim 750\,\mathrm{cm}^{-1}$) and OH-stretching peak ($\sim 3,500\,\mathrm{cm}^{-1}$) peaks exhibit dispersion.

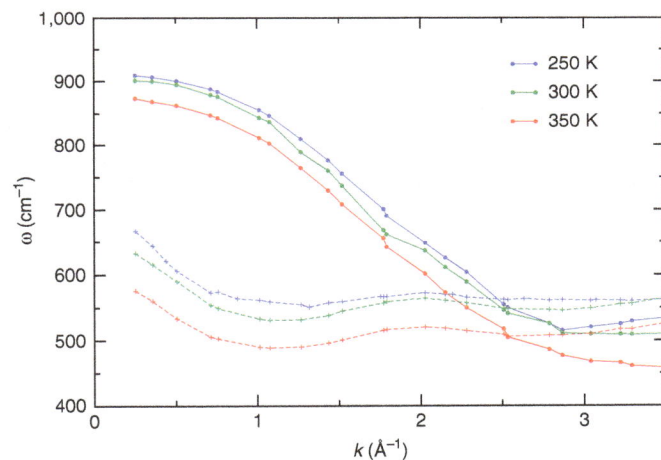

Figure 4 | Dispersion relations for the propagating librational modes. For TIP4P/2005f at three different temperatures (squares = longutudinal, pluses = transverse). A similar plot was found for TTM3F, but with lower frequencies.

longitudinal case the relaxation time $\tau(k)$ of the underlying exponential relaxation exhibits non-monotonic behaviour with k, reaching a maximum at $k \approx 3\,\text{Å}^{-1}$ (Supplementary Fig. 2). At wavenumbers greater than $k \approx 2.5\,\text{Å}$ only intramolecular motions contribute.

Dispersion of the librational peak. Figure 3 shows the imaginary part of the longitudinal and transverse susceptibility for TTM3F. In the longitudinal case the librational peak is clearly seen to shift with k. In the transverse case, the lower frequency portion of the band is seen to shift slightly with k. Dispersion relations for the longitudinal and transverse librational peaks are shown in Fig. 4 for three different temperatures, using one peak fits. The dispersion relations appear to be that of optical phonons. In both the longitudinal and transverse case the dampening factors remain less than the resonance frequencies, indicating an underdamped oscillation (Supplementary Fig. 3). The longitudinal dispersion

relation for TIP4P/2005f agrees with that found by Bopp and Kornyshev (who used the flexible BJH model)[21]. Resat *et al.*[36] also obtained a similar dispersion relation (but at a higher frequency), using the reference memory function approximation for TIP4P instead of molecular dynamics.

Resonance frequencies and lifetimes for the smallest k are shown in Table 1. The speed of propagation of these modes was computed by finding the slope $d\omega/dk$ in the regime of linear dispersion. For TIP4P/2005f we found speeds of $\approx 2,700$ and $\approx 1,800\,\mathrm{m\,s}^{-1}$ for the longitudinal and transverse modes. These propagation speeds are above the speed of sound in water ($1,500\,\mathrm{m\,s}^{-1}$) but below the speed of sound in ice ($4,000\,\mathrm{m\,s}^{-1}$). The temperature dependence of the propagation speed was found to be very small.

Table 1 | Resonance frequencies and lifetimes.

Model	Temp	ω_{LO}	τ_{LO}	ω_{TO}	τ_{TO}	$\omega_{LO} - \omega_{TO}$
TIP4P/2005f	250	905	0.38	667	0.23	233
	300	900	0.44	632	0.18	268
	350	871	0.34	574	0.18	297
	400	826	0.25	423	0.17	400
3*TTM3F	250	757	0.49	496		261
	300	721	0.44	410		311
	350	710	0.20	380		330
expt[34]	253	820		641		179
expt[35]	300	759		556		203

Frequencies are given in cm^{-1} and lifetimes in ps. The values from simulation were computed at the smallest k in the system. The experimental values are based on the position of the max of the band and therefore only approximate.

In both the longitudinal and transverse cases, the residual of the peak fitting shows features not captured by our Debye + one resonance fit of the librational peak. In both the longitudinal and transverse cases there is a non-dispersive peak at higher frequency, located at $\approx 900\,cm^{-1}$ for TIP4P/2005f and at $\approx 650\,cm^{-1}$ in TTM3F. This peak is negligibly small in the $k = 0$ longitudinal susceptibility but appears as a shoulder as k increases. In the transverse case the overlapping peak persists at $k = 0$, so we found that the $k = 0$ transverse spectra is best fit with two peaks, in agreement with experimental spectra. As we describe later, the higher frequency transverse peak is largely due to the self part of the response and is associated with the wagging librations of single molecules.

Importance of polarizability. There are several notable differences between TTM3F and the non-polarizable model TIP4P/2005f. First of all, the librational band of TIP4P/2005f is at higher frequency, in worse agreement with experiment. This difference in frequency is likely related to the parameters of TIP4P/2005f and not its lack of polarization. More importantly, we find that TTM3F exhibits dispersion in the OH-stretching band ($\approx 3,500\,cm^{-1}$) in the longitudinal case, while TIP4P/2005f does not (Fig. 5). The transverse susceptibility of TTM3F does not exhibit such dispersion but the magnitude of the OH-stretching band increases at small k, indicating long-range intermolecular correlations. TIP4P/2005f does not exhibit this behaviour. Similarly, at $k = 0$ TTM3F exhibits significant LO–TO splitting in the OH-stretching band, while TIP4P/2005f does not (Fig. 6). These findings are consistent with Heyden et al.'s results for the k-resolved infrared spectra from ab initio simulation, where they concluded that polarization allows for intermolecular correlations at the OH-stretch frequency[37].

These findings can be understood from the dipole derivative in equation 3. In the librational band the derivative of the dipole moment with respect to normal coordinate is purely due to rotation, while in the OH-stretching band it is due to changes in the geometry of the molecule and electronic polarization of the molecule during the OH-stretching. In principle, there may be coupling between the librational and stretching motions, but typically such rotational–vibrational coupling effects are negligibly small[38]. The dipole moment surface (fluctuating charges) and polarization dipole incorporated in TTM3F account for the changes in polarization that occur during OH-stretching motion. These results confirm the significance of polarization in capturing the OH-stretching response of water[37].

Figure 6 shows a comparison of TTM3F, TIP4P/2005f and experiment at $k = 0$. While the location of the peaks in TTM3F are in good agreement with the experimental data at 298 K, the magnitude of the longitudinal response is greatly overestimated in

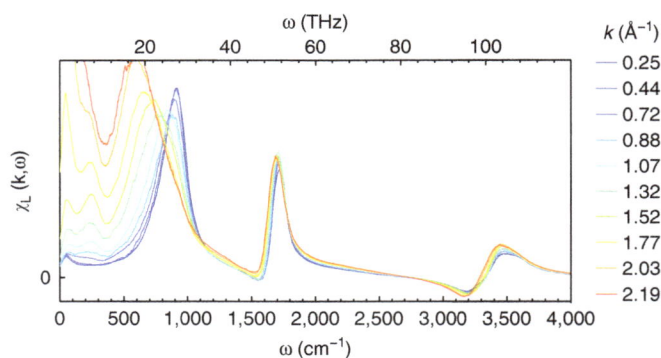

Figure 5 | Imaginary part of the longitudinal susceptibility. For TIP4P/2005f at 300 K. No significant dispersion is observed in the OH-stretching peak.

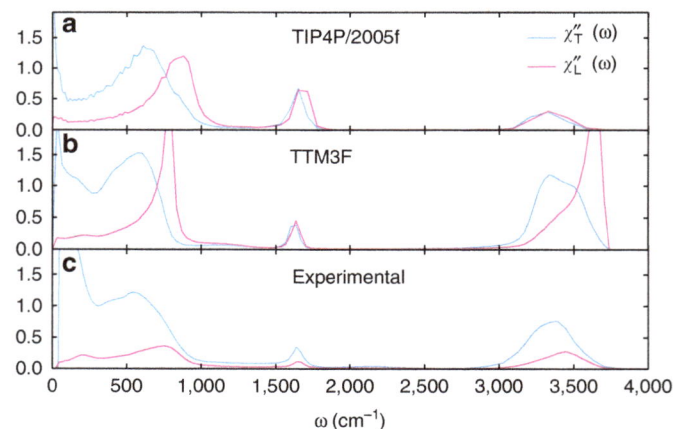

Figure 6 | Imaginary parts of dielectric susceptibility. We compare (**a**) the non-polarizable model TIP4P/2005f, (**b**) the polarizable model TTM3F, and (**c**) experimental data[35] at 298 K. The effects of polarization can be seen in the LO–TO splitting of the stretching mode and in the low-frequency features.

TTM3F. The degree of LO–TO splitting in the OH-stretching peak is also overestimated in TTM3F. In general it appears that TTM3F overestimates the dipole derivative in equation 3, while TIP4P/2005f underestimates it. Figure 6 also shows the effect of polarization at low frequencies, in particular the appearance of an H-bond stretching response at $\approx 250\,cm^{-1}$ in TTM3F, which is absent in TIP4P/2005f (ref. 31).

LO–TO splitting versus temperature. The frequencies of the librational and stretching modes are shown in Table 1. Once again we compare our results to experimental data[34,35,39]. The comparison is imperfect since the TIP4P/2005f and TTM3F data come from data at finite k (the smallest k in the system). For all three systems (TIP4P/2005f, TTM3F and experiment) the increase in the LO–TO splitting of the librational band is puzzling, since the right hand side of the LST relation predicts a decrease in splitting, corresponding to a smaller dielectric constant and weaker dipole–dipole interactions. We found verifying the generalized LST equation is difficult because water contains either two or three Debye relaxations which must be taken into account[40,41]. Uncertainties in how to fit the region of 1–300 cm^{-1} (0.2–9 THz), which includes contributions from many H-bonding modes, precludes a direct application of the generalized LST relation to water. By ignoring this region, however, we were able to achieve an approximate validation of the generalized LST equation for TIP4P/2005f. A more detailed analysis of how to fit the low-frequency region will be the focus of future work. Since the generalized LST equation is an exact sum rule it can be used to assist in testing the validity of different fit functions.

Relation to phonons in ice. Naturally we would like to find corresponding optical phonon modes in ice. As shown in Fig. 1 the dielectric spectra and LO–TO splitting of supercooled water resembles that of ice. Recently evidence has been presented for propagating librational phonon modes in ice XI (refs 42,43). Three of the twelve librational modes of ice XI are infrared active (labelled WR1, RW1 and RW2) and all three exhibit LO–TO splitting. The splittings have been found from Raman scattering to be 255, 135 and 35 cm^{-1} (ref. 43). These modes all consist of coupled wagging and rocking motions. The WR1 mode, which has the largest infrared intensity, most closely matches our results. WR1 and RW2 have the same transverse frequency and RW1 has a smaller infrared intensity, which may help explain why the librational band is well fit by a single optical mode. LO–TO splitting in the OH-stretching modes of hexagonal ice has been discussed previously[17].

Range of propagation. The range of propagation of these modes can be calculated as $R = \tau v_g$ where τ is the lifetime and $v_g = d\omega/dk$ is the group velocity. For TIP4P/2005f we find a range of propagation of ≈ 1.1 nm for the longitudinal librational mode and ≈ 0.3 nm for the transverse mode. Similar results hold for TTM3F.

To verify that the modes we observe are actually propagating and to further quantify the range of propagation we study the spatial extent of polarization dipole correlations as a function of frequency. We investigated several different methodologies that can be used to decompose a spectra into distance-dependent components (Supplementary Note 1). We choose to start with the polarization correlation function:

$$\phi(k,t) = \left\langle \sum_i \mathbf{p}_i(k,0) \cdot \sum_j \mathbf{p}_j(k,t) \right\rangle \quad (8)$$

Here $\mathbf{p}_i(k,t)$ represents either the longitudinal or transverse molecular polarization vector of molecule i. We now limit the molecules in the second sum to those in a sphere of radius R around each molecule i:

$$\phi(k,t,R) = \left\langle \sum_i \mathbf{p}_i(k,0) \cdot \sum_{j \in R_i} \mathbf{p}_j(k,t) \right\rangle \quad (9)$$

The resulting function exhibits the expected $R \to 0$ limit, yielding only the self-contribution. R can be increased to the largest R in the system $(\sqrt{3}L/2)$, where the full response function for the simulation box is recovered. As R increases, the contributions of the distinct term add constructively and destructively to the self term, illustrating the contributions from molecules at different distances.

Figure 7 shows the distance decomposed longitudinal and transverse susceptibilities for TIP4P/2005f in a 4-nm box at the smallest k available in the system. Similar results are obtained for TTM3F, but with additional contributions in the stretching band. Figure 8 shows the distance decomposed longitudinal susceptibility for TTM3F in a 1.97 nm box. The entire region between 0–1,000 cm^{-1} contains significant cancellation between the self and distinct parts, in qualitative agreement with a previous study[44] (Fig. 8). In the longitudinal susceptibility, the self component has two peaks (at 500 and 900 cm^{-1} for TIP4P/2005f) representing the two infrared active librational motions (rocking and wagging, respectively). The self part is the same in both the longitudinal and transverse cases, reflecting an underlying isotropy, which is only broken when dipole–dipole correlations are introduced. Further insight into the self-distinct cancellation comes from the results of Bopp, *et al.*, who project the hydrogen currents into a local molecular frame, allowing them to study the cross correlations between the rocking and wagging librations[21]. They find that in the longitudinal case cross

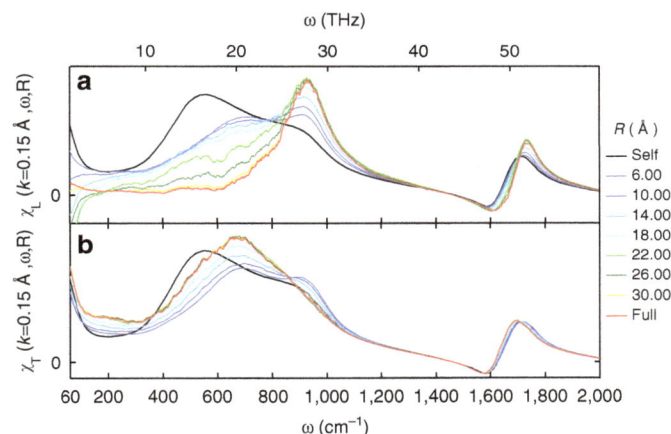

Figure 7 | Imaginary part of the distance decomposed susceptibility for TIP4P/2005f. Transverse (**a**) and longitudinal (**b**) susceptibilities, calculated with a 4 nm box at 300 K, using the smallest **k** vector in the system. Gaussian smoothing was applied. Long-range contributions to the librational peak extending to $R = 2$ nm are observed.

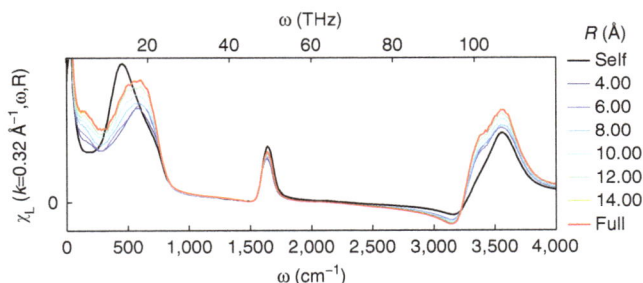

Figure 8 | Imaginary part of the distance decomposed longitudinal susceptibility for TTM3F at 300 K. Long-range contributions are observed in the OH-stretching band.

correlations between rocking and wagging contribute negatively in the region of $480\,cm^{-1}$ and positively in the region of $740\,cm^{-1}$, suppressing the lower frequency peak to zero and enhancing the higher frequency peak.

In both the transverse and longitudinal cases as R increases a new peak emerges, corresponding to the propagating mode. Incidentally, the shift in the peak between the self and distinct parts rules out the possibility that the propagating mode is the proposed dipolar plasmon resonance, since the dipolar plasmon must be a resonance of both the of single molecule and collective motion[45–47]. Interestingly, there are very long-range contributions to this peak. In our simulations with a 4-nm box of TIP4P/2005f contributions persist up to 3 nm in the longitudinal case and 2 nm in the transverse case. As noted, recent studies of ice XI suggest that the propagating modes consist of coupled wagging and rocking librations[42,43]. The results for the transverse mode seem to confirm this hypothesis for liquid water, since the propagating mode peak lies between the single-molecule rocking and wagging peaks. In the longitudinal case the propagating mode overlaps more with the wagging peak, suggesting a greater role for these type of librations in the longitudinal phonon.

Methanol and acentonitrile. To provide further evidence the aforementioned optical modes propagate through the hydrogen-bond network of water we decided to repeat our analysis for other polar liquids, both H-bonding and non H-bonding. As an H-bonding liquid we choose methanol, which is known to contain winding H-bonded chains. According to results from MD simulation, most of these chains have around 5–6 molecules[48,49], with a small percentage of chains containing 10–20 molecules[50]. Chain lifetimes have been estimated to be about 0.5 ps (ref. 50). Therefore, we expect methanol can also support a librational phonon mode that propagates along hydrogen bonds, but perhaps with a shorter lifetime and range than water. As a non H-bonding polar liquid we choose acetonitrile, because it has a structure similar to methanol, but with the hydroxyl group replaced by a carbon atom. We find that the OH librational band of methanol ($\approx 700\,cm^{-1}$ (ref. 51)) is indeed dispersive (Supplementary Fig. 4). As with water, the transverse spectra also exhibits dispersion, but to a much lesser extent. LO–TO splitting of about $100\,cm^{-1}$ is observed in the $700\,cm^{-1}$ librational peak. The results for acetonitrile are more ambiguous—we observe dispersion in the broad peak at $\approx 100\,cm^{-1}$, however, this peak contains contributions from translational and (free) rotational modes, as well as the CH_3 torsion mode, and it is not clear, which modes are responsible for the dispersion (Supplementary Fig. 5).

Discussion

In this work, we have presented several lines of evidence for short-lived optical phonons that propagate along the H-bond network of water. The longitudinal and transverse nonlocal susceptibility exhibit dispersive peaks with dispersion relations resembling optical phonons. As the temperature is lowered, the resonance frequencies and LO–TO splittings of these modes converge towards the values for phonons in ice Ih. By comparing our results with a recent study of ice XI we believe both modes likely consist of coupled wagging and rocking librations[42,43].

This work fundamentally changes our understanding of the librational band in the Raman spectra of water by assigning the lower and higher frequency peaks to transverse and longitudinal optical modes. Our analysis of the self-distinct cancellation indicates that the middle Raman peak ($\approx 600\,cm^{-1}$) belongs to the remnant of the single-molecule wagging response, which

remains after the cancellation. We are also led to a new interpretation the librational region of the real part of the dielectric function. In the case of a lossless optical phonon the transverse phonon occurs where $\varepsilon'(\omega) = \infty$, while the longitudinal phonon occurs where $\varepsilon'(\omega) = 0$. The presence of dampening smooths the divergence leading to a peak followed by a sharp dip. This is what is observed in the real part of the dielectric function of water between 300 and $500\,cm^{-1}$ (the features are shifted to lower frequencies by the tail of the low-frequency Debye relaxation).

One might wonder how our work relates to existing work on acoustic modes in water, in particular, the controversial 'fast sound' mode[52,53]. Acoustic modes, which are observable through the dynamic structure factor, have been explored as means of understanding the hydrogen-bond structure and low-temperature anomalies of water[5]. In this work, we have argued that optical modes can also provide insight into water's structure and dynamics. The fast sound mode lies at much lower frequencies than the librational and OH stretch modes that we studied. The H-bond bending and stretching modes also primarily lie at at frequencies below the librational region. However, normal mode analysis of liquid water and clusters shows that the H-bond stretching modes have a wide distribution of frequencies, which overlaps with the librational modes, so some coupling between these modes is possible[54,55]. Recently, it was shown that there is coupling between the acoustic and optic modes in water—that is, between fluctuations in mass density and fluctuations in charge density[56].

The large spatial range and coherent propagation of these modes is surprising and implies the existence of an extended hydrogen-bond network, in contrast to earlier ideas about the structure of water which emphasize dynamics as being confined within small clusters[57]. Simulations with larger simulation boxes are needed to fully quantify the extent of the longitudinal modes. The ability of water to transmit phonon modes may be relevant to biophysics, where such modes could lead to dynamical coupling between biomolecules, a phenomena that is currently only being considered at much lower frequencies[58–60]. The methodology used in this paper to analyse LO–TO splitting opens up a new avenue to understanding the structure and dynamics of water. The fact that the librational LO–TO splitting increases with temperature instead of the expected decrease is likely due to significant changes in the structure of the liquid. One likely possibility is that the volume per 'unit cell' term in equation 3 decreases with temperature. This could be caused by the local quasi-structure determined by H-bonding changing from a more ice-like structure (four molecules per unit cell) to a more cubic structure (1 molecule per unit cell). More research is needed to understand the microscopic origin of the LO–TO splitting in water, both in the librational and stretching modes.

Methods

Theory of the nonlocal susceptibility. If the external field is sufficiently small, then the relation between the polarization response of a medium and the electric displacement field D for a spatially homogeneous system is given by:

$$\mathbf{P}(r, t) = \epsilon_0 \int_V \int_{-\infty}^{t} dr' dt' \,\overleftrightarrow{\chi}(r - r', t - t') \mathbf{D}(r', t'). \tag{10}$$

This expression Fourier transforms to:

$$\mathbf{P}(\mathbf{k}, \omega) = \epsilon_0 \overleftrightarrow{\chi}(\mathbf{k}, \omega) \mathbf{D}(\mathbf{k}, \omega). \tag{11}$$

For isotropic systems, the tensor $\overleftrightarrow{\chi}$ can be decomposed into longitudinal and transverse components:

$$\overleftrightarrow{\chi}(\mathbf{k}, \omega) = \chi_L(k, \omega)\hat{\mathbf{k}}\hat{\mathbf{k}} + \chi_T(k, \omega)\left(\mathbf{I} - \hat{\mathbf{k}}\hat{\mathbf{k}}\right). \tag{12}$$

The easiest starting point for deriving microscopic expressions for $\chi_L(k, \omega)$ and $\chi_T(k, \omega)$ is the classical Kubo formula[61]:

$$\chi_{L/T}(\mathbf{k}, \omega) = \frac{\beta}{\epsilon_0} \int_0^\infty dt \frac{d}{dt} \left\langle \mathbf{P}_{L/T}(\mathbf{k}, t) \cdot \mathbf{P}_{L/T}^*(\mathbf{k}, 0) \right\rangle e^{i\omega t}. \quad (13)$$

This expression relates the susceptibility to the time correlation function of the polarization in equilibrium. The longitudinal part of the polarization can be calculated by Fourier transforming the defining expression for the polarization $\nabla \cdot \mathbf{P}(\mathbf{r}, t) = -\rho(r,t)$, leading to $\hat{\mathbf{k}} \cdot \mathbf{P} = i\rho(\mathbf{k}, t)/k = P_L$. To calculate the transverse part of the polarization we use the method of Raineri and Friedman to find the polarization vector of each molecule (Supplementary Note 2)[62]. We can rewrite equation 13 in terms of the normalized polarization correlation function (equation 7), and taking into account the isotropy of water:

$$\chi_{L/T}(k, \omega) = \chi_{L/T}(k, 0) \int_0^\infty \dot{\Phi}_{L/T}(k, t) e^{i\omega t} dt. \quad (14)$$

Computational methods. The three water models we used were TIP4P/ε (ref. 63), TIP4P/2005f (ref. 64) and TTM3F (ref. 65). To simulate methanol and acetonitrile we used the General AMBER Forcefield (GAFF)[66], a forcefield with full intramolecular flexibility, which has been shown to satisfactory reproduce key properties of both liquids[67]. Our TTM3F simulations were performed using an in-house code that uses the TTM3F force calculation routine of Fanourgakis and Xantheas. All other simulations were ran using the GROMACS package (ver. 4.6.5)[68]. We used particle–mesh Ewald summation for the long-range electrostatics with a Coloumb cutoff of 2 nm for our 4 + nm simulations and a cutoff of 1.2 nm for our simulations with 512 molecules. Our TTM3F simulations had 256 molecules and used Ewald summation with a Coulomb cutoff of 0.9 nm. The principle TIP4P/2005f simulations contained 512 molecules and were 8 ns long ($\Delta t_{out} = 8$ fs) and 0.6–1.2 ns long ($\Delta t_{out} = 4$ fs). Other simulations were 1–2 ns long. Simulations with MeOH and ACN contained 1,000 molecules and were 1-ns long. All simulations were equilibrated for at least 50 ps before outputting the data.

Because of periodic boundary conditions, the possible $\mathbf{k}$ vectors are limited to the form $\mathbf{k} = 2\pi n_x \hat{\mathbf{i}}/L_x + 2\pi n_y \hat{\mathbf{j}}/L_y + 2\pi n_z \hat{\mathbf{k}}/L_z$, where n_x, n_y and n_z are integers. We calculated correlation functions separately for each $\mathbf{k}$ and then average over the results for $\mathbf{k}$ vectors with the same magnitude, a process we found reduced random noise.

One can question whether a purely classical treatment is justified here because the librational dynamics we are interested have frequencies of 700–900 cm^{-1} for which $\hbar\omega \approx 3$–$4k_BT$ at 300 K. Previously it was shown that the widely used harmonic correction does not change the spectrum[21]. Furthermore, comparison of k-resolved infrared spectra taken from molecular dynamics and ab inito density functional theory (DFT) simulation show that they give qualitatively similar results for all frequencies below 800 cm^{-1} (ref. 37). For the OH-stretching peak, however, quantum effects are known to be very important.

Fitting the librational band. To obtain resonance frequencies and lifetimes for the librational peak in the imaginary part of the response we used a damped oscillator model. A Debye peak overlaps significantly with the librational band in both the longitudinal and transverse cases and must be included in the peak fitting. Equation 14 can be used to relate the form of the time correlation function to the absorption peak lineshape. For Debye response one has the following expressions:

$$\begin{aligned} \Phi(k, t) &= A e^{-t/\tau_D} \\ \frac{\text{Im}\{\chi(k, \omega)\}}{\chi(k, 0)} &= \frac{A\omega\tau_D}{1 + \omega^2\tau_D^2} \end{aligned} \quad (15)$$

For resonant response with resonance frequency $\omega_0(k)$ and dampening factor $\gamma \equiv 1/\tau$ we have:

$$\begin{aligned} \Phi(k, t) &= B e^{-t/\tau} \cos(\omega_0 t) \\ \frac{\text{Im}\{\chi(k, \omega)\}}{\chi(k, 0)} &= \frac{B}{2} \left(\frac{\omega\tau}{1 + (\omega + \omega_0)^2\tau^2} + \frac{\omega\tau}{1 + (\omega - \omega_0)^2\tau^2} \right) \end{aligned}$$

We find this lineshape (the Van Vleck–Weisskopf lineshape[69,70]) yields results identical to the standard damped harmonic oscillator response for the range of τ, ω_0 values we are interested in. We found a two function (Debye + resonant) fit worked very well for fitting the librational peak in the longitudinal case (Supplementary Figs 6 and 7). The H-bond stretching peak at ≈ 200 cm^{-1} overlaps with the librational band for $2 < k < 2.5$, and we found that it can be included in the fit using an additional damped harmonic oscillator, but usually this was not necessary. Because of this overlap and due to the broad nature of the transverse band, the fitting in the transverse case is only approximate. We found this was especially true for TTM3F and the experimental data, so we do not report lifetimes for such cases.

References

1. Santra, Jr. B., DiStasio, Jr. R. A., Martelli, F. & Car, R. Local structure analysis in ab initio liquid water. *Mol. Phys.* 1–13 in the press (2015).
2. Errington, J. R. & Debenedetti, P. G. Relationship between structural order and the anomalies of liquid water. *Nature* **409**, 318–321 (2001).
3. English, N. J. & Tse, J. S. Density fluctuations in liquid water. *Phys. Rev. Lett.* **106**, 037801 (2011).
4. Huang, C. *et al.* The inhomogeneous structure of water at ambient conditions. *Proc. Natl Acad. Sci. USA* **106**, 15214–15218 (2009).
5. Mallamace, F., Corsaro, C. & Stanley, H. E. Possible relation of water structural relaxation to water anomalies. *Proc. Natl Acad. Sci. USA* **110**, 4899–4904 (2013).
6. Sahle, C. J. *et al.* Microscopic structure of water at elevated pressures and temperatures. *Proc. Natl Acad. Sci. USA* **110**, 6301–6306 (2013).
7. Walrafen, G. E. Raman spectral studies of water structure. *J. Phys. Chem.* **40**, 3249 (1964).
8. Carey, D. M. & Korenowski, G. M. Measurement of the Raman spectrum of liquid water. *J. Chem. Phys.* **108**, 2669–2675 (1998).
9. Walrafen, G. E. Raman spectrum of water: transverse and longitudinal acoustic modes below ≈ 300 cm^{-1} and optic modes above ≈ 300 cm^{-1}. *J. Phys. Chem.* **94**, 2237–2239 (1990).
10. Walrafen, G. E. Raman spectral studies of the effects of temperature on water structure. *J. Phys. Chem.* **47**, 114–126 (1967).
11. Walrafen, G. E., Fisher, M. R., Hokmabadi, M. S. & Yang, W. Temperature dependence of the low and high frequency Raman scattering from liquid water. *J. Chem. Phys.* **85**, 6970–6982 (1986).
12. Castner, E. W., Chang, Y. J., Chu, Y. C. & Walrafen, G. E. The intermolecular dynamics of liquid water. *J. Chem. Phys.* **102**, 653–659 (1995).
13. Zelsmann, H. R. Temperature dependence of the optical constants for liquid H$_2$O and D$_2$O in the far IR region. *J. Mol. Struct.* **350**, 95–114 (1995).
14. Fukasawa, T. *et al.* Relation between dielectric and low-frequency Raman spectra of hydrogen-bond liquids. *Phys. Rev. Lett.* **95**, 197802 (2005).
15. Aure, P. & Chosson, A. The translational lattice-vibration Raman spectrum of single crystal ice 1 h. *J. Glaciol.* **21**, 65–71 (1978).
16. Klug, D. D., Tse, J. S. & Whalley, E. The longitudinalopticotranverseoptic mode splitting in ice Ih. *J. Chem. Phys.* **95**, 7011–7012 (1991).
17. Whalley, E. A detailed assignment of the oh stretching bands of ice i. *Can. J. Chem.* **55**, 3429–3441 (1977).
18. Klug, D. D. & Whalley, E. Origin of the high-frequency translational bands of ice i*. *J. Glaciol.* **21**, 55–63 (1978).
19. Galeener, F. L. & Lucovsky, G. Longitudinal optical vibrations in glasses: GeO$_2$ and SiO$_2$. *Phys. Rev. Lett.* **37**, 1474–1478 (1976).
20. Resat, H., Raineri, F. O. & Friedman, H. L. Studies of the optical like high frequency dispersion mode in liquid water. *J. Chem. Phys.* **98**, 7277 (1993).
21. Bopp, P. A., Kornyshev, A. A. & Sutmann, G. Frequency and wave-vector dependent dielectric function of water: Collective modes and relaxation spectra. *J. Chem. Phys.* **109**, 1939 (1998).
22. Ricci, M. A., Rocca, D., Ruocco, G. & Vallauri, R. Theoretical and computer-simulation study of the density fluctuations in liquid water. *Phys. Rev. A* **40**, 7226–7238 (1989).
23. Lyddane, R. H., Sachs, R. G. & Teller, E. On the polar vibrations of alkali halides. *Phys. Rev.* **59**, 673–676 (1941).
24. Barker, A. S. Long-wavelength soft modes, central peaks, and the Lyddane-Sachs-Teller relation. *Phys. Rev. B* **12**, 4071–4084 (1975).
25. Sievers, A. J. & Page, J. B. Generalized Lyddane-Sachs-Teller relation and disordered solids. *Phys. Rev. B* **41**, 3455–3459 (1990).
26. Payne, M. & Inkson, J. Longitudinal-optic-transverse-optic vibrational mode splittings in tetrahedral network glasses. *J. Non-Cryst. Solids* **68**, 351–360 (1984).
27. Sekimoto, K. & Matsubara, T. To-lo splittings of glassy dielectrics. *Phys. Rev. B* **26**, 3411 (1982).
28. Decius, J. C. & Hexter, R. M. *Molecular Vibrations in Crystals* (McGraw-Hill, 1977).
29. Decius, J. C. Dipolar coupling and molecular vibration in crystals. i. general theory. *J. Chem. Phys.* **49**, 1387–1392 (1968).
30. Burba, C. M. & Frech, R. Existence of optical phonons in the room temperature ionic liquid 1-ethyl-3-methylimidazolium trifluoromethanesulfonate. *J. Chem. Phys.* **134**, 134503 (2011).
31. Elton, D. C. & Fernández-Serra, M.-V. Polar nanoregions in water: a study of the dielectric properties of TIP4P/2005, TIP4P/2005f and TTM3F. *J. Chem. Phys.* **140**, 124504 (2014).
32. Madden, P. & Kivelson, D. In *Advances in Chemical Physics* 467 (John Wiley & Sons, Inc., 2007).
33. Hansen, J.-P. & McDonald, I. R. In *Theory of Simple Liquids* 3rd edn 341 (Academic Press, 2006).
34. Zasetsky, A. Y., Khalizov, A. F., Earle, M. E. & Sloan, J. J. Frequency dependent complex refractive indices of supercooled liquid water and ice determined from aerosol extinction spectra. *J. Phys. Chem. A* **109**, 2760 (2005).
35. Hale, G. & Querry, M. Optical constants of water in the 200-nm to 200- mu m wavelength region. *Appl. Opt.* **12**, 555 (1973).
36. Resat, H., Raineri, F. O. & Friedman, H. L. A dielectric theory of the optical like high frequency mode in liquid water. *J. Chem. Phys.* **97**, 2618 (1992).
37. Heyden, M. *et al.* Understanding the origins of dipolar couplings and correlated motion in the vibrational spectrum of water. *J. Phys. Chem. Lett.* **3**, 2135–2140 (2012).

38. Woodward, L. *Introduction to the Theory Of Molecular Vibrations and Vibrational Spectroscopy* (Clarendon Press, 1972).

39. Wagner, R. *et al.* Mid-infrared extinction spectra and optical constants of supercooled water droplets. *J. Phys. Chem. A* **109**, 7099–7112 (2005).

40. Vinh, N. Q. *et al.* High-precision gigahertz-to-terahertz spectroscopy of aqueous salt solutions as a probe of the femtosecond-to-picosecond dynamics of liquid water. *J. Chem. Phys.* **142**, 164502 (2015).

41. Ellison, W. J. Permittivity of pure water, at standard atmospheric pressure, over the frequency range 0-25 THz and the temperature range 0100c. *J. Phys. Chem. Ref. Data* **36**, 1–18 (2007).

42. Iwano, K., Yokoo, T., Oguro, M. & Ikeda, S. Propagating librations in ice xi: Model analysis and coherent inelastic neutron scattering experiment. *J. Phys. Soc. Jpn* **79**, 063601 (2010).

43. Shigenari, T. & Abe, K. Vibrational modes of hydrogens in the proton ordered phase XI of ice: Raman spectra above 400 cm^1. *J. Chem. Phys.* **136**, 174504 (2012).

44. Wan, Q., Spanu, L., Galli, G. A. & Gygi, F. Raman spectra of liquid water from ab initio molecular dynamics: vibrational signatures of charge fluctuations in the hydrogen bond network. *J. Chem. Theory. Comput.* **9**, 4124–4130 (2013).

45. Lobo, R., Robinson, J. E. & Rodriguez, S. High frequency dielectric response of dipolar liquids. *J. Chem. Phys.* **59**, 5992–6008 (1973).

46. Pollock, E. L. & Alder, B. J. Frequency-dependent dielectric response in polar liquids. *Phys. Rev. Lett.* **46**, 950–953 (1981).

47. Chandra, A. & Bagchi, B. Collective excitations in a dense dipolar liquid: How important are dipolarons in the polarization relaxation of common dipolar liquids? *J. Chem. Phys.* **92**, 6833 (1990).

48. Haughney, M., Ferrario, M. & McDonald, I. R. Molecular-dynamics simulation of liquid methanol. *J. Phys. Chem.* **91**, 4934–4940 (1987).

49. Yamaguchi, T., Benmore, C. J. & Soper, A. K. The structure of subcritical and supercritical methanol by neutron diffraction, empirical potential structure refinement, and spherical harmonic analysis. *J. Chem. Phys.* **112**, 8976–8987 (2000).

50. Matsumoto, M. & Gubbins, K. E. Hydrogen bonding in liquid methanol. *J. Chem. Phys.* **93**, 1981–1994 (1990).

51. Crowder, G. A. & Cook, B. R. Acetonitrile: far-infrared spectra and chemical thermodynamic properties. discussion of an entropy discrepancy. *J. Phys. Chem.* **71**, 914–916 (1967).

52. Santucci, S. C., Fioretto, D., Comez, L., Gessini, A. & Masciovecchio, C. Is there any fast sound in water? *Phys. Rev. Lett.* **97**, 225701 (2006).

53. Sampoli, M., Ruocco, G. & Sette, F. Mixing of longitudinal and transverse dynamics in liquid water. *Phys. Rev. Lett.* **79**, 1678–1681 (1997).

54. Cho, M., Fleming, G. R., Saito, S., Ohmine, I. & Stratt, R. M. Instantaneous normal mode analysis of liquid water. *J. Chem. Phys.* **100**, 6672–6683 (1994).

55. Garberoglio, G., Vallauri, R. & Sutmann, G. Instantaneous normal mode analysis of correlated cluster motions in hydrogen bonded liquids. *J. Chem. Phys.* **117**, 3278–3288 (2002).

56. Sedlmeier, F., Shadkhoo, S., Bruinsma, R. & Netz, R. R. Charge/mass dynamic structure factors of water and applications to dielectric friction and electroacoustic conversion. *J. Chem. Phys.* **140**, 054512 (2014).

57. Bosma, W. B., Fried, L. E. & Mukamel, S. Simulation of the intermolecular vibrational spectra of liquid water and water clusters. *J. Chem. Phys.* **98**, 4413–4421 (1993).

58. Conti Nibali, V. & Havenith, M. New insights into the role of water in biological function: Studying solvated biomolecules using terahertz absorption spectroscopy in conjunction with molecular dynamics simulations. *J. Am. Chem. Soc.* **136**, 12800 (2014).

59. Ebbinghaus, S. *et al.* An extended dynamical hydration shell around proteins. *Proc. Natl Acad. Sci. USA* **104**, 20749 (2007).

60. Kim, S., Born, B., Havenith, M. & Gruebele, M. Real-time detection of protein-water dynamics upon protein folding by terahertz absorption spectroscopy. *Angew. Chem. Int. Ed.* **47**, 6486 (2008).

61. Kubo, R. Statistical-mechanical theory of irreversible processes. i. general theory and simple applications to magnetic and conduction problems. *J. Phys. Soc. Jpn* **12**, 570–586 (1957).

62. Raineri, F. O. & Friedman, H. L. Static transverse dielectric function of model molecular fluids. *J. Chem. Phys.* **98**, 8910–8918 (1993).

63. Fuentes-Azcatl, R. & Alejandre, J. Non-polarizable force field of water based on the dielectric constant: TIP4P/ε. *J. Phys. Chem. B* **118**, 1263–1272 (2014).

64. Gonzalez, M. A. & Abascal, J. L. A flexible model for water based on TIP4P/2005. *J. Chem. Phys.* **135**, 224516 (2011).

65. Fanourgakis, G. S. & Xantheas, S. S. Development of transferable interaction potentials for water. v. extension of the flexible, polarizable, thole-type model potential (TTM3-F, v. 3.0) to describe the vibrational spectra of water clusters and liquid water. *J. Chem. Phys.* **128**, 074506 (2008).

66. Wang, J. *et al.* Development and testing of a general AMBER force field. *J. Compt. Chem.* **25**, 1157 (2004).

67. Caleman, C. *et al.* Force field benchmark of organic liquids: Density, enthalpy of vaporization, heat capacities, surface tension, isothermal compressibility, volumetric expansion coefficient, and dielectric constant. *J. Chem. Theory Comput.* **8**, 61 (2012).

68. Hess, B., Kutzner, C., van der Spoel, D. & Lindahl, E. GROMACS 4: Algorithms for highly efficient, load-balanced, and scalable molecular simulation. *J. Chem. Theory Comput.* **4**, 435–447 (2008).

69. Toda, M., Kubo, R., Saitō, N. & Hashitsume, N. In *Statistical Phys. II: Nonequilibrium Statistical Mechanics, Series C* (Springer, 1991).

70. Van Vleck, J. H. & Weisskopf, V. F. On the shape of collision-broadened lines. *Rev. Mod. Phys.* **17**, 227–236 (1945).

Acknowledgements

This work was partially supported by DOE Award No. DE-FG02-09ER16052 (D.C.E.) and by DOE Early Career Award No. DE-SC0003871 (M.V.F.S.).

Author contributions

D.C.E. Performed the simulations and helped prepare the manuscript. M.V.F.S supervised the project and helped prepare the manuscript.

Additional information

17

Two distinctive energy migration pathways of monolayer molecules on metal nanoparticle surfaces

Jiebo Li[1,2], Huifeng Qian[3], Hailong Chen[2], Zhun Zhao[3], Kaijun Yuan[4], Guangxu Chen[5], Andrea Miranda[2], Xunmin Guo[2], Yajing Chen[4], Nanfeng Zheng[5], Michael S. Wong[3,2] & Junrong Zheng[1,2]

Energy migrations at metal nanomaterial surfaces are fundamentally important to heterogeneous reactions. Here we report two distinctive energy migration pathways of monolayer adsorbate molecules on differently sized metal nanoparticle surfaces investigated with ultrafast vibrational spectroscopy. On a 5 nm platinum particle, within a few picoseconds the vibrational energy of a carbon monoxide adsorbate rapidly dissipates into the particle through electron/hole pair excitations, generating heat that quickly migrates on surface. In contrast, the lack of vibration-electron coupling on approximately 1 nm particles results in vibrational energy migration among adsorbates that occurs on a twenty times slower timescale. Further investigations reveal that the rapid carbon monoxide energy relaxation is also affected by the adsorption sites and the nature of the metal but to a lesser extent. These findings reflect the dependence of electron/vibration coupling on the metallic nature, size and surface site of nanoparticles and its significance in mediating energy relaxations and migrations on nanoparticle surfaces.

[1] College of Chemistry and Molecular Engineering, Beijing National Laboratory for Molecular Sciences, Peking University, Beijing 100871, China. [2] Department of Chemistry, Rice University, 6100 Main Street, Houston, Texas 77005, USA. [3] Department of Chemical and Biomolecular Engineering, Rice University, 6100 Main Street, Houston, Texas 77005, USA. [4] State Key Laboratory of Molecular Reaction Dynamics, Dalian Institute of the Chemical Physics, Chinese Academy of Sciences, Dalian, Liaoning 116023, China. [5] College of Chemistry and Chemical Engineering, Xiamen University, Xiamen, Fujian 361005, China. Correspondence and requests for materials should be addressed to J.Z. (email: zhengjunrong@gmail.com).

On the surfaces of heterogeneous catalysts, energy dissipation and migration dynamics determine the reaction local temperature and the ability of an adsorbate to cross the reaction barrier. Knowledge about the energy dynamics is essential for a thorough understanding of the selectivity and kinetics of catalytic reactions—and indispensable for developing rational guiding theories of catalyst design[1-3]. However, to experimentally investigate and theoretically describe energy dissipation and migration mechanisms on the surface of a metal nanoparticle, a key component of many important heterogeneous catalysts, remains a grand challenge[1,4-10]. Typically, theoretically modelling a reaction requires modifications of the nuclear potential energy surface with the assumption of separated nuclear and electronic motions during the nuclear movements across its transition state[11,12]. This approach, known as the Born–Oppenheimer approximation, which serves as the cornerstone of chemical kinetic theory, allows the electronic energy to be considered independent of the nuclear kinetic energy. Based on the Born–Oppenheimer approximation, the energy (< 0.5 eV) released in chemisorption or physisorption of molecules on metal surfaces was usually believed to be dissipated by surface vibrations (phonons), and electronic excitations through electron/vibration coupling were neglected[13-16]. However, some experimental evidence[17-22] and theoretical studies[23] have shown the nonadiabatic effects of vibration/electron coupling on flat metal and some metal nanoparticle surfaces[10,24], questioning the general applicability of the approach. This problem is even more complicated on the surfaces of catalytic metal nanoparticles, as the electronic properties can vary with particle size[25] and surface sites[26].

In this work, we demonstrate that the real-time energy migration dynamics between molecules on different surface sites of metal nanoparticles can be monitored with an ultrafast multiple dimensional vibrational spectroscopy. Two distinctive energy pathways are observed for a monolayer of carbon monoxide (CO) molecules adsorbed on different surface sites of a series of platinum (Pt) nanoparticles. On the surface of a 5 nm particle, which is metallic and traditionally theorized to exhibit electron/vibrational coupling[27], the energy migrations among different sites are very fast, occurring within a few picoseconds. In contrast, the energy migration dynamics between surface sites of small, semi-conductive Pt nanoparticles (~ 1 nm clusters) with a much smaller degree of electron/vibrational coupling are more than 1 order of magnitude slower. Further investigations reveal a similar particle-size dependence for the vibrational energy relaxations of the CO molecules: the vibrational relaxations on larger particles (2–11 nm) are also more than 1 order of magnitude faster than those on the 1 nm clusters. Changing the surface adsorption sites and the nature of the metal from Pt to Palladium (Pd) also alters the fast energy relaxation but to a lesser extent compared with the size effect.

Results

CO molecules on different surface sites of Pt particles.
The samples investigated in this work are 1 (Fig. 1a), 2, 5 (Fig. 1b) and 11 nm Pt clusters and nanoparticles and ~ 1 and 5 nm Pd clusters and particles, all of which are coated with a monolayer of CO and ligand molecules (particle characterizations are in Supplementary Figs 1–5). Pt and Pd nanoparticles are important metals in the field of catalysis. They have been used to catalyse the oxidation of CO, aiding in the removal of automobile exhausts, and also serve as classical model systems for mechanistic studies of heterogeneous catalysis for many years[28,29]. On the surface of the triangular prismatic 1 nm particle, CO molecules can reside on either the atop site (I) on the top of a Pt atom or the bridge site

(II) between two Pt atoms (Fig. 1a). The CO molecules on the atop sites have a CO stretch 0–1 transition frequency at $\sim 2,047$ cm^{-1}, and those at the bridge sites have a frequency at $\sim 1,860$ cm^{-1} (Fig. 1c (black curve))[30]. On the surface of 5 nm polyhedral-like Pt nanoparticles (Fig. 1b), CO molecules can occupy the terrace atop site (III), the step atop site (I) and the bridge site (II)[31,32]. As displayed in the two-dimensional infrared (2D IR) spectrum in Fig. 1d, CO molecules on site III have a CO stretch 0–1 transition frequency (ω_3, probe frequency) at $\sim 2,113$ cm^{-1}. The frequency on site I is $\sim 2,053$ cm^{-1}, and that on site II is $\sim 1,880$ cm^{-1} (Fig. 1c). The blue peaks underneath the red peaks along the ω_3 axis in Fig. 1d are the CO stretch 1–2 transition peaks. The blue peaks have lower ω_3 frequencies than the corresponding red peaks because of vibrational anharmonicities[33-35]. On the surfaces of other Pt particles (2 and 11 nm), CO molecules can also occupy the atop, bridge and terrace sites. Because the surface structures and the relative surface-site numbers differ on particles of different sizes, the frequency and relative intensity of the CO vibrational peak on each surface site vary with the particle size (Fig. 1c).

Two distinctive energy migration pathways. We employed two experimental strategies to investigate the energy migration pathways on the surfaces of these Pt nanoparticles by the real-time detection of (a) heat/vibrational energy migration on the nanoparticle surfaces and (b) direct measurements of the CO stretch first excited state population relaxations at different surface sites under various conditions. In doing so, we applied an ultrafast frequency-resolved mid-IR pulse to excite CO molecules on each site to the first vibrational excited state. Another ultrafast mid-IR pulse that covers a frequency range[36] from $\sim 1,000$ to $\sim 3,500$ cm^{-1} simultaneously detects the optical responses at the exciting frequencies, as well as frequencies associated with molecules on other surface sites. Tuning the waiting time between the excitation and detection pulses, (a) the heat generation was monitored by recording the heat-induced CO stretch 0–1 transition bleaching signals of CO molecules and (b) the direct CO stretch excitation relaxations were monitored in real-time by measuring the decays of the CO stretch 0–1 and 1–2 transition signals. (Signal origins are presented in Supplementary Method).

As displayed in Fig. 2a (where only the 0–1 transition signal is plotted), on the surface of the 5 nm Pt, after the CO molecules on the bridge site II are initially excited by a laser pulse, at zero waiting time, there is only one diagonal peak (peak 1) in the 0 ps spectrum. At 2 ps, a cross peak (peak 2) grows in at 2,062 cm^{-1}, ~ 9 cm^{-1} higher in frequency than that of the CO 0–1 transition on site I (Fig. 2b). At 5 ps, peak 2 grows in further and surpasses peak 1 in intensity. Depiction of 2D IR approach of the 5 nm Pt sample is shown in Supplementary Fig. 6. In contrast, the growth of the cross peak 2' on the surface of the 1 nm Pt particles is significantly slower (Fig. 2c,d). At a waiting time of 50 ps, its intensity is still less than 20% of the diagonal peak 1' (Fig. 2d). The growth of peaks 2 and 3 (Fig. 2b) is caused by the heat generated from the vibrational relaxation of the CO stretch on site II, similar to those in liquid samples described in great detail in our previous work[36,37]. The growth of peaks 2' and 6' (Fig. 2d) is caused by vibrational energy exchanging between two sites. Illustration of 2D IR approach on the 1 nm Pt sample is shown in Supplementary Fig. 7.

The particle size-dependent responses of the step site CO indicate different energy migration pathways. Recent work[25,38] suggests that the electronic properties of metal nanoparticles are dictated by the number of atoms and thus the particle size. For example, 5 nm polyhedral particle is metallic[39,40] and exhibits a

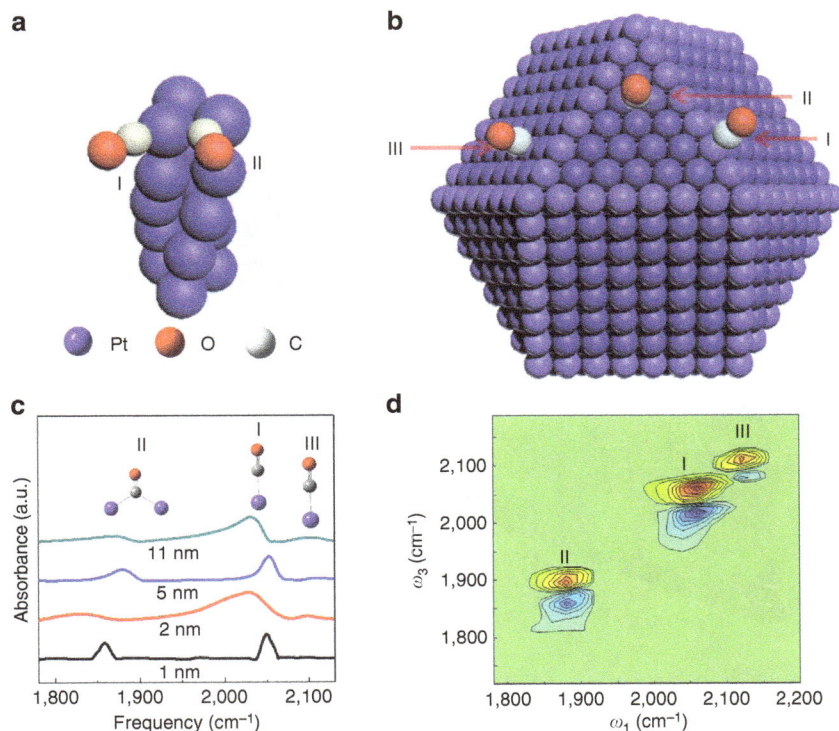

Figure 1 | Illustration and spectrums of CO molecules on Pt particles. (**a**) Illustrated CO molecules on the atop (I) and bridge (II) sites of a 1 nm Pt nanoparticle (Pt_{15}). In the sample, the actual CO coverage on both surface sites is 100%. (**b**) Illustrated CO molecules on the step (I), bridge (II) and terrace (III) sites of a 5 nm Pt nanoparticle. In the sample, the actual CO coverage on both surface sites is saturated. (**c**) FTIR spectra of CO molecules on Pt nanoparticles in the frequency range of $C \equiv O$ stretch 0–1 transition. (**d**) 2D IR spectrum of CO molecules on 5 nm Pt nanoparticles at zero waiting time, clearly showing the three different CO species (y-axis is probe frequency, and x-axis is pump frequency). The intensities of peaks II and III and their corresponding blue peaks along the y-axis are magnified by 13 and 22 times, respectively, to be comparable to that of peak I.

continuum of electronic states. As a result, theoretical studies predict that the conduction electrons on these metallic Pt nanoparticle surfaces can be excited from just below to just above the Fermi level (excitation of an electron/hole pair) by the resonant energy transfer from the CO stretch vibrational excitation[27]. The electron/hole pairs excited by the vibrational relaxation can quickly recombine and release their energy to the lattice motions (phonons, or 'heat') within $1 \sim 2$ ps, and quickly migrate on the surface, raising the metal surface temperature[41,42]. As a result of the temperature increase, the CO stretch 0–1 transition frequency ($2,053 \, cm^{-1}$) on site I red-shifts, which is corroborated by the temperature-dependent Fourier transform infrared spectroscopy (FTIR) spectra (in Supplementary Fig. 8). Such a red shift produces a bleaching signal at a frequency ($2,062 \, cm^{-1}$) higher than $2,053 \, cm^{-1}$ and a new absorption at a frequency ($2,047 \, cm^{-1}$) lower than $2,053 \, cm^{-1}$ in the 2D IR spectra (Fig. 2b). The excitation frequency ($\omega_1 = 1,880 \, cm^{-1}$) of peaks 2 and 3 reflects the origin (site II) of the energy dissipation process, and the respective detection frequencies ($\omega_3 = 2,062$ and $2,047 \, cm^{-1}$) demonstrate the spectral changes of the CO stretch on site I caused by the process. Therefore, how the intensities of the two peaks change with the waiting time is indicative of thermal energy migration dynamics from site II to site I. As displayed in Fig. 3a, peak 2 grows very fast and reaches a maximum at $2 \sim 3$ ps and then decays slowly, suggesting that the thermal energy migrates from site II to site I very rapidly within $2 \sim 3$ ps and then dissipate to environment slowly. By comparing the 2D IR results to the temperature difference FTIR spectra (Fig. 3b), the temperature increase on site II within the initial $2 \sim 3$ ps is found to be >57 K (Fig. 3b), and the increase drops to about 25 K at a delay of 25 ps after the thermal energy has further

dissipated to the environment. The spectral changes caused by the temperature increase are very small in the frequency range of bridge site CO stretch ($\sim 1,880 \, cm^{-1}$) (Supplementary Fig. 8b). The additional overlap, occurring with the tail of a very strong 1–2 absorption signal of the step atop site CO stretch after 2 ps (Fig. 2b), makes the heat-induced cross peaks around 6 and 7 ($\omega_1 = 2,053 \, cm^{-1}$, $\omega_3 = 1,880 \, cm^{-1}$) too weak to be observed.

In contrast to peaks 2 and 3 of the 5 nm Pt particle, the growth of peaks 2' and 3' are not caused by the migration of thermal energy generated from vibrational relaxation, but by the direct vibrational energy exchange between sites I and II CO as their frequencies are identical to those of the diagonal peaks 4' and 5' (Fig. 2d). The growth of the cross peaks is not caused by chemical exchange, in which case the peak growth rate constant ratio of 6' over 2' would be equal to 1 (ref. 35). The 1 nm cluster ($H_2[Pt_3(CO)_3(\mu_2\text{-}CO)_3]_5$) is a semiconductor with a HOMO-LUMO energy gap of ~ 1.7 eV (see visible spectrum in Supplementary Fig. 1d)[43,44]. This energy splitting is considerably larger than the CO vibrational 0–1 transition energy of 0.25 eV, and thus the CO stretch vibrational energy cannot excite the Pt electrons from the HOMO to the LUMO. Therefore, the CO vibrational energy on 1 nm sample cannot be mediated by electronic excitation, it can only transfer directly to the bridge site (II) C–O stretch (peaks 2' and 6') via vibration/vibration coupling. When direct vibrational energy exchange occurs, the population ratio is determined by detailed balance (this system is $e^{\frac{2,047cm^{-1} - 1,860cm^{-1}}{200cm^{-1}}} = 2.5$)[45]. Kinetic analyses show that the peak growth rate ratio (Fig. 3c) is also 2.5 (detailed kinetic analyses are found in Supplementary Note 1). No heat signal is observed up to 50 ps (Fig. 2d) (The control

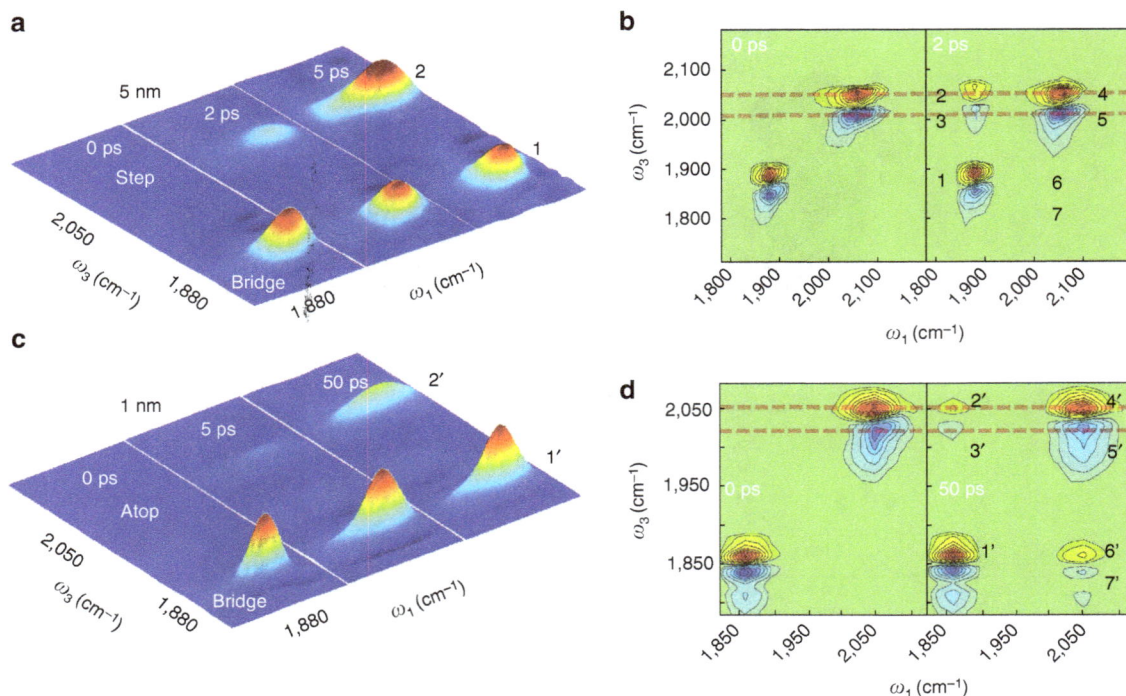

Figure 2 | Two distinctive pathways for vibrational energy generation. (a) 2D IR spectra of CO stretch (0–1 transition) on 5 nm Pt particles at three waiting times after exciting CO on the bridge site and detecting responses on both bridge and step atop sites. The growth of peak 2 at the step site indicates the generation of heat. **(b)** 2D IR spectra of CO stretch (both 0–1 and 1–2 transitions) on 5 nm Pt particles at two waiting times. The frequencies of the cross peak pairs (peaks 2 and 3) are different from those of the diagonal peak pair (peaks 4 and 5), indicating that peaks 2 and 3 arise because of heat generation. Red dashed lines are drawn to highlight the frequency differences. **(c)** 2D IR spectra of CO stretch (0–1 transition) on 1 nm Pt particles at three waiting times by exciting CO on the bridge site and detecting responses on both bridge and step atop sites. The growth of the peak 2′ at the step site indicates the direct vibrational exchange between CO molecules on the two surface sites. **(d)** 2D IR spectra of CO stretch (both 0–1 and 1–2 transitions) on 1 nm Pt particles at two waiting times. The frequencies of the cross peak pairs (peaks 2′ and 3′) are the same as those of the diagonal peak pair (peaks 4′ and 5′), indicating that peaks 2′ and 3′ are due to direct vibrational transfer between CO molecules on the two sites. The doublet of the blue peak of bridge CO is attributed to a Fermi resonance[57]. (The diagonal red peaks 1 and 1′ are normalized to 1). For the 2D IR spectra, y-axis is probe frequency, and x-axis is pump frequency.

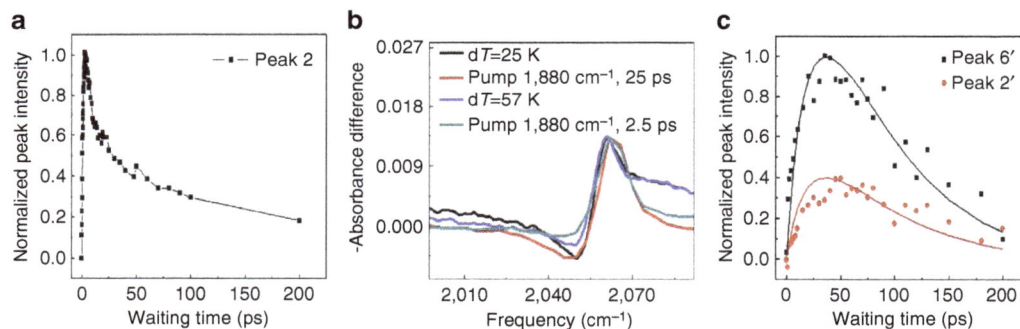

Figure 3 | Energy migration induced time-dependent spectrum dynamics. (a) The time-dependent intensities of peak 2 ($\omega_1 = 1{,}880$ cm^{-1}, $\omega_3 = 2{,}062$ cm^{-1}) on the 5 nm Pt nanoparticle. **(b)** Comparison of temperature difference FTIR (black and blue curves) and 2D IR data at waiting times 2.5 ps (red) and 25 ps (cyan). Intensities were normalized to similar values for better illustration. **(c)** The time-dependent intensities of the cross peak signals 6′ ($\omega_1 = 2{,}047$ cm^{-1}, $\omega_3 = 1{,}860$ cm^{-1}) and 2′ ($\omega_1 = 1{,}860$ cm^{-1}, $\omega_3 = 2{,}047$ cm^{-1}) on the 1 nm Pt nanoparticle.

temperature-dependent FITR of 1 nm sample is shown in Supplementary Fig. 9). This result indicates that the heat generation induced by CO vibrational relaxation on the 1 nm Pt particle is significantly slower than 50 ps, which is already more than 20 times slower than on the 5 nm Pt particle.

Two distinctive energy dissipation pathways. The distinct energy migration pathways following the surface molecular vibrational excitations on the two sizes particles are further supported by direct measurements of the CO stretch first excited

state population relaxations at different surface sites under various conditions. Figure 4 displays the waiting time-dependent CO stretch 0–1 (red) and 1–2 (blue) transition signals of CO molecules on the step atop site of 1, 2, 5 and 11 nm Pt particles after excitation to the first excited state of the CO stretch. On the surface of 1 nm Pt particles (Fig. 4a), after 20 ps, both 0–1 and 1–2 transition signals still remain > 50% of the initial intensity. However, on larger particles, the signals decay much faster. Within 4 ps more than 80% of the signal has decayed (Fig. 4b–d). This demonstrates that the CO vibrational energy dissipates

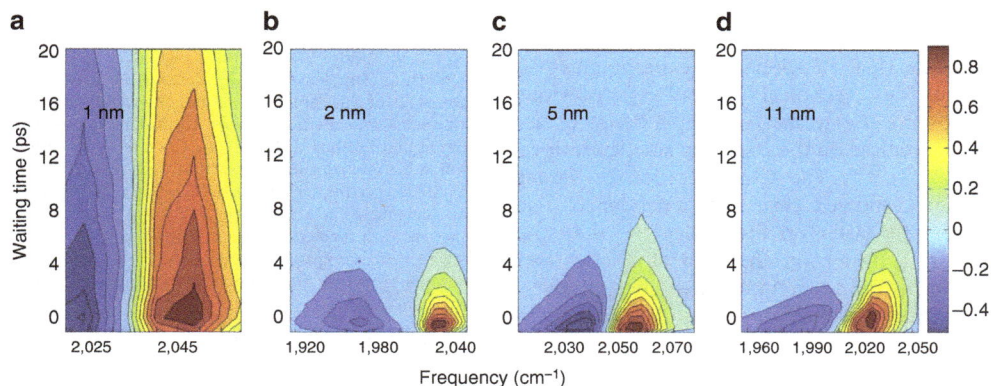

Figure 4 | Time-dependent spectrum of CO on Pt nanoparticles. The waiting time-dependent CO stretch 0-1 (red) and 1-2 (blue) transition signals of CO molecules on the atop step site of (**a**) 1 nm Pt particles, (**b**) 2 nm Pt particles, (**c**) 5 nm Pt nanoparticles and (**d**) 11 nm Pt nanoparticles after these molecules are excited to the CO stretch first excited state. The signal from the 1 nm particle lasts significantly longer than the signal from larger particles.

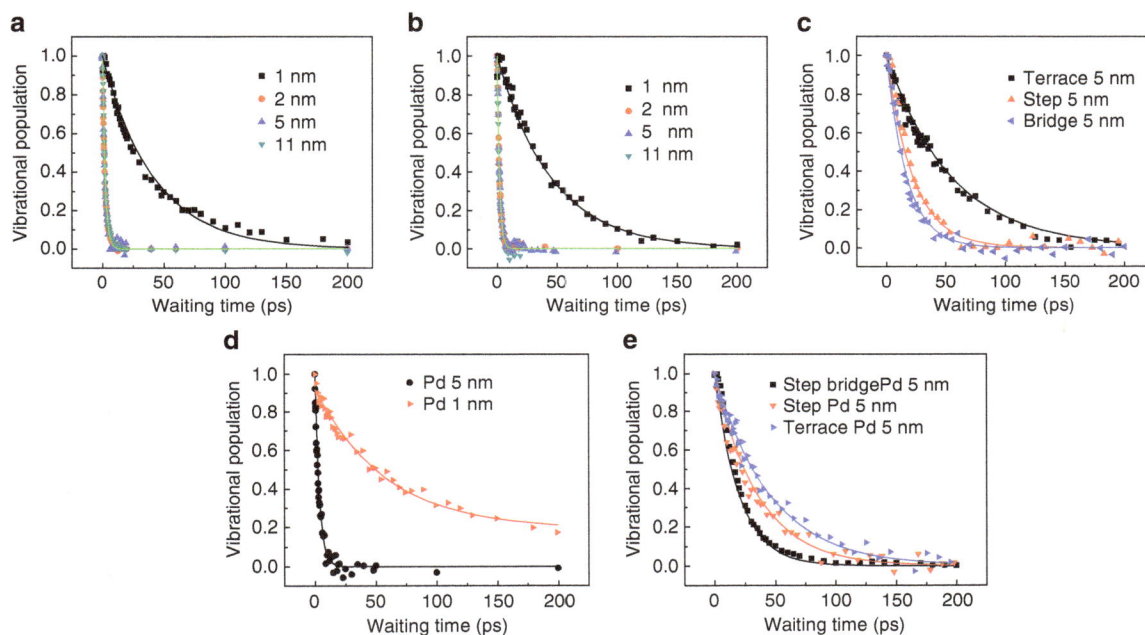

Figure 5 | Vibrational relaxation of CO in different conditions. (**a**) Vibrational relaxations of CO at the step site (I) on differently sized Pt particles; (**b**) Vibrational relaxations of CO at the bridge site (II) on differently sized Pt particles; (**c**) Vibrational relaxations of CO on the 5 nm Pt particle at different sites; (**d**) Vibrational relaxations of CO on the step atop site of 5 nm Pd particles and the atop site of 1 nm Pd (Pd/Ag) particles; (**e**) Vibrational relaxations of CO on 5 nm Pd particles at different sites. Experimental data points are fit with curves.

much faster on the surfaces of larger nanoparticles. This fast relaxation is consistent with the observed rapid heat generation from CO relaxation on step sites of metallic nanoparticles. From Fig. 4b–d, we can also see that the 0–1 transition signals slightly blue-shift with the increase of waiting time and last a little longer than the 1–2 transition signals, which is similar to Fig. 2c.

Quantitative analyses (Fig. 5a) of the data show that the CO stretch first excited state vibrational energy decays with a time constant of 42 ± 5 ps on site I of the 1 nm particle. In contrast, the relaxations time constants are more than 15 times faster and identical on the same site of the 3 larger-sized Pt particles, 2.2 ± 0.2 ps. Similarly, particle size dependence of the CO vibrational relaxations is observed on the bridge site (II) (Fig. 5b). The CO stretch first vibrational excited state population decays with a time constant of 47 ± 5 ps on the bridge site of 1 nm particles. On the same surface site of the larger (2–11 nm) Pt particles, the decay is significantly faster with a time constant of 1.8 ± 0.3 ps, which is consistent with fast temperature rise in

Fig. 3a. In addition, the fast relaxation dynamics on step site CO are independent (within the experimental uncertainty $\sim 10\%$) of isotopic labelling $^{13}C^{18}O$ (Supplementary Fig. 10a), temperature variance (293–80 K) (Supplementary Fig. 10b) and environmental modifications (Supplementary Fig. 10c). In contrast, without the surface electronic excitation, the CO relaxation on 1 nm Pt particle surfaces follows a similar mechanism (vibration/vibration coupling) as the metal carbonyl small molecules[46], 1–2 orders slower than the dynamics observed on metallic nanoparticle surfaces.

The fast CO stretch vibrational decays on metallic Pt particles surfaces (≥ 2 nm) were also observed on other surface sites besides sites I and II. Close examination of the rates reveals that different kinds of binding at the surface sites result in different dynamics. As shown in Fig. 5c, the CO vibrational energy relaxation on the terrace site (III) has a time constant 5.5 ± 0.6 ps. The relaxations (1.8 ± 0.3, 2.2 ± 0.2 and 5.5 ± 0.6 ps) on all three surface sites are of the same order time scale and all significantly faster than the dynamics observed on the 1 nm Pt particle surface (40–50 ps). The observed metal/nonmetal dependence of CO relaxations is similar to that of previous experiments[10,24]. However, probably due to technical limitations, previous experiments did not observe the surface-site dependence of the CO lifetimes and the CO lifetime on the step atop site determined then was too long ($\sim$7 ps)[10,24]. The CO relaxation on larger particle sizes in this work showed clear site dependence. The energy dissipation is slightly faster on the bridge site than the terrace atop site, and it is faster on the step atop site than the terrace atop site. According to previous theoretical study[27], vibrational relaxations induced by electron/vibration coupling are determined not only by the available electronic energy levels but also by the coupling strength between the surface electrons and the chemical bond that vibrates. The available energy levels of surface electrons are identical for each surface site. And, thus, the experimental results suggest that the CO/surface electron coupling strengths vary on different surface sites of nanoparticle, consistent with the different binding strength of CO molecules on distinct surface sites from previous studies on Pt flat surface[26,47].

Particle-size and surface-site dependences on Pd surfaces. Similar trends are also observed on the surfaces of Pd nanoparticles. As displayed in Fig. 5d, the CO stretch vibrational relaxation (3.5 ± 0.5 ps) on the step atop site of 5 nm Pd metallic particles is about 20 times faster than that (67 ± 15 ps) on the atop site of $\sim$1 nm Pd semi-conductive clusters (Pd/Ag bimetallic particle <2 nm). The relaxation times follow the same relative order as those of metallic Pd particles: bridge (2.0 ± 0.2 ps) <step atop (3.5 ± 0.5 ps)< terrace atop (6.1 ± 0.5 ps) ($\ll$1 nm (67 ± 15 ps)) (Fig. 5e). Comparison of the 5 nm Pd to the 5 nm Pt shows the CO stretch vibrational relaxation on Pd is slightly slower than that on Pt at each surface site. This is probably because CO/Pt interactions are slightly stronger than CO/Pd interactions[48].

Discussion

Two distinct energy migration pathways on nanoparticle surfaces were observed. On a large metallic nanoparticle, the vibrational energy of a surface molecule can rapidly dissipate into the particle through electron/hole pair excitations. The electron/hole pairs can then quickly recombine and release their energy to the lattice motions (phonons, or 'heat'), and quickly migrate on the surface, raising the surface temperature. All the processes occur within a few picoseconds. On a small and semiconducting particle, because of the lack of electron/vibration coupling, the surface vibrational relaxation and the energy migration between surface molecules are mainly through the vibration/vibration coupling mechanism and are >1 order of magnitude slower. Systematic studies also reveal that the electron/vibration coupling is little dependent on temperature (from 80 to 293 K) or the particle size only if the particle is metallic, but is dependent on the particle surface site and the nature of the metal. The findings may stimulate further application and development of existing nonadiabatic theory to explore the dependences[23].

Methods

Materials and synthesis procedures. The chemicals used for synthesizing the nanoparticles were purchased from Sigma-Aldrich and used without further purification and the CO gas was purchased from Matheson and Sigma.

Preparation and characterization of 1 nm Pt nanoparticle. About 0.5 ml (0.1 M) aqueous solution of chloroplatinic acid ($H_2PtCl_6 \cdot 6H_2O$) dissolved in 10 ml dimethylformamide (DMF) can be sealed into 1.5 atm CO glass vessel for 28 h to obtain the blue–green sample. The structure of the cluster was characterized with mass spectrometry[43]. In the DMF solution (Fig. 1b and Supplementary Fig. 1a),

the atop site CO absorption is at 2,047 cm^{-1}, and the bridge site CO absorption is around 1,860 cm^{-1}. The FTIR spectrum in Fig. 1c multiplies the intensities of the atop CO and bridge CO absorption regions by 25 times for better illustration. The IR spectrum (Supplementary Fig. 1c) showed in solid phase also confirmed two main peaks of this sample. The 1 nm Pt particles are unstable in the solid state. All ultrafast measurements presented in this paper were conducted in the DMF solution and finished within 2–3 h. The ultraviolet–visible spectra of the reaction solution before and after the reaction were recorded using a Varian Cary 5000 UV-visible spectrometer. Compared with the H_2PtCl_6 solution, two new peaks (Supplementary Fig. 1d) at 400 and 630 nm appear in the cluster solution, consistent with those of the $[Pt_3(CO)_3(\mu_2\text{-}CO)_3]_5^{2-}$. The 1.7 eV energy gap is calculated according to the ultraviolet–visible spectrum because this sample's electronic properties are comparable to a quantum dot.

Preparation and characterization of 2 nm Pt nanoparticle solid sample. K_2PtCl_4 (3 mM ethylene glycol solution) and 10 mM sodium acetate were mixed in hot 1, 2-ethanediol at 80 °C, stirred for 30 min. Thus, $\sim$2 nm Pt nanoparticles were synthesized following the reported method[49]. The hot solution was then cooled down. CO gas was bubbled at the rate of 200 ml min^{-1} within 30 min in a hood. The solution was then mixed with water at a v/v ratio 3:1. The mixture was centrifuged for three times with 26,000 r.p.m. for 38 h or 40,000 r.p.m. for 20 h. The precipitate was suspended in several drops of 1,2-ethanediol. The mixture was dropped onto CaF$_2$ windows and transferred into the vacuum oven overnight to remove solvent. The sample-covered CaF$_2$ window was placed into vacuum chamber and oil-free pumped overnight. We applied the same procedures to bubble Pt nanoparticles (2, 5 and 11 nm) and Pd nanoparticles (1 and 5 nm) samples, prepared the solid samples in vacuum chamber, measure the solid samples FTIR spectrum and collect ultrafast data. The frequency ranges (1,830–1,890 cm^{-1}) and (2,090–2,130 cm^{-1}) of FTIR Fig. 1c of 2 nm sample in the result part were zoomed in three times for better illustration. The sample was characterized by JEM 2100F TEM and Rigaku D/Max Ultima XRD. The transmission electron microscopy (TEM) and XRD data are shown in Supplementary Fig. 2a,b (high resolution) and Supplementary Fig. 3a. The temperature control measurements were performed with JANIS model ST-100.

Preparation and characterization of 2 nm Pt nanoparticle solution sample. K_2PtCl_4 (0.0174 g, 4.2 × 10^{-5} mol), poly(acrylic acid sodium salt) (molecular weight (MW) $\sim$5,100, 0.0394 g) and 10 ml ethylene glycol were mixed in a 50 ml flask. It is noted that poly(acrylic acid sodium salt) shows poor solubility in ethylene glycol at room temperature. Under vigorously stirring, the solution was heated to 180 °C and kept at 180 °C for 10 min. The colour of solution would become dark and clear with increasing temperature, which indicates the formation of $\sim$2–3 nm Pt NPs. Then, the sample was dropped on the CaF$_2$ window. Vacuum pump was used to remove main parts of ethylene glycol solvents in the vacuum oven. Only two drops of high-concentration solution were left on the CaF$_2$ window. Another CaF$_2$ was used to quickly seal the sample between two CaF$_2$. This solution sample did not need to be transferred into vacuum chamber for measurement. The FTIR is shown in Supplementary Fig. 3b. The method to prepare isotope-labelled 2 nm Pt was exactly same as that for the powder 2 nm Pt NP sample. K_2PtCl_4 (3 mM ethylene glycol solution) and 10 mM sodium acetate were mixed in hot 1, 2-ethanediol at 80 °C, stirred for 30 min. The isotope $^{13}C^{18}O$ gas 0.5 l (purchased from Sigma) was bubbled into the solution. Samples were then prepared according to the procedure described above.

FTIR spectra of CO and $^{13}C^{18}O$ on the 2 nm Pt nanoparticles were shown in Supplementary Fig. 3c. The vibrational decay data of the $^{13}C^{18}O$ stretch excitation in this paper were obtained by exciting the shoulder at $\sim$1,940 cm^{-1}. The 2D IR of this sample is shown in Supplementary Fig. 3d. The blue peak around (1940 cm^{-1}, 1870 cm^{-1}) stems from the $^{13}C^{18}O$ 1-2 transition and indicates its binding formation on the step site.

Preparation and characterization of 5 nm Pt nanoparticle. Pt(II) acetylacetonate (80 mg, $\sim$0.2 mmol) and polyvinylpyrrolidone (PVP, 55 mg, MW = 55,000) were dissolved in 5 ml of ethylene glycol. The solution was then heated to 200 °C in a microwave reactor (Anton Paar Monowave 300) under a stirring rate of 1200 r.p.m. min^{-1} and held at that temperature for 5 min and then cooled to room temperature. The as-prepared Pt nanoparticles were precipitated with 45 ml of acetone and re-dispersed in 10 ml of ethylene glycol to remove excess PVP. Thus, the 5 nm Pt nanoparticles were synthesized according to the reported method[50]. The TEM image is shown in Supplementary Fig. 2c. The original FTIR intensities of the bridge and terrace are weak. In the Results part, FTIR Fig. 1c bridge area (1,830–1,890 cm^{-1}) and terrace area (2,090–2,130 cm^{-1}) of 5 nm sample were zoomed in 10 times for better illustration.

Preparation and characterization of 11 nm Pt nanoparticle. First, 250 ml 1 × 10^{-4} M K_2PtCl_4 solution in DI water was prepared, followed by adding 0.2 ml of 0.1 M sodium polyacrylate. Then N$_2$ gas (>99.99%, Matheson) was bubbled into the mixture at a flow rate of 200 ml min^{-1} for 20 min. Reduction of Pt ions was started by switching the bubbling gas to H$_2$ (>99.99%, Matheson) for 10 min at a flow rate of 200 ml min^{-1}. The reaction vessel was then sealed and left under room temperature for 48 h to complete the reduction. Thus, the preparation method of the 11 nm Pt nanoparticles in this study followed the method by Ahmadi et al.[51] The TEM picture is shown in Supplementary Fig. 2d.

Preparation and characterization of 5 nm Pd nanoparticle. About 0.449 (2.00 mmol) of palladium(II) acetate and 0.85 g of PVP (MW = 55,000) were dissolved in 20 ml of 2-ethoxyethanol. The solution was then heated to 110 °C in a microwave reactor (Anton Paar Monowave 300) under a stirring rate of 1,200 r.p.m. min^{-1} and held at that temperature for 60 min, and then cooled to room temperature. The as-prepared Pd NPs are used for spectroscopic measurement. The TEM picture is shown in Supplementary Fig. 2e. The FTIR of this sample is shown in Supplementary Fig. 4.

Preparation and characterization of 1 nm Pd–Ag nanoparticle. The preparation method was a revised method for the preparation of Ag$_{44}$ (ref. 52). Supplementary Fig. 5a briefly presented the synthesis procedure. In detail, 0.25 mmol of silver nitrate (AgNO$_3$) and 0.0625 mmol of palladium acetate (Pd(O$_2$CCH3)$_2$) were dissolved in 21 ml water and 1.25 mmol of para-mercaptobenzoic acid (p-MPA) were dissolved in 12 ml ethanol. The aqueous solution and ethanolic solution were mixed to form a yellowish insoluble precursor. The pH was then raised to 12 to deprotonate p-MPA by using aqueous CsOH (50% w/v). The colour of solution immediately became red with the increase of pH value. After that 9 ml of aqueous NaBH$_4$ (3.125 mmol) was slowly added to the precursor solution with vigorously stirring. The colour of solution turned to dark after adding NaBH$_4$ solution. After 1 h reaction, the obtained product was cleaned first by centrifuging to remove any solid and then precipitating the clusters with DMF to remove salts and other left-over soluble materials from the reaction. TEM picture is shown in Supplementary Fig. 5b and X-ray photoelectron spectroscopy (XPS) is shown in Supplementary Fig. 5c. The ratio of Pd/Ag/S equals 1:6.2:6.8, calculated from the peak area/relatively sensitivity factor with the software Multipak of the XPS machine. Thus, the as-prepared nanoparticle is the Pd–Ag bimetallic structure. The low metal/ligand ratio suggested the small size of this particle. The multi-absorbance peaks in ultraviolet–visible spectrum also indicate that the size of Pd–Ag bimetallic nanoparticles is <2 nm—otherwise, a strong surface Plasmon resonance peak from Ag nanoparticles would be observed at ∼420 nm (ref. 53). The TEM image (Supplementary fig. 5b) also suggests that the size is very small (<2 nm) and results in poor image resolution. Because the binding strength of CO with Pd is much stronger than that of CO with Ag (refs 54,55), the CO molecules on the Ag–Pd bimetallic nanoparticle dominantly reside on Pd atoms.

Optical methods. Briefly, the same seed pulse is employed to synchronize a picosecond amplifier and a femtosecond amplifier. The two amplifiers are then used to produce pump and probe IR beams, respectively. The ∼0.8 ps mid-IR pulses (vary from 0.7 to 0.9 ps according to the frequency) were generated by the ps amplifier pumping an optical parametric amplifier. The bandwidth of the 1 kHz mid-IR beam as pumping beam in measurement was around 10–35 cm^{-1} in a tunable frequency range of 400 to 4,000 cm^{-1} with energy ranging from 1 ∼ 40 µJ per pulse (>10 µJ per pulse for 1,000–4,000 cm^{-1}). The probing beam (a high-intensity mid-IR and terahertz super-continuum pulse with a duration of <100 fs between 400 to 3,000 cm^{-1} at 1 KHz) was generated from the fematosecond amplifier. Two polarizers are inserted into the probe beam pathway before and after the sample, respectively, to selectively measure the parallel or perpendicular polarized signal relative to the excitation beam. In IR pump/IR probe experiments, the picosecond IR pulse is the pump beam and the super-continuum pulse is the probe beam. Two polarizers are inserted in the probe beam pathway to selectively measure the parallel or perpendicular polarized signal relative to the excitation beam. To avoid the scattering signal (especially for the weak signals at the bridge site and terrace atop site), we used data from the perpendicular configuration rather than the usual rotation-free data to describe the vibrational decay[56]. This approach can introduce some uncertainty in the vibrational lifetimes as resonant vibrational energy transfers and rotation can occur. To address this issue, for both 1 and 5 nm Pt samples, we measured both rotation free (parallel signal plus two perpendicular signals) and perpendicular signals. The data are plotted in Supplementary Fig. 11 and show that the decays from the perpendicular measurements are about 10% slower than the rotation-free data. Because our experimental data have ± 10% uncertainty, the data of the isotope labelling ^{13}C^{18}O, the data do not present a clear difference with the un-labelled data. In all our 2D IR spectra in this paper, the probe frequency resolution is around ± 4 cm^{-1} and the pump frequency resolution is ∼ 20 cm^{-1}. The FTIR spectral resolution is around 2 cm^{-1}. The control experiment of heat generation of peak 2 in Figs 2 and 3 is introduced in Supplementary Fig. 12. Our signal origins of both diagonal peaks and cross peaks are presented in Supplementary Method.

References

1. Wodtke, A. M., Tully, J. C. & Auerbach, D. J. Electronically non-adiabatic interactions of molecules at metal surfaces: Can we trust the Born-Oppenheimer approximation for surface chemistry? *Inter. Rev. Phys. Chem.* **23**, 513–539 (2004).

2. Tully, J. C. Chemical dynamics at metal surfaces. *Ann. Rev. Phys. Chem.* **51**, 153–178 (2000).

3. Wodtke, A. M., Matsiev, D. & Auerbach, D. J. Energy transfer and chemical dynamics at solid surfaces: The special role of charge transfer. *Prog. Surf. Sci.* **83**, 167–214 (2008).

4. Jaeger, N. I. Bridging gaps and opening windows. *Science* **293**, 1601–1602 (2001).

5. Rosenfeld, D. E., Gengeliczki, Z., Smith, B. J., Stack, T. & Fayer, M. Structural dynamics of a catalytic monolayer probed by ultrafast 2D IR vibrational echoes. *Science* **334**, 634–639 (2011).

6. Forsblom, M. & Persson, M. Vibrational lifetimes of cyanide and carbon monoxide on noble and transition metal surfaces. *J. Chem. Phys.* **127**, 154303 (2007).

7. Nieto, P. et al. Reactive and nonreactive scattering of H2 from a metal surface is electronically adiabatic. *Science* **312**, 86–89 (2006).

8. Díaz, C. et al. Multidimensional effects on dissociation of N$_2$ on Ru (0001). *Phys. Rev. Lett.* **96**, 096102 (2006).

9. Harris, A. & Rothberg, L. Surface vibrational energy relaxation by sum frequency generation: five-wave mixing and coherent transients. *J. Chem. Phys.* **94**, 2449–2457 (1991).

10. Berkerle, J., Casassa, M., Cavanagh, R., Heilweil, E. & Stephenson, J. Time resolved studies of vibrational relaxation dynamics of CO (v = 1) on metal particle surfaces. *J. Chem. Phys.* **90**, 4619–4620 (1989).

11. Hasselbrink, E. Capturing the complexities of molecule-surface interactions. *Science* **326**, 809–810 (2009).

12. Wodtke, A. M. Chemistry in a computer: advancing the *in silico* dream. *Science* **312**, 64–65 (2006).

13. Hurst, J. E., Wharton, L., Janda, K. C. & Auerbach, D. J. Direct inelastic scattering Ar from Pt (111). *J. Chem. Phys.* **78**, 1559–1581 (1983).

14. Hurst, J. E. et al. Observation of direct inelastic scattering in the presence of trapping-desorption scattering: Xe on Pt (111). *Phys. Rev. Lett.* **43**, 1175 (1979).

15. Tully, J. C. Theories of the dynamics of inelastic and reactive processes at surfaces. *Annu. Rev. Phys. Chem.* **31**, 319–343 (1980).

16. Auerbach, D. J. Hitting the surface--softly. *Science* **294**, 2488–2489 (2001).

17. Gergen, B., Nienhaus, H., Weinberg, W. H. & McFarland, E. W. Chemically induced electronic excitations at metal surfaces. *Science* **294**, 2521–2523 (2001).

18. Huang, Y., Rettner, C. T., Auerbach, D. J. & Wodtke, A. M. Vibrational promotion of electron transfer. *Science* **290**, 111–114 (2000).

19. Hou, H. et al. Enhanced reactivity of highly vibrationally excited molecules on metal surfaces. *Science* **284**, 1647–1650 (1999).

20. White, J. D., Chen, J., Matsiev, D., Auerbach, D. J. & Wodtke, A. M. Conversion of large-amplitude vibration to electron excitation at a metal surface. *Nature* **433**, 503–505 (2005).

21. Kroes, G.-J., Díaz, C., Pijper, E., Olsen, R. A. & Auerbach, D. J. Apparent failure of the Born-Oppenheimer static surface model for vibrational excitation of molecular hydrogen on copper. *Proc. Natl Acad. Sci. USA* **107**, 20881–20886 (2010).

22. Beckerle, J. D., Casassa, M. P., Cavanagh, R. R., Heilweil, E. J. & Stephenson, J. C. Ultrafast infrared response of adsorbates on metal surfaces: vibrational lifetime of CO/Pt(111). *Phys. Rev. Lett.* **64**, 2090–2093 (1990).

23. Shenvi, N., Roy, S. & Tully, J. C. Dynamical steering and electronic excitation in NO scattering from a gold surface. *Science* **326**, 829–832 (2009).

24. Heilweil, E., Cavanagh, R. & Stephenson, J. Picosecond study of the population lifetime of CO (v = 1) chemisorbed on SiO$_2$-supported rhodium particles. *J. Chem. Phys.* **89**, 5342–5343 (1988).

25. Guasco, T. L., Elliott, B. M., Johnson, M. A., Ding, J. & Jordan, K. D. Isolating the spectral signatures of individual sites in water networks using vibrational double-resonance spectroscopy of cluster isotopomers. *J. Phys. Chem. Lett.* **1**, 2396–2401 (2010).

26. Backus, E. H. G., Forsblom, M., Persson, M. & Bonn, M. Highly efficient ultrafast energy transfer into molecules at surface step sites. *J. Phys. Chem. C* **111**, 6149–6153 (2007).

27. Persson, B. & Persson, M. Vibrational lifetime for CO adsorbed on Cu (100). *Solid State Commun.* **36**, 175–179 (1980).

28. Wang, Z., Ahmad, T. & El-Sayed, M. Steps, ledges and kinks on the surfaces of platinum nanoparticles of different shapes. *Surf. Sci.* **380**, 302–310 (1997).

29. Tripa, C. E. & Yates, J. T. Surface-aligned reaction of photogenerated oxygen atoms with carbon monoxide targets. *Nature* **398**, 591–593 (1999).

30. Longoni, G. & Chini, P. Synthesis and chemical characterization of platinum carbonyl dianions [Pt3 (CO) 6] n2-(n = . apprx. 10, 6, 5, 4, 3, 2, 1). A new series of inorganic oligomers. *J. Am. Chem. Soc.* **98**, 7225–7231 (1976).

31. Hayden, B. E. & Bradshaw, A. M. The adsorption of CO on Pt(111) studied by infrared reflection--Absorption spectroscopy. *Surf. Sci.* **125**, 787–802 (1983).

32. Lundwall, M. J., McClure, S. M. & Goodman, D. W. Probing terrace and step sites on Pt nanoparticles using CO and ethylene. *J. Phys. Chem. C* **114**, 7904–7912 (2010).

33. Golonzka, O., Khalil, M., Demirdoven, N. & Tokmakoff, A. Vibrational anharmonicities revealed by coherent two-dimensional infrared spectroscopy. *Phys. Rev. Lett.* **86**, 2154–2157 (2001).

34. Cervetto, V., Helbing, J., Bredenbeck, J. & Hamm, P. Double-resonance versus pulsed Fourier transform two-dimensional infrared spectroscopy: an experimental and theoretical comparison. *J. Chem. Phys.* **121**, 5935–5942 (2004).

35. Zheng, J. *et al.* Ultrafast dynamics of solute-solvent complexation observed at thermal equilibrium in real time. *Science* **309**, 1338–1343 (2005).

36. Chen, H., Bian, H., Li, J., Wen, X. & Zheng, J. Relative intermolecular orientation probed via molecular heat transport. *J. Phys. Chem. A* **117**, 6052–6065 (2013).

37. Chen, H., Bian, H., Li, J., Wen, X. & Zheng, J. Ultrafast multiple-mode multiple-dimensional vibrational spectroscopy. *Inter. Rev. Phys. Chem.* **31**, 469–565 (2012).

38. Yau, S. H., Varnavski, O. & Goodson, III T. An ultrafast look at Au nanoclusters. *Acc. Chem. Res.* **46**, 1506–1516 (2013).

39. Schmid, G. Large clusters and colloids. Metals in the embryonic state. *Chem. Rev.* **92**, 1709–1727 (1992).

40. Li, L. *et al.* Investigation of catalytic finite-size-effects of platinum metal clusters. *J. Phys. Chem. Lett.* **4**, 222–226 (2012).

41. Wang, Z. *et al.* Ultrafast flash thermal conductance of molecular chains. *Science* **317**, 787–790 (2007).

42. Frischkorn, C. & Wolf, M. Femtochemistry at metal surfaces: nonadiabatic reaction dynamics. *Chem. Rev.* **106**, 4207–4233 (2006).

43. Chen, G., Yang, H., Wu, B., Zheng, Y. & Zheng, N. Supported monodisperse Pt nanoparticles from [Pt3 (CO)$_3$ (μ2-CO)$_3$] 52-clusters for investigating support-Pt interface effect in catalysis. *Dalton Trans.* **42**, 12699–12705 (2013).

44. Selvakannan, P., Lampre, I., Erard, M. & Remita, H. Platinum carbonyl clusters: double emitting quantum dots. *J. Phys. Chem. C* **112**, 18722–18726 (2008).

45. Chen, H., Wen, X., Guo, X. & Zheng, J. Intermolecular vibrational energy transfers in liquids and solids. *Phys. Chem. Chem. Phys.* **16**, 13995–14014 (2014).

46. Heilweil, E., Stephenson, J. & Cavanagh, R. Measurements of carbonyl (v = 1) population lifetimes: metal-carbonyl cluster compounds supported on silica. *J. Phys. Chem.* **92**, 6099–6103 (1988).

47. Backus, E. H., Eichler, A., Kleyn, A. W. & Bonn, M. Real-time observation of molecular motion on a surface. *Science* **310**, 1790–1793 (2005).

48. Hammer, B., Morikawa, Y. & Nørskov, J. K. CO chemisorption at metal surfaces and overlayers. *Phys. Rev. Lett.* **76**, 2141 (1996).

49. Dablemont, C. *et al.* FTIR and XPS study of Pt nanoparticle functionalization and interaction with alumina. *Langmuir* **24**, 5832–5841 (2008).

50. Wang, H. *et al.* Influence of size-induced oxidation state of platinum nanoparticles on selectivity and activity in catalytic methanol oxidation in the gas phase. *Nano Lett.* **13**, 2976–2979 (2013).

51. Ahmadi, T. S., Wang, Z. L., Green, T. C., Henglein, A. & El-Sayed, M. A. Shape-controlled synthesis of colloidal platinum nanoparticles. *Science* **272**, 1924–1925 (1996).

52. Desireddy, A. *et al.* Ultrastable silver nanoparticles. *Nature* **501**, 399–402 (2013).

53. Peng, S., McMahon, J. M., Schatz, G. C., Gray, S. K. & Sun, Y. Reversing the size-dependence of surface plasmon resonances. *Proc. Natl Acad. Sci. USA* **107**, 14530–14534 (2010).

54. Zeinalipour-Yazdi, C. D., Cooksy, A. L. & Efstathiou, A. M. CO adsorption on transition metal clusters: trends from density functional theory. *Surf. Sci.* **602**, 1858–1862 (2008).

55. Zhao, S., Ren, Y., Ren, Y., Wang, J. & Yin, W. Density functional study of CO binding on small Ag<i>n</i>Pd<i>m</i> clusters. *J. Mol. Struct.* **955**, 66–70 (2010).

56. Bian, H. *et al.* Molecular conformations and dynamics on surfaces of gold nanoparticles probed with multiple-mode multiple-dimensional infrared spectroscopy. *J. Phys. Chem. C* **116**, 7913–7924 (2012).

57. Zheng, J. *et al.* Accidental vibrational degeneracy in vibrational excited states observed with ultrafast Two-Dimentional IR vibrational echo spectroscopy. *J. Chem. Phys.* **123**, 164301 (2005).

Acknowledgements

This material is based on the work supported by the AFOSR Award No. FA9550-11-1-0070 and the Welch foundation under Award No. C-1752. J.Z. also thanks supports from PKU and the David and Lucile Packard Foundation for a Packard fellowship and the Alfred P. Sloan Foundation for a Sloan fellowship. N.Z. appreciates the funding from the National Nature Science Foundation of China (21131005, 21420102001). M.S.W. acknowledges J. Evans Attwell-Welch Postdoctoral Fellowship Program of the Smalley Institute of Rice University (to H.Q.) and the National Science Foundation (CBET-1134535).

Author contributions

J.Z. and J.L. designed the experiments. J.Z. supervised the project. J.L., H.C., K.Y., X.G. and A.M. performed ultrafast experiments. H.Q., J.L., Z.Z., X.G., N.Z. and M.S.W. conducted the synthesis experiments. Y.C. performed 1 nm Pt sample temperature-dependent FTIR. J.Z., J.L., H.Q. and A.M. prepared and revised the manuscript.

Additional information

Hierarchy of bond stiffnesses within icosahedral-based gold clusters protected by thiolates

Seiji Yamazoe[1,2], Shinjiro Takano[1], Wataru Kurashige[3], Toshihiko Yokoyama[4], Kiyofumi Nitta[5], Yuichi Negishi[3] & Tatsuya Tsukuda[1,2]

Unique thermal properties of metal clusters are believed to originate from the hierarchy of the bonding. However, an atomic-level understanding of how the bond stiffnesses are affected by the atomic packing of a metal cluster and the interfacial structure with the surrounding environment has not been attained to date. Here we elucidate the hierarchy in the bond stiffness in thiolate-protected, icosahedral-based gold clusters $Au_{25}(SC_2H_4Ph)_{18}$, $Au_{38}(SC_2H_4Ph)_{24}$ and $Au_{144}(SC_2H_4Ph)_{60}$ by analysing Au L_3-edge extended X-ray absorption fine structure data. The Au–Au bonds have different stiffnesses depending on their lengths. The long Au–Au bonds, which are more flexible than those in the bulk metal, are located at the icosahedral-based gold core surface. The short Au–Au bonds, which are stiffer than those in the bulk metal, are mainly distributed along the radial direction and form a cyclic structural backbone with the rigid Au–SR oligomers.

[1] Department of Chemistry, School of Science, The University of Tokyo, 7-3-1 Hongo, Bunkyo-ku, Tokyo 113-0033, Japan. [2] Elements Strategy Initiative for Catalysts and Batteries (ESICB), Kyoto University, Katsura, Kyoto 615-8520, Japan. [3] Department of Applied Chemistry, Faculty of Science, Tokyo University of Science, 1-3 Kagurazaka, Shinjuku-ku, Tokyo 162-8601, Japan. [4] Department of Materials Molecular Science, Institute for Molecular Science, Myodaiji, Okazaki, Aichi 444-8585, Japan. [5] Japan Synchrotron Radiation Research Institute, SPring-8, 1-1-1 Koto, Sayo, Hyogo 679-5198, Japan. Correspondence and requests for materials should be addressed to T.T. (email: tsukuda@chem.s.u-tokyo.ac.jp).

etal nanoparticles (NPs) exhibit specific thermal properties and phase transition behaviour that are quite different from the corresponding bulk metal[1-4]. For example, the melting point of a metal NP is significantly depressed with respect to that of the bulk metal[5]. The melting of a metal NP starts with premelting of the surface layer, which then expands towards the inner core and leads to a complete transition to the liquid[6]. These size-specific thermal behaviours are explained in terms of the large surface energy of a metal NP. In other words, a crucial factor that governs thermal behaviours is the hierarchy of the bonding within NPs; metal–metal bonds on the surface are generally softer than those within the core[3,7,8]. The thermal behaviours of metal NPs are not only affected by their size, but also by the atomic packing of the core[3,7-16] and the interaction with the environment, such as an organic ligand[17] or solid support[3,18,19]. For example, the vibrational spectrum of a metal NP with an icosahedral (Ih) structure has a component with a vibrational frequency that is higher than those of other structures such as cuboctahedra and decahedra, which suggests the formation of stiff bonds within the NPs[7]. The influence of surface adsorbates on the thermal properties of metal NPs has been ascribed to a change in the bond stiffness; Pt–Pt bonds of small Pt NPs are softened by hydrogen adsorption, but stiffened by oxidation[19] or capping with a polymer[17]. However, an atomic-level understanding of how the bond stiffnesses are affected by a variety of structural parameters has not been attained to date because of the experimental difficulties in defining the atomic packing of a metal NP and the interfacial structure with the surrounding environment.

Recently, a series of thiolate (RS)-protected gold clusters $Au_n(SR)_m$ with well-defined compositions have gained attention as ideal platforms to study the structure–property correlation and the size-dependent evolution of properties[20]. Among others, $Au_n(SR)_m$ with $(n, m) = (25, 18), (38, 24)$ and $(144, 60)$ have been studied most extensively as prototypical systems. Single-crystal X-ray diffraction (XRD) analysis revealed that $Au_{25}(SR)_{18}$ and $Au_{38}(SR)_{24}$ have icosahedral Au_{13} and bi-icosahedral Au_{23} cores, respectively, which are protected by combinations of $-SR-(Au-SR)_2-$ and $-SR-Au-SR-$ oligomers (Fig. 1)[21-23]. The atomic structure of $Au_{144}(SR)_{60}$ has not yet been determined by X-ray crystallography, but has been theoretically predicted as composed of a hollow icosahedral Au_{114} core protected by 30 $-SR-Au-SR-$ oligomers (Fig. 1)[24,25]. In addition, it has been identified that there are distinct Au–Au bonds with different lengths in the range of 2.7–3.3 Å in these clusters (Supplementary Fig. 1). The $Au_n(SR)_m$ clusters ($n = 25, 38, 144$) provide an ideal opportunity to study the hierarchy of the bond stiffness within Au clusters with well-defined atomic structures and surface modification.

In this study, the stiffnesses of the Au–Au and Au–S bonds in $Au_n(PET)_m$ (phenylethanethiolate (PET) = PhC_2H_4S) are examined using Au L_3-edge extended X-ray absorption fine structure (EXAFS) analysis, which is a powerful tool to study the local structure of the ligand-protected gold clusters[26,27]. The temperature dependence of the Debye–Waller (DW) factors of individual bonds are analysed in the framework of the Einstein model, in which a metal cluster is treated as an ensemble of quantum harmonic oscillators with individual Einstein temperatures that depend on the bonding. The hierarchy in the bond stiffness is elucidated. The Au–S bonds are much stiffer than Au–Au bonds in all the clusters and there are two types of Au–Au bonds; one is stiffer and the other is softer than those in the bulk Au. A major portion of the shorter and stiffer Au–Au bonds is distributed within the Ih-based Au core as radial bonds. The stiff Au–Au bonds distributed on the surface of the core are connected by the rigid Au–S oligomers to form a ring structure in

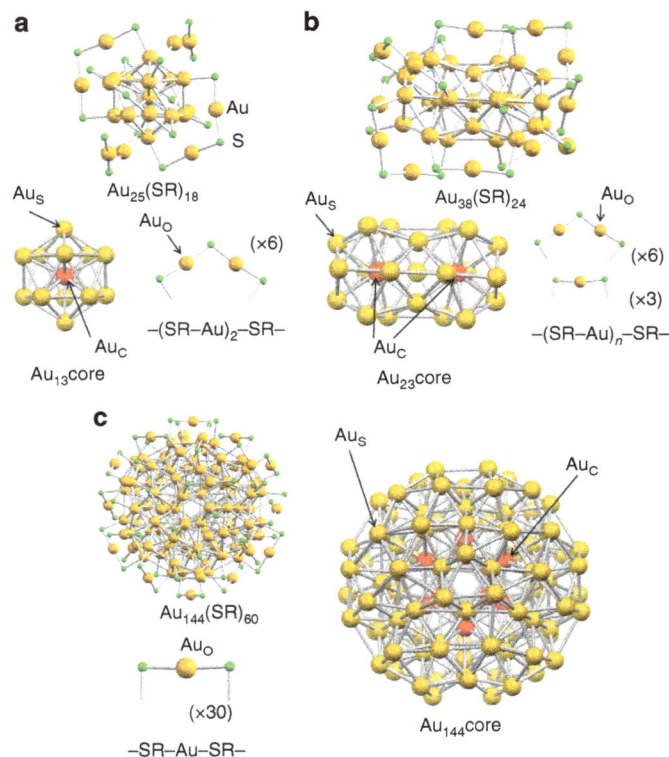

Figure 1 | Geometric structures of $Au_n(SR)_m$ samples. (a) Au–S framework of $Au_{25}(SR)_{18}$ determined by single crystallography[22], **(b)** Au–S framework of $Au_{38}(SR)_{24}$ determined by single crystallography[23] and **(c)** Au–S framework of $Au_{144}(SR)_{60}$ predicted by density functional theory calculation (ref. 24). Au_C, Au_S and Au_O represent Au atoms bonded only to Au atoms, those bonded both to Au and S atoms, and those bridged by two S atoms, respectively for $Au_{25}(SR)_{18}$ and $Au_{38}(SR)_{24}$. The definition of AuC, Au_S, and Au_O for $Au_{144}(SR)_{60}$ is given in the text. R is omitted for simplicity.

all the clusters. These ring structures may act as a rigid framework to enhance the thermal stability of the thiolate-protected Au clusters.

Results

Structural analysis. Figure 2a shows EXAFS oscillations of $Au_{25}(PET)_{18}$ measured at 300 and 8 K, respectively. EXAFS oscillation is clearly observed up to the k range of 21 Å^{-1} at 8 K, whereas it is damped in the k range of > 14 Å^{-1} at 300 K. Given that the EXAFS oscillation in the high k region mainly originates from heavy atoms, the large amplitude in Fig. 2a suggests that thermal fluctuation of the Au–Au bonds in $Au_{25}(PET)_{18}$ is significantly suppressed at 8 K. Figure 2b represents Fourier transformed (FT)-EXAFS spectra of $Au_{25}(PET)_{18}$ at 300 and 8 K, respectively. Most notably, the peak due to Au–Au bonds is clearly observed in the bond length (r) range of 2.1–3.0 Å at 8 K, in addition to that for Au–S bonds in the r range of 1.5–2.0 Å, whereas the Au–Au peak is hardly discernible at 300 K (Fig. 2b). Similar temperature dependence has been reported in the literature[26] and is attributed to suppression of the thermal fluctuation of Au–Au bonds at low temperature. Previous single crystallographic study revealed that the Au–Au bonds of $Au_{25}(PET)_{18}$ are classified into three groups according to their lengths: the Au–Au(x) bonds ($x = 1-3$) assigned mainly to the Au_C–Au_S, Au_S–Au_S and Au_S–Au_O bonds, respectively (see Fig. 1 and Supplementary Note 1; Supplementary Fig. 1a). The average coordination number (CN) and r values calculated for the

Au–Au(x) bonds (x = 1–3) are summarized in Table 1. Based on this information, curve-fitting analysis of the FT-EXAFS data for $Au_{25}(PET)_{18}$ at 8 K was conducted by assuming a single Au–S and three types of Au–Au bonds with different r values. However, we found that contribution from the longest Au–Au bonds is significantly smaller than the other two Au bonds due to larger DW factor (see Supplementary Discussion, Supplementary Fig. 2, and Supplementary Table 1). Thus, the FT-EXAFS data was analysed by assuming a single Au–S and two types of Au–Au bonds, Au–Au(S) and Au–Au(L). Their CN and r values thus obtained are summarized in Table 1. The CN and r values for the Au–S, Au–Au(S) and Au–Au(L) bonds obtained

by EXAFS are in good agreement with those of the Au–S, Au–Au(1) and Au–Au(2) bonds determined by single crystal XRD data, respectively. This agreement indicates that low-temperature EXAFS measurements allow quantitative probing of Au–Au bonds within the Au_{13} core: Au_C–Au_S and Au_S–Au_S bonds. The FT-EXAFS spectrum simulated assuming DW factors of 0.0027, 0.0040 and 0.0037 $Å^2$ (Supplementary Fig. 3) for Au_C, Au_S and Au_O, respectively, corresponded well with the experimental spectrum (Supplementary Fig. 4).

Similar temperature dependence was observed in the Au L_3-edge EXAFS spectrum for $Au_{38}(PET)_{24}$ measured at 8 K (Fig. 2a), where a clear oscillation was observed in the k range of 3.0–21.0 $Å^{-1}$. The FT-EXAFS spectrum (Fig. 2b) at 8 K has peaks for the Au–S (1.5–2.0 Å) and Au–Au (2.1–3.0 Å) bonds. Single-crystal XRD measurements of $Au_{38}(PET)_{24}$ show that 38 Au atoms are categorized into three different sites, Au_C, Au_S and Au_O, as in the case of $Au_{25}(PET)_{18}$ (Fig. 1b). The length distribution of Au–Au bonds obtained from the single crystal structure[23] is shown in Supplementary Fig. 1b. The Au–Au bonds were classified into three groups with border distances of 2.90 and 3.05 Å: Au–Au(x) bonds (x = 1–3) although there is larger ambiguity in the determination of the border distances than in the case of $Au_{25}(PET)_{18}$ (Supplementary Fig. 1a). The Au–Au(x) bonds (x = 1–3) are composed of mainly Au_C–Au_S, Au_S–Au_S and Au_S–Au_O bonds, respectively. The average CN and r values calculated for the Au–Au(x) bonds (x = 1–3) are summarized in Table 1. The FT-EXAFS spectrum was analysed by assuming a single Au–S and two types of Au–Au bonds with different lengths, Au–Au(S) and Au–Au(L), as in the case of $Au_{25}(PET)_{18}$. The CN and r values for the Au–S, Au–Au(S) and Au–Au(L) bonds obtained by curve-fitting analysis (Table 1) quantitatively agree with those for the Au–S, Au–Au(1) and

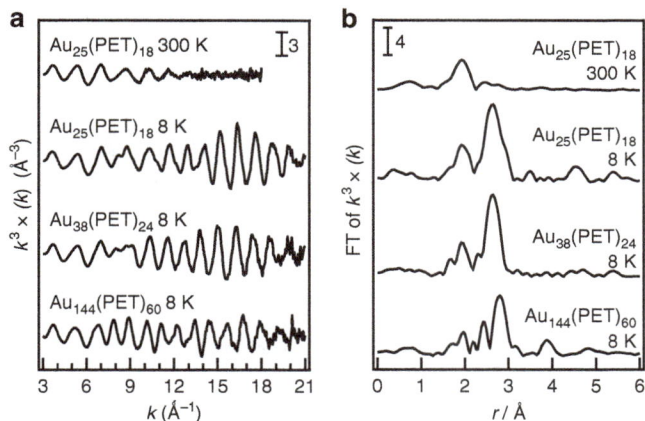

Figure 2 | EXAFS oscillations and FT-EXAFS spectra. (a) EXAFS oscillations for $Au_{25}(PET)_{18}$ measured at 300 and 8 K, $Au_{38}(PET)_{24}$ at 8 K, and $Au_{144}(PET)_{60}$ at 8 K and (b) the corresponding FT-EXAFS spectra.

Table 1 | Structural parameters of $Au_n(PET)_m$ obtained by curve-fitting analysis of Au L_3-edge FT-EXAFS and single-crystal XRD data.

n, m	Method	Bonds	CN*	r (Å)†	σ² (Å²)‡	R (%)§	σ²ₛ (Å²)‖	θ_E (K)¶
25, 18	EXAFS at 8 K	Au–S	1.6 (2)	2.319 (4)	0.0037 (18)	10.6	0.0017 (3)	429 (38)
		Au–Au (S)	1.5 (4)	2.770 (3)	0.0027 (11)		0.0008 (6)	137 (10)
		Au–Au (L)	1.5 (6)	2.936 (6)	0.0040 (24)		0.0020 (6)	101 (4)
	Single crystal XRD#	Au–S	1.4	2.33				
		Au–Au (1)	1.4	2.78				
		Au–Au (2)	1.9	2.95				
		Au–Au (3)	2.9	3.16				
38, 24	EXAFS at 8 K	Au–S	1.2 (2)	2.315 (4)	0.0030 (10)	10.7	0.0012 (5)	416 (57)
		Au–Au (S)	2.8 (4)	2.788 (1)	0.0041 (4)		0.0028 (12)	153 (11)
		Au–Au (L)	1.6 (9)	2.982 (15)	0.0090 (49)		0.0066 (5)	106 (15)
	Single crystal XRD**	Au–S	1.3	2.33				
		Au–Au (1)	2.7	2.82				
		Au–Au (2)	1.8	2.98				
		Au–Au (3)	2.2	3.21				
144, 60	EXAFS at 8 K	Au–S	0.9 (2)	2.326 (8)	0.0048 (29)	12.7	0.0024 (9)	381 (45)
		Au–Au (S)	1.2 (5)	2.733 (9)	0.0035 (20)		0.0021 (5)	148 (11)
		Au–Au (L)	6.0 (7)	2.870 (4)	0.0059 (13)		0.0041 (6)	128 (9)
	DFT††	Au–S	0.8	2.34				
		Au–Au (1)	1.5	2.77				
		Au–Au (2)	5.9	2.92				
		Au–Au (3)	1.5	3.24				

*Coordination number.
†Bond length.
‡Debye–Waller factor.
§$R = (Σ(k^3χ^{data}(k)−k^3χ^{fit}(k))^2)^{1/2}/(Σ(k^3χ^{data}(k))^2)^{1/2}$.
‖Static component of Debye–Waller factor at 8 K.
¶Einstein temperature.
#ref. 22.
**ref. 23.
††ref. 24.

Au–Au(2) bonds, respectively. The FT-EXAFS spectrum simulated using DW factors for each of the Au sites shown in Supplementary Fig. 3 reproduced the experimental spectrum (Supplementary Fig. 5).

These analyses demonstrate that low-temperature EXAFS data provides quantitative structural information on the Au bonds within the $Au_n(SR)_m$. Following this successful characterization, we next conducted Au L_3-edge EXAFS measurement of $Au_{144}(PET)_{60}$ at 8 K, of which the structure has not yet been determined by single-crystal XRD. Au L_3-edge EXAFS oscillation was clearly observed in the k range as large as 21.0 Å^{-1}, as shown in Fig. 2a. The FT-EXAFS spectrum of $Au_{144}(PET)_{60}$ (Fig. 2b) shows a small peak for the Au–S bond in the range of 1.5–2.0 Å and a large peak for the Au–Au bond in the range of 2.1–3.0 Å. The curve-fitting analysis indicates that the data can be reproduced by assuming Au–S (CN = 0.9, r = 2.326 Å), Au–Au(S) (CN = 1.2, r = 2.733 Å) and Au–Au(L) (CN = 6.0, r = 2.870 Å) bonds, as shown in Table 1. The Au–Au bonds are slightly shorter than those of $Au_{25}(PET)_{18}$ and $Au_{38}(PET)_{24}$, whereas the Au–S bond length is comparable to those of $Au_{25}(PET)_{18}$ and $Au_{38}(PET)_{24}$. According to density functional theory (DFT) calculations[24], $Au_{144}(SCH_3)_{60}$ has a hollow icosahedral Au_{12} core. The absence of the central Au atom may be a cause of the reduction in the Au–Au bond lengths compared with those of $Au_{25}(PET)_{18}$ and $Au_{38}(PET)_{24}$. Supplementary Figure 1c plots the bond lengths between the nearest neighbour Au atoms in $Au_{144}(SCH_3)_{60}$ obtained by DFT calculations[24]. When Au–Au bonds are divided into three groups Au–Au(x) (x = 1–3) with border distances of 2.83 and 3.10 Å, the experimentally obtained CN and r values of the Au–S, Au–Au(S) and Au–Au(L) bonds are in quantitative agreement with those of the Au–S, Au–Au(1) and Au–Au(2) bonds, respectively, as shown in Table 1. The EXAFS analysis strongly supports the $Au_{144}(SCH_3)_{60}$ model structure[24]. We then examined which Au–Au bonds contribute mainly to the Au–Au(1) and Au–Au(2) bonds. When the Au atoms of $Au_{144}(SCH_3)_{60}$ are divided into three groups: 12 Au atoms at the hollow icosahedral core (Au_C), 102 Au atoms located at the middle and the outermost layers (Au_S) and 30 Au atoms at the oligomer (Au_O) (Fig. 1), the Au–Au(1) and Au–Au(2) bonds are assigned mainly to the Au_C–Au_C and Au_S–Au_S bonds, respectively. The EXAFS oscillation is again dominated by the Au–Au bonds in the core and the shell, whereas the contribution from the bonds with Au_O is negligibly small. The FT-EXAFS spectrum simulated using DW factors for each of the Au sites shown in Supplementary Fig. 3

corresponded well with the experimental spectrum (Supplementary Fig. 6).

Einstein temperatures for Au–S and Au–Au bonds. The DW factors for the Au–S, Au–Au(L) and Au–Au(S) bonds of $Au_n(PET)_m$ (n = 25, 38, 144) were determined by least-squares fit analysis while keeping the CN values the same as those in Table 1, and the results are plotted as a function of temperature in Fig. 3. Details of the analysis are given in Methods section. Figure 3 shows that the DW factors for both the Au–Au(S) and Au–Au(L) bonds increase monotonically with the temperature, whereas those of Au–S bonds remain almost constant in the temperature range of 8–300 K in all clusters. The DW factors for the Au–Au(S) bonds are smaller than those for the Au–Au(L) bonds and are comparable to that for the Au–Au bond in the bulk Au (Supplementary Fig. 7).

The stiffnesses of the Au–S bonds and two types of Au–Au bonds of $Au_n(PET)_m$ (n = 25, 38, 144) were evaluated within the framework of the Einstein model. In the framework of the Einstein model, we assume three independent quantum oscillators with different Einstein frequencies (ω_E) for the Au–S, Au–Au(L) and Au–Au(S) bonds. The Einstein temperatures (θ_E) were determined by fitting the temperature dependence of the DW factors for the Au–S, Au–Au(L) and Au–Au(S) bonds. Details of the fitting procedure are given in Methods section. The best-fit results are shown in Fig. 3, and the optimized θ_E and σ_S^2 values are listed in Table 1.

Discussion

In this study, the hierarchy of the bond stiffnesses within $Au_n(PET)_m$ (n = 25, 38, 144) was examined using Au L_3-edge EXAFS. EXAFS results provided quantitative structural information (r and CN values) of two types of Au–Au bonds (Au–Au(S) and Au–Au(L)) within the Ih-based Au cores in addition to the Au–S bonds. The temperature dependence of the DW factors of the individual bonds was analysed within the framework of the Einstein model. Einstein temperatures for the two types of Au–Au bonds were determined to be much lower than that of the Au–S bonds in all clusters, which suggests that thiolate-protected Au clusters can be viewed as soft Au cores capped by rigid Au–SR staples. This is in agreement with the results of molecular dynamic (MD) simulation on a model system $Au_{25}(SH)_{18}^-$ that the –SR–(Au–SR)$_2$– units are rigid and confine the elastic Au core internally[28].

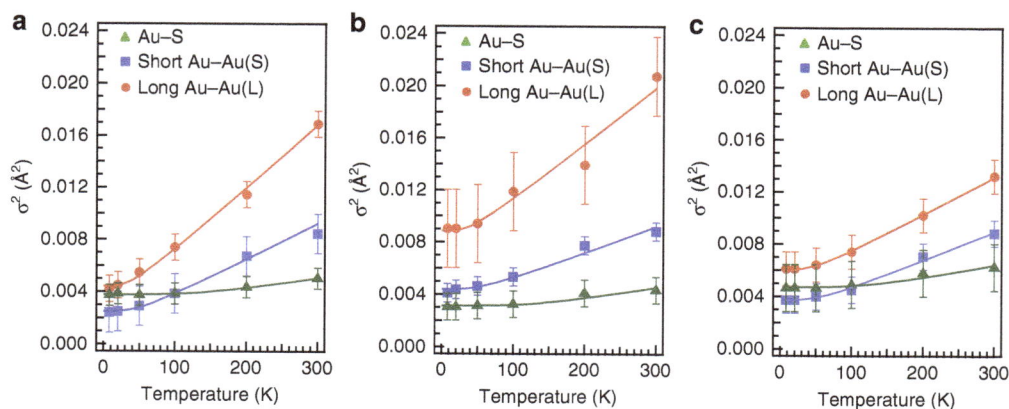

Figure 3 | Temperature dependence of the DW factors. Temperature dependence of the DW factors (σ^2) for Au–S, Au–Au(S) and Au–Au(L) bonds. (**a**) $Au_{25}(PET)_{18}$, (**b**) $Au_{38}(PET)_{24}$ and (**c**) $Au_{144}(PET)_{60}$. Solid lines represent the fitting results. Error bars represent the s.d. in the curve-fitting procedure of FT-EXAFS.

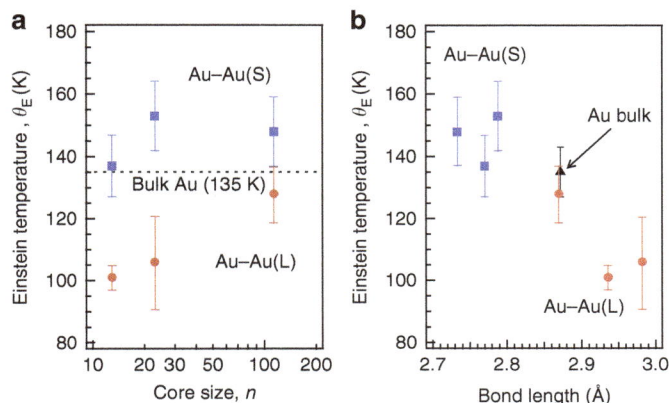

Figure 4 | Einstein temperature. (a) Einstein temperature (θ_E) as a function of the core size and **(b)** bond length for Au-Au(S) (square) and Au-Au(L) (circle) bonds for core size of $n = 13$, 23 and 114. θ_E and Au-Au bond distance for bulk Au are 135 K and 2.872 Å, respectively. Error bars represent the s.d. in the fitting procedure of θ_E.

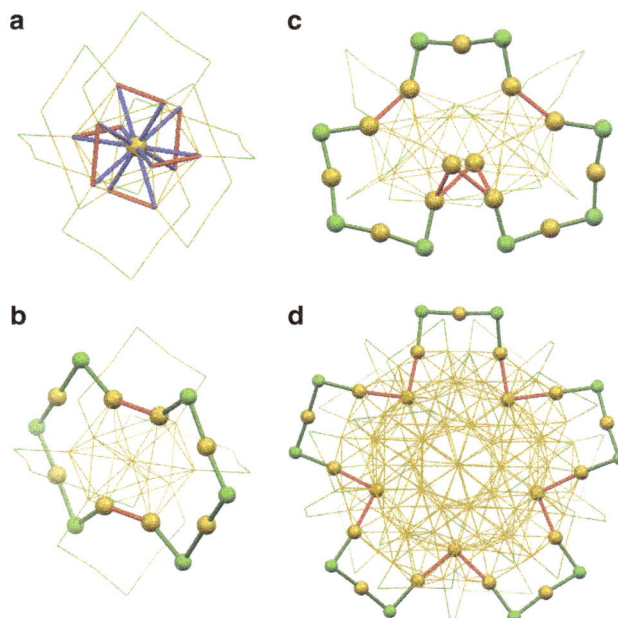

Figure 5 | Network structures of stiff bonds. (a) Distribution of stiff Au-Au bonds in $Au_{25}(PET)_{18}$. Red and blue bonds represent Au_S-Au_S and Au_C-Au_S bonds, respectively. Ring structures in **(b)** $Au_{25}(PET)_{18}$, **(c)** $Au_{38}(PET)_{24}$ and **(d)** $Au_{144}(PET)_{60}$.

More interestingly, two classes of Au–Au bonds with different stiffnesses were identified. Figure 4a shows the Einstein temperatures of the Au–Au(S) and Au–Au(L) bonds as a function of the Au core sizes (13, 23, 114) of $Au_n(PET)_m$ and that of Au–Au bonds in the bulk Au (135 ± 8 K)[29]. This plot indicates that the Au–Au(S) and Au–Au(L) bonds are respectively stiffer and more flexible than the Au–Au bonds in the face-centred cubic bulk Au, regardless of the cluster size. To the best of our knowledge, this result is the first experimental verification of hierarchy of Au–Au bond stiffnesses in thiolate-protected Au clusters. The correlation suggests that the bond stiffness is related to the bond length. To confirm this hypothesis, Einstein temperatures of the Au–Au bonds were plotted as a function of their lengths in Fig. 4b. The Au–Au bond becomes stiffer with a reduction of the bond length. The Au–Au(S) bonds that are shorter and stiffer than Au–Au bonds in the bulk Au are specific to the Ih core because Au–Au(S) bonds are formed along the radial direction in the icosahedral Au core. The formation of Au–Au bonds in $Au_n(PET)_m$ that are stiffer than those in the bulk Au is consistent with the theoretical prediction that the vibrational density of states distribution of the Ih cluster has a tail towards a higher frequency than that of bulk Au[4,7,8]. The Au–Au(L) bonds between the surface Au atoms are more flexible on average than those in the bulk Au, even though they are bonded with the rigid Au–SR oligomers. The above consideration suggests that the Au core of the thiolate-protected Au clusters tends to deform more easily along the lateral direction than in the radial direction. This is consistent with theoretical prediction based on the MD simulation of $Au_{25}(SH)_{18}^-$ that Au core atoms prefer vibrating in the tangential directions as opposed to the radial direction[28].

Close inspection of the distribution of Au–Au(S) bonds reveals that the stiff Au–Au bonds are distributed not only in the centre of the core, but also on the surface of the core. For example, 6 Au_S–Au_S bonds and 12 Au_C–Au_S bonds in $Au_{25}(PET)_{18}$ are categorized as the Au–Au(S) bonds, as shown in Fig. 5a. By bridging these stiff Au–Au surface bonds with the –SR–(Au–SR)$_2$– oligomers, a ring structure is formed in $Au_{25}(PET)_{18}$ (Fig. 5b). Similar ring structures can be found in $Au_{38}(PET)_{24}$ and $Au_{144}(PET)_{60}$ (Fig. 5c,d). Tlahuice-Flores et al.[30] also suggested the formation of circular networks ($Au_{20}Cl_{10}$) with short Au–Au and Au–Cl bonds as subunits in $Au_{144}Cl_{60}^{2+}$, which is modelled by replacement of the SR ligands of $Au_{144}(SR)_{60}$ with Cl. These results suggest that thiolate-protected Au clusters contain rigid

ring subunits that consist of short Au–Au bonds at the surface and Au–SR staples. The formation of these rigid ring structures may act as frameworks to enhance the thermal stability of the thiolate-protected Au clusters.

Methods

Chemicals. Hydrogen tetrachloroaurate tetrahydrate ($HAuCl_4 \cdot 4H_2O$) was purchased from Tanaka Kikinzoku. phenylethanethiol, tetraoctylammonium bromide (TOABr), sodium tetrahydroborate ($NaBH_4$), methanol, acetone, dichloromethane and toluene were obtained from Wako Pure Chemical Industries. All chemicals were used without further purification. Deionized water with a resistivity above 18.2 MΩ cm was used.

Synthesis and characterization. $Au_{25}(PET)_{18}$ was synthesized by the methods similar to that in the literature[31]. First, $HAuCl_4 \cdot 4H_2O$ (0.75 mmol) was dissolved in 25 ml tetrahydrofuran solution containing TOABr (0.76 mmol) at room temperature. After stirring for 15 min, phenylethanethiol (4.7 mmol) was added to this solution and the solution was stirred for 15 min. A cold aqueous solution (5.8 ml) containing $NaBH_4$ (8.7 mmol) was then rapidly added to the solution and then the solution was stirred at room temperature. After 12 h, tetrahydrofuran solvent was evaporated and the remaining red brown powder was washed with methanol to remove excess thiol and other byproducts. The $Au_{25}(PET)_{18}$ cluster was extracted from the dried sample using acetonitrile.

$Au_{38}(PET)_{24}$ was synthesized by the methods similar to that in the literature[32]. First, $HAuCl_4 \cdot 4H_2O$ (0.50 mmol) and glutathione (0.25 mmol) were dissolved in acetone (20 ml). This solution was stirred for 15 min at room temperature and for another 15 min at 0 °C. A cold aqueous solution (6.0 ml) containing $NaBH_4$ (5.0 mmol) was then rapidly added to this solution and the solution was stirred in ice bath. After 15 min, the acetone was discarded and the residue was dissolved in water (6.0 ml). This solution was then added to the mixture of toluene (2.0 ml) and ethanol (0.30 ml) containing phenylethanethiol (15 mmol). The resulting solution was stirred at 80 °C. After 12 h, the organic solution was evaporated and the resulting black powder was washed by methanol to remove byproducts. The $Au_{38}(PET)_{24}$ thus obtained was further purified by the gel permeation chromatography.

$Au_{144}(PET)_{60}$ was synthesized by the methods similar to that in the literature[33]. First, $HAuCl_4 \cdot 4H_2O$ (0.60 mmol) and TOABr (0.69 mmol) were dissolved in methanol (30 ml) at room temperature. After stirring for 5 min, phenylethanethiol (3.7 mmol) was added to this solution and the solution was stirred for 15 min. A cold aqueous solution (10 ml) containing $NaBH_4$ (6.0 mmol) was then rapidly added to this solution and the solution was stirred at room temperature. After 4 h, the black precipitate was separated by centrifugation and washed repeatedly with methanol or acetonitrile to remove excess thiol, byproducts and the other-sized clusters. The $Au_{144}(PET)_{60}$ thus obtained was further purified by gel permeation chromatography.

Chemical compositions and purities of the $Au_n(PET)_m$ samples were confirmed by a high-resolution Fourier-transform mass spectrometer with an electrospray ionization source (Bruker, Solarix). A 1-mg ml^{-1} dispersion of $Au_n(SR)_m$ in toluene/acetonitrile (1:1, v-v) was electrosprayed at a flow rate of 800 μl per hour. Ultraviolet-visible-near infrared (UV-Vis-NIR) spectra of $Au_n(PET)_m$ in toluene were recorded with a spectrometer (Jasco, V-670).

EXAFS measurements. Au L$_3$-edge EXAFS measurements were conducted using the BL01B1 beamline at the SPring-8 facility of the Japan Synchrotron Radiation Research Institute. The incident X-ray beam was monochromatized by a Si(311) double crystal monochromator. A solid sample of each $Au_n(PET)_m$ cluster was diluted with boron nitride powder, pressed into a pellet, and then mounted on a copper holder attached to a cryostat. All EXAFS spectra were measured in the transmission mode using ionization chambers at 8-300 K. The X-ray energy was calibrated using Cu foil. The EXAFS analysis was conducted using REX2000 Ver. 2.5.9 program (Rigaku Co.) as follows. The χ spectra were extracted by subtracting the atomic absorption background using cubic spline interpolation and were normalized to the edge height. The k^3-weighted χ spectra within the k ranges of 3.0–21.0 Å^{-1} for structural analysis and of 3.0–16.0 Å^{-1} for analysis of temperature dependence of the DW factor were Fourier-transformed into r space. The curve-fitting analysis was performed for Au–S and Au–Au bonds over the r range of 1.5–3.1 Å. In the curve-fitting analysis, the phase shifts and back-scattering amplitude functions for Au–S and Au–Au were extracted from Au_2S (ICSD#78718) and Au metal (ICSD#44362), respectively, using the FEFF8 program[34] by setting σ^2 to 0.0036. This σ^2 value did not significantly affect the phase shift and back-scattering amplitude functions. EXAFS spectra for $Au_n(PET)_m$ were also simulated using the FEFF8 (ref. 34) programs. Reported crystal structures for $Au_{25}(PET)_{18}$ (ref. 22) and $Au_{38}(PET)_{24}$ (ref. 23) and the calculated structure for $Au_{144}(SCH_3)_{60}$ (ref. 24) were used for the EXAFS simulations. The theoretical bond lengths for the $Au_{144}(SCH_3)_{60}$ structure were scaled by 0.974 because the Perdew, Burke and Ernzerhof approximation used to the exchange-correlation functional overestimates the Au–Au bond lengths[24]. Different DW factors for Au atoms at different sites, as shown in Supplementary Fig. 3, were used for EXAFS simulations because the DW factors for Au atoms are dependent on the sites shown in Table 1.

Evaluation of DW factors. The EXAFS equation is generally expressed as:

$$\chi(k) = S_0^2 \sum \frac{CN}{kr^2} f(k;\pi) \exp\left(-2\sigma^2 k^2\right) \sin\left(2kr + \delta(k) - \frac{4}{3}C_3 k^3\right), \quad (1)$$

where S_0^2, CN, k, r, $f(k;\pi)$, σ^2, $\delta(k)$ and C_3 are passive electron reduction factor, coordination number, photoelectron energy, bond distance, back-scattering amplitude function, DW factor, phase shift, and the third cumulant, respectively[35]. Supplementary Figure 9 shows FT-EXAFS spectra for $Au_n(PET)_m$ at various temperatures obtained from the temperature-dependent EXAFS data ($3.0 \leq k \leq 16.0$ Å^{-1}) (Supplementary Fig. 8). It was confirmed by the analysis of bulk Au that the temperature dependence of C_3 must be taken into account in addition to the σ^2 and r values (Supplementary Fig. 7)[14] to reproduce the FT-EXAFS spectra. C_3, r and σ^2 values for individual Au–S and Au–Au bonds were determined by least-squares fit analysis while keeping the CN values the same as those in Table 1, and the results are plotted as a function of temperature in Supplementary Fig. 10. In all clusters, the DW factors for both the Au–Au(S) and Au–Au(L) bonds increase monotonically with the temperature, whereas those of Au–S bonds remain almost constant in the temperature range of 8–300 K (Supplementary Fig. 10). The DW factors for the Au–Au(S) bonds are smaller than those for the Au–Au(L) bonds and are comparable to that for the Au–Au bond in the bulk Au (Supplementary Figs 7 and 10).

Evaluation of Einstein temperatures. The DW factor σ^2 is composed of a dynamic component σ_D^2 and a static component σ_S^2, which arises from the thermal atomic oscillation and the temperature-independent structural disorder, respectively:

$$\sigma^2 = \sigma_S^2 + \sigma_D^2, \quad (2)$$

We assume within the framework of the Einstein model three independent quantum oscillators with different Einstein frequencies (ω_E) for the Au–S and Au–Au bonds. The Einstein temperature (θ_E) is defined as:

$$\theta_E = \frac{h\omega_E}{2\pi k_B}, \quad (3)$$

where h and k_B are the Planck and Boltzmann constants, respectively. According to the Einstein model, the dynamic DW factor σ_D^2 can be fitted using the equation:

$$\sigma_D^2 = \frac{h^2}{8\pi^2 \mu k_B \theta_E} \coth\frac{\theta_E}{2T}, \quad (4)$$

where μ and T represent the reduced mass of adjacent atoms and the temperature, respectively. The θ_E values were determined by fitting the temperature dependence

of the DW factors for the Au–S, Au–Au(L) and Au–Au(S) bonds using equations (2)–(4). The best-fit results are shown in Fig. 3, and the optimized θ_E and σ_S^2 values are listed in Table 1.

References

1. Bai, H. Y., Luo, J. L. & Jin, D. Particle size and interfacial effect on the specific heat of nanocrystalline Fe. *J. Appl. Phys.* **79**, 361–364 (1996).
2. Schmidt, M., Kusche, R., Issendorff, B. V. & Haberland, H. Irregular variations in the melting point of size-selected atomic clusters. *Nature* **393**, 238–240 (1998).
3. Narvaez, G. A., Kim, J. & Wilkins, W. Effects of morphology on phonons of nanoscopic silver grains. *Phys. Rev. B* **72**, 155411 (2005).
4. Araujo, L. L., Kluth, P., Azevedo, G. M. & Ridgway, M. C. Vibrational properties of Ge nanocrystals determined by EXAFS. *Phys. Rev. B* **74**, 184102 (2006).
5. Buffat, P. & Borel, J.-P. Size effect on the melting temperature of gold particles. *Phys. Rev. A* **13**, 2287–2298 (1976).
6. Cleveland, C. L., Luedtke, W. D. & Landman, U. Melting of gold clusters: icosahedral precursors. *Phys. Rev. Lett.* **81**, 2036–2039 (1998).
7. Sauceda, H. E., Salazar, F., Pérez, L. A. & Garzón, I. L. Size and shape dependence of the vibrational spectrum and low-temperature specific heat of Au nanoparticles. *J. Phys. Chem. C* **117**, 25160–25168 (2013).
8. Ortigoza, M. A., Heid, R., Bohnen, K.-P. & Rahman, T. S. Anomalously soft and stiff modes of transition-metal nanoparticles. *J. Phys. Chem. C* **118**, 10335–10347 (2014).
9. Yang, C. C., Xiao, M. X., Li, W. & Jiang, Q. Size effects on Debye temperature, Einstein temperature, and volume thermal expansion coefficient of nanocrystals. *Solid State Commun.* **139**, 148–152 (2006).
10. Cuenya, B. R. *et al.* Size-dependent evolution of the atomic vibrational density of states and thermodynamic properties of isolated Fe nanoparticles. *Phys. Rev. B* **86**, 165406 (2012).
11. Yokoyama, T., Kimoto, S. & Ohta, T. Temperature-dependent EXAFS study on supported silver and palladium clusters. *Jpn J. Appl. Phys.* **28**, L851–L853 (1989).
12. Cuenya, B. R. *et al.* Thermodynamic properties of Pt nanoparticles: size, shape, support, and adsorbate effects. *Phys. Rev. B* **84**, 245438 (2011).
13. Cuenya, B. R. *et al.* Anomalous lattice dynamics and thermal properties of supported size- and shape-selected Pt nanoparticles. *Phys. Rev. B* **82**, 155450 (2010).
14. Comaschi, T., Balerna, A. & Mobilio, S. Temperature dependence of the structural parameters of gold nanoparticles investigated with EXAFS. *Phys. Rev. B* **77**, 075432 (2008).
15. Dubiel, M. *et al.* Temperature dependence of EXAFS cumulants of Ag nanoparticles in glass. *J. Phys. Conf. Ser.* **190**, 012123 (2009).
16. Sauceda, H. E., Pelayo, J. J., Salazar, F., Pérez, L. A. & Garzón, I. L. Vibrational spectrum, caloric curve, low-temperature het capacity, and Debye temperature of sodium clusters: the Na_{139}^+ case. *J. Phys. Chem. C* **117**, 11393–11398 (2013).
17. Giovanetti, L. J. *et al.* Anomalous vibrational properties induced by surface effects in capped Pt nanoparticles. *J. Phys. Chem. C* **111**, 7599–7604 (2009).
18. Kang, J. H., Menard, L. D., Nuzzo, R. G. & Frenkel, A. I. Unusual non-bulk properties in nanoscale materials: thermal metal-metal bond contraction of γ-alumina-supported Pt catalysts. *J. Am. Chem. Soc.* **128**, 12068–12069 (2006).
19. Sanchez, S. I. *et al.* The emergence of nonbulk properties in supported metal clusters: negative thermal expansion and atomic disorder in Pt nanoclusters supported on γ-Al$_2$O$_3$. *J. Am. Chem. Soc.* **131**, 7040–7054 (2009).
20. Tsukuda, T. & Häkkinen, H. *Protected Metal Clusters: from Fundamentals to Applications* (Elsevier, 2015).
21. Heaven, M. W., Dass, A., White, P. S., Holt, K. M. & Murray, R. W. Crystal structure of the gold nanoparticle [N(C$_8$H$_{17}$)$_4$][Au$_{25}$(SCH$_2$CH$_2$Ph)$_{18}$]. *J. Am. Chem. Soc.* **130**, 3754–3755 (2008).
22. Zhu, M., Aikens, C. M., Hollander, F. J., Schatz, G. C. & Jin, R. Correlating the crystal structure of a thiol-protected Au$_{25}$ cluster and optical properties. *J. Am. Chem. Soc.* **130**, 5883–5885 (2008).
23. Qian, H., Echenhoff, W. T., Zhu, Y., Pintauer, T. & Jin, R. Total structure determination of thiolate-protected Au$_{38}$ nanoparticles. *J. Am. Chem. Soc.* **132**, 8280–8281 (2010).
24. Lopez-Acevedo, O., Akola, J., Whetten, R. L., Grönbeck, H. & Häkkinen, H. Structure and bonding in the ubiquitous icosahedral metallic gold cluster Au$_{144}$(SR)$_{60}$. *J. Phys. Chem. C* **113**, 5035–5038 (2009).
25. Weissker, H.-Ch. *et al.* Information on quantum states pervades the visible spectrum of the ubiquitous Au$_{144}$(SR)$_{60}$ gold nanocluster. *Nat. Commun.* **5**, 3785 (2014).
26. Zhang, P. X-ray spectroscopy of gold-thiolate nanoclusters. *J. Phys. Chem. C* **118**, 25291–25299 (2014).
27. Chevrier, D. M., Yang, R., Chatt, A. & Zhang, P. Bonding properties of thiolate-protected gold nanoclusters and structural analogs from X-ray absorption spectroscopy. *Nanotechnol. Rev.* **4**, 193–206 (2015).

28. Mäkinen, V. & Häkkinen, H. Density functional theory molecular dynamics study of the $Au_{25}(SH)^{-}_{18}$ cluster. *Eur. Phys. J. D* **66**, 310 (2011).
29. Kluth, P. *et al.* Vibrational properties of Au and Cu nanocrystals formed by ion implantation. *AIP Conf. Proc.* **882**, 731–733 (2007).
30. Tlahuice-Flores, A., Black, D. M., Bach, S. B. H., Yose-Yacamán, M. & Whetten, R. L. Structure & bonding of the gold-subhalide cluster I-$Au_{144}Cl^{[z]}_{60}$. *Phys. Chem. Chem. Phys.* **15**, 19191–19195 (2013).
31. Dharmaratne, A. C., Krick, T. & Dass, A. Nanocluster size evolution studied by mass spectrometry in room temperature $Au_{25}(SR)_{18}$ synthesis. *J. Am. Chem. Soc.* **131**, 13604–13605 (2009).
32. Qian, H., Zhu, Y. & Jin, R. Size-focusing synthesis, optical and electrochemical properties of monodisperse $Au_{38}(SC_2H_4Ph)_{24}$ nanoclusters. *ACS Nano* **3**, 3795–3803 (2009).
33. Qian, H. & Jin, R. Ambient synthesis of $Au_{144}(SR)_{60}$ nanoclusters in methanol. *Chem. Mater.* **23**, 2209–2217 (2011).
34. Ankudinov, A. L., Ravel, B., Rehr, J. J. & Conradson, S. D. Real-space multiple-scattering calculation and interpretation of X-ray-absorption near-edge structure. *Phys. Rev. B* **58**, 7565–7576 (1998).
35. Tranquada, J. M. & Ingalls, R. Extended X-ray-absorption fine-structure study of anharmonicity in CuBr. *Phys. Rev. B* **28**, 3520–3528 (1983).

Acknowledgements

The EXAFS measurements were conducted under an approval of JASRI (Proposal Nos 2015A1590, 2014B1430, 2013B1659, 2013B1421, 2012B1421 and 2012B1986). We thank Professor Hannu Häkkinen (University of Jyväskylä) for providing atomic coordinates of $Au_{144}(SCH_3)_{60}$ obtained by DFT calculations. This research was financially supported by the Funding Program for Next Generation World-leading Researchers (NEXT Program, GR-003), 'Elements Strategy Initiative to Form Core Research Center', a Grant-in-Aid for Scientific Research (No. 26248003) from the MEXT (Ministry of Education, Culture, Sports, Science and Technology) and a Grant for Basic Science Research Projects from The Sumitomo Foundation.

Author contributions

S.Y. and T.T. designed this study. Y.N. and W.K. synthesized and characterized the gold cluster samples. K.N. developed a sample holder for low-temperature XAFS measurements. S.Y. and S.T. conducted the XAFS measurements. S.Y. analysed the experimental and theoretical spectra with the help of T.Y. All authors took part in the writing of this manuscript.

Additional information

19

Silicon oxycarbide glass-graphene composite paper electrode for long-cycle lithium-ion batteries

Lamuel David[1], Romil Bhandavat[1], Uriel Barrera[1] & Gurpreet Singh[1]

Silicon and graphene are promising anode materials for lithium-ion batteries because of their high theoretical capacity; however, low volumetric energy density, poor efficiency and instability in high loading electrodes limit their practical application. Here we report a large area (approximately 15 cm × 2.5 cm) self-standing anode material consisting of molecular precursor-derived silicon oxycarbide glass particles embedded in a chemically-modified reduced graphene oxide matrix. The porous reduced graphene oxide matrix serves as an effective electron conductor and current collector with a stable mechanical structure, and the amorphous silicon oxycarbide particles cycle lithium-ions with high Coulombic efficiency. The paper electrode (mass loading of $2\,mg\,cm^{-2}$) delivers a charge capacity of $\sim 588\,mAh\,g^{-1}_{electrode}$ ($\sim 393\,mAh\,cm^{-3}_{electrode}$) at 1,020th cycle and shows no evidence of mechanical failure. Elimination of inactive ingredients such as metal current collector and polymeric binder reduces the total electrode weight and may provide the means to produce efficient lightweight batteries.

[1] Mechanical and Nuclear Engineering Department, Kansas State University, 3002 Rathbone Hall, Kansas, Manhattan, Kansas 66506, USA. Correspondence and requests for materials should be addressed to G.S. (email: gurpreet@ksu.edu).

Concentrated efforts are currently employed to discover a practical replacement for traditional Li-ion battery electrodes that is, graphite anode and LiCoO$_2$ cathode with materials that continuously deliver high power and energy densities at high cycling efficiencies without damage[1-5]. Alloying reaction electrodes such as silicon that can deliver as much as 5–10 times higher discharge capacity than traditional graphite, are at the forefront of this research. High capacity electrodes, however, are prone to enormous volume changes ($\sim$300%) that generally lead to structural collapse and capacity fading during successive lithiation/delithiation[6-12]. Recent work has shown that decreasing particle size or electrode nanostructuring allows the electrode to withstand high volumetric strains associated with repeated Li alloying and de-alloying. Pomegranate-inspired carbon-coated Si nanoparticles, yoke shell-structured SiC nanocomposites and Si/C core/shell composites (prepared at low mass loading) have proven to survive several hundred cycles without damage[9-13]. Yet, electrode nanostructuring has lead to new fundamental challenges such as low volumetric capacity (low tap density), increased electrical resistance between the nanoparticles, increased manufacturing costs and lower Coulombic efficiency due to side reactions with the electrolyte. These challenges have not been fully addressed. What's more, a particle-based electrode's long-term cyclability hinges on the inter-particle electrical connection and particle adhesion to the metallic substrate, which decreases rapidly with increasing charge/discharge cycles, particularly for thick high-loading electrodes[9].

In this context, the graphene-based multicomponent composite anodes are an attractive alternative to traditional (binder and carbon-black) designs, chiefly because of graphene's superior electronic conductivity, mechanical strength and ability to be interfaced with Li active redox components, such as particles of Si, Ge, and transition metals sulfides/oxides resulting in electrodes that are intrinsically conducting and promote faster ion diffusion[14-38]. Additional advantages include weight savings of up to 10% of the total battery weight[7], if the electrode is prepared in the freestanding form, improved corrosion resistance (elimination of metal foil), and enhanced flexibility, particularly for bendable, implantable, and roll-up electronics.

In spite of these advantages, graphene-paper electrodes do not offer an absolute solution because of the following associative disadvantages: (a) potential limiting of overall battery capacity due to insufficient active mass (thickness generally limited to submicrometers), (b) expensive techniques required for synthesis of Li-redox components and (c) more important, paper anodes generally show very high first cycle loss (50–60%), low cycling efficiency (95–98%) and poor capacity retention at high current densities (damage at high C-rates)[23-39], making graphene-paper electrodes somewhat impractical for use in an Li-ion battery full-cell. Here again, very few studies have been performed to investigate the mechanical and fracture properties of composite paper-based electrodes.

Continued search for better anodes has brought attention to unique, rarely studied molecular precursor-derived Si-based glass-ceramics (such as silicon oxycarbide or SiOC and silicon carbonitride or SiCN) materials[40-50]. SiOC is a high-temperature glass-ceramic with an open polymer-like network structure consisting of two interpenetrating amorphous phases of SiOC (Si bonded to O and C) and disordered carbon[42]. Its low weight density ($\sim$2.1 g cm^{-3}) and open structure enables high charge and discharge rates with a gravimetric capacity more than twice that of commercial graphite electrode. More important, major portion of the electrochemical capacity in SiOC is due to reversible Li-adsorption in the disordered carbon phase and not the conventional alloying reaction with Si, ensuing relatively

lower volumetric changes[43,44]. Regrettably, the glass-ceramics that show high lithiation capacity are poor conductors of electronic/ionic current and consequently the electrode preparation involves incorporation of conducting agents and binders in order to hold the particles on a metal current collector, a method known as screen printing[45-47]. Such foil-based electrodes carry the dead weight of conducting agents, polymeric binders, and the metal foil that do not contribute towards the battery capacity.

As an attractive solution to screen printed electrodes, we present our results related to fabrication of a well-organized, interleaved, freestanding, large-area composite anode consisting of SiOC particles supported by crumpled reduced graphene oxide matrix. The electrode delivers higher volumetric capacity than the recently reported pomegranate Si/carbon nanotube (310 mAh cm^{-3}) paper-electrode[9]. Large micrometer size reduced graphene oxide (rGO) sheets serve as host material to SiOC particles, providing the necessary electronic path and consistent cycling performance at high current densities along with high structural stability. Because of their unique nanodomain amorphous structure, SiOC particles offer required chemical and thermodynamic stability and high Li intercalation capacity for the electrode. As a result the electrode (at least 2 mg cm^{-2} weight loading) has first cycle charge capacity of 702 mAh g^{-1}$_{electrode}$ (total weight of electrode considered) and $\sim$470 mAh cm^{-3}$_{electrode}$ (total volume of electrode considered) at 100 mA g^{-1}$_{electrode}$ and stable charge capacity of 543 mAh g^{-1}$_{electrode}$ ($\sim$363 mAh cm^{-3}$_{electrode}$) at charge current density of 2,400 mA g^{-1}$_{electrode}$. The capacity is $\sim$200 mAh g^{-1}$_{electrode}$ when cycled at $\sim$-15 °C. Further, the composite electrode has exceptionally high strain-to-failure (exceeds 2%) as measured in a uniaxial tensile test and the mode of failure differ significantly from pristine rGO papers.

Results

Material synthesis and electrode fabrication. Polymer-derived SiOC ceramic particles were prepared by controlled thermolysis of 1,3,5,7-tetramethyl-1,3,5,7-tetravinylcyclotetrasiloxane (TTCS) polymeric precursor while graphene oxide (GO) was prepared by the modified Hummer's method[51] (for details, see Methods section). The polymer-to-ceramic transformation was complete at 1,000 °C[41]. Detailed characterization of cross-linked polymer and resulting SiOC material is presented in Fig. 1a–g. SEM images of SiOC particles in Fig. 1a confirmed average particle size to be $\sim$4 μm (with s.d. = 1.8 μm). X-ray photoelectron spectroscopy (XPS) showed O 1s, C 1s, Si 2s, Si 2p and O 2s peaks for both cross-linked and pyrolyzed SiOC ceramic (Fig. 1b). Close analysis of the deconvoluted silicon band (for Si 2p photoelectrons) in SiOC revealed the emergence of peaks at 103.5 and 102.2 eV, corresponding to SiO$_4$ and CSiO$_3$ phases, respectively (Fig. 1c). In addition, peaks at 534.5, 533.1 and 532.4 eV corresponding to C=O, SiO$_2$ and Si–O phases, respectively, were observed in O 1s band (Fig. 1d), whereas the C 1s band (Fig. 1e) was fitted with 3 peaks at 286.5, 284.5 and 284.7 eV corresponding to C=O, C–C and C–Si phases, respectively. Surface elemental composition from XPS was measured to be C=62.55 at% (50.35 wt%), O=25.73 at% (27.57 wt%) and Si=11.72 at% (22.06 wt%). XPS composition after 80 min of depth profiling (with 5 keV Ar-ion) showed lower carbon and oxygen content of 50.78 at% (34.47 wt%) and 18.44 at% (16.66 wt%), respectively with Si at 30.78 at% (48.85 wt%) (see Supplementary Fig. 1). Bulk composition of SiOC particles was also determined from combustion and inert gas fusion techniques (see Supplementary Fig. 2a,b and Methods section for details). The composition was found to be C=51.24 at% (38.3 wt%), O=19.79 at% (19.7 wt%),

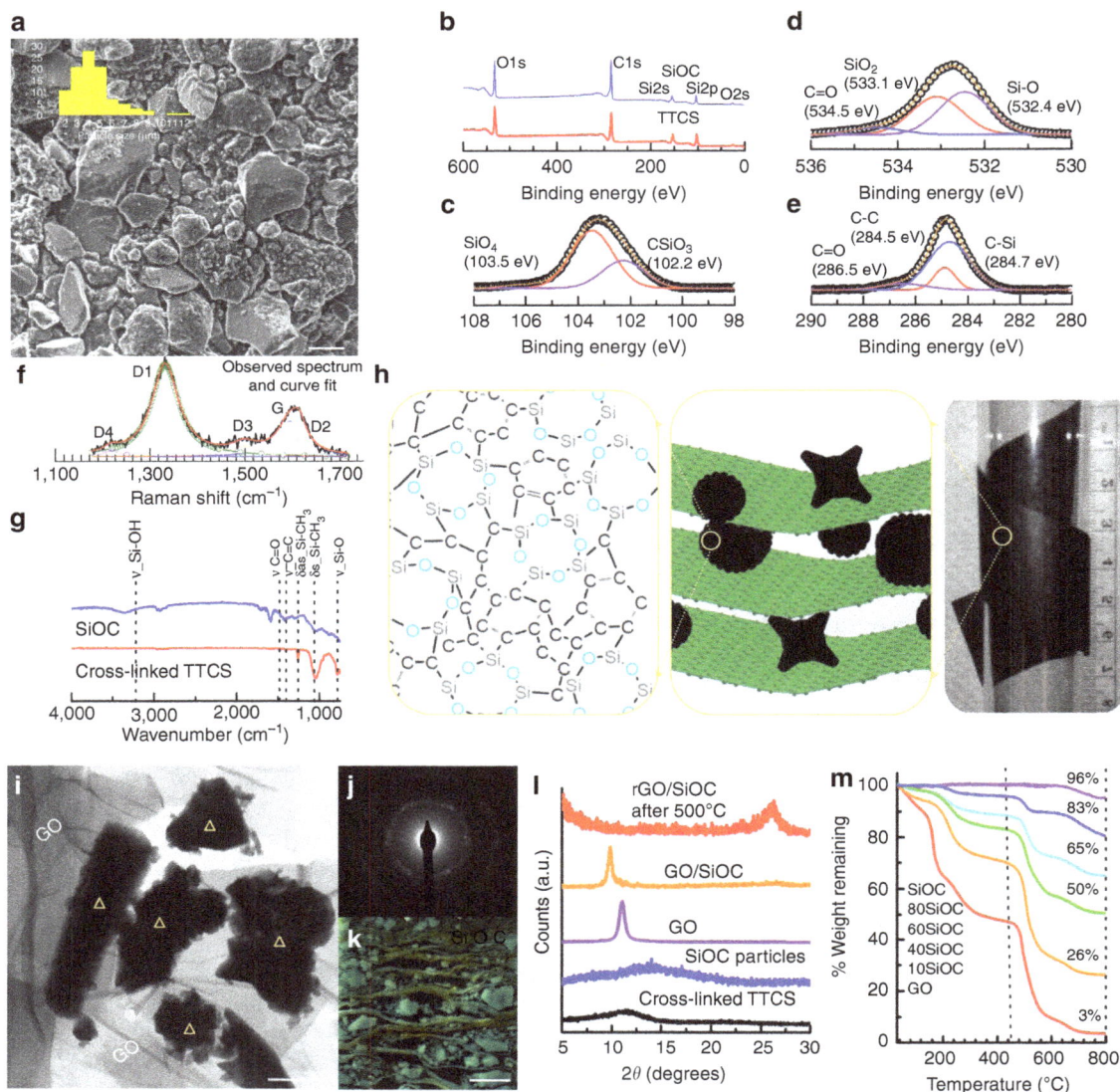

Figure 1 | Characterization of SiOC ceramic and SiOC/graphene composite papers. (**a**) SEM image of SiOC particles after pyrolysis of the polymeric TTCS particles. Sharp glass-like particles decorated with sub-micron size particles were observed. Scale bar is 5 µm. (**b**) XPS survey scans for cross-linked TTCS and pyrolyzed SiOC. High resolution XPS spectrum of pyrolyzed SiOC particles in the (**c**) Si 2p region, (**d**) O 1s region and (**e**) C 1s region were consistent with the polymer-derived SiOC nanodomain model. Deconvoluted peaks indicate the various bonds between Si, C and O atoms that are distinct to pyrolyzed SiOC. (**f**) Raman spectrum of SiOC showed peaks that are characteristic of graphite-like carbon (D1-peak: 1,350 cm^{-1} and G-peak: 1,590 cm^{-1}). (**g**) Fourier Transform Infrared Spectroscopy spectra of SiOC and cross-linked TTCS (v: stretching vibration mode and δ: bending vibration mode). (**h**) Digital camera picture and schematic illustration of proposed hybrid structure of the freestanding paper along with the atomic structure of pyrolyzed SiOC particle. (**i**) TEM image of SiOC/GO composite material. Large GO flakes covering SiOC particles (Δ) were observed. Scale bar is 500 nm. (**j**) Corresponding TEM selected area electron diffraction pattern showed multiple spot pattern due to polycrystallinity of restacked GO sheets with faint ring pattern attributed to amorphous SiOC material. (**k**) FIB cross-sectional EDX elemental map of 60SiOC paper in which Si, C and O are indicated by blue, red and green, respectively. The scale bar is 5 µm. Additional TEM and SEM images are presented in Supplementary Figs. 6–8. (**l**) XRD of cross-linked TTCS, SiOC particles, GO and composite papers before and after thermal reduction (annealing). Complete reduction of GO to rGO is illustrated in the plot. (**m**) TGA curves of GO paper and unannealed composite paper measured from 30 to 800 °C (10 °C min^{-1}) in flowing air (20 ml min^{-1}). The weight percentage of SiOC in the unannealed composite is as indicated in the figure.

H = 5.09 at% (0.31 wt%) and Si = 23.85 at% (41.68 wt%). The elemental composition obtained from various techniques is summarized in Supplementary Table 1. Raman spectroscopy of SiOC particles was performed to further confirm the existence of the free or excess carbon domains. As shown in Fig. 1f, five peaks could be fitted into the spectrum: D1 or D-band ($\sim$1,330 cm^{-1}), D2 ($\sim$1,615 cm^{-1}), D3 ($\sim$1,500 cm^{-1}), D4 ($\sim$1,220 cm^{-1}) and the G-band ($\sim$1,590 cm^{-1}) [52]. D1, D2 and D4 originate from disordered graphitic lattice (graphene layer edges, surface layers and polyenes and so on) while D3 is associated

with amorphous carbon soot. G-band corresponds to the ideal graphitic lattice. In addition, two bumps centered at $\sim$2,640 (2*D overtone) and $\sim$2,915 cm^{-1} (D + G combination) were also observed (Supplementary Fig. 3). Similarly, Fourier Transform Infrared Spectroscopy (FTIR) analysis also confirmed transformation of TTCS polymer to ceramic SiOC (Fig. 1g)[41]. Based on spectroscopic evidence, the predicted chemical structure of the cross-linked polymer and resultant ceramic is presented in Supplementary Fig. 4, which is in agreement with previous work on polymer-derived SiOC[42].

The composite papers were prepared following a vacuum filtration technique (see Materials section for details and schematic in Supplementary Fig. 5). Samples were labeled as rGO, 10SiOC, 40SiOC, 60SiOC and 80SiOC for rGO paper and GO with 10, 40, 60 and 80 wt% of SiOC in the paper, respectively. The digital camera image and schematic in Fig. 1h highlights the flexibility and structure of the composite paper, respectively.

Morphology of the composite and thermally reduced (annealed) freestanding papers was studied by electron and focused ion beam (FIB) microscopy. The transmission electron microscope (TEM) image (Fig. 1i) showed large micrometer-sized thin GO sheets along with random shape glass-like SiOC particles (also see Supplementary Fig. 6a–e). Large SiOC particles were seen to be covered with smaller nanometer size particles. The graphene platelets seem to ocassionaly fold and cover individual SiOC particles and other instances show GO being interlayered by SiOC. EDX elemental mapping performed in scanning-TEM mode (Supplementary Figs. 6a–e) confirmed the uniform distribution of Si, O, C in the particles with higher concentration of C observed near the edges possibly due to graphene platelets. For the selected area electron diffraction pattern in Fig. 1j, the multiple spot pattern is a result of polycrystallinity of restacked GO sheets and the faint ring pattern is attributed to amorphous SiOC material. The SEM images of the freestanding papers showed a sheet-like structure with a relatively smooth top surface for rGO paper[53–56], which became increasingly rough and porous with higher loading of SiOC particles in the composite (Supplementary Fig. 7a–d). Cross-sectional SEM of the fractured samples revealed ordered stacks of rGO with SiOC particles interlayered between the sheets (Supplementary Fig. 7e–h). Several micrometer sized particles could be seen for 60SiOC specimen along with clumped nanometer sized particles. Nonetheless, mechanically fractured composite papers were largely uneven and showed signs of damage to the interface. To obtain a smooth and defect-free cross-section, the 60SiOC paper was sectioned by means of a FIB milling (see Methods section and Supplementary Fig. 8a for details regarding specimen preparation). The uniform distribution of SiOC particles and wrapping by large-area graphene platelets could be clearly observed in the electron beam (Supplementary Fig. 8b) and ion-beam images (Supplementary Fig. 8c). Elemental mapping by means of EDX (Fig. 1k and Supplementary Fig. 8d–f) further established the inter-layered morphology of the composite. Depending up on the SiOC content, the average thickness of the papers varied between $\sim$20 and 30 μm.

The reduction of GO (non-conducting) to rGO (conducting) was confirmed by use of X-ray diffraction (XRD). As shown in Fig. 1l, both GO and unannealed composite papers, had peaks at 11.05 and 9.8°, corresponding to interlayer spacing of 8 and 12 Å, respectively. Interlayer spacing was large compared with that of graphite (with major peak (002) at 26.53°, corresponding to 3.36 Å) because of oxygen functional groups present in GO and water molecules held between the layers. After thermal annealing at 500 °C for 2h, the paper showed a broad peak at $2\theta = 26°$, typical of reduced GO material[55,56]. The broad peak observed in the spectra suggests inhomogeneous spacing between the layers. XRD spectra of cross-linked TTCS and SiOC particles were both featureless, confirming the amorphous nature of these ceramics (hallmark of these materials). Raman spectrum (I_d/I_g) pre and post thermal reduction showed a slight change in accordance with previous reports (Supplementary Fig. 9)[39]. Reduction of GO to rGO was verified by the disappearance of oxide peaks in the high resolution XPS analysis of C 1s peak (Supplementary Fig. 10).

Thermogravimetric analysis (TGA) was performed to ascertain the mass loading of SiOC in the composite papers. Figure 1m shows the percentage composition of filtered composite paper prior to their thermal reduction. Significant weight loss was observed in the 50–100 °C and 100–400 °C temperature ranges, which is attributed to evaporation of trapped water molecules in the GO and oxygen functionalities, respectively[57–59]. The weight loss was highest for GO and lowest for 80SiOC (see Supplementary Table 2). Final weight loss in the 400–800 °C range is due to burning of carbon material. Comparatively, the initial weight loss was not observed in thermally reduced samples (mere 1.2% for rGO at 400 °C, Supplementary Fig. 11) that suggests high degree of water removal and oxygen groups by thermal annealing. Approximately 3% and 6–10% residue was noted for GO and rGO material at $\sim$800 °C. As a result SiOC content (or percentage weight remaining) in the thermally reduced composite was higher than unannealed specimens; SiOC content in 10SiOC, 40SiOC, 60SiOC and 80SiOC increased from $\sim$10–30%, $\sim$50–65%, $\sim$65–78% and $\sim$83–92%, respectively. In the traditional method of electrode preparation, active material (including recently reported graphene embedded PDC material) is mixed with polymeric binder and conductive agent in an $\sim$80:10:10 ratio, followed by slurry coating on metal current collector foil[47]. However, using the present method we have made a freestanding and lightweight electrode, containing up to $\sim$78% SiOC as active material and $\sim$22% of rGO (acting as binder and conductive agent). Paper electrodes were directly utilized as the working electrodes. Electrochemical performance is presented in the following section.

Electrochemical performance. Figure 2a shows charge capacities and columbic efficiency of rGO, 10SiOC, 40SiOC, 60SiOC electrodes asymmetrically cycled at varying charge current densities. For rGO, the first-cycle charge capacity at 100 mA g^{-1}$_{electrode}$ was $\sim$210 mAh g^{-1}$_{electrode}$, it dropped to $\sim$200 mAh g^{-1}$_{electrode}$ in the second cycle, and then the charge capacity stabilized at $\sim$180 mAh g^{-1}$_{electrode}$ after five cycles. When charge current density increased to 2,400 mA g^{-1}$_{electrode}$, charge capacity was retained at $\sim$175 mAh g^{-1}$_{electrode}$. Returning the current density back to 100 mA g^{-1}$_{electrode}$ led to the return of higher capacity of 192 mAh g^{-1}$_{electrode}$. High irreversible first-cycle capacity results from electrochemical reaction contributed to solid-electrolyte interphase (SEI) layer formation. For the composite electrode, the first-cycle charge capacity increased in correspondence to the percentage of SiOC in the electrode. For example, 10SiOC showed 376 mAh g^{-1}$_{electrode}$, while 40SiOC and 60SiOC showed 546 mAh g^{-1}$_{electrode}$ and 702 mAh g^{-1}$_{electrode}$ (volumetric capacity of $\sim$470 mAh cm^{-3}$_{electrode}$), respectively. The 60SiOC capacity was lower than the capacity calculation based on a 'rule of mixture' approach ($\sim$793 mAh g^{-1}) with constituent rGO (first cycle reversible capacity $\sim$210 mAh g^{-1}) at $\sim$22 wt% as lower bound and SiOC (highest first cycle reversible capacity $\sim$958 mAh g^{-1} from ref. 46) at $\sim$78 wt% as upper bound. Similar to rGO electrode, when charge current density increased to 2,400 mA g^{-1}$_{electrode}$, composites 10SiOC, 40SiOC and 60SiOC showed high reversible capacity at 296, 417 and 543 mAh g^{-1}$_{electrode}$, respectively. Capacity retention at 2,400 mA g^{-1}$_{electrode}$ of 83.5% (compared with cycle number 5 at 100 mA g^{-1}$_{electrode}$) and first-cycle efficiency of 68% for 60SiOC is among the highest reported performances for a freestanding graphene-based electrode (see Supplementary Table 3 and Supplementary Table 4 for summary and comparison, respectively)[14–19,23,25,32,38]. When charge current density was lowered again to 100 mA g^{-1}$_{electrode}$ at cycle number 31, charge capacity increased to stable values of 304 mAh g^{-1}$_{electrode}$ ($\sim$80% retained), 471 mAh g^{-1}$_{electrode}$ ($\sim$96% retained) and

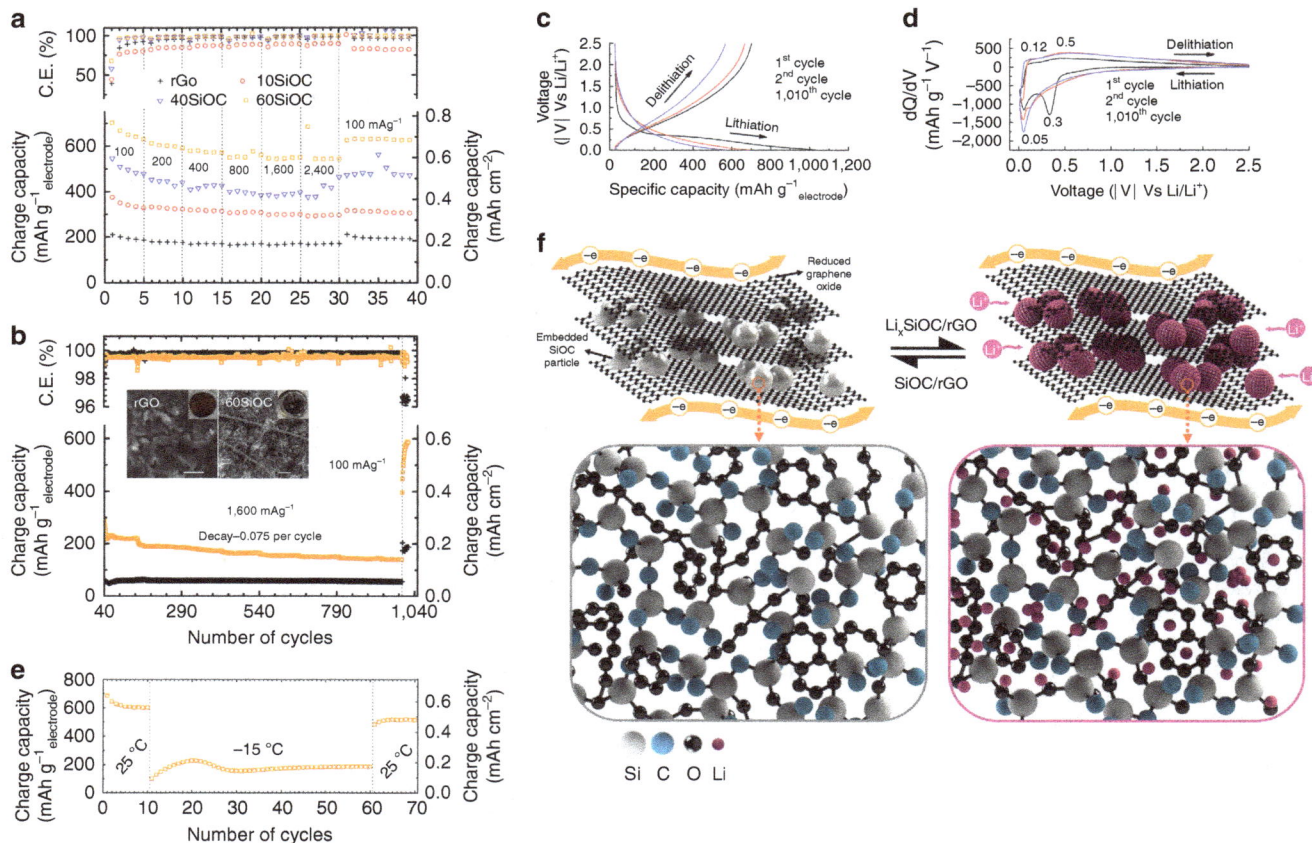

Figure 2 | Electrochemical characteristics and proposed lithium storage mechanism. (a) Charge capacity and cycling efficiency of various paper electrodes when cycled asymmetrically at increasing charge current densities. **(b)** Extended cycling behavior of rGO and 60SiOC electrodes cycled symmetrically at 1,600 mA g^{-1}$_{electrode}$. After 970 cycles, the electrodes showed good recovery when the current density was lowered back to 100 mA g^{-1}$_{electrode}$. Insets show the post-cycling digital and SEM images of the dissembled rGO and 60SiOC electrodes. Scale bar is 10 µm. **(c)** Voltage profile of 60SiOC electrode and corresponding **(d)** differential capacity curves for 1st, 2nd and 1,010th cycle. **(e)** Cycling behavior of 60SiOC at sub-zero temperature. After cooling down to ~-15 °C, the cell demonstrated a stable charge capacity of ~200 mAh g^{-1}$_{electrode}$ at 100 mA g^{-1}$_{electrode}$. The cell regained ~86% of its initial capacity when returned to cycling at room temperature (~25 °C). **(f)** Schematic representing the mechanism of lithiation/delithiation in SiOC particles. Majority of lithiation occurs *via* adsorption at disordered carbon phase, which is uniformly distributed in the SiOC amorphous matrix. Large rGO sheets serve as an efficient electron conductor and elastic support.

626 mAhg^{-1}$_{electrode}$ (~97% retained) for 10SiOC, 40SiOC and 60SiOC, respectively.

In order to test cyclic stability of the electrodes, the same cells were subjected to symmetric cycling at a current density of 1,600 mA g^{-1}$_{electrode}$. Charge capacity for this test is shown in Fig. 2b. Charge capacity of 60SiOC showed some decline as the cells were subjected to prolong symmetric cycling at 1,600 mA g^{-1}$_{electrode}$. The capacity decay over the 970-cycle range was observed to be approximately 0.075 mAh g^{-1}$_{electrode}$ per cycle. This decline was not observed in the rGO specimen, thereby demonstrating the importance of graphene in the composite material. Nonetheless, the average composite paper capacity in this range was approximately three times higher than pristine rGO electrode (~170 versus ~58 mAh g^{-1}$_{electrode}$). Most significantly, the cell capacities were ~185 (rGO) and 568 mAh g^{-1}$_{electrode}$ (60SiOC) at 1,010th cycle when the current density was brought back to 100 mA g^{-1}$_{electrode}$ and stabilized to 186 and 588 mAh g^{-1}$_{electrode}$, respectively at 1,020th cycle before the tests were stopped for post-cycling analysis. This represents ~94% capacity retention for 60SiOC when compared with capacity value at the 40th cycle prior to beginning of the long-term cycling test (see Supplementary Table 3). No measureable change in cycling efficiency of 60SiOC (~99.6%) was observed during this period. This shows that, even after 1,020 cycles, the

composite electrode was robust and continued to function without appreciable degradation.

Supplementary Fig. 12a shows voltage profiles of rGO for the 1st, 2nd and 1,010th cycle. Differential capacity profiles in Supplementary Fig. 12b were similar to previous reports on rGO electrodes, with a primary reduction peak at ~50 mV, a secondary reduction peak at ~(520–560) mV, and an oxidation peak at ~(120–130) mV[39]. The peak at ~50 mV, present in all subsequent cycles, is associated with lithiation of graphitic carbon, whereas the peak at ~560 mV signifies formation of SEI, which exists only in the first cycle. Supplementary Fig. 12c and d show the voltage profile and differential capacity curves of 1st and 2nd cycle of 10SiOC, respectively. The first cycle contained three reduction peaks at around ~50, ~240 and ~520 mV, attributed to rGO lithiation, irreversible Li$_x$SiOC formation, and SEI formation, respectively[39,41,45]. In contrast, only one subtle extraction peak at ~110 mV is observed, which represents rGO de-lithiation with an extended bulge at ~500 mV that represents Li$_x$SiOC de-lithiation[38,45–47]. As the SiOC content increased to 40% (Supplementary Fig. 12e,f) and 60% (see Fig. 2c,d), domination of SiOC lithiation increased, as proven by increased intensity of the irreversible Li$_x$SiOC formation peak at ~(270–300) mV. Peak intensity of rGO de-lithiation at ~120 mV diminished with respect to Li$_x$SiOC de-lithiation bulge at ~ 500 mV. In addition, the 2nd and the 1,010th cycle charge/discharge and differential capacity curves of the electrodes had similar profiles, showing that no new phases formed even after more than 1,000 cycles. More importantly, the efficiency of 60SiOC remained high throughout the cycling test.

Additional rate capability test involving extreme symmetric cycling were performed on freshly prepared 60SiOC paper electrode with even higher mass loading (approximately 3 mg cm^{-2}). The data is presented in Supplementary Fig. 13. Stable capacity of ~ 700 mAh g^{-1}electrode was observed at 100 mA g^{-1}electrode which decreased to ~ 100 mAh g^{-1}electrode at 2,400 mA g^{-1}electrode and showed complete recovery when the current density was brought back to 100 mA g^{-1}electrode. Such stable performance is rarely reported for precursor-derived ceramic materials even on traditionally prepared electrode on copper foil where the current density and capacity are reported with respect to the active material only[46–48]. Tests were also conducted on 80SiOC specimen to ascertain if the charge capacity of the freestanding paper-based electrodes can be improved even further due to higher SiOC content. These attempts, however, were not successful because electrodes prepared at 80% SiOC loading were brittle and showed erratic behavior after only a few initial cycles. First-cycle charge capacity for 80SiOC was ~ 762 mAh g^{-1}electrode and showed domination of Li$_x$SiOC lithiation (~ 330 mV) and delithiation (~ 500 mV) over rGO peaks, similar to other composite electrodes (Supplementary Fig. 14a,b). The 80SiOC electrode began to demonstrate random spikes in charge capacity and efficiency with increased cycle number at high C-rate possibly due to mechanical disintegration and loss of electrical contact due to insufficient rGO loading (Supplementary Fig. 15a). Crack could be observed in the post-cycling SEM images (see Supplementary Fig. 15b–e).

Four-point electrical conductivity measurements were performed and compared for all specimens (for details, see Supplementary Note 1 and Supplementary Fig. 16). Data is summarized in Supplementary Table 5. Although average four-point resistance for 60SiOC (580 Ω) was higher than rGO paper (40 Ω), it still represents an important achievement because TTCS derived SiOC (under present pyrolysis conditions and for the given composition) was observed to be poor electrical conductor and the improved conductivity of the composite paper (5×10^{-2} S cm^{-1} versus $\sim 10^{-12}$ S cm^{-1} for SiOC powder[41]) is key to better C − rate characteristics. This is more evident when we compare the C − rate data for SiOC particle electrode prepared on traditional copper current collectors[46], where the electrochemical capacity was observed to be near zero for cycling current density of 1,600 mA g^{-1}.

In addition to room temperature testing, the best performing specimen (that is, 60SiOC) was subjected to electrochemical cycling at sub-zero temperature at ~ -15 °C (for details, see Supplementary Note 2). When initially cycled at room temperature (Fig. 2e), the cell had a stable charge capacity of ~ 600 mAh g^{-1}electrode that then reduced to a stable charge capacity of ~ 200 mAh g^{-1}electrode when cycled at low temperature. The cell regained $\sim 86\%$ of its initial capacity when it returned to cycling at room temperature.

In order to verify electrode integrity, the cells were dissembled in their lithiated state and the electrode was recovered for additional characterization. The inset in Fig. 2b and Supplementary Fig. 17 show the digital photograph and SEM image of the cycled electrodes. Post-cycling Raman spectroscopy data is presented in Supplementary Fig. 18 and Supplementary Table 6. No evidence of surface cracks, volume change, or physical imperfections were observed in the SEM images, suggesting high mechanical/structural strength of the composite paper towards continuous Li-cycling which could be attributed to

unique structure of the electrode as shown in Fig. 2f. In all cases, evidence of SEI formation due to repeated cycling of Li-ions was observed. Contamination in the specimen, indicated by arrows, was a result of residue of glass separator fibers. The electrodes were briefly exposed to air during the transfer process, resulting in oxidation of Li, which appeared as bright spots in the images due to non-conducting nature.

To illustrate the kinetics of charge/discharge of the composite paper, Galvanostatic intermittent titration cycling was performed for the 60SiOC electrode at room and low temperature (for details, see Supplementary Note 3). Acquired D$_{Li+}$ varied between $\sim 10^{-14}$ and $\sim 10^{-15}$ m^2 s^{-1} during insertion and extraction (Supplementary Fig. 19). These values are comparable with values reported for polymer-derived SiOC (Kasper et al. 10^{-13} to 10^{-15} m^2 s^{-1})[44]. In addition, total polarization potential and time dependent change in open-circuit voltage (OCV) at various states of charge were inferred for these experiments, as shown in Supplementary Fig. 20a–d. Reaction resistance to Li insertion and extraction from the 60SiOC electrode was calculated by taking a ratio of OCV to the current density (Supplementary Fig. 20e,f). Reaction resistance was fairly constant at 2 Ohm g during room temperature insertion. However, it increased exponentially to 8 Ohm g during Li extraction in the 1.5–2.0 V range, which highlights the difficulty in extracting the very last Li atoms from amorphous SiOC structure (Fig. 2f). Density of state calculations (Supplementary Fig. 21) show that Li is stored at several energy levels in the amorphous SiOC structure, with majority of the insertion occurring in the 0–0.5 V range. Further, a voltage hysteresis of ~ 0.5 V exists during the extraction half, which could be attributed to the hydrogen (H-terminated edges of free carbon phase) that are generally present in the SiOC derived from thermal decomposition of organosilicon polymers. H content in pyrolyzed ceramic particles was measured to be $\sim 0.25–0.3$ wt% (for details, see Methods section, Supplementary Fig. 2, Supplementary Table 1). Galvanostatic intermittent titration performed at low temperature (~ -15 °C) showed D$_{Li+}$ values in the $\sim (10^{-15}$ to $10^{-13})$ m^2 s^{-1} range during Li-ion insertion and extraction (Supplementary Fig. 22). The total polarization potential, time dependent change in OCV at various states of charge performed at ~ -15 °C and corresponding reaction resistance plots are included in Supplementary Fig. 23.

Mechanical strength of the electrode. Static uniaxial tensile tests were conducted to quantify the strength and strain-to-failure for the freestanding composite papers by use of a custom-built set-up. Figure 3a shows a schematic of the test setup, in which the load cell is attached to a digital meter, connected to a transducer electronic data sheet in order to transfer the data to host computer through an RS232 serial port using a program written in MATLAB. Engineering stress–strain plots and tensile modulus, derived from load–displacement curves for various paper electrodes are compared in Fig. 3b,c, respectively. The rGO sample showed average tensile strength of ~ 10.7 MPa at a failure strain of 2.8%, while 60SiOC sample had tensile strength of ~ 2.7 MPa at a strain of 1.1%. Low tensile strength of the 60SiOC specimen was expected considering that it contained only $\sim 20\%$ rGO. Overall, strength and modulus for these crumpled composite papers was lower than GO and rGO papers prepared from techniques other than high temperature reduction[53,54]. However, the strain-to-failure was almost 5 to 10 times higher than a typical GO, rGO or rGO-composite paper, suggesting that crumpled composite papers may be able to sustain larger volume changes. Surface analysis using SEM of rGO (Fig. 3d) showed occurrence of micro features after tensile test, which we suggest, are due to

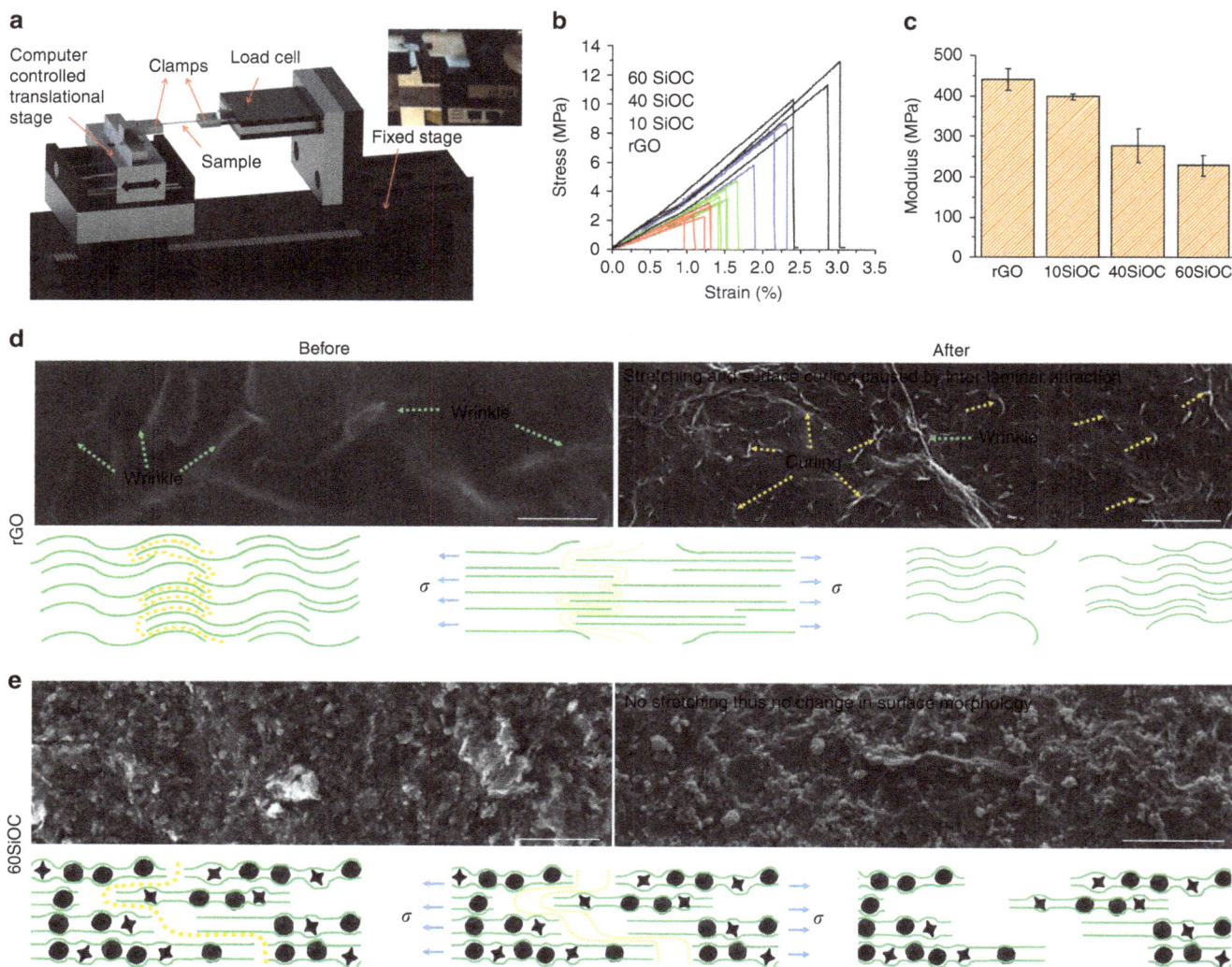

Figure 3 | Mechanical testing data. (a) Schematic of the tensile testing setup with a photograph of rGO paper immediately after the fracture. Scale reading in the photograph indicate the change in length to be ~0.28 mm. (b) Engineering stress versus strain plots of various freestanding papers derived from load versus displacement data, and (c) their corresponding modulus values. Error bars are 26.8, 7.6, 41.5, 24.1 MPa for rGO, 10SiOC, 40SiOC, and 60SiOC, respectively. The SEM images and schematic illustration to show the predicted mechanism of fracture in rGO and 60SiOC freestanding papers: (d) The rGO paper experienced stretching and rearrangement of graphene sheets before failure. (e) For 60SiOC paper, negligible stretching or rearrangement occurred. Fracture line follows SiOC particles embedded in rGO flakes, resulting in gradual separation/tearing of the paper. The scale bar is 20 µm in all images. Tensile test videos are included as Supplementary Movies 1 and 2.

rearrangement of rGO sheets under tensile load. These micro features are assumed to be due to curling of individual graphene sheets on the top surface when they lose contact with the sheets below them. However, for 60SiOC in Fig. 3e, ceramic particles acted as the point of fracture and caused rGO sheets to separate without stretching, as proven by SEM images that show no distinguishable changes before and after tensile test. Supplementary Fig. 24a–h are the top and cross-sectional view SEM images of fractured surface. The rGO because of higher elasticity had an irregular crumpled appearance, but composite papers were more brittle and had sharper cross-section. Mode of fracture in rGO and 60SiOC papers differed significantly, as presented in Supplementary Movies 1 and 2. A loud distinct sound indicated almost instantaneous fracture of the rGO specimen, accompanied by curling of both ends of the fractured paper. Fracture of 60SiOC specimen was similar to a thin plate with an edge crack, the crack propagation could be clearly observed. In addition, stress lines could be observed only in the rGO specimen, radiating from one clamp to another and indicating distribution of stress throughout the length of the

specimen. These observations are explained with the help of a schematic in Fig. 3d,e. *Ex situ* Raman analysis (Supplementary Fig. 25) from the top surface of the specimens before and after tests showed increase in average intensity ratio of the I_d and I_g peaks for rGO (0.88 versus 1.02) while the ratio was largely unaffected for composite specimen.

Discussion

Electrochemical characterization shows that 60SiOC is best long-term cycling electrode with reversible capacities of ~702 mAh g^{-1}$_{electrode}$ at 1st cycle and ~588 mAh g^{-1}$_{electrode}$ at 1,020th cycle, respectively. Although 80SiOC offers highest first reversible capacity of ~762 mAh g^{-1}$_{electrode}$, it undergoes capacity fading and mechanical damage after few initial cycles at high currents. Hence, the capacity and cycling stability are affected by the relative amounts of SiOC and graphene in the composite, respectively. We ascribe the superior electrochemical performance of 60SiOC electrode to remarkable physical and chemical properties of its constituents and the unique morphological features of the paper. Because graphene sheets in 60SiOC occupy larger volume in the composite, well-dispersed GO sheets during the layer − by − layer filtration process arrange themselves around the SiOC particles to form a flexible composite paper. TEM (Supplementary Fig. 6), SEM (Supplementary Fig. 7) and FIB (Supplementary Fig. 8) characterization shows that morphology of the composite paper is planar and porous. The porous design therefore facilitated liquid electrolyte to reach the very interior of the electrode thereby providing easy path for solvated ions to be transported on to the surface of SiOC particles. Further, rGO because of its high electrical conductivity and mechanical flexibility provided an electrically conducting (see Supplementary Table 5) and mechanically robust (see Fig. 3b) matrix for the Li-active SiOC particles thereby buffering volume changes in the electrode and maintaining inter particle connection during long-term cycling. Microscopy (Fig. 2b, Supplementary Fig. 17) and Raman spectroscopy (Supplementary Fig. 18) of the disassembled cell reveal formation of stable SEI on a completely integral electrode, which could explain the high cycling efficiency observed in these composites.

We attribute high reversible capacity of molecular precursor derived SiOC to its amorphous structure, which is comprised of silica domains, $-sp^2$ carbon chains (or the free carbon phase), nano-voids and silicon/carbon open bonds (Fig. 2f and Supplementary Fig. 4 for proposed SiOC structure), that offer large number of sites, in which Li-ion can be reversibly stored. We notice that even the composite electrodes are not free from charge–discharge voltage hysteresis (or energy inefficiency) that is generally observed in precursor derived ceramics during the extraction half[46,49,50]. Lowering hydrogen content[60] and doping of silica domains (such as B) in SiOC could be a useful strategy for improving electrical properties and lowering of voltage hysteresis in these ceramics[40,46]. Another important area for future investigation could be to tailor the rGO flakes for residual oxygen and hydrogen surface groups and edge defects so that lithium irreversibility and voltage hysteresis[60] that arises from active defect sites could be minimized without compromising Li-ions' mobility and access to the SiOC particles[4].

In summary we have demonstrated fabrication of a freestanding multi-component composite paper consisting of SiOC glass-ceramic particles supported in rGO matrix as a stable and durable battery electrode. The porous 3-D rGO matrix served as an effective current collector and electron conductor with a stable chemical and mechanical structure while, embedded amorphous SiOC particles actively cycled Li-ions with high efficiency. Elimination of inactive ingredients such as metal current collector, non-conducting polymeric binder and conducting agent reduces the total electrode weight and provides the means to produce highly efficient lightweight batteries.

Methods

Preparation of polymer derived SiOC ceramic.
SiOC was prepared through the polymer pyrolysis route[41], liquid 1,3,5,7-tetramethyl-1,3,5,7-tetravinylcyclotetrasiloxane (TTCS, Gelest, PA) precursor (with 1 wt% dicumyl peroxide added as the cross-linking agent) was cross-linked at 380 °C in argon for 5 h, which resulted in a white infusible mass. The infusible polymer was ball-milled in to fine powder and pyrolyzed at 1,000 °C for 10 h in flowing argon resulting in a fine black SiOC ceramic powder.

Chemicals.
Sodium nitrate (99.2%), potassium permanganate (99.4%), sulfuric acid (96.4%), hydrogen peroxide (31.3% solution in water), hydrochloric acid (30% solution in water) and methanol (99.9%) were purchased from Fisher Scientific. All materials were used as received without further purification.

Preparation of GO and SiOC composite paper.
Modified Hummer's method was used to make GO[51]. A total of, 20 ml colloidal suspension of GO in 1:1 (v/v) water

and isopropanol was made by sonication. Varying weight percentages of SiOC particles (with respect to GO) were added to the solution and the solution was sonicated for 1 h and stirred for ∼6 h for homogenous mixing. The composite suspension was then filtered by vacuum filtration through a 10 µm filter membrane (HPLC grade, Millipore). The GO/SiOC composite paper obtained was carefully removed from the filter paper, dried, and thermally reduced at 500 °C under argon atmosphere for 2 h. The large-area paper with 60SiOC composition (with an ∼6.25 inch diameter, cut into rectangular strip) was similarly prepared by use of a Büchner funnel with a polypropylene filter paper (Celgard). The heat-treated paper was then punched (cut) into small circles and used as working electrode material for Li-ion battery half-cells.

Coin cell assembly and electrochemical measurements.
Li-ion battery coin cells were assembled in an argon-filled glove box. 1 M LiPF$_6$ (Alfa Aesar) in (1:1 v/v) dimethyl carbonate:ethylene carbonate (ionic conductivity 10.7 mS cm^{-1}) was used as the electrolyte. A 25 µm thick (19 mm diameter) glass separator soaked in electrolyte was placed between the working electrode and pure Li foil (14.3 mm diameter, 75 µm thick) as the counter electrode. Washer, spring, and a top casing were placed to complete the assembly before crimping.

Electrochemical performance of the assembled coin cells was tested using a multichannel BT2000 Arbin test unit sweeping between 2.5 V to 10 mV versus Li/Li$^+$ that followed a cycle schedule: (a) Asymmetric mode: Li was inserted at 100 mA g^{-1}$_{electrode}$ while the extraction was performed at increasing current densities of 100, 200, 400, 800, 1,600 and 2,400 mA g^{-1}$_{electrode}$ for 5 cycles each, and returned to 100 mA g^{-1}$_{electrode}$ for the next 10 cycles. (b) Symmetric mode: later, all the cells were subjected to symmetric cycling at a current density of 1,600 mA g^{-1}$_{electrode}$ for up to 1,000 cycles, returning to 100 mA g^{-1}$_{electrode}$ for the last 20 cycles.

Instrumentation and characterization.
SEM of SiOC powder was carried out on a Carl Zeiss EVO MA10 system with incident voltage of 5–30 kV. TEM images were digitally acquired by use of a Phillips CM100 operated at 100 kV. TEM elemental mapping was performed by using a 200 kV S/TEM system (FEI Osiris) equipped with chemiSTEM technology, a high angle annular dark field (HAADF) and Super-X windowless EDX detector. Super-X windowless EDX detector system with silicon drift detector technology allowed fast EDX data collection (a factor of more than 50 enhancement in acquisition speed of EDX chemical mapping) and large field of view elemental mapping. Acceleration voltage was 200 kV and acquisition time was 10 min.

A FIB system (FEI Versa 3D Dual Beam) was used for milling and imaging cross-section of the paper electrodes following standard procedures. Briefly, a platinum protective layer (∼25 µm × 10 µm × 5 µm in x, y and z axes, respectively) was first deposited at an ion beam current of ∼5 nA. Milling was then performed using regular cross section at an ion beam current of ∼65 nA to create trenches on either side and bottom face of platinum-coated area. Followed by cleaning cross-section feature (∼20 µm × 1 µm × 6 µm in x, y and z axes, respectively) to fine mill contamination at the bottom face of platinum coated area. The acceleration voltage of Ga$^+$ was 30 kV. An ion-beam current of ∼40 pA was used for imaging purposes. In-column detector for secondary electrons in beam deceleration mode was used for SEM imaging of the milled cross-section. Elemental mapping (EDS) was performed by use of an inbuilt energy dispersive spectroscopy silicon drift detector (Oxford Instruments).

Raman spectra were collected using a confocal Raman imaging system (Horiba Jobin Yvon LabRam ARAMIS) with 633 nm HeNe laser (laser power of 17 mW) as the light source with a × 100 microscope objective. Data acquisition was performed at an exposure time of 20 s with at least four accumulations at each point. D1 filter (10% transparency) was employed for the ceramic powder samples. Additional material characterization was made using XRD operating at room temperature, with nickel-filtered CuKα radiation ($\lambda = 1.5418$ Å). The surface chemical composition was studied by XPS (PHI Quantera SXM-03 Scanning XPS Microprobe) using monochromatic Al Kα radiation. For XPS depth profiling, sputtering was performed with a 5 keV Argon ion gun for 20 min followed by survey scan. The sputtered area was set to ∼2 mm × 2 mm. The process was repeated four times with total sputtering time reaching 80 min.

Further, bulk elemental composition of the pyrolyzed SiOC ceramic was measured following procedures similar to as described in the literature[46]. Analysis was done for carbon, oxygen and hydrogen content. Silicon content was calculated as a difference to 100%. The carbon content was measured by use of LECO Analyzer Model CS844 (LECO Corp. Analytical Bus, St Joseph, MI) by the combustion method and IR detection. Approximately 50 mg of SiOC powder mixed with accelerants as Iron chips and Lecocel II HP was used for this test. The oxygen and hydrogen contents were measured by use of LECO Analyzer Model No. ONH-836 (LECO Corp. Analytical Bus, St Joseph, MI) based on inert gas fusion thermal conductivity/infrared detection method. Specimen preparation involved mixing ∼34 mg of SiOC ceramic powder with graphite powder (LECO Corp.) as an accelerant in a nickel capsule (LECO Corp.) followed by placement in graphite crucible. The crucible was then heated to ∼3,000 °C in the chamber and gaseous products transferred to IR/thermal conductivity detectors for analysis. The mass per cent of carbon and oxygen were quantified in reference to the IR spectrum generated from graphite and tungsten oxide powders, respectively.

Hydrogen content in SiOC ceramic was also confirmed by use of another equipment based on combustion/thermal conductivity detector method, CE-440 Elemental Analyser (Exeter Analytical, UK). Combustion of the weighed sample (1.8056 mg of fine powder) was carried out in the instrument chamber in pure oxygen under static conditions. Helium carried the combustion products through the analytical system to atmosphere. Between the thermal conductivity cells absorption trap removed water from the sample gas. The differential signal read before and after the trap reflected the water concentration and, therefore, the amount of hydrogen in the original sample. The hydrogen content by this method was observed to be 0.25 wt% with an error of 0.06%. TGA was performed using Shimadzu 50 TGA (limited to 800 °C). Samples weighing, ~ 2.5 mg, were heated in a platinum pan at a rate of $10\,°C\,min^{-1}$ in air flowing at $20\,ml\,min^{-1}$. Electrical conductivity measurements were carried out by use of a four-point probe setup and Keithley 2636A (Cleveland, OH) dual channel sourcemeter in the Ohmic region. Electrochemical cycling of assembled cells was carried out using multichannel Battery Test Equipment (Arbin-BT2000, Austin, TX) at atmospheric conditions.

Mechanical testing. Static uniaxial in-plane tensile tests were conducted in a custom-built test setup. One end of the setup was connected to a 1N load cell (ULC-1N Interface) and the other end was clamped to a computer-controlled translation stage (M-111.2DG from PI). The entire setup was located on a bench with self-adjusting feet. All tensile tests were conducted in controlled strain rate mode with a strain rate of $0.2\%\,min^{-1}$. Paper electrodes were cut (punched out) into rectangular strips of $\sim 5 \times 15\,mm^2$ for testing without any further modification.

References

1. Xiong, P. et al. Chemically integrated two-dimensional hybrid zinc manganate/graphene nanosheets with enhanced lithium storage capability. *ACS Nano* **8**, 8610–8616 (2014).
2. Ko, M., Chae, S., Jeong, S., Oh, P. & Cho, J. Elastic a-silicon nanoparticle backboned graphene hybrid as a self-compacting anode for high-rate lithium ion batteries. *ACS Nano* **8**, 8591–8599 (2014).
3. David, L., Bhandavat, R., Barrera, U. & Singh, G. Polymer-derived ceramic functionalized MoS_2 composite paper as a stable lithium-ion battery electrode. *Sci. Rep* **5**, 9792 (2015).
4. Raccichini, R., Varzi, A., Passerini, S. & Scrosati, B. The role of graphene for electrochemical energy storage. *Nat. Mater.* **14**, 271–279 (2015).
5. Bhandavat, R., David, L. & Singh, G. Synthesis of surface-functionalized WS_2 nanosheets and performance as Li-ion battery anodes. *J. Phys. Chem. Lett.* **3**, 1523–1530 (2012).
6. Manthiram, A. Materials challenges and opportunities of lithium ion batteries. *J. Phys. Chem. Lett.* **2**, 176–184 (2011).
7. Cui, L.-F., Hu, L., Choi, J. W. & Cui, Y. Light-weight free-standing carbon nanotube-silicon films for anodes of lithium ion batteries. *ACS Nano* **4**, 3671–3678 (2010).
8. Luo, J. Y. et al. Crumpled graphene-encapsulated Si nanoparticles for lithium ion battery anodes. *J. Phys. Chem. Lett.* **3**, 1824–1829 (2012).
9. Liu, N. et al. A pomegranate-inspired nanoscale design for large-volume-change lithium battery anodes. *Nat. Nanotechnol* **9**, 187–192 (2014).
10. Hwang, T. H., Lee, Y. M., Kong, B. S., Seo, J. S. & Choi, J. W. Electrospun core-shell fibers for robust silicon nanoparticle-based lithium ion battery anodes. *Nano Lett.* **12**, 802–807 (2012).
11. Pan, L. et al. Facile synthesis of yolk-shell structured Si-C nanocomposites as anodes for lithium-ion batteries. *Chem. Commun.* **50**, 5878–5880 (2014).
12. Cui, L. F., Yang, Y., Hsu, C. M. & Cui, Y. Carbon-silicon core-shell nanowires as high capacity electrode for lithium ion batteries. *Nano Lett.* **9**, 3370–3374 (2009).
13. Hertzberg, B., Alexeev, A. & Yushin, G. Deformations in Si-Li anodes upon electrochemical alloying in nano-confined space. *J. Am. Chem. Soc.* **132**, 8548–8549 (2010).
14. Wang, B. et al. Adaptable silicon-carbon nanocables sandwiched between reduced graphene oxide sheets as lithium ion battery anodes. *ACS Nano* **7**, 1437–1445 (2013).
15. Hu, T. et al. Flexible free-standing graphene-TiO_2 hybrid paper for use as lithium ion battery anode materials. *Carbon* **51**, 322–326 (2013).
16. Liang, J., Zhao, Y., Guo, L. & Li, L. Flexible free-standing graphene/SnO_2 nanocomposites paper for Li-ion battery. *ACS Appl. Mater. Interfaces* **4**, 5742–5748 (2012).
17. Jia, X. et al. Building robust architectures of carbon and metal oxide nanocrystals toward high-performance anodes for lithium-ion batteries. *ACS Nano* **6**, 9911–9919 (2012).
18. Noerochim, L., Wang, J. Z., Chou, S. L., Wexler, D. & Liu, H. K. Free-standing single-walled carbon nanotube/SnO_2 anode paper for flexible lithium-ion batteries. *Carbon* **50**, 1289–1297 (2012).
19. Chockla, A. M. et al. Electrochemical lithiation of graphene-supported silicon and germanium for rechargeable batteries. *J. Phys. Chem. C* **116**, 11917–11923 (2012).
20. Yue, L., Zhong, H. & Zhang, L. Enhanced reversible lithium storage in a nano-Si/MWCNT free-standing paper electrode prepared by a simple filtration and post sintering process. *Electrochim. Acta* **76**, 326–332 (2012).
21. Chen, Z. et al. In situ generation of few-layer graphene coatings on SnO_2-SiC core-shell nanoparticles for high-performance lithium-ion storage. *Adv. Eng. Mater.* **2**, 95–102 (2012).
22. Ji, L. et al. Graphene/Si multilayer structure anodes for advanced half and full lithium-ion cells. *Nano Energy* **1**, 164–171 (2012).
23. Zhang, B., Zheng, Q. B., Huang, Z. D., Oh, S. W. & Kim, J. K. SnO_2-graphene-carbon nanotube mixture for anode material with improved rate capacities. *Carbon* **49**, 4524–4534 (2011).
24. Ji, L. et al. Fe_3O_4 Nanoparticle-integrated graphene sheets for high-performance half and full lithium ion cells. *Phys. Chem. Chem. Phys.* **13**, 7170–7177 (2011).
25. Yu, A. et al. Free-standing layer-by-layer hybrid thin film of graphene-MnO_2 nanotube as anode for lithium ion batteries. *J. Phys. Chem. Lett.* **2**, 1855–1860 (2011).
26. Xiao, J. et al. Electrochemically induced high capacity displacement reaction of PEO/MoS_2/graphene nanocomposites with lithium. *Adv. Funct. Mater.* **21**, 2840–2846 (2011).
27. Lee, J. K., Smith, K. B., Hayner, C. M. & Kung, H. H. Silicon nanoparticles-graphene paper composites for Li ion battery anodes. *Chem. Commun.* **46**, 2025–2027 (2010).
28. Wang, H. et al. Mn_3O_4 − graphene hybrid as a high-capacity anode material for lithium ion batteries. *J. Am. Chem. Soc.* **132**, 13978–13980 (2010).
29. Wu, Z. S. et al. Graphene anchored with Co_3O_4 nanoparticles as anode of lithium ion batteries with enhanced reversible capacity and cyclic performance. *ACS Nano* **4**, 3187–3194 (2010).
30. Yang, S., Feng, X., Ivanovici, S. & Müllen, K. Fabrication of graphene-encapsulated oxide nanoparticles: Towards high-performance anode materials for lithium storage. *Angew. Chem. Int. Ed.* **49**, 8408–8411 (2010).
31. Wang, D. et al. Ternary self-assembly of ordered metal oxide − graphene nanocomposites for electrochemical energy storage. *ACS Nano* **4**, 1587–1595 (2010).
32. Wang, J. Z., Zhong, C., Chou, S. L. & Liu, H. K. Flexible free-standing graphene-silicon composite film for lithium-ion batteries. *Electrochem. Commun.* **12**, 1467–1470 (2010).
33. Zhou, G. et al. Graphene-wrapped Fe_3O_4 anode material with improved reversible capacity and cyclic stability for lithium ion batteries. *Chem. Mater.* **22**, 5306–5313 (2010).
34. Wang, D. et al. Self-assembled TiO_2−graphene hybrid nanostructures for enhanced Li-ion insertion. *ACS Nano* **3**, 907–914 (2009).
35. Paek, S. M., Yoo, E. & Honma, I. Enhanced cyclic performance and lithium storage capacity of SnO_2/graphene nanoporous electrodes with three-dimensionally delaminated flexible structure. *Nano Lett.* **9**, 72–75 (2008).
36. Kumar, A. et al. Direct synthesis of lithium-intercalated graphene for electrochemical energy storage application. *ACS Nano* **5**, 4345–4349 (2011).
37. Li, X. et al. Superior cycle stability of nitrogen-doped graphene nanosheets as anodes for lithium ion batteries. *Electrochem. Commun.* **13**, 822–825 (2011).
38. Zhao, X., Hayner, C. M., Kung, M. C. & Kung, H. H. Flexible holey graphene paper electrodes with enhanced rate capability for energy storage applications. *ACS Nano* **5**, 8739–8749 (2011).
39. David, L. & Singh, G. Reduced graphene oxide paper electrode: opposing effect of thermal annealing on Li and Na cyclability. *J. Phys. Chem. C* **118**, 28401–28408 (2014).
40. Bhandavat, R. & Singh, G. Improved electrochemical capacity of precursor-derived Si(B)CN-carbon nanotube composite as Li-ion battery anode. *ACS Appl. Mater. Interfaces* **4**, 5092–5097 (2012).
41. Bhandavat, R. & Singh, G. Stable and efficient Li-ion battery anodes prepared from polymer-derived silicon oxycarbide-carbon nanotube shell/core composites. *J. Phys. Chem. C* **117**, 11899–11905 (2013).
42. Saha, A., Raj, R. & Williamson, D. L. A model for the nanodomains in polymer-derived SiCO. *J. Am. Ceram. Soc.* **89**, 2188–2195 (2006).
43. Sanchez-Jimenez, P. E. & Raj, R. Lithium insertion in polymer-derived silicon oxycarbide ceramics. *J. Am. Ceram. Soc.* **93**, 1127–1135 (2010).
44. Kaspar, J., Graczyk-Zajac, M. & Riedel, R. Determination of the chemical diffusion coefficient of Li-ions in carbon-rich silicon oxycarbide anodes by electro-analytical methods. *Electrochim. Acta* **115**, 665–670 (2014).
45. Konno, H. et al. Si-C-O glass-like compound/exfoliated graphite composites for negative electrode of lithium ion battery. *Carbon* **45**, 477–483 (2007).
46. Ahn, D. & Raj, R. Cyclic stability and C-rate performance of amorphous silicon and carbon based anodes for electrochemical storage of lithium. *J. Power Sources* **196**, 2179–2186 (2011).
47. Ahn, D., Lee, M. & Shah, S. R. (2009). U.S. Patent Application 12/483,631.
48. Feng, Y., Feng, N., Wei, Y. & Bai, Y. Preparation and improved electrochemical performance of SiCN-graphene composite derived from poly(silylcarbondiimide) as Li-ion battery anode. *J. Mater. Chem. A* **2**, 4168–4177 (2014).

49. Graczyk-Zajac, M., Fasel, C. & Riedel, R. Polymer-derived-SiCN ceramic/graphite composite as anode material with enhanced rate capability for lithium ion batteries. *J. Power Sources* **196**, 6412–6418 (2011).

50. Xing, W., Wilson, A. M., Eguchi, K., Zank, G. & Dahn, J. R. Pyrolyzed polysiloxanes for use as anode materials in lithium- ion batteries. *J. Electrochem. Soc.* **144**, 2410–2416 (1997).

51. Marcano, D. C. *et al.* Improved synthesis of graphene oxide. *ACS Nano* **4**, 4806–4814 (2010).

52. Sadezky, A., Muckenhuber, H., Grothe, H., Niessner, R. & Poschl, U. Raman micro spectroscopy of soot and related carbonaceous materials: Spectral analysis and structural information. *Carbon* **43**, 1731–1742 (2005).

53. Compton, O. C., Dikin, D. A., Putz, K. W., Brinson, L. C. & Nguyen, S. T. Electrically conductive "alkylated" graphene paper via chemical reduction of amine-functionalized graphene oxide paper. *Adv. Mater.* **22**, 892–896 (2010).

54. Compton, O. C. & Nguyen, S. T. Graphene oxide, highly reduced graphene oxide, and graphene: versatile building blocks for carbon-based materials. *Small* **6**, 711–723 (2010).

55. Park, S. *et al.* Aqueous suspension and characterization of chemically modified graphene sheets. *Chem. Mater.* **20**, 6592–6594 (2008).

56. Park, S. *et al.* Colloidal suspensions of highly reduced graphene oxide in a wide variety of organic solvents. *Nano Lett.* **9**, 1593–1597 (2009).

57. Ganguly, A., Sharma, S., Papakonstantinou, P. & Hamilton, J. Probing the thermal deoxygenation of graphene oxide using high-resolution in situ X-ray-based spectroscopies. *J. Phys. Chem. C* **115**, 17009–17019 (2011).

58. Haubner, K. *et al.* The route to functional graphene oxide. *ChemPhysChem* **11**, 2131–2139 (2010).

59. Park, S. *et al.* Hydrazine-reduction of graphite- and graphene oxide. *Carbon* **49**, 3019–3023 (2011).

60. Zheng, T., McKinnon, W. R. & Dahn, J. R. Hysteresis during lithium insertion in hydrogen-containing carbons. *J. Electrochem. Soc.* **143**, 2137–2145 (1996).

Acknowledgements

We thank Professor Rishi Raj (University of Colorado) for introducing us to the field of polymer-derived ceramics. We also thank Professor Kevin Lease, Dr Nasim Rahmani, and Dr Dan Boyle (all K-State) for access to equipment their lab. G.S. would like to thank Dr Xingzhong Li (University of Nebraska-Lincoln) and Ms. Heather Shinogle (University of Kansas) for allowing access to and training on FEI Osiris and FEI Versa 3D Dual Beam, respectively for elemental mapping. We sincerely thank Mr Andy Flores, Mr Bruce and Ms. Mary Landis (all LECO Corporation, St Joseph, MI) for access to elemental analysis instruments on LECO Analytical Bus, Ms. Elisabeth Eves (UIUC) for analysis using CE-440 Exeter Analytical instrument and Dr Jerry Hunter (UW-Madison) and Andrew Gio (Virginia Tech) for access to XPS. Publication of this article was funded in part ($\sim 35\%$) by the K-State Open Access Publishing Fund. Tasks related to synthesis of SiOC and flexible paper electrodes are supported by the United States National Science Foundation grant number NSF CBET-1335862 and NSF CAREER CMMI-1454151, respectively.

Author contribution

L.D. prepared all composite specimens, performed electrochemical testing, raman spectroscopy, low magnification TEM and mechanical testing. U.B. assisted L.D. with cell assembly. R.B. synthesized SiOC particles. G.S. conceived the idea, designed the experiments, performed elemental analysis/mapping and wrote the manuscript with inputs from L.D. All authors discussed the results and commented or revised the manuscript.

Additional information

Competing financial interests: The authors declare the following competing financial interest(s): G. S., R. B. and L. D. have filed for a provisional patent: U.S. Provisional Patent Application 61/817,626—Flexible Silicon oxycarbide Graphene Composite Electrodes for High Rate Performance Lithium-ion Batteries; Filed 30 April 2013.

20

Electron–phonon coupling in hybrid lead halide perovskites

Adam D. Wright[1], Carla Verdi[2], Rebecca L. Milot[1], Giles E. Eperon[1], Miguel A. Pérez-Osorio[2], Henry J. Snaith[1], Feliciano Giustino[2], Michael B. Johnston[1] & Laura M. Herz[1]

Phonon scattering limits charge-carrier mobilities and governs emission line broadening in hybrid metal halide perovskites. Establishing how charge carriers interact with phonons in these materials is therefore essential for the development of high-efficiency perovskite photovoltaics and low-cost lasers. Here we investigate the temperature dependence of emission line broadening in the four commonly studied formamidinium and methylammonium perovskites, $HC(NH_2)_2PbI_3$, $HC(NH_2)_2PbBr_3$, $CH_3NH_3PbI_3$ and $CH_3NH_3PbBr_3$, and discover that scattering from longitudinal optical phonons via the Fröhlich interaction is the dominant source of electron–phonon coupling near room temperature, with scattering off acoustic phonons negligible. We determine energies for the interacting longitudinal optical phonon modes to be 11.5 and 15.3 meV, and Fröhlich coupling constants of ~ 40 and 60 meV for the lead iodide and bromide perovskites, respectively. Our findings correlate well with first-principles calculations based on many-body perturbation theory, which underlines the suitability of an electronic band-structure picture for describing charge carriers in hybrid perovskites.

[1] Clarendon Laboratory, Department of Physics, University of Oxford, Parks Road, Oxford OX1 3PU, UK. [2] Department of Materials, University of Oxford, Parks Road, Oxford OX1 3PH, UK. Correspondence and requests for materials should be addressed to L.M.H. (email: laura.herz@physics.ox.ac.uk).

Hybrid lead halide perovskites have attracted intense research activity following their first implementation as light absorbers in thin-film solar cells[1] that now reach power conversion efficiencies (PCEs) in excess of 20% (refs 2,3). These compounds are described by the general formula ABX$_3$, where A is typically an organic cation such as methylammonium (CH$_3$NH$_3^+$ or MA$^+$) or formamidinium (HC(NH$_2$)$_2^+$ or FA$^+$), B is a divalent metal cation (usually Pb^{2+}) and X is a halide anion (I$^-$, Br$^-$ or Cl$^-$)[4]. Such hybrid organic–inorganic materials straddle the divide between organic and inorganic semiconductors, facilitating photovoltaic devices that combine the low processing costs of the former with the high PCEs of the latter[2,3,5]. In addition, their structural flexibility allows a wide compositional parameter space to be explored. Although MAPbI$_3$ is the most commonly investigated perovskite, the currently most efficient perovskite solar cells replace all[2] or most[3] of the MA$^+$ with FA$^+$. FAPbI$_3$ has the advantage of a smaller bandgap than MAPbI$_3$ (1.48 versus 1.57 eV), making it more suited for use in single-junction solar cells[6], and furthermore shows greater resistance to heat stress[6]. High-performing devices have used formulations containing both FA and MA cations, to counteract the thermodynamic instability of FAPbI$_3$ in its perovskite phase at room temperature[7–9]. Meanwhile, mixed-halide perovskites incorporating both I$^-$ and Br$^-$ have been investigated for use in tandem solar cells[5,6,10], as these systems allow for bandgap optimization across a wide tuning range.

The success of hybrid perovskites in photovoltaic applications has been widely attributed to their high absorption coefficients across the visible spectrum[11], their low exciton binding energies[4,12] facilitating charge formation and their long charge-carrier diffusion lengths enabling efficient charge extraction[13,14]. However, although much recent attention has been devoted towards unravelling the charge-carrier recombination mechanisms underlying these properties[4,15], the interaction of charge carriers with lattice vibrations (phonons) is currently still a subject of intense debate[16,17]. Such electron–phonon interactions matter, because they set a fundamental intrinsic limit to charge-carrier mobilities in the absence of extrinsic scattering off impurities or interfaces[18]. In addition, charge-carrier cooling following non-resonant (above bandgap) photon absorption is governed by interactions between charges and phonons[4]. Slow charge-carrier cooling components (compared with GaAs) have been postulated for hybrid perovskites[19], which may open the possibility for PCEs beyond the Shockley–Queisser limit. Furthermore, electron–phonon coupling has been shown to yield predominantly homogeneous emission line broadening in hybrid lead iodide perovskite at room temperature, making it suitable as a gain medium for short-pulse lasers[20].

Despite the importance of electron–phonon interactions to the optoelectronic properties of these materials, currently no clear picture has emerged of which mechanisms are active. To address this issue, a number of studies have examined the temperature dependence of the charge-carrier mobility μ, which was found[21–24] to scale with T^m with m in the range between -1.4 and -1.6. Several groups[16,17,24] therefore proposed that electron–phonon coupling at room temperature is almost solely governed by deformation potential scattering with acoustic phonons, which is known[18,25] to theoretically result in $\mu \propto T^{-3/2}$. Although such behaviour may be adopted by non-polar inorganic semiconductors such as silicon or germanium[18,26], it would be extremely unusual for perovskites that exhibit polar[27,28] lead–iodide bonds. These findings have therefore raised the puzzling question of why such hybrid perovskites appear to evade the Fröhlich interactions between charge carriers and polar longitudinal optical (LO) phonon modes that normally govern polar inorganic semiconductors, such as GaAs[18,29], at room temperature.

Here we clarify the relative activity of different charge-carrier scattering mechanisms in hybrid lead halide perovskites by investigating charge-carrier scattering through an analysis of the photoluminescence (PL) linewidth as a function of temperature between 10 and 370 K. By carefully examining the low-temperature regime in which thermal energies fall below those of high-energy LO phonon modes, we are able to clearly separate competing contributions from charge-carrier interactions with acoustic and optical phonons. We are therefore able to show unambiguously that Fröhlich coupling to LO phonons is the predominant cause of linewidth broadening in these materials at room temperature, with scattering from acoustic phonons and impurities being a minor component. We further demonstrate excellent agreement between the experimentally determined temperature dependence of the PL linewidth and theoretical values derived from ab initio calculations for MAPbI$_3$. To elucidate how charge-carrier–phonon interaction strengths depend on perovskite composition, we examine FAPbI$_3$, FAPbBr$_3$, MAPbI$_3$ and MAPbBr$_3$, which represent a comprehensive set of the most commonly implemented organic and halide ingredients in hybrid perovskites. We show that although the choice of organic cation has relatively little effect on the Fröhlich interactions, bromide perovskites exhibit higher Fröhlich coupling than iodide perovskites as a result of their smaller high-frequency values of the dielectric function. Overall, our results conclusively demonstrate that electron–phonon coupling in hybrid lead halide perovskites follows a classic bandstructure picture for polar inorganic semiconductors, which are dominated by Fröhlich coupling between charge carriers and LO phonon modes in the high-temperature regime.

Results

Temperature-dependent PL spectra. To conduct our analysis of electron–phonon coupling in hybrid lead halide perovskites, we recorded steady-state PL spectra of solution-processed FAPbI$_3$, FAPbBr$_3$, MAPbI$_3$ and MAPbBr$_3$ thin films over temperatures from 10 to 370 K in increments of 5 K (Fig. 1). The observed PL peak positions at room temperature are consistent with those reported previously for these materials[6,30–33]. The colour plots in Fig. 1 exhibit abrupt shifts in PL peak energies at various temperatures that are associated with phase transitions commonly found in these relatively soft materials. For example, MAPbI$_3$, MAPbBr$_3$ and FAPbI$_3$ have been reported to undergo a phase transition from an orthorhombic to a tetragonal structure between 130 and 160 K (refs 30,33,34). A further phase transition to the cubic phase follows at higher temperatures for the MA perovskites (at ~330 K for MAPbI$_3$ (refs 30,34) and 240 K for MAPbBr$_3$ (ref. 34)), whereas a transition to a trigonal phase occurs at around 200 K for FAPbI$_3$ (ref. 30). The lower-temperature phase transition to the orthorhombic phase below 130–160 K is generally associated with larger energetic shifts as it marks a strong reduction in the extent of rotational freedom of the organic cation[34–36]. Higher-temperature structural changes are more subtle in this regard and therefore much harder to discern[21]. Apart from these discontinuities, the PL peak in all four materials shifts continuously towards higher energy with increasing temperature. This general trend is in contrast to that of typical semiconductors such as Si, Ge and GaAs, for which the bandgap decreases with temperature as a result of lattice dilation[26,37]. The atypical positive bandgap deformation potential of hybrid lead halide perovskites has been attributed to a stabilization of out-of-phase band-edge states as the lattice expands[38].

In addition to these well-understood temperature trends, the PL spectra of MA-containing perovskites exhibit strong inhomogeneous broadening and multi-peak emission in the

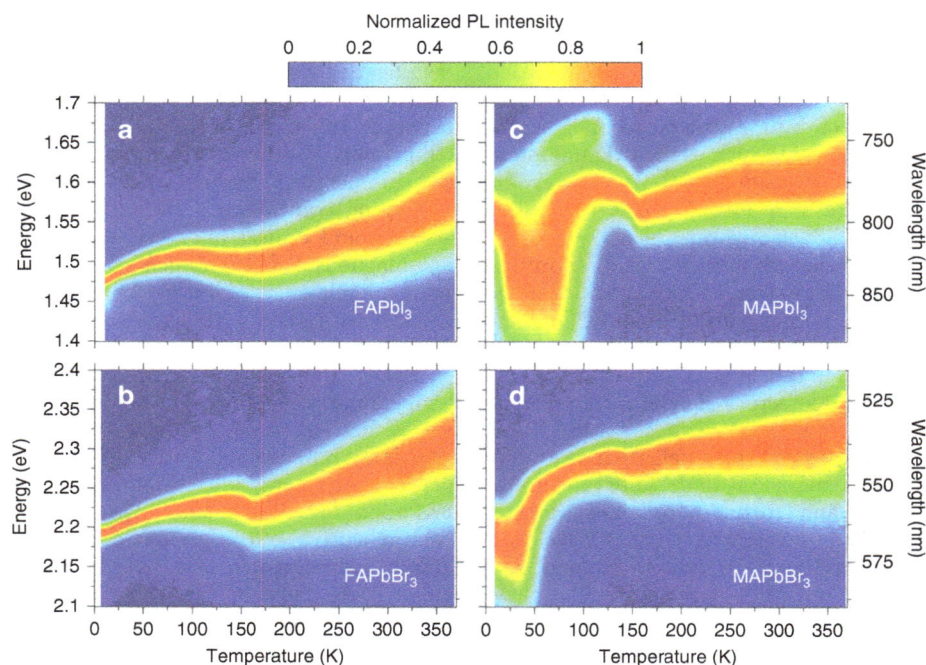

Figure 1 | Temperature dependence of steady-state PL. Colour plots of normalized steady-state photoluminescence spectra of (**a**) FAPbI₃, (**b**) FAPbBr₃, (**c**) MAPbI₃ and (**d**) MAPbBr₃ thin films at temperatures between 10 and 370 K.

low-temperature orthorhombic phase. Such behaviour has been reported by ourselves[21,39] and others[24,31,40,41] on many occasions, yet a precise explanation is still outstanding. For example, the PL spectra of MAPbI₃ develop additional peaks at temperatures below 150 K, as can be seen clearly in Supplementary Fig. 1, which shows spectra from the colour plots in Fig. 1 at selected temperatures. The emerging consensus is that these are caused by additional charge or exciton trap states that are only active in the low-temperature orthorhombic phase[21,24,31,39–41]. It has been proposed that these traps could derive from a small fraction of inclusions of the room-temperature tetragonal phase that are populated through charge or exciton transfer from the majority orthorhombic phase[39]. These inclusions could be a result of strain or the proposed impossibility of a continuous structural transition from the tetragonal to orthorhombic phase in MA perovskites, as reported by Baikie et al.[35]. We further note that as these complicating features do not appear in the PL spectra of the equivalent FA perovskites, they are probably not intrinsic to hybrid lead halide perovskites, which re-affirms the prevailing view that they originate from trap states. Importantly, the lack of uncomplicated low-temperature spectra has to date prevented proper analysis of the linewidth broadening of MA perovskites and the associated phonon coupling to charge carriers. As we show below, such analysis requires access to a low-temperature range over which PL spectra are dominated by intrinsic phonon broadening, rather than trap-related PL, for the contributions from acoustic and optical phonons to the clearly separated. Therefore, the discovery that the equivalent FA lead halide perovskites do not exhibit extrinsic defect-related PL in the low temperature range allows us to carry out such analysis unhindered.

Analysis of PL linewidth broadening. Analysis of temperature-dependent emission broadening has long been used to assess the mechanisms of electron–phonon coupling in a wide range of inorganic semiconductors[42] (see Supplementary Table 1 for a

literature overview of the results). We here apply these methods to hybrid lead halide perovskites by first extracting the full width at half-maximum (FWHM) of the PL spectra shown in Fig. 1 and then analysing its temperature dependence (plotted in Fig. 2). For most inorganic semiconductors, different mechanisms of scattering between charge carriers and phonons or impurities are associated with different functional dependencies of the PL linewidth $\Gamma(T)$ on temperature, which can be expressed as the sum over the various contributions[42,43]:

$$\begin{aligned} \Gamma(T) &= \Gamma_0 + \Gamma_{ac} + \Gamma_{LO} + \Gamma_{imp} \\ &= \Gamma_0 + \gamma_{ac}T + \gamma_{LO}N_{LO}(T) + \gamma_{imp}e^{-E_b/k_BT}. \end{aligned} \quad (1)$$

Here, Γ_0 is a temperature-independent inhomogeneous broadening term, which arises from scattering due to disorder and imperfections[42,44]. The second and third terms (Γ_{ac} and Γ_{LO}) are homogeneous broadening terms, which result from acoustic and LO phonon (Fröhlich) scattering[28,42,44] with charge-carrier–phonon coupling strengths of γ_{ac} and γ_{LO}, respectively. Electron–phonon coupling is in general proportional to the occupation numbers of the respective phonons, as given by the Bose–Einstein distribution function[45,46], taken as $N_{LO}(T) = 1/[e^{E_{LO}/k_BT} - 1]$ for LO phonons, where E_{LO} is an energy representative of the frequency for the weakly dispersive LO phonon branch[18,47]. For acoustic phonons whose energy is much smaller than k_BT over the typical observation range, a linear dependence on temperature is generally assumed[46,48]. Ab initio calculations of the relevant phonon energies and occupation numbers are shown in Supplementary Fig. 4, which confirms that the linear approximation to the acoustic phonon population used in equation (1) is appropriate. The final term, Γ_{imp}, phenomenologically accounts for scattering from ionized impurities with an average binding energy E_b (ref. 43). These impurities contribute γ_{imp} of inhomogeneous broadening to the width when fully ionized[43,45].

In general, the two major mechanisms governing the electron–phonon coupling in inorganic semiconductors are deformation potential scattering, in which distortions of the lattice change the electronic band structure, and electromechanical or piezoelectric

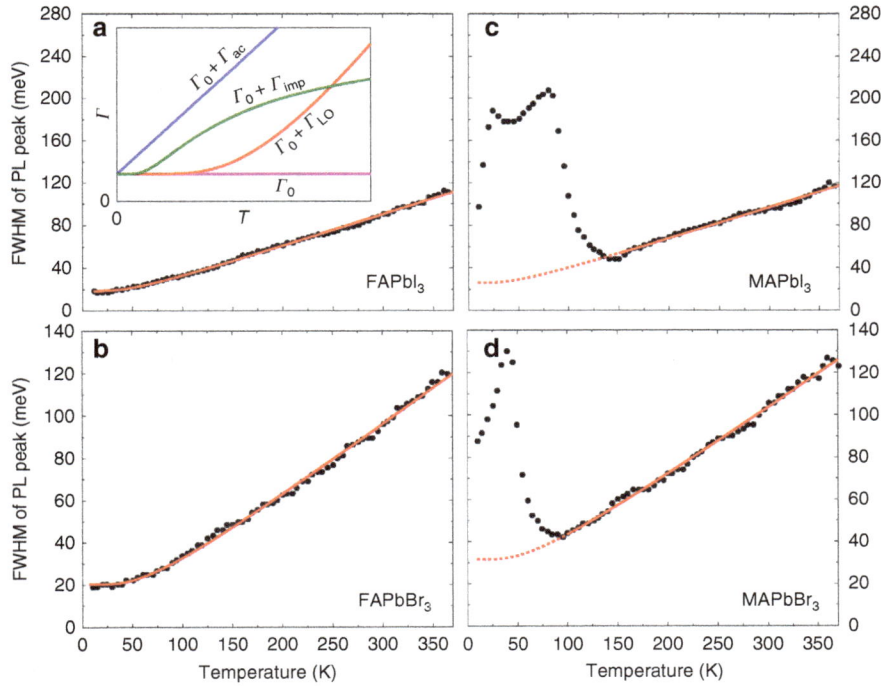

Figure 2 | Temperature dependence of linewidth. FWHM of the steady-state PL spectra as a function of temperature for (**a**) FAPbI$_3$, (**b**) FAPbBr$_3$, (**c**) MAPbI$_3$ and (**d**) MAPbBr$_3$ thin films plotted as black dots. The solid red lines are fits of $\Gamma(T) = \Gamma_0 + \Gamma_{LO}$, which account for contributions from inhomogeneous broadening and Fröhlich coupling with LO phonons. For the perovskites containing MA, the fits are extrapolated into the low-temperature region in which the model does not hold, as indicated by dashed red lines (actual fits were carried out between 150 and 370 K for MAPbI$_3$, and between 100 and 370 K for MAPbBr$_3$). The inset shows the functional form of the temperature dependence of the contributions to PL linewidth in semiconductors from inhomogeneous broadening (Γ_0, magenta), Fröhlich coupling between charge carriers and LO phonons (Γ_{LO}, red) and acoustic phonons (Γ_{ac}, blue), and scattering from ionized impurities (Γ_{imp}, green), as given by the terms of equation (1). An alternative presentation of the linewidths as a mutliple of the thermal energy is given in Supplementary Fig. 2.

interactions, in which lattice-related electric fields modify the electronic Hamiltonian[18]. Specifically, the LO phonon term in equation (1) accounts for the Fröhlich interaction between LO phonons and electrons, which arises from the Coulomb interaction between the electrons and the macroscopic electric field induced by the out-of-phase displacements of oppositely charged atoms caused by the LO phonon mode[18]. Although both transverse optical and LO phonons interact with electrons via non-polar deformation potentials, equation (1) only accounts for LO phonons because of the dominant influence of their Fröhlich interaction with electrons in polar crystals at higher temperatures[49]. As optical phonons in semiconductors typically have energies of the order of tens of meV (ref. 18), their population at low temperatures ($T < 100$ K) is very small; thus, homogeneous broadening in this regime predominantly results from acoustic phonons[18,49]. Therefore, careful examination of the low-temperature regime allows separation of the contributions from optical and acoustic modes. Long-wavelength acoustic phonons induce atomic displacements, which can correspond to macroscopic crystal deformation, affecting electronic energies via either the resultant deformation potential or a piezoelectrically induced electric field[18].

To establish qualitatively which electron–phonon scattering mechanisms contribute in hybrid lead halide perovskites, we compare the temperature-dependent PL linewidth plotted in Fig. 2 with the functional form of the terms in equation (1). To aid comparison, the inset to Fig. 2a shows example functions for the separate components. First, we assess the possibility of electron scattering with ionized impurities playing a significant role. Comparison of the curves in the inset with the data in the main Fig. 2 makes it apparent that the shape of the ionized

impurity scattering term Γ_{imp} could not produce the observed linear variation with T of the linewidths at high temperatures. Therefore, we conclude that scattering with ionized impurities does not play any major role here, in agreement with findings based on recent analyses of the temperature-dependence of the charge-carrier mobility in this regime[21–24]. We therefore assume $\Gamma_{imp} \approx 0$ for the rest of the analysis.

To separate the contributions from acoustic and optical phonon modes, we first focus on an analysis of the PL linewidth for perovskites containing FA as the organic cation. As Fig. 2a,b show, these materials exhibit smooth variation of the linewidth, whereas for MA-containing perovskites the presence of the impurity emission discussed above leads to additional emission broadening in the low-temperature phase (Fig. 2c,d). Both FAPbI$_3$ and FAPbBr$_3$ approach a PL linewidth of the order of 20 meV towards $T = 0$, which can therefore be identified as the temperature-independent inhomogeneous broadening term Γ_0 arising from disorder. To qualitatively assess the relative importance of acoustic versus optical phonon contributions, an inspection of the gradient of these curves in the low-temperature regime is essential. Although the optical phonon terms lead to a gradient of zero in the regime for which $E_{LO} < k_B T$, the smaller energies of acoustic phonons should result in a non-zero gradient given by γ_{ac} here. However, visual inspection of the graphs in Fig. 2a,b shows that the gradient of the FWHM versus temperature approaches zero at low temperature, suggesting negligible acoustic phonon contribution ($\gamma_{ac} \approx 0$). Indeed, we find that fits of equation (1) to these curves converge with $\gamma_{ac} \to 0$. This result is not surprising, given that in polar inorganic semiconductors the contribution of acoustic phonons to the broadening at room temperature is typically dwarfed by that of the LO phonons and indeed several studies ignore the contribution of acoustic phonons in such systems[49]. However, we may obtain an upper limit to γ_{ac} by careful examination of the data in the low-temperature regime in which acoustic phonons are still expected to contribute significantly. Here we may fit $\Gamma(T) = \Gamma_0 + \gamma_{ac}T$ to the data in the low-temperature ($T < 60$ K) region[50] or, as an alternative method, obtain the gradient of the data near $T = 0$ K from differentiation. Both methods yield upper limits around $\gamma_{ac} = 60 \pm 20$ μeV K^{-1} for FAPbI$_3$ and FAPbBr$_3$; therefore, acoustic phonons will only contribute up to ~ 18 meV to the linewidth broadening at 300 K. Hence, our analysis illustrates that the majority of broadening in the room-temperature regime arises from charge-carrier interactions with optical phonons, as would be expected for a polar semiconductor.

To further quantify the dominant Fröhlich coupling in these systems, we proceed by fitting all data including only the mechanisms based on temperature-independent inhomogeneous broadening and Fröhlich coupling to LO phonon modes. For perovskites containing FA cations, fits of $\Gamma(T) = \Gamma_0 + \Gamma_{LO}$ to the PL linewidth data are plotted in Fig. 2a,b (red lines) and the extracted fitting parameters are presented in Table 1. Apart from Fröhlich coupling strengths, we are also able to determine the energy of LO phonon modes that play the dominant role in electron–LO-phonon coupling. We find $E_{LO} = 11.5$ meV for FAPbI$_3$, with the value for FAPbBr$_3$ (15.3 meV) being 1.3 times larger, which is only slightly greater than the factor of 1.2 expected from a crude model of the frequency of a diatomic harmonic oscillator. These LO phonon energies of hybrid lead halide perovskites are somewhat lower than those typically measured for a range of inorganic semiconductors (see Supplementary Table 1 for an overview); however, they agree well with a recent combined experimental and density functional theory (DFT) study assigning LO phonon modes of the Pb-I lattice in MAPbI$_3$ with energies near 10 meV (ref. 51) and with our present calculations (see next section).

As noted above, the situation is more complex for MA perovskites, owing to the additional trap-related emission in the low-temperature orthorhombic phase that gives rise to sizeable (>100 meV) additional broadening (see Fig. 2c,d). However, in the high-temperature regime, PL linewidths of the MA perovskites vary much in the same manner as that of their FA counterparts, suggesting very similar mechanisms. This may be expected, as the organic cation has relatively little influence on the vibrations of the lead-halide lattice. Hence, we model the linewidth broadening of MA perovskites in the high-temperature regime again using $\Gamma(T) = \Gamma_0 + \Gamma_{LO}$ (solid red lines in Fig. 2c,d), using the LO phonon energies determined previously for FA perovskites. The resultant γ_{LO} values are, as expected, very

similar to those for the corresponding FA perovskites (Table 1). We may also extrapolate these fits down through the low-temperature regime (dashed red lines), where they do not reflect experimental reality but rather show the broadening that would be present if the additional defects in the orthorhombic phase were absent. As such, values extracted for the parameter Γ_0 here mostly reflect inhomogeneous disorder present in the high-temperature phases of MAPbI$_3$ and MAPbBr$_3$ at temperatures above 150 and 100 K, respectively.

Finally, we comment on the extent to which excitonic effects may influence the coupling between charge carriers and phonons. Although the exact values for the exciton-binding energies in these systems are still a matter of debate, most reported values fall into the range of a few to a few tens of milli-electronvolts (see ref. 4 for a review). These values are compatible with numerous studies demonstrating that at room temperature, following non-resonant excitation, hybrid lead halide perovskites sustain free charge carriers as the predominant species[13,14,52]. Excitonic effects in the generated charge-carrier population are expected to increase as the temperature is lowered. However, we have recently examined infrared photoinduced transmission spectra for methylammonium lead iodide perovskite[21] and found these to be predominantly governed by free-charge (Drude-like) features, with localization effects (for example, from excitons) only contributing at low temperature and not more than around 23% even at 8 K. This suggests that over the temperature window we examine here, emission broadening is mostly governed by interactions between phonons and free charge carriers rather than excitons. Such predominantly free-charge behaviour of the photogenerated species may be understood, for example, in terms of the Saha equation or a low Mott density for these systems, and considering that the initial highly non-resonant excitation generates predominantly free electron–hole pairs[12]. We may further inspect in more detail the temperature-dependent line shapes of the emission spectra and find these to exhibit high-energy Boltzmann tails corresponding to a thermalized electron–hole density near the lattice temperature (see Supplementary Fig. 3). These overall observations are therefore compatible with the presence of a thermalized free electron–hole charge-carrier density that scatters off mostly LO phonons whose occupancy is governed by the Bose–Einstein distribution function. Analysis of how the energies of such thermalized free electrons and holes influence the precise lineshape of the PL spectra as a function of temperature could provide additional insight into the scattering mechanisms for free charge carriers. Such additional analysis is however beyond the scope of the present investigation, which is limited to considering only the extent of linewidth broadening.

Table 1 | Extracted linewidth parameters.

Sample	Γ_0/meV	γ_{LO}/meV	E_{LO}/meV
FAPbI$_3$	19 ± 1	40 ± 5	11.5 ± 1.2
FAPbBr$_3$	20 ± 1	61 ± 7	15.3 ± 1.4
MAPbI$_3$	26 ± 2	40 ± 2	[11.5]
MAPbBr$_3$	32 ± 2	58 ± 2	[15.3]

FA, formamidinium; LO, longitudinal optical; MA, methylammonium.
Linewidth broadening parameters extracted from fits of $\Gamma(T) = \Gamma_0 + \Gamma_{LO}$ to the PL linewidth data for the four hybrid perovskite films. Γ_0 is the inhomogeneous broadening (the linewidth at 0 K), γ_{LO} is the strength of the LO phonon-charge-carrier Fröhlich coupling and E_{LO} is the relevant LO phonon energy. For fits to the data from MA-containing perovskites, the values of E_{LO} extracted previously for the FA-containing perovskites were used (hence, are italicized and enclosed in square brackets). Because of the additional defect luminescence present in the orthorhombic phase of the MA-containing perovskites, fits were only carried out between 150 and 370 K for MAPbI$_3$, and between 100 and 370 K for MAPbBr$_3$ (see solid lines in Fig. 2c,d). Therefore, the extracted parameters do not reflect the lineshape broadening in the low-temperature phase of the MA-containing perovskites.

First-principles calculations. We further corroborate our analysis with first-principles calculations of the electron–phonon coupling in MAPbI$_3$ (see details in Methods section and in the Supplementary Note 1). In accordance with the above discussion, we separately consider the broadening arising from the interaction of phonons with free conduction-band electrons and free valence-band holes[53]. The combined broadening arising from both types of charge carrier is then compared with the experimentally determined emission broadening. In Fig. 3a, we present a heat-map of the imaginary part of the electron–phonon self-energy, Im(Σ), projected on the quasiparticle band structure of MAPbI$_3$; 2 Im(Σ) represents the linewidth of electrons and holes arising from the electron–phonon interaction before accounting for many-body effects (see Methods) and is therefore directly comparable to the experimentally determined FWHM of the PL emission linewidth. Figure 3b shows $Im(\Sigma)$ as a function of electron energy, together with the density of electronic states. This figure indicates that the increase in the linewidth is linked to the phase-space availability for electronic transitions, that is, $Im(\Sigma)$ increases with increasing density of electronic states, because each state can scatter into a higher number of states by absorbing or emitting a phonon. Our calculations show that the dominant contribution to the electron–phonon self-energy arises from the coupling with the LO mode at $\omega_{LO} \approx 13$ meV, which is shown schematically in Fig. 3c. This observation is compatible with the analysis of the temperature dependence of the PL broadening, as presented in Fig. 2, yielding similar energy for the predominantly coupling LO phonon mode in MAPbI$_3$.

Figure 3d shows that our calculated temperature dependence of the PL broadening is in good agreement with experiment. In this figure we compared our calculations with the experimental trends obtained from the fit shown in Fig. 2 (which does not account for the anomalous broadening below 150 K, as discussed above). The blue triangles represent the data calculated based on Fermi's golden rule, which is equivalent to using the imaginary part of the electron–phonon self-energy, $2 Im(\Sigma)$, whereas the red triangles were obtained by using Brillouin–Wigner perturbation theory[54], which corresponds to scaling the self-energy by the quasiparticle renormalization factor Z, that is $2 Z Im(\Sigma)$ (see Methods). The

comparison between calculations and experiments shows that the experimental data are most accurately described by fully taking into account the many-body renormalization of the electron lifetime. Taken together, the impressive agreement between (1) our measured and calculated characteristic phonon energy scale (11.5 and 13 meV, respectively), (2) the magnitude of our measured and calculated linewidth broadening at room temperature (90 and 75 meV, respectively), and (3) the phonon energy scale identified here and the LO phonon identified between 10 and 13 meV in our previous study[51], strongly support the notion that the broadening of the PL spectra reflects the interaction between free carriers and LO phonons.

In addition, we may use first-principle calculations to elucidate why the lead bromide perovskites exhibit Fröhlich coupling constants that are larger than those of the lead iodide system by a factor of 1.5. To investigate this trend, we performed DFT calculations for MAPbBr$_3$ (see Supplementary Note 1) to compare the associated Born effective charges between the two systems. Our results indicate that the electron–phonon coupling in the bromide perovskite is 40% stronger than in the iodide perovskite, in excellent agreement with our measurements. We find that the increased Fröhlich coupling in MAPbBr$_3$ is primarily connected with the smaller high-frequency value of the dielectric function compared with MAPbI$_3$.

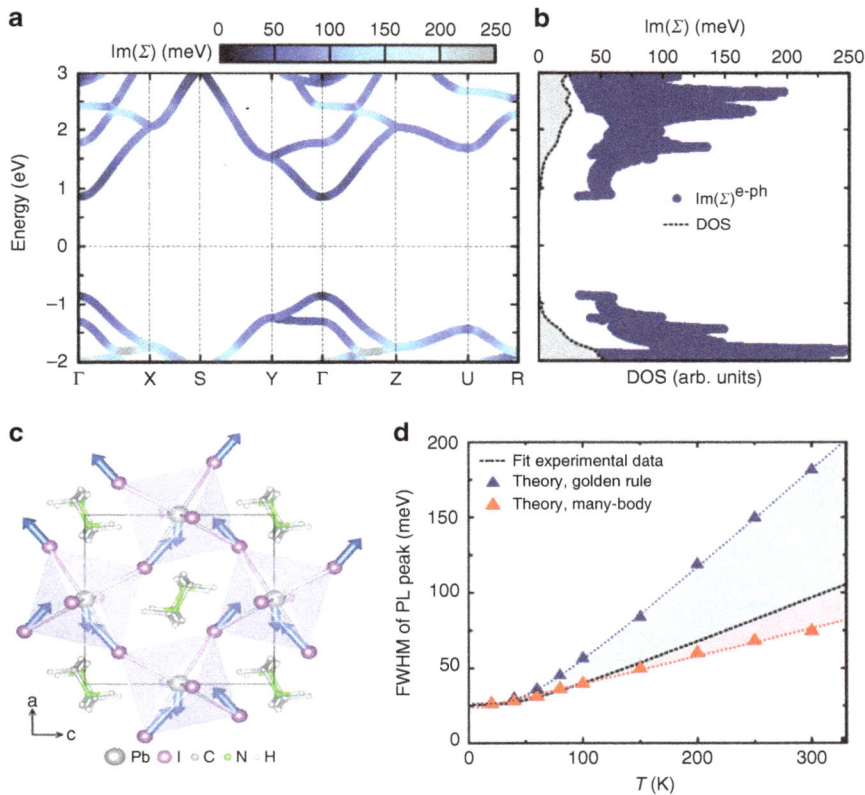

Figure 3 | *Ab initio* calculations of electron–phonon coupling and PL broadening in MAPbI$_3$. (a) Electronic band structure of orthorhombic MAPbI$_3$, calculated within the GW approximation as in ref. 55, combined with a heat map of the imaginary part of the electron–phonon self-energy ($Im(\Sigma)$) at $T = 200$ K. The zero of the energy is placed in the middle of the bandgap. In **b**, the imaginary part of the electron–phonon self-energy is shown together with the electronic density of states (DOS). $2 Im(\Sigma)$ corresponds to the electron/hole linewidth arising from electron–phonon coupling (apart from the quasiparticle renormalization factor Z). (**c**) Ball-and-stick representation of the LO vibration responsible for the broadening of the PL peaks. The blue arrows indicate the displacements of Pb and I atoms, for the phonon wavevector $\mathbf{q} \rightarrow 0$ along the [100] direction. This is a Pb–I stretching mode with B_{3u} symmetry[51]. (**d**) Temperature dependence of the FWHM of the PL peak in MAPbI$_3$: fit to the experimental data (dashed black line) and theoretical calculations using Fermi's golden rule (blue triangles) and the more accurate Brillouin–Wigner perturbation theory (red triangles). The theoretical broadening is obtained as the sum of $2 Im(\Sigma)$ at the valence and conduction band edges in the case of Fermi's golden rule, and the sum of $2 Z Im(\Sigma)$ when including many-body quasiparticle renormalization, rigidly shifted by the FWHM at $T = 0$ K (25.72 meV), to account for inhomogenous broadening. The lines are guides to the eye.

Discussion

Our analysis of the experimentally determined emission linewidth broadening and our first-principles calculations strongly support the notion that Fröhlich coupling to LO phonons is the predominant charge-carrier scattering mechanism in hybrid lead halide perovskites. As already discussed above, such behaviour is in many ways to be expected for these materials, because the lead-halide bond is sufficiently polar (see our calculated Born effective charges in Supplementary Table 2) to lead to macroscopic polarizations from LO phonon modes that modify the electronic energies, causing electron–phonon scattering. Coupling of charge carriers to phonons with energies in the meV range has also been postulated from the signature of such modes in the room- and low-temperature photoconductivity spectra[21,56]. However, the predominance of Fröhlich coupling appears at first sight to contradict the measured[21–24] temperature dependence of the charge-carrier mobility, which has been stated[16,17,24] to approach the expected form for acoustic deformation potential scattering ($\mu \propto T^{-3/2}$). We note, however, that although electron–phonon coupling generally leads to charge-carrier mobilities that increase with decreasing temperature, the exact functional dependence is usually a composite of many different scattering mechanisms that can be hard to attribute uniquely[18]. Charge-carrier scattering with low-energy acoustic phonons can relatively easily be quantified by considering the distribution function of a nondegenerate electron gas approximated by a Boltzmann distribution, which gives the probability that a particular state with energy E_k is occupied at any temperature T. Such calculations result in predicted variations in mobility following $\mu \propto T^{-3/2}$ for acoustic phonon deformation potential scattering[25,26] and $\mu \propto T^{-1/2}$ for acoustic phonon piezoelectric scattering[18]. For Fröhlich interactions between charges and LO phonons, analytical solutions are harder to establish[57], but the reduction in LO phonon occupancy with decreasing temperature similarly leads to an increase in charge-carrier mobility in polar semiconductors[18]. Even for non-polar semiconductors such as silicon, electron–phonon interactions are found to be complex, for example, involving higher-order phonon terms and intervalley scattering[58]. Therefore, we conclude that although acoustic phonon deformation potential scattering may result in $\mu \propto T^{-3/2}$, the converse may not necessarily also hold. In addition, the rapid energy loss of electrons observed following non-resonant excitation[19,59] can only sensibly be explained by a succession of high-energy optical phonon emissions[60], as is typically observed in inorganic semiconductors[4,61]. Therefore, we conclude that Fröhlich coupling to LO phonons, rather than acoustic phonon deformation potential coupling, is the dominant charge-carrier scattering mechanism at room temperature in these hybrid lead halide perovskites. Although our findings themselves do not explain the temperature dependence of the charge-carrier mobility in these materials, they support the hypothesis that the observed $\mu \propto T^{-3/2}$ relationship is not wholly attributable to acoustic deformation potential scattering. It is also clear that for these high-quality materials, scattering from ionized impurities is a negligible component at room temperature, with both our PL linewidth data and earlier charge-carrier mobility measurements[21–23], indicating a complete absence of such contributions.

In addition, our findings give early answers to the question of how perovskite composition affects Fröhlich interactions between charge carriers and phonons. Such interactions determine the maximum charge-carrier mobilities intrinsically attainable, which in turn affects charge-carrier extraction in solar cells. We show that Fröhlich coupling in hybrid lead bromide perovskites appears to be stronger because of the lower dielectric function in the high-frequency regime. Indeed, the THz charge-carrier mobility for FAPbBr$_3$ thin films has recently been shown to be somewhat lower than that for FAPbI$_3$ films[10], which could be related to decreased momentum scattering time resulting from increased scattering with LO phonons. However, it may also be partly related to a higher propensity towards disorder in the bromide perovskites, as the inhomogeneous broadening parameter Γ_0 appears to be somewhat higher for bromide than iodide perovskites here. Similarly, higher Urbach energies have previously been reported for MAPbBr$_3$ compared with MAPbI$_3$, in accordance with larger energetic disorder in the former[62]. We also find that Γ_0 tends to be lower for the perovskites containing FA as the organic cation, which points to larger material uniformity as one reason behind this material's recent success in the highest efficiency perovskite solar cells[2,7].

We may also compare these findings with those of other polar inorganic semiconductors for which Fröhlich coupling is known to be active. Supplementary Table 1 provides a detailed literature survey of values established for γ_{LO} in other inorganic semiconductors. It has been pointed out[17] that charge-carrier mobilities established for lead halide perovskites (typically ≤ 100 cm^2 (V s)$^{-1}$)[15,17] are relatively modest compared with those achieved in high-quality GaAs despite the effective charge-carrier masses in perovskites being only slightly elevated above those in GaAs. The values of $\gamma_{LO} \approx 40$ meV and $\gamma_{LO} \approx 60$ meV we extract from our data for the respective iodide and bromide perovskites (see Table 1) are somewhat higher than the range reported for GaAs (see comparison in Supplementary Table 1), which may partly explain these discrepancies. However, they are significantly lower than those typically found in highly polar materials such as GaN and ZnO where they can be over an order of magnitude higher[28]. Further theoretical modelling of charge-carrier mobility in these systems based on our findings will most probably allow more quantitative explanations and predictions to be made.

In summary, we have conducted an in-depth analysis of charge-carrier–phonon interactions in hybrid lead halide perovskites, considering the four currently most implemented organic and halide components in hybrid perovskite photovoltaics, which are FAPbI$_3$, FAPbBr$_3$, MAPbI$_3$ and MAPbBr$_3$. Our analysis of the temperature-dependent emission linewidth of FAPbI$_3$ and FAPbBr$_3$ allowed us to establish that the Fröhlich interaction between charge carriers and LO phonons provides the dominant contribution to the predominantly homogeneous linewidth broadening in these hybrid perovskites at room temperature. We successfully corroborated our findings with DFT and many-body perturbation theory calculations, which underline the suitability of an electronic bandstructure picture for describing charge carriers in perovskites. We furthermore obtained experimentally measured energies of LO phonon modes responsible for Fröhlich interactions in these materials and showed that Fröhlich interactions are higher for bromide perovskites than iodide perovskites, providing a link between composition and electron–phonon scattering that fundamentally limits charge-carrier motion. These results lay the groundwork for more quantitative models of charge-carrier mobility values and cooling dynamics that underpin photovoltaic device operation.

Methods

Sample preparation. All materials, unless otherwise stated, were purchased from Sigma-Aldrich and were used as received. Methylammonium iodide, methylammonium bromide, formamidinium iodide and formamidinium bromide were purchased from Dyesol. Thin films were prepared on Z-cut quartz substrates. These were initially cleaned sequentially with acetone followed by propan-2-ol, then treated with oxygen plasma for 10 min.

MA perovskite films were deposited in a nitrogen-filled glovebox using a solvent quenching method wherein an excess of antisolvent is deposited onto the wet substrate while spin-coating[63]. A 1:1 molar ratio solution of MAX and PbX_2 ($X = I$, Br) was dissolved in anhydrous N,N-dimethylformamide at 1 M. This was then spin-coated onto the quartz substrates at 5,000 r.p.m. for 25 s. During spin coating, after 7 s an excess of anhydrous chlorobenzene was rapidly deposited onto the spinning film. After spin-coating, films were annealed at 100 °C for 10 min.

FA perovskite films were deposited using an acid-addition method to produce smooth and uniform pinhole-free films[6]. FAX and PbX_2 ($X = I$, Br) were dissolved in anhydrous N,N-dimethylformamide in a 1:1 molar ratio at 0.55 M. Immediately before film formation, small amounts of acid were added to the precursor solutions, to enhance the solubility of the precursors and allow smooth and uniform film formation. Thirty-eight microlitres of hydroiodic acid (57% mass/mass) was added to 1 ml of the 0.55 M $FAPbI_3$ precursor solution and 32 μl of hydrobromic acid (48% mass/mass) was added to 1 ml of the 0.55 M $FAPbBr_3$ precursor solution. Films were then spin coated from the precursor plus acid solution on warm (85 °C) oxygen plasma-cleaned substrates at 2,000 r.p.m. in a nitrogen-filled glovebox and subsequently annealed in air at 170 °C for 10 min.

PL spectroscopy. Each sample was photoexcited by a 398-nm picosecond pulsed diode laser (PicoHarp, LDH-D-C-405M) with a repetition rate of 10 kHz and a fluence of 490 nJ cm^{-2}. The resultant PL was collected and focused into a grating spectrometer (Princeton Instruments, SP-2558), which directed the spectrally dispersed PL onto an iCCD (PI-MAX4, Princeton Instruments). The sample was mounted under vacuum ($P < 10^{-6}$ mbar) in a cold-finger liquid helium cryostat (Oxford Instruments, MicrostatHe). An associated temperature controller (Oxford Instruments, ITC503) monitored the temperature at two sensors mounted on the heat exchanger of the cryostat and the end of the sample holder, respectively; the reading from the latter was taken as the sample temperature. PL measurements were taken as the sample was heated in increments of 5 K between 10 and 370 K.

Computational methods. We carried out *ab initio* calculations on $MAPbI_3$ in the orthorhombic phase using the crystallographic data in ref. 35. The ground-state electronic structure was computed within the local density approximation to DFT including spin–orbit coupling, as implemented in the Quantum ESPRESSO package[64]. We used norm-conserving pseudopotentials to describe the core–valence interaction, with the semicore d states taken explicitly into account in the case of Pb and I. Ground-state calculations were converged with a plane-wave cutoff of 100 Ry and a $6 \times 6 \times 6$ unshifted Brillouin-zone grid. The electronic quasiparticle energies were calculated with the SS-GW method described in ref. 65 using the Yambo code[66] and interpolated by means of Wannier functions as in ref. 55, using wannier90 (ref. 67). This yields bandgap and effective masses in good agreement with experiment (see Supplementary Note 1). The lattice dynamical properties were computed within density functional perturbation theory at the Γ point, as in ref. 51. The LO–TO splitting was included through the evaluation of the non-analytic contribution to the dynamical matrix. The electron–phonon coupling was calculated using the EPW code[68,69], v.4. The electron–phonon self-energy, $\Sigma_{n\mathbf{k}}$, was calculated as:

$$\Sigma_{n\mathbf{k}} = \sum_{m\nu\mathbf{q}} \left| g_{mn}^{\nu}(\mathbf{k},\mathbf{q}) \right|^2 \left[\frac{n_{\mathbf{q}\nu} + f_{m\mathbf{k}+\mathbf{q}}}{\epsilon_{n\mathbf{k}} - \epsilon_{m\mathbf{k}+\mathbf{q}} + \omega_{\mathbf{q}\nu} - i\eta} + \frac{n_{\mathbf{q}\nu} + 1 - f_{m\mathbf{k}+\mathbf{q}}}{\epsilon_{n\mathbf{k}} - \epsilon_{m\mathbf{k}+\mathbf{q}} - \omega_{\mathbf{q}\nu} - i\eta} \right].$$

(2)

Here, $f_{m\mathbf{k}+\mathbf{q}}$ and $n_{\mathbf{q}\nu}$ are the Fermi–Dirac and the Bose–Einstein occupations, respectively, $\epsilon_{n\mathbf{k}}$ and $\epsilon_{m\mathbf{k}+\mathbf{q}}$ are electron energies, $\hbar\omega_{\mathbf{q}\nu}$ is the energy of a phonon with wavevector $\mathbf{q}$ and polarization ν, and η is a small broadening (10 meV in Fig. 3a,b; 1 meV in Fig. 3d). Only the interaction with the LO phonons was taken into account, by calculating the electron–phonon matrix element $g_{mn}^{\nu}(\mathbf{k},\mathbf{q})$ as in ref. 70 (see Supplementary equation (1)). The self-energy in equation (2) was converged by using up to two million random $\mathbf{q}$ points in the Brillouin zone. The quasiparticle renormalization factor Z is defined as the frequency derivative:

$$Z_{n\mathbf{k}} = \left[1 - \frac{1}{\hbar} \frac{\partial \, \mathrm{Re}(\Sigma_{n\mathbf{k}})}{\partial \omega} \right]^{-1},$$

(3)

and is evaluated at the band edges. In the case of the conduction band edge, we find $Z = 0.54$ at zero temperature.

References

1. Kojima, A., Teshima, K., Shirai, Y. & Miyasaka, T. Organometal halide perovskites as visible-light sensitizers for photovoltaic cells. *J. Am. Chem. Soc.* **131**, 6050–6051 (2009).
2. Yang, W. S. *et al.* High-performance photovoltaic perovskite layers fabricated through intramolecular exchange. *Science* **348**, 1234–1237 (2015).
3. Bi, D. *et al.* Efficient luminescent solar cells based on tailored mixed-cation perovskites. *Sci. Adv.* **2**, e1501170 (2016).
4. Herz, L. M. Charge carrier dynamics in organic-inorganic metal halide perovskites. *Annu. Rev. Phys. Chem.* **67**, doi:10.1146/annurev-physchem-040215-112222 (2016).
5. McMeekin, D. P. *et al.* A mixed-cation lead mixed-halide perovskite absorber for tandem solar cells. *Science* **351**, 151–155 (2016).
6. Eperon, G. E. *et al.* Formamidinium lead trihalide: a broadly tunable perovskite for efficient planar heterojunction solar cells. *Energy Environ. Sci.* **7**, 982–988 (2014).
7. Jeon, N. J. *et al.* Compositional engineering of perovskite materials for high-performance solar cells. *Nature* **517**, 476–480 (2015).
8. Binek, A., Hanusch, F. C., Docampo, P. & Bein, T. Stabilization of the trigonal high temperature phase of formamidinium lead iodide. *J. Phys. Chem. Lett.* **6**, 1249–1253 (2015).
9. Eperon, G. E., Beck, C. E. & Snaith, H. J. Cation exchange for thin film lead iodide perovskite interconversion. *Mater. Horiz.* **3**, 63–71 (2015).
10. Rehman, W. *et al.* Charge-carrier dynamics and mobilities in formamidinium lead mixed-halide perovskites. *Adv. Mater.* **27**, 7938–7944 (2015).
11. Lee, M. M., Teuscher, J., Miyasaka, T., Murakami, T. N. & Snaith, H. J. Efficient hybrid solar cells based on meso-superstructured organometal halide perovskites. *Science* **338**, 643–647 (2012).
12. D'Innocenzo, V. *et al.* Excitons versus free charges in organo-lead tri-halide perovskites. *Nat. Commun.* **5**, 3586 (2014).
13. Stranks, S., Eperon, G. & Grancini, G. Electron-hole diffusion lengths exceeding 1 micrometer in an organometal trihalide perovskite absorber. *Science* **342**, 341–344 (2013).
14. Wehrenfennig, C., Eperon, G. E., Johnston, M. B., Snaith, H. J. & Herz, L. M. High charge carrier mobilities and lifetimes in organolead trihalide perovskites. *Adv. Mater.* **26**, 1584–1589 (2014).
15. Johnston, M. B. & Herz, L. M. Hybrid Perovskites for photovoltaics: charge-carrier recombination, diffusion, and radiative efficiencies. *Acc. Chem. Res.* **49**, 146–154 (2016).
16. Zhu, X.-Y. & Podzorov, V. Charge carriers in hybrid organic-inorganic lead halide perovskites might be protected as large polarons. *J. Phys. Chem. Lett.* **6**, 4758–4761 (2015).
17. Brenner, T. M. *et al.* Are mobilities in hybrid organic-inorganic halide perovskites actually 'high'? *J. Phys. Chem. Lett.* **6**, 4754–4757 (2015).
18. Yu, P. Y. & Cardona, M. *Fundamentals of Semiconductors. Graduate Texts in Physics* (Springer-Verlag, 2010).
19. Yang, Y. *et al.* Observation of a hot-phonon bottleneck in lead-iodide perovskites. *Nature Photon.* **10**, 53–59 (2015).
20. Wehrenfennig, C., Liu, M., Snaith, H. J., Johnston, M. B. & Herz, L. M. Homogeneous emission line broadening in the organo lead halide perovskite $CH_3NH_3PbI_{3-x}Cl_x$. *J. Phys. Chem. Lett.* **5**, 1300–1306 (2014).
21. Milot, R. L., Eperon, G. E., Snaith, H. J., Johnston, M. B. & Herz, L. M. Temperature-dependent charge-carrier dynamics in $CH_3NH_3PbI_3$ perovskite thin films. *Adv. Funct. Mater.* **25**, 6218–6227 (2015).
22. Oga, H., Saeki, A., Ogomi, Y., Hayase, S. & Seki, S. Improved understanding of the electronic and energetic landscapes of perovskite solar cells: high local charge carrier mobility, reduced recombination, and extremely shallow traps. *J. Am. Chem. Soc.* **136**, 13818–13825 (2014).
23. Savenije, T. *et al.* Thermally activated exciton dissociation and recombination control the carrier dynamics in organometal halide perovskite. *J. Phys. Chem. Lett.* **5**, 2189–2194 (2014).
24. Karakus, M. *et al.* Phonon-electron scattering limits free charge mobility in methylammonium lead iodide perovskites. *J. Phys. Chem. Lett.* **6**, 4991–4996 (2015).
25. Seitz, F. On the mobility of electrons in pure non-polar insulators. *Phys. Rev.* **73**, 549–564 (1948).
26. Bardeen, J. & Shockley, W. Deformation potentials and mobilities in non-polar crystals. *Phys. Rev.* **80**, 72–80 (1950).
27. Benavides-Garcia, M. & Balasubramanian, K. Bond energies, ionization potentials, and the singlet-triplet energy separations of $SnCl_2$, $SnBr_2$, SnI_2, $PbCl_2$, $PbBr_2$, PbI_2, and their positive ions. *J. Chem. Phys.* **100**, 2821–2830 (1994).
28. Viswanath, A., Lee, J., Kim, D., Lee, C. & Leem, J. Exciton-phonon interactions, exciton binding energy, and their importance in the realization of room-temperature semiconductor lasers based on GaN. *Phys. Rev. B* **58**, 16333–16339 (1998).
29. Stillman, G., Wolfe, C. & Dimmock, J. Hall coefficient factor for polar mode scattering in n-type GaAs. *J. Phys. Chem. Solids* **31**, 1199–1204 (1970).
30. Stoumpos, C., Malliakas, C. & Kanatzidis, M. Semiconducting tin and lead iodide perovskites with organic cations: phase transitions, high mobilities, and near-infrared photoluminescent properties. *Inorg. Chem.* **52**, 9019–9038 (2013).
31. Fang, H.-H. *et al.* Photophysics of organic-inorganic hybrid lead iodide perovskite single crystals. *Adv. Funct. Mater.* **25**, 2378–2385 (2015).

32. Ha, S. T. *et al.* Synthesis of organic-inorganic lead halide perovskite nanoplatelets: towards high-performance perovskite solar cells and optoelectronic devices. *Adv. Opt. Mater.* **2**, 838–844 (2014).

33. Fang, H. H. *et al.* Photoexcitation dynamics in solution-processed formamidinium lead iodide perovskite thin films for solar cell applications. *Light Sci. Appl.* **5**, e16056 (2016).

34. Onoda-Yamamuro, N., Matsuo, T. & Suga, H. Calorimetric and IR spectroscopic studies of phase transitions in methylammonium trihalogenoplumbates (II). *J. Phys. Chem. Solids* **51**, 1383–1395 (1990).

35. Baikie, T. *et al.* Synthesis and crystal chemistry of the hybrid perovskite (CH_3NH_3)PbI_3 for solid-state sensitised solar cell applications. *J. Mater. Chem. A* **1**, 5628–5641 (2013).

36. Wasylishen, R., Knop, O. & Macdonald, J. Cation rotation in methylammonium lead halides. *Solid State Commun.* **56**, 581–582 (1985).

37. Varshni, Y. Temperature dependence of the energy gap in semiconductors. *Physica* **34**, 149–154 (1967).

38. Frost, J. M. J. *et al.* Atomistic origins of high-performance in hybrid halide perovskite solar cells. *Nano Lett.* **14**, 2584–2590 (2014).

39. Wehrenfennig, C., Liu, M., Snaith, H. J., Johnston, M. B. & Herz, L. M. Charge carrier recombination channels in the low-temperature phase of organic-inorganic lead halide perovskite thin films. *APL Mater* **2**, 081513 (2014).

40. Wu, X. *et al.* Trap states in lead iodide perovskites. *J. Am. Chem. Soc.* **137**, 2089–2096 (2015).

41. Priante, D. *et al.* The recombination mechanisms leading to amplified spontaneous emission at the true-green wavelength in $CH_3NH_3PbBr_3$ perovskites. *Appl. Phys. Lett.* **106**, 081902 (2015).

42. Rudin, S., Reinecke, T. L. & Segall, B. Temperature-dependent exciton linewidths in semiconductors. *Phys. Rev. B* **42**, 11218–11231 (1990).

43. Lee, J., Koteles, E. S. & Vassell, M. O. Luminescence linewidths of excitons in GaAs quantum wells below 150 K. *Phys. Rev. B* **33**, 5512–5516 (1986).

44. Malikova, L. *et al.* Temperature dependence of the direct gaps of ZnSe and $Zn_{0.56}Cd_{0.44}$Se. *Phys. Rev. B* **54**, 1819–1824 (1996).

45. Chen, Y., Kothiyal, G., Singh, J. & Bhattacharya, P. Absorption and photoluminescence studies of the temperature dependence of exciton life time in lattice-matched and strained quantum well systems. *Superlattice Microst.* **3**, 657–664 (1987).

46. Bartolo, B. D. & Chen, X. *Advances in Energy Transfer Processes* (World Scientific, 2001).

47. Selci, S. *et al.* Evaluation of electron-phonon coupling of $Al_{0.27}Ga_{0.73}$As/GaAs quantum wells by normal incidence reflectance. *Solid State Commun.* **79**, 561–565 (1991).

48. Masumoto, Y. & Takagahara, T. *Semiconductor Quantum Dots* (Springer-Verlag, 2002).

49. Zhang, X. B., Taliercio, T., Kolliakos, S. & Lefebvre, P. Influence of electron-phonon interaction on the optical properties of III nitride semiconductors. *J. Phys.: Condens. Matter* **13**, 7053–7074 (2001).

50. Gammon, D., Rudin, S., Reinecke, T. L., Katzer, D. S. & Kyono, C. S. Phonon broadening of excitons in GaAs/AlGaAs quantum wells. *Phys. Rev. B* **51**, 16785–16789 (1995).

51. Pérez-Osorio, M. A. *et al.* Vibrational properties of the organic-inorganic halide perovskite $CH_3NH_3PbI_3$ from theory and experiment: factor group analysis, first-principles calculations, and low-temperature infrared spectra. *J. Phys. Chem. C* **119**, 25703–25718 (2015).

52. Yamada, Y. *et al.* Photocarrier recombination dynamics in perovskite $CH_3NH_3PbI_3$ for solar cell applications. *J. Am. Chem. Soc.* **136**, 11610–11613 (2014).

53. Gopalan, S., Lautenschlager, P. & Cardona, M. Temperature dependence of the shifts and broadenings of the critical points in GaAs. *Phys. Rev. B* **35**, 5577–5584 (1987).

54. Grimvall, G. *The Electron-Phonon Interaction in Metals* (North-Holland, 1981).

55. Filip, M. R., Verdi, C. & Giustino, F. *GW* band structures and carrier effective masses of $CH_3NH_3PbI_3$ and hypothetical perovskites of the type APbI$_3$: A = NH_4, PH_4, AsH_4, and SbH_4. *J. Phys. Chem. C* **119**, 25209–25219 (2015).

56. Wehrenfennig, C., Liu, M., Snaith, H. J., Johnston, M. B. & Herz, L. M. Charge-carrier dynamics in vapour-deposited films of the organolead halide perovskite $CH_3NH_3PbI_{3-x}Cl_x$. *Energy Environ. Sci.* **7**, 2269–2275 (2014).

57. Petritz, R. & Scanlon, W. Mobility of electrons and holes in the polar crystal, PbS. *Phys. Rev.* **97**, 1620–1626 (1955).

58. Ferry, D. First-order optical and intervalley scattering in semiconductors. *Phys. Rev. B* **14**, 1605–1609 (1976).

59. Price, M. *et al.* Hot carrier cooling and photo-induced refractive index changes in organic-inorganic lead halide perovskites. *Nat. Commun.* **6**, 8420 (2015).

60. Kawai, H., Giorgi, G., Marini, A. & Yamashita, K. The mechanism of slow hot-hole cooling in lead-iodide perovskite: first-principle calculation on carrier lifetime from electron-phonon interaction. *Nano Lett.* **15**, 3103–3108 (2015).

61. von der Linde, D. & Lambrich, R. Direct measurement of hot-electron relaxation by picosecond spectroscopy. *Phys. Rev. Lett.* **42**, 1090–1093 (1979).

62. Sadhanala, A. *et al.* Preparation of single phase films of $CH_3NH_3Pb(I_{1-x}Br_x)_3$ with sharp optical band edges. *J. Phys. Chem. Lett.* **5**, 2501–2505 (2014).

63. Xiao, M. *et al.* A fast deposition-crystallization procedure for highly efficient lead iodide perovskite thin-film solar cells. *Angew. Chem. Int. Ed.* **126**, 10056–10061 (2014).

64. Giannozzi, P. *et al.* QUANTUM ESPRESSO: a modular and open-source software project for quantum simulations of materials. *J. Phys.: Condens. Matter* **21**, 395502 (2009).

65. Filip, M. R. & Giustino, F. GW quasiparticle band gap of the hybrid organic-inorganic perovskite $CH_3NH_3PbI_3$: Effect of spin-orbit interaction, semicore electrons, and self-consistency. *Phys. Rev. B* **90**, 245145 (2014).

66. Marini, A., Hogan, C., Grüning, M. & Varsano, D. yambo: An *ab initio* tool for excited state calculations. *Comp. Phys. Commun.* **180**, 1392–1403 (2009).

67. Mostofi, A. A. *et al.* wannier90: A tool for obtaining maximally-localised Wannier functions. *Comp. Phys. Commun.* **178**, 685–699 (2008).

68. Giustino, F., Cohen, M. L. & Louie, S. G. Electron-phonon interaction using Wannier functions. *Phys. Rev. B* **76**, 165108 (2007).

69. Poncé, S., Margine, E. R., Verdi, C. & Giustino, F. EPW: electron-phonon coupling, transport and superconducting properties using maximally localized Wannier functions. Preprint at http://arxiv.org/abs/1604.03525 (2016).

70. Verdi, C. & Giustino, F. Fröhlich electron-phonon vertex from first principles. *Phys. Rev. Lett.* **115**, 176401 (2015).

Acknowledgements

This work was supported by the Leverhulme Trust (Grant RL-2012-001), the UK Engineering and Physical Sciences Research Council (grant numbers EP/J009857/1, EP/L024667, EP/L016702/1 and EP/M020517/1) and the Graphene Flagship (EU FP7 grant number 604391). This work used the ARCHER UK National Supercomputing Service via the AMSEC Leadership project and the Advanced Research Computing (ARC) facility of the University of Oxford.

Author contributions

A.D.W. performed the PL experiments, did the data analysis and participated in the experimental planning. C.V. carried out the first-principles calculations. R.L.M. gave guidance for the PL experiments. G.E.E. prepared the samples. H.J.S. gave guidance on sample preparation. M.A.P.-O. provided support with the first-principles calculations. The project was conceived, planned and supervised by F.G., M.B.J. and L.M.H. A.D.W. wrote the first version of the manuscript. All authors contributed to the discussion and preparation of the final version of the paper.

Additional information

Insertion compounds and composites made by ball milling for advanced sodium-ion batteries

Biao Zhang[1,2,3], Romain Dugas[1,2,3], Gwenaelle Rousse[1,2,3,4], Patrick Rozier[2,3,5], Artem M. Abakumov[6,7] & Jean-Marie Tarascon[1,2,3,4]

Sodium-ion batteries have been considered as potential candidates for stationary energy storage because of the low cost and wide availability of Na sources. However, their future commercialization depends critically on control over the solid electrolyte interface formation, as well as the degree of sodiation at the positive electrode. Here we report an easily scalable ball milling approach, which relies on the use of metallic sodium, to prepare a variety of sodium-based alloys, insertion layered oxides and polyanionic compounds having sodium in excess such as the $Na_4V_2(PO_4)_2F_3$ phase. The practical benefits of preparing sodium-enriched positive electrodes as reservoirs to compensate for sodium loss during solid electrolyte interphase formation are demonstrated by assembling full $C/P'2-Na_1[Fe_{0.5}Mn_{0.5}]O_2$ and $C/'Na_{3+x}V_2(PO_4)_2F_3'$ sodium-ion cells that show substantial increases ($>10\%$) in energy storage density. Our findings may offer electrode design principles for accelerating the development of the sodium-ion technology.

[1] Chimie du Solide-Energie, FRE 3677, Collège de France, 11 Place Marcelin Berthelot, Paris, Cedex 05 75231, France. [2] Réseau sur le Stockage Electrochimique de l'Energie (RS2E), FR CNRS 3459, Amiens 80039, France. [3] ALISTORE-European Research Institute, Amiens 80039, France. [4] Sorbonne Universités—UPMC Univ Paris 06, 4 Place Jussieu, Paris F-75005, France. [5] University of Toulouse III Paul Sabatier, CIRIMAT CNRS UMR 5085, 118 route de Narbonne, Toulouse, Cedex 09 31062, France. [6] EMAT, University of Antwerp, Groenenborgerlaan 171, Antwerp B-2020, Belgium. [7] Skolkovo Institute of Science and Technology, Moscow 143025, Russia. Correspondence and requests for materials should be addressed to J.-M.T. (email: jean-marie.tarascon@college-de-france.fr).

Considering elemental abundance, the most appealing alternative to Li-based battery technology is undoubtedly sodium. This is reflected in the revival of Na-ion battery (NIB) research, a field in which intense efforts are currently devoted to the search for high-performance electrode materials. Progress in sodium intercalation chemistry is primarily inherited from work on Li-ion materials, with the negative and positive electrodes for sodium systems based on similar structural types[1-3]. For negative electrodes comprising sodium metal, Na_xSb is the most attractive[4]. Antimony, however, is not the most desirable component considering its scarcity, moderate toxicity and large mass. There is thus great interest in carbon negative electrodes, whose reversible capacity can reach $300\,mAh\,g^{-1}$ after accounting for a $\sim25\%$ irreversibility penalty associated with formation of the solid electrolyte interface (SEI) in the first cycle[5]. Among the positive electrode candidates, polyanionic compounds such as $Na_3V_2(PO_4)_2F_3$ (NVPF) (refs 6,7) and $Na_2Fe_2(SO_4)_3$ (ref. 8), and layered compounds like the $O3-NaNi_{0.5}Mn_{0.5}O_2$ (ref. 9) and $P2-Na_{0.67}[Fe_{0.5}Mn_{0.5}]O_2$ (ref. 10) phases are presently the most studied. Nevertheless, the performances of NIB prototypes based on the aforementioned materials suffer from the large irreversibility of initial Na-uptake–removal processes at the carbon negative electrode.

The sodium loss in the formation of the SEI is similar to what is observed in the case of Li-ion batteries (LIBs), which has been widely studied. Because the only source of Li in the Li-ion cell is the Li-bearing electrode, compensating for lithium loss in the formation of the SEI at the negative electrode has been of great importance for achieving high energy densities in LIBs, and it is becoming more urgent with the emerging Si anodes, which show larger initial irreversibility than carbon electrodes. Through the years various routes have been explored. By adding to the positive electrode either sacrificial lithium species or additional intercalation compounds, the full capacity of a cathode material can be realized with modest weight penalties, For example, $Li_2C_2O_4$ is used as a sacrificial source that decomposes during oxidation to provide extra lithium[11]. Alternately, in compounds such as $Li_{1+x}Mn_2O_4$, which contains two redox voltages, the low-voltage $\sim3\,V$ plateau associated with xLi^+ can be used as a Li reservoir[12]. Another approach, mainly applied to Si, is *in situ* or *ex situ* pre-lithiation. This can be accomplished either by Li ball milling[13], or by placing Si in contact with Li so that when electrolyte is added formation of the SEI occurs conjointly with the uptake of Li by the negative Si electrode (Li_xSi) (ref. 14).

Considering the lower coulombic efficiency of hard carbon in NIBs, that is, $<80\%$ (ref. 15), a strategy to compensate for Na loss to the SEI is sorely needed. It is also equally important to tune the degree of sodiation in cathodes to be used in NIBs, especially in the case of 'sodium deficient' P2-type layered oxide electrodes of formula $Na_{2/3}MO_2$ (M = Mn, Fe, Co...)[10,16], which only contain $\sim0.67\,Na^+$ per formula unit. These materials can achieve performances as high as $200\,mAh\,g^{-1}$ but only in half cells against sodium metal which, in excess, enables $NaMO_2$ compositions to be reached upon cycling. In contrast, the capacity drops by 30% in Na-ion cells because of the lack of a Na reservoir to compensate for the missing $0.33\,Na^+$ in the pristine P2 phase. Solving these two issues constitutes the driving force of the present study.

Surprisingly few studies have been conducted before now to address the development of Na reservoir sources and pre-sodiation reagents for NIBs while minimizing energy penalty. Recently, NaN_3 was proposed as a sacrificial Na source, but the generation of N_2 gas during its non-reversible electrochemically driven decomposition upon oxidation is detrimental to the battery. Moreover, the low Na content in NaN_3 (35% Na by weight) imposes a penalty in energy density[17]. Similar to LIBs, Na metal would obviously be the best pre-sodiation reagent for NIBs. However, it is one of the most difficult metals to handle because of its ductility and tendency to stick to metallic surfaces. Following our previous work on preparing Li-alloy through ball milling[18], we report herein the synthesis of Na-based alloys and other insertion compounds via the ball milling of Na-metal with the appropriate chemical elements. We find this approach equally suitable for tuning the amount of Na in insertion electrodes. In particular, we find that Na_3P-the Na-based compound showing the closest capacity ($804\,mAh\,g^{-1}$) to Na metal ($1{,}165\,mAh\,g^{-1}$) and hence offering the minimum weight penalty-can be used as a sacrificial Na source, increasing for instance the reversible capacity of $C/P'2-Na[Fe_{0.5}Mn_{0.5}]O_2$ cells by 20%. We begin by describing the ball milling synthesis of Na-alloys, then present the use of the Na-ball milling approach for pre-sodiation of insertion positive electrodes ($P2-Na_{0.67}[Fe_{0.5}Mn_{0.5}]O_2$, NVPF), and end with the implementation of both approaches towards the optimization of full Na-ion cells.

Results
Ball milling-driven formation of Na-based alloys. Na_3P powders were prepared by adding stoichiometric amounts of Na-lumps and red phosphorous powder into a hardened steel ball milling jar, loaded in an Ar-filled glove box, with a ball/powder weight ratio of 35. Strikingly, room temperature continuous ball milling for 2 h, using a SPEX 8000 milling apparatus, was found to be sufficient to produce well crystalline and single-phased Na_3P powders as deduced by X-ray diffraction (XRD) (Fig. 1a,b), while shear grinding for 27 h in planetary ball mill was shown to give a mixture of Na_3P and amorphous phase[19]. These powders were mixed with carbon SP (30% by weight) and further ball milled for 20 min to obtain homogeneous composite electrodes. When cycled versus Na^+/Na^0, these composites showed reversible capacity (Fig. 1c) of $\sim600\,mAh\,g^{-1}$ based on the weight of Na_3P.

To highlight the simplicity of this ball milling synthesis we recall the two approaches previously adopted to prepare Na_3P powders: (i) electrochemical sodiation of P by making a battery using P and Na metal as working and counter electrode[20], respectively; and (ii) chemical alloying of Na and P, either through a solvothermal reaction at $150\,°C$ (ref. 21), or high temperature annealing ($480\,°C$) in evacuated silica ampoules[22]. Encouraged by these results, we successfully extended the approach to the synthesis of Na_xM alloys, where M is selected from the group consisting of Sn, Sb and Pb. For reasons of conciseness, the preparation of single-phase Na_3Sb powders (Fig. 1d) whose voltage-composition curves fully mirror previous literature data (Fig. 1e)[23] is solely reported herein.

Although speculative, we believe that the key to success for such unexpected reactions depends on the large free energy (ΔG) of the alloy formation reaction. As depicted in the schematic representations of Fig. 1a, a monolayer of alloy is rapidly created by contact of metal particles with Na lumps, hence separating the Na from the metal container and balls. Ball milling will continuously break the alloy shell to expose fresh surfaces, which rapidly react in a progressive alloying of all the Na, most likely facilitated by local heating. The repeated alloying/fracturing sequences are accompanied by a continuous peeling and breaking of the alloy shell into loose powders.

Ball milling-driven pre-sodiation of insertion electrodes. The successful synthesis of Na-based alloys by room temperature Na ball milling encouraged us to exploit the reducing power of both Na and Na_3P to simulate electrochemical reduction of positive electrode materials so as to increase their sodium content.

We tested this possibility using the P2-type layered oxide phase, $Na_{0.67}[Fe_{0.5}Mn_{0.5}]O_2$, which can be electrochemically reduced at a potential near 1.5 V to form the P'2-$Na_1[Fe_{0.5}Mn_{0.5}]O_2$ phase that delivers a capacity of ~190 mAh g^{-1} (ref. 24). Attempts to synthesize P'2 type $Na_1[Fe_{0.5}Mn_{0.5}]O_2$ through direct solid-state reaction have so far failed, resulting in an O3 type phase with poor capacity[10]. Stoichiometric amounts of $Na_{0.67}[Fe_{0.5}Mn_{0.5}]O_2$ (XRD shown in Fig. 2b) and Na metal were added under argon into a hardened steel ball milling jar using a ball to powder weight ratio of 35. The XRD pattern of the obtained powder (Fig. 2a) is consistent with the reflections reported for the electrochemically produced P'2 phase[24], which confirms the successful production of fully sodiated P'2-$Na_1[Fe_{0.5}Mn_{0.5}]O_2$ after only 2 h of ball milling. At shorter milling time or when a sub-stoichiometric amount of Na is used, the ball-milled samples are a mixture of the P2 and P'2 phases.

Fig. 2c–e compare the electrochemical behaviour of a Na/P2 cell with that of a Na/P'2 cell with the P'2 phase made by ball milling with Na. Note that only 0.45 Na$^+$ can be removed from the P2-based cell, as opposed to nearly 0.8 Na$^+$ for the P'2 cell, clearly confirming the success of the ball milling-induced sodiation process. Aside from this difference, the cells behave identically in terms of reversible and sustained capacity upon cycling, independently of whether pristine P2 or P'2 obtained by Na ball milling was initially used.

In addition to layered oxides, the polyanionic compound NVPF ($Na_3V_2(PO_4)_2F_3$) is also of great interest as a positive electrode in Na-ion cells[25,26] since it shows high-voltage plateaus near 3.6 and 4.2 V, whose equal amplitudes provide a cumulative capacity of ~110 mAh g^{-1}. To generalize our approach, we explored the ball milling-driven reactivity of Na against NVPF.

$Na_3V_2(PO_4)_2F_3$ powders were mixed with various amounts of Na and various ball milling times (Supplementary Fig. 1). To our surprise, ball milling 1 molar equivalent of NVPF with 2 equivalents of Na metal lumps for 30 min (ball to powder weight ratio of 35) results in loose composite powders whose XRD pattern differs from that of NVPF with namely the onset of extra peaks corresponding to the onset of a second phase. By increasing the milling time we progressively increase the amount of this extra phase, which was obtained as a single phase after 3 h of ball milling (Fig. 3a–c). The elemental distribution and chemical composition of this as-prepared powder was studied by high-angle annular dark field scanning transmission electron microscopy. The images demonstrate that crystallites of the main phase are surrounded by Na nanoparticles with sizes of 20–50 nm, as deduced from compositional energy-dispersive X-ray spectroscopy mapping (Supplementary Fig. 2a,b). Note that this is in agreement with the residual amount of Na metal, as deduced by differential scanning calorimetry (DSC) (Supplementary Fig. 3). Moreover, the analysis of the main phase crystallites provides a Na:V:P = 4.0(2):2.0(1):2.1(2) atomic ratio, consistent with a $Na_4V_2(PO_4)_2F_3$ formulae.

The Synchrotron XRD patterns of the pristine NVPF and $Na_4V_2(PO_4)_2F_3$ phase, and of a mixture of both phases are shown in Fig. 3a–c. The diffraction peaks of the phase formed upon ball milling with Na can be indexed in the same orthorhombic cell as for the pristine NVPF[26], but with different lattice parameters, that is, $a = 9.2208(2)$ Å, $b = 9.2641(2)$ Å and $c = 10.6036(2)$ Å. This corresponds to an increase of the unit cell by 3.2% ($V = 905.79(3)$ Å^3) relative to the pristine NVPF phase ($V = 878.05(3)$ Å^3), which is consistent with the uptake of extra sodium upon reduction.

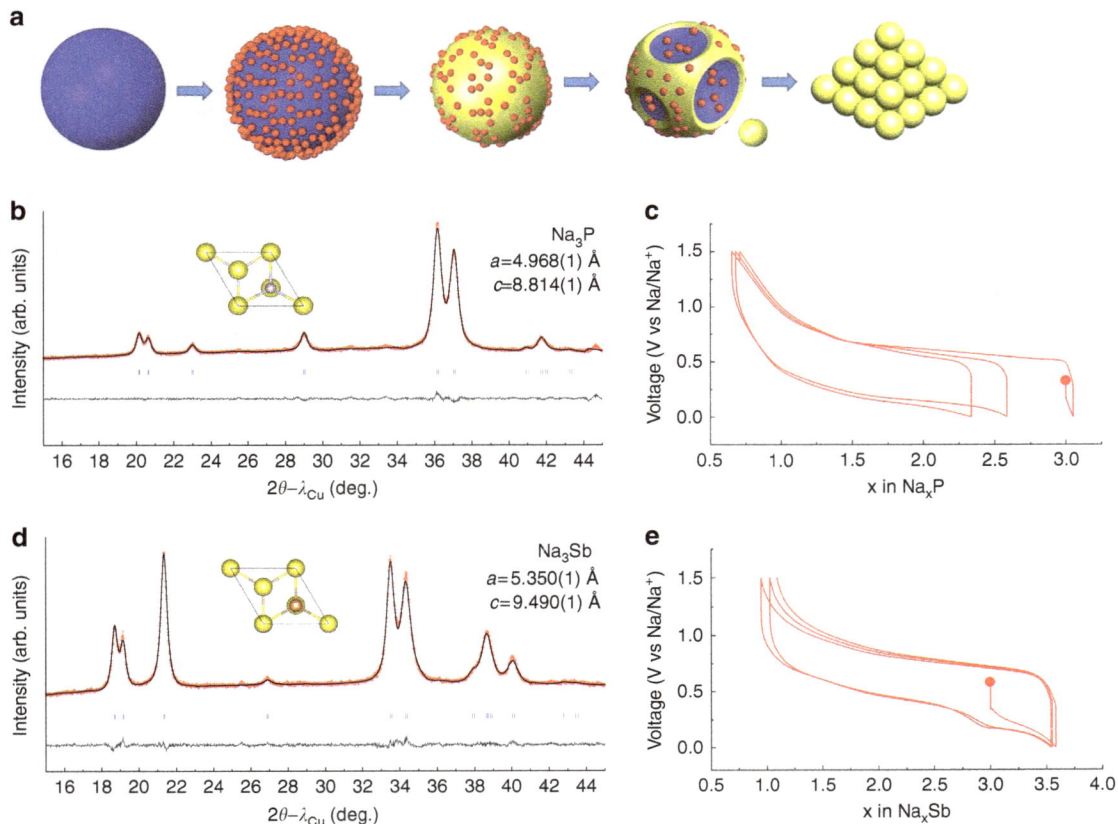

Figure 1 | Synthesis of Na-based alloys. (a) Schematic diagram showing the ball milling process exemplified for the formation of Na$_x$M alloys (yellow) by reacting Na (blue) with metal M (M = P or Sb) (red). In **b,c** and **d,e**, the Rietveld refined X-ray powder patterns of Na$_3$P and Na$_3$Sb powders are respectively shown together with their corresponding electrochemical voltage-profile in Na-half cells.

Refinement of the synchrotron XRD pattern (Fig. 3c) of the obtained $Na_4V_2(PO_4)_2F_3$ was undertaken based on the structural model reported for pristine NVPF by Bianchini et al.[26] The VO_6 octahedra and PO_4 tetrahedra arrangement of NVPF is kept, and the best agreement between the observed and calculated patterns was found for the Na positions as listed in Supplementary Tables 1 and 2 and Supplementary Note 1. The Na environments are shown in Fig. 3e. The three distinct Na sites are all seven-fold coordinated with four oxygen and three fluorine atoms, which is analogous to the coordination of Na1 in NVPF (Fig. 3g). The structural analysis fully confirms the chemical composition $(Na_4V_2(PO_4)_2F_3)$ and indicates that there is apparently no further space for Na insertion. Lastly, the synchrotron XRD patterns of the sample prepared by ball milling NVPF with Na for 30 min can be perfectly refined with a two phases model: NVPF and $Na_4V_2(PO_4)_2F_3$, as shown in Fig. 3b.

The occurrence of $Na_4V_2(PO_4)_2F_3$ comes as a total surprise, as no extra capacity has ever been reported for NVPF at low potential. It is however worth noting that the reversible insertion of 1 Na^+ at 0.3 V was just recently reported for $Na_3V_2(PO_4)_3$ (ref. 27). Knowing the existence of the two $Na_3V_2(PO_4)_2F_3$ and $Na_4V_2(PO_4)_2F_3$ end-member phases, we deliberately prepared composites of nominal compositions having different amounts of NVPF and $Na_4V_2(PO_4)_2F_3$ that will be denoted hereafter '$Na_{3+x}V_2(PO_4)_2F_3$'. Such as-prepared composites show similar electrochemical performance as NVPF, except x more Na is removed during the first oxidation (Fig. 3h). They deliver a

stable capacity of $\sim 110\,mAh\,g^{-1}$ when cycled between 4.4 and 3 V.

In light of our finding on the $Na_4V_2(PO_4)_2F_3$ phase, we explored electrochemical intercalation in NVPF down to low voltages. In situ XRD measurements were conducted on a Na/NVPF cell, with XRD patterns collected for every 90 min. As the cell was being discharged at 0.15 C, we observed the progressive appearance of an additional set of peaks (Fig. 4a,b) corresponding to the $Na_4V_2(PO_4)_2F_3$ phase obtained by ball milling. They appear at the expense of the NVPF reflections, which barely change in position but decrease in intensity. By careful refinement of the XRD patterns, we quantified (Fig. 4c) the growing amounts of $Na_4V_2(PO_4)_2F_3$, promoted by continuously lowering the reduction voltage; $Na_4V_2(PO_4)_2F_3$ constitutes nearly 60% of the composite when the cell potential reaches 0 V. This corresponds to a x value of 0.6 (in '$Na_{3+x}V_2(PO_4)_2F_3$'), which cannot be determined accurately from coulometric titration because of side reactions (Supplementary Note 2 and Supplementary Figs 4 and 5). Turning to the charging process, a reverse trend is observed, with the recovery at 3.6 V of an XRD pattern similar to that of the pristine phase indicating the full reversibility of the Na-uptake–removal process (Fig. 4a). However, it is worth noting the drastic difference in the charge and discharge profiles, highlighted in the derivative plot (Supplementary Fig. 4a,b), which is indicative of a different reacting pathway. Moreover, we note the continuous growing of $Na_4V_2(PO_4)_2F_3$ when the cell is switched back to

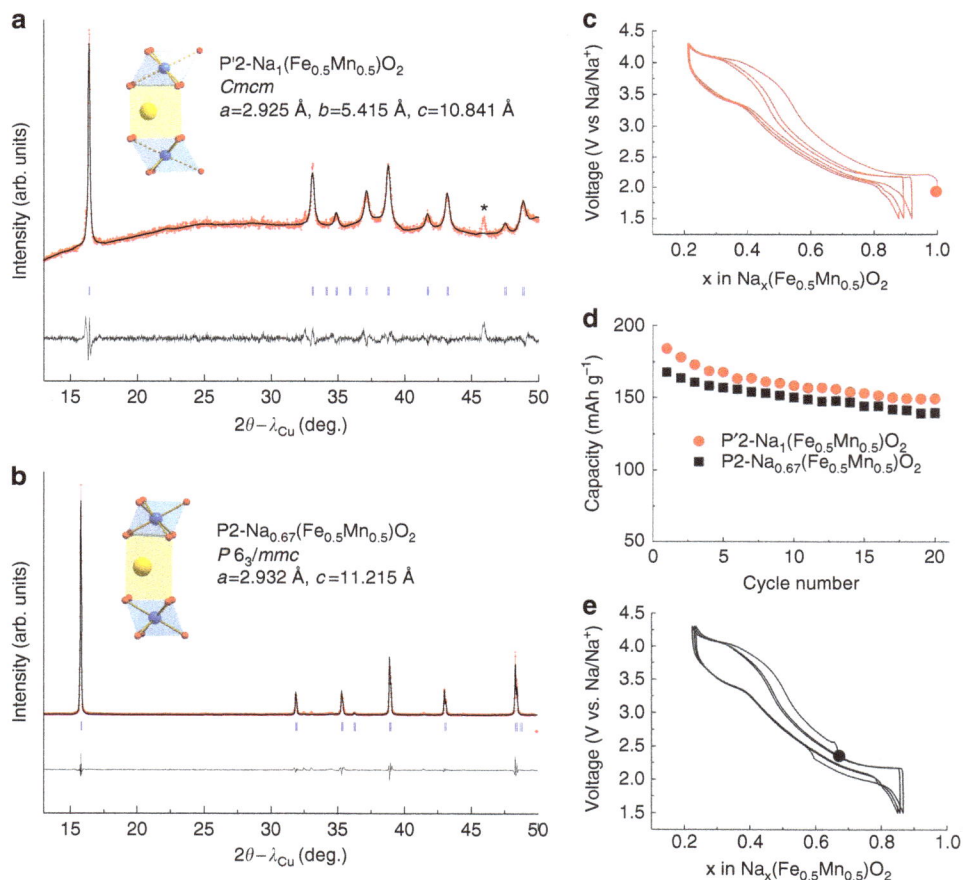

Figure 2 | Synthesis of the P′2 phase by Na ball milling. XRD powder pattern profile matching and schematic representation of the structure of (**a**) P′2-$Na_1[Fe_{0.5}Mn_{0.5}]O_2$ obtained using ball milling with Na and (**b**) pristine P2-$Na_{0.67}[Fe_{0.5}Mn_{0.5}]O_2$. The refined cell parameters are in agreement with reported ones for both P2 and P′2 phases. The change of symmetry from P2 to P′2 accounts for the distortion of the MnO_6 octahedral due to the Jahn–Teller effect of reduced Mn^{3+}; voltage profiles of (**c**) P′2-$Na_1[Fe_{0.5}Mn_{0.5}]O_2$, (**e**) P2-$Na_{0.67}[Fe_{0.5}Mn_{0.5}]O_2$ and their corresponding cycling performances in Na-half cells (**d**). The peak marked with an asterisk (*) in **a** is attributed to Be window used to measure such a moisture sensitive sample.

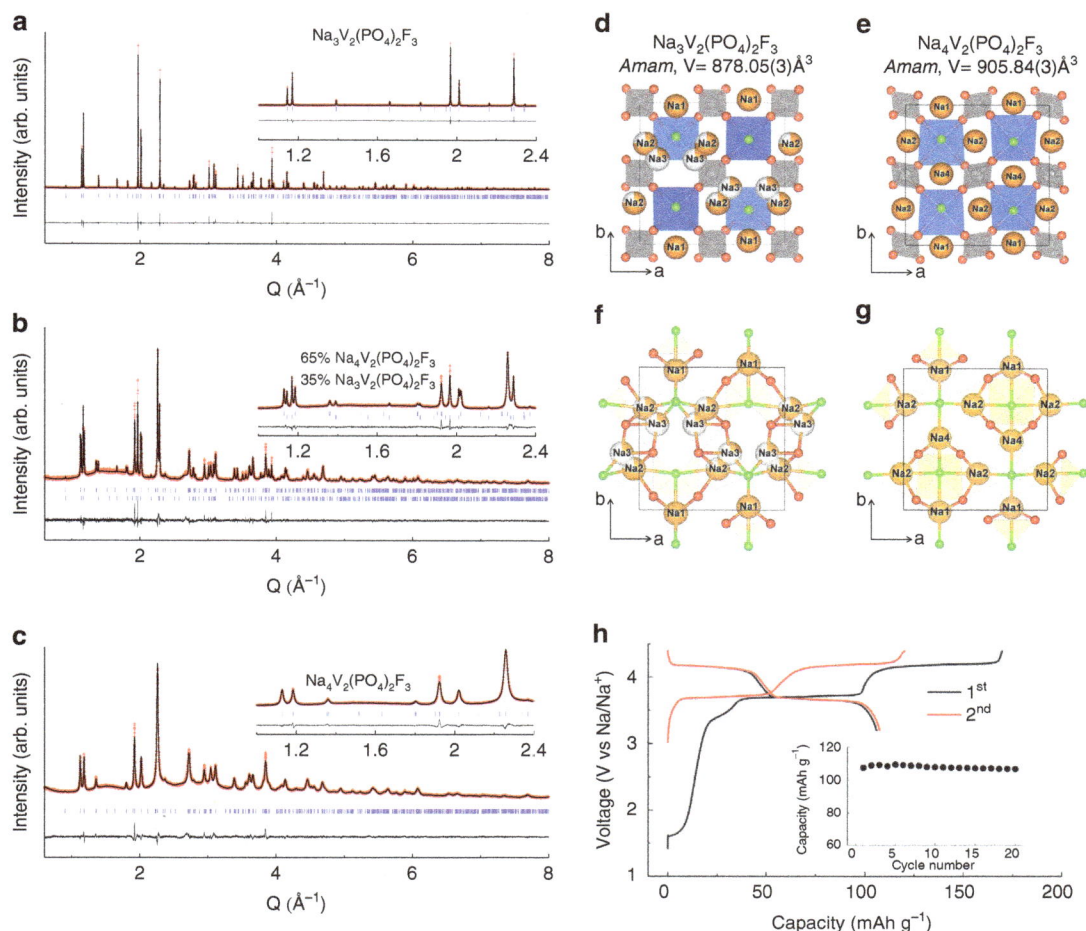

Figure 3 | Synthesis and electrochemistry of Na₄V₂(PO₄)₂F₃ by Na reductive ball milling. Rietveld refinements of pristine $Na_3V_2(PO_4)_2F_3$ (**a**), a mixture of $Na_3V_2(PO_4)_2F_3$ and $Na_4V_2(PO_4)_2F_3$ (**b**) and pure $Na_4V_2(PO_4)_2F_3$ (**c**). The red crosses, black continuous line and bottom grey line represent the observed, calculated and difference patterns, respectively. Vertical blue tick bars mark the Bragg reflections arising from the *Amam* space group. Patterns are given in Q-space for allowing a direct comparison. Structure of $Na_3V_2(PO_4)_2F_3$ (**d,f**) and of $Na_4V_2(PO_4)_2F_3$ (**e,g**) as deduced from the refinement of the synchrotron patterns. The Na environment of $Na_3V_2(PO_4)_2F_3$ and $Na_4V_2(PO_4)_2F_3$ is respectively highlighted in (**f,g**). VO₄F₂ octahedral and PO₄ tetrahedral are coloured in blue and grey, respectively. Na and F atoms are shown as orange and green balls, respectively. Vacancies on the Na1 and Na2 sites on $Na_3V_2(PO_4)_2F_3$ are coloured in white. The electrochemical behaviour of composite '$Na_{3.5}V_2(PO_4)_2F_3$' (that is, with equal amounts of $Na_3V_2(PO_4)_2F_3$ and $Na_4V_2(PO_4)_2F_3$) in Na-half cell is shown in **h**.

oxidation, further implying a complex Na-uptake–removal process. We believe this partial reduction to be nested in kinetics blockages that are most likely due to the growth of a thick insulating SEI layer due to copious electrolyte decomposition and/or the formation of peculiar self-limiting core—shell-like '$Na_{3+x}V_2(PO_4)_2F_3$' particles upon reduction. Besides, attempts to modify the SEI with the use of fluoroethylene carbonate (FEC) did not result in subsequent changes in increasing the amount of the $Na_4V_2(PO_4)_2F_3$ phase (Supplementary Note 2). In comparison, there are two specific aspects that facilitate the production of pure $Na_4V_2(PO_4)_2F_3$ in the case of ball milling synthesis. They enlist the absence of SEI formation because of the lack of electrolyte, and the continuous formation of highly reacting fresh surfaces due to repeated fracturing.

These examples highlight the benefits of ball milling-driven Na-reduction reactions, which are free of the complexities associated with handling reactive solutions or using mild temperature processing to prepare fully reduced materials. The synthesized P′2-$Na_1[Fe_{0.5}Mn_{0.5}]O_2$ and '$Na_{3+x}V_2(PO_4)_2F_3$' show similar cyclic and rate performance (Supplementary Fig. 6) as the pristine P2-$Na_{0.67}[Fe_{0.5}Mn_{0.5}]O_2$ and NVPF, respectively, demonstrating that the short ball milling time does not bring any

detrimental effect to its electrochemical performance. Such reactions, leading to the transformation of P2 to P′2, or NVPF to $Na_4V_2(PO_4)_2F_3$, are topotatic since the host structural frameworks are unaltered, and can simply be viewed as insertion reactions. Thus, the reactivity is simply dictated by the redox potential associated with the insertion of Na in various compounds. As shown in the electrochemical energy scale schematic representation in Fig. 5b, although Na_3P is a milder reducing reagent than Na (0.5 V versus Na^+/Na^0), it is also expected to reduce P2 and NVPF. To test this point, as we have done for Na, a survey of various ball milling times together with various amounts of Na_3P were conducted and the obtained composites were periodically checked for phase purity. We find the possibility to successfully prepare the $Na_4V_2(PO_4)_2F_3$ phase (Fig. 5a) as well as the P′2 phase (Fig. 5c) by ball milling powdered mixtures of NVPF and P2 with 1 and 0.2 molar amounts of Na_3P, respectively. The reaction between Na_3P and NVPF could be classified as chemical sodiation. The successful sodiation of NVPF by Na_3P suggests that the reduction potential of NVPF should be higher than the potential for Na_3P oxidation, that is, ~0.5 V. The advantage of Na_3P as compared with Na metal is the fact it is a powder, although there is a compromise for

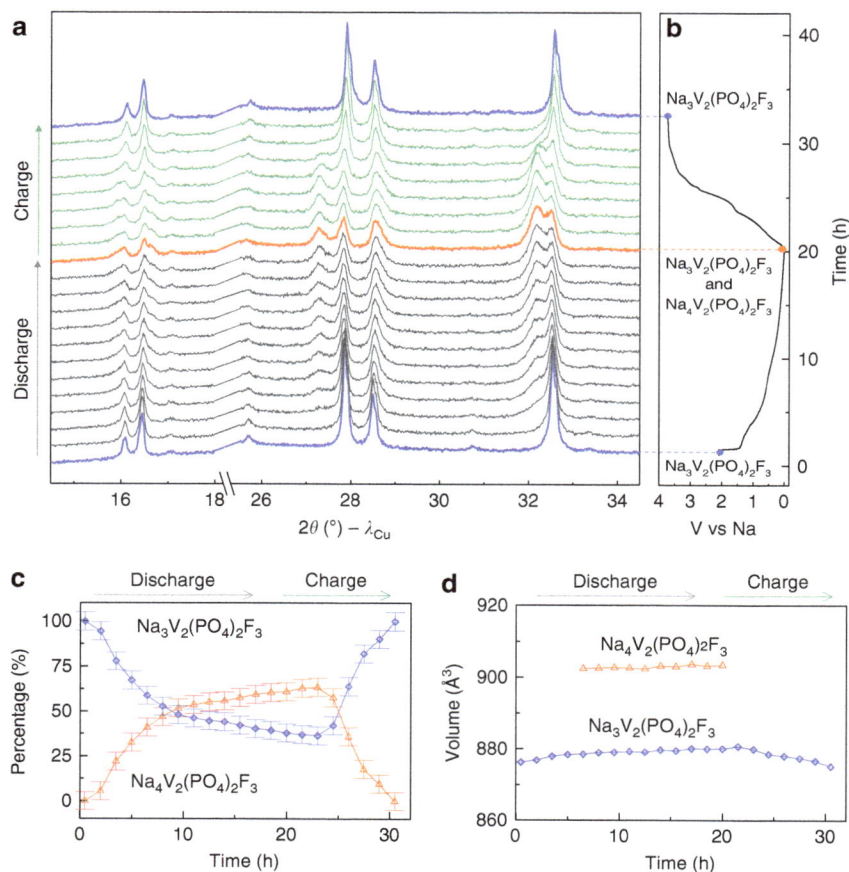

Figure 4 | *In situ* XRD patterns collected for a Na$_3$V$_2$(PO$_4$)$_2$F$_3$/Na cell. XRD patterns (**a**) and the corresponding electrochemical curve (**b**) is shown. The percentage of each phase and their unit cell volumes are shown in **c,d**, respectively. Error bars represent standard deviation obtained from the Rietveld refinement. Note that the Na$_4$V$_2$(PO$_4$)$_2$F$_3$ phase mainly forms at the beginning of the process (within the 1.4–0.8 V domain) and then the Na-uptake appears self-limiting for reasons explained with the text. For time >20 h the amount of Na$_4$V$_2$(PO$_4$)$_2$F$_3$ within the accuracy of the measurements still increases, while the cell is already in its oxidation mode.

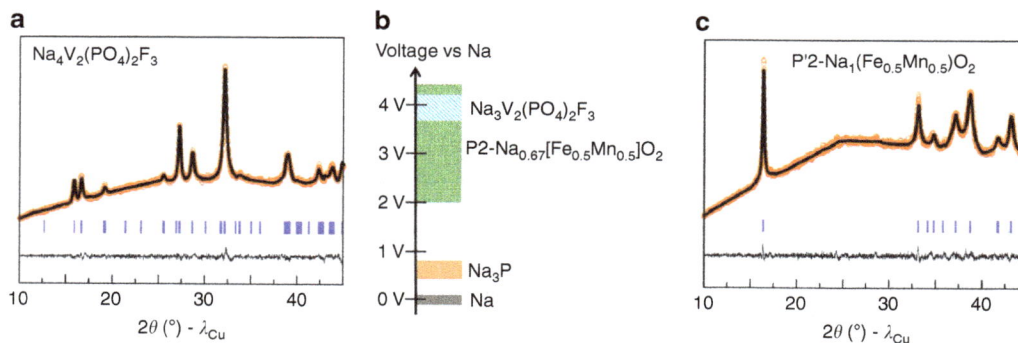

Figure 5 | Synthesis of Na$_4$V$_2$(PO$_4$)$_2$F$_3$ and P′2 phase by Na$_3$P ball milling. The XRD Rietveld refinement of (**a**) Na$_4$V$_2$(PO$_4$)$_2$F$_3$ and the profile matching of (**c**) P′2-Na$_1$[Fe$_{0.5}$Mn$_{0.5}$]O$_2$ phase obtained by ball milling using Na$_3$P as a reducing reagent are shown together with in the middle an electrochemical scale ranking the various used compounds in terms of redox potentials (**b**). In **a,c** the orange open circles, black continuous line and bottom grey line represent the observed, calculated, and difference patterns, respectively. Vertical blue tick bars mark the Bragg reflections.

this convenience with the requirements of excess amounts of Na$_3$P and longer ball milling time.

Na-enriched phases for highly efficient Na-ion batteries. The successful preparation of Na-rich Na$_4$V$_2$(PO$_4$)$_2$F$_3$ or composites with known amounts of NVPF and Na$_4$V$_2$(PO$_4$)$_2$F$_3$ enables control over the extra Na content by playing with ball milling times, and is of great importance for enhancing the performances

of C/NVPF Na-ion cells. This situation closely mirrors the use of Li$_{1+x}$Mn$_2$O$_4$ as a Li source in C/LiMn$_2$O$_4$ Li-ion cells[12], since the extra Na$^+$ ions present in NVPF can be removed at low potential without an added weight penalty with exception to the weight of the x added Na. As a proof of concept, we assembled various electrochemical cells having carbon as the negative electrode and either the pristine NVPF phase or the 'Na$_{3+x}$V$_2$(PO$_4$)$_2$F$_3$' composites as the positive electrode. For conciseness, we focus

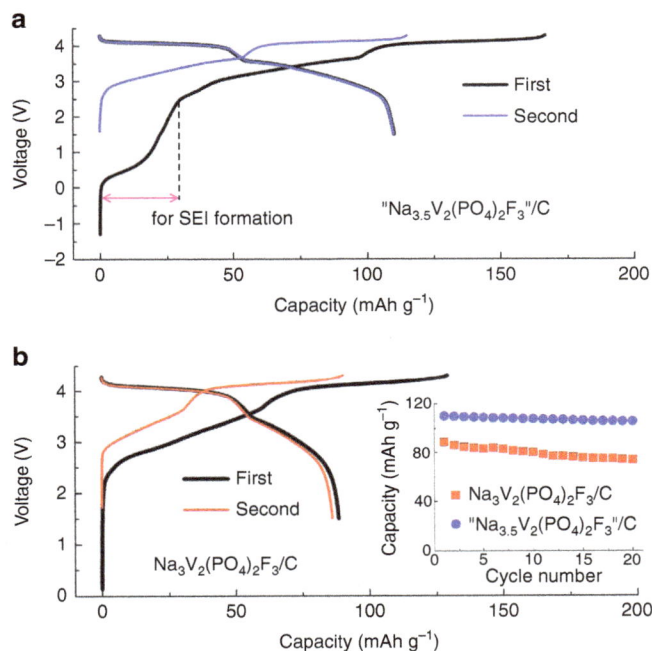

Figure 6 | Performances of C/'Na$_{3+x}$V$_2$(PO$_4$)$_2$F$_3$' Na-ion cells. The voltage-composition profiles are reported for two first cycles for cell C1 using Na$_3$V$_2$(PO$_4$)$_2$F$_3$ (**b**) and cell C2 using 'Na$_{3.5}$V$_2$(PO$_4$)$_2$F$_3$' (**a**). The weight ratio of cathode to anode is 2.7 and 1.9 for C1 and C2 cell, respectively. The cells were cycled at a 0.2C rate (C rate being defined as the extraction of 2 Na from NVPF). For both cells a 1 M NaClO$_4$ in EC/DMC (50/50 by volume) electrolyte was used. The capacity is calculated based on the weight of positive materials. Inset in (**b**) shows the capacity retention in the first 20 cycles.

here only on the performance for the optimum x value. This value was estimated to be equal to 0.5 for compensating the ~25% irreversible capacity related to the SEI. Details about the balancing of cathode and anode are described in Supplementary Fig. 7 and Supplementary Note 3. Figure 6 shows the electrochemical performances of cells C1 and C2, having pristine NVPF and composite 'Na$_{3.5}$V$_2$(PO$_4$)$_2$F$_3$' as positive electrodes, respectively. The cells were tested electrochemically between 1.5 and 4.3 V. The voltage trace for C1 mirrors reports in the literature for similar cells with a charging capacity of 129 mAh g^{-1} and a discharge capacity of 89 mAh g^{-1}, which remains stable upon cycling. Providing sacrificial Na in the form of Na-rich NVPF (C2) strongly modifies the voltage profile: an initial capacity near 0.5 V corresponds to the removal of Na from 'Na$_{3.5}$V$_2$(PO$_4$)$_2$F$_3$' to compensate for the SEI formation at the negative electrode; afterwards the potential rises, associated with removal of Na from NVPF. The C2 cell exhibits an overall charging capacity of 167 mAh g^{-1} and a discharge capacity of 110 mAh g^{-1}, which is a 24% enhancement compared with cell C1. This corresponds to an overall 10% increase in energy density as described in Supplementary Note 4. Lastly, there is no evidence to suggest that use of the Na-rich phase jeopardizes the cycle life, with the capacity remaining nearly constant over 20 cycles (the maximum we have tried). Needless to say that further optimization of cell balancing via the use of three electrodes is being pursued to fine tune the proper value of x for achieving optimum performance.

We next implemented a similar approach in the optimization of C/P2-type Na$_{0.67}$[Fe$_{0.5}$Mn$_{0.5}$]O$_2$ Na-ion systems. In contrast to NVPF, the positive P′2-Na$_1$[Fe$_{0.5}$Mn$_{0.5}$]O$_2$ phase cannot act as an extra source of Na since Na-rich Na$_{1+x}$[Fe$_{0.5}$Mn$_{0.5}$]O$_2$ does

not exist. This is easily overcome by first preparing the P′2-Na$_1$[Fe$_{0.5}$Mn$_{0.5}$]O$_2$ phase by Na-ball milling and then homogeneously mixing the P′2 phase with the proper amount of Na$_3$P to compensate for the carbon irreversible capacity. We here use extra Na$_3$P, rather than Na metal, as it is easily added as powders to positive electrode materials to make homogeneous composite electrodes. Upon charging the cell, the added Na$_3$P will oxidize to compensate for the Na consumed in SEI formation during the first cycle. This leaves behind elemental phosphorous, which remains as an electrochemical spectator within the cell upon subsequent cycles (Supplementary Fig. 8), but which also has the capability to act as a safety buffer by reinserting Na at a constant voltage in case of cell over-discharge.

A series of full Na-ion cells were assembled using carbon negative electrodes and composite positive electrodes made of P2-Na$_{0.67}$[Fe$_{0.5}$Mn$_{0.5}$]O$_2$ (cell D1), P′2-Na$_1$[Fe$_{0.5}$Mn$_{0.5}$]O$_2$ (cell D2) and P′2-Na$_1$[Fe$_{0.5}$Mn$_{0.5}$]O$_2$ + 10 wt.% Na$_3$P (cell D3). The corresponding voltage profiles for the cells, collected upon cycling between 0 and 4.3 V at a current rate of 0.1C, are shown in Fig. 7a–c. The D1 cell shows a charge capacity of 112 mAh g^{-1} and a discharge capacity of only 71 mAh g^{-1} that is maintained upon subsequent cycling. These capacities are considerably lower than those obtained for Na/Na$_{0.67}$[Fe$_{0.5}$Mn$_{0.5}$]O$_2$ half cells (168 mAh g^{-1}), due to the replacement of the Na anode by carbon, and hence the absence of a Na source to enable formation of the P′2 phase and to compensate for losses to the SEI at the carbon electrode. In contrast, the D2 and D3 cells show charge capacities of 185 and 247 mAh g^{-1}, respectively, with corresponding discharge capacities of 128 and 155 mAh g^{-1} that stabilized to 110 and 131 mAh g^{-1} after 20 cycles (Fig. 7d), the maximum we have tried so far. Note that cell D3, as opposed to cell D2, presents an extra voltage feature below 1 V in the first charge. The feature mirrors the voltage charge profile of a Na$_3$P/C cell (Fig. 7e), which nicely demonstrates the way Na$_3$P works as a sacrificial Na source in a full Na-ion cell. Owing to its 0.5 V redox potential versus Na$^+$/Na^0, Na$_3$P initially clamps the positive electrode voltage at a lower potential than that of the negative electrode-carbon starts to uptake Na$^+$ at solely 1.5 V versus Na$^+$/Na, thus resulting in a negative output voltage of the cell to start with. Once the cell is charged, Na$^+$ ions are released from Na$_3$P to compensate for the irreversible SEI formation on carbon with the cell voltage climbing to 1 V; afterwards Na$^+$ is released from the P′2 part of the positive electrode so that the voltage profiles of both cells become nearly identical. Application wise the observed increase in reversible capacities from 71 to 128 and 155 mAh g^{-1} for the D1, D2, and D3 cells, respectively, which occurs without any sacrifice to the capacity retention behaviour upon cycling (Fig. 7d), clearly demonstrates the benefits of pre-producing the P′2 phase and incorporating Na$_3$P as a Na reservoir. Such capacity improvements increase the energy density by 30% between C/P2- Na$_{0.67}$[Fe$_{0.5}$Mn$_{0.5}$]O$_2$ and C/P′2-Na$_1$[Fe$_{0.5}$Mn$_{0.5}$]O$_2$ Na-ion cells, and by an additional of 7% using P′2-Na$_1$[Fe$_{0.5}$Mn$_{0.5}$]O$_2$/Na$_3$P composites rather than solely P′2-Na$_1$[Fe$_{0.5}$Mn$_{0.5}$]O$_2$ powders as positive electrodes (Supplementary Note 4).

Discussion

We have reported the preparation of Na-based alloys (Na$_3$Sb, Na$_3$P and so on) via an approach that relies on the ball milling of Na metal and have demonstrated that such an approach, using either Na or Na$_3$P as reducing agents, can equally be used to easily prepare P′2-Na$_1$[Fe$_{0.5}$Mn$_{0.5}$]O$_2$ layered oxide insertion compound, and to produce 'Na$_{3+x}$V$_2$(PO$_4$)$_2$F$_3$' composites with the $x = 1$ composition being a single phase never reported so far. Such findings provide new insights for combating the irreversible capacity of NIBs, thereby significantly enhancing their

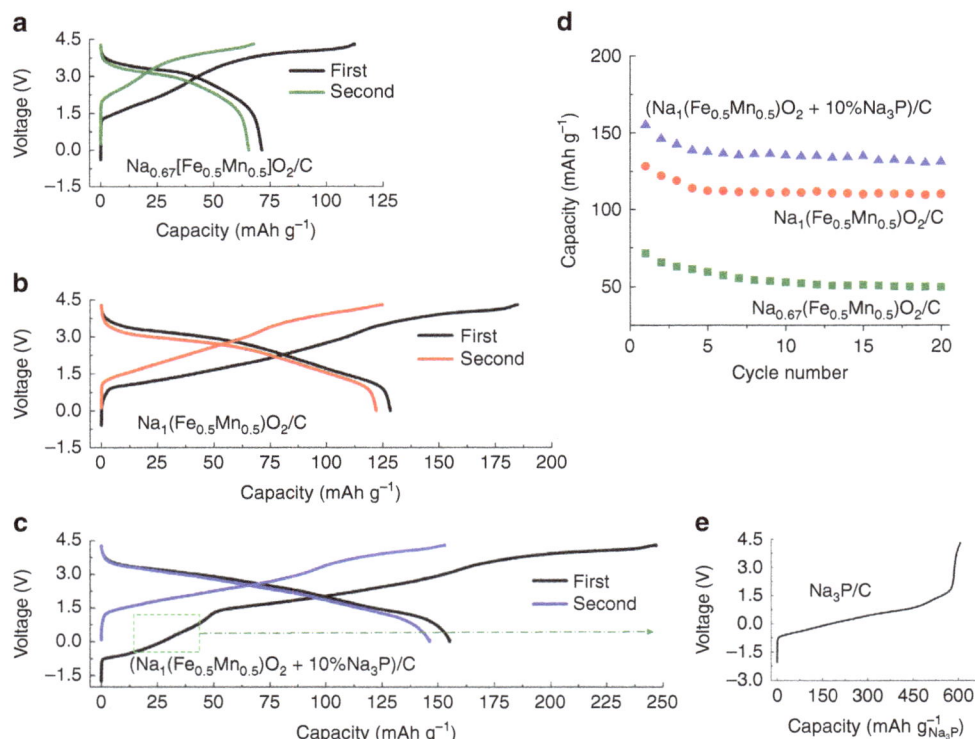

Figure 7 | Performances of C/P2-Na$_{0.67}$[Fe$_{0.5}$Mn$_{0.5}$]O$_2$ Na-ion cells. The voltage-composition profiles are reported for the two first cycles for cells using (**a**) as-made P2-Na$_{0.67}$[Fe$_{0.5}$Mn$_{0.5}$]O$_2$ (cell D1), (**b**) P'2-Na$_1$[Fe$_{0.5}$Mn$_{0.5}$]O$_2$ made by Na ball milling (cell D2) and (**c**) for a P'2-Na$_1$[Fe$_{0.5}$Mn$_{0.5}$]O$_2$/Na$_3$P composite (cell D3) as positive electrode. The negative electrode is C. The weight ratio of positive to negative electrode is 2.6, 1.6 and 1.4 for D1, D2 and D3 cell, respectively. The cells were cycled at 0.1 C rate. The capacity is based on the weight of positive materials, including Na$_3$P for D3 cell. The capacity retention of D1, D2 and D3 cell is given (**d**), and **e** shows the first charge profile of a C/Na$_3$P cell. For all cells a 1 M NaClO$_4$ in EC/DMC (50/50 by volume) electrolyte was used.

performances. We demonstrate the feasibility of assembling full Na-ions cells showing marked enhancements in energy storage density (10–30%) by using 'Na$_{3+x}$V$_2$(PO$_4$)$_2$F$_3$' and P'2-Na$_1$[Fe$_{0.5}$Mn$_{0.5}$]O$_2$ as positive electrodes, respectively, with an extra 7% been achievable for the latter by adding proper amounts of Na$_3$P sacrificial salt to P'2-Na$_1$[Fe$_{0.5}$Mn$_{0.5}$]O$_2$ powders. The improvement in energy density associated to the addition of Na$_3$P is here limited due to the low voltage of cathode at the end of the full discharge of the Na-ion cell. Since N-methyl-2-pyrrolidone (NMP) is used in today's electrode formulation technology we have checked the reactivity of Na$_3$P towards NMP by directly immersing Na$_3$P/C composite in NMP overnight. No change has been observed in the Na$_3$P crystallinity and its electrochemical activity was preserved with a capacity of ~600 mAh g^{-1} when oxidized to 4.3 V. Nevertheless, an inherent difficulty, application wise, with such fully sodiated electrodes is their reactivity towards moisture, hence the need to design coating-grafting techniques to minimize such moisture sensitivity[28] as it is being eagerly pursued in our group. Despite this practical pending issue, we anticipate that the here discussed means for enhancing the performances of NIBs will have strong implications towards their upcoming commercialization.

Methods

Synthesis of Na$_3$P. Stoichiometric amounts of metallic sodium as bulk (Sigma) and red phosphorus (Alfa, 325 mesh) were filled into a hard steel ball-milled jar (30 cm^3) of a Spex 8000M ball-miller in an Ar-filled glove box and equipped with seven hard steel balls each having a weight of 7 g and a diameter of 12 mm. These solid materials were ball-milled for 2 h to obtain Na$_3$P particles. The mass ratio of balls to Na$_3$P was maintained at 35. A similar protocol was used to prepare the reported Na$_3$Sb powders.

Synthesis of P'2 type Na$_1$[Fe$_{0.5}$Mn$_{0.5}$]O$_2$. P2 phase of Na$_{0.67}$[Fe$_{0.5}$Mn$_{0.5}$]O$_2$ was produced by solid-state reaction under 900 °C for 12 h in air[10]. It was ball milled with stoichiometric amount of Na to produce Na$_1$[Fe$_{0.5}$Mn$_{0.5}$]O$_2$ for durations from 20 min to 2 h. Once the obtained powders were single phased, 20 wt% carbon SP was added and the mixture was ball milled for additional 10 min for making electrodes. Excess amounts of Na$_3$P (0.2 mole Na$_3$P per Na$_{0.67}$[Fe$_{0.5}$Mn$_{0.5}$]O$_2$) was also used to transfer P2 into P'2 phase without noting any detrimental effect on the phase formation.

Synthesis of 'Na$_{3+x}$V$_2$(PO$_4$)$_2$F$_3$' composites. Pure NVPF was obtained from CEA via a recently patented process. It was ball milled with increasing amounts of Na ranging from 0.5 Na to 2 Na per mole of NVPF and times ranging from 20 min to 3 h. We find the formation of single phased materials for a stoichiometry of 2 Na and for ball milling time > 3 h (Supplementary Fig. 1). Such samples were shown to contain tiny amounts of remaining Na as deduced by DSC experiments (Supplementary Fig. 3). Equally, we could produce the fully sodiated Na$_4$V$_2$(PO$_4$)$_2$F$_3$ phase by ball milling for 3 h of NVPF with 1 M of Na$_3$P. Composites with adjusted values of x in 'Na$_{3+x}$V$_2$(PO$_4$)$_2$F$_3$' to compensate for the carbon SEI negative electrode were made as above by properly adjusting amount of Na and ball milling time. Once the desired composite obtained, it was ball milled for 10 min with 20% additional carbon SP to make the electrode.

X-ray diffraction. XRD patterns were collected on a Bruker D8-Advance diffractometer equipped with Cu Kα radiation source. Additional synchrotron XRD patterns were collected on powders put in sealed glass capillaries (diameter 0.7 mm) either at the European Synchrotron Radiation Facility on ID22 with $\lambda = 0.3543$ Å (Fig. 3a,c) or at 11BM-Argonne National Lab with $\lambda = 0.4142$ Å (Fig. 3b). The in situ XRD patterns were recorded using electrochemical cells, assembled similarly to our Swagelok cell, but equipped with a beryllium window as current collector on the positive side. These cells were placed on the Bruker D8-Advance diffractometer (Cu Kα radiation) and connected to the VMP2 system. All patterns were analysed using the Rietveld method as implemented in the FullProf program[29]. Phase quantification was performed on the in situ patterns by applying a overall correction on the patterns to account for the absorption from the Be window.

Electrochemical tests. Coin cells were used to study the electrochemical performances. 1 M $NaClO_4$ solution in a mixture of EC/DMC in 1:1 ratio was used as electrolyte for the full cells with C as negative electrode. 5% FEC was systematically added in half cells when Na metal was used as counter electrode. The only exception regards the cells made of NVPF as positive electrode that were initially discharged because of experimented interferences between the decomposition potential of FEC and the insertion reduction potential of Na into NVPF (Supplementary Figs 4 and 5). Two pieces of Whatman glass fibres soaked with the electrolyte were used as separator between the positive and negative electrode. The powders of active materials were mixed with 20% carbon SP using Spex 8000M mixer mill. A typical weight of 5 mg of active electrode material was used per cell whatever Swagelok's or coin cells. The cells were galvanostatic charged/discharged with a VMP automatic cycling/data recording system (Biologic Co., Claix, France) using various ranges of scanning potential and C rates with 1 C corresponding to the uptake or removal of 1 Na^+ and 2 Na^+ per formula unit in 1 h for P2- $Na_{0.67}[Fe_{0.5}Mn_{0.5}]O_2$ and NVPF, respectively. Note: While the present manuscript was under review, the $Na_4V_2(PO_4)_2F_3$ phase was predicted, but not synthesized, through computational calculation by Ceder's group[30].

References

1. Larcher, D. & Tarascon, J. Towards greener and more sustainable batteries for electrical energy storage. *Nat. Chem.* **7**, 19–29 (2014).
2. Kundu, D., Talaie, E., Duffort, V. & Nazar, L. F. The emerging chemistry of sodium ion batteries for electrochemical energy storage. *Angew. Chem. Int. Ed. Engl.* **54**, 3431–3448 (2015).
3. Palomares, V. *et al.* Na-ion batteries, recent advances and present challenges to become low cost energy storage systems. *Energy Environ. Sci.* **5**, 5884–5901 (2012).
4. Kim, Y., Ha, K.-H., Oh, S. M. & Lee, K. T. High-capacity anode materials for sodium-ion batteries. *Chem. Eur. J.* **20**, 11980–11992 (2014).
5. Stevens, D. A. & Dahn, J. R. High capacity anode materials for rechargeable sodium-ion batteries. *J. Electrochem. Soc.* **147**, 1271–1273 (2000).
6. Ponrouch, A. *et al.* Towards high energy density sodium ion batteries through electrolyte optimization. *Energy Environ. Sci.* **6**, 2361–2369 (2013).
7. Park, Y. U. *et al.* A family of high-performance cathode materials for Na-ion batteries, $Na_3(VO_{1-x}PO_4)_2F_{1+2x}$ ($0 \leq x \leq 1$): combined first-principles and experimental study. *Adv. Funct. Mater.* **24**, 4603–4614 (2014).
8. Barpanda, P., Oyama, G., Nishimura, S.-I., Chung, S.-C. & Yamada, A. A 3.8-V earth-abundant sodium battery electrode. *Nat. Commun.* **5**, 4358 (2014).
9. Komaba, S. *et al.* Study on the reversible electrode reaction of $Na_{1-x}Ni_{0.5}Mn_{0.5}O_2$ for a rechargeable sodium-ion battery. *Inorg. Chem.* **51**, 6211–6220 (2012).
10. Yabuuchi, N. *et al.* P2-type $Na_x[Fe_{1/2}Mn_{1/2}]O_2$ made from earth-abundant elements for rechargeable Na batteries. *Nat. Mater.* **11**, 512–517 (2012).
11. Shanmukaraj, D. *et al.* Sacrificial salts: compensating the initial charge irreversibility in lithium batteries. *Electrochem. Commun.* **12**, 1344–1347 (2010).
12. Tarascon, J. M. & Guyomard, D. The $Li_{1+x}Mn_2O_4$/C rocking-chair system: a review. *Electrochim. Acta* **38**, 1221–1231 (1993).
13. Tang, W. S., Chotard, J.-N. & Janot, R. Synthesis of single-phase LiSi by ball-milling: electrochemical behavior and hydrogenation properties. *J. Electrochem. Soc.* **160**, A1232–A1240 (2013).
14. Liu, N., Hu, L., Mcdowell, M. T., Jackson, A. & Cui, Y. Prelithiated silicon nanowires as an anode for lithium ion batteries. *ACS Nano* **5**, 6487–6493 (2011).
15. Ponrouch, A., Goñi, A. R. & Palacín, M. R. High capacity hard carbon anodes for sodium ion batteries in additive free electrolyte. *Electrochem. Commun.* **27**, 85–88 (2013).
16. Billaud, J. *et al.* $Na_{0.67}Mn_{1-x}Mg_xO_2$ ($0 \leq x \leq 0.2$): a high capacity cathode for sodium-ion batteries. *Energy Environ. Sci.* **7**, 1387–1391 (2014).
17. Singh, G. *et al.* An approach to overcome first cycle irreversible capacity in P2-$Na_{2/3}[Fe_{1/2}Mn_{1/2}]O_2$. *Electrochem. Commun.* **37**, 61–63 (2013).
18. Morcrette, M., Gillot, F., Monconduit, L. & Tarascon, J.-M. Ballmilling elaboration of Li-based negative electrode materials. *Electrochem. Solid-State Lett.* **6**, A59 (2003).
19. Fullenwarth, J., Darwiche, A., Soares, A., Donnadieu, B. & Monconduit, L. NiP_3: a promising negative electrode for Li- and Na-ion batteries. *J. Mater. Chem. A* **2**, 2050–2059 (2014).
20. Qian, J., Wu, X., Cao, Y., Ai, X. & Yang, H. High capacity and rate capability of amorphous phosphorus for sodium ion batteries. *Angew. Chem. Int. Ed. Engl.* **52**, 4633–4636 (2013).
21. Xie, Y., Su, H., Li, B. & Qian, Y. Solvothermal preparation of tin phosphide nanorods. *Mater. Res. Bull.* **35**, 675–680 (2000).
22. Jarvis, R. F., Jacubinas, R. M. & Kaner, R. B. Self-propagating metathesis routes to metastable group 4 phosphides. *Inorg. Chem.* **39**, 3243–3246 (2000).
23. Darwiche, A. *et al.* Better cycling performances of bulk Sb in Na-ion batteries compared to Li-ion systems: an unexpected electrochemical mechanism. *J. Am. Chem. Soc.* **134**, 20805–20811 (2012).
24. de Boisse, B. M., Carlier, D., Guignard, M., Bourgeois, L. & Delmas, C. P2 phase used as positive electrode in Na batteries. *Inorg. Chem.* **53**, 11197–11205 (2014).
25. Bianchini, M. *et al.* Comprehensive investigation of the $Na_3V_2(PO_4)_2F_3$-$NaV_2(PO_4)_2F_3$ system by operando high resolution synchrotron X-ray diffraction. *Chem. Mater.* **27**, 3009–3020 (2015).
26. Bianchini, M. *et al.* $Na_3V_2(PO_4)_2F_3$ revisited: a high-resolution diffraction study. *Chem. Mater.* **26**, 4238–4247 (2014).
27. Jian, Z., Sun, Y. & Ji, X. A new low-voltage plateau of $Na_3V_2(PO_4)_3$ as an anode for Na-ion batteries. *Chem. Commun.* **51**, 6381–6383 (2015).
28. Zhao, J. *et al.* Artificial solid electrolyte interphase protected LixSi nanoparticles: an efficient and stable prelithiation reagent for lithium-ion batteries. *J. Am. Chem. Soc.* **137**, 8372–8375 (2015).
29. Rodríguez-Carvajal, J. Recent advances in magnetic structure determination by neutron powder diffraction. *Physica B Condens. Matter* **192**, 55–69 (1993).
30. Matts, I. L., Dacek, S., Pietrzak, T. K., Malik, R. & Ceder, G. Explaining Performance-Limiting Mechanisms in Fluorophosphate Na-Ion Battery Cathodes through Inactive Transition-Metal Mixing and First-Principles Mobility Calculations. *Chem. Mater.* **27**, 6008–6015 (2015).

Acknowledgements

We thank L. Simonin and Y. Chatillon from CEA for sending us the pristine $Na_3V_2(PO_4)_2F_3$ materials that we have been using for the present study; Antonella Iadecola and Carlotta Giacobbe for the collection of Synchrotron X-ray diffraction patterns at European Synchrotron Radiation Facility on ID22; and the RS2E' Na-ion task force (C. Masquelier, L. Croguennec) for helpful discussions and J. Kurzman for his critical reading of the paper. Use of the Advanced Photon Source at Argonne National Laboratory was supported by the US Department of Energy, Office of Science, Office of Basic Energy Sciences, under Contract No. DE-AC02–06CH11357.

Author contributions

B.Z. carried out the synthesis; B.Z., R.D. and J.-M.T. conducted the electrochemical work and designed the research approach; and P.R. produced P2 phase. G.R. analysed the crystal structure and diffraction patterns; A.M.A. carried out the transmission electron microscopy test; and B.Z., G.R. and J.-M.T. wrote the manuscript. All authors discussed the experiments and final manuscript.

Additional information

Putting pressure on aromaticity along with *in situ* experimental electron density of a molecular crystal

Nicola Casati[1], Annette Kleppe[2], Andrew P. Jephcoat[3] & Piero Macchi[4]

When pressure is applied, the molecules inside a crystal undergo significant changes of their stereoelectronic properties. The most interesting are those enhancing the reactivity of systems that would be otherwise rather inert at ambient conditions. Before a reaction can occur, however, a molecule must be activated, which means destabilized. In aromatic compounds, molecular stability originates from the resonance between two electronic configurations. Here we show how the resonance energy can be decreased in molecular crystals on application of pressure. The focus is on *syn*-1,6:8,13-Biscarbonyl[14]annulene, an aromatic compound at ambient conditions that gradually localizes one of the resonant configurations on compression. This phenomenon is evident from the molecular geometries measured at several pressures and from the experimentally determined electron density distribution at 7.7 GPa; the observations presented in this work are validated by periodic DFT calculations.

[1] Paul Scherrer Institute, WLGA/229, CH-5232 Villigen, Switzerland. [2] Diamond light source Ltd., Harwell Science and innovation Campus, Didcot OX11ODE, UK. [3] Institute for Study of the Earth's interior, Okayama University, Yamada 827, Misasa, Tottori 682-0193, Japan. [4] Department of Chemistry and Biochemistry, University of Bern, Freiestrasse 3, Bern CH-3012, Switzerland. Correspondence and requests for materials should be addressed to P.M. (email: piero.macchi@dcb.unibe.ch) or to N.C. (email: nicola.casati@psi.ch).

One of the most important and famous classes of chemical compounds is that of aromatic molecules. According to Hückel[1,2], the conjugation of an odd number of electron pairs in a ring is stabilized by the resonance between two equivalent electronic configurations. The implications for the chemistry of these compounds are enormous: the extra stability of aromatic hydrocarbons implies more severe conditions to induce reactions, compared with non-conjugated poly-olefins. Aromaticity affects also the structure of a molecule and its response to external magnetic fields, in particular, the nuclear resonant frequency. In general, a planar geometry formed by equally distant C atoms, an induced diamagnetic current in the ring and a scarce tendency to react are the main clues of aromaticity[3]. Nevertheless, an unbiased and universal criterion for quantifying aromaticity remains elusive. Over the years, chemists have challenged the very concept of aromaticity at times, synthesizing ever more exotic molecular systems and using different investigating methods (diffraction, NMR, reactivity tests, molecular orbital calculations and so on), with the aim of finding universal criteria based on structural, energetic and magnetic parameters[4-8]. Insight often came by comparing different molecular species, mimicking a continuous variation of the more relevant parameters, to solve the aromatic riddle[3,9]. In this work, instead, we adopt a different strategy to investigate aromaticity, which is probing its variation in a single species, while modifying continuously the molecular geometry through compression.

Reducing the aromaticity of a species requires significant external energy, to stabilize one of the two electronic configurations over the other, breaking the resonance. Typically, chemists make use of heat, light or electrochemical potential to attack an aromatic molecule. An alternative source of external energy is pressure, which rises the internal energy and the enthalpy of a system. This may perturb the molecular conformation and/or the electronic state stable at the ambient conditions, in favour of an otherwise inaccessible configuration.

Recent research in high-pressure (HP) solid-state chemistry led to the discovery of very peculiar phenomena, such as the polymerization of molecules like N_2 (ref. 10), CO (ref. 11) and CO_2 (ref. 12), or the transformation of metals into non-metals, for example, Na (ref. 13). Studies on organic crystals are less abundant and have appeared only more recently[14,15]. Benzene—that is, the prototype of aromaticity—is an exception, because the first, seminal investigations of its HP forms date back to 1960s (ref. 16). While its phase diagram remains controversial, there is consensus on the occurrence of an irreversible polymerization above 24 GPa, but the structure of this phase is known only from theoretical predictions[17], without experimental confirmation yet. Ciabini et al.[18] estimated that below a critical intermolecular distance (C–C $\approx$ 2.6 Å) lattice phonons are able to bring atoms of neighbouring molecules sufficiently close to induce an intermolecular addition reaction, which eventually leads to one or more polymeric products[19]. On crystals of s-triazine, a progressive destabilization of the π bonding orbitals has been reported, based on two-photon induced fluorescence[20], and the enhanced reactivity of the species was ascribed to this process. In both cases, however, details are missing for the pressure-induced distortions of the molecular geometry that could favour reactivity. In fact, the HP crystalline phases of benzene that anticipate the polymerization[21] are not known with sufficient precision to enable fine speculations and the high-pressure structures of triazine are poorly characterized. It is also worth noting that most of related studies are conducted in non-hydrostatic conditions.

To obtain relevant information on the aromaticity of species under pressure, the structural, electronic and energetic changes

should be monitored. The elective method for investigating molecular geometries is single crystal X-ray diffraction, which maps the electron density (ED, $\rho(\mathbf{r})$) distribution in a crystal. Measurements of particular accuracy and completeness are able to reveal not only the maxima of $\rho(\mathbf{r})$ (coinciding with the nuclear sites) but also the smaller fraction of electrons present in between the atoms and responsible of the chemical bonding[22]. The ED mapping from X-ray diffraction is nowadays a well-established technique, but it requires quenching the atomic motion at low temperature for a sufficient deconvolution of the ED from the atomic displacements. In addition, a more sophisticated modelling based on atomic multipolar expansion[23] is necessary to extract the desired information from the diffraction data. In this respect, benzene may not be the perfect test case, because its solidification must occur in the HP apparatus and the crystal sample cannot be of the highest quality. Experimental ED determinations of molecular crystals at HP are not known so far. A few examples reported ED maps of simple inorganic compounds[24] or pure elements[25], obtained by maximum entropy method. Models using multipoles restricted to theoretical values have been tested against X-ray diffraction data for propionamide[26] and piperazinium hydrogen oxalate[27], but no full refinement was so far reported.

The main obstacles to ED mapping from HP X-ray diffraction are due to the pressure apparatus (the Diamond Anvil Cell, DAC), which reduces drastically the resolution, completeness and quality of available data[28]. However, such pitfalls may be overcome by a combination of higher pressure (which significantly attenuates thermal motion even at ambient temperature), modern synchrotron sources (easily providing high-intensity and very short wavelength radiation) and careful experimental strategies; these are discussed in details below.

A complementary approach to study the HP forms of molecules in crystals is first principle calculations, in particular density functional theory (DFT) with periodic boundary conditions[29]. This not only allows to validate the experimental observations and predict the occurrence of new phases, but also to calculate quantities otherwise not available or too difficult to measure, such as electron correlation, current density and electronic energy of a system. Moreover, a theoretical analysis enables extending the pressure range achievable with experiments.

In the following, we report on the experimental and theoretical investigations of an aromatic molecule in its crystal form and we analyse how its aromatic character is reduced by the pressure-induced modifications of the molecular geometry.

Results

HP single crystal X-ray diffraction. Our investigation focused on a doubly bridged annulene, namely syn-1,6:8,13-Biscarbonyl[14] annulene (BCA), Fig. 1, for which high-quality crystals are available. Initially studied within the debate on aromaticity of annulenes[30], BCA was also the subject of a very detailed ED study at ambient P and low T (19 K) (ref. 31), which is an excellent benchmark. Due to the strain in the ring, the aromaticity of BCA is quite smaller than for benzene, therefore, a stronger response to perturbation is expected.

Single crystals of BCA were investigated with multi-temperature and multi-pressure X-ray diffraction, to determine experimentally the structural changes on varying the thermo-dynamic conditions. HP diffraction experiments were carried out at the I15 beamline of Diamond light source using a pinhole defined monochromatic beam with 0.31 Å wavelength and an Atlas CCD detector. Six simple data collections, based on perpendicular scans (ω and φ at $\chi = 90$) with frontal detector (resulting in a resolution up to 0.8 Å), were performed in the

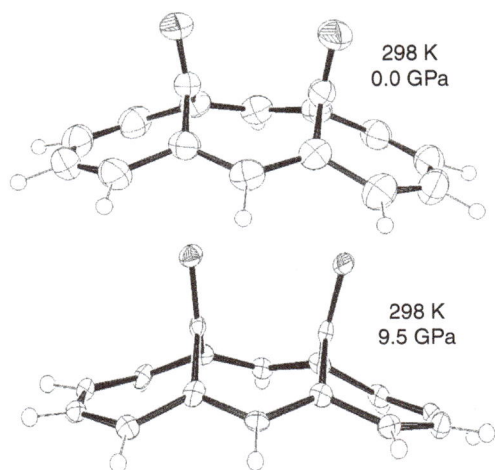

Figure 1 | BCA molecule. The molecular geometry and atomic displacement parameters of BCA at ambient pressure and at 9.5 GPa.

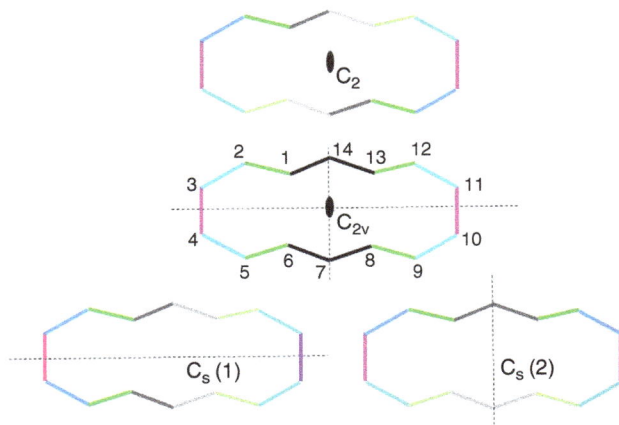

Figure 2 | The possible configurations of BCA. BCA symmetries (only the annulene skeleton is drawn): C_{2v} is the most symmetric isomer, stable in the gas phase and very close to the molecule in the crystal at ambient pressure; $C_s(1)$ is the sub-symmetry which is close to the configuration found at HP; C_2 and $C_s(2)$ are the other two possible sub-symmetries. Note that only $C_s(1)$ implies breaking the resonance between ψ_1 and ψ_2.

range from 0.0001 to 9.5 GPa. These were intended to determine the main structural changes occurring to the molecule as a function of pressure, up to the hydrostatic limit of the pressure-transmitting medium that we adopted (methanol:ethanol 4:1 mixture; *ca.* 10 GPa). The data collection at 7.7 GPa (experiment EE7741-1) was more extensive, involving several ω and φ scans at four different χ positions (0°, 30°, 60° and 90°), using different θ positions for the detector, as it was intended to measure with the best accuracy the diffraction intensities and to reach the highest resolution ($d = 0.5$ Å). Accuracy here means: (a) high data completeness (that is, the portion of reciprocal space that is measurable), obtained using two crystals in the DAC; (b) redundant measures of the diffracted intensities (with every ω and φ scan re-performed with a 2° offset of φ and ω, respectively) to minimize random errors and therefore enable a multipolar expansion of the ED. The high-energy radiation chosen reduces absorption effects and allows collection of higher-order reflections. A beam smaller than the crystal was selected by a 30-μm pinhole, to probe one crystal at a time and to maximize the sample/diamond diffraction intensity ratio. Periodic and molecular DFT calculations were used to simulate the structures, validate the experimental results, compute the geometry at pressures above the experimentally available range and compute those properties that are not directly accessible through experiments, like the current densities and the electron delocalization indices. Details of all experiments and calculations are in Methods section and in Supplementary Methods.

Theoretical calculations and molecular geometries. With its 14 carbon atom ring, BCA is a $4n + 2$ Hückel system, therefore potentially aromatic. However, the two carbonyl bridges distort the planarity of the ring, reducing the conjugation of C–C bonds and therefore decreasing the aromaticity of the system. The C–C distances are quite important indicators: in an aromatic molecule, the ring skeleton bonds should have homogeneous distances, quite shorter than single bonds but longer than double bonds. In the gas phase of a molecule like benzene, all C–C distances are equal by symmetry (1.389 Å), whereas in BCA the lower molecular symmetry (C_{2v}) does not imply equalization of C–C distances and the two bridges produce an heterogeneous distribution in the range 1.37–1.41 Å. Four C–C bonds are symmetry independent, Fig. 2. Gas phase DFT calculations on the isolated molecule (at B3LYP/6–31 + G(2d,2p) level) indicate that C1–C14, C2–C3 and all their symmetry equivalents, are shorter whereas C1–C2, C3–C4 and all their symmetry equivalents, are

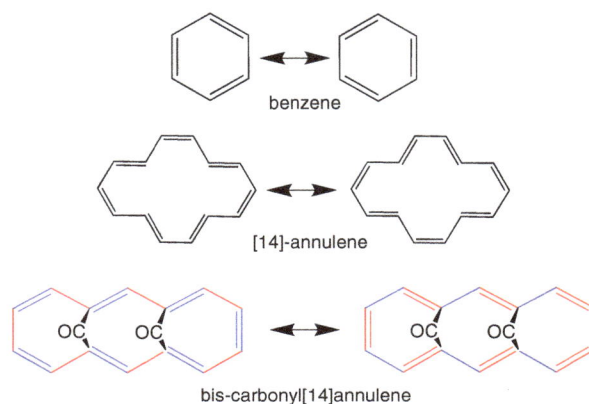

Figure 3 | Resonant configurations in aromatic molecules. Skeletal formulae of the resonant configuration in selected aromatic molecules.

longer. Nevertheless, a C_{2v} symmetry still implies a perfect resonance between the two electronic configurations ψ_1 and ψ_2 (Figs 2 and 3). In the crystal phase, the BCA molecule sits on a general position (therefore, without any intramolecular symmetry element), although its geometry remains close to the gas phase C_{2v} isomer, as confirmed by X-ray diffraction. The small root mean square deviation from C_{2v} progressively increases on lowering the temperature (from 0.031 Å at 298 K to 0.036 Å at 19 K). It is worth noting that, even in the absence of carbonyl bridges, a ring strain significantly affects the aromaticity of the 'parent', unbridged [14]-annulene (Fig. 3), which is in fact not planar[32].

The aromaticity is not only reflected by the C–C distances, hence by the position of the $\rho(\mathbf{r})$ maxima, but also by the amount of ED in the chemical bonds, correlated with the bond strength. In this respect, the detailed experimental study by Destro and Merati[31] on BCA went beyond a routine geometrical analysis, providing also an accurate determination of $\rho(\mathbf{r})$ at 19 K. At the bond critical points, (that is, saddle points of the three-dimensional ED function, according to Bader's QTAIM[33]), $\rho(\mathbf{r})$ parallels the bond distances and confirms the perfect resonance between the two electronic configurations. Interestingly, this analysis revealed an unexpected bond critical point interconnecting the two bridging carbons (C15–C16 in Supplementary Fig. 1). Although the ED at the critical point is small, the feature is not anticipated from a simple Lewis structural formula of BCA. Moreover, the C–C distance of 2.593 Å is more than 1 Å longer than a typical covalent bond. A Møller-Plesset[34] perturbation calculation on the isolated molecule, indicates a small population of two virtual molecular orbitals containing in-phase combinations of the π^*-type C=O orbitals. Nevertheless, this interaction is principally of closed-shell type.

As anticipated, the aromaticity also affects the magnetically induced current density in the ring. This modifies substantially the interaction of an external magnetic field with the spin active nuclei, as it can be monitored by nuclear magnetic resonance. However, the atomic chemical shift may depend on the position of the spin active nuclei (most often the protons, revealed by H^1 NMR). Therefore, geometries of cyclic molecules that differ for conformation may introduce biases because of shielding/deshielding effects not directly related to the aromatic behaviour. For this reason, a nuclear-independent chemical shift (NICS) indicator has been introduced[35] to provide an unbiased indication of the ring current. NICS is available only from theoretical simulations, because it requires calculating the shielding due to the ring current in the centre of the ring, assuming a virtual atom in that site. The NICS is the negative of the shielding tensor, for sake of consistency with traditional NMR chemical shifts. We calculated a negative NICS at the centre of the BCA ring, which implies the diatropic ring current produced by aromaticity[3].

As mentioned in the introduction, our goal was observing how all parameters, correlated with aromaticity, vary when the molecule is compressed. The HP experiments revealed a quite large compressibility of the crystal, which exceeds 25% from ambient pressure up to 9.5 GPa, in keeping with theoretical predictions (Fig. 4): the experimental K_0, calculated using a fourth order Birch–Murnagham equation, is only 7.4 GPa (compared, for example, to 37.1 GPa for quartz)[36]. As the crystal shrinks, the electric field experienced by a molecule and generated by all other molecules in the crystal, increases significantly (Fig. 5). Because the molecule sits on a general position, inside the monoclinic unit cell, atoms which would be equivalent under the ideal C_{2v} symmetry of the molecule, experience a different crystal electric field and this asymmetry increases with pressure. Therefore, an external stress gradually, but not uniformly, perturbs all the covalent bonds of the annulene skeleton.

Good indicators of this phenomenon are the average C–C distances of the hypothetical double bonds for each of the two resonant configurations (Figs 3 and 6). At ambient P, the observed and calculated pseudo-C_{2v} symmetry implies that the two sets of bonds have coincident average distances (1.396 Å). As P increases, however, one of the two resonant configurations of Fig. 3 ($\psi 1$) becomes progressively dominant. This is evident, because all the double bonds of this configuration shorten with respect to the ambient pressure geometry, whereas all the hypothetical double bonds of the alternative configuration (ψ_2) are unaltered or even slightly elongated (see also Supplementary

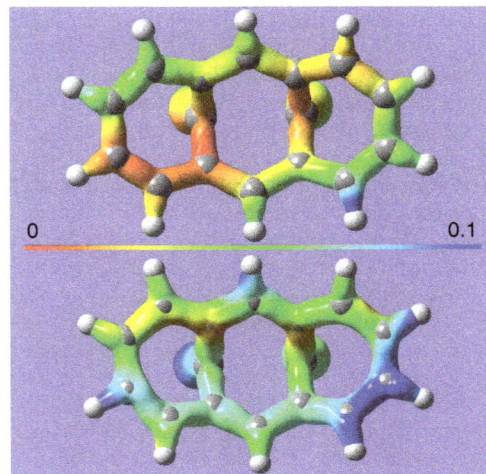

Figure 5 | BCA electric field. The magnitude of the electric field acting on a BCA molecule, plotted on a electron density isovalue surface of 0.2 a.u. of BCA at 0.0 GPa (top) and 7.7 GPa (bottom). The colour scale for the electric field is in atomic units (= 5.1 × 10^11 Vm^−1, blue). a.u., arbitrary units.

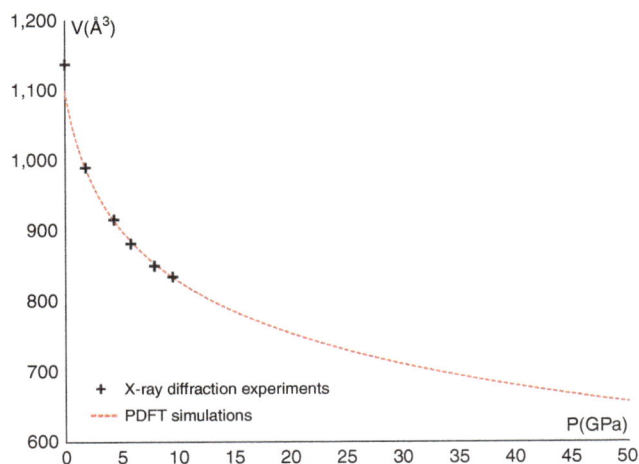

Figure 4 | BCA compressibility. The unit cell volume as a function of pressure, from single crystal diffraction experiments and from periodic DFT calculations.

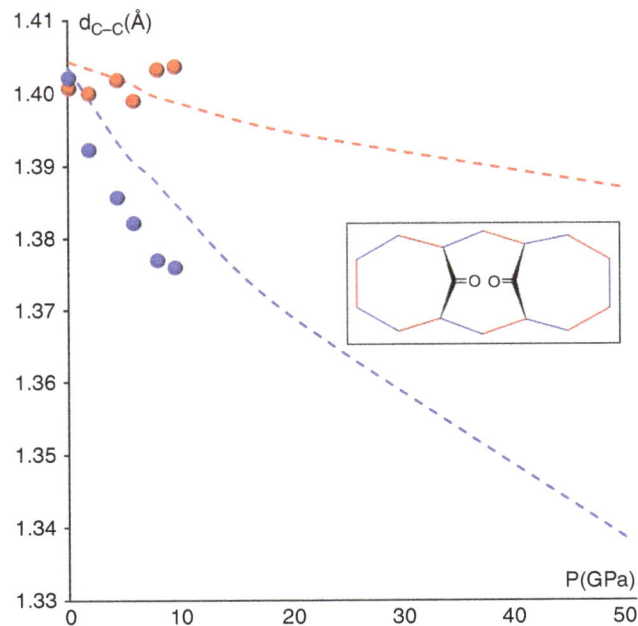

Figure 6 | Resonant configurations bond evolution. The average bond distances for hypothetical C–C double bonds of ψ_1 and ψ_2 (blue and red, following Fig. 3) as a function of pressure from the single crystal measurements (circles) or the theoretical calculations (dashed lines). Experimental data are corrected for thermal libration effects.

Fig. 7 and Table 4). From the experimental structure at 9.5 GPa, the average distances of the two sets of bonds are 1.375 and 1.405 Å, respectively. The gap is even larger at 50 GPa, from the theoretical simulations (1.338 vs 1.387 Å, see also Supplementary Fig. 8) This shows that BCA clearly distorts from C_{2v} symmetry, but the C–C bonds in the annulene ring respect one of the two mirror symmetries, namely the $C_s(1)$ configuration represented in Fig. 2. This distortion destabilizes the molecule (by ca. 65 kcal mol^{-1}) not only because of breaking the aromaticity, but also because of other geometrical distortions imposed by the reduced volume available for each molecule, including a significant bending of the bridging CO's, Fig. 1. The distance between them shortens from 2.591 (ambient pressure) to 2.525 Å (at 9.5 GPa, from experiment) or even 2.46 Å (at 50 GPa, from calculations), at the expense of the C = O bonds that experience a small but continuous elongation under pressure. This observation confirms the weak bonding proposed by Destro and Merati[31], and it suggests a slight strengthening with pressure.

Electron and current density distribution. The partial localization of one electronic configuration should be visible also from the analysis of the ED distribution. For this reason, we determined the accurate $\rho(\mathbf{r})$ from the X-ray diffraction intensities measured at 7.7 GPa. We used two methods to reconstruct $\rho(\mathbf{r})$: the traditional multipolar refinement[22] and the X-ray constrained wave function[37]. Details of the model refinements are in SupplementaryMethods, whereas the technical details on the necessities and pitfalls of ED determinations at HP will be subject of a forthcoming paper. Here we report only on the most relevant results of the $\rho(\mathbf{r})$ analysis. To validate the experimental models, the ED was also computed at various pressure points with periodic DFT calculations.

In Fig. 7, we plot $\rho(\mathbf{r})$ at isosurface values (0.305 arbitrary units) chosen to visualize the ED level of the mixed single–double bonds of the molecular skeleton. At 0.0001 GPa, the theoretical calculations and the experimental model[31] show an almost perfectly symmetric distribution. The bonds C1–C14 and C2–C3 (and all pseudosymmetry related ones, Fig. 2) display larger amount of ED, in agreement with their shorter distances, whereas C1–C2 and C3–C4 and all pseudosymmetry equivalents, have lower density. This scenario respects the resonance scheme, that,

however, partially breaks at 7.7 GPa. In fact, both the experimental models and the periodic DFT calculations demonstrate that bonds C2-C3, C4-C5 and C6-C7 gain ED whereas C1-C2, C3-C4 and C5-C6 loose it (Fig. 7). In the left part of the molecule, the scenario is now closer to a localized configuration because the ED accumulations are clearly associated only with the hypothetical double bonds of ψ_1 in Fig. 3. On the other hand, in the right part of the molecule a larger delocalization persists. To explain this, one should consider that localizing ψ_1 implies strengthening bonds C8-C9, C10-C11 and C12-C13 (originally weaker at ambient conditions) at the expense of C7-C8, C9-C10, C11-C12 and C13-C14 (originally stronger). Anyway, C8-C9, C10-C11 and C12-C13 have increased the ED amount compared with their pseudo-symmetric counter parts C5-C6, C3-C4 and C1-C2. A complete localization of ψ_1 would eventually occur at 50 GPa, where only periodic DFT calculations are available without confirmation from the experiment. At this pressure, all the double bonds of ψ_1 are associated with larger amounts of ED peaks, compared with all single bonds. The molecule has therefore become closer to a cyclic non-aromatic poly-ene.

The qualitative agreement between the experimental models (multipole or X-ray constrained wave function) and the first principle calculations (periodic DFT) at 7.7 GPa is remarkable, which excludes potential biases in the analysis and it allows to thrust the theoretical values computed at 50 GPa.

From the multipolar model or the X-ray constrained wave function, we could also determine the molecular graph (Fig. 8), that is, the set of lines of maximum ED (bond paths) that interconnect bonded atoms, following QTAIM. The partial localization of one of the two resonant configurations is visible also from the increased ED at the critical points (saddle points along the bond paths). On average, $\rho(\mathbf{r})$ is 2.29 and 2.15 e Å^{-3} in the partial double or single bonds, respectively, which varies from what observed at ambient pressure at 19 K (on average, 2.05 e Å^{-3} for both configurations).

As observed at the ambient P and low T by Destro and Merati[31], a bond path links the two carbonyl carbons, a feature confirmed also by the periodic DFT calculations.

Having an experimental wave function available, we can compute other quantities, otherwise not available from the

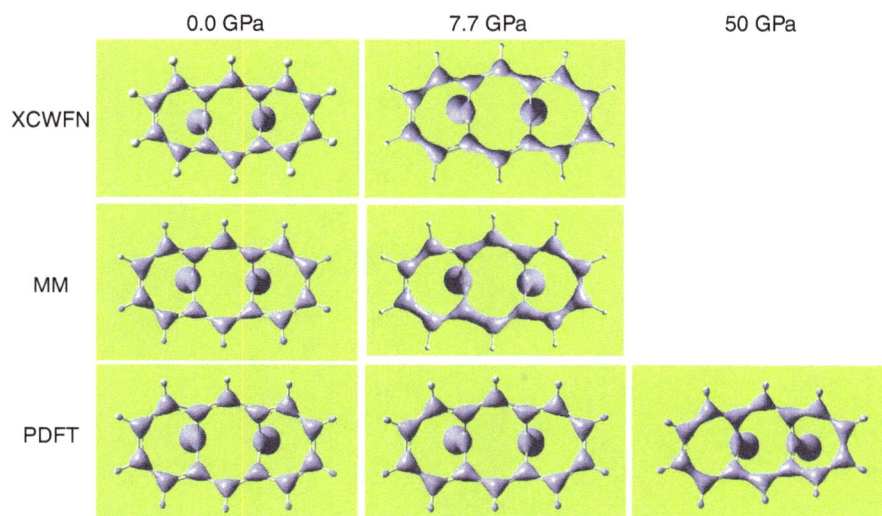

Figure 7 | ED distribution of BCA. Plots are shown at various pressure and from various sources: PDFT are periodic DFT calculations at B3LYP level of theory; XCWFN are X-ray constrained wave functions computed at Hartree–Fock level, but constrained against the experimentally measured diffraction intensities; MM is the electron density derived from multipolar expansion, with coefficients refined against experimentally measured intensities. Experimental data at ambient pressure are taken from Destro and Merati[31], collected at 19 K; the 7.7 GPa data are from this work.

multipolar model only; in particular, the delocalization indexes (δ) (ref. 38) that address the amount of electron pairs shared between two atoms. While the theoretical calculations for the C_{2v} symmetric geometry give on average $\delta_{C-C} = 1.34e$ for both configurations, at 7.7 GPa the partially localized double bonds have $\delta_{C-C} = 1.42e$ against $\delta_{C-C} = 1.28e$ of the partially localized single bonds. For the theoretically calculated structure at 50 GPa, $\delta_{C-C} = 1.47e$ and $1.22e$, respectively, in keeping with the further shortening of the double bonds and lengthening of the single bonds. Noteworthy, the non-planar and highly strained geometry of BCA would hamper a full localization of double bonds and the ideal $\delta_{C-C} = 2.0e$.

As anticipated above, the aromaticity of a molecule also affects the current density in the ring and the shielding experienced by the atomic nuclei. Current density can only be calculated, with gauge invariant orbitals for molecules, after geometry optimizations in the crystal at various pressure points. These results are illustrated in Fig. 9, where one can easily see the modified orientation of the current density vectors at 50 GPa. The NICS can be also used to analyse the pressure-induced changes. Recent works suggest to scan the out of plane components along a direction perpendicular to the ring[39]. In BCA, we can scan only along the direction opposed to the two CO bridges, otherwise the shielding of $C=O$ would severely interfere. In agreement with standard aromatic systems, NICS is a bit larger (in absolute value) at ca. 0.5–1.0 Å out of the plane, and then it decreases (Fig. 9c). NICS decreases from the ambient pressure to the 50 GPa structure. This is a further proof, based on magnetic criteria, that the aromaticity of BCA decreases as a function of pressure, in keeping with the structural, electronic and energetic criteria above discussed.

Discussion

We have analysed the pressure-induced loss of aromaticity of a carbonyl annulene. The molecule activates by partially localizing one of the two resonant configurations. This mechanism may occur also in other aromatic systems under pressure and could be

representative of the steps that anticipate addition reactions leading to polymerization, like in benzene.

A fortunate circumstance for BCA is that the activation mechanism is induced by a smaller pressure and it does not damage the crystal quality, thus the molecular geometries as well as the ED could be determined with sufficient accuracy to elucidate many important details.

This observation of progressive aromatic loss in BCA is extremely useful to test how different indicators of aromaticity respond to drastic changes of the molecular geometry. Criteria based on C–C distances appear to be extremely sensitive. Moreover, they are quite easy to determine whether single crystal X-ray diffraction is available. Deviations from a uniform distribution of C–C bond distances directly correlate with the raise of molecular energy that can be determined with theoretical methods but not experimentally. While magnetic criteria are also sensitive to an increased or decreased aromaticity, NICS are only available from theoretical calculations, whereas the ^{1}H NMR chemical shifts, although measurable even at HP, may not reveal so directly the ongoing changes of electronic configuration. In fact, at each pressure point, the protons would probe the ring current in different positions, due to the geometrical distortions that involve themselves as well. Therefore, they would be unable to reveal the actual modifications of the current density due only to the breaking of aromaticity.

In this study, we have also presented a way to determine experimentally the charge density of a molecular crystal under pressure, which provides additional indicators of the aromaticity based on the topological analysis of the ED. We believe the use of a synchrotron is presently mandatory to achieve the necessary quality and quantity of unique reflections; potential improvements on our setup include the use of poly-nanocrystalline diamonds, which should eliminate the problem of diamond dips (though introducing a significant background), the use of photon counting detectors (possibly with a high Z material such as GaAs), which would reduce the noise and an improved usage of panoramic cells, presently limited in the literature. The ongoing research on new liquid jet microsources of relatively high energy may also enable in the near future such experiments with laboratory sources, nevertheless the present limitations are difficult to overcome and limit the possibilities of such studies, though not their potential. As we have shown for BCA at 7.7 GPa, it was possible to successfully refine a full multipolar model and an X-ray constrained wave function. This enabled us to visualize the partial localization of double and single bonds, which eventually becomes more complete at higher pressure, according to theoretical predictions. ED criteria are seamlessly replicating the geometric criteria that, in this case, provide easily accessible and sufficiently reliable indicators of even minor changes of aromaticity. This cross-validation opens up the possibility of using this method for far more complex observation of ED localization in molecular systems.

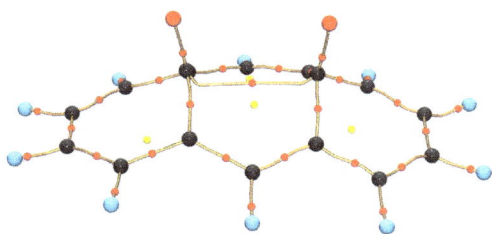

Figure 8 | BCA bonding scheme. Molecular graph for BCA at 7.7 GPa from the experimentally refined multipolar model. Atoms are represented by black (carbon), red (oxygen) and blue (hydrogen) spheres, critical points are shown as small red dots. Noteworthy, depending on the refinement model, a bond path interconnecting the two O atoms can also be localized.

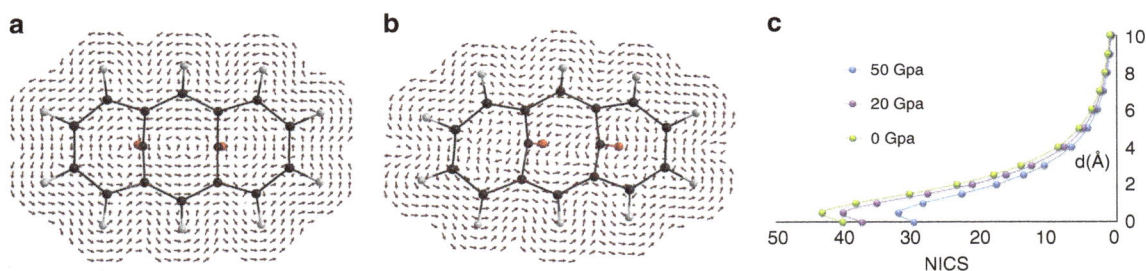

Figure 9 | BCA ring current. The calculated current density $J(\mathbf{r})$ for the molecule of BCA at 0.0001 GPa (**a**) and 50 GPa (**b**). (**c**) NICS scan perpendicularly to the average plane of the BCA ring (anti with respect to the bridging carbonyls).

Methods

HP diffraction measurements. The two separate kinds of experiments were conducted using a Betsa and an own University of Oxford screw driven-type DAC equipped with 0.5 mm culet diamonds. In the first setup a single crystal of BCA was loaded, using a methanol;ethanol 4:1 mixture as pressure medium and ruby fluorescence for pressure measurement. At $P = 7.7$ GPa, a second setup was adopted, using two crystals pre-oriented with crystallographic axis almost normal to each. A sufficiently high redundancy was also sought to correct for problems such as diamond dips, which were also separately identified by recording transmission scans through the cell using a diode immediately after the cell itself. With this setup, the overall data redundancy was 4.7 (6.0 for data up to 0.8 Å resolution) and 70% of the unique reflections were measured (88% for data up to 0.8 Å resolution). This guarantees a sufficient sampling of the reciprocal space, especially in the region where the valence electrons are mostly scattering.

For both setups, a monochromatic radiation of 40 keV was focused down to about $90 \times 90\,\mu m^2$ and then collimated by pinhole of 30 μm in diameter. In the 7.7 GPa experiment the full beam size was always probing only one of the two crystals, which were separately centred and measured. The beam was notably smaller than the crystal, which means all the beam was effectively used for diffraction by the crystal itself. In the case of beams larger than the crystal, problems may arise from the significant diffraction of diamonds, saturating the detector and of the other elements of the cell, which contribute to background. Data were treated using the dedicated HP routines present in the package of CrysalisPro[40] for shading areas, carefully assigning a well-describing vector and opening angle to the cell. Diamond reflections were individually masked in a similar way to already described procedures[41], which proved also important in obtaining a smooth background subtraction from the programme itself. Suspiciously badly fitting reflections were investigated and manually rejected when: (a) their intensity was significantly lower than their equivalents and on the border of masks, (b) their intensity was significantly higher than their equivalents and on the tails of a diamond reflection and (c) they were on images collected at angles were an obvious diamond dip occurred, as revealed by the mentioned transmission scans. No change of space group was detected up to the maximum pressure; therefore, standard refinements were carried on the known structure using ShelXL[42] as included in the WingX package[43]. The non-standard P2$_1$/n space group was used for all determinations for sake of consistency with previous studies[31].

Experimental determination and modelling of the ED distribution at 7.7 GPa. To obtain an accurate ED mapping, it is necessary to collect accurately and extensively the X-ray data, up to a sufficient resolution. High resolution is necessary because the large number of parameters of a multipolar model requires many more intensities to match a sufficient observation/parameter ratio. The data set should be sufficiently complete to avoid systematic effects in the refinement. This is particularly cogent for the low angle data, because the valence electrons, which are mainly responsible for aspherical scattering, are not contributing to reflection intensities at higher resolution. To reach the goal of determination of the ED in a crystal, it is also necessary a very accurate measurement of the reflection intensities, which means minimizing the effect of experimental errors such as absorption by the sample, extinction and, very important here, absorption by diamonds and metal gasket. Apart from accurate correction of the data, the modern area detector technologies offer exceptional possibility to improve the precision of the measurement by repeated collection of the same intensities.

The data set collected at 7.7 GPa was an ideal candidate to attempt a determination of accurate ED in BCA, because the pressure is sufficient to reduce the thermal parameters by a factor of *ca.* 4 (making the atomic displacement parameters comparable to those measured at 120 K) and because both crystals were still sufficiently free from damages. Below this pressure, the atomic motion is still too large and above it the hydrostaticity of the medium decreases and therefore some damage could occur to the samples, which does not affect a conventional structure determination but hampers any accurate ED mapping. Data from the two difference crystals were linearly scaled and merged without any weighting scheme using the appropriate routine in WingX.

A full multipolar model could be refined based on the 7.7 GPa data, expanding each C and O atom up to an octupole level and each H atom up to a dipole level (refining only the bond directed dipole). H positions were fixed at values calculated from the periodic DFT calculations. The final R factor is larger than what one could obtain from low temperature experiments at ambient pressure (*ca.* 2% versus 6%), but an accurate analysis of the residuals reveal that they obey a normal distribution and the larger peaks do not occur in important regions of the molecules. This implies that, despite being noisy, the data do not contain systematic errors and the model is therefore not significantly biased. The program XD2006[44] was used for the multipolar modelling and for the calculation of the ED and the molecular graph. MolCoolQT[45] and Gaussview[46] were used for plotting isosurfaces.

Theoretical calculations. Calculations on molecules (structural optimization, ED, NICS and wing current) were carried out with Gaussian09 (ref. 47), using B3LYP/6-31(2d,2p) level of theory. AimAll[48] was used to compute and visualize the ED, the theoretical molecular graph and the ring current. Calculations on periodic systems (geometry optimization and ED at various pressure points)

were carried out at the same level of theory, including empirical corrections for dispersion effects. The program CRYSTAL14 (ref. 29) was used.

X-ray constrained wave function calculations were carried out using the program TONTO[49]. This method is a calculation of molecular orbitals by means of a modified variational approach, which minimizes a function that couples the Hartree–Fock energy of a molecule and the experimentally observed structure factors of its crystal. The same basis set of previous calculations was used. X-ray constrained wave functions enable to exploit the experimental information, though avoiding a dangerous over-fitting of noisy data that could occur using the multipolar model. This is especially cogent when data are not of the highest quality, such as the X-ray diffraction measured from crystal in a DAC.

References

1. Hückel, E. *Grundzüge der Theorie ungesättigter und aromatischer Verbindungen* 71–85 (Verlag Chemie, 1938).
2. Von, W., Doering, E. & Detert, F. L. Cycloheptatrienylium Oxide. *J. Am. Chem. Soc.* **73**, 876–877 (1951).
3. Stanger, A. What is aromaticity: a critique of the concept of aromaticity—can it really be defined? *Chem. Commun.* 1939–1947 (2009).
4. Krygowski, T. M. & Cyranski, M. K. Structural aspects of aromaticity. *Chem. Rev.* **101**, 1385–1419 (2009).
5. Mitchell, R. H. Measuring aromaticity by NMR. *Chem. Rev.* **101**, 1301–1315 (2001).
6. Gomes, J. A. N. F. & Mallion, R. B. Aromaticity and ring currents. *Chem. Rev.* **101**, 1349–1383 (2001).
7. Bürgi, H. B. Getting more out of crystal-structure analyses. *Helv. Chim. Acta* **86**, 1625–1640 (2003).
8. Abersfelder, K., White, A. J. P., Rzepa, H. S. & Scheschkewitz, D. A tricyclic aromatic isomer of hexasilabenzene. *Science* **327**, 564–566 (2010).
9. Hey, J. *et al.* Heteroaromaticity approached by charge density investigations and electronic structure calculations. *Phys. Chem. Chem. Phys.* **15**, 20600 (2013).
10. Eremets, M. I., Gavriliuk, A. G., Trojan, I. A., Dzivenko, D. A. & Boehler, R. Single-bonded cubic form of nitrogen. *Nat. Mater.* **3**, 558–563 (2004).
11. Lipp, M. J., Evans, W. J., Baer, B. J. & Yoo, C. S. High-energy density extended CO solid. *Nat. Mater.* **4**, 211–215 (2005).
12. Santoro, M. & Gorelli, F. A. High pressure solid state chemistry of carbon dioxide. *Chem. Soc. Rev.* **35**, 918–931 (2006).
13. Ma, Y. *et al.* Transparent dense sodium. *Nature* **458**, 182–185 (2009).
14. Boldyreva, E. High-pressure diffraction studies of molecular organic solids. A personal view. *Acta Crystallogr. A* **A64**, 218–231 (2008).
15. Katrusiak, A. High-pressure crystallography. *Acta Crystallogr. A* **A64**, 131–148 (2007).
16. Piermarini, G. J., Mighell, A. D., Weir, C. E. & Block, S. Crystal structure of benzene II at 25 kilobars. *Science* **165**, 1250–1255 (1969).
17. Wen, X.-D., Hoffmann, R-. & Achcroft, N. W. Benzene under High Pressure: a Story of Molecular Crystals Transforming to Saturated Networks, with a Possible Intermediate Metallic Phase. *J. Am. Chem. Soc.* **133**, 9023–9035 (2011).
18. Ciabini, L. *et al.* Triggering dynamics of the high-pressure benzene amorphization. *Nat. Mater.* **6**, 39–43 (2006).
19. Fitzgibbons, T. C. *et al.* Benzene-derived carbon nanothreads. *Nat. Mater.* **15**, 43–47 (2015).
20. Fanetti, S., Citroni, M. & Bini, R. Tuning the aromaticity of s-triazine in the crystal phase by pressure. *J. Phys. Chem. C* **118**, 13764–13768 (2014).
21. Budzianowski, A. & Katrusiak, A. Pressure-frozen benzene I revisited. *Acta Crystallogr. A* **B62**, 94–101 (2006).
22. Coppens, P. *X-ray Charge Density and Chemical Bonding* (Oxford University, 1997).
23. Hansen, N. K. & Coppens, P. Testing aspherical atom refinements on small-molecule data sets. *Acta Crystallogr. A* **A34**, 909–921 (1978).
24. Yamanaka, T., Okada, T. & Nakamoto, Y. Electron density distribution and static dipole moment of KNbO3 at high pressure. *Phys. Rev. B* **80**, 094108 (2009).
25. Tse, J. S., Klug, D. D., Patchkovskii, S. & Dewhurst, J. K. Chemical bonding, electron-phonon coupling, and structural transformations in high-pressure phases of Si. *J. Phys. Chem. B* **110**, 3721–3726 (2006).
26. Fabbiani, F. P. A., Dittrich, B., Pulham, C. R. & Warren, J. E. Towards charge-density analysis of high-pressure molecular crystal structures. *Acta Crystallogr. A.* **67**, C376 (2011).
27. Macchi, P. & Casati, N. Strong hydrogen bonds in crystals under high pressure. *Acta Crystallogr. A* **67**, C163–C164 (2011).
28. Katrusiak, A. High-pressure crystallography. *Acta Crystallogr. A* **A64**, 135–148 (2008).
29. Dovesi, R. *et al.* CRYSTAL14: a program for the ab initio investigation of crystalline solids. *Int. J. Quantum Chem.* **114**, 1287–1317 (2014).
30. Destro, R. & Simonetta, M. Syn-1,6 : 8,13-Bisearbonyl[14]annulene. *Acta Crystallogr. A* **B33**, 3219–3221 (1977).
31. Destro, R. & Merati, F. Bond lengths, and beyond. *Acta Crystallogr. A* **B51**, 559–570 (1995).
32. Chiang, C. C. & Paul, I. C. Crystal and molecular structure of [14]annulene. *J. Am. Chem. Soc.* **94**, 4741–4743 (1972).

33. Bader, R. W. F. *A Quantum Theory* (Oxford University Press, 1990).
34. Møller, C. & Plesset, M. S. Note on an approximation treatment for many-electron systems. *Phys. Rev.* **46**, 618–622 (1934).
35. Merino, G., Heine, T. h. & Seifert, G. The induced magnetic field in cyclic molecules. *Chemistry* **10**, 4367–4371 (2004).
36. Angel, R. J., Allan, D. R., Miletich, R. & Finger, W. The use of quartz as an internal pressure standard in high-pressure crystallography. *J. Appl. Crystallogr.* **30**, 461–466 (1997).
37. Jayatilaka, D. & Grimwood, D. J. Wavefunctions derived from experiment. I. Motivation and theory. *Acta Crystallogr. A* **A57**, 76–86 (2001).
38. Bader, R. F. W. & Stephens, M. E. Spatial localization of the electronic pair and number distributions in molecules. *J. Am. Chem. Soc.* **97**, 7391–7399 (1975).
39. Stanger, A. Nucleus-independent chemical shifts (NICS): distance dependence and revised criteria for aromaticity and antiaromaticity. *J. Org. Chem.* **71**, 883–893 (2006).
40. CrysAlis PRO. Agilent Technologies UK Ltd, Yarnton, England, 2014.
41. Casati, N., Macchi, P. & Sironi, A. Improving the quality of diamond anvil cell data collected on an area detector by shading individual diamond overlay. *J. Appl. Crystallogr.* **40**, 628–630 (2007).
42. Sheldrick, G. M. A short history of SHELX. *Acta Crystallogr. A* **64**, 112–122 (2008).
43. Farrugia, L. J. WinGX suite for small-molecule single-crystal crystallography. *J. Appl. Crystallogr* **32**, 837–838 (1999).
44. Volkov, A. *et al.* XD2006 - A Computer Program Package for Multipole Refinement, Topological Analysis of Charge Densities and Evaluation of Intermolecular Energies from Experimental and Theoretical Structure Factors (2006).
45. Hübschle, C. B. & Dittrich, B. MoleCoolQt - a molecule viewer for charge-density research. *J. Appl. Cryst.* **44**, 238–240 (2011).
46. Dennington, R., Keith, T. & Millam, J. GaussView, Version 5, Semichem Inc. (Shawnee Mission, KS, USA, 2009).
47. Frisch, M. J. *et al.* Gaussian 09, Revision D.01, Gaussian, Inc. (Wallingford, CT, USA, 2009).
48. Keith, T. A. TK Gristmill Software, AIMAll (Version 14.11.23) Software (Overland Park, KS, USA, 2014) (aim.tkgristmill.com).
49. Jayatilaka, D. & Grimwood, D. J. Tonto: a fortran based object-oriented system for quantum chemistry and crystallography. *Lect. Notes Comput. Sci.* **2660**, 142–151 (2003).

Acknowledgements

We thank the Swiss National Science foundation (project 144534 and 162861) for financial support. We thank Dr Anna Krawczuk, Dr Shaun Evans and Dr Heribert Whilelm for assistance during one of the experiments at Diamond Light Source, Professor Riccardo Destro for providing the crystal samples.

Author contributions

N.C. and P.M. conceived the project. N.C., A.P.J. and A.K. set-up and carried out the HP experiments and N.C. analysed the data refined the models. P.M. carried out the multi-temperature experiments, the multipolar refinements and the theoretical calculations. N.C. and P.M. wrote the paper.

Additional information

Accession codes: The X-ray crystallographic coordinates for structures reported in this study have been deposited at the Cambridge Crystallographic Data Centre (CCDC), under deposition numbers: 1438912-1438922. These data can be obtained free of charge from The Cambridge Crystallographic Data Centre via www.ccdc.cam.ac.uk/data_request/cif.

Competing financial interests: The authors declare no competing financial interests.

Structural complexity of simple Fe$_2$O$_3$ at high pressures and temperatures

E. Bykova[1,2], L. Dubrovinsky[1], N. Dubrovinskaia[2], M. Bykov[1,2], C. McCammon[1], S.V. Ovsyannikov[1], H.-P. Liermann[3], I. Kupenko[1,4], A.I. Chumakov[4], R. Rüffer[4], M. Hanfland[4] & V. Prakapenka[5]

Although chemically very simple, Fe$_2$O$_3$ is known to undergo a series of enigmatic structural, electronic and magnetic transformations at high pressures and high temperatures. So far, these transformations have neither been correctly described nor understood because of the lack of structural data. Here we report a systematic investigation of the behaviour of Fe$_2$O$_3$ at pressures over 100 GPa and temperatures above 2,500 K employing single crystal X-ray diffraction and synchrotron Mössbauer source spectroscopy. Crystal chemical analysis of structures presented here and known Fe(II, III) oxides shows their fundamental relationships and that they can be described by the homologous series nFeO · mFe$_2$O$_3$. Decomposition of Fe$_2$O$_3$ and Fe$_3$O$_4$ observed at pressures above 60 GPa and temperatures of 2,000 K leads to crystallization of unusual Fe$_5$O$_7$ and Fe$_{25}$O$_{32}$ phases with release of oxygen. Our findings suggest that mixed-valence iron oxides may play a significant role in oxygen cycling between earth reservoirs.

[1] Bayerisches Geoinstitut, University of Bayreuth, Universitaetsstrasse 30, D-95447 Bayreuth, Germany. [2] Laboratory of Crystallography, University of Bayreuth, Universitaetsstrasse 30, D-95447 Bayreuth, Germany. [3] Photon Sciences, Deutsches Elektronen-Synchrotron, Notkestrasse 85, D-22607 Hamburg, Germany. [4] European Synchrotron Radiation Facility, 71 avenue des Martyrs, Grenoble F-38000, France. [5] Center for Advanced Radiation Sources, University of Chicago, 9700 South Cass Avenue, Illinois, Argonne 60437, USA. Correspondence and requests for materials should be addressed to E.B. (email: elena.bykova@uni-bayreuth.de).

The structures, properties and high-pressure behavior of corundum-type oxides have been extensively investigated because of their wide variety of elastic, electrical and magnetic properties and importance in earth sciences and technology[1-3]. High-pressure studies of hematite, α-Fe_2O_3 (Fig. 1a), have attracted special attention due to their geophysical interest and the unclear role of Fe^{3+} in the nature and dynamics of the earth's lower mantle[1,2]. Particular attention has been focused on elucidating the nature of phase transition(s) and the structure of the high-pressure phase of hematite observed above $\sim 50\,GPa$ (refs. 1,4–12). For this phase two structures have been proposed by different groups: Rh_2O_3-II-type (space group *Pbcn*, no. #60) and $GdFeO_3$-perovskite-type (space group *Pbnm*, no. #62) structures[4,7]. While Mössbauer spectroscopic and resistivity measurements clearly demonstrate the importance of electronic changes in Fe^{3+} and seem to support the Rh_2O_3-II-type structure[5], powder diffraction data collected by various groups over several decades did not allow an unambiguous assignment of the structural type (see refs 4,5,7,8 and references therein). Only recent single-crystal high-P,T diffraction data[12] were able to solve this challenge; they demonstrated that the Rh_2O_3-II-type phase of Fe_2O_3 (which we refer to below as ι-Fe_2O_3, Fig. 1b) forms upon laser heating at pressures above $\sim 40\,GPa$; whereas, compression of hematite at ambient temperature to over $\sim 50\,GPa$ results in the formation of a phase with distorted $GdFeO_3$-perovskite-type, dPv ζ-Fe_2O_3, structure (Fig. 1c). Experiments in laser-heated diamond anvil cells (DACs) revealed the formation of a $CaIrO_3$-type phase ('post-perovskite' (PPv) η-Fe_2O_3, Fig. 1d) at pressures above $\sim 60\,GPa$ (refs. 1,9,12,13). However, the behaviour of this phase under compression is not well studied. The phase diagram of Fe_2O_3 in the megabar pressure range is incomplete and the data are often conflicting[1,9,10,13].

In order to study the high-pressure high-temperature (HPHT) behaviour of ferric iron (Fe^{3+}) oxide we apply the complementary methods of single crystal X-ray diffraction in laser-heated DACs and synchrotron Mössbauer source (SMS) spectroscopy (see Methods section). We observe hitherto unknown Fe–O phases, show the results of their structure solution and refinement, and characterize the pressure–temperature conditions, at which different Fe_2O_3 polymorphs occur. Crystal chemical analysis of the new structures and known Fe(II, III) oxides reveals their fundamental relationships as members of the homologous series $nFeO \cdot mFe_2O_3$. We observe that at pressures above 60 GPa and at high temperatures (that is, at conditions of the earth's lower mantle), Fe_2O_3 decomposes with release of oxygen. The same phenomenon is observed for Fe_3O_4. Our results indicate that mixed-valence iron oxides may play a significant role in oxygen cycling between the earth's atmosphere and mantle.

Results

Structural transformations in Fe_2O_3. In agreement with previous studies[4,5,7,8,12], our cold compression experiments on hematite single crystals to 54(1) GPa result in a transition to the ζ-Fe_2O_3 phase manifested by a $\sim 8.4\%$ volume discontinuity (Supplementary Fig. 1). Although in earlier work we indexed the diffraction pattern of ζ-Fe_2O_3 in a monoclinic unit cell[12], the new extended dataset acquired in the present work showed that the structure is in fact triclinic (see Supplementary Note 1 for details), similar to Mn_2O_3 (ref. 14). An insufficient number of independent reflections prevented structural refinement of ζ-Fe_2O_3 in triclinic symmetry, so we used a monoclinic model[12] to qualitatively constrain the atomic arrangement in ζ-Fe_2O_3. Upon

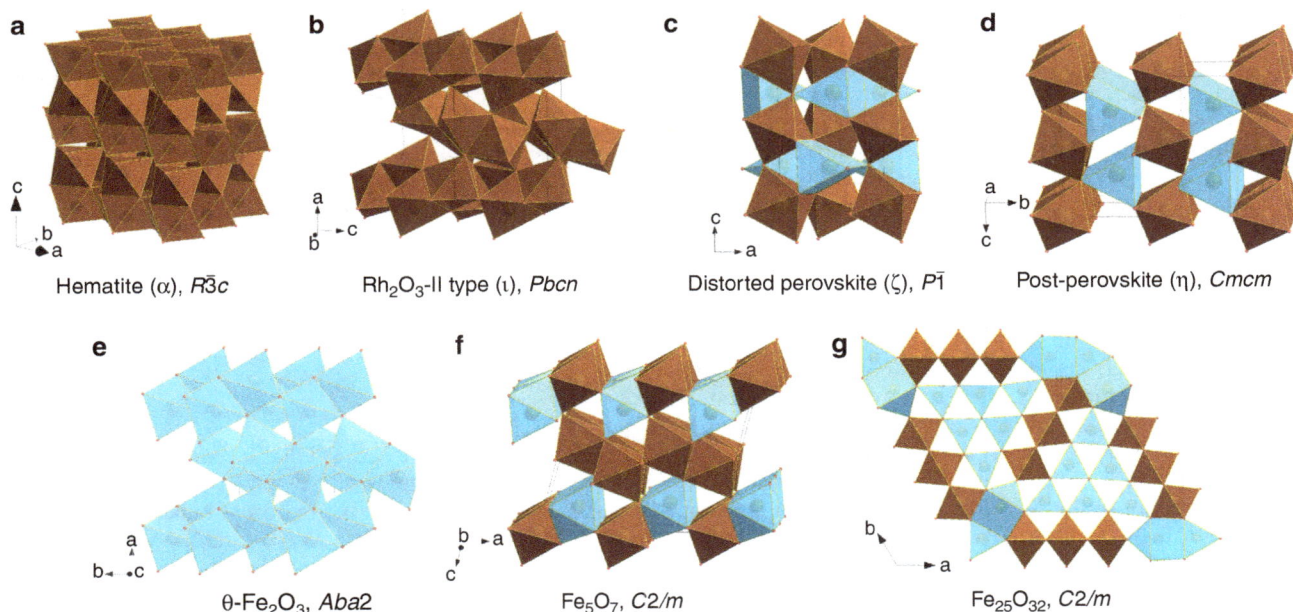

Figure 1 | Crystal structures of iron oxide phases studied in the present work. Building blocks are octahedra (brown) and trigonal prisms (blue). The prisms in Fe_5O_7, $Fe_{25}O_{32}$ and η-Fe_2O_3 have one or two additional apices. Hematite (**a**) consists of FeO_6 octahedra connected in a corundum-like motif, namely each octahedron connects with three neighbours via edges in honeycomb layers, and layers are interconnected through common triangular faces of octahedra. The ι-Fe_2O_3 structure (**b**) is built of only FeO_6 octahedra but each two octahedra are connected through a common triangular face; such units pack in a herringbone pattern and layers pack with a shift along the *c*-direction having common edges. In distorted perovskite ζ-Fe_2O_3 (**c**) octahedra connect through common vertices and prisms share only common edges. θ-Fe_2O_3 (**e**) adopts the packing motif from ι-Fe_2O_3 but instead of octahedra it consists of FeO_6 prisms. Post-perovskite (**d**) and Fe_5O_7 (**f**) are members of the homologous series $nFeO \cdot mFe_2O_3$ (see also Fig. 4), where prisms are connected through common triangular faces, while octahedra connect only via shared edges. In addition to triangular face-shared prisms and edge-shared octahedra, $Fe_{25}O_{32}$ (**g**) has edge-shared one-capped prisms; therefore it belongs neither to the homologous series nor adopts any other known structural motif.

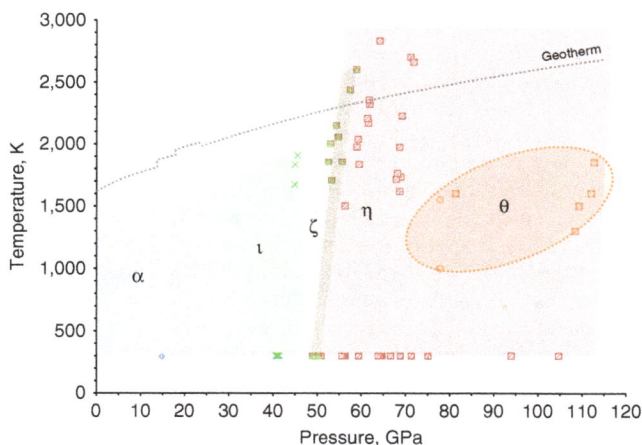

Figure 2 | Transformational phase diagram of Fe₂O₃. (a) ◇-$R\bar{3}c$ hematite (α-Fe_2O_3), △-$P\bar{1}$ distorted perovskite (ζ-Fe_2O_3), ○-$Aba2$ (θ-Fe_2O_3, probably metastable), □-$Cmcm$ post-perovskite (η-Fe_2O_3) and ×-Rh_2O_3-II type phase (ι-Fe_2O_3). The boundary between hematite α-Fe_2O_3 and ι-Fe_2O_3 is defined according to ref. 10. The geotherm is defined according to refs 39,40.

further pressure increase from 54(1) to 67(1) GPa, we observed a reduction in the splitting of reflections, indicating an increase in symmetry. The structure of ζ-Fe_2O_3 thus becomes closer to that of $GdFeO_3$-type-perovskite (Supplementary Fig. 2b). At 67(1) GPa a small drop in the unit cell volume (~1.7%) manifests the next transformation to the θ-Fe_2O_3 phase (Fig. 1e) with orthorhombic symmetry (space group $Aba2$, no. #41, a = 4.608(7), b = 4.730(4), c = 6.682(18) Å (Supplementary Table 1)). On compression at ambient temperature θ-Fe_2O_3 can be observed to at least 100 GPa (Supplementary Fig. 1). The transformational P–T diagram for Fe_2O_3 is given in Fig. 2.

During *in situ* laser heating of θ-Fe_2O_3 between ~1,000 and 1,550(50) K at 78(2) GPa, we observed no evidence of a phase transformation. The absence of transformations may either be evidence that θ-Fe_2O_3 is stable at these conditions or an indication that higher temperatures are required to overcome kinetic barriers to further structural transitions. Indeed, heating at 1,600(50) K results in the formation of post-perovskite type η-Fe_2O_3 coexisting with θ-Fe_2O_3. Both phases (θ-Fe_2O_3 and η-Fe_2O_3) were observed *in situ* simultaneously upon heating to 1,850(50) K at pressures up to 113(1) GPa. However, temperature-quenched products contained only η-Fe_2O_3 (Fig. 2). Once synthesized, η-Fe_2O_3 may be preserved at ambient temperature down to at least 26 GPa. At lower pressures it transforms back to hematite (see Figs 1 and 2 for structures and phase relations). Moderate heating to ~2,000 K at pressures of about 50 GPa provokes a transition to the dPv ζ-Fe_2O_3 phase. Decompression of ζ-Fe_2O_3 or η-Fe_2O_3 to 41(1) GPa with heating at 1,800(100) K results in growth of Rh_2O_3-II type ι-Fe_2O_3 (Supplementary Fig. 1, Supplementary Table 1). Interestingly, ι-Fe_2O_3 was synthesized earlier[10,11] from hematite, thus bracketing the possible P–T stability field of the phase (Fig. 2).

Electronic transformations in Fe₂O₃. The sequence of phase transitions in Fe_2O_3 in the megabar pressure range and temperatures up to about 2,500 K (Fig. 2) can be neatly rationalized through the variation of molar volumes of the phases observed as a function of pressure (Supplementary Fig. 1), complemented by the corresponding SMS spectroscopy data (Fig. 3). The bulk modulus of hematite, 219(7) GPa, is in good agreement with

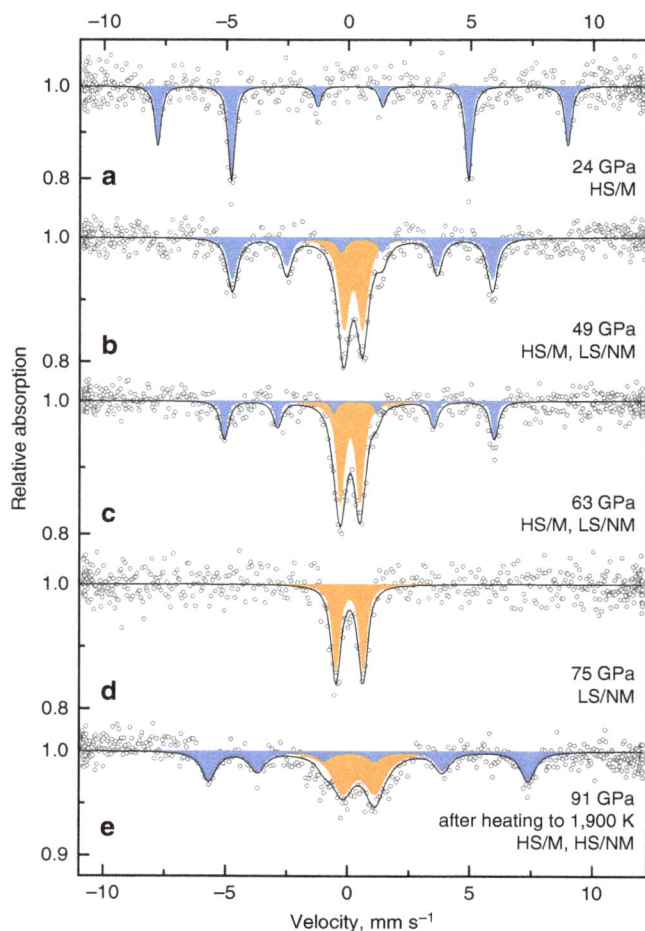

Figure 3 | Evolution of SMS spectra of Fe₂O₃. Spectra collected during compression (**a–d**) and after heating (**e**). In hematite (**a**) iron atoms have a HS state (at ~24 GPa CS = 0.306(4) mm s⁻¹), and spectra are split due to magnetic ordering (M). After the first transition at 49 GPa (**b**) a new non-magnetic (NM) component appears with CS of 0.074(5) mm s⁻¹ corresponding to a LS state. During further compression a fraction of the magnetic component decreases (**c**) and it disappears completely after the second transition to the θ-Fe_2O_3 phase (**d**) that has only one non-magnetic position of LS iron atoms in the crystal structure. After heating above 1,600(50) K (**e**) a transformation to η-Fe_2O_3 occurs. The crystal structure has two HS-iron positions (both CS are ~0.45 mm s⁻¹), where one position is magnetically ordered and the other is non-magnetic.

previous studies[15] and at 67 GPa it reaches ~392(10) GPa, whereas the bulk modulus of ζ-Fe_2O_3 at 54 GPa is substantially lower, 320(18) GPa. Such a large drop of bulk modulus (~18%) associated with a large reduction of molar volume (~8.4%) is very unusual and is likely caused by changes in the electronic state of Fe^{3+}. The Mössbauer spectrum of ζ-Fe_2O_3 collected immediately after the transition at ~50 GPa shows two components (Fig. 3b), a magnetic sextet having centre shift (CS) of 0.424(7) mm s⁻¹ corresponding to the high-spin (HS) state of Fe^{3+}, and a doublet (CS = 0.074(5) mm s⁻¹) with hyperfine parameters characteristic for low-spin (LS) Fe^{3+} in an octahedral oxygen environment[16]. The relative abundance of the components is ~1:1, as expected for the perovskite-type structure of ζ-Fe_2O_3 with HS-Fe^{3+} located in large bipolar prisms and LS-Fe^{3+} in smaller octahedra (Fig. 1c). Upon further compression of ζ-Fe_2O_3 the amount of HS-Fe^{3+} decreases (Fig. 3c), which explains the anomalously high compressibility of this phase.

Transformation to θ-Fe_2O_3 is associated with a small decrease of molar volume ($\sim 1.7\%$) and an increase of bulk modulus as expected (418(11) GPa for θ-Fe_2O_3 at 67 GPa compared with 371(20) GPa for ζ-Fe_2O_3 at 70 GPa) (Supplementary Fig. 1). The Mössbauer spectrum of θ-Fe_2O_3 (Fig. 3d) shows that all Fe^{3+} is in the LS state and there is only one type of iron atom in the crystal structure in accordance with the single crystal X-ray diffraction data (Fig. 1e).

Heating of θ-Fe_2O_3 above 1,600 K at pressures above 70 GPa resulted in partial or complete transformation into $CaIrO_3$-PPv-type η-Fe_2O_3 (Fig. 2). The Mössbauer spectrum of pure η-Fe_2O_3 at 91(2) GPa (Fig. 3e) contains two components (a magnetically ordered sextet and a paramagnetic doublet) with equal abundances and almost equal CS (~ 0.45 mm/s) corresponding to HS-Fe^{3+}. Within the accuracy of our X-ray diffraction data the molar volumes of θ-Fe_2O_3 and as-synthesized η-Fe_2O_3 are indistinguishable (Supplementary Fig. 1), suggesting that the atomic packing density increase in the $CaIrO_3$-PPv-type η-Fe_2O_3 structure compensates the difference in ionic radii of HS and LS Fe^{3+} ions in the ζ-Fe_2O_3 structure. Note that Shim et al.[1] also reported magnetic ordering in η-Fe_2O_3 based on nuclear forward scattering measurements. One of the magnetic sites described by the authors[1] has hyperfine parameters close to those that we observed; however, the second non-magnetic component in the nuclear forward scattering spectra was not identified in ref. 1.

Thermal stability of Fe_2O_3. The behaviour of η-Fe_2O_3 under heating is rather remarkable. First, we noted that its unit cell volume increases by up to 1% upon laser heating to about 2,000 K at ~ 56 and 64 GPa. (Supplementary Fig. 3). Second, after heating for a few seconds to 2,700–3,000 K and 71 GPa we observed the immediate appearance of new sharp spots in the diffraction pattern. The peaks were indexed in the $C2/m$ space group and the structure solution using direct methods identified the phase as a novel mixed-valence iron oxide with stoichiometry Fe_5O_7 ($FeO \cdot 2Fe_2O_3$) (Fig. 1f, Supplementary Table 2). Visual observations (particularly preservation of the shape of the samples upon heating) and careful analysis of diffraction patterns (absence of diffuse scattering) verify that samples were not melted in experiments where Fe_5O_7 was synthesized. The phase is preserved on decompression down to at least 41(1) GPa. Thus, we explain our observations as a continuous loss of oxygen by η-Fe_2O_3 upon

heating at moderate temperatures and pressures above ~ 60 GPa according to the reaction η-$Fe_2O_3 \rightarrow$ η-$Fe_2O_{3-\delta} + 0.5\delta \cdot O_2$. Note that a similar process is well known for perovskites[17] and other oxides[18]. The reaction is accompanied by a partial reduction of Fe^{3+} to larger-sized Fe^{2+} that consequently increases the unit cell volume. Upon heating at sufficiently high temperature (above $\sim 2,700$ K), the oxygen deficiency in η-Fe_2O_3 reaches a critical limit and provokes a reconstructive phase transition resulting in the formation of the mixed-valence iron oxide Fe_5O_7: 5η-$Fe_2O_3 \rightarrow 2Fe_5O_7 + 0.5O_2$. From both X-ray diffraction and Raman spectroscopy we did not find any evidence to suggest involvement of carbon from the diamond anvils in the chemical reactions. Indeed this was not expected, because at the HPHT conditions of our experiments carbon and oxygen do not react[19]. Mössbauer experiments show that laser heating of Fe_2O_3 at pressures above 80 GPa leads to formation of phases containing iron with hyperfine parameters characteristic of a mixed valence state (Supplementary Fig. 4).

Similarities in the crystal structures of η-Fe_2O_3, Fe_5O_7, high-pressure polymorph of Fe_3O_4 (HP-Fe_3O_4, space group $Bbmm$, no. #63, Supplementary Tables 1 and 2), and the recently discovered Fe_4O_5 (ref. 20) and Fe_5O_6 (ref. 21) (Fig. 4) demonstrate[22] that iron oxide phases form a homologous series $nFeO \cdot mFe_2O_3$ (with wüstite, FeO and η-Fe_2O_3 as the end-members) and indicate that a mixed-valence state of iron may become crystal chemically important at high pressures and temperatures.

Discussion

Our results demonstrate clearly the complex behaviour of iron oxide subjected to high pressures and temperatures and may have significant consequences for modelling of the earth's interior. Hematite is one of the major components of banded iron formations (BIFs) and ironstones, and these huge sedimentary rock formations occurring on all continents may reach up to several hundred meters in thickness and hundreds of kilometres in length. Deposited in the world's oceans, BIFs as part of the ocean floor are recycled into the earth's interior by subduction[2,23] to depths extending possibly to the core–mantle boundary region[2]. Available experimental data[2,13,24] suggest that iron oxides melt above the geotherm in the entire mantle and thus remain solid in slabs that are colder than the surrounding mantle. Even assuming a slow subduction rate of 1 cm per year with slabs

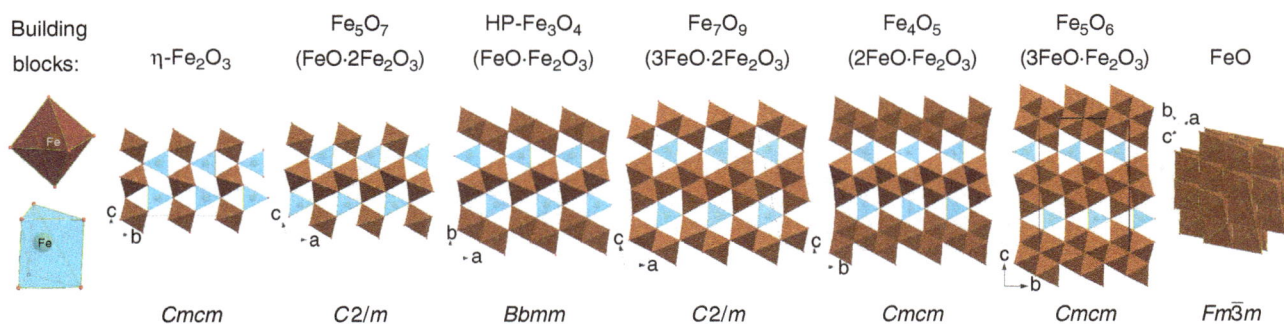

Figure 4 | Homologous series of iron oxides described by the common formula $nFeO \cdot mFe_2O_3$. The structures may be described as assembled from two building blocks, FeO6 octahedra and trigonal prisms (prisms could be two-capped but they are not shown for simplicity). Prisms connect to each other through triangular faces, while octahedra share edges, so that they form parallel columns of face-shared prisms and edge-shared octahedra arranged in different motifs as seen in the figure with structures viewed from the top of the columns. Increasing Fe^{2+} content favours octahedral packing over mixed octahedral and prismatic packing. This requires denser packing of FeO6 octahedra and as a result columns of octahedra condense in slabs by sharing common edges. In particular, η-Fe_2O_3 has ordinary columns of prisms and octahedra with a chequerboard-like arrangement; Fe_5O_7 has ordinary and doubled columns of octahedra; the HP-Fe_3O_4 (ref. 41) possesses only doubled columns; Fe_7O_9 (ICSD reference number CSD-430601)[42] has doubled columns and tripled columns organized in zigzag slabs; Fe_4O_5 (ref. 20) possesses only tripled and Fe_5O_6 (ref. 21) only quadruple zigzag slabs. The end-member of the homologous series wüstite (FeO) consists of octahedra with a maximum (12) number of edge-shared neighbours.

reaching a depth of about 2,000 km in ∼200 Ma, this geological time is sufficient to influence only a few tens of meters of rocks beneath the BIF's surface. Thus the fate of iron oxides, a major component of subducted BIFs, depends on the pressures and temperatures (P–T), to which they are exposed. Upon subduction of BIFs into the lower mantle, hematite undergoes numerous phase transformations. At pressures above ∼60 GPa the HP phase η-Fe_2O_3 starts to decompose, producing oxygen. Moreover, experiments on Fe_3O_4, the second major component of BIFs, show that it also decomposes upon heating at pressures above ∼70 GPa, forming in particular the phase $Fe_{25}O_{32}$ (Fig. 1g, see also Supplementary Note 2). Based on estimates of the amount of BIFs subducted into the earth's mantle[2] and that BIFs may consist of ∼50% Fe_2O_3 by volume, the amount of oxygen produced by the formation of Fe_5O_7 alone can be as high as 8–10 times the mass of oxygen in the modern atmosphere. Extrapolation of available data[25] indicates that oxygen would be in the liquid state at geotherm temperatures. Since the oxygen fugacity of the lower mantle is expected to be low through equilibrium with metallic iron, an oxygen-rich fluid could locally oxidize surrounding material (particularly Fe^{2+} in ferropericlase as well as bridgmanite, and metallic iron in a (Fe, Ni)-metal phase[26]). Seismic tomography reveals pronounced complex heterogeneities in the lower mantle at depths of 1,500–2,000 km associated with subducted slabs[27–29] and the presence of oxidized material may be a reason for these observations[30]. On the other hand, a low oxygen chemical activity at high pressure[19,31,32] could prevent the immediate reaction of oxygen in the lower mantle or even in the transition zone, and instead allow an oxygen-rich fluid to pass to the upper mantle, thus shifting Fe^{2+}/Fe^{3+} equilibria in silicate minerals and greatly raising the oxygen fugacity in this region. In any case, our study suggests the presence of an oxygen-rich fluid in the deep earth's interior that can significantly affect geochemical processes by changing oxidation states and mobilizing trace elements.

Methods

Sample preparation. Single crystals of α-Fe_2O_3 enriched with ^{57}Fe ($^{57}Fe_2O_3$) were grown by means of HPHT technique at 7 GPa and 800 °C in a 1,200-t Sumitomo press at Bayerisches Geoinstitut (Bayreuth, Germany). As a precursor, a 1:1 mixture of a powder of non-enriched hematite (α-Fe_2O_3) of 99.998% purity and a pure powder of $^{57}Fe_2O_3$ (96.64%-enriched) was used. Magnetite synthesis was performed in the same way at 9.5 GPa and 1,100 °C as described in ref. 33. Synthesis of non-enriched hematite single crystals was described in ref. 15.

Single crystals with an average size of $0.03 \times 0.03 \times 0.005$ mm^3 were preselected on a three-circle Bruker diffractometer equipped with a SMART APEX CCD detector and a high-brilliance Rigaku rotating anode (Rotor Flex FR-D, Mo-Kα radiation) with Osmic focusing X-ray optics.

Selected crystals together with small ruby chips (for pressure estimation) were loaded into BX90-type DACs[34]. Neon was used as a pressure transmitting medium loaded at Bayerisches Geoinstitut.

X-ray diffraction. The single-crystal X-ray diffraction experiments were conducted on the ID09A beamline at the European synchrotron radiation facility (ESRF), Grenoble, France (MAR555 detector, $\lambda = 0.4126$–0.4130 Å); on the 13-IDD beamline at the advanced photon source (APS), Chicago, USA (MAR165 CCD detector, $\lambda = 0.3344$ Å); and on the extreme conditions beamline P02.2 at PETRA III, Hamburg, Germany (PerkinElmer XRD1621 flat panel detector, $\lambda = 0.2898$–0.2902 Å). The X-ray spot size depended on the beamline settings and varied from 4 to 30 μm, where typically a smaller beam was used for laser heating experiments. A portable double-sided laser heating[35] system was used for experiments on ID09A (ESRF) to collect *in situ* single-crystal X-ray diffraction. State-of-the art stationary double-side laser-heating set-up at IDD-13 (APS) has been used for temperature-quenched single-crystal X-ray diffraction. Crystals were completely 'surrounded' by laser light and there were no measurable temperature gradients within the samples. In the case of prolonged heating experiments the temperature variation during the heating did not exceed ± 100 K. Pressures were calculated from the positions of the X-ray diffraction lines of Ne (http://kantor.50webs.com/diffraction.htm). X-ray diffraction images were collected during continuous rotation of DACs typically from −40 to +40 on ω; while data collection experiments were performed by narrow 0.5–1° scanning of the same ω range. The crystallographic information is also available as Supplementary Data 1-7.

Data analysis. Integration of the reflection intensities and absorption corrections were performed using CrysAlisPro software[36]. The structures were solved by the direct method and refined in the isotropic approximation by full matrix least-squares refinements using SHELXS and SHELXL software[37], respectively.

SMS spectroscopy. Energy-domain Mössbauer measurements were carried out at the nuclear resonance beamline ID18 at ESRF (see ref. 38 for more details).

References

1. Shim, S.-H. *et al.* Electronic and magnetic structurses of the postperovskite-type Fe_2O_3 and implications for planetary magnetic records and deep interiors. *Proc. Natl Acad. Sci. USA* **106**, 5508–5512 (2009).
2. Dobson, D. P. & Brodholt, J. P. Subducted banded iron formations as a source of ultralow-velocity zones at the core-mantle boundary. *Nature* **434**, 371–374 (2005).
3. Tuček, J. *et al.* Zeta-Fe_2O_3–A new stable polymorph in iron(III) oxide family. *Sci. Rep.* **5**, 15091 (2015).
4. Olsen, J. S., Cousins, C. S. G., Gerward, L., Jhans, H. & Sheldon, B. J. A study of the crystal structure of Fe_2O_3 in the pressure range up to 65 GPa using synchrotron radiation. *Phys. Scr.* **43**, 327–330 (1991).
5. Pasternak, M. *et al.* Breakdown of the Mott-Hubbard state in Fe_2O_3: A first-order insulator-metaltransition with collapse of magnetism at 50 GPa. *Phys. Rev. Lett.* **82**, 4663–4666 (1999).
6. Badro, J. *et al.* Nature of the high-pressure transition in Fe_2O_3 hematite. *Phys. Rev. Lett.* **89**, 205504 (2002).
7. Rozenberg, G. *et al.* High-pressure structural studies of hematite Fe_2O_3. *Phys. Rev. B* **65**, 064112 (2002).
8. Liu, H., Caldwell, W. A., Benedetti, L. R., Panero, W. & Jeanloz, R. Static compression of α-Fe_2O_3: linear incompressibility of lattice parameters and high-pressure transformations. *Phys. Chem. Miner.* **30**, 582–588 (2003).
9. Ono, S., Kikegawa, T. & Ohishi, Y. High-pressure phase transition of hematite, Fe_2O_3. *J. Phys. Chem. Solids* **65**, 1527–1530 (2004).
10. Ito, E. *et al.* Determination of high-pressure phase equilibria of Fe_2O_3 using the Kawai-type apparatus equipped with sintered diamond anvils. *Am. Mineral.* **94**, 205–209 (2009).
11. Dubrovinsky, L. *et al.* Single-crystal X-ray diffraction at megabar pressures and temperatures of thousands of degrees. *High Pressure Res.* **30**, 620–633 (2010).
12. Bykova, E. *et al.* Novel high pressure monoclinic Fe_2O_3 polymorph revealed by single-crystal synchrotron X-ray diffraction studies. *High Pressure Res.* **33**, 534–545 (2013).
13. Ono, S. & Ohishi, Y. In situ X-ray observation of phase transformation in Fe_2O_3 at high pressures and high temperatures. *J. Phys. Chem. Solids* **66**, 1714–1720 (2005).
14. Ovsyannikov, S. V. *et al.* Perovskite-like Mn_2O_3: a path to new manganites. *Angew. Chem. Int. Ed. Engl.* **52**, 1494–1498 (2013).
15. Schouwink, P. *et al.* High-pressure structural behavior of α-Fe_2O_3 studied by single-crystal X-ray diffraction and synchrotron radiation up to 25 GPa. *Am. Mineral.* **96**, 1781–1786 (2011).
16. Xu, W. *et al.* Pressure-induced hydrogen bond symmetrization in iron oxyhydroxide. *Phys. Rev. Lett.* **111**, 175501 (2013).
17. Mizusaki, J., Yamauchi, S., Fueki, K. & Ishikawa, A. Nonstoichiometry of the perovskite-type oxide $La_{1-x}Sr_xCrO_{3-\delta}$. *Solid State Ionics* **12**, 119–124 (1984).
18. Brazhkin, V. V., Voloshin, R. N., Lyapin, A. G. & Popova, S. V. Phase equilibria in partially open systems under pressure: the decomposition of stoichiometric GeO_2 oxide. *Physics-Uspekhi* **46**, 1283–1289 (2003).
19. Litasov, K. D., Goncharov, A. F. & Hemley, R. J. Crossover from melting to dissociation of CO_2 under pressure: Implications for the lower mantle. *Earth Planet. Sci. Lett.* **309**, 318–323 (2011).
20. Lavina, B. *et al.* Discovery of the recoverable high-pressure iron oxide Fe_4O_5. *Proc. Natl Acad. Sci. USA* **108**, 17281–17285 (2011).
21. Lavina, B. & Meng, Y. Unraveling the complexity of iron oxides at high pressure and temperature: Synthesis of Fe_5O_6. *Sci. Adv.* **1**, e1400260 (2015).
22. Guignard, J. & Crichton, W. A. Synthesis and recovery of bulk Fe_4O_5 from magnetite, Fe_3O_4. A member of a self-similar series of structures for the lower mantle and transition zone. *Mineral. Mag.* **78**, 361–371 (2014).
23. Polat, A., Hofmann, A. W. & Rosing, M. T. Boninite-like volcanic rocks in the 3.7-3.8 Ga Isua greenstone belt, West Greenland: Geochemical evidence for intra-oceanic subduction zone processes in the early Earth. *Chem. Geol.* **184**, 231–254 (2002).
24. Ozawa, H., Hirose, K., Tateno, S., Sata, N. & Ohishi, Y. Phase transition boundary between B1 and B8 structures of FeO up to 210 GPa. *Phys. Earth Planet. Inter.* **179**, 157–163 (2010).
25. Freiman, Y. A. & Jodl, H. J. Solid oxygen. *Phys. Rep.* **401**, 1–228 (2004).
26. Frost, D. J. *et al.* Experimental evidence for the existence of iron-rich metal in the Earth's lower mantle. *Nature* **428**, 409–412 (2004).
27. Li, C., Van Der Hilst, R. D., Engdahl, E. R. & Burdick, S. A new global model for P wave speed variations in Earth's mantle. *Geochemistry, Geophys. Geosystems* **9**, Q05018 (2008).

28. Van der Hilst, R. D., Widiyantoro, S. & Engdahl, E. R. Evidence for deep mantle circulation from global tomography. *Nature* **386**, 578–584 (1997).
29. Zhao, D. Global tomographic images of mantle plumes and subducting slabs: Insight into deep Earth dynamics. *Phys. Earth Planet. Inter.* **146**, 3–34 (2004).
30. Glazyrin, K. *et al.* Magnesium silicate perovskite and effect of iron oxidation state on its bulk sound velocity at the conditions of the lower mantle. *Earth Planet. Sci. Lett.* **393**, 182–186 (2014).
31. Rohrbach, A. & Schmidt, M. W. Redox freezing and melting in the Earth's deep mantle resulting from carbon-iron redox coupling. *Nature* **472**, 209–212 (2011).
32. Stagno, V., Ojwang, D. O., McCammon, C. A. & Frost, D. J. The oxidation state of the mantle and the extraction of carbon from Earth's interior. *Nature* **493**, 84–88 (2013).
33. Glazyrin, K. *et al.* Effect of high pressure on the crystal structure and electronic properties of magnetite below 25 GPa. *Am. Mineral.* **97**, 128–133 (2012).
34. Kantor, I. *et al.* BX90: A new diamond anvil cell design for X-ray diffraction and optical measurements. *Rev. Sci. Instrum.* **83**, 125102 (2012).
35. Kupenko, I. *et al.* Portable double-sided laser-heating system for Mössbauer spectroscopy and X-ray diffraction experiments at synchrotron facilities with diamond anvil cells. *Rev. Sci. Instrum.* **83**, 124501 (2012).
36. CrysAlisPro Software system. Version 1.171.37.35. (Agilent Technologies UK Ltd., Oxford, UK, 2014).
37. Sheldrick, G. M. A short history of SHELX. *Acta Cryst.* **64**, 112–122 (2008).
38. Potapkin, V. *et al.* The ^{57}Fe Synchrotron Mössbauer Source at the ESRF *J. Synchrotron Radiat.* **19**, 559–569 (2012).
39. Dziewonski, A. M. & Anderson, D. L. Preliminary reference Earth model. *Phys. Earth Planet. Inter.* **25**, 297–356 (1981).
40. Katsura, T., Yoneda, A., Yamazaki, D., Yoshino, T. & Ito, E. Adiabatic temperature profile in the mantle. *Phys. Earth Planet. Inter.* **183**, 212–218 (2010).
41. Dubrovinsky, L. S. *et al.* The structure of the metallic high-pressure Fe3O4 polymorph: Experimental and theoretical study. *J. Phys. Condens. Matter* **15**, 7697–7706 (2003).
42. Belsky, A., Hellenbrandt, M., Karen, V. L. & Luksch, P. New developments in the Inorganic Crystal Structure Database (ICSD): Accessibility in support of materials research and design. *Acta Crystallogr. B* **58**, 364–369 (2002).

Acknowledgements

We thank K. Glazyrin for synthesis of non-enriched Fe$_2$O$_3$ and V. Cerantola for synthesis of FeCO$_3$ and for assistance with HPHT experiments at ESRF and APS. We appreciate the technical assistance of S. Linhardt and S. Übelhack. N.D. and L.D. thank the German Research Foundation (Deutsche Forschungsgemeinschaft, DFG) and the Federal Ministry of Education and Research (BMBF, Germany) for funding. N.D. thanks the DFG for financial support through the Heisenberg Program and project no. DU 954-8/1, and BMBF for the grant no. 5K13WC3 (Verbundprojekt O5K2013, Teilprojekt 2, PT-DESY).

We acknowledge the European Synchrotron Radiation Facility for provision of synchrotron radiation facilities. Portions of this work were performed at GeoSoilEnviroCARS (sector 13), Advanced Photon Source (APS), Argonne National Laboratory. GeoSoilEnviroCARS is supported by the National Science Foundation-Earth Sciences (EAR-1128799) and Department of Energy-GeoSciences (DE-FG02-94ER14466). This research used resources of the Advanced Photon Source, a US Department of Energy (DOE) Office of Science User Facility operated for the DOE Office of Science by Argonne National Laboratory under contract no. DE-AC02-06CH11357.

Author contributions

L.D. and S.V.O. provided the sample. E.B. selected the single-crystals and analysed all X-ray diffraction data. M.B., E.B., L.D., V.P., M.H. and H.-P.L. conducted the HPHT single-crystal X-ray diffraction experiments. The SMS were collected by I.K., L.D., A.I.C., R.R. and analysed by L.D., I.K. and C.M. E.B., N.D., C.M. and L.D. interpreted the results and wrote the manuscript with contributions of all authors.

Additional information

Accession codes: The X-ray crystallographic coordinates for structures reported in this article have been deposited at the Inorganic Crystal Structure Database (ICSD) under deposition number CSD 430557–430563. These data can be obtained free of charge from FIZ Karlsruhe, 76344 Eggenstein-Leopoldshafen, Germany (fax: (+ 49)7247-808-666; e-mail: crysdata@fiz-karlsruhe.de) through the hyperlink 'https://www.fiz-karlsruhe.de/en/leistungen/kristallographie/kristallstrukturdepot/order-form-request-for-deposited-data.html'.

Competing financial interests: The authors declare no competing financial interests.

Direct observation of mineral–organic composite formation reveals occlusion mechanism

Kang Rae Cho[1,2], Yi-Yeoun Kim[3], Pengcheng Yang[4], Wei Cai[5], Haihua Pan[1,6], Alexander N. Kulak[3], Jolene L. Lau[1], Prashant Kulshreshtha[1], Steven P. Armes[4], Fiona C. Meldrum[3] & James J. De Yoreo[1,7]

Manipulation of inorganic materials with organic macromolecules enables organisms to create biominerals such as bones and seashells, where occlusion of biomacromolecules within individual crystals generates superior mechanical properties. Current understanding of this process largely comes from studying the entrapment of micron-size particles in cooling melts. Here, by investigating micelle incorporation in calcite with atomic force microscopy and micromechanical simulations, we show that different mechanisms govern nanoscale occlusion. By simultaneously visualizing the micelles and propagating step edges, we demonstrate that the micelles experience significant compression during occlusion, which is accompanied by cavity formation. This generates local lattice strain, leading to enhanced mechanical properties. These results give new insight into the formation of occlusions in natural and synthetic crystals, and will facilitate the synthesis of multifunctional nano-composite crystals.

[1] The Molecular Foundry, Lawrence Berkeley National Laboratory, Berkeley, California 94720, USA. [2] Bioscience and Biotechnology Division, Physical and Life Sciences Directorate, Lawrence Livermore National Laboratory, Livermore, California 94550, USA. [3] School of Chemistry, University of Leeds, Leeds LS2 9JT, UK. [4] Department of Chemistry, University of Sheffield, Brook Hill, Sheffield S3 7HF, UK. [5] Department of Mechanical Engineering, Stanford University, Stanford, California 94305, USA. [6] Department of Chemistry, Zhejiang University, Hangzhou 310027, China. [7] Physical Sciences Division, Pacific Northwest National Laboratory, Richland, Washington 99352, USA. Correspondence and requests for materials should be addressed to K.R.C. (email: kangraecho@lbl.gov) or to J.J.D.Y. (email: james.deyoreo@pnnl.gov) or to F.C.M. (email: F.Meldrum@leeds.ac.uk).

oluble additives are widely used to control crystallization processes, providing an experimentally simple and versatile strategy to generate crystals with defined polymorphs, morphologies and sizes[1]. Significant insights into additive-directed crystallization, and the effects of these additives on crystal properties, have come from *in vitro* studies of biomineralization processes[2,3], where the use of advanced characterization methods have demonstrated that the macromolecules active in controlling crystallization become occluded within the crystal structure, modifying crystal texture and lattice strain[4–8], causing local disorder[9] and enhancing mechanical properties[10]. Tomography-based studies have even succeeded in revealing atomic-scale structural[11] and chemical[12] characteristics of interfaces between the crystal lattice and occluded organic materials.

Translation of this strategy to synthetic systems promises the ability to control crystal properties and to generate composite structures[5,13–16]. However, implementation of this approach is currently limited by a poor understanding of the mechanisms by which nanoscale additives are incorporated within single crystals. Much of our knowledge comes from *in situ* atomic force microscopy (AFM) studies supported by computational models, in which changes in crystal shape and step speed (growth rate) are used as indirect probes of additive/crystal interactions[17]. Without the ability to image the individual additives, this process cannot be fully characterized or understood. In this study, we profit from recent demonstrations that high densities of functionalized nanoparticles can be occluded within inorganic crystals[5,15,16] to reveal the mechanisms that govern the adsorption and occlusion of organic additives within single calcite crystals.

Results

Adsorption of copolymer micelles during calcite growth.
Our approach employed carboxylated (Fig. 1a,b) and sulfonated (Supplementary Figs 1 and 2) block copolymer micelles as additives that act as mimics of the biomacromolecules occluded within biominerals. Incorporation of these micelles within calcite generates 'artificial biominerals' with microstructures and mechanical properties comparable to those of biogenic calcite[5]. Crucially, they are sufficiently large enough to be visualized simultaneously with atomic steps on crystal surfaces using *in situ* AFM.

Characterization of the carboxylated micelles in bulk $CaCO_3$ solution using dynamic light scattering (DLS) revealed a monomodal population with an average diameter of ~ 75 nm (Fig. 1d), while two distinct populations were observed on mica (Fig. 1c,e) and calcite surfaces, together with occasional micellar aggregates (Fig. 1e, right inset). The adsorbed micelles on mica were ellipsoidal in morphology, exhibiting heights and diameters of 23.1 (average) $\pm$ 4.6 nm (s.d.) and 41.9 $\pm$ 5.8 nm for the large micelles and 5.2 $\pm$ 1.4 and 15.6 $\pm$ 3.8 nm for the small ones, respectively. The calcite surfaces in pure $CaCO_3$ solutions consisted of terraces separated by 0.31 nm high 'acute' and 'obtuse' atomic steps[18] emanating from screw dislocations (Fig. 1f,g). Under continuous flow of micelle-containing solution at or near equilibrium with respect to calcite (Fig. 1g), both small and large micelles were immobilized at the step edges rather than on the terraces (Fig. 1h,i). Adsorption was complete within a few minutes and adsorption was slightly preferred on the acute over the obtuse steps (63 $\pm$ 3.1% versus 37 $\pm$ 2.6%) (Fig. 1h and Supplementary Movie 1). Identical behaviour was observed at all flow rates (including zero) and for all supersaturations investigated, indicating that neither binding to terraces nor crystallization speed played a direct role in micelle incorporation.

Incorporation of the micelles into calcite single crystals.
Micelle incorporation was then studied over a range of supersaturations

($\sigma = \sim 0$–2.66; see the Methods section for definition of σ). Both the micelle diameter and height decreased with the passage of each step as the micelles were gradually entrapped (Fig. 2a–c) (Supplementary Fig. 3a–c and Supplementary Movie 2). As incorporation continued, a cavity began to form around the perimeter of each micelle and persisted—often for significant time periods—after complete burial of the micelle (A–C in Fig. 2c,d–g). Cavities associated with large micelles had greater diameters and persisted for much longer times than those of small ones (compare Fig. 2g,h with Supplementary Fig. 4b,c), but all were eventually either covered or possibly filled in by advancing steps (Fig. 2h). Significantly, cavities also appeared on dissolution of the composite crystals in undersaturated solutions. Dissolution first generated etchpits and then cavities, and ultimately exposed the buried micelles (Fig. 2k and Supplementary Fig. 4d–i).

Higher supersaturations led to a greater number density of incorporated micelles because the density of steps generated at a dislocation source—and therefore the probability of capturing a micelle from solution—increases quadratically with supersaturation[18] (Fig. 2i,j). Close examination of the micelle-step boundary also showed that steps experienced little or no inhibition by adsorbed micelles (inset, Fig. 2c and Supplementary Fig. 3d–i), as demonstrated by their similar growth rates and straight edges in both pure and micelle-containing solutions (compare Fig. 2a–c with Fig. 1f). This behaviour contrasts with that of other macromolecular adsorbates that bind to kink sites at steps and effectively block addition of solute ions at similar additive concentrations and solution supersaturations[19], presumably because the effective copolymer concentration is greatly reduced by micelle formation.

Careful measurements of micelle heights revealed a further key aspect of incorporation, which is best illustrated by comparing the relative heights of individual large micelles above the calcite surface to the absolute position of the growing crystal face (Fig. 3a). Here, a trajectory parallel to the thick dashed line, which has a slope of unity, represents a micelle that is simply buried by the growing face without any change in its absolute height. Our data show that the actual micelle trajectories had slopes < -1, which demonstrates that the micelles continuously retracted vertically during burial (Supplementary Fig. 5e for detailed analysis). At the same time, the micelles also underwent simultaneous lateral compression (Fig. 2d,e), suggesting loss of associated water molecules and/or collapse of polymer chains during occlusion. In all cases, the ellipsoidal shapes of the large micelles were maintained during crystal growth (Fig. 3b), with the micelle diameter and height decreasing from 94.1 $\pm$ 9.4 and 19.9 $\pm$ 3.1 nm upon adsorption to 38.2 $\pm$ 5.9 and 13.1 $\pm$ 1.9 nm after burial, respectively.

The small micelles with initial heights of about 3.2 $\pm$ 1.1 nm, exhibited similar behaviour with one distinct difference; slopes were initially slightly < -1, but often became > -1 during the last 2 nm of burial (Fig. 3c). These micelles therefore retracted downward slightly during the initial phase of compression, but once buried roughly halfway, they began to extend upward with continued growth. Based on the analysis from Fig. 3c and AFM images, the diameters and heights of the small micelles were estimated to change from 32.0 $\pm$ 6.0 and 3.2 $\pm$ 1.1 nm, respectively, upon adsorption to 9.0 $\pm$ 3.1 and 4.6 $\pm$ 0.7 nm, respectively, after burial.

Micromechanical simulations of micelle incorporation.
Simulations of micelle incorporation into a crystal growing through layer-by-layer addition were also performed to understand the morphological behaviour observed for the small micelles, as well as the mechanical consequences of burial. For

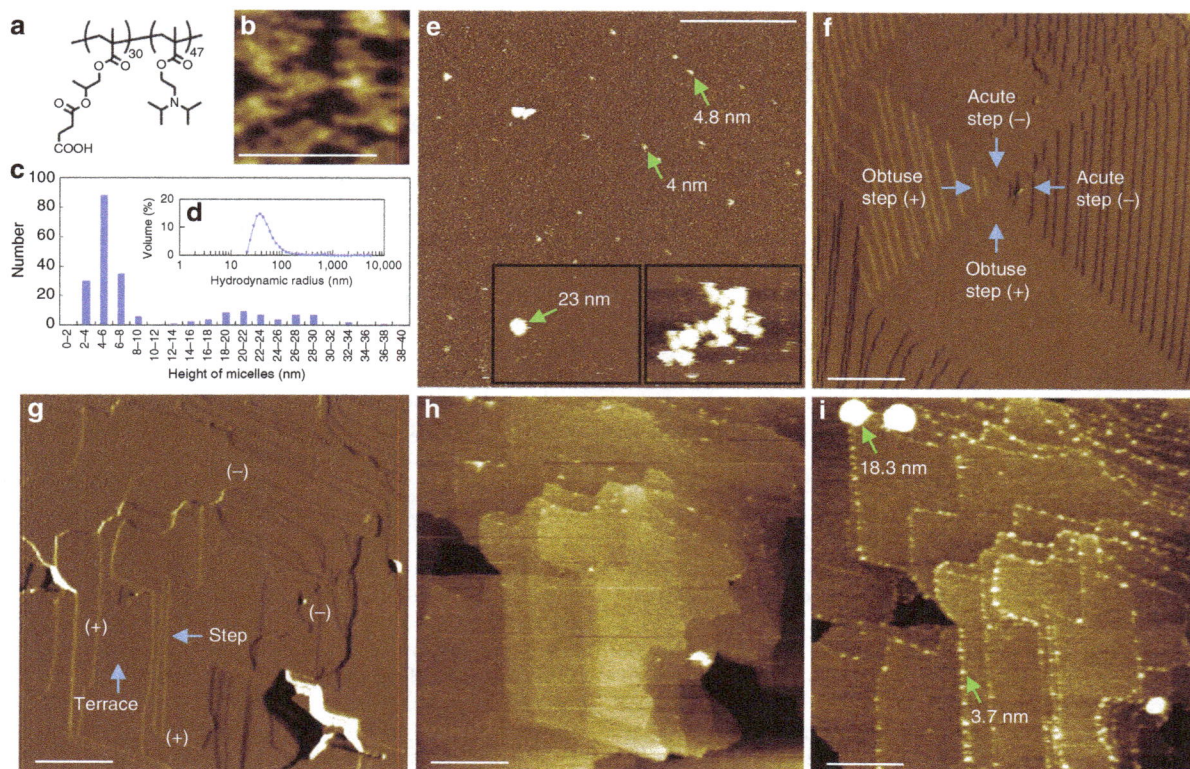

Figure 1 | Structure and adsorption dynamics of carboxylated micelles on calcite. (**a**) Chemical structure and (**b**) morphologies of copolymers observed on positive (poly-L-lysine-treated) mica surface in water. (**c**) Height distribution measured by AFM and (**d**) hydrodynamic radii measured by DLS of micelles in calcium carbonate solution ($\sigma = \sim 1.81$, pH $= \sim 8.5$). The heights of the micelles were determined from images such as (**e**), which is a positive mica surface showing micelles belonging to two populations, one ~ 3.5–6.5 nm and another ~ 20–30 nm in height (left inset). Some aggregates were also observed (right inset). The main image and insets are at the same scale. (**f**) Morphology of {104} calcite growth surface showing growth hillocks composed of two crystallographically distinct types of atomic steps (acute ($-$) and obtuse ($+$)) generated at screw dislocations[18]. (**g**) Calcite surface in (near)—equilibrium solution (step speed = 0). (**h**) Same surface as in **g** after exposure for ~ 2 min to a micelle-containing solution at near-equilibrium conditions (step speed = 0), demonstrating that micelles were immobilized at the step edges rather than on the terraces, with slightly greater affinity for the acute steps over obtuse steps. (**i**) Same surface as in **h** after exposure for another 48 min to flowing micelle-containing solution showing overwhelming dominance of steps edge adsorption and that micelles adsorbed on calcite formed ellipsoids with the short axis perpendicular to the plane of the crystal face. Scale bars, (**b**) 50 nm, (**e**) 300 nm, (**f**) 600 nm and (**g-i**) 300 nm. All images were collected in tapping mode except for **f** and **g** which collected in contact mode.

simplicity, we assumed that the micelles were initially spherical and used a two-dimensional model, in which the initial micelle shape is represented as a circle (Fig. 4). (Note that while this simple model correctly explains the general pattern of shape change of the small micelles, it predicts a higher final aspect ratio than observed experimentally, because the actual initial micelle shape is a flattened ellipsoid (Fig. 3d, stage 1)). The bottom of the circle was fixed to the initial crystal surface to mimic binding to the step edge. As the height of the crystal was increased, it was allowed to compress the micelles laterally until the pressure exerted by the crystal was balanced by the bending resistance of the micelle. Some representative results are given in Supplementary Figs 6–8.

Two conditions were considered for the boundary between the crystal and the buried part of the micelle. In the first, the buried boundary was allowed to continuously relax, thereby mimicking continued communication with the solution. Under this condition the crystal exerts an isotropic pressure field on the micelle. Thus, the buried micelle returns either to a spherical shape (under low pressure) or to some non-convex but symmetrical shape (under high pressure). Hence, the final shape of the micelle cannot be ellipsoidal, because the bending energy of the micelle can be reduced without changing the work done by the pressure field if the micelle is transformed to a sphere of the

same volume. In the second condition, the buried boundary was assumed to be static, which simulates no ion exchange with the solution. The crystal-micelle boundary therefore becomes fixed shortly after burial beneath the top surface of the crystal. This model predicts that the top of the micelle retracts vertically at the beginning of incorporation, before eventually extending upwards, as observed experimentally for small micelles.

Discussion
Based on our experimental data, micromechanical simulations and previous transmission electron microscopy (TEM) data, which show that the occluded micelles—and any associated cavities that may be retained—are ellipsoidal in form, we can now reconstruct the entire incorporation process for both large (Fig. 3b) and small micelles (Fig. 3d). Following micelle binding to step edges (Fig. 3b,d, stage 1), the passing steps begin to compress the micelles (Fig. 3b,d, stage 2). For each layer, the steps stop when the increase in bending energy of the micelle balances the chemical potential that drives step advancement. The bending resistance of the micelle thus strains the surrounding crystal, which reduces the driving force for crystallization around the micelle. This then contributes to the formation of a gap (Fig. 3b,d, stage 3) in a manner similar to the creation of hollow

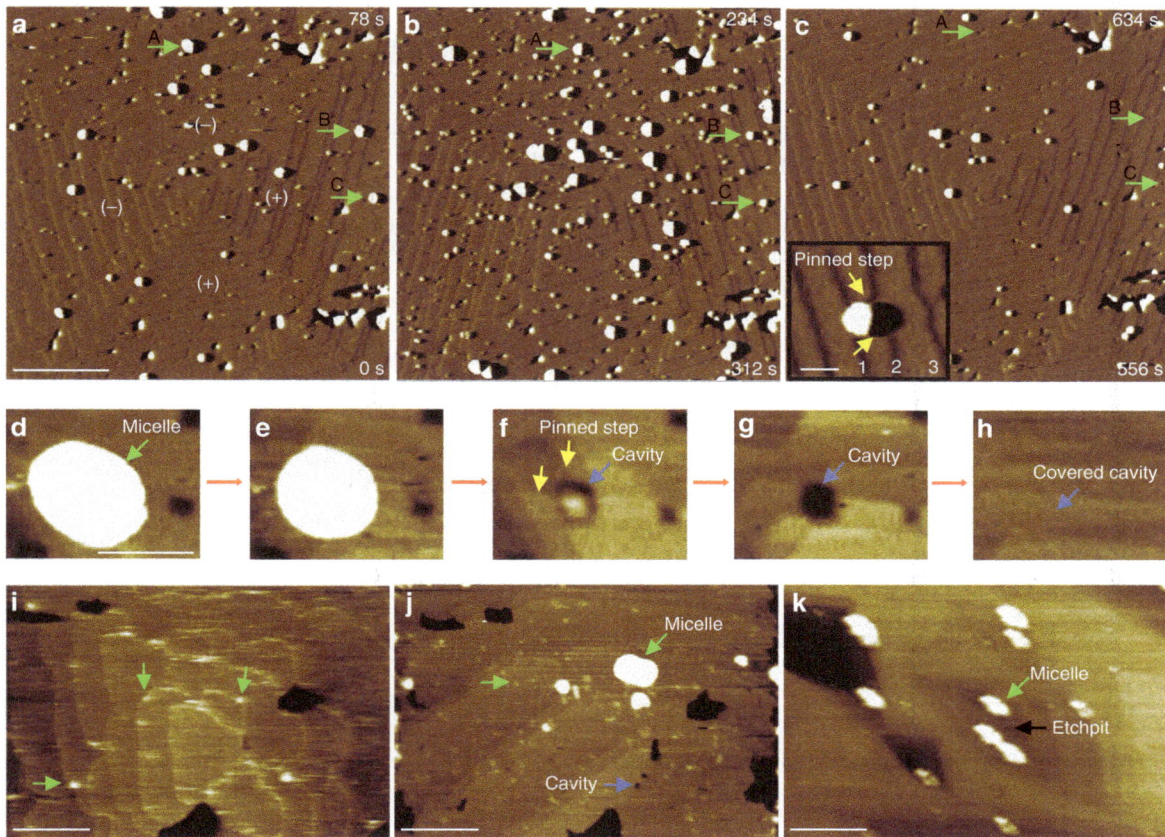

Figure 2 | Incorporation of carboxylated micelles and associated cavity generation. (a–c) Sequential *in situ* $3 \times 3\,\mu$m AFM images of growing calcite surface at $\sigma = 1.49$. Particles indicated by arrows in (**a**) subsequently show a decrease in height (**b**), before undergoing complete burial (**c**). The inset in **c** shows one step (1) that has just reached a micelle, one (2) that has just closed around it and another (3) that has nearly recovered to a straight morphology with little overall inhibition. The times (*t*) at which the bottom and top of the images were collected are given in lower and upper right corners of images, respectively, where the bottom of (**a**) was arbitrarily set to $t = 0$ s. The copolymer concentration was (**a,b**) 127 nM and (**c**) zero. (**d–h**) Detailed view of the incorporation of a large micelle ($\sigma = 2.66$). (**d**; 0 s) An ellipsoidal micelle with a height of 19.6 nm and in-plane diameters of 93 nm is adsorbed to an acute step (see Supplementary Fig. 5f for determination of true micelle diameter). (**e**; 34 s) As incorporation proceeds, the micelle decreases in height (13.3 nm) and its shape evolves to that of an ellipsoid with a circular cross-section and reduced diameter of 84 nm. (**f**; 2 min 35 s) Upon further burial, a gap (blue arrow) forms around the micelle with the diameter further reduced to 33 nm (height: 0.8 nm). (**g**; 3 min 10 s) The cavity remains (blue arrow) after complete micelle burial before eventually closing over, (**h**; 31 min 58 s). (**i**) Calcite surface with adsorbed micelles (green arrows) in micelle-containing solution at (near-) equilibrium (step speed = 0, copolymer concentration: 69 nM). (**j**) Upon introduction of highly supersaturated solution ($\sigma = 2.66$) and consequent step advancement, many micelles (green arrows) appear because of increased capture rate by steps. Many have already become incorporated and formed cavities (blue arrow). (**k**) Reappearance of incorporated micelles in etchpits upon dissolution. Scale bars; (**a**) 750 nm, (**d**, inset to **c**) 100 nm and (**i–k**) 300 nm. Images **a–h** and **i–k** were collected in contact and tapping mode, respectively.

cores in the strain field of a dislocation[20]. As the micelle becomes approximately half buried, the difficulty of extending the crystal steps over the surface of the underlying micelle—and thus creating an overhang—may also contribute to gap formation. As the growth front passes beyond the micelle, the gap develops into a channel (Fig. 3b,d, stage 4), which ultimately closes over as the strain field of the compressed micelle recedes behind the advancing growth front (Fig. 3b,d, stages 5 and 6).

These data do not enable us to determine whether cavities are retained or eliminated during the growth process. Either scenario should be possible depending on the supersaturation and particle size. A simple extrapolation to biominerals—in which similar sized occlusions are observed by TEM—suggests they may be a combination of cavities and proteins. Indeed, this is consistent with a previous TEM study of the aragonite plates in nacre[21], and would also explain why occlusions can far exceed the sizes expected for individual proteins[22]. However, previous studies of a wide range of calcite/particle composite crystals using scanning electron microscopy and TEM have not reported evidence for

retained cavities[14,16]. This is particularly clear for occluded gold nanoparticles, which should be easily distinguished from an associated cavity using TEM by virtue of their high electron density. The cavities observed upon dissolution of the composite crystals suggest that the cavities may remain; however, these might also be attributable to the high lattice strain present in the vicinity of the micelles, which renders these regions more soluble than strain-free regions (Fig. 2k and Supplementary Fig. 4d–i).

This study provides significant new insights into additive-controlled crystallization processes. Considering first the calcite/micelle system, the previous study of Kim *et al.*[5] concluded from a purely structural analysis that the ellipsoidal occlusions seen in TEM corresponded to occluded micelles that were incorporated as a result of their affinity for adsorption to the crystal face with a morphology determined by that interaction. The mechanistic understanding obtained from our current *in situ* observations refines this picture. Micelles actually bind to advancing steps and undergo a morphological change during occlusion as a result of compression by the steps. Kim *et al.*[5] also showed that the calcite/

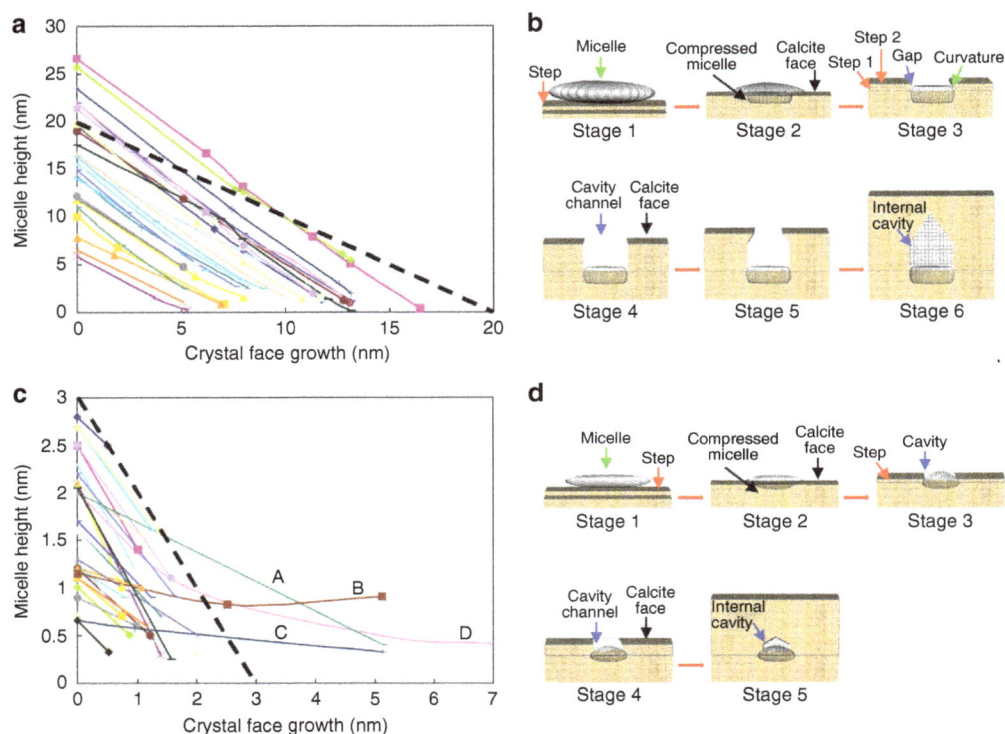

Figure 3 | Micelle height trajectories and schematic of micelle incorporation. (**a**) Plot showing changes in heights of individual large micelles relative to crystal face versus absolute height of the growing crystal face. (Crystal face growth (nm) = 0 marks time when each micelle is first imaged and end point of each trajectory marks last micelle image before complete incorporation.) Errors in height caused by tip compression of the micelles under these conditions were < 3 nm for the largest micelles and < 0.5 nm for the smallest at crystal face growth = 0, leading to maximum errors in slope of 8–11 %. (**b**) Schematic diagram illustrating the incorporation of the large micelles. A micelle adsorbs to a step as an ellipsoid (stage 1) and contracts vertically, while it is compressed laterally (stage 2) as the calcite face advances. A gap then begins to form around the micelle periphery (stage 3). Successive steps do not advance beyond this gap, creating a cavity. As growth continues, the gap develops into a cylindrical channel (stage 4), whose width decreases (stage 5) until it closes to form an internal cavity (stage 6). (**c**) Plot of height changes for small micelles determined as in **a** showing shallower slopes than for large micelles, including slopes > -1 (for example, A–D). (**d**) Schematic representation showing that the incorporation process of small micelles is similar to that of large micelles, except that micelles elongate vertically between stages 2 and 3 and channel lengths is much smaller. Face growth rate $= 0.033\,\mathrm{nm\,s^{-1}}$ ($\sigma = 1.49$).

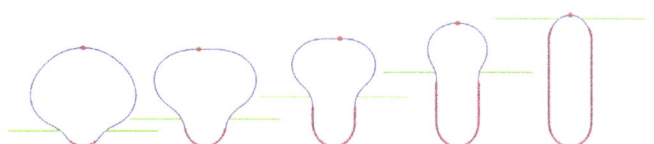

Figure 4 | Micromechanical simulation of spherical micelle incorporation. During burial, lateral micelle compression is accompanied by an initial downward contraction, followed by upward extension. The bottom dashed line indicates the surface on which the micelle is initially attached, while the green solid lines represent the top surface of the crystal. The red crosses mark the fixed portion of the (buried) micelle. The red circle marks the highest point of the micelle. While the true initial micelle shape is that of a flattened ellipsoid rather than a sphere, the mechanical response to lateral compression resembles that observed in Fig. 3d.

micelle crystals are harder than pure calcite, where this change was associated with a large compressive strain gradient in the crystal. The current work implies this local strain is a direct result of the chemical driving force behind the step advancement, which ensures that the steps will continue to compress the micelle until the resulting lattice strain energy equals the solution chemical potential.

Much of our current understanding of additive occlusion derives from optical and theoretical studies on incorporation of large particulates within crystals during their freezing transitions[23,24]. Under those conditions, particle entrapment occurs only above a critical growth velocity of the solid–liquid interface[23,24], and depends on many factors including the thermal diffusivity of the particle and melt, wetting of the crystal face by the particles, the size and buoyancy of the particles, and the melt viscosity. Particle entrapment during growth from solution, in contrast, has received considerably less attention[14], although it has been suggested that similar factors may be important.

The model that emerges here stands in stark contrast to those previously proposed[23–25,] in which particles are incorporated either because of the slow diffusive velocity of the liquid towards the crystal surface relative to the interface growth velocity or because of hydrodynamic forces that correlate with crystallization speed. Instead the particles become entrapped because they bind specifically to steps, enabling successive steps to close around them. Further, the increase in incorporation efficiency at higher growth rates is a consequence of the greater rate of step generation with increasing supersaturation[18], rather than the enhanced ·growth front velocity. These results therefore provide a new understanding of the dynamics of additive/crystal interactions. Combined with the insight they provide concerning evidence on the strain-induced toughening of minerals, these findings will inform the synthesis of novel composite crystals through the optimization of variables, such as adsorbate–step interactions, bending rigidity and supersaturation.

Methods

Micelle stock solution preparation. PSPMA$_{30}$-PDPA$_{47}$ block copolymer powder was dissolved in MilliQ water for which the pH was initially adjusted to $\sim$4.8 with 1 M HCl solution. The prepared stock solution volumes were typically 100 ml and included 5.76 μM (fixed concentration) copolymer. (See Supplementary Information of ref. 5 for the copolymer synthesis.) After the copolymers were completely dissolved, rapid micelle formation was then induced by increasing the solution pH to $\sim$9.5–$\sim$10 by adding 1 M NaOH solution. The pH of the stored stock solutions decreased over time from the initial value to $\sim$8.0. However, the micelles kept their micellar structure, because the pK$_a$ values[26] of PSPMA and PDPA are $\sim$5.5 and 6.3, respectively. Micelle stock solutions of sulfonated SBA: PHPMA$_{30}$-PDPA$_{47}$ copolymer were prepared in the same manner (see Supplementary Methods and Supplementary Fig. 2 for the synthesis and characterization of the sulfonated copolymer).

Micelle-containing calcium carbonate solution preparation. Calcium carbonate (CaCO$_3$) solutions (volume 100 ml) at supersaturations (σ) = $\sim$0–2.66 with a calcium to typical carbonate ratio of $\sim$0.5 were prepared. Here, σ is defined by $\sigma = \ln a(\text{Ca}^{2+})\, a(\text{CO}_3^{2-})/K_{sp}$ where $a(\text{Ca}^{2+})$ and $a(\text{CO}_3^{2-})$ are the activities of the calcium and carbonate ions, respectively, and K_{sp} is the solubility constant for calcite. The activities of calcium and carbonate ions were obtained using the multicomponent speciation program Visual Minteq[27], which uses the Davies equation to give the ion activity coefficient. The desired amounts of micelle stock solution (typically <2.2 ml) were first introduced into $\sim$46 ml CaCl$_2$.2H$_2$O (reagent grade) solutions at pH $\sim$10 (adjusted using NaOH). The solution volume was then adjusted to 50 ml using MilliQ water. These solutions were mixed with 50 ml NaHCO$_3$ (reagent grade) solutions containing the desired amounts of NaCl to give CaCO$_3$ solutions with ionic strength (I) of $\sim$0.05 M. In this way, micelle-containing CaCO$_3$ solutions with a typical copolymer concentration <127 nM and I = $\sim$0.05 at pH $\sim$8.5 were prepared.

In situ AFM imaging. Freshly cleaved geologic calcite (Iceland spar, Ward's Natural Science, Rochester, NY) was glued to the AFM specimen disk to expose a fresh {104} surface for investigation. A commercial fluid cell (MTFML, Veeco Probes) with O-ring was placed on the cleaved calcite face inside an AFM. (Multimode Nanoscope IIIa or VIII from Digital Instruments, Santa Barbara, CA). A syringe pump was used to flow solution through the fluid cell and imaging was performed at room temperature using commercially available SiN cantilevers (Bruker, NP-S with spring constant of 0.12 N m^{-1} for imaging on calcite surface, and Olympus, TR400PSA with spring constant of 0.08 N m^{-1} for imaging on mica surface) with a nominal radius of 15 nm (as reported by the manufacturer).

The *in situ* images of the calcite surface were obtained in either contact mode or tapping mode at solution flow rates of 0.3 ml min^{-1}. Images were typically collected at a scan rate 3.3 Hz (lines per second) with 256 scan lines per image. We found that applying the minimum possible scan force, a suitable scan rate, and scan size were critical to being able to record quality, reproducible images of the micelles on the calcite surface. Tapping mode images were collected at $\geq$ 60% of cantilever free amplitude. The above scan parameters were appropriate for imaging areas between 1 × 1 μm and 3 × 3 μm and for obtaining reasonable acquisition rates. To account for the distortion in step orientations because of finite scan rates and non-zero step speeds[28] in images collected *in situ*, step speeds were obtained by orienting the fast scan direction perpendicular to true step directions and measuring the change in the apparent angle of the steps in the images collected during upward and downward scans[28]. To image the negatively charged micelles on mica surfaces, the surface charge of the mica was adjusted from negative to positive by depositing poly-L-lysine on freshly cleaved mica before introducing the micelle solutions ($\sim$50 μl) into the fluid cell.

Dynamic light scattering. Dynamic light scattering measurements of carboxylated PSPMA$_{30}$-PDPA$_{47}$ micelles were obtained on a Malvern Zetasizer Nano Series ZS (Malvern Instruments) and analysed with the provided Zetasizer software. The samples came from the same calcium carbonate solution as those used for AFM imaging. Measurements were recorded at 25 °C, the refractive index of the buffer was estimated to be 1.330 and the viscosity 0.8894 cP.

Determination of micelle and step heights. Heights of micelles and steps were obtained by analysing height profiles from images such as those in Fig. 1e,i using standard Veeco Nanoscope image analysis software. In addition, the scanning probe image processor (SPIP 5.1.4) was used for measuring heights of micelles from images obtained on mica surfaces. The height distribution of micelles in Fig. 1c was based on measurements from three 1 × 1 μm *in situ* images, including Fig. 1e.

Simulations of micelle incorporation. A two-dimensional model was developed in which the shape of the micelle was represented as a continuous and closed line (that is, loop). The line is discretized into a set of nodes {$\mathbf{r}_i$}, where $\mathbf{r}_i = (x_i, y_i)$, $i = 1$, ..., N. The nodes are connected to their neighbours: $\mathbf{r}_i$ is connected to $\mathbf{r}_{i+1}$ for $i = 1, ..., N-1$ and $\mathbf{r}_N$ is connected to $\mathbf{r}_1$. A total free energy F is defined as a function of nodal positions.

$$F(\{\mathbf{r}_i\}) = \sum_{\langle i,j \rangle} \frac{1}{2} K \left(|\mathbf{r}_i - \mathbf{r}_j| - L_0 \right)^2 + \sum_{\langle j,i,k \rangle} \frac{1}{2} \kappa |\mathbf{r}_i - \mathbf{r}_j|^2 |\mathbf{r}_i - \mathbf{r}_k|^2 (1 + \cos \theta_{jik})^2 + p \cdot A_{sub} \tag{1}$$

where K is the stretching stiffness and L_0 is the equilibrium 'bond length' between neighbouring nodes, κ is the bending stiffness and θ_{jik} is the angle formed between 'bonds' i-j and i-k, p is the pressure exerted by the crystal and A_{sub} is the part area enclosed by the loop below the top surface of the crystal. The three terms in the free energy function F represent the contributions from the stretching of the membrane at the surface, the bending of the membrane and the interaction with the crystal, respectively. The free energy contribution from the volume change of the micelle (that is, the bulk contribution) is not included in the above expression. We have assumed the stretching stiffness to be sufficiently large so that the perimeter of the micelle remains essentially constant during the deformation. Therefore, the shape change of the micelle is mainly determined by the competition between the bending energy (second term) and the interaction with the crystal (third term). In an extended model, we have also considered the effect of the bulk free energy contribution due to the volume change of the micelle (Supplementary Note 1). However, we found that if the bulk contribution is too large, the micelle shape remains essentially circular throughout the entire burial process, which is inconsistent with the experimental observation. Deformation of the micelle is only obtained when the bulk term is sufficiently small, but in this case the predicted behaviour is qualitatively the same as that for the bulk term equal to zero. Therefore, we have excluded the bulk term in the above free energy expression.

For a given position of the top surface of the crystal, y_{sub}, the nodes evolve in the steepest direction, that is

$$\frac{d\mathbf{r}_i}{dt} = -\frac{\partial F}{\partial \mathbf{r}_i} \tag{2}$$

until equilibrium positions are reached. y_{sub} is then incremented by a small step and the nodes are allowed to evolve into their new equilibrium positions. The simulation shown in Fig. 4 was performed using the following (dimensionless) parameters: $K = 2$, $L_0 = 0.63$, $\kappa = 1$, $P = 10^{-4}$. Nodes with $y_i < y_{sub} - 2$ are fixed (red crosses in Fig. 4) to simulate the condition of no communication between the solution and the buried part of the micelle.

References

1. Song, R.-Q. & Cölfen, H. Additive controlled crystallization. *CrystEngComm* **13**, 1249–1276 (2011).
2. Weiner, S. & Addadi, L. Design strategies in mineralized biological materials. *J. Mater. Chem.* **7**, 689–702 (1997).
3. Meldrum, F. C. & Colfen, H. Controlling mineral morphologies and structures in biological and synthetic systems. *Chem. Rev* **108**, 4332–4432 (2008).
4. Berman, A. *et al.* Intercalation of sea-urchin proteins in calcite: study of a crystalline composite material. *Science* **250**, 664–667 (1990).
5. Kim, Y. Y. *et al.* An artificial biomineral formed by incorporation of copolymer micelles in calcite crystals. *Nat. Mater.* **10**, 890–896 (2011).
6. Berman, A., Addadi, L. & Weiner, S. Interactions of sea-urchin skeleton macromolecules with growing calcite crystals—a study of intracrystalline proteins. *Nature* **331**, 546–548 (1988).
7. Pokroy, B., Fitch, A. N. & Zolotoyabko, E. The microstructure of biogenic calcite: a view by high-resolution synchrotron powder diffraction. *Adv. Mater.* **18**, 2363–2368 (2006).
8. Kim, Y. Y. *et al.* A critical analysis of calcium carbonate mesocrystals. *Nat. Commun.* **5**, 4341 (2014).
9. Metzler, R. A. *et al.* Nacre protein fragment templates lamellar aragonite growth. *J. Am. Chem. Soc.* **132**, 6329–6334 (2010).
10. Weiner, S., Addadi, L. & Wagner, H. D. Materials design in biology. *Mater. Sci. Eng. C* **11**, 1–8 (2000).
11. Li, H., Xin, H. L., Muller, D. A. & Estroff, L. A. Visualizing the 3D internal structure of calcite single crystals grown in agarose hydrogels. *Science* **326**, 1244–1247 (2009).
12. Gordon, L. M. & Joester, D. Nanoscale chemical tomography of buried organic-inorganic interfaces in the chiton tooth. *Nature* **469**, 194–197 (2011).
13. Brif, A., Ankonina, G., Drathen, C. & Pokroy, B. Bio-inspired band gap engineering of zinc oxide by intracrystalline incorporation of amino acids. *Adv. Mater.* **26**, 477–481 (2014).
14. Kim, Y.-Y. *et al.* Bio-inspired synthesis and mechanical properties of calcite-polymer particle composites. *Adv. Mater.* **22**, 2082–2086 (2010).
15. Kulak, A. N. *et al.* One-pot synthesis of an inorganic heterostructure: uniform occlusion of magnetite nanoparticles within calcite single crystals. *Chem. Sci* **5**, 738–743 (2014).
16. Kulak, A. N., Yang, P. C., Kim, Y. Y., Armes, S. P. & Meldrum, F. C. Colouring crystals with inorganic nanoparticles. *Chem. Commun.* **50**, 67–69 (2014).

17. De Yoreo, J. J., Wierzbicki, A. & Dove, P. M. New insights into mechanisms of biomolecular control on growth of inorganic crystals. *CrystEngComm* **9**, 1144–1152 (2007).
18. Teng, H. H., Dove, P. M. & Orme, C. A. Thermodynamics of calcite growth: baseline for understanding biomineral formation. *Science* **282**, 724–727 (1998).
19. Elhadj, S., De Yoreo, J. J., Hoyer, J. R. & Dove, P. M. Role of molecular charge and hydrophilicity in regulating the kinetics of crystal growth. *Proc. Natl Acad. Sci. USA* **103**, 19237–19242 (2006).
20. De Yoreo, J. J., Land, T. A. & Lee, J. D. Limits on surface vicinality and growth rate due to hollow dislocation cores on KDP {101}. *Phys. Rev. Lett.* **78**, 4462–4465 (1997).
21. Gries, K., Kroger, R., Kubel, C., Fritz, M. & Rosenauer, A. Investigations of voids in the aragonite platelets of nacre. *Acta Biomater.* **5**, 3038–3044 (2009).
22. Li, H. *et al.* Calcite prisms from mollusk shells (Atrina Rigida): swiss-cheese-like organic-inorganic single-crystal composites. *Adv. Funct. Mater.* **21**, 2028–2034 (2011).
23. Uhlmann, D. R., Chalmers, B. & Jackson, K. A. Interaction between particles and a solid-liquid interface. *J. Appl. Phys.* **35**, 2986–2993 (1964).
24. Rempel, A. W. & Worster, M. G. The interaction between a particle and an advancing solidification front. *J. Cryst. Growth* **205**, 427–440 (1999).
25. Li, H. Y. & Estroff, L. A. Calcite growth in hydrogels: assessing the mechanism of polymer-network incorporation into single crystals. *Adv. Mater.* **21**, 470–473 (2009).
26. Vo, C. D., Armes, S. P., Randall, D. P., Sakai, K. & Biggs, S. Synthesis of zwitterionic diblock copolymers without protecting group chemistry. *Macromolecules* **40**, 157–167 (2007).
27. Gustaffson, J. P. Visual Minteq 2.30 edn. http://vminteq.lwr.kth.se/ (2004).
28. Land, T. A., De Yoreo, J. J. & Lee, J. D. An *in-situ* AFM investigation of canavalin crystallization kinetics. *Surf. Sci.* **384**, 136–155 (1997).

Acknowledgements

We thank Drs Qiaona Hu and Raymond Friddle for help with the AFM experimental setup, Dr Debin Wang for help with the AFM analysis software, Drs Dominik Ziegler and Paul Ashby for discussion about AFM and Matthew Rames for help with editing the manuscript. Research on micelle incorporation and deformation was performed under the auspices of the US Department of Energy, Office of Basic Energy Sciences, Division of Chemical Sciences, Geosciences and Biosciences at Lawrence Berkeley National Laboratory (LBNL) under contract DE-AC02-05CH11231 and the Pacific Northwest National Laboratory (PNNL), which is operated by Battelle under Contract DE-AC05-76RL01830. Analysis of solution micelle formation was supported by grant DC011614 from the National Institutes of Health. AFM and DLS measurements were performed at the Molecular Foundry, a National User Facility operated by LBNL on behalf of the US Department of Energy, Office of Basic Energy Sciences. K.R.C. acknowledges support from the Postdoctoral Program at Lawrence Livermore National Laboratory, which is operated for the US Department of Energy under Contract DE-AC52-07NA27344. We thank the Engineering and Physical Sciences Research Council (EPSRC) for financial support via grants EP/G00868X/1 and EP/K006304/1 (A.K. and F.C.M.) and EP/J018589/1 (Y-Y.K. and F.C.M.). This work was also supported by an EPSRC Leadership Fellowship (EP/H005374/1; F.C.M. and Y.Y.K.). S.P.A. acknowledges support from EPSRC (EP/K006290/1 and EP/J018589/1) and also a 5-year ERC Advanced Investigator grant (PISA 320372).

Author Contributions

K.R.C., W.C., S.P.A., F.C.M. and J.J.D.Y. designed the research. K.R.C. and H.P. performed the AFM measurements. K.R.C. analysed the AFM data. W.C. developed and performed the micromechanical simulations. P.Y. synthesized and characterized the block copolymers. J.L.L. performed DLS measurements. K.R.C., Y-Y.K., P.Y., W.C., A.N.K., P.K., S.P.A., F.C.M. and J.J.D.Y. wrote the paper.

Additional information

Water electrolysis on $La_{1-x}Sr_xCoO_{3-\delta}$ perovskite electrocatalysts

J. Tyler Mefford[1], Xi Rong[2], Artem M. Abakumov[3,4], William G. Hardin[5], Sheng Dai[6], Alexie M. Kolpak[2], Keith P. Johnston[5,7] & Keith J. Stevenson[1,3,5,8]

Perovskite oxides are attractive candidates as catalysts for the electrolysis of water in alkaline energy storage and conversion systems. However, the rational design of active catalysts has been hampered by the lack of understanding of the mechanism of water electrolysis on perovskite surfaces. Key parameters that have been overlooked include the role of oxygen vacancies, B–O bond covalency, and redox activity of lattice oxygen species. Here we present a series of cobaltite perovskites where the covalency of the Co–O bond and the concentration of oxygen vacancies are controlled through Sr^{2+} substitution into $La_{1-x}Sr_xCoO_{3-\delta}$. We attempt to rationalize the high activities of $La_{1-x}Sr_xCoO_{3-\delta}$ through the electronic structure and participation of lattice oxygen in the mechanism of water electrolysis as revealed through *ab initio* modelling. Using this approach, we report a material, $SrCoO_{2.7}$, with a high, room temperature-specific activity and mass activity towards alkaline water electrolysis.

[1] Department of Chemistry, The University of Texas at Austin, Austin, Texas 78712, USA. [2] Department of Mechanical Engineering, Massachusetts Institute of Technology, Cambridge, Massachusetts 02139, USA. [3] Center for Electrochemical Energy Storage, Skolkovo Institute of Science and Technology, 143026 Moscow, Russia. [4] Electron Microscopy for Material Science, University of Antwerp, Groenenborgerlaan 171, B-2020 Antwerp, Belgium. [5] Texas Materials Institute, The University of Texas at Austin, Austin, Texas 78712, USA. [6] Chemical Sciences Division, Oak Ridge National Laboratory, Oak Ridge, Tennessee 37831, USA. [7] Department of Chemical Engineering, The University of Texas at Austin, Austin, Texas 78712 , USA. [8] Center for Nano and Molecular Science and Technology, The University of Texas at Austin, Austin, Texas 78712, USA. Correspondence and requests for materials should be addressed to K.J.S. (email: K.Stevenson@skoltech.ru).

The scarcity of fossil fuels and the increasing awareness of the environmental and geopolitical problems associated with their use have encouraged significant efforts towards the development of advanced energy storage and conversion systems using materials that are cheap, abundant and environmentally benign. A major thrust in the field of renewable energy has been to develop higher power and more energy-dense storage devices, including low-temperature regenerative fuel cells and rechargeable metal–air batteries that function through the electrocatalysis of oxygen. Inherent to these systems are the electrolysis of water $(2H_2O \rightarrow O_2 + 4H^+ + 4e^-$; oxygen evolution reaction (OER)) and the reduction of molecular oxygen $(O_2 + 4H^+ + 4e^- \rightarrow 2H_2O$; oxygen reduction reaction (ORR)), both of which require the use of an electrocatalyst due to their slow reaction kinetics. The most active catalysts for the ORR are Pt-alloys and other precious metals, Ir, Ru and Pd[1–3]. However, while the Pt group metals perform well for the ORR, the formation of an oxide surface film at high potentials, especially in the case of Pt, decreases their ability to catalyse the OER[4]. This problem, coupled with the Pt group metal scarcity and restrictive cost represent major roadblocks to mass adoption of fuel cells and metal–air batteries in renewable energy technologies.

Using alkaline electrolytes opens up the possibility to use transition metal oxides as catalysts due to their structural stability, resistance to electrolytic corrosion and their high activities for both the OER and ORR[5–7]. Among the wide variety of metal oxides available, the crystal family of perovskite oxides $ABO_{3 \pm \delta}$, of which A is a rare-earth or alkaline earth element and B is a transition metal, are attractive candidates due to their high ionic and electronic conductivities, structural stability, and the ability to substitute into the A and B sites elements of varying valency, electronegativity or ionic size to tune the structural, physical and electronic properties of the catalyst. Even though the electrolysis of water to oxygen is one of the most extensively studied reactions, predating even the fields of catalysis and electrochemistry, the lack of a conclusive mechanism for metal oxides in alkaline electrolyte remains a significant limitation in the rational design of electrocatalysts for the OER[8]. Thus, much of the research on perovskites for the OER and ORR has been focused on identifying descriptors for the activities of perovskites based on the electronic and structural properties of the surface or bulk[9–11]. Since the initial discovery of $La_{0.8}Sr_{0.2}CoO_3$ as an active ORR catalyst, many mechanistic theories have been put forward over the past 40 years[12]. A recent review summarizes the current understanding of mechanistic processes for the OER, specifically highlighting correlations between bulk and surface properties of metal oxides and their electrocatalytic activities[13]. Notably, the idea that the e_g filling of the transition metal in the ABO_3 perovskite controls the intermediates binding strength and thus the electrocatalytic activity has recently gained significant credence[14,15]. However, we have observed that among a series of perovskites with a nominal e_g filling of ~ 1 ($LaBO_3$, where $B = Mn$, Co, Ni, or $Ni_{0.75}Fe_{0.25}$), there exists significant differences in their activities for both the ORR and the OER, indicating that the surface chemistry may not be adequately rationalized by bulk electronic descriptions[16,17].

A previously overlooked parameter concerns the role of oxygen vacancy defects, which allows for crystalline oxygen to be mobile at the surface of perovskites. It is well-known that the stoichiometry of oxygen in the crystal structure of perovskites often differs from the nominal value of 3 for the formula ABO_3, affecting both the lability of surface oxygen and reflecting the underlying electronic structure of these materials[18–20]. The degree of vacancy formation reflects the relative positions of the transition metal 3d bands compared with the oxygen 2p band

in the crystal, with more covalent systems exhibiting higher vacancy concentrations as shown in Fig. 1. In addition, it is well-documented that the concentrations of oxygen vacancies in perovskite electrodes can be controlled through an applied electrical potential, with room temperature diffusion coefficients of lattice oxygen for a number of perovskites in the range of 10^{-14} to 10^{-11} cm^2 s^{-1} (refs 21–26). In a previous paper, we demonstrated that this effect could be used as a means of pseudocapacitive energy storage in an oxygen-deficient $LaMnO_{2.91}$ electrode[27]. We have previously hypothesized the role of lattice oxygen and vacancy exchange in the OER mechanism on $LaNiO_3$ refs 16,17. We now revisit this idea to investigate the role of mobile lattice oxygen in the electrolysis of water by examining the system $La_{1-x}Sr_xCoO_{3-\delta}$, $0 \leq x \leq 1$. Through substitution of the lower valence Sr^{2+} ion for La^{3+}, the amount of oxygen vacancy defects and the oxidation state of cobalt can be tuned through the relation[28]:

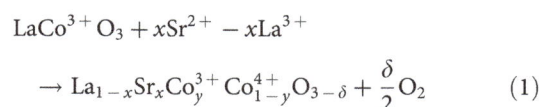

$$LaCo^{3+}O_3 + xSr^{2+} - xLa^{3+}$$

$$\rightarrow La_{1-x}Sr_xCo_y^{3+}Co_{1-y}^{4+}O_{3-\delta} + \frac{\delta}{2}O_2 \quad (1)$$

where, δ is the oxygen non-stoichiometry parameter, x is the amount of Sr^{2+}, and y is the amount of Co^{3+} in $La_{1-x}Sr_xCoO_{3-\delta}$, hereafter referred to as $LSCO(1-x)x$ (that is, LSCO28 for $La_{0.2}Sr_{0.8}CoO_{3-\delta}$).

Herein, we describe the intrinsic activities of $La_{1-x}Sr_xCoO_{3-\delta}$ for the OER across the full series from $0 \leq x \leq 1$, including the previously unreported perovskite phase $SrCoO_{2.7}$ with the layered ordering of oxygen vacancies. The controlled substitution of Sr^{2+} for La^{3+} across the full phase space of the LSCO system while maintaining the perovskite structure allows us to probe the effects of covalency, vacancy defects and oxygen exchange on the electrocatalysis of the OER. The high activities for materials with $x > 0.4$ are rationalized through the high oxygen ion diffusivity and the covalency of the Co 3d and O 2p bonding in these materials allowing access to a newly hypothesized lattice oxygen-based mechanism as predicted through DFT modelling.

Results

Crystallographic characterization. LCO, LSCO and SCO samples were synthesized using our previously developed reverse-phase hydrolysis scheme, using a 950 °C calcination temperature

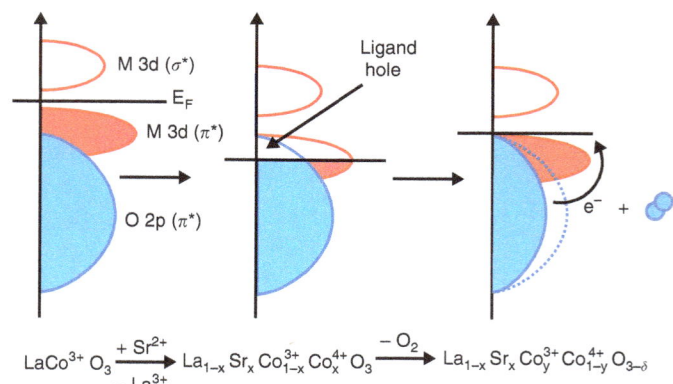

Figure 1 | Relationship between oxygen vacancy concentration and Co-O bond covalency. As the oxidation state of Co is increased through Sr^{2+} substitution, the Co 3d/O 2p band overlap is increased (covalency increases) and the Fermi level decreases into the Co 3d/O 2p π^* band, creating ligand holes. Oxygen is released from the system resulting in oxygen vacancies and pinning the Fermi level at the top of the Co 3d/O 2p π^* band[61].

instead of 700 °C to ensure that the correct phase was synthesized[16,17,27]. Figure 2a shows the powder X-ray diffraction patterns for the system, demonstrating the successful synthesis of the perovskite phases across the whole-composition range. The only minor admixture found in the LCO and LSCO samples was Co_3O_4. The crystal structures of all compositions have been verified using a combination of powder X-ray diffraction and transmission electron microscopy. The unit cell parameters and space groups of the respective materials are given in Supplementary Table 1. The powder X-ray diffraction and selected area electron diffraction (SAED) patterns of the $x = 0$–0.4 compositions are characteristic of the perovskite $R\bar{3}c$ structure with the $a^- a^- a^-$ tilting distortion of the octahedral framework (Fig. 2b,e). The monoclinic distortion due to orbital ordering reported for this compositional range was not detected being beyond resolution of our powder X-ray diffraction experiment[29–31]. The LSCO46 composition crystallizes in a cubic $Pm\bar{3}m$ perovskite structure. In the crystal structures of LSCO28 and SCO ordering of oxygen vacancies becomes obvious from both SAED patterns and high-angle annular dark-field scanning transmission electron microcopy (HAADF-STEM) images (Fig. 2c,d,f,g). Oxygen vacancies reside in the $(CoO_{2-\delta})$ anion-deficient perovskite layers alternating with the complete (CoO_2) layers that results in a tetragonal $a_p \times a_p \times 2a_p$

(a_p indicates the parameter of the perovskite subcell) supercell in LSCO28. The anion-deficient layers manifest themselves as faintly darker stripes in the HAADF-STEM images (marked with arrowheads in Fig. 2f,g), which according to Kim et al.[32] is related to the structural relaxation in these planes. The anion-deficient layers form nanoscale-twinned patterns in both the LSCO28 and SCO samples (Fig. 2f,g). In general, the crystallographic observations on the LCO and LSCO samples are in agreement with the $La_{1-x}Sr_xCoO_{3-\delta}$ phase diagram[33]. However, in contrast to the earlier reported $Sr_2Co_2O_5$ brownmillerite or hexagonal $Sr_6Co_5O_{15}$ phases[34,35], the SCO sample demonstrates another type of oxygen vacancy ordering. The $[010]_p$ SAED pattern of SCO (Fig. 2d, top) is strongly reminiscent to that of the $Ln_{1-x}Sr_xCoO_{3-\delta}$ ($Ln = $ Sm-Yb, Y) perovskites with the $I4/mmm$ $2a_p \times 2a_p \times 4a_p$ supercell[33,36,37]. A detailed deconvolution of this SAED pattern into contributions from the twinned domains is presented in Supplementary Fig. 1. This supercell also allows complete indexing of the powder X-ray diffraction pattern of SCO (Supplementary Fig. 2). The layered ordering of the oxygen vacancies in the LSCO28 and SCO samples was directly visualized using annular bright-field STEM (ABF-STEM) imaging (Fig. 3a,b). In both structures the anion-complete (CoO_2) and anion-deficient ($CoO_{2-\delta}$) layers can be clearly distinguished, alternating along the c-axis of the tetragonal supercells. However,

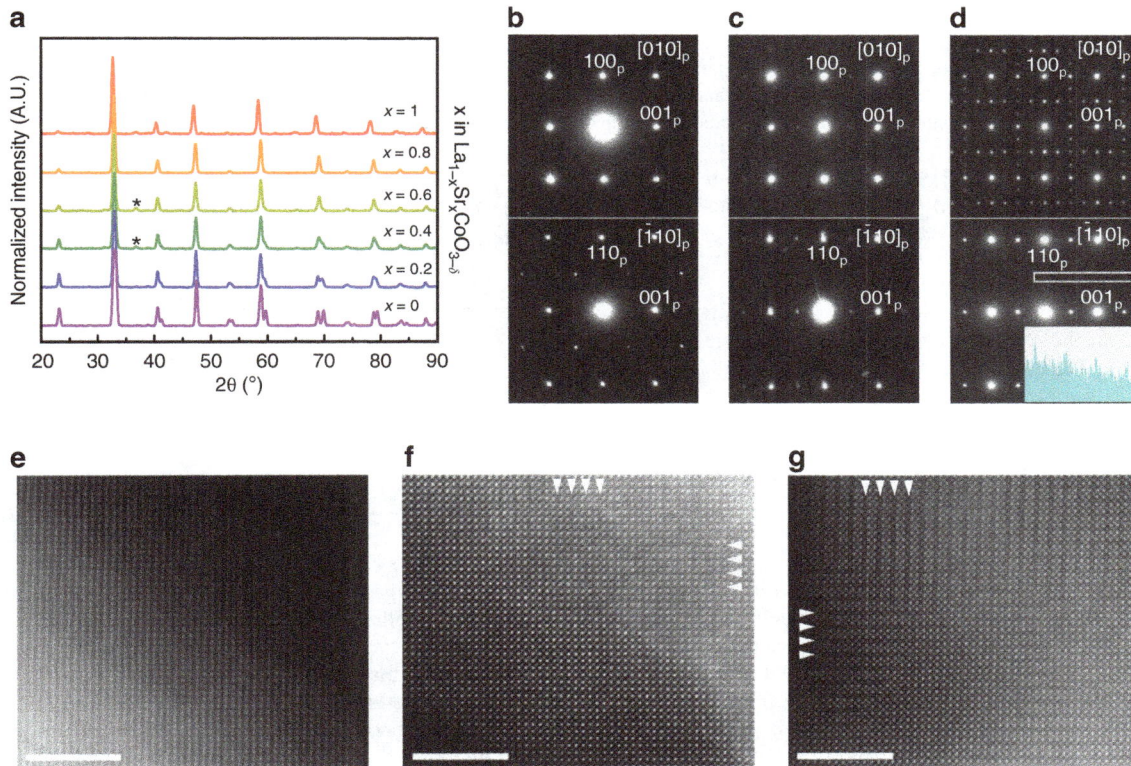

Figure 2 | Structural characterization of $La_{1-x}Sr_xCoO_{3-\delta}$. (**a**) Powder X-ray diffraction patterns for $La_{1-x}Sr_xCoO_{3-\delta}$ ($0 \leq x \leq 1$). The reflection from Co_3O_4 is marked with an asterisk. (**b–d**) SAED patterns of LSCO82 (**b**), LSCO28 (**c**) and SCO (**d**). The reflections of the basic perovskite structure are indexed. The $[\bar{1}10]_p$ SAED pattern of LSCO82 shows weak $G_p \pm 1/2 <111>_p$-type reflections (G_p—reciprocal lattice vector of the perovskite structure) characteristic of the $a^- a^- a^-$ octahedral tilting distortion of the perovskite structure. The $[010]_p$ SAED pattern of LSCO28 demonstrates the orientationally twinned $G_p \pm 1/2 <001>_p$ superlattice reflections resulting in the $P4/mmm$ $a_p \times a_p \times 2a_p$ supercell. The superstructure in the $[010]_p$ SAED pattern of SCO can be described with the $G_p \pm n/4 <201>_p$ (n—integer) and $G_p \pm 1/2 <110>_p$ superstructure vectors corresponding to the orientationally twinned $I4/mmm$ $2a_p \times 2a_p \times 4a_p$ supercell (see details in Supplementary Fig. 1). Note that the $G_p \pm 1/2 <110>_p$ superlattice reflections are barely visible in the $[\bar{1}10]_p$ SAED patterns of SCO, but the intensity profile (shown as insert in **d**) along the area marked with the white rectangle demonstrates their presence undoubtedly. (**e–g**) $[010]_p$ HAADF-STEM images of LSCO82 (**e**), LSCO28 (**f**) and SCO (**g**). The image of LSCO82 shows uniform perovskite structure, whereas the images of LSCO28 and SCO show faint darker stripes spaced by $2a_p$ (marked by arrowheads) indicating nanoscale-twinned arrangement of the alternating (CoO_2) perovskite layers and ($CoO_{2-\delta}$) anion-deficient layers. Scale bars are 5 nm.

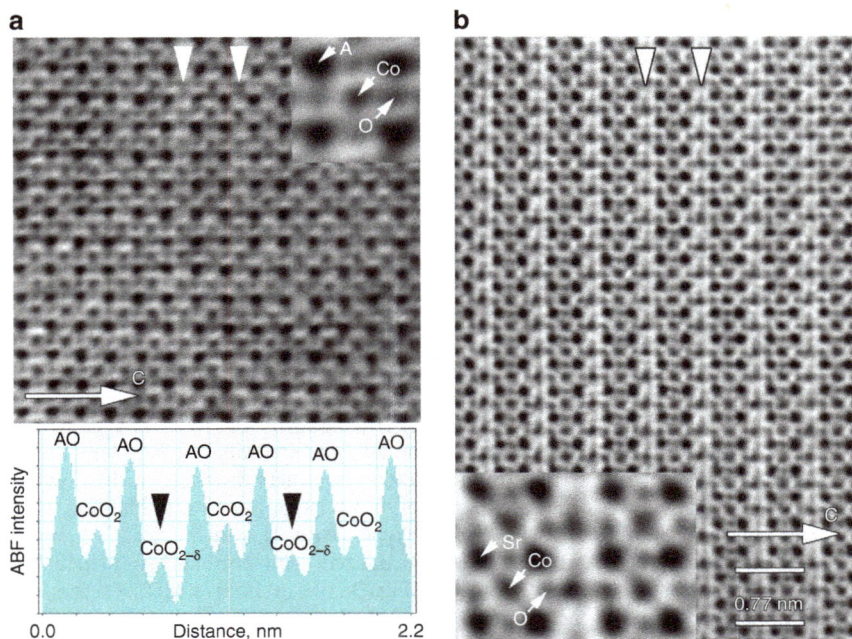

Figure 3 | ABF-STEM imaging of oxygen vacancy ordering in La$_{1-x}$Sr$_x$CoO$_{3-\delta}$ ($x = 0.8, 1.0$). (a) [001]$_p$ ABF-STEM image of LSCO28 showing the cation and anion sublattices. The contrast is inverted in comparison with the HAADF-STEM images. The assignment of the atomic columns is shown in the enlargement at the top right corner. Half of the perovskite (CoO$_2$) layers appear brighter indicating oxygen deficiency (marked with white arrowheads). The complete (CoO$_2$) layers and anion-deficient (CoO$_{2-\delta}$) layer alternate (see the ABF intensity profile below, the anion-deficient layers are marked with black arrowheads) resulting in doubling of the perovskite lattice parameter in the direction perpendicular to the layers. **(b)** [001]$_p$ ABF-STEM image of SCO showing layered anion-vacancy ordering. The (CoO$_{2-\delta}$) layers are marked with the white arrowheads and demonstrate the contrast clearly distinct from that of the (CoO$_2$) layers. The assignment of the atomic columns is shown in the enlarged part at the bottom left.

establishing the exact ordering patterns of the oxygen atoms and vacancies in these (CoO$_{2-\delta}$) layers requires more detailed neutron powder diffraction investigation.

In order to understand the effects of Sr^{2+} substitution on oxygen vacancy concentrations in La$_{1-x}$Sr$_x$CoO$_{3-\delta}$, iodometric titrations were performed. It should be noted that processing conditions affect the oxygen content and oxidation state of cobalt significantly through equation 1. The results of the iodometric titrations are presented in Table 1. As can be seen, there is both an increase in the bulk oxidation state of Co as well as an increase in the concentration of oxygen vacancies as lower valence Sr^{2+} is substituted for La^{3+}. The high concentration of oxygen vacancies in SrCoO$_{2.7}$ corroborates their pronounced layered ordering.

Microstructural characterization. The overall morphology of the LSCO series was investigated with bright-field TEM images, presented in Supplementary Fig. 3. The samples consist of highly agglomerated and partially sintered nanoparticles with size ranging from 20–50 nm to few hundred nanometres. The LCO and SCO materials demonstrate somewhat larger and more sintered crystallites compared with those of the mixed LSCO samples. HAADF-STEM and ABF-STEM images of the surface structure of LCO and SCO are shown in Supplementary Fig. 4, where the particles remain crystalline at the surface and for SCO the anion-deficient layers, evident through the nanoscale-twinned domain columns, extend to the surface. Brunauer–Emmett–Teller surface areas measured through N$_2$ adsorption showed similar surface areas for all samples of 3.1–4.5 m^2 g^{-1} (Supplementary Table 2). This surface area is approximately half the surface area of the materials reported in our previous studies, which results from the higher calcination temperatures used for the LSCO series than the previously investigated LaCoO$_3$, LaNiO$_3$, LaMnO$_3$ and LaNi$_{0.75}$Fe$_{0.25}$O$_3$.

Table 1 | Oxygen vacancy concentration, δ, and cobalt oxidation state, y.

x in La$_{1-x}$Sr$_x$CoO$_{3-\delta}$	δ	y
0	-0.01 ± 0.01	3.01 ± 0.01
0.2	0.01 ± 0.01	3.18 ± 0.02
0.4	0.05 ± 0.04	3.30 ± 0.08
0.6	0.09 ± 0.01	3.43 ± 0.01
0.8	0.16 ± 0.01	3.48 ± 0.02
1.0	0.30 ± 0.03	3.40 ± 0.06

Error is based on the s.d. of triplicate measurements.

Electrochemical characterization. In order to better understand the role of oxygen vacancies in La$_{1-x}$Sr$_x$CoO$_{3-\delta}$ during electrochemical applications, the intercalation of oxygen in LSCO was studied using cyclic voltammetry in Ar saturated 1 M KOH solutions. The insertion and removal of oxygen ions appear as redox peaks in Fig. 4a. It is apparent that an increase in the oxygen vacancy concentration as Sr^{2+} is substituted for La^{3+} in LSCO increases the tendency for oxygen intercalation as indicated through the high current densities measured in the intercalation region. In addition, it is interesting to note that the position of the intercalation redox peaks shifts to higher potentials with increased oxygen vacancies which can be described through the common pseudocapacitive Nernst Equation:

$$E = E^0 + \frac{RT}{nF} \ln \frac{\sigma}{1-\sigma} \qquad (2)$$

where, E represents the measured potential for oxygen intercalation, E^0 represents the standard potential for oxygen intercalation, R is the universal gas constant (8.3145 J K^{-1} mol^{-1}), T is the temperature during the measurement, F is

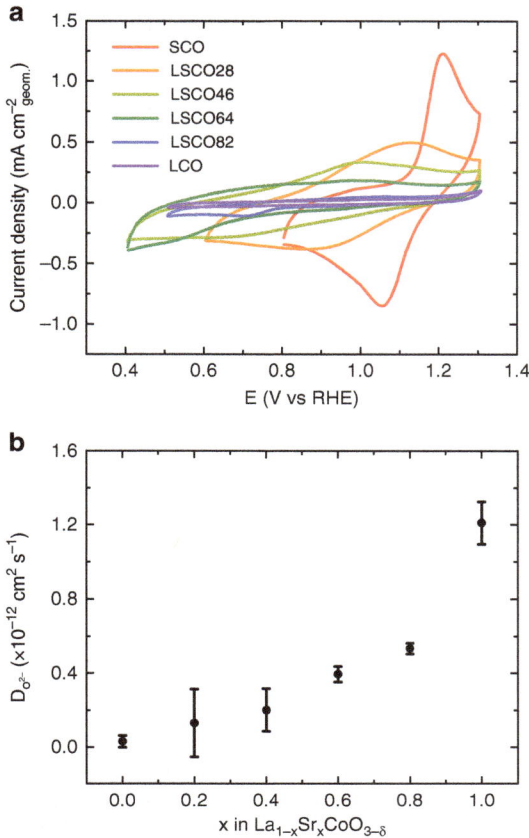

Figure 4 | Electrochemical oxygen intercalation into $La_{1-x}Sr_xCoO_{3-\delta}$. (**a**) Cyclic voltammetry at $20\,mV\,s^{-1}$ for each member of LSCO in Ar saturated 1 M KOH. The redox peaks, indicative of the insertion and removal of oxygen from the crystal, shift to higher potentials with increasing Sr^{2+} and oxygen vacancy concentrations. (**b**) Oxygen diffusion rates measured at 25 °C chronoamperometrically. The diffusion rate increases with Sr^{2+} and oxygen vacancy concentrations as well. Error bars represent the standard deviation of triplicate measurements.

Faraday's constant $(96,485\,C\,mol^{-1})$, and σ is the occupancy fraction of accessible lattice vacancy sites[38] for the reaction:

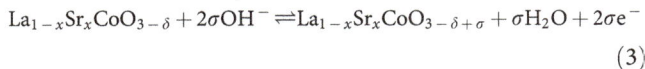

$$La_{1-x}Sr_xCoO_{3-\delta} + 2\sigma OH^- \rightleftharpoons La_{1-x}Sr_xCoO_{3-\delta+\sigma} + \sigma H_2O + 2\sigma e^-$$

$$(3)$$

This type of Nernst Equation is commonly associated with pseudocapacitive-type intercalation mechanisms, indicative of facile oxygen ion diffusion.

The diffusion rates of oxygen ions in LSCO were measured chronoamperometrically based on a bounded 3D solid-state diffusion model with a rotating disk electrode (RRDE) rotating at 1,600 r.p.m. in Ar saturated 1 M KOH[39–41]. These results are presented in Fig. 4b, and a more detailed description of the theory behind the model is included as Supplementary Fig. 5. It was found that SCO, with a vacancy concentration of $\delta = 0.30 \pm 0.03$, had a diffusion rate of $D = 1.2 \pm 0.1 \times 10^{-12}\,cm^2\,s^{-1}$ at room temperature, which is ~40 × faster than for LCO, with a complete oxygen sublattice and a diffusion rate of $D = 3 \pm 1 \times 10^{-14}\,cm^2\,s^{-1}$. As a general comment, diffusion coefficients in the range of 10^{-9} to $10^{-14}\,cm^2\,s^{-1}$ have been found as usual values for the short circuit diffusion of oxygen along high-diffusivity pathways, including grain boundaries[24]. Although it is unclear whether the measured diffusion rates are from bulk diffusion or along

grain boundaries, isotope tracer studies have shown that diffusion rates trend in the order of surface oxygen > oxygen at grain boundaries > bulk oxygen in perovskite systems, and thus the fast diffusion rates found in this study represent the lower boundary on the mobility of oxygen at the surface[42]. Further, the crystallite size and density of grain boundaries is relatively consistent across the LSCO series due to the similar synthetic conditions, indicating that the diffusion rates can at least be compared against each other. The results indicate that the diffusion rates scale with Sr concentration because of the correlation with vacancies and Sr content. The results highlight the benefit of substitution of a lower valence ion into the A-site as an effective means of increasing the mobility of oxygen in perovskite oxide electrodes.

The electrolysis of water. The OER activities for LSCO and for a commercial IrO_2 sample were quantified through cyclic voltammetry in O_2 saturated 0.1 M KOH at 1,600 r.p.m., as shown in Fig. 5a. Each material was mixed at a mass loading of 30 wt% perovskite on a mesoporous nitrogen-doped carbon (NC) or onto Vulcan Carbon XC-72 (VC) for stability measurements. An evaluation of the carbon loading and total mass loading is presented in Supplementary Fig. 6, Supplementary Table 5 and the Supplementary Discussion. There is a shift towards more active Tafel slopes with increasing Sr content, with LCO and IrO_2 having similar Tafel slopes of $\partial V/\partial \ln i = 58\,mV\,dec^{-1}$ ($\approx 2RT/F$) which decreases towards SCO with a Tafel slope of $\partial V/\partial \ln i = 31\,mV\,dec^{-1}$ ($\approx RT/F$). This shift of Tafel slope for the OER may be indicative of the facile surface kinetics for oxygen exchange with increasing vacancy content, whereby OER kinetics that are limited by high-coverage Langmuir like behaviour where surface oxygen is not exchanged rapidly ($\theta \rightarrow 1$) show Tafel slopes of $2RT/F$. In contrast, those materials showing more rapid surface oxygen exchange in the intermediate coverage Temkin condition range ($0.2 < \theta < 0.8$) have slopes of RT/F[9]. The specific activities at an overpotential of 400 mV, based on perovskite surface area from BET, are presented in Fig. 5b. It is clear that substitution of Sr^{2+} for La^{3+} in LSCO, and thereby the creation of oxygen vacancies, is beneficial to the OER, with the fully substituted $SrCoO_{2.7}$ at $28.4\,mA\,cm^{-2}_{ox}$ which is ~6 × more active than $LaCoO_{3.005}$ $(4.3\,mA\,cm^{-2}_{ox})$, ~23 × more active than the commercial IrO_2 sample $(1.2\,mA\,cm^{-2}_{ox})$, and ~1.5 × more active than previously reported high-vacancy concentration cobaltite perovskites $(Ba_{0.5}Sr_{0.5}Co_{0.8}Fe_{0.2}O_{2.6}: \sim 20\,mA\,cm^{-2}_{ox};$ $Pr_{0.5}Ba_{0.5}CoO_{2.85}: \sim 20\,mA\,cm^{-2}_{ox})$ (refs 14,43). In addition, due to the small particle size from the reverse-phase hydrolysis synthesis, $SrCoO_{2.7}$ $(3.6\,m^2\,g^{-1})$ had a mass activity of $1,020 \pm 20\,mA\,mg^{-1}_{ox}$ at $+1.63\,V$ versus the reversible hydrogen electrode (RHE), which is ~2 × more active than BSCF with a similar surface area ($\sim 500\,mA\,mg^{-1}_{ox}, 3.9\,m^2\,g^{-1}$) (ref. 14). To verify that the measured current was due only to the OER, and not to side-reactions or corrosion of the electrode material, rotating-ring-disk (RRDE) cyclic voltammetry was performed with a Pt ring poised at $+0.4\,V$ versus RHE, whereby O_2 generated at the disk from the OER is collected and reduced at the ring. The results for $SrCoO_{2.7}/NC$ and IrO_2/NC are shown in Fig. 5c. The collection efficiency for both $SrCoO_{2.7}/NC$ and IrO_2/NC was 37%, which was equal to the collection efficiency measured during calibration of the RRDE for the oxidation of 0.3 mM ferrocene-methanol in 0.1 M KCl. Therefore, we can confirm that the current is exclusively due to the generation of oxygen on the SCO or the IrO_2 surface within the precision of the RRDE measurements.

The stability of $SrCoO_{2.7}$ and of the carbon supports under OER conditions were tested galvanostatically at $10\,A\,g^{-1}_{ox}$ and 1,600 r.p.m., shown in Fig. 5d. As is readily apparent, both the

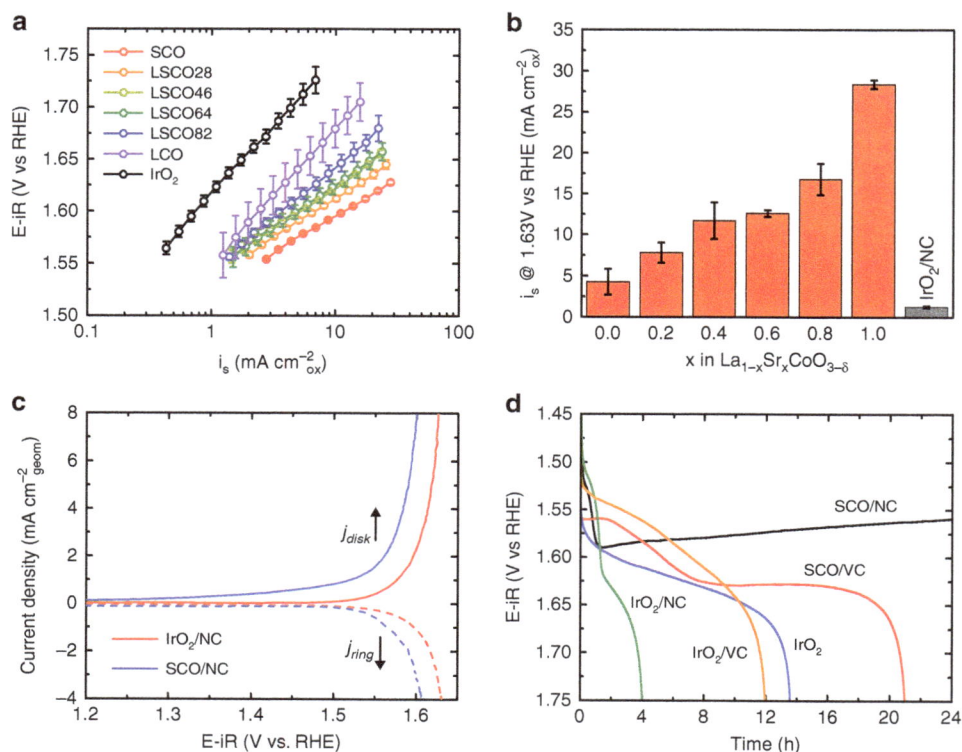

Figure 5 | Electrochemical characterization of La$_{1-x}$Sr$_x$CoO$_{3-\delta}$ for the OER. (**a**) Capacitance corrected specific OER current densities in O$_2$ saturated 0.1 M KOH, a scan rate of 10 mV s^{-1}, and ω = 1,600 r.p.m., for 30 wt% La$_{1-x}$Sr$_x$CoO$_{3-\delta}$ supported on 2 at. % NC. The performance of 30 wt% IrO$_2$ supported on 2 at. % NC is included as a reference. (**b**) Specific activities of La$_{1-x}$Sr$_x$CoO$_{3-\delta}$ and IrO$_2$ at a 400 mV overpotential for the OER (1.63 V versus RHE). (**c**) Confirmation of oxygen generation using a RRDE. The disk has a thin layer of either 30 wt% SrCoO$_{2.7}$/NC or 30 wt% IrO$_2$/NC and the ring is Pt. O$_2$ is generated at the disk then reduced back to OH$^-$ at the ring which is poised at +0.4 V versus RHE. The collection efficiency of the RRDE was found to be 37%. (**d**) Galvanostatic stability at 10 A g$^{-1}_{ox}$ and ω = 1,600 r.p.m. of SrCoO$_{2.7}$ and IrO$_2$ supported on two different carbons, 2 at % nitrogen-doped NC and non-nitrogen doped VC. It is evident that both carbons are unstable at the anodic potentials of the OER, with rapid degradation occurring for all samples once the potential is >1.65 V versus RHE. The high activity and stability of SrCoO$_{2.7}$ on NC allows the electrode to generate 10 A g$^{-1}_{ox}$ of current without reaching this potential, which results in a relatively stable catalyst for 24 h of operation. For all electrochemical studies the mass loading of the electrode was 51 μg$_{tot}$ cm$^{-2}_{geom}$. Error bars represent the standard deviation of triplicate measurements.

NC and VC are not stable carbon supports for the OER, and we hypothesize that this dominates the mechanism of failure for the composite electrodes at potentials > +1.65 V versus RHE. However, other variables may be responsible for the failure of the electrodes, including the degradation of the Nafion binder due to the oxidative conditions and the rapid rotation of the electrode. SrCoO$_{2.7}$, however, appears to be active enough to sustain the OER for 24 h at 10 A g$^{-1}_{ox}$ without reaching the potential where rapid carbon corrosion occurs. Further studies are needed in order to better understand the variables that influence catalyst stability, however, it is clear that carbon may not be the optimal catalyst support under the OER conditions. In addition, it should be noted that IrO$_2$ which has become the benchmark comparison for OER catalysts is not stable under the anodic conditions of the OER, forming the soluble complex anion IrO$_4^{2-}$ in alkaline environments[44]. This is demonstrated in the stability plot in Fig. 5d, where even the unsupported IrO$_2$ electrode failed after ~14 h.

The catalytic activity towards the OER was found to strongly correlate with the oxygen diffusion rate and the vacancy concentration, δ, presented in Fig. 6c,d. On the basis of these correlations, we hypothesize a new OER mechanism in Fig. 6a based on the exchange of lattice oxygen species that takes into account the role of surface oxygen vacancies and B–O bond covalency (lattice oxygen-mediated OER, LOM). In contrast to the general adsorbate evolution mechanism (AEM) which considers only the redox activity of the transition metal B-site,

we find a better electronic explanation arises when the covalency of the M–O bond is considered, indicative of the overlap of the Co 3d and O 2p bands in the crystal, as first proposed by Matsumoto et al.[14,45]. As the oxidation state of Co is increased, the d orbitals of the Co ion have a greater overlap with the s, p orbitals of the O^{2-} ion, leading to the formation of π^* and σ^* bands, as described through Fig. 1 and in the partial density of states (PDOS) diagrams in Fig. 6a and refs 11,13,22,43. When the overlap is great enough, ligand holes (oxygen vacancies) are formed and the metal 3d π^* band can no longer be treated as isolated in energy from the oxygen O 2p π^* band. At this point, the surface of the crystal and bound intermediates can be treated as a single energy surface, where the Fermi energy can be modulated through the hybridized Co 3d–O 2p π^* band with applied electrical potential, opening up the possibility for lattice oxygen redox activity[46]. A recent in situ ambient pressure XPS study has confirmed the validity of this model in perovskites and other oxides[47,48]. In addition, oxygen redox activity has been observed in LSCO with high Sr^{2+} content in the regime of oxygen intercalation, which occurs approximately at the onset potentials of the OER in these materials[49–51].

To test the validity of this lattice oxygen-mediated mechanism (LOM) and identify the rate-determining step, we modelled the reaction pathway using density functional theory[52]. Supplementary Fig. 7a shows that OH$^-_{(aq)}$ tends to electrochemically fill the surface O vacancies of LSCO under the operational electrode potential of OER, as described through

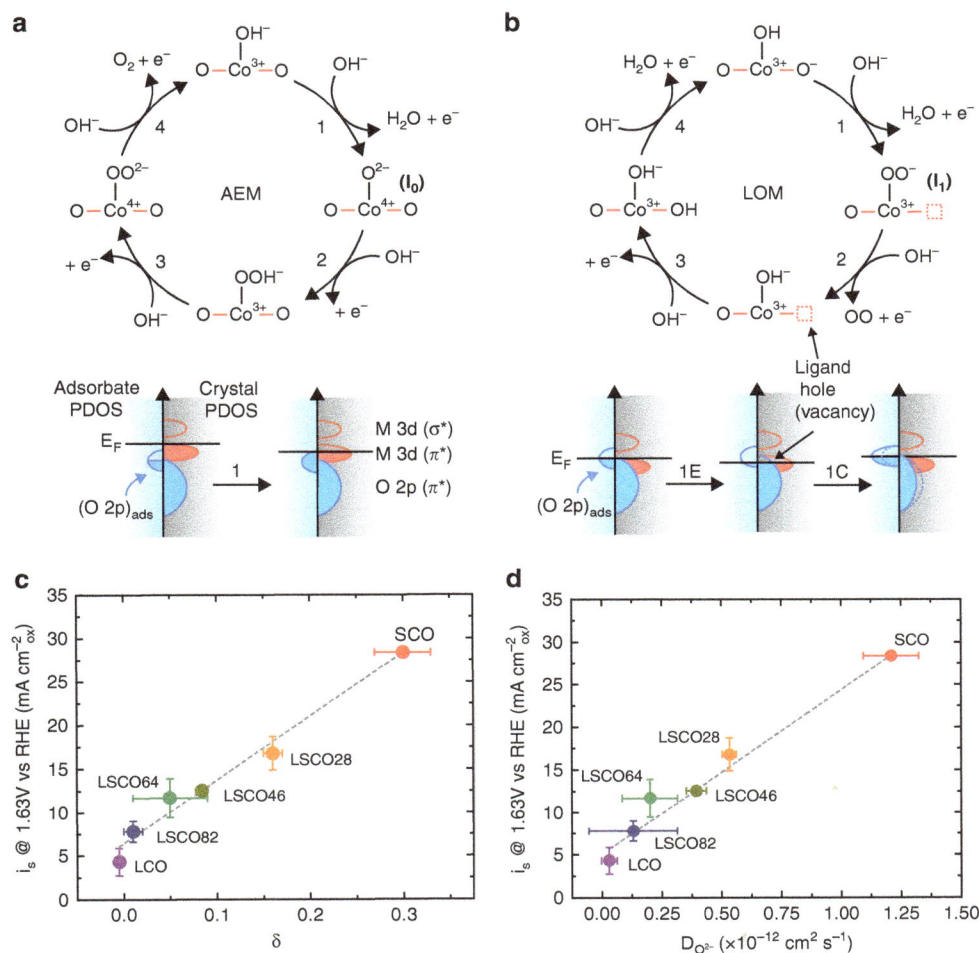

Figure 6 | Oxygen evolution mechanisms on $La_{1-x}Sr_xCoO_{3-\delta}$ and activity correlations. (a) AEM[14,62]. In the AEM, the transition metal 3d bands are significantly higher in energy than the O 2p band in the lattice as shown qualitatively in the PDOS diagram below the mechanism. Because of this, all intermediates during the reaction originate from the electrolyte and Co in the active-site undergoes the catalytic redox reactions. This allows Co to access a higher oxidation state of Co^{4+} in Step 1 **(a)** AEM. As the covalency of the material increases, the transition metal 3d bands are lowered into the O 2p band in the lattice, where the Fermi energy is pinned at the top of the O 2p band through generation of oxygen vacancies[61]. In contrast, in Step 1 **(b)** of the LOM, applying an anodic potential oxidizes a ligand hole in the O 2p band allowing for exchange of lattice oxygen to the adsorbed intermediate to yield the superoxide ion O_2^- rather than oxidizing Co to Co^{4+}. This is shown qualitatively in the PDOS diagram below the mechanism where Step 1 of the LOM is separated into an electrochemical (1E) step in which the ligand hole is generated and a chemical step (1C) in which the lattice oxygen is exchanged into the adsorbed intermediate. For both **(a,b)** lattice species are shown in red and electrolyte species are shown in blue. In the PDOS diagrams, the electrolyte species are shown to the left of the energy axis and the crystal PDOS are shown to the right. **(c)** Correlation of oxygen evolution activity with the vacancy parameter δ. The vacancy parameter is indicative of the underlying electronic structure where vacancies are generated when there is significant Co 3d and O 2p band overlap. **(d)** Correlation of oxygen evolution activity with the oxygen ion diffusion rate, indicating that increased surface exchange kinetics trend with increased OER activity. Error bars represent standard deviation of triplicate measurements.

reaction 3 and LOM 1 in Fig. 6a, leading to an *in situ* surface–layer stoichiometry close to that of stoichiometric bulk ABO_3. Consequently, we begin by constructing the [001] BO_2 terminated surfaces (Supplementary Fig. 8a) with $\frac{1}{4}$ ML OER intermediate adsorbates[52] based on the $2 \times 2 \times 2$ cubic stoichiometric bulk LSCO for the initial identification of the reactivity trend and reaction mechanism[53]. We subsequently investigated more realistic bulk phases with oxygen vacancies and various surface structures, which we find do not alter the preference of LOM over AEM; further details of these computations are provided in the Supplementary Methods.

Our results show that Step 1 differentiates the LOM, involving the intermediate with adsorbed –OO and lattice O vacancies (I_1 in Figs 6a and 7a), from the AEM, involving the generally proposed adsorbed –O (I_0 in Figs 6a and 7a). Therefore, the relative stabilities (free energy difference, ΔG) between these two

isomeric intermediates are key to identifying if OER proceeds via the LOM or AEM for a given LSCO composition. This identification approach has been successfully used to demonstrate the preference of LOM on $LaNiO_3$ (ref. 52). The computed values of ΔG are shown as a function of LSCO composition in Fig. 7b, which illustrates two key points. First, increasing x in $La_{1-x}Sr_xCoO_{3-\delta}$ reduces the O vacancy formation energy and therefore bulk stability. Second, ΔG decreases with the decreased bulk stability, becoming negative between $0.25 < x < 0.5$. Therefore, OER on perovskites with low stability such as $La_{0.5}Sr_{0.5}CoO_{3-\delta}$, $La_{0.25}Sr_{0.75}CoO_{3-\delta}$ and $SrCoO_{3-\delta}$ is predicted to occur via the LOM, whereas $LaCoO_3$ and $La_{0.75}Sr_{0.25}CoO_{3-\delta}$ are expected to follow the AEM.

The transition from the AEM to the LOM is related to the ineffectiveness of the surface Co as electron donors. The double bond of the adsorbed O formed in reaction 1 of the AEM

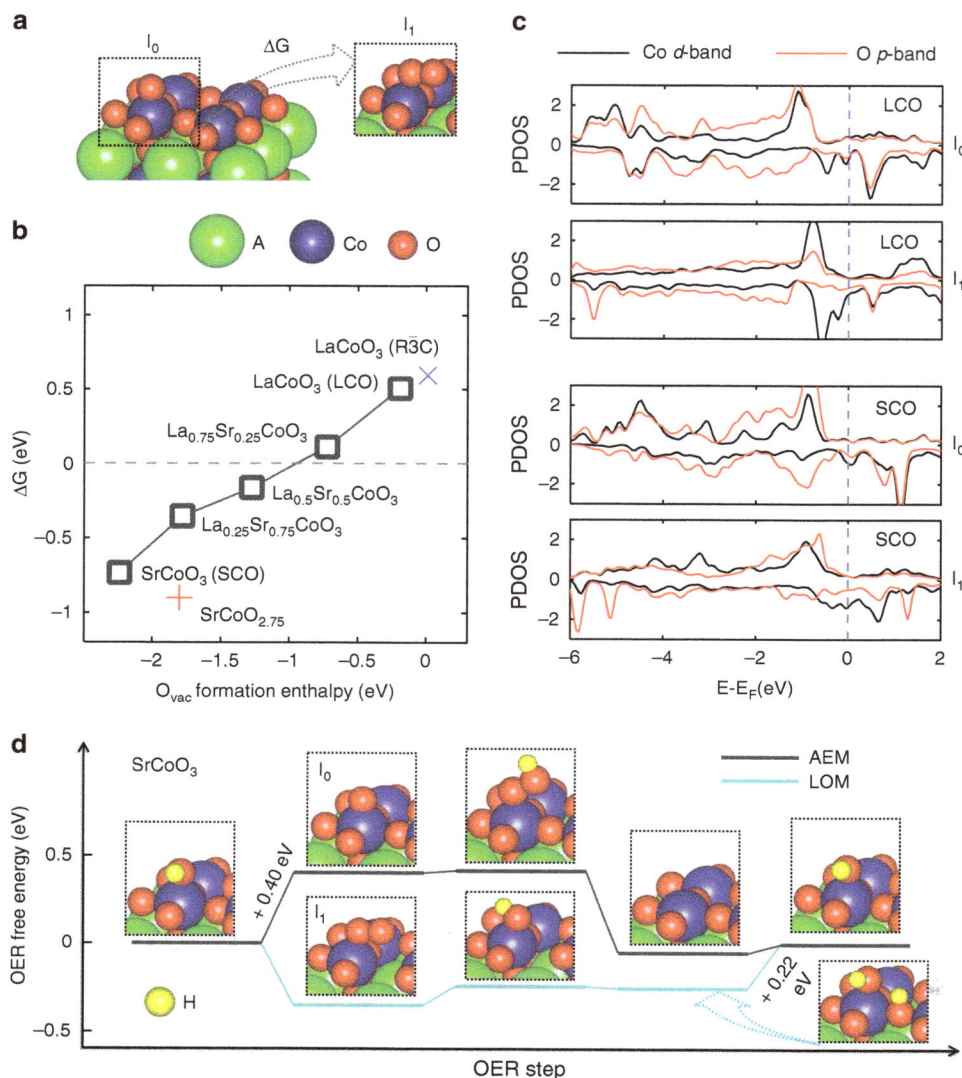

Figure 7 | Density functional theory modelling of vacancy-mediated oxygen evolution on La$_{1-x}$Sr$_x$CoO$_{3-\delta}$. (a) Surface configurations of the intermediate after AEM Step 1 (I$_0$) and the one after LOM Step 1 (I$_1$). (b) The free energy change of I$_1$ over I$_0$ versus the O vacancy formation enthalpy in the bulk, for the cubic La$_{1-x}$Sr$_x$CoO$_{3-\delta}$ (black mark), where $x = 0$, 0.25, 0.5, 0.75 and 1, with the rhombohedral LaCoO$_3$ and optimized SrCoO$_{2.75}$ phases; for $x = 0.25$ and 0.75, the most energetic favourable vacancy site is selected; the O vacancy formation energy is calculated at the concentration of 1 per $2 \times 2 \times 2$ unit cell with respect to H$_2$O(g) and H$_2$(g) at standard condition; using O$_2$(g) as the reference will shift the O vacancy formation enthalpy around $+ 2.5$ eV larger. (c) The density of states of d-band for the active surface Co and the overall p-band for its ligand O, for LaCoO$_3$ and SrCoO$_3$ before and after the lattice oxygen exchange. (d) The OER free energy changes of LOM and AEM on SrCoO$_3$ at the concentration of $\frac{1}{4}$ ML, with indicated intermediates structures and potential-determining steps.

significantly increases the oxidation state of surface Co to $> 3 +$. In the LOM step 1, the transfer of a surface O to form a surface O vacancy and the single-bonded –OO adsorbate decreases the nominal valence charge on the Co to $3 +$. Thus, the LOM pathway has higher stability than the AEM pathway, particularly for those LSCO with large x. The relative stability of I$_1$ to I$_0$ is also apparent in the projected density of states of the d-band for the active surface Co and the overall p-band for its ligand O (Fig. 7c). The overlap of the peaks in these two bands indicates the orbital hybridization and Co–O binding. For the AEM intermediate on LaCoO$_3$ (I$_0$), the strong overlap of peaks in the spin-up (down) bands centred around $- 1$ eV (0.5 eV) indicates the strong Co–O covalent bonding state. These overlaps, however, are significantly weakened for I$_1$, consistent with the stability. The reverse is true for the LSCO with low stability. Compared with I$_0$ for SrCoO$_3$, I$_1$ preserves a significant overlap of spin-up state around $- 1$ eV,

but has negligible overlap of the unoccupied spin-down states, which are anti-bonding in character, indicating the greater stability of I$_1$.

To understand the phase and stoichiometry effects on the relative stability of I$_1$ to I$_0$, we perform the analogous calculations on the rhombohedral LaCoO$_3$ and the nonstoichiometric SrCoO$_{2.7}$ phases. The rhombohedral LaCoO$_3$ phase is modelled by optimizing an initial $2 \times 2 \times 2$ orthorhombic cell with octahedral rotation; the optimized structure exhibits a Co–O–Co angle of 162° and a Co–O distance of 1.96 Å, consistent with experimental measurements. The SrCoO$_{2.7}$ phase is approximated as SrCoO$_{2.75}$, which can be modelled by relaxing the cubic $2 \times 2 \times 2$ SrCoO$_3$ structure with two oxygen vacancies. By comprehensively searching the vacancy ordering, we identify the most stable configuration as the presence of the two vacancies surrounding one Co, which therefore leads to the formation of a tetrahedral CoO_4 linked to two tetragonal pyramidal CoO_5 units (Supplementary Fig. 8b). The lattice constant of this optimized $SrCoO_{2.75}$ is within 1.1% difference from that of the derived pseudocubic $SrCoO_{2.7}$ (Supplementary Table 1). This configuration is further validated by introducing two more vacancies to form $SrCoO_{2.5}$ in the same way, so as to maximize the number of tetrahedral CoO_4. The relaxed $SrCoO_{2.5}$ shows alternating octahedral (CoO_2) and tetrahedral zigzag-like (CoO) layers with respect to the (001) direction of the reference cubic phase (Supplementary Fig. 8b), in full agreement with experimental observations. The $SrCoO_{2.75}$ slab is subsequently constructed by exposing the ($CoO_{2-\delta}$) layer (Supplementary Fig. 8c), but with added oxygen anions to attain the correct stoichiometry (Supplementary Fig. 8d) to simulate the intercalation phenomenon as described by Supplementary Fig. 7a and LOM Step 3. As Fig. 7b shows, the octahedral rotation stabilizes the rhombohedral $LaCoO_3$, leading to a slight increase in the oxygen vacancy formation energy and ΔG. In the case of $SrCoO_{2.75}$, the existing oxygen deficiency increases the oxygen vacancy formation energy by 0.33 eV, while slightly stabilizing I_1 relative to I_0, compared with $SrCoO_3$. The lattice constant of the predicted $SrCoO_{2.75}$ is 0.7% larger than that of $SrCoO_3$, leading to the slightly weaker adsorption strength and lower stability of I_0 (ref. 54). However, the small magnitude of this change indicates the similar reactivity of $SrCoO_3$ to that of the intercalated $SrCoO_{2.7}$ surface under OER conditions. From the above analysis, we conclude that neither the phase nor the non-stoichiometry alters the qualitative stability of I_1 to I_0, although it leads to a horizontal shift in the overall trend of bulk vacancy formation to higher energetic cost.

We also compute the free energy of electrochemical OER on $SrCoO_3$ to demonstrate the switch in the reaction mechanism due to the relative change in I_1-to-I_0 stability on $SrCoO_{2.7}$. In accordance with the procedure in ref. 53 the free energy of each reaction step is determined by $\Delta G_R = E + \Delta ZPE - T\Delta S - eU_{RHE}$ at $U_{RHE} = +1.23$ V, where ΔE is the DFT-computed enthalpy change for $\frac{1}{4}$ ML of intermediates relative to H_2O and H_2 molecules (Supplementary Table 3) and $\Delta ZPE - T\Delta S$ gives the corrections for zero-point energy and entropy of both adsorbates and $H_2(g)$ and H_2O (l) under OER conditions (Supplementary Table 4) (refs 53,54). The largest free energy is the estimated overpotential, η. As ΔG_R is independent of the initial OER intermediate considered, we—in practice—start from the stoichiometric hydroxylated surface (the surface before LOM 1). Figure 7d shows that the first step ($-OH$ to I_0) of AEM is the potential-determining step, with $\eta = 0.4$ V. However, it becomes remarkably energetically favourable to follow LOM 1, forming the superoxide-like $-OO$ (V_O) adsorbates (I_1) with an O-to-O bond length of 1.28 Å. Therefore, LOM is the relevant mechanism for $SrCoO_{2.7}$. Once I_1 forms, it requires small energetically uphill and downhill reactions, respectively, to evolve back to $-OH$ (V_O) and electrochemically fill the vacancy by OH^- (aq) in Step 2 and 3 of the LOM. This electrochemical surface hydroxylation during Step 3 occurs concomitantly with an electron transfer to leave the surface in a neutral state. The subsequent step of electrochemical deprotonation is identified as the potential-determining step, similar to the results for $LaNiO_3$ (ref. 52). Further, the computed overpotential of 0.22 V is fully consistent with experiments.

We note that consideration of modified surface configurations, which may occur under operating conditions, could lead to different values of ΔG. For example, full surface hydroxylation can further decrease the value of ΔG due to oxidation of the surface Co, making the LOM more favourable, while moderate protonation of surface oxygen can increase ΔG by donating electrons to the surface. In addition, the O-to-O overbinding effects in the superoxide formation (I_1) by RPBE can increase ΔG by <0.3 eV

(ref. 55), while the use of GGA + U_{eff} can lead to the weaker adsorption strength of I_0, decreasing ΔG by >0.3 eV (ref. 56). Nevertheless, the behaviour of the model surfaces expected to be qualitatively correct for these systems, and also independent of exchange correlation functional, as demonstrated for $LaNiO_3$.

Interestingly, significant oxygen deficiencies of the LSCO series begin to appear at $x = 0.4$, matching well with the predicted transition from the AEM to LOM at $0.25 < x < 0.50$. These oxygen deficiencies reveal the saturated charge states of Co, which become unable to donate enough electrons to attain oxygen stoichiometry as described through Fig. 1. The bulk oxygen deficiency is consequently indicative of the LOM, since the double bonded $-O$ (AEM) induces a higher oxidation state of the surface Co than that in the bulk. The transition is further demonstrated by the experimental observation that the current density at $U_{RHE} = +1.63$ V increases on a very different scale with increasing δ when $x > 0.4$ from that when $x < 0.4$. Our work thus provides a strong theoretical framework, consistent with experiments, to describe the transition of the OER mechanism as a function of bulk stability. Further discussion about the applicability of this mechanism to other metal oxide catalysts is included in Supplementary Fig. 9 and in the Supplementary Discussion.

Discussion

We have demonstrated that oxygen vacancy defects are a crucial parameter in improving the electrocatalysis of oxygen on metal oxide surfaces, whereby they may control the physical parameters of ionic diffusion rates and reflect the underlying electronic structure of the catalyst. The vacancy-mediated mechanism proposed offers insight into the design of highly active OER catalysts, and allows for the rationalization of the electrolysis of water using surface chemistry parameters, as described through the modulation of the Fermi energy through transition metal 3d and oxygen 2p partial density of states at the surface. As such, the role of oxygen vacancy defects cannot be ignored, and should be a critical component in the benchmarking of metal oxide oxygen electrocatalysts and the advancement of the mechanistic theory behind the OER.

Methods

General. All chemicals were used as received. Anhydrous ethanol and 5 wt% Nafion solution in lower alcohols were purchased from Sigma-Aldrich. Lanthanum (III) nitrate hexahydrate (99.999%), strontium (II) nitrate hexahydrate (99.9%), cobalt (II) nitrate hexahydrate (99.9%), tetrapropylammonium bromide (98%), tetramethylammonium hydroxide (TMAH) pentahydrate (99%), 2-propanol, potassium hydroxide, potassium iodide (≥99%), sodium thiosulfate (0.1 N), potassium iodate (0.1 N) and hydrochloric acid were obtained from Fisher Scientific. Absolute ethanol (200 proof) was obtained from Aaper alcohol. The commercial IrO_2 sample was obtained from Strem Chemicals. Oxygen (99.999%) and argon (99.999%) gases were obtained from Praxair. VC was obtained from Cabot Corporation and the NC was prepared as reported elsewhere[57].

Synthesis of La$_{1-x}$Sr$_x$CoO$_{3-\delta}$, $0 \leq x \leq 0.8$. La$_{1-x}$Sr$_x$CoO$_{3-\delta}$ was synthesized following our previously reported reverse-phase hydrolysis approach[16,17,27]. Mixed metal hydroxides were prepared by reverse-phase hydrolysis of La, Sr and Co nitrates in the presence of an equimolar amount of tetrapropylammonium bromide (TPAB) dissolved in 1 wt% TMAH. An ~10 mM solution of mixed metal nitrates of the appropriate stoichiometry was added dropwise at ~1–2 ml min^{-1} to 200 ml of the 1 wt% TMAH solution containing TPAB. The resulting precipitated mixed metal hydroxide nanoparticles were collected by centrifugation and washed with deionized water, followed by re-suspension in deionized water through probe sonication. The solution was frozen as a thin film on a rotating steel drum at cryogenic temperatures (-79 °C), and then lyophilized at -10 °C at a fixed pressure of ~50 mTorr for 20 h. The lyophilized powder was calcined in a tube furnace under dehumidified air at a flow rate of 150 ml min^{-1} for 5 h at 950 °C. The resulting perovskites are then washed with ethanol followed by water and allowed to dry in an oven at 80 °C overnight.

Synthesis of SrCoO$_{2.7}$. Synthesis of $SrCoO_{2.7}$ followed a similar procedure to the one used above, but used a slower addition rate of metal nitrate solution to TMAH/ TPAB of <0.5 ml min^{-1}. In addition, the hydrolysis reaction was allowed to proceed for 5 days before collection by centrifugation. Finally, the flow rate of dehumidified air during calcination was adjusted to 20 ml min^{-1}.

Materials characterization. Bulk crystal structures were determined through wide-angle X-ray diffraction (Rigaku Spider, Cu Kα radiation, $\lambda = 1.5418$ Å) and analysed with JANA2006 software[58]. The TEM samples were prepared by crushing the crystals in a mortar in ethanol and depositing drops of suspension onto holey carbon grid. Electron diffraction patterns, TEM images, HAADF-STEM images, ABF-STEM images and energy dispersive X-ray spectra were obtained with an aberration-corrected Titan G^3 electron microscope operated at 200 kV using a convergence semi-angle of 21.6 mrad. The HAADF and ABF inner collection semi-angles were 70 mrad and 10 mrad, respectively. Iodometric titrations were performed according to the referenced procedure[19]. In short, 3 ml of de-oxygenated 2 M KI solution was added to a flask containing 15–20 mg of perovskite under an Ar atmosphere and allowed to disperse for three minutes. After a few minutes 25 ml of 1 M HCl is added and the perovskite is allowed to dissolve. This solution is then titrated to a faint golden colour with a solution of ~40 μM solution of Na$_2$S$_2$O$_3$ that has been pre-standardized with 0.1 N KIO$_3$. Starch indicator is then added and the solution is titrated until clear, marking the end point. BET surface area measurements were performed through nitrogen sorption on a Quantachrome Instruments NOVA 2000 high-speed surface area BET analyser at a temperature of 77 K, using 7 points from the linear region of the adsorption isotherm to determine the surface area.

Electrode preparation. All La$_{1-x}$Sr$_x$CoO$_{3-\delta}$ nanopowders and the commercial IrO$_2$ sample were loaded onto carbon through ball milling with a Wig-L-Bug ball mill. For rotating disk electrode (RDE) and for the RRDE measurements the LSCO nanopowders were loaded at a mass loading of ~30 wt% onto NC. For the galvanostatic stability tests, LSCO nanopowders and IrO$_2$ were also loaded onto VC (XC-72, Cabot Corporation) at a mass loading of ~30 wt%. The LSCO/carbon mixtures were dispersed in ethanol containing 0.05 wt% Na-substituted Nafion at a ratio of 1 mg ml^{-1} and sonicated for 45 min. This solution was spuncast onto a glassy carbon RDE (0.196 cm$^2_{geom}$, Pine Instruments) and for the RRDE (Glassy Carbon Disk: 0.2472 cm$^2_{geom}$; Pt ring: 0.1859 cm$^2_{geom}$, Pine Instruments) at a total mass loading of 51.0 μg cm$^{-2}_{geom,disk}$ (LSCO loading: 15.3 μg cm$^{-2}_{geom}$). The synthesis of the NC is described elsewhere[57]. For the oxygen intercalation cyclic voltammetry studies the LSCO nanopowders were loaded at a mass loading of 85 wt% on VC (Cabot Corporation). The LSCO/carbon mixtures were dispersed in ethanol containing 0.1 wt% Na-substituted Nafion at a ratio of 2 mg ml^{-1} and sonicated for 45 min. This solution was spun cast onto the glassy carbon RDE at a total mass loading of 102.0 μg cm$^{-2}_{geom}$ (LSCO loading 86.7 μg cm$^{-2}_{geom}$). The electrodes were cleaned before spin casting by sonication in a 1:1 deionized water:ethanol solution. The electrodes were then polished using 50 nm alumina powder, sonicated in a fresh deionized water:ethanol solution and dried under a scintillation vial in ambient air.

Electrochemical testing. Electrochemical testing was performed on a CH Instruments CHI832a potentiostat or a Metrohm Autolab PGSTAT302N potentiostat, both equipped with high-speed rotators from Pine Instruments. For the OER studies, the testing was done at room temperature in O$_2$ saturated 0.1 M KOH (measured pH ≈ 12.6). The current interrupt and positive-feedback methods were used to determine electrolyte resistance (50 Ω) and all data was *iR* compensated after testing. Each measurement was performed in a standard three-electrode cell using a Hg/HgO (1 M KOH) reference electrode, a Pt wire counter electrode, and a film of catalyst ink on the glassy carbon working electrode. All OER testing was performed on a new electrode that had not undergone previous testing. Cyclic voltammetry was performed from +0.9 to +1.943 V at 10 mV s^{-1} with a rotation rate of 1,600 r.p.m. To compensate for capacitive effects, the currents were averaged for the forward and backwards scans (Supplementary Fig. 10) The current at +1.63 V was selected from the polarization curves to compare the OER activities. For the rotating-ring-disk studies, the same parameters were used for the disk and the Pt ring electrode was held at a constant potential of +0.4 V versus RHE for the reduction of O$_2$ to OH$^-$. The Pt ring of the RRDE was electrochemically cleaned before testing by cyclic voltammetry on only the polished electrode in 0.1 M KOH through the hydrogen reduction potential regime at 5 mV s^{-1} for 20 cycles. The collection efficiency of the RRDE was measured as $N = 0.37$ through calibration in 0.3 mM Ferrocene-methanol in 0.1 M KCl electrolyte (Supplementary Fig. 11). Stability tests were performed galvanostatically at a current density of 10 A g$^{-1}_{ox}$ and a rotation rate of 1,600 r.p.m. for 24 h for SrCoO$_{2.7}$ and IrO$_2$ supported on either NC or on VC. A cutoff potential of +1.75 V versus RHE was used to stop the test to preserve the integrity of the glassy carbon electrode supports. All potential are reported versus the RHE, which was measured as $E_{RHE} = E_{Hg/HgO} + 0.8456$ V through the reduction of hydrogen in 1 atm H$_2$ saturated 0.1 M KOH (Supplementary Fig. 12).

Oxygen intercalation and diffusion rate measurements. The reversible intercalation of oxygen into LSCO was measured using cyclic voltammetry in an Ar saturated 1 M KOH electrolyte at 20 mV s^{-1} in a standard 3-electrode cell, using a Hg/HgO (1 M KOH) reference electrode, a Pt wire counter electrode, and a

working electrode of a thin film of LSCO/VC on a glassy carbon electrode as described above. The electrodes were stationary during testing and cycled twice. The data shown is from the second cycle. Following this, the diffusion rates of oxygen in the crystal were measured based on an adaptation of the procedure given in refs 39–41. In short, following the cyclic voltammetry oxygen intercalation measurements, the $E_{1/2}$ of the intercalation redox peaks was determined as the potential half way between the peak currents for intercalation and de-intercalation. The same electrodes were tested chronoamperometrically by applying a potential 50 mV more anodic of the $E_{1/2}$. The electrodes were rotated at 1,600 r.p.m. to get rid of electrolyte based mass-transfer effects, and the current was measured as a function of time for 4 h. The current was plotted versus $t^{-1/2}$ and the linear section of the curve was fit to find the intercept with the $t^{-1/2}$ axis. Using a bounded 3-dimensional solid-state diffusion model, this intersect is indicative of the diffusion rate of oxygen according to the relation $\lambda = \frac{a}{\sqrt{Dt}}$, where, λ is a shape factor for the particles (in this case $\lambda = 2$ for rounded paralelipipeds), a is the radius of the particle (in this case 150 nm was used for all LSCO samples), $t^{-1/2}$ is determined from the intersection with the $t^{-1/2}$ axis, and D is the diffusion rate of oxygen ions in the crystal measured at room temperature.

Density function theory calculations and surface models. DFT calculations[55,59,60] are performed using VASP with PAW pseudopotentials and the RPBE-GGA functional. More details are provided in the Supplementary Methods.

References

1. Cui, C., Gan, L., Heggen, M., Rudi, S. & Strasser, P. Compositional segregation in shaped Pt alloy nanoparticles and their structural behaviour during electrocatalysis. *Nat. Mater.* **12,** 765–771 (2013).
2. Slanac, D. A., Hardin, W. G., Johnston, K. P. & Stevenson, K. J. Atomic ensemble and electronic effects in Ag-rich AgPd nanoalloy catalysts for oxygen reduction in alkaline media. *J. Am. Chem. Soc.* **134,** 9812–9819 (2012).
3. Gupta, G. *et al.* Highly stable and active Pt − Cu oxygen reduction electrocatalysts based on mesoporous graphitic carbon supports. *Chem. Mater.* **21,** 4515–4526 (2009).
4. James Patrick, H. *The Electrochemistry of Oxygen* (Interscience Publishers, 1968).
5. McCrory, C. C. L., Jung, S., Peters, J. C. & Jaramillo, T. F. Benchmarking heterogeneous electrocatalysts for the oxygen evolution reaction. *J. Am. Chem. Soc.* **135,** 16977–16987 (2013).
6. Gorlin, Y. & Jaramillo, T. F. A bifunctional nonprecious metal catalyst for oxygen reduction and water oxidation. *J. Am. Chem. Soc.* **132,** 13612–13614 (2010).
7. Jasem, S. M. & Tseung, A. C. C. A potentiostatic pulse study of oxygen evolution on teflon–bonded nickel–cobalt oxide electrodes. *J. Electrochem. Soc.* **126,** 1353–1360 (1979).
8. Dau, H. *et al.* The mechanism of water oxidation: from electrolysis via homogeneous to biological catalysis. *ChemCatChem* **2,** 724–761 (2010).
9. Otagawa, T. & Bockris, J. O. Oxygen evolution on perovskites. *J. Phys. Chem.* **87,** 2960–2971 (1983).
10. Bockris, J. O. & Otagawa, T. The electrocatalysis of oxygen evolution on perovskites. *J. Electrochem. Soc.* **131,** 290–302 (1984).
11. Matsumoto, Y. & Sato, E. Electrocatalytic properties of transition metal oxides for oxygen evolution reaction. *Mater. Chem. Phys.* **14,** 397–426 (1986).
12. Meadowcroft, D. B. Low-cost Oxygen electrode material. *Nature* **226,** 847–848 (1970).
13. Hong, W. T. *et al.* Toward the rational design of non-precious transition metal oxides for oxygen electrocatalysis. *Energy Env. Sci.* **8,** 1404–1427 (2015).
14. Suntivich, J., May, K. J., Gasteiger, H. A., Goodenough, J. B. & Shao-Horn, Y. A perovskite oxide optimized for oxygen evolution catalysis from molecular orbital principles. *Science* **334,** 1383–1385 (2011).
15. Suntivich, J. *et al.* Design principles for oxygen-reduction activity on perovskite oxide catalysts for fuel cells and metal-air batteries. *Nat. Chem.* **3,** 546–550 (2011).
16. Hardin, W. G. *et al.* Highly active, nonprecious metal perovskite electrocatalysts for bifunctional metal–air battery electrodes. *J. Phys. Chem. Lett.* **4,** 1254–1259 (2013).
17. Hardin, W. G. *et al.* Tuning the electrocatalytic activity of perovskites through active site variation and support interactions. *Chem. Mater.* **26,** 3368–3376 (2014).
18. Peña, M. a. & Fierro, J. L. Chemical structures and performance of perovskite oxides. *Chem. Rev.* **101,** 1981–2017 (2001).
19. Conder, K., Pomjakushina, E., Soldatov, A. & Mitberg, E. Oxygen content determination in perovskite-type cobaltates. *Mater. Res. Bull.* **40,** 257–263 (2005).
20. Takeda, Y. *et al.* Phase relation and oxygen-non-stoichiometry of perovskite-like compound SrCoO$_x$ (2.29 < x < 2.80). *Z. Anorg. Allg. Chem.* **540/541,** 259–270 (1986).

21. Kudo, T., Obayashi, H. & Gejo, T. Electrochemical behavior of the perovskite-type $Nd_{1-x}Sr_xCoO_3$ in an aqueous alkaline solution. *J. Electrochem. Soc.* **122**, 159–163 (1975).

22. Wattiaux, A., Grenier, J. C., Pouchard, M. & Hagenmuller, P. Electrolytic oxygen evolution in alkaline medium on $La_{1-x}Sr_xFeO_{3-y}$ perovskite-related ferrites II. Influence of bulk properties. *J. Electrochem. Soc.* **134**, 1718–1724 (1987).

23. Nemudry, A., Rudolf, P. & Schöllhorn, R. Topotactic electrochemical redox reactions of the defect perovskite $SrCoO_{2.5+x}$. *Chem. Mater.* **8**, 2232–2238 (1996).

24. Nemudry, A., Goldberg, E. L., Aguirre, M. & Alario-Franco, M. Á. Electrochemical topotactic oxidation of nonstoichiometric perovskites at ambient temperature. *Solid State Sci.* **4**, 677–690 (2002).

25. Jeen, H. *et al.* Reversible redox reactions in an epitaxially stabilized $SrCoO_x$ oxygen sponge. *Nat. Mater.* **12**, 1057–1063 (2013).

26. Jeen, H. *et al.* Topotactic phase transformation of the brownmillerite $SrCoO_{2.5}$ to the perovskite $SrCoO_{3-\delta}$. *Adv. Mater.* **25**, 3651–3656 (2013).

27. Mefford, J. T., Hardin, W. G., Dai, S., Johnston, K. P. & Stevenson, K. J. Anion charge storage through oxygen intercalation in $LaMnO_3$ perovskite pseudocapacitor electrodes. *Nat. Mater.* **13**, 726–732 (2014).

28. Mastin, J., Einarsrud, M.-A. & Grande, T. Structural and thermal properties of $La_{1-x}Sr_xCoO_{3-\delta}$. *Chem. Mater.* **18**, 6047–6053 (2006).

29. Maris, G. *et al.* Evidence for orbital ordering in $LaCoO_3$. *Phys. Rev. B* **67**, 224423 (2003).

30. Takami, T., Zhou, J.-S., Goodenough, J. B. & Ikuta, H. Correlation between the structure and the spin state in $R_{1-x}Sr_xCoO_3$ (R = La, Pr, and Nd). *Phys. Rev. B* **76**, 144116 (2007).

31. Wang, Y. *et al.* Correlation between the structural distortions and thermoelectric characteristics in $La_{1-x}A_xCoO_3$ (A = Ca and Sr). *Inorg. Chem.* **49**, 3216–3223 (2010).

32. Kim, Y.-M. *et al.* Probing oxygen vacancy concentration and homogeneity in solid-oxide fuel-cell cathode materials on the subunit-cell level. *Nat. Mater.* **11**, 888–894 (2012).

33. James, M. *et al.* Orthorhombic superstructures within the rare earth strontium-doped cobaltate perovskites: $Ln_{1-x}Sr_xCoO_{3-\delta}$ (Ln = Y^{3+}, Dy^{3+}–Yb^{3+}; $0.750 \leqslant x \leqslant 0.875$). *J. Solid State Chem.* **180**, 2233–2247 (2007).

34. Grimaud, A. *et al.* Oxygen evolution activity and stability of $Ba_6Mn_5O_{16}$, $Sr_4Mn_2CoO_9$, and $Sr_6Co_5O_{15}$: the influence of transition metal coordination. *J. Phys. Chem. C* **117**, 25926–25932 (2013).

35. Takeda, Y. *et al.* Properties of $SrMO_{3-d}$ (M = Fe, Co) as oxygen electrodes in alkaline solution. *J. Appl. Electrochem.* **12**, 275–280 (1982).

36. Lindberg, F. *et al.* Synthesis and characterization of $Sr_{0.75}Y_{0.25}Co_{1-x}M_xO_{2.625+\delta}$ (M = Ga, $0.125 \leqslant x \leqslant 0.500$ and M = Fe, $0.125 \leqslant x \leqslant 0.875$). *J. Solid State Chem.* **179**, 1434–1444 (2006).

37. Withers, R. L., James, M. & Goossens, D. J. Atomic ordering in the doped rare earth cobaltates $Ln_{0.33}Sr_{0.67}CoO_{3-\delta}$ (Ln = Y^{3+}, Ho^{3+} and Dy^{3+}). *J. Solid State Chem.* **174**, 198–208 (2003).

38. Conway, B. E. *Electrochemical Supercapacitors* (Springer, 1999).

39. Van Buren, F. R., Broers, G. H. J., Bouman, A. J. & Boesveld, C. An electrochemical method for the determination of oxygen ion diffusion coefficients in $La_{1-x}Sr_xCoO_{3-y}$ compounds: theoretical aspects. *J. Electroanal. Chem. Interfacial Electrochem.* **87**, 389–394 (1978).

40. Van Buren, F. R., Broers, G. H. J., Bouman, A. J. & Boesveld, C. The electrochemical determination of oxygen ion diffusion coefficients in $La_{0.50}Sr_{0.50}CoO_{3-y}$: experimental results and related properties. *J. Electroanal. Chem. Interfacial Electrochem.* **88**, 353–361 (1978).

41. Kobussen, A. G. C., van Buren, F. R. & Broers, G. H. J. The influence of the particle size distribution on the measurement of oxygen ion diffusion coefficients in $La_{0.50}Sr_{0.50}CoO_{3-y}$. *J. Electroanal. Chem. Interfacial Electrochem.* **91**, 211–217 (1978).

42. Royer, S., Duprez, D. & Kaliaguine, S. Oxygen mobility in $LaCoO_3$ Perovskites. *Catal. Today* **112**, 99–102 (2006).

43. Grimaud, A. *et al.* Double perovskites as a family of highly active catalysts for oxygen evolution in alkaline solution. *Nat. Commun.* **4**, 2439 (2013).

44. Minguzzi, A., Fan, F.-R. F., Vertova, A., Rondinini, S. & Bard, A. J. Dynamic potential–pH diagrams application to electrocatalysts for water oxidation. *Chem. Sci.* **3**, 217–229 (2011).

45. Matsumoto, Y., Yoneyama, H. & Tamura, H. Influence of the nature of the conduction band of transition metal oxides on catalytic activity for oxygen reduction. *J. Electroanal. Chem. Interfacial Electrochem.* **83**, 237–243 (1977).

46. Goodenough, J. B. Covalency criterion for localized vs collective electrons in oxides with the perovskite structure. *J. Appl. Phys.* **37**, 1415–1422 (1966).

47. Mueller, D. N., Machala, M. L., Bluhm, H. & Chueh, W. C. Redox activity of surface oxygen anions in oxygen-deficient perovskite oxides during electrochemical reactions. *Nat. Commun.* **6**, 6097 (2015).

48. Sathiya, M. *et al.* Reversible anionic redox chemistry in high-capacity layered-oxide electrodes. *Nat. Mater.* **12**, 827–835 (2013).

49. Imamura, M., Matsubayashi, N. & Shimada, H. Catalytically active oxygen species in $La_{1-x}Sr_xCoO_{3-\delta}$ studied by XPS and XAFS spectroscopy. *J. Phys. Chem. B* **104**, 7348–7353 (2000).

50. Sunstrom, IV J. E., Ramanujachary, K. V., Greenblatt, M. & Croft, M. The synthesis and properties of the chemically oxidized perovskite, $La_{1-x}Sr_xCoO_{3-\delta}$ ($0.5 \leqslant x \leqslant 0.9$). *J. Solid State Chem.* **139**, 388–397 (1998).

51. Le Toquin, R., Paulus, W., Cousson, A., Prestipino, C. & Lamberti, C. Time-resolved *in situ* studies of oxygen intercalation into $SrCoO_{2.5}$, performed by neutron diffraction and X-ray absorption spectroscopy. *J. Am. Chem. Soc.* **128**, 13161–13174 (2006).

52. Rong, X., Parolin, J. & Kolpak, A. M. A fundamental relationship between reaction mechanism and stability in metal oxide catalysts for oxygen evolution. *ACS Catal.* **6**, 1153–1158 (2016).

53. Man, I. C. *et al.* Universality in oxygen evolution electrocatalysis on oxide surfaces. *ChemCatChem* **3**, 1159–1165 (2011).

54. Akhade, S. A. & Kitchin, J. R. Effects of strain, d-band filling, and oxidation state on the surface electronic structure and reactivity of 3d perovskite surfaces. *J. Chem. Phys.* **137**, 084703 (2012).

55. Hammer, B., Hansen, L. B. & Nørskov, J. K. Improved adsorption energetics within density-functional theory using revised Perdew-Burke-Ernzerhof functionals. *Phys. Rev. B* **59**, 7413–7421 (1999).

56. Lee, Y.-L., Kleis, J., Rossmeisl, J. & Morgan, D. *Ab initio* energetics of $LaBO_3$ (B = Mn, Fe, Co, and Ni) for solid oxide fuel cell cathodes. *Phys. Rev. B* **80**, 224101 (2009).

57. Wang, X. *et al.* Ammonia-treated ordered mesoporous carbons as catalytic materials for oxygen reduction reaction. *Chem. Mater.* **22**, 2178–2180 (2010).

58. Petříček, V., Dušek, M. & Palatinus, L. Crystallographic computing system JANA2006: general features. *Z. Für. Krist.* **229**, 345–352 (2014).

59. Kresse, G. & Furthmüller, J. Efficient iterative schemes for *ab initio* total-energy calculations using a plane-wave basis set. *Phys. Rev. B* **54**, 11169–11186 (1996).

60. Blöchl, P. E. Projector augmented-wave method. *Phys. Rev. B* **50**, 17953–17979 (1994).

61. Maiyalagan, T., Jarvis, K. A., Therese, S., Ferreira, P. J. & Manthiram, A. Spinel-type lithium cobalt oxide as a bifunctional electrocatalyst for the oxygen evolution and oxygen reduction reactions. *Nat. Commun.* **5**, 3949 (2014).

62. Goodenough, J. B. & Cushing, B. L. in *Handbook of Fuel Cells—Fundamentals, Technology and Applications* 2, 520–533 (Wiley, 2003).

Acknowledgements

Financial support for this work was provided by the R.A. Welch Foundation (grants F-1529 and F-1319). X.R. and A.M.K. acknowledge support from the Skoltech-MIT Center for Electrochemical Energy Storage. Computations were performed using computational resources from XSEDE and NERSC. S.D. was supported as part of the Fluid Interface Reactions, Structures and Transport (FIRST) Center, an Energy Frontier Research Center funded by the U.S. Department of Energy, Office of Science, and Office of Basic Energy Sciences. We thank D.W. Redman for help with the RHE measurements.

Author contributions

J.T.M. and W.G.H. performed the synthesis. J.T.M. performed the X-ray diffraction, and electrochemical characterization. X.R. and A.M.K. performed the DFT modelling. A.M.A. performed the SAED, HAAFD-STEM, ABF-STEM, energy dispersive X-ray measurements and crystallographic analysis. S.D. contributed the carbon support. J.T.M., K.P.J. and K.J.S. planned the experiment and analysed the data. All authors contributed to the writing of the paper.

Additional information

Light-enhanced liquid-phase exfoliation and current photoswitching in graphene–azobenzene composites

Markus Döbbelin[1], Artur Ciesielski[1], Sébastien Haar[1], Silvio Osella[2], Matteo Bruna[3], Andrea Minoia[2], Luca Grisanti[2], Thomas Mosciatti[1], Fanny Richard[1], Eko Adi Prasetyanto[4], Luisa De Cola[4], Vincenzo Palermo[5], Raffaello Mazzaro[6,7], Vittorio Morandi[6], Roberto Lazzaroni[2], Andrea C. Ferrari[3], David Beljonne[2] & Paolo Samorì[1]

Multifunctional materials can be engineered by combining multiple chemical components, each conferring a well-defined function to the ensemble. Graphene is at the centre of an ever-growing research effort due to its combination of unique properties. Here we show that the large conformational change associated with the *trans–cis* photochemical isomerization of alkyl-substituted azobenzenes can be used to improve the efficiency of liquid-phase exfoliation of graphite, with the photochromic molecules acting as dispersion-stabilizing agents. We also demonstrate reversible photo-modulated current in two-terminal devices based on graphene–azobenzene composites. We assign this tuneable electrical characteristics to the intercalation of the azobenzene between adjacent graphene layers and the resulting increase in the interlayer distance on (photo)switching from the linear *trans*-form to the bulky *cis*-form of the photochromes. These findings pave the way to the development of new optically controlled memories for light-assisted programming and high-sensitive photosensors.

[1] Nanochemistry Laboratory, ISIS & icFRC, Université de Strasbourg & CNRS, 8 allée Gaspard Monge, 67000 Strasbourg, France. [2] Laboratory for Chemistry of Novel Materials, University of Mons, Place du Parc 20, B-7000 Mons, Belgium. [3] Cambridge Graphene Centre, University of Cambridge, 9 JJ Thomson Avenue, Cambridge CB3 OFA, UK. [4] Laboratory of Supramolecular Biomaterials and Chemistry, ISIS & icFRC, Université de Strasbourg & CNRS, 8 allée Gaspard Monge, 67000 Strasbourg, France. [5] ISOF-CNR, via Gobetti 101, 40129 Bologna, Italy. [6] IMM-CNR Sezione di Bologna, via Gobetti, 101, 40129 Bologna, Italy. [7] Dipartimento di Chimica 'G. Ciamician', Università di Bologna, via Selmi 2, 40126 Bologna, Italy. Correspondence and requests for materials should be addressed to A.C. (email:ciesielski@unistra.fr) or to P.S. (email: samori@unistra.fr).

Graphene is a one-atom-thick two-dimensional material with unique mechanical, optical, thermal and electrical properties[1], promising future emerging technologies, including flexible and wearable electronics[2]. Two main approaches are being followed for graphene production: bottom-up and top-down[3]. The former relies on the assembly of suitably designed molecular building blocks, undergoing chemical reaction to form covalently linked networks[4]. The latter occurs via exfoliation of graphite into graphene[3,5]. Bottom-up techniques, in particular those based on organic syntheses starting from small molecular modules[6,7], when performed in liquid media, are both size limited, because macromolecules become less soluble with increasing size[6], and suffer from the occurrence of side reactions with increasing molecular weight[6]. The growth on solid (ideally catalytically active) surfaces allows one to circumvent these issues. Substrate-based growth can also be achieved by chemical vapour deposition (CVD)[8,9] or via silicon evaporation from silicon carbide[10], which rely on the ability to follow a narrow thermodynamic path. Top-down approaches can be accomplished under different environmental conditions. Among them, liquid-phase exfoliation (LPE)[2,3,5,11–15] is an attractive strategy, being extremely versatile and applicable to a variety of environments and on different substrates. While bottom-up methods, in particular CVD, can yield large size, LPE gives limited sheet sizes[11,16–18]. Nevertheless, LPE has several advantages. It is a viable low-cost process, which can be easily upscaled to mass-produced dispersions processable by well-established techniques such as spin-coating, drop-casting, screen-printing and ink-jet printing[17]. Increasing research efforts are being devoted to the production of graphene via LPE to improve the material's physicochemical and electrical properties.

Small organic molecules, such as surfactants or dispersion-stabilizing agents, can promote the exfoliation of graphite into graphene in organic solvents[5,11], in particular when such molecules have higher adsorption energies on the basal plane of graphene than those involved in solvent–graphene interaction[11]. We recently found that long alkanes are particularly suitable for enhancing the yield of exfoliation[19,20].

Photochromic molecules, in particular azobenzene-based molecules, covalently linked to reduced graphene oxide[21], physisorbed directly on graphene[22,23] or through pyrene anchoring groups[24] on the graphene surface, or even when graphene is adsorbed on an azobenzene self-assembled monolayer[25], can be used to reversibly modulate the graphene's electronic properties. Moreover, it was shown that covalent functionalization of gold surface[26] or carbon nanotubes[27] with azobenzene molecules can be used as a route for fabrication of light-driven electronic switches. Nonetheless, the use of additional responsive functions provided by a suitably designed molecule to assist the LPE of graphite into functional inks is still unexplored. This approach could lead to the production of functional hybrid materials and nanocomposites in a one-pot process.

Here we combine photochromic systems and LPE graphene exploiting the properties of these materials. We focus on a commercial alkoxy-substituted azobenzene, i.e., 4-(decyloxy) azobenzene. The presence of a long alkoxy side chain is expected to enhance the molecular interaction with graphene[28], and promote the exfoliation of graphite towards graphene. Here we demonstrate that the use of 4-(decyloxy)azobenzene has two major advantages: first, the amount of graphene dispersed in N-methyl-2-pyrrolidone (NMP) can be enhanced by exploiting the photoisomerization of 4-(decyloxy)azobenzene molecules during exfoliation. Second, the isomerization of 4-(decyloxy)azobenzene from trans to cis and vice versa, when physisorbed between adjacent graphene layers leads to a reversible modulation of the inter-flake distance on the sub-Ångström scale. This is reflected in a light response of the electrical properties of the hybrid material.

Results

Liquid-phase exfoliation. To test the ability of alkoxy-substituted azobenzenes to increase Y_W (%), which is the yield of graphene exfoliation, defined as the ratio between the weight of dispersed graphitic material and that of the starting graphite flakes, as well as to exploit the photochromic nature of such molecules when interacting with graphene, vials containing NMP, graphite powder and azobenzene are exposed to tip sonication for 3 h (see Methods and Supplementary Note 1 for details). This experiment is carried out either in dark or under ultraviolet irradiation, using a portable ultraviolet lamp, at two temperatures: 20 and 40 °C (see experimental set-up in Supplementary Fig. 1). Subsequently, the dispersions are allowed to settle for 15 min, then decanted and centrifuged for 1 h at 10,000 r.p.m. Control samples are also prepared, consisting of dispersions prepared in the absence of 4-(decyloxy)azobenzene both in dark and irradiated at 365 nm at either 20 or 40 °C. To quantify the concentration of graphene after centrifugation, a mixture of graphene dispersion and 2-propanol (IPA) is first heated to 50 °C for 30 min and then passed through polytetrafluoroethylene membrane filters. The remaining solvent and weakly interacting 4-(decyloxy) azobenzene molecules are washed out several times with diethyl ether and IPA. Measurements of the filtered mass are performed on a microbalance to infer the concentration of graphitic material in dispersion, needed to quantify Y_W (%) (see Fig. 1b). The presence of adsorbed molecules may affect the mass measurements and Y_W (%). Thus, the heating step is necessary to completely desorb the 4-(decyloxy)azobenzene molecules, as proven with X-ray photoelectron spectroscopy (Supplementary Fig. 2 and Supplementary Note 2). By analysing 20 independent experiments, we find that when the exfoliation is performed in the absence of 4-(decyloxy)azobenzene (control experiments), comparable Y_W are observed when dispersions are kept at 20 °C (0.64%) or heated at 40 °C (0.62%). The irradiation at 365 nm does not affect Y_W, neither at 20 nor at 40 °C (0.63% in both cases).

LPE in the presence of alkoxy-substituted azobenzene acting as dispersion-stabilizing agent can undergo different mechanisms depending on the isomeric form of the photochromic molecule (Fig. 1c). In our case, the presence of 4-(decyloxy)azobenzene has no major influence on Y_W of dispersions prepared in dark (0.71% and 0.72% at 20 and 40 °C, respectively). As previously reported[11], the use of small organic molecules, in particular those based on aromatic cores[11] and alkyl functionalization[19,20], can promote graphite exfoliation because of their high affinity for the basal plane of graphene hindering interflake stacking. The relation between interaction energy and exfoliation yield is not straightforward. In some cases, molecules with lower adsorption energy on graphene give higher exfoliation yield, as demonstrated for graphite exfoliation[29], as well as for exfoliation of a wide range of other layered materials[30]. This is because a critical step of the process is the adsorption of the molecule on the graphene surface and the displacement of the adsorbed solvent layer, overcoming an activation energy barrier that depends significantly on molecular structure and conformation. Therefore, even a minor increase of Y_W (0.06–0.09%) on addition of a small amount of 4-(decyloxy)azobenzene (5 wt%) highlights the importance of these molecules during LPE.

In contrast to the experiments performed in dark, irradiation with ultraviolet light at 365 nm increases the concentration of exfoliated material. At 20 °C we obtain $\sim 81 \pm 5\,\mu g\,ml^{-1}$,

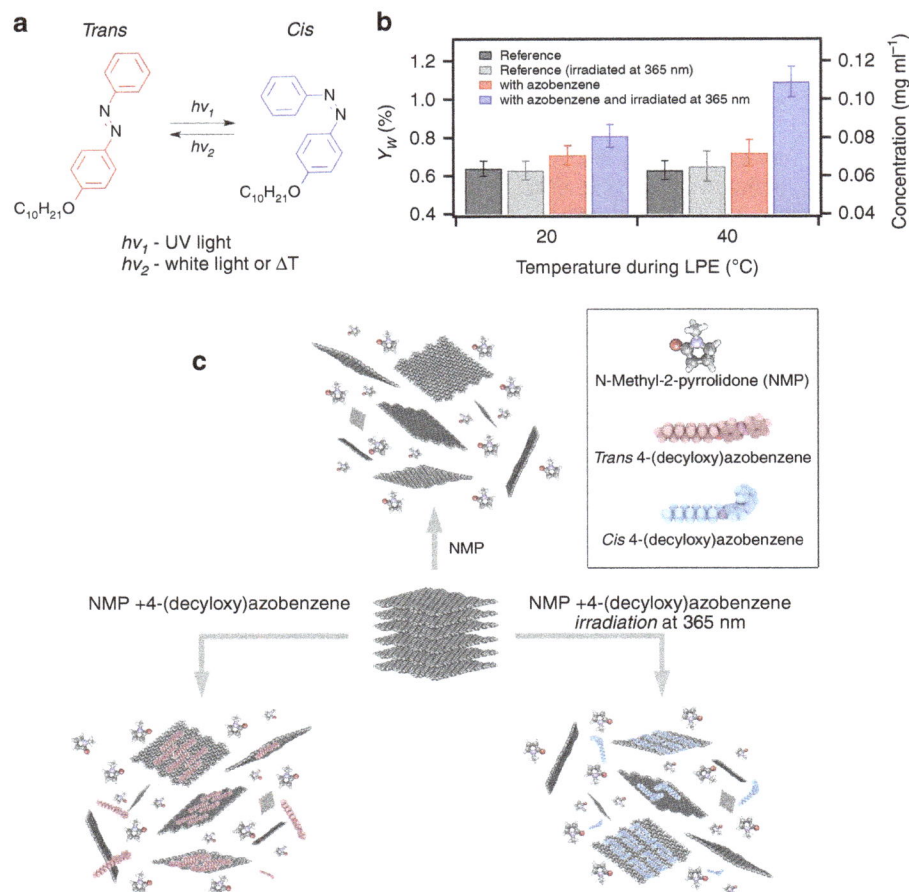

Figure 1 | Light-assisted graphene's LPE process in the presence of 4-(decyloxy)azobenzene and yield of exfoliation. (**a**) Chemical structure of 4-(decyloxy)azobenzene and reversible *trans*–*cis* photoisomerization under UV and visible light. (**b**) Y_W and concentration after filtration. The error bars reflect a statistical analysis on 20 independent experiments. (**c**) Schematic representation of the LPE process under different experimental conditions: in NMP (reference experiment); and in NMP with azobenzene in dark and under UV irradiation.

corresponding to $\sim 30\%$ increase in Y_W, when compared with reference samples ($63\,\mu g\,ml^{-1}$). Yet, the largest increase in concentration is obtained by irradiating the dispersions heated at $40\,°C$, reaching $110 \pm 8\,\mu g\,ml^{-1}$, which corresponds to a 75% increase in Y_W. These results demonstrate that the addition of 4-(decyloxy)azobenzene during LPE leads to a significant increase in Y_W only in samples containing 4-(decyloxy)azobenzene irradiated at $365\,nm$. Molecules rapidly switching conformation, such as azobenzenes on ultraviolet (UV) illumination, heating and sonication, can overcome the activation barriers to adsorption, acting as 'nano-spinners' that disrupt the adsorbed solvent layer more effectively than molecules having a static conformation. Therefore, the increase in concentration can be attributed to the motion resulting by the *trans*-to-*cis* photoisomerization. The Y_W temperature dependence may be due to differences in molecule–graphene interactions at 20 and $40\,°C$, as well as different kinetics of the *trans*-to-*cis* photoisomerization.

Trans-to-cis photoisomerization and cis-to-trans relaxation. The transition from the thermodynamically stable *trans* to the metastable *cis* form can be induced by irradiation with UV light, and takes place on the picosecond timescale[31,32], whereas the *cis*-to-*trans* isomerization is typically triggered by irradiation with visible light or heat[33]. The latter can also be activated by the use of ultrasound-induced mechanical force[33]. Ref. 34 claimed

that mechanoisomerization of azobenzene typically results in the cleavage of the $N = N$ azo bond. However, more recent theoretical calculations[35] and experiments[36] have challenged this view.

To gain a quantitative understanding of the photoisomerization of 4-(decyloxy)azobenzene, we perform a kinetic analysis of both *trans*-to-*cis* and *cis*-to-*trans* isomerization. The UV–visible spectrum of 4-(decyloxy)azobenzene dissolved in NMP (Fig. 2a, black line) shows an absorption maximum at $353\,nm$, arising from the π–$\pi^\star$ transition[33], and a peak at $442\,nm$, due to n–$\pi^\star$ transitions[33]. The UV–visible spectrum of the above solution when irradiated at $365\,nm$ shows a reduction of the π–$\pi^\star$ transition band at $353\,nm$, and an increase in absorbance of the $442\,nm$ (n–$\pi^\star$ transition) peak, which is a characteristic feature of the *trans*-to-*cis* photoisomerization (Fig. 2a, red line; and Supplementary Fig. 3). The photostationary state can be reversibly converted back to the *trans* isomer on irradiation at $450\,nm$ (see Methods for details). The yield of the conversion back to the thermodynamically stable *trans*-4-(decyloxy)azobenzene is higher than 95% (Fig. 2a, black dashed line).

To probe the *cis*-to-*trans* isomerization of the azobenzene molecules under experimental conditions corresponding to those used during the LPE process, i.e., at 20 and $40\,°C$, in the presence of ultrasound-induced mechanical forces, a stock solution of 4-(decyloxy)azobenzene in NMP ($16.2\,mM$) is prepared, and then exposed to $365\,nm$ light and sonicated for $3\,h$ (see Supplementary Note 1 for details). Every $30\,min$

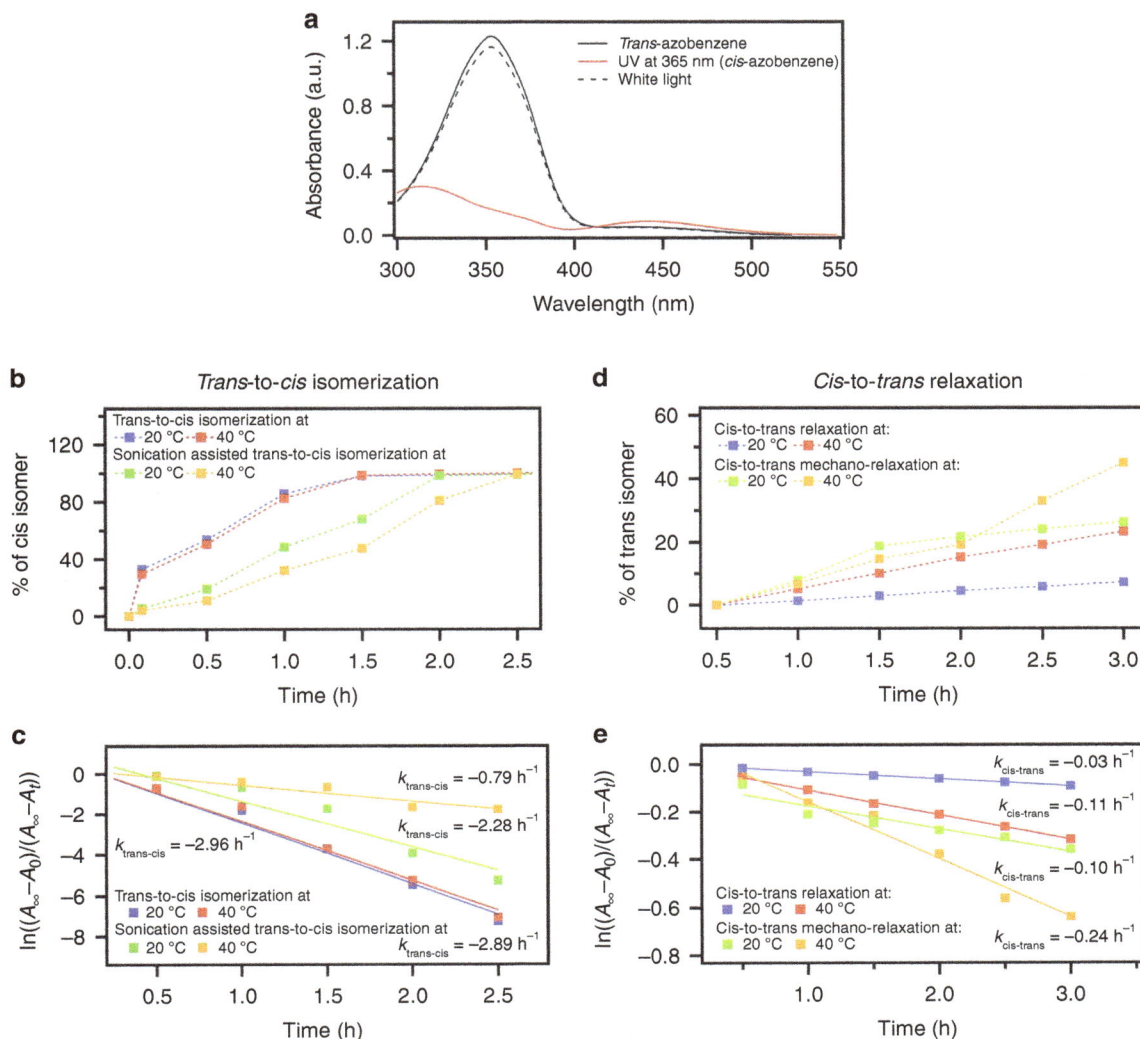

Figure 2 | Spectroscopic characterization of *trans*-to-*cis* isomerization and *cis*-to-*trans* relaxation. (a) UV-ndash;visible spectra of 4-(decyloxy) azobenzene in NMP (1.62 mM) showing the disappearance of the 353-nm band and an increase of the 442-nm band for 365-nm irradiation. The solid black line corresponds to the parent *trans*-4-(decyloxy)azobenzene spectrum, the red line represents a typical spectrum of a solution irradiated with UV light (*cis*-4-(decyloxy)azobenzene), whereas the black dashed line corresponds to a *cis*-4-(decyloxy)azobenzene solution irradiated with visible light (thus re-transformed into *trans*-4-(decyloxy)azobenzene). (b) Conversion percentage from *trans* to *cis* as a function of UV irradiation time at 20 and 40 °C. (c) First-order kinetic plots and rate constants ($k_{trans-cis}$) for *trans*-to-*cis* isomerization at 20 and 40 °C. (d) Conversion percentage from *cis* to *trans* as a function of time at 20 and 40 °C. (e) First-order kinetic plots and rate constants ($k_{cis-trans}$) for the *cis*-to-*trans* relaxation at 20 and 40 °C. (b–e) Reference samples prepared without sonication: blue (20 °C) and red (40 °C); sonicated samples: green (20 °C) and orange (40 °C).

(except for a first measurement performed after 5 min) 0.2 ml of UV-irradiated solution is transferred (in the dark) into a quartz cuvette, diluted 10 times with NMP. The optical absorption spectrum is then recorded (shown in Supplementary Fig. 4). Finally, to ensure the stability of the azo bond during sonication, a solution sonicated for 3 h is exposed to white light and an absorption spectrum is recorded. A similar procedure is applied for blank experiments, i.e., in the absence of ultrasound-induced mechanical forces.

Analysis of the absorption and proton nuclear magnetic resonance spectra (Supplementary Fig. 5) acquired for different experimental conditions allows us to quantify the percentage of conversion from *trans*- to *cis*-4-(decyloxy)azobenzene, as well as the rate constants for *trans*-to-*cis* isomerization. Both nuclear magnetic resonance and high-performance liquid chromatography data (Supplementary Figs 5 and 6, and Supplementary Note 3) show that the decyloxy chains are not

being chopped from the azobenzene core. The slow kinetics of the *trans*-to-*cis* isomerization can be explained by the fact that both spectroscopic experiments and LPE process are carried out under unusual isomerization conditions, including a low UV light power density (0.34 mW cm^{-2}). A different time (30 min) is needed for reaching a photostationary state, in the reference and sonicated samples, as displayed in Fig. 2b. Previous studies[36,37] on the *trans*-to-*cis* isomerization of azobenzene in the presence of UV light showed that the reaction follows first-order kinetics and the rate constant k can be written as follows:

$$\ln\frac{(A_\infty - A_t)}{(A_\infty - A_0)} = -kt \qquad (1)$$

where A_0, A_t and A_∞ are the absorbances before irradiation, at irradiation time t and after irradiation for a prolonged time (ca. 6 h). Applying equation (1), we find that the isomerization constant $k_{trans-cis}$ at 20 °C for the 4-(decyloxy)azobenzene linked to graphene is similar to $k_{trans-cis}$ at 40 °C: 2.96 and 2.89 h^{-1}, respectively. Significantly different $k_{trans-cis}$ kinetics are observed for the samples exposed to ultrasound-induced mechanical forces, i.e., 2.28 and 0.79 h^{-1} for dispersions prepared at 20 and 40 °C, respectively. To ensure the stability of the azo bond during sonication, the two samples are exposed to white light, and the absorption spectra recorded. The absorption at 365 nm is consistent with the *trans*-4-(decyloxy)azobenzene spectra in both cases (Supplementary Figs 3 and 4), indicating that the molecules, in particular the azo bond, are stable during sonication at various temperatures.

Since the *trans*-to-*cis* isomerization and the LPE processes are performed under UV light, the possible *cis*-to-*trans* back-conversion can be considered promoted by thermal or mechano-thermal relaxation, for experiments performed in the absence or presence of ultrasound-induced forces, respectively.

To probe both thermal and mechano-thermal isomerization, four solutions (16.2 mM) are prepared in NMP and the UV–visible spectra recorded. The solutions are then irradiated with UV light (365 nm) for 3 h to photoisomerize the 4-(decyloxy)azobenzene from its *trans* to *cis* configuration (Supplementary Fig. 3). The solutions are then kept for 3 h at 20 and/or 40 °C, in the presence and/or absence of ultrasound-induced mechanical forces. Every 30 min, 0.2 ml of solution is transferred (in dark) onto a quartz cuvette and diluted 10 times with NMP. Absorption spectra are then recorded. Owing to the lack of white light during the 3 h relaxation, 4-(decyloxy)azobenzene molecules do not convert fully from their *cis*- to *trans*-isomers. As shown in Fig. 2c, the percentage of *cis*-to-*trans* conversion in the blank samples (without sonication) at 20 and 40 °C is ~ 8.6% and 27.6%, respectively. In contrast, the ultrasound-induced conversion is much higher, i.e., 29.8% and 51.2% for samples kept at 20 and 40 °C, respectively. Similar to the *trans*-to-*cis* isomerization, the *cis*-to-*trans* relaxation of 4-(decyloxy)azobenzene follows first-order kinetics, and k can be obtained from equation (1). $k_{cis-trans}$ for 4-(decyloxy)azobenzene kept at 20 °C is lower than $k_{cis-trans}$ for 4-(decyloxy)azobenzene relaxed at 40 °C, i.e., 0.03 and 0.11 h^{-1}, respectively (Fig. 2d). Notably different $k_{cis-trans}$ are also observed for samples exposed to ultrasound-induced mechanical forces, i.e., 0.10 and 0.24 h^{-1} for dispersions kept at 20 and 40 °C, respectively.

These results indicate that the *trans*-to-*cis* isomerization of 4-(decyloxy)azobenzene strongly depends on the experimental

conditions. In particular, it is hindered by ultrasound-induced mechanical forces, which most likely cause the occurrence of the competing *cis*-to-*trans* isomerization. Furthermore, comparison of $k_{trans-cis}$ and $k_{cis-trans}$ at 20 and 40 °C reveals that, at higher temperature, the process is more dynamic. As a result, the higher concentration of flakes (110 µg ml^{-1}) obtained at 40 °C can be attributed to the dynamic conformational *trans*-to-*cis* and *cis*-to-*trans* changes of 4-(decyloxy)azobenzene molecules triggered simultaneously by the competing effect activated by UV light, temperature and ultrasounds.

Analysis of dispersions. To fully characterize the exfoliated flakes, both qualitative and quantitative information are required. While quantitative insights can be assessed by providing the Y_W, a qualitative analysis must give more relevant details, such as the percentage of single-layer graphene (SLG) and multi-layer graphene (MLG) flakes, the lateral size of the flakes and the presence/absence of defects. The number of graphene layers (N) can be quantified using high resolution-transmission electron microscopy (HR-TEM)[3,5] and Raman spectroscopy[17,18]. Together with the information coming from electron diffraction patterns, in HR-TEM N can be directly counted by analysing the folded edges[38]. We focus on the samples prepared by LPE at 40 °C in the presence of azobenzene, while irradiating with UV light (highest increase in Y_W), and compare them with those exfoliated in NMP at 40 °C under UV irradiation (in the absence of azobenzene). The statistical analyses of N and the lateral flake size are reported in Supplementary Fig. 7a,b, respectively (Supplementary Note 4). In terms of N, minor differences are observed between the reference sample and flakes exfoliated in the presence of azobenzene molecules, where the percentage of SLG amounts to 20% and 14%, respectively. The lateral size of the flakes is not affected by the presence of azobenzene molecules during LPE (Supplementary Fig. 7b). Representative HR-TEM micrographs are reported in Supplementary Figs 8, 9 and 10.

The Raman spectrum of graphite and MLG consists of two fundamentally different sets of peaks. Those, such as D, G and 2D, present also in SLG, due to in-plane vibrations[39–41], and others such as the shear (C) modes[42] and the layer-breathing modes (LBM)[40,43], due to the relative motions of the planes themselves, are either perpendicular or parallel to their normal. The G peak corresponds to the high-frequency E_{2g} phonon at Γ.

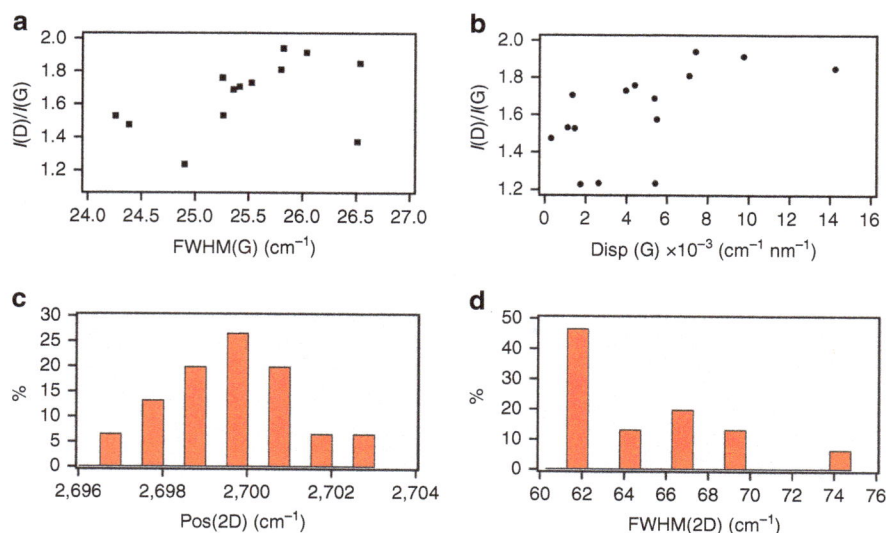

Figure 3 | Raman analysis of graphene/4-(decyloxy)azobenzene films. *I*(D)/*I*(G) as a function of (**a**) FWHM(G) and (**b**) Disp(G). Distribution of (**c**) Pos(2D) and (**d**) FWGM(2D).

The D peak is due to the breathing modes of six-atom rings and requires a defect for its activation[41,44,45]. It comes from transverse optical phonons around the Brillouin Zone edge **K** (refs 41,45), is active by double resonance[46] and is strongly dispersive with excitation energy due to a Kohn anomaly at **K** (ref. 47). Double resonance can also happen as intra-valley process, i.e., connecting two points belonging to the same cone around **K** or **K′**. This gives the so-called D′ peak. The 2D peak is the D peak overtone while the 2D′ peak is the D′ overtone. Since 2D and 2D′ originate from a process where momentum conservation is satisfied by two phonons with opposite wave vectors, no defects are required for their activation, and are thus always present[38]. The 2D peak is composed of a single Lorentzian in SLG, whereas it splits into several components as N increases, reflecting the evolution of the electronic band structure[38].

In disordered carbons, the G peak position, Pos(G), increases as the excitation wavelength decreases from infrared to ultraviolet[44]. Therefore, the dispersion of the G peak, Disp(G), i.e. the rate of change of Pos(G) with the laser excitation wavelength, increases with disorder[44]. Similar to Disp(G), also the full width at half maximum of the G peak, FWHM(G), always increases with disorder[44]. The analysis of the intensity ratio of the D and G peaks, $I(D)/I(G)$, combined with that of FWHM(G) and Disp(G) allows us to discriminate between disorder localized at the edges and disorder in the bulk of the samples. In the latter case, a higher $I(D)/I(G)$ would correspond to higher FWHM(G) and Disp(G).

Raman measurements are done on the same set of samples to characterize the quality of the graphitic material. Figure 3a,b show a small correlation between these parameters. This implies that the D peak is mostly due to edges, as well as the presence of some defects in the samples. FWHM(2D) (Fig. 3d) is larger with respect to that of graphene flakes produced by micromechanical cleavage, but still has a single Lorentzian line shape. This is consistent with the presence of MLGs composed of folded or juxtaposed SLGs, but still electronically decoupled[17].

Figure 4a plots the Raman spectra of a graphene–azobenzene film after visible and UV light illumination, acquired at 514.5 nm with power $<250\,\mu W$ to avoid any Raman laser-induced isomerization of azobenzene. None of the graphene Raman

parameters show significant change on UV or visible light irradiation indicating negligible change in the defects and doping.

The *trans*-to-*cis* isomerization can also be monitored by Raman spectroscopy[24,48]. Figure 4b plots the Raman spectrum of azobenzene after both visible and UV light illumination. The spectrum is composed of five main peaks, labelled C1–C5 (ref. 48). These can also be seen on the spectra of the graphene–azobenzene hybrids (Fig. 4c). The intensity ratio of peaks C3 and C4 can be used to monitor the isomerization of azobenzene as described in ref. 48. On illumination with visible light, the photochromic molecules fall into the *trans* configuration and $I(C3)/I(C4)$ approaches ~ 1. When the sample is illuminated with UV light, azobenzene molecules undergo *trans*-to-*cis* isomerization and $I(C3)/I(C4)$ decreases markedly. The effect is fully reversible and the *trans* form is recovered by a further visible light exposure as shown in Fig. 4d.

X-ray analysis. Powder X-ray diffraction is further used to characterize the structure of the graphene–azobenzene hybrid powder and to compare it with the control samples (Supplementary Note 5). Similarly to graphite, the powder prepared from LPE has a sharp peak at 26.7°, corresponding to an interlayer spacing of ~ 0.33 nm (Supplementary Fig. 11a). The X-ray diffraction patterns of powders prepared from graphene/*trans*-azobenzene and graphene/*cis*-azobenzene show new peaks at $2\theta = 12.3°$ ($d_{\text{spacing}} \sim 0.72$ nm) and 9.9° ($d_{\text{spacing}} \sim 0.89$ nm), respectively (Supplementary Fig. 11b,c). This confirms that *trans* and/or *cis*-azobenzenes can sandwich between graphene layers, thus increasing the overall spacing.

Electrical characterization. To probe the electrical properties of the graphene–azobenzene hybrid and, in particular, to explore the potential light-responsive nature of the material, we drop-cast a ~ 100-nm-thick graphene–azobenzene film on a $n^{++}Si/SiO_2$ substrates exposing pre-patterned interdigitated gold electrodes (channel length = 10 μm; Fig. 6a,b).

Typical I–V curves exhibit a nonlinear resistive behaviour, as shown in Fig. 5c. The conductivity of the graphene–azobenzene hybrid is lower than that measured in control devices, i.e.,

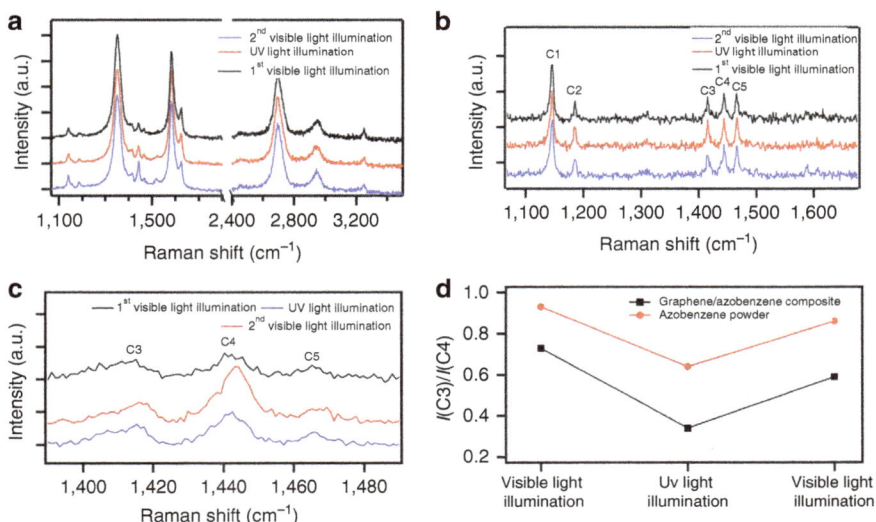

Figure 4 | Evolution of Raman spectra on illumination. (**a**) Comparison of the Raman spectrum of a graphene/4-(decyloxy) azobenzene hybrid film after 90-min illumination with visible light, UV light and visible light again to probe the effect of isomerization. (**b**) The same experiment on azobenzene powder. (**c**) 4-(decyloxy)azobenzene Raman peaks C3, C4 and C5 extracted from graphs in **a** showing the change of peaks ratio on illumination, which confirms the isomerization. (**d**) Dependence upon illumination of $I(C3)/I(C4)$ for both hybrid material and powders.

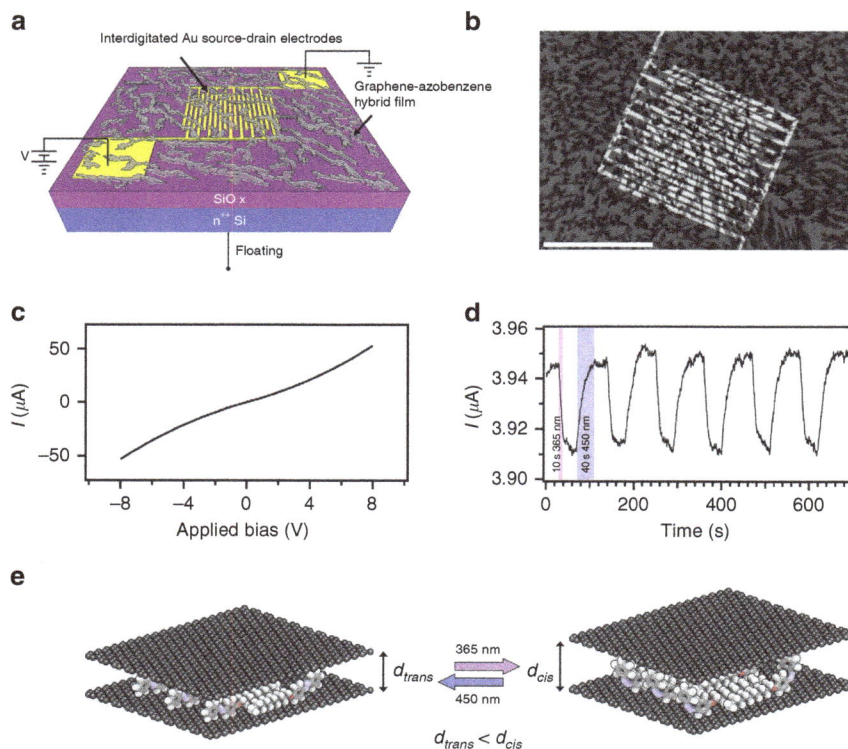

Figure 5 | Electrical characteristics of hybrid materials. (**a**) Scheme of the two-terminal device configuration, (**b**) scanning electron microscopy image of interdigitated Au electrodes covered with a hybrid film. Scale bar, 500 μm. (**c**) *I–V* characteristics, (**d**) optical modulation of current response for a static bias and dynamic alternative UV and visible light irradiation cycles. (**e**) Schematic of graphene-azobenzene hybrid undergoing UV and visible irradiation cycles.

comprising as active layer graphene exfoliated in NMP in the absence of azobenzene (Supplementary Fig. 12). Figure 6d shows the current modulation in a two-terminal device with graphene–azobenzene on cycles of UV and visible light irradiation, for 1 V applied between the two electrodes. One cycle consists of 10-s UV light (365 nm), 1-min rest in dark followed by 40-s visible light (450 nm) irradiation. Under UV exposure (magenta background) a current decrease is detected, which reversibly increases to about the initial value under visible light irradiation (violet background) over various cycles. No fatigue is observed during six cycles of photoisomerization. Conversely, the control devices show no response to light (Supplementary Fig. 12). Noteworthy, in some cases the increase of current in the hybrid films can be the result of π-electron interactions between the molecules and graphene[49]; however, such phenomenon, known as increased photo-conductivity, is irreversible and can only be modulated by varying the ratio between graphene and *ad hoc* molecules.

In the azobenzene–graphene film (Supplementary Figs 13 and 14), the molecules are physisorbed on top and in between the graphene sheets. Under UV irradiation, the azobenzene molecules undergo a conformational change from the less-bulky *trans* isomer to the bulkier *cis* isomer. Such a process is therefore accompanied by an increase in the few-layer graphene (FLG) inter-sheet distance, thereby hindering the charge transport via hopping between the sheets, resulting in lower conductivity of the hybrid film. The conductivity of the hybrid material can be fully restored by irradiating with visible light, causing the *cis*-to-*trans* photoisomerization.

To gain greater insight into the conduction mechanism and the effect of isomerization, we then perform Kelvin probe measurements (Supplementary Note 6). We find that the work function is reversibly photo-modulated over various cycles (Supplementary Fig. 15), providing further evidence for the isomerization of the

azobenzene molecules. This difference in work function has a direct effect on charge injection from the electrodes to the graphene/azobenzene conductive material, which is less favourable in the presence of the *cis* isomer. Light modulation can be applied multiple times to the systems, with no evidence of decrease in performance. We assign this to graphene's shielding of the azobenzene molecules from the outer environment. An ideal case consists of stacks of alternating graphene and azobenzenes, the latter having a thickness, which can be photo-modulated (Fig. 5e). In reality, the flakes (with average lateral sizes of 200 nm) form aggregates possessing a poor degree of order. Within these aggregates the azobenzenes are intercalated in between adjacent layers of graphene. The control samples show no change in work function on UV and visible light cycling (Supplementary Fig. 15).

Molecular modelling simulations. To gain a better understanding of the interactions of the azobenzene molecules with graphene, we perform molecular dynamics simulations (see Supplementary Note 7 for details). All calculations are done using the Groningen Machine for Chemical Simulations (GROMACS)[50] package and a modified version of the all-atom Optimized Potentials for Liquid Simulations (OPLSAA) force field developed for azobenzenes[51,52]. To assess the relative adsorption affinity of the *trans* versus *cis* conformations, we first compute the interaction energy for a single azobenzene molecule by running a 1-ns simulation in vacuum. The graphene layer is taken as an infinite rigid body in these calculations. The computed interaction energies of -130.1 ± 7.5 and $-115.7 \pm 8.0\,\mathrm{kJ\,mol^{-1}}$ for *trans*- and *cis*-azobenzene, respectively, are consistent with the expected stronger adsorption of *trans*-azobenzene, resulting from its planar geometry and the concomitant maximized π–π interactions between the phenyl

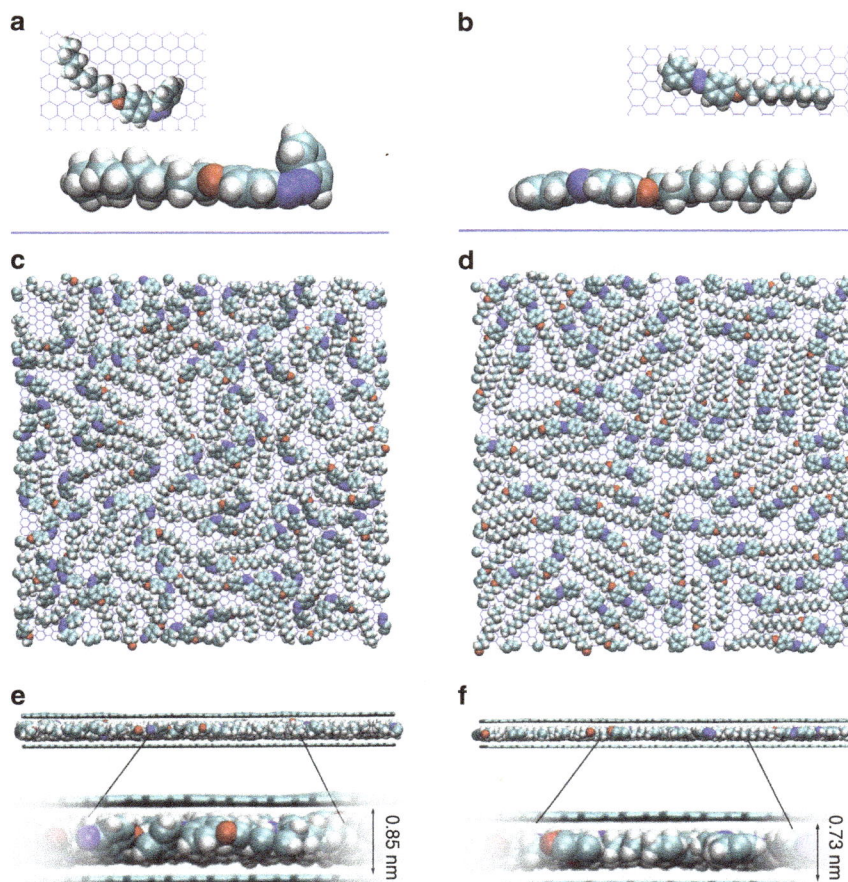

Figure 6 | Molecular modelling of graphene interaction with *cis* and *trans* azobenzene. Side and top views of a single (**a**) *cis*- and (**b**) *trans*-azobenzene molecule adsorbed on SLG. Top view of the (**c**) *cis* and (**d**) *trans*-monolayer, and side view of the molecular system consisting of the (**e**) *cis* and (**f**) *trans*-monolayer sandwiched between two SLGs, with the indication of the SLGs separation.

groups and graphene. The average adsorption distance for the *trans* molecule is ∼0.07 nm shorter than *cis* (0.34 versus 0.41 nm). Figure 6a,b plots top and side views of the adsorbed *cis* and *trans* molecules from the last snapshots along the molecular dynamics trajectories.

It should be noted that the conformation giving the higher exfoliation yield is the *cis* one, with lower interaction energy with graphene (14.4 kJ mol^{-1} difference). This counterintuitive finding highlights the importance of kinetic factors on exfoliation yield rather than thermodynamic factors, in good agreement with previous results obtained with rigid molecules[29,30]. The 70% increase in exfoliation yield could be due to the rapid, dynamic change of conformation between *cis* and *trans* that the molecules undergo during the exfoliation, which is more rapid at high temperature and under sonication (Fig. 2).

We then study the supramolecular organization of *trans*- and *cis*-azobenzenes on SLG in vacuum. In the absence of experimental structural data, we consider highly regular self-assembled monolayers (Supplementary Fig. 16) and test their stability by performing simulations of the assemblies at room temperature. The regular assemblies are unstable at 300 K and disassemble after few tens of picoseconds, suggesting that the azobenzene molecules form disordered layers at the SLG surface. To investigate the formation and structure of such disordered layers, we use the following protocol. Azobenzene molecules are introduced one by one in the simulation box at a distance of 1.8 nm above the plane of the infinite SLG with a random orientation. The initial atomic velocities are assigned so that the resulting vectors point towards the SLG, thereby prompting

physical adsorption on the surface. Room-temperature molecular dynamics simulations of a few hundreds of picoseconds are then performed to explore the configurational space after each molecular deposition, and the last saved configuration is used as input for the landing of the subsequent molecule. The scheme is iterated multiple times until the desired surface coverage is reached. A 5-ns simulation is then performed on the final structure to equilibrate the monolayer at room temperature. A more detailed description of the methodology is provided in Supplementary Note 8. Full coverage of the graphene periodic layer of 115.8 nm^2 is reached when 64 (72) molecules of *trans* (*cis*)-azobenzene are adsorbed, which translates into a surface contact area on graphene of ∼1.8 nm^2 for the *trans* and ∼1.6 nm^2 for the *cis* conformation. Figure 6c,d plots a top view of the azobenzene monolayers obtained for the two conformers in the case of full coverage. Supplementary Figs 17 and 18 show the formation of those monolayers as a function of the number of molecules adsorbed on SLG.

To validate the hypothesis that the insertion of a *trans*-/*cis*-monolayer sandwiched between two SLGs can result in different interlayer spacing according to the ideal model, a second SLG is placed on top of the azobenzene monolayers prepared using the deposition protocol described above. The 1-ns simulations are then done at room temperature. The results show that *cis*-azobenzenes maintain the SLG at a distance ∼0.85 nm, while the layer separation decreases to ∼0.73 nm for *trans* (Fig. 6e,f), in excellent agreement with the X-ray data. This supports the view that photoisomerization of the molecules from *trans* to *cis* can favour the exfoliation process. To investigate the effect of this switchable separation on charge transport, we perform quantum chemical calculations (at the intermediate neglect of differential overlap level[53]) on two square-shaped nanographenes ($\sim 20\,nm^2$) interacting either through-space or in the presence of the sandwiched (*trans* or *cis*) azobenzenes (Supplementary Fig. 19). In the first case (no sandwiched layer), the electronic couplings mediating positive and negative charge carrier tunnelling from one nanographene to the other, as obtained from the diabatization scheme reported in ref. 54, decrease by a factor ~ 100 when the interlayer distance is increased from 0.73 (corresponding to the azobenzenes in *trans* conformation) to 0.85 nm (*cis* conformation). In both cases, these interactions are very small due to the fast fall-off decay of the nanographene wavefunction overlap. Interestingly, the corresponding couplings calculated in the presence of the sandwiched azobenzenes are several orders of magnitude larger and show a less pronounced change (by a factor ~ 10) on photoisomerization, suggesting that the loss in electrical transmission associated with the larger inter-graphene distances is at least partly compensated by conformational-dependent superexchange effects involving the molecular orbitals of the azobenzenes. Our experimental current modulation on photoisomerization of azobenzenes (Fig. 5d) is much smaller than the calculated one (factor ~ 100). This can be understood considering that the simulations are run for two SLG intercalated by a tightly packed, physisorbed, self-assembled layer of azobenzene, whereas the experiments are done for hybrid films with a distribution of layer thicknesses (Supplementary Fig. 7a) separated by a small quantity of azobenzene (below 5% of the area of graphene).

As aforementioned the ideal model does not fully describe our real system that consists of poorly ordered aggregates comprising flakes intercalated with azobenzenes, and charges hopping between adjacent graphenes. The distance between the adjacent graphene sheets can be varied on photoisomerization of the azobenzenes.

Discussion

We demonstrated that alkoxy-substituted photochromic molecules can act as photo-addressable surfactant and as dispersion-stabilizing agents to enhance the yield of exfoliation in an upscalable molecule-assisted LPE-based method. The simultaneous use of UV light, promoting the *trans*-to-*cis* isomerization, as well as thermal annealing at 40 °C and mechanical forces generated by sonication, both favouring *cis*-to-*trans* isomerization, promotes the exfoliation in liquid media. The most effective exfoliation is obtained with azobenzene molecules irradiated with UV light in NMP at 40 °C, with a concentration of exfoliated graphene of 110 µg ml^{-1}. This corresponds to an $\sim 80\%$ increase in exfoliation yield when compared with pure NMP (63 µg ml^{-1}). By depositing the hybrid film onto Au pre-patterned SiO$_2$ substrates, light-responsive thin hybrid films, formed in a one-step co-deposition process, can be realized, whose conductivity can be reversibly modulated by the *trans*–*cis* photoisomerization of the azobenzenes. By combining this approach with cost-effective techniques, such as ink-jet printing, more complex responsive device designs and architectures may be realized. This paves the way to future applications such as optically controllable memory switch elements for light-assisted programming and photosensors.

Methods

Materials. Graphite synthetic flakes (Product No. 332461), 4-(decyloxy) azobenzene (Product No. S931950) and NMP (Product No. 270458) are sourced from Sigma-Aldrich.

Liquid-phase exfoliation. Graphene dispersions are prepared by adding graphite powder (100 mg) to NMP (10 ml), and tip-sonication (Labsonic M, Platinum tip, diameter 2 mm, sound rating density 300 W cm^{-2}) for 3 h either in dark or under UV light irradiation, using a portable laboratory ultraviolet lamp (8 W, 365 nm, 0.34 mW cm^{-2}, Herolab GmbH), in glass vials (Pyrex), in the presence of azobenzene molecules (5.5 mg). The dispersions are then allowed to settle for 15 min, then decanted and centrifuged (Eppendorf, centrifuge 5804) for 1 hour at 10,000 r.p.m. From the centrifuged dispersions 70 vol% are pipetted off the top for characterization and film deposition. To quantify the concentration of flakes after centrifugation, the dispersions are passed through polytetrafluoroethylene membrane filters (pore size 100 nm). Measurements of the filtered mass are performed with a microbalance (Sartorius MSA2.75)

Device fabrication. n^{++}Si substrates with a thermally grown SiO$_2$ layer (230 nm) and pre-patterned interdigitated gold source and drain electrodes (IPMS Fraunhofer) with different channel length ($L = 2.5, 5, 10$ and $20\,\mu m$) and constant channel width ($W = 10\,mm$) are used. The substrates are cleaned before device fabrication in an ultrasonic bath (FB 15047, Fisher Scientific) of acetone and isopropanol, 15 min in each solvent, and treated 5 min ($+30$ min incubation) with an UV surface decontamination system (PSD-UV, Novascan) to improve wetting of the solvent. The dispersions (50 µl) are drop-cast on clean substrates and dried for 48 h in vacuum at 30 °C.

Characterization. The electrical characterization is carried out in inert atmosphere (glovebox) with an electrometer (Keithley 2636A) interfaced with LabTracer software. For the light-induced switch, a Polychrome V monochromator (Till Photonics) is used as ultraviolet light (350 nm, 5.64 mW cm^{-2}) and visible light source (450 nm, 4.83 mW cm^{-2}). For the Raman experiments, graphene dispersions are drop-cast onto pre-cleaned n^{++}Si substrates with a thermally grown SiO$_2$ layer (300 nm) and dried for 48 h. Raman spectra are collected with a Renishaw InVia spectrometer at 457, 514.5 and 633 nm. The excitation power is kept below 1 mW to avoid the effects of local heating. The scattered light is collected with a $\times 100$ objective. HR-TEM is done in a FEI Tecnai F20 equipped with a Schottky emitter and operated at 120 keV primary beam energy. Scanning electron microscopy images are recorded with a Quanta FEG 250 from FEI.

The thickness of the hybrid films is determined with an Alpha-Step IQ Surface Profiler from KLA Tencor. X-ray photoelectron spectroscopy analyses are carried out on a Thermo Scientific K-Alpha X-ray photoelectron spectrometer, with a basic chamber pressure $\sim 10^{-8}$ mbar and an Al anode as the X-ray source (X-ray radiation of 1,486 eV). Spot sizes of 400 µm, pass energies of 200 eV for survey scans and 50 eV for high-resolution scans are used. A volume of 150 µl of the dispersions are spin-coated on Au substrate for 1 min at 1,000 r.p.m. and substrates are annealed for 1 day at 100 °C in a oven under vacuum. Samples for X-ray diffraction are prepared by precipitating graphene and/or graphene/azobenzene by adding water:ethanol (1:1, vol:vol) into NMP. The collected precipitate is dried under vacuum for 24 h. The powder X-ray diffraction patterns are obtained using a Bruker AXS D2 Phaser (LYNXEYE detector) with Ni-filtrated Cu-Kα radiation ($\lambda = 1.5406$ Å) with a 1-mm air-scattering slit and 0.1-mm equatorial slit. Samples are deposited on the surface of a single crystal Si wafer (cut of (911)). X-ray diffraction patterns are collected with 0.016° steps and 10 s per step increments from 8° to 80°.

References

1. Geim, A. K. & Novoselov, K. S. The rise of graphene. *Nat. Mater.* **6,** 183–191 (2007).
2. Ferrari, A. C. *et al.* Science and technology roadmap for graphene, related two-dimensional crystals, and hybrid systems. *Nanoscale* **7,** 4598–4810 (2015).
3. Bonaccorso, F. *et al.* Production and processing of graphene and 2d crystals. *Mater. Today* **15,** 564–589 (2012).
4. Narita, A. *et al.* Synthesis of structurally well-defined and liquid-phase-processable graphene nanoribbons. *Nat. Chem.* **6,** 126–132 (2014).
5. Hernandez, Y. *et al.* High-yield production of graphene by liquid-phase exfoliation of graphite. *Nat. Nanotechnol.* **3,** 563–568 (2008).
6. Chen, L., Hernandez, Y., Feng, X. L. & Müllen, K. From nanographene and graphene nanoribbons to graphene sheets: chemical synthesis. *Angew. Chem. Int. Ed.* **51,** 7640–7654 (2012).
7. Palma, C.-A. & Samorì, P. Blueprinting macromolecular electronics. *Nat. Chem.* **3,** 431–436 (2011).
8. Kim, K. S. *et al.* Large-scale pattern growth of graphene films for stretchable transparent electrodes. *Nature* **457,** 706–710 (2009).
9. Li, X. S. *et al.* Large-area synthesis of high-quality and uniform graphene films on copper foils. *Science* **324,** 1312–1314 (2009).
10. Berger, C. *et al.* Electronic confinement and coherence in patterned epitaxial graphene. *Science* **312,** 1191–1196 (2006).
11. Ciesielski, A. & Samorì, P. Graphene via sonication assisted liquid-phase exfoliation. *Chem. Soc. Rev.* **43,** 381–398 (2014).

12. Parvez, K. *et al.* Electrochemically exfoliated graphene as solution-processable, highly conductive electrodes for organic electronics. *ACS Nano* **7**, 3598–3606 (2013).

13. Parvez, K. *et al.* Exfoliation of graphite into graphene in aqueous solutions of inorganic salts. *J. Am. Chem. Soc.* **136**, 6083–6091 (2014).

14. León, V. *et al.* Few-layer graphenes from ball-milling of graphite with melamine. *Chem. Commun.* **47**, 10936–10938 (2011).

15. Sampath, S. *et al.* Direct exfoliation of graphite to graphene in aqueous media with diazaperopyrenium dications. *Adv. Mater.* **25**, 2740–2745 (2013).

16. Coleman, J. N. Liquid exfoliation of defect-free graphene. *Acc. Chem. Res.* **46**, 14–22 (2013).

17. Torrisi, F. *et al.* Inkjet-printed graphene electronics. *ACS Nano* **6**, 2992–3006 (2012).

18. Marago, O. M. *et al.* Brownian motion of graphene. *ACS Nano* **4**, 7515–7523 (2010).

19. Ciesielski, A. *et al.* Liquid-phase exfoliation of graphene using intercalating compounds: a supramolecular approach. *Angew. Chem. Int. Ed.* **53**, 10355–10361 (2014).

20. Haar, S. *et al.* A supramolecular strategy to leverage the liquid-phase exfoliation of graphene in the presence of surfactants: unraveling the role of the length of fatty acids. *Small* **11**, 1691–1702 (2014).

21. Luo, W. *et al.* A high energy density azobenzene/graphene hybrid: a nano-templated platform for solar thermal storage. *J. Mater. Chem. A* **3**, 11787–11795 (2015).

22. Peimyoo, N. *et al.* Photocontrolled molecular structural transition and doping in graphene. *ACS Nano* **6**, 8878–8886 (2012).

23. Seo, S., Min, M., Lee, S. M. & Lee, H. Photo-switchable molecular monolayer anchored between highly transparent and flexible graphene electrodes. *Nat. Commun.* **4**, 1920 (2013).

24. Kim, M., Safron, N. S., Huang, C. H., Arnold, M. S. & Gopalan, P. Light-driven reversible modulation of doping in graphene. *Nano Lett.* **12**, 182–187 (2012).

25. Margapoti, E. *et al.* Emergence of photoswitchable states in a graphene–azobenzene–Au platform. *Nano Lett.* **14**, 6823–6827 (2014).

26. Mativetsky, J. M. *et al.* Azobenzenes as light-controlled molecular electronic switches in nanoscale metal-molecule-metal junctions. *J. Am. Chem. Soc.* **130**, 9192–9193 (2008).

27. Feng, Y. Y., Zhang, X. Q., Ding, X. S. & Feng, W. A light-driven reversible conductance switch based on a few-walled carbon nanotube/azobenzene hybrid linked by a flexible spacer. *Carbon* **48**, 3091–3096 (2010).

28. Rabe, J. P. & Buchholz, S. Commensurability and mobility in 2-dimensional molecular-patterns on graphite. *Science* **253**, 424–427 (1991).

29. Schlierf, A. *et al.* Nanoscale insight into the exfoliation mechanism of graphene with organic dyes: effect of charge, dipole and molecular structure. *Nanoscale* **5**, 4205–4216 (2013).

30. Yang, H. F. *et al.* Dielectric nanosheets made by liquid-phase exfoliation in water and their use in graphene-based electronics. *2D Mater.* **1**, 011012 (2014).

31. Griffiths, J. Photochemistry of azobenzene and its derivatives. *Chem. Soc. Rev.* **1**, 481–493 (1972).

32. Tamai, N. & Miyasaka, H. Ultrafast dynamics of photochromic systems. *Chem. Rev.* **100**, 1875–1890 (2000).

33. Bandara, H. M. D. & Burdette, S. C. Photoisomerization in different classes of azobenzene. *Chem. Soc. Rev.* **41**, 1809–1825 (2012).

34. Joseph, J. M., Destaillats, H., Hung, H. M. & Hoffmann, M. R. The sonochemical degradation of azobenzene and related azo dyes: rate enhancements via Fenton's reactions. *J. Phys. Chem. A* **104**, 301–307 (2000).

35. Turansky, R., Konopka, M., Doltsinis, N. L., Stich, I. & Marx, D. Switching of functionalized azobenzene suspended between gold tips by mechanochemical, photochemical, and opto-mechanical means. *Phys. Chem. Chem. Phys.* **12**, 13922–13932 (2010).

36. Surampudi, S. K., Patel, H. R., Nagarjuna, G. & Venkataraman, D. Mechano-isomerization of azobenzene. *Chem. Commun.* **49**, 7519–7521 (2013).

37. Shin, K. H. & Shin, E. J. Photoresponsive azobenzene-modified gold nanoparticle. *Bull. Korean Chem. Soc.* **29**, 1259–1262 (2008).

38. Ferrari, A. C. *et al.* Raman spectrum of graphene and graphene layers. *Phys. Rev. Lett.* **97**, 187401–187405 (2006).

39. Ferrari, A. C. & Robertson, J. Raman spectroscopy of amorphous, nanostructured, diamond-like carbon, and nanodiamond. *Philos. Trans. R. Soc. Lond. Ser. A* **362**, 2477–2512 (2004).

40. Ferrari, A. C. & Basko, D. M. Raman spectroscopy as a versatile tool for studying the properties of graphene. *Nat. Nanotechnol.* **8**, 235–246 (2013).

41. Tuinstra, F. & Koenig, J. L. Raman spectrum of graphite. *J. Chem. Phys.* **53**, 1126–1130 (1970).

42. Tan, P. H. *et al.* The shear mode of multilayer graphene. *Nat. Mater.* **11**, 294–300 (2012).

43. Lui, C. H. *et al.* Observation of layer-breathing mode vibrations in few-layer graphene through combination Raman scattering. *Nano Lett.* **12**, 5539–5544 (2012).

44. Ferrari, A. C. & Robertson, J. Resonant Raman spectroscopy of disordered, amorphous, and diamondlike carbon. *Phys. Rev. B* **64**, 075414–075426 (2001).

45. Ferrari, A. C. & Robertson, J. Interpretation of Raman spectra of disordered and amorphous carbon. *Phys. Rev. B* **61**, 14095–14107 (2000).

46. Thomsen, C. & Reich, S. Double resonant Raman scattering in graphite. *Phys. Rev. Lett.* **85**, 5214–5217 (2000).

47. Piscanec, S., Lazzeri, M., Mauri, F., Ferrari, A. C. & Robertson, J. Kohn anomalies and electron-phonon interactions in graphite. *Phys. Rev. Lett.* **93**, 185503 (2004).

48. Zheng, Y. B. *et al.* Surface-enhanced Raman spectroscopy to probe reversibly photoswitchable azobenzene in controlled nanoscale environments. *Nano Lett.* **11**, 3447–3452 (2011).

49. Chunder, A., Pal, T., Khondaker, S. I. & Zhai, L. Reduced graphene oxide/copper phthalocyanine composite and its optoelectrical properties. *J. Phys. Chem. C* **114**, 15129–15135 (2010).

50. Abraham, M. J. *et al.* GROMACS: high performance molecular simulations through multi-level parallelism from laptops to supercomputers. *SoftwareX* **1**, 19–25 (2015).

51. Schafer, L. V., Muller, E. M., Gaub, H. E. & Grubmuller, H. Elastic properties of photoswitchable azobenzene polymers from molecular dynamics simulations. *Angew. Chem. Int. Ed.* **46**, 2232–2237 (2007).

52. Pipolo, S. *et al.* First-principle-based MD description of azobenzene molecular rods. *Theor. Chem. Acc.* **131**, 1274–1287 (2012).

53. Ridley, J. & Zerner, M. An intermediate neglect of differential overlap technique for spectroscopy: pyrrole and the azines. *Theor. Chem. Acc.* **32**, 111–134 (1973).

54. Kondov, I., Cizek, M., Benesch, C., Wang, H. B. & Thoss, M. Quantum dynamics of photoinduced electron-transfer reactions in dye-semiconductor systems: first-principles description and application to coumarin 343-TiO$_2$. *J. Phys. Chem. C* **111**, 11970–11981 (2007).

Acknowledgements

We acknowledge funding from the European Commission through the Graphene Flagship (GA-604391), the Marie Curie project GREAT (PIEF-2011-GA-298246), the FP7-NMP-2012 project SACS (GA-310651), the Agence Nationale de la Recherche through the LabEx project Nanostructures in Interaction with their Environment (ANR-11-LABX-0058_NIE), the International Center for Frontier Research in Chemistry (icFRC), the Belgian National Fund for Scientific Research (FNRS), the FNRS/FRFC 'PHOTOGRAPH' PDR project, the Belgian Science Policy Office (IAP 7/05), the ERC synergy grant Hetero2D, ERC PoC HiGRAPHINK, and the Royal Society Wolfson Research Merit Award, EP/K01711X/1, EP/K017144/1 and EP/L016087/1. D.B. is a FNRS Research Director. M. Eredia, M. El Garah and A. Galanti are acknowledged for their assistance in X-ray photoelectron spectroscopy and high-performance liquid chromatography measurements.

Author contributions

P.S. and A.C. conceived the experiments and designed the study; M.D. and S.H. participated in the planning of the study and carried out the exfoliation (LPE) and characterization in solution (ultraviolet); M.D. performed the electrical characterization; S.H. and A.C. performed spectroscopic characterization; F.R. and S.H. performed the X-ray photoelectron spectroscopy experiments; M.D. and T.M. performed Kelvin probe measurements; M.B. and A.C.F. performed and analysed Raman experiments; S.O., A.M., R.L. and D.B. conducted theoretical studies; R.M. and V.M. analysed the samples with TEM; E.A.P. and L.D.C. performed and analysed X-ray diffraction experiments; all authors discussed results and contributed to the interpretation of data; D.B., A.C.F., A.C. and P.S. co-wrote the paper; all authors contributed to editing the manuscript.

Additional information

Creating single-atom Pt-ceria catalysts by surface step decoration

Filip Dvořák[1,*], Matteo Farnesi Camellone[2,*], Andrii Tovt[1], Nguyen-Dung Tran[2,3], Fabio R. Negreiros[2,†], Mykhailo Vorokhta[1], Tomáš Skála[1], Iva Matolínová[1], Josef Mysliveček[1], Vladimír Matolín[1] & Stefano Fabris[2,3]

Single-atom catalysts maximize the utilization of supported precious metals by exposing every single metal atom to reactants. To avoid sintering and deactivation at realistic reaction conditions, single metal atoms are stabilized by specific adsorption sites on catalyst substrates. Here we show by combining photoelectron spectroscopy, scanning tunnelling microscopy and density functional theory calculations that Pt single atoms on ceria are stabilized by the most ubiquitous defects on solid surfaces—monoatomic step edges. Pt segregation at steps leads to stable dispersions of single Pt^{2+} ions in planar PtO_4 moieties incorporating excess O atoms and contributing to oxygen storage capacity of ceria. We experimentally control the step density on our samples, to maximize the coverage of monodispersed Pt^{2+} and demonstrate that step engineering and step decoration represent effective strategies for understanding and design of new single-atom catalysts.

[1]Charles University in Prague, Faculty of Mathematics and Physics, V Holešovičkách 2, Prague 18000, Czech Republic. [2]CNR-IOM DEMOCRITOS, Istituto Officina dei Materiali, Consiglio Nazionale delle Ricerche, Via Bonomea 265, Trieste 34136, Italy. [3]SISSA, Scuola Internazionale Superiore di Studi Avanzati, Via Bonomea 265, Trieste 34136, Italy. * These authors contributed equally to this work. † Present address: Universidade Federal do ABC. Av. dos Estados, 5001 Bairro Bangu, Santo André SP CEP 09210-580, Brasil. Correspondence and requests for materials should be addressed to J.M. (email: josef.myslivecek@mff.cuni.cz) or to S.F. (email: fabris@democritos.it).

Single-atom catalysts represent the limiting realization of supported metal catalysts with metal load ultimately dispersed as single atoms[1,2]. This maximizes the utilization of supported metals and helps development of sustainable catalytic technologies for renewable energies and environmental applications with reduced precious metal contents[3,4]. A central prerequisite for understanding and knowledge-based design of single-atom catalysts is the identification of specific adsorption sites on catalyst supports that provide the stabilization of single metal atoms under reaction conditions at elevated temperatures and pressures. For oxide supports, understanding specific adsorption sites presently concentrates on low-index oxide facets[5-9]. Single-atom catalysts are, however, nanostructured large-area materials; thus, a question arises whether single supported atoms can be stabilized at defect sites of nanostructured oxide supports.

Highly dispersed platinum (Pt) ions on ceria qualify as single-atom catalysts[2] and hold a promise of radical reduction of Pt load in critical large-scale catalytic applications—hydrogen production[3], three-way catalytic converters[10] and fuel cells[11]. Ceria surfaces provide a limited amount of low coordinated surface sites where Pt^{2+} ions can adsorb and remain stable in real applications[2,3,10,11]. Recent studies on large-area ceria samples identify the necessity of nanostructuring the ceria substrates for obtaining supported Pt^{2+} ions[2,4] and propose a square-planar PtO_4 unit as a Pt^{2+}-containing surface moiety[4]. In the present model study on the single crystalline $CeO_2(111)$ surface, we demonstrate

that single-ion dispersions of Pt^{2+} are stabilized at monolayer (ML)-high ceria step edges. Pt^{2+} ions at step edges are located in PtO_4 units that can be considered the elementary building blocks of Pt^{2+}/ceria single-atom catalysts. The PtO_4 units incorporate excess O and can act as oxygen source for redox reactions. Besides clarifying the nature of Pt^{2+} stabilization on ceria, our study demonstrates the importance of step edges—the most common surface defects on oxide supports[12]—for single-atom catalyst stabilization. We experimentally adjust the step density on the ceria supports for maximizing the load of monodispersed Pt^{2+} ions. This identifies step engineering[13] and step decoration[14,15] as advanced techniques for designing new single-atom catalysts.

Results

Pt deposits on highly defined $CeO_2(111)$ surfaces. The experiments were performed on model $CeO_2(111)$ surfaces prepared as 20 to 40 Å thick ceria films on Cu(111) using procedures that allow adjusting the density of ML-high steps[16] and the density of surface oxygen vacancies on ceria surface[17]. On these highly defined surfaces, we deposit 0.06 ML of Pt, anneal at 700 K in ultra-high vacuum (UHV) and observe stabilization of Pt^{2+} species and/or nucleation of Pt clusters with scanning tunnelling microscopy (STM) and with photoelectron spectroscopy (PES). Deposition and annealing of Pt on $CeO_2(111)$ surfaces containing low concentrations of defects— ML-high steps and surface oxygen vacancies (Fig. 1a)—yield

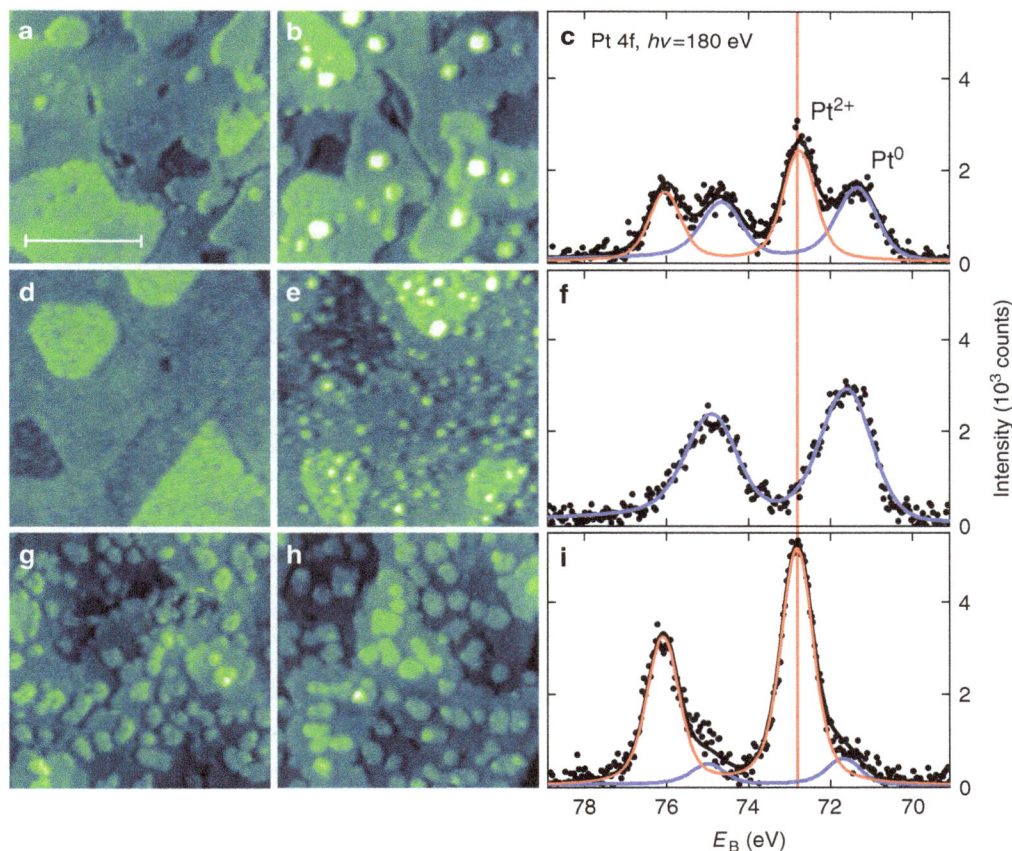

Figure 1 | Nucleation of Pt and stabilization of Pt^{2+} on ceria surfaces containing controlled amount of surface defects. (a-c) $CeO_2(111)$ surface with low density of surface oxygen vacancies and ML-high steps. **(d-f)** $CeO_{1.7}$ surface with increased density of surface oxygen vacancies. **(g-i)** $CeO_2(111)$ surface with increased density of ML-high steps. **(a,d,g)** STM images of clean surfaces before deposition of Pt. **(b,e,h)** STM images after deposition of 0.06 ML Pt and annealing at 700 K in UHV. All STM images 45 × 45 nm², tunnelling current 25–75 pA, sample bias voltage 2.5–3.5 V. Scale bar, 20 nm **(a)**. **(c,f,i)** PES spectra of the Pt deposit after annealing. All PES spectra were acquired with photon energy $hv = 180$ eV (black points). Fits indicate metallic (Pt^0, blue line) and ionic (Pt^{2+}, red line) contributions to Pt 4f signal. E_B is the photoelectron binding energy.

metallic Pt^0 clusters (Fig. 1b,c) coexisting with ionic Pt^{2+} species (Fig. 1c). To determine whether the charge of the supported Pt species is selectively induced by a specific defect type, we repeat the experiment varying independently the amount of surface O vacancies—up to 0.16 ML, creating $CeO_{1.7}$ surface (Fig. 1d–f)—and the amount of ML-high steps on the $CeO_2(111)$ surface—up to 0.15 ML (Fig. 1g–i). We observe that surface oxygen vacancies do not promote the dispersion of Pt^{2+} species but lead to small metallic Pt^0 clusters (Fig. 1e,f)[18]. On the other hand, the increased step density leads to almost complete oxidation of the Pt deposit to Pt^{2+} (Fig. 1h,i) proving that step edges selectively promote the stabilization of Pt^{2+} species. Detailed STM images allow to exclude formation of three-dimensional and two-dimensional PtO_x clusters (Supplementary Fig. 1), and allow to conclude that Pt^{2+} species are incorporated in the ceria step edges. Nucleation of Pt^0 clusters and stabilization of Pt^{2+} species represent concurrent processes. Differently to Pt^0 clusters (Fig. 1b,e), Pt^{2+} species at the step edges are not discernible in empty states STM imaging (cf. Fig. 1g,h without and with Pt deposit) because of their electronic structure. STM imaging in occupied states on metal-supported ceria is unavailable[19].

The possibility to adjust the density of ML-high steps on the model $CeO_2(111)$ surfaces[16] allows us to obtain a quantitative correlation between the step density and the amount of Pt^{2+} species. We prepare $CeO_2(111)$ samples with step density between 0.06 and 0.20 ML[16] and deposit 0.06 or 0.18 ML Pt at 300 K. Parameters of the prepared samples are summarized in Supplementary Table 1. After annealing at 700 K the amount of Pt stabilized in the form of Pt^{2+} is determined by PES. For quantification, all relevant parameters—the density of ceria steps, deposited amount of Pt and amount of stabilized Pt^{2+}—are expressed in ML where 1 ML corresponds to the density of Ce atoms on the $CeO_2(111)$ surface, that is, $7.9 \times 10^{14} \, cm^{-2}$. The density of steps is defined as the density of Ce atoms located at the ceria step edges[20].

The amount of Pt^{2+} ions as a function of the step density is plotted in Fig. 2a. For the higher Pt coverage 0.18 ML, the analysis reveals a linear dependence between the amount of stabilized Pt^{2+} ions and the step density (Fig. 2a, blue symbols), confirming the activation of Pt oxidation to Pt^{2+} and the localization of Pt^{2+} at the surface steps. The highest step density 0.20 ML allows converting up to 80% of the Pt deposit to Pt^{2+}. The degree of oxidation of the Pt deposit increases with decreasing the amount of deposited Pt to 0.06 ML (Fig. 2a, black symbols). In this case, up to 90% of Pt converts to Pt^{2+}. The concentrations of ceria step edges and the amount of Pt^{2+} stabilized on the surface obey a classical supply-and-demand scenario characteristic for single-atom catalysts[1,4,10]: when sufficient step edges, the amount of oxidized Pt^{2+} is limited by the amount of deposited Pt. Otherwise, the amount of oxidized Pt^{2+} is limited by the step density regardless of the amount of deposited Pt. The Pt deposit exceeding the available step sites cannot be oxidized and nucleates as metallic Pt^0 clusters on the surface. Besides the high oxidative power of the step edges towards Pt, our quantitative analysis reveals also the capacity of ceria step edges to accommodate a high density of Pt^{2+} ions. Up to 0.16 ML of Pt^{2+} ions can be stabilized by the sample containing 0.18 ML of Pt deposit and 0.20 ML of steps (Fig. 2a, blue symbols). This corresponds to 80% of the step-edge sites being occupied by Pt^{2+}. In the whole range of the step densities between 0.06 and 0.20 ML, the occupation of the step-edge sites by Pt^{2+} varies between 50 and 80%.

Stability and charge state of $Pt^{2+}/CeO_2(111)$ samples. The necessity of annealing the Pt deposit on ceria in UHV at 700 K for obtaining Pt^{2+} stabilization in our experiment indicates the

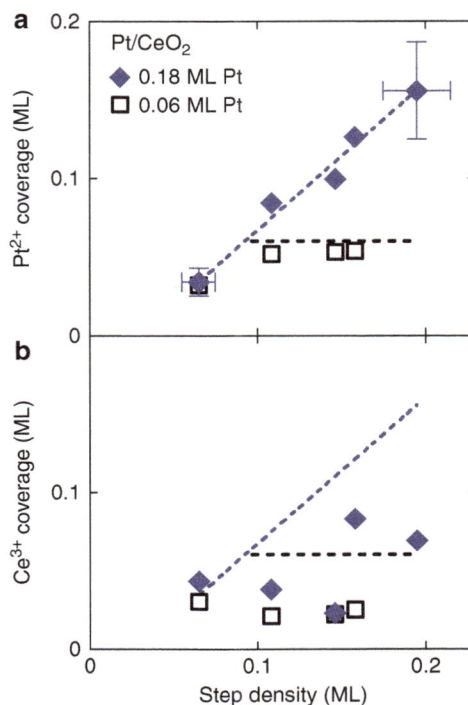

Figure 2 | Capacity of stepped $CeO_2(111)$ surface to accommodate Pt^{2+}. (a) Amount of Pt^{2+} stabilized on $CeO_2(111)$ substrates with different density of steps for 0.18 ML (blue symbols) and 0.06 ML (black symbols) of deposited platinum. Pt not stabilized in the form of Pt^{2+} remains metallic. Lines represent guides to the eyes. Blue line is a linear fit of 0.18 ML Pt data. Black line represents the maximum achievable amount of Pt^{2+} in the case of 100% oxidation of Pt for 0.06 ML of deposited Pt. (b) Reduction of the ceria surface accompanying the stabilization of Pt^{2+} ions determined by resonant PES expressed as a coverage of the surface by Ce^{3+} ions. Lines represent guides to the eyes from a and indicate the Pt^{2+} concentration. The Ce^{3+} concentration is lower or equal to the Pt^{2+} concentration on all samples.

activated nature of Pt segregation at the ceria steps and oxidation, and implies considerable thermal stability of Pt^{2+} ions on ceria. High-temperature annealing represents a prerequisite for obtaining Pt^{2+} ions also in the experiments on large-area nanostructured ceria samples[3,4]. Once created, Pt^{2+} ions remain stable on repeated annealing at 700 K in UHV. The Pt^{2+} ions in our experiment also remain stable on adsorption and thermal desorption of CO in UHV (Supplementary Fig. 2), or on exposure to air at ambient conditions (Supplementary Fig. 3).

Parallel to the charge state of the Pt deposit we determine the charge state of the CeO_2 support, in particular the concentration of surface Ce^{3+} ions that is indicative of reduction of the ceria surface. Contrary to the case of stabilizing Ni^{2+} ions on ceria[21], we observe that Pt oxidation during annealing is not accompanied by a corresponding reduction of $CeO_2(111)$ surface (Fig. 2b). This rules out the direct participation of ceria into the observed Pt oxidation at steps and indicates the involvement of other oxidizing agents in the Pt^{2+} stabilization, such as excess oxygen atoms. In the UHV environment of our experiments, the eligible source of excess oxygen can be water adsorbing in sub-ML amounts from background atmosphere (Supplementary Fig. 4) and undergoing dissociation on reduced ceria and Pt/ceria substrates[22,23]. In the large-area Pt^{2+}/CeO_2 catalysts displaying high concentration of Pt^{2+} ions and exceptional redox reactivity, excess O atoms may be incorporated during the synthesis that proceeds in air[3,10].

Segregation of Pt at CeO₂(111) step edges. *Ab initio* density functional theory (DFT) calculations allow to interpret the above experimental results. We calculate the segregation thermodynamics and the atomic and electronic structures of Pt atoms in representative adsorption sites on $CeO_2(111)$ surfaces. The results

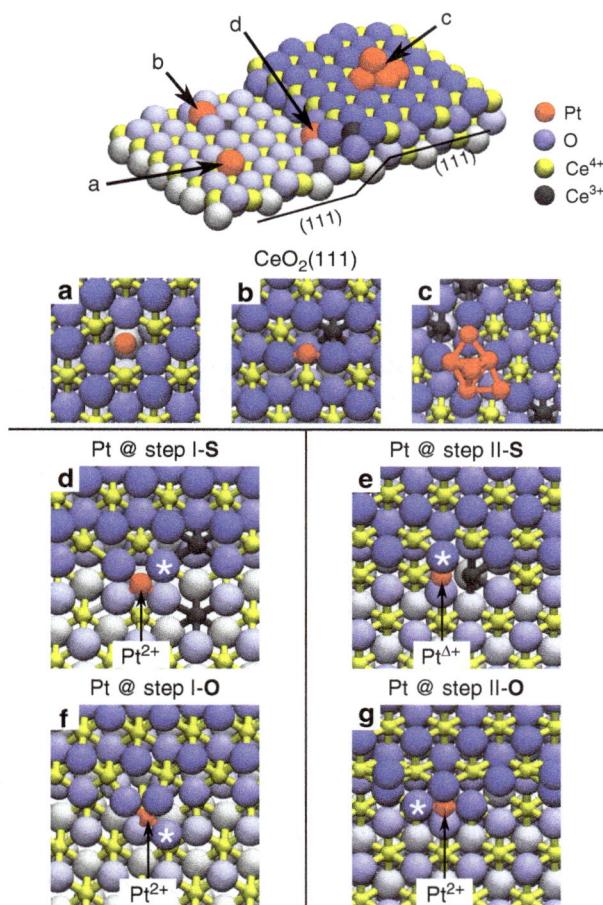

Pt
O
Ce⁴⁺
Ce³⁺

$CeO_2(111)$

Pt @ step I-S   Pt @ step II-S

Pt @ step I-O   Pt @ step II-O

Figure 3 | Pt adsorption sites on the CeO₂(111) surface obtained from DFT calculations. (**a**) Pt adatom in a surface O vacancy, (**b**) on the stoichiometric CeO₂(111) terrace and (**c**) supported Pt₆ cluster. (**d**) Pt adatom at the stoichiometric step I (step I-S) and (**e**) at the stoichiometric step II (step II-S). (**f**) Pt adatom at the step I with excess O (step I-O) and (**g**) at the step II with excess O (step II-O). Binding energies and Bader charges are summarized in Table 1. (**d-g**) The * symbol denotes the O atom removed to calculate the O vacancy formation energy reported in Table 2.

for the lowest-energy configurations are summarized in Fig. 3 and Table 1. The model adsorption sites include oxygen vacancies (Fig. 3a), regular sites (Fig. 3b) and Pt clusters[24] (Fig. 3c) on the $CeO_2(111)$ terrace, as well as two low-energy ML-high steps, which we label following ref. 25 as step I (Fig. 3d) and step II (Fig. 3e). A detailed list of the systems considered in the DFT analysis is reported in the Supplementary Note 1. The steps I and II represent the preferred types of steps at the $CeO_2(111)$ surfaces at temperatures <1,000 K (ref. 25). On our experimental samples, the steps I and II appear in equal proportion as evidenced from the absence of triangularly shaped islands in Fig. 1a,g[25,26].

In agreement with the experiment, our calculations predict the preferential segregation of Pt adatoms at the steps I and II, independently on the local step geometry and stoichiometry. The binding energies of Pt at the steps are 1.6–3.4 eV higher than at stoichiometric or defective (111) terraces (Table 1). This driving force for Pt segregation at the steps is in qualitative agreement with recent calorimetric studies of other metal clusters on $CeO_2(111)$[27,28], see the Supplementary Discussion. The particular binding energy and the charge state of Pt atoms at the steps depend on both the local step geometry and stoichiometry (Table 1). In the following, we show that a good agreement with all the experimental observations can only be achieved when considering segregation of Pt at steps in the presence of excess O (calculations denoted as O), while Pt segregation at stoichiometric steps (denoted S) exhibits significant discrepancies.

Pt at stoichiometric CeO₂(111) step edges. Pt segregation on step I-S yields Pt^{2+} species that are coordinated by four lattice O atoms in a characteristic PtO_4 planar unit (Fig. 3d). The PtO_4 unit is remarkably similar to that one proposed for Pt-doped ceria nanoparticles[4] and for surface reconstructions of Pd–ceria systems[29]. Instead, the different atomic structure of the step II-S edge prevents the formation of PtO_4 units, hinders the full Pt oxidation to Pt^{2+} and yields weakly oxidized $Pt^{\Delta+}$ species (Fig. 3e). Calculation results presented in Fig. 3d–g correspond to the Pt coverage at the steps 1/3 (1 Pt atom per 3 Ce step-edge atoms). For interpreting the capacity of the ceria step edges to accommodate a high density of Pt^{2+} ions, we calculate the adsorption of Pt at the ceria steps with increasing Pt coverage at the steps (Fig. 4), ranging from 1/3 to 1 (1 Pt atom per 1 Ce step-edge atom). On the step I-S, the maximum coverage of Pt^{2+} species is 2/3 (Fig. 4a). Higher Pt^{2+} coverages are unattainable and lead to nucleation of metallic Pt clusters, due to the large strain buildup resulting from long sequences of interconnected PtO_4 step units (Fig. 4b). On the step II-S, metallic Pt^0 species appear already for a coverage higher than 1/3 (Fig. 4c). Thus, on

Table 1 | Properties of Pt on CeO₂(111) obtained from DFT calculations.

Pt adsorption site	Binding energy (eV)	Formal charge	Bader charge (e)
Pt @ O vacancy	2.8	$Pt^{\Delta-}$	10.9
Pt @ CeO₂(111)	3.3	$Pt^{\Delta+}$	9.7
Pt₆ @ CeO₂(111)	4.4	Pt^0	9.9
Pt @ step I-S	5.0	Pt^{2+}	9.2
Pt @ step II-S	5.1	$Pt^{\Delta+}$	9.7
Pt @ step I-O	6.6	Pt^{2+}	8.6
Pt @ step II-O	6.7	Pt^{2+}	9.0
Pt bulk	5.5*	Pt^0	10.0

DFT, density functional theory; Pt, platinum; I-S and II-S, stoichiometric step types I and II; I-O and II-O, step types I and II with excess O atoms.
Results for the binding energies per atom, formal charges and Bader charges of Pt adatoms at the adsorption sites displayed in Fig. 3a–g. The binding energy for the Pt₆ cluster is the total binding energy divided by the number of Pt atoms. For reference, Pt bulk values are given in the last line.
*Bulk cohesive energy.

Pt step coverage

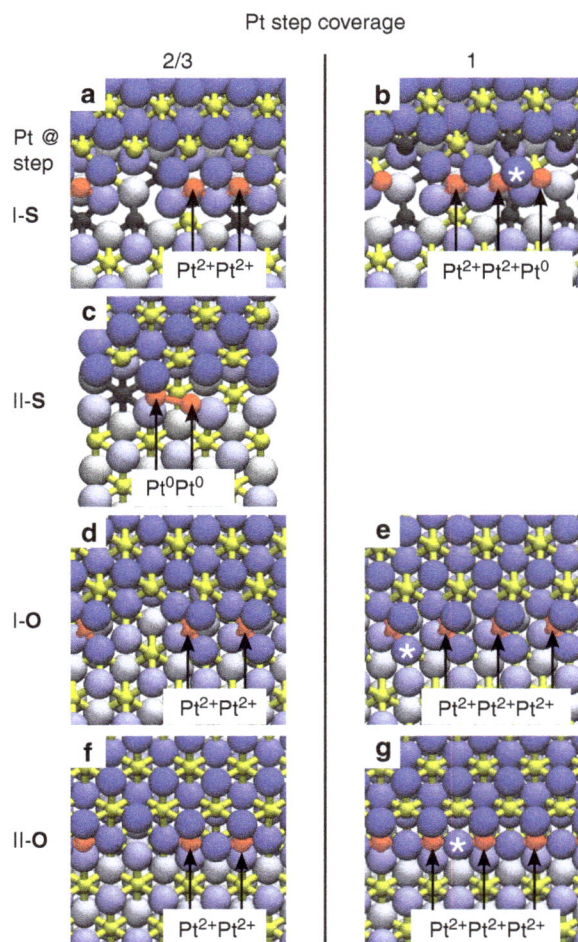

Figure 4 | Capacity of the CeO₂(111) step edges to accommodate Pt²⁺ ions obtained from DFT calculations. Calculated top views of the Pt binding to the steps I-S (**a,b**), step II-S (**c**), step I-O (**d,e**) and the step II-O (**f,g**) for Pt step coverage 2/3 (**a,c,d,f**) and 1 (**b,e,g**). At the step I-S, the limiting coverage of Pt²⁺ is 2/3 (**a**), additional Pt attaches to step edge as Pt⁰ (**b**). At the step II-S, the Pt²⁺ coverage is 0. Pt atoms attach as weakly ionized Pt^Δ+ and readily form metallic dimers (**c**) and clusters. On both steps I-O and II-O, excess oxygen can stabilize ionic Pt²⁺ at step edges as single ions appearing isolated or in groups up to 100% step coverage (**d–g**). The * symbol denotes the O atom removed to calculate the O vacancy formation energy reported in Table 2.

Table 2 | Minimum energy to remove an O atom from the CeO₂(111) surface obtained from DFT calculations.

O vacancy site	Pt step coverage	O vacancy formation energy (eV)
Pt @ step II-S	1/3	3.8
Pt @ step I-S	1/3	3.3
CeO₂(111)		2.5
step II-S	0	2.0
Pt @ step I-O	1/3	1.9
Pt @ step II-O	1/3	1.9
step I-S	0	1.8
Pt @ step I-O	1	1.8
Pt @ step II-O	1	1.7

DFT, density functional theory.
Results for steps I and II, and for Pt at steps I and II in the presence and absence of excess O. The energies for Pt @ steps were calculated by removing the O atom marked by the * symbol in Figs 3 and 4.

Pt²⁺ at CeO₂(111) step edges with excess O. Agreement between the theory and the experiment can be achieved when taking into account the step edges in the presence of an excess of O atoms. Irrespective of the local step geometry and Pt coverage at the steps, we find that excess O atoms readily bind to Pt at the ceria steps and drive a rearrangement of the step morphology forming ionized Pt²⁺ species incorporated in the planar PtO₄ moieties on both steps I and II (Fig. 3f,g). In the presence of excess of oxygen, Pt atoms bind stronger to the ceria step edges, with calculated binding energies up to 6.7 eV, which are higher than at the stoichiometric steps edges by ∼1.6 eV, and which are also higher than the cohesive energy of bulk metallic Pt (Table 1). This condition, which determines the stability of the Pt²⁺ species at steps with respect to metallic Pt clusters, is fulfilled only in the presence of excess oxygen at the steps. The computed electronic structure and density of states of the PtO₄ moieties at the steps I-O and II-O (Supplementary Figs 5 and 6) confirm that the Pt⁰ → Pt²⁺ oxidation results from the ionic Pt-O bond in the PtO₄ planar units, and that Ce³⁺ ions do not form in agreement with the experimental evidence (Fig. 2b). The calculated maximum coverage of Pt²⁺ at the steps I-O and II-O is 100% (Fig. 4e,g and Supplementary Table 2), as interconnected assemblies of the PtO₄ units can optimally fit the periodicities of both steps I and II at calculated Pt coverages at the step edges 1/3, 2/3 and 1 (Figs 3f,g and 4d–g). The presence of excess oxygen at steps therefore explains also the maximal Pt²⁺ ionization experimentally measured on the ceria-supported catalysts.

The stabilization of excess oxygen in the PtO₄ moieties by the Pt²⁺ ions suggests an oxygen source for redox reactions and hence provides a link between the presence of highly dispersed ionic Pt species on ceria and the increased redox reactivity of Pt²⁺/CeO₂ single-atom catalysts[3]. The oxygen buffering capacity of ceria-based catalysts is associated with easy oxygen vacancy formation. We calculate the vacancy formation energy on the clean CeO₂(111) terrace, on the stoichiometric steps and on the Pt-decorated steps (Table 2). Compared with the energy of 2.5 eV calculated on the CeO₂(111) terrace, the energies required to remove an oxygen atom bound to Pt at the stoichiometric steps are 3.3 eV (from the PtO₄ unit at the step I-S) and 3.8 eV (from the PtO₄ unit at the step II-S; see Table 2). Pt segregation at the stoichiometric step edges yields the formation of strong Pt-O bonds and therefore hinders ceria O-buffering. Much lower energies are instead needed to remove the excess O incorporated in the PtO₄ units at steps I-O and II-O, where the O vacancy formation energy can be as low as 1.7 eV, lower or comparable to the values for the stoichiometric steps without Pt (2.0–1.8 eV;

samples with equal proportion of the stoichiometric steps I and II, *ab initio* calculations predict maximum Pt²⁺ coverage at the steps (≤33% of the step-edge sites) and maximum conversion of the Pt deposit to Pt²⁺ (≤33% of deposited Pt) that are well below the experimental values (50–80% of step-edge sites, up to 90% of deposited Pt, *cf.* Fig. 2a).

Most importantly, the calculations on the stoichiometric steps predict that Pt segregation, oxidation and the formation of the Pt²⁺ species are always accompanied by the reduction of surface Ce atoms from Ce⁴⁺ to Ce³⁺ (denoted in gray in Figs 3 and 4). The resulting concentration of the Ce³⁺ ions exceeds that of the Pt²⁺ ions by a factor of 2. This is in stark contrast with the resonant PES measurements on our samples showing that the concentration of Ce³⁺ is considerably lower than the concentration of Pt²⁺ after annealing the samples (Fig. 2b). This indicates that Pt is preferentially oxidized by other mechanisms than the Pt⁰/Ce⁴⁺ redox couple.

Table 2)[30]. This indicates that the dispersed Pt^{2+} ions can enhance the oxygen storage capacity of ceria-based catalysts by assisting the reversible storage of excess O atoms.

The single-ion nature of Pt^{2+} in the PtO_4 units is preserved at all coverages, even when densely packed at the ceria step edges as interconnected PtO_4 units. Indeed, the Pt charge state, its local electronic structure and the O vacancy formation energy of the densely packed PtO_4 units are comparable to that of the isolated PtO_4 units at the steps (Supplementary Fig. 7 and Table 2). The PtO_4 units exhibit a large adaptability in stabilizing at different types of surface step edges, resulting in high effectiveness and capacity of the ceria surface to accommodate the Pt^{2+} ions. Regardless of the particular organization on the surface, the PtO_4 units are also always accessible to the reactants. Thus, the square-planar PtO_4 units carrying monodispersed Pt^{2+} ions can be considered elementary building blocks of single-atom Pt-ceria catalysts.

Discussion

On large-area samples, Pt^{2+} ions on ceria show exceptional reactivity with minimum Pt load in important applications: water–gas shift reaction[3], hydrogen oxidation on the anode of proton-exchange membrane fuel cell[11] and in the three-way catalyst converter[10]. In these applications, Pt^{2+} ions on ceria exhibit long-term stability under realistic reaction conditions of elevated temperatures and ambient pressure of reactant gases[3,31,32]. Pt^{2+} in large-area samples is routinely identified with PES. Complementary measurements with extended X-ray absorption fine structure (EXAFS)[33] and high-resolution transmission electron microscopy[3,4,31,32] confirm the absence of three-dimensional Pt or PtO_x clusters and, in agreement with the advanced PES measurements[34], identify Pt^{2+} as highly dispersed surface species on ceria[32].

Our present study identifies the stabilization of monodispersed Pt^{2+} ions with one particular defect site on the ceria surface—the monoatomic step edge—and excludes the stabilization of Pt^{2+} on the oxygen vacancies. Monodispersed Pt on ceria is observed to be effective in incorporating excess oxygen even in the unfavourable conditions of UHV experiment. Excess oxygen and Pt^{2+} arrange in the square-planar PtO_4 moieties decorating different types of the surface steps at coverages up to one PtO_4 per one step-edge Ce atom. The excess oxygen can be easily detached, indicating enhancement of the redox properties of ceria loaded with the Pt^{2+} ions. Adjusting the step density and the Pt load on the model $CeO_2(111)$ surface allows maximizing the coverage of Pt^{2+}, while suppressing the nucleation of metallic Pt^0 clusters. In the present experiment, we achieve surface coverage of Pt^{2+} 0.05 ML. A further increase of the completely monodispersed ionized Pt^{2+} coverage to 0.1 ML can be expected.

Step edges on ceria have been previously identified as preferred nucleation sites for supported metal clusters[20,21,27,28,35–37]. Our present study highlights the property of the step edges on ceria to provide specific structural and electronic environments for selective formation of monodispersed, thermally and chemically stable Pt^{2+} ions. The step edges represent intrinsic defects ubiquitously present on nanostructured ceria surfaces[4,38]; our results are thus applicable for the interpretation of the properties and the optimization of the Pt^{2+} load on large-area ceria supports[3,10,11]. More generally, the step edges may represent a common type of adsorption sites providing stabilization for monodispersed metal atoms and ions in any oxide-supported single-atom catalysts[15,39]. Our results therefore introduce important concepts of step reactivity[40] and step engineering[13,14] in understanding the stability, the activity and in designing new single-atom catalysts.

Methods

Experiment. The experiments were performed on surface science apparatuses in Surface Science Laboratory in Prague (STM, laboratory X-ray PES (XPS) with $hv = 1,487$ eV (Al Kα), low-energy electron diffraction) and at the Materials Science Beamline in Trieste (PES with $hv = 22$–$1,000$ eV (synchrotron), laboratory XPS with $hv = 1,487$ eV (Al Kα) and low-energy electron diffraction).

Preparation of the ceria substrates. The ceria layers and their Pt loading were prepared using the same procedures and parameters in both laboratories, and investigated by surface science methods in situ without exposing to air. The procedures and parameters of all samples are summarized in Supplementary Table 1. The ceria layers were prepared by deposition of Ce metal (Ce wire 99.9%, Goodfellow Cambridge Ltd) from Ta or Mo crucible heated by electron bombardment on clean Cu(111) substrate (MaTecK GmbH) in a background pressure of 5×10^{-5} Pa of O_2 (5.0, Linde AG). The growth rate of CeO_2 was 6 ML per hour. Varying densities of 1 ML high steps on the prepared CeO_2 layers were obtained by growth of CeO_2 at constant substrate temperature 423 or 523 K (Method I in Supplementary Table 1) or linearly increasing substrate temperature from room temperature to 723 K (Method II in Supplementary Table 1)[16]. For experiments in Fig. 1, the ordered, fully oxidized layer of CeO_2 (Fig. 1a,b) and the ordered reduced layer of $CeO_{1.7}$ (Fig. 1d,e) were obtained by approach published in ref. 17 that yields the lowest step density. In this approach, first, fully reduced Ce_2O_3 layer is prepared by depositing metallic Ce on a CeO_2 layer and annealing in vacuum. Subsequently, the Ce_2O_3 layer is exposed to a controlled dose of O_2 at 5×10^{-5} Pa and annealed to obtain desired stoichiometry $CeO_{1.7}$ or CeO_2 (Method III in Supplementary Table 1). The CeO_2 surface imaged in Fig. 1g,h was obtained by depositing 0.3 ML CeO_2 on the $CeO_2(111)$ substrate as in Fig. 1a, forming small ML-high islands. This homoepitaxy of CeO_2 on CeO_2 yields clearly arranged samples with high step density (Method IV in Supplementary Table 1).

Characterization of the ceria substrates. The thickness of the ceria layers was determined from the attenuation of the substrate Cu $2p_{3/2}$ XPS signal measured at $hv = 1,487$ eV. For calculations, we used inelastic mean free path of electrons in CeO_2 11.2 Å. The thickness of the ceria layers was set between 20 and 40 Å or 7 and 12 ML with 1 ML corresponding to 3.1 Å, the distance between Ce(111) atomic planes of CeO_2. In this range of thickness, the coverage of the Cu substrate by ceria ranges between 97 and 100 % (ref. 17). For determining the density of 1 ML high steps, we use a semi-automated procedure when the first step outlines are marked in STM images manually. Step outlines are then mapped onto a properly scaled and rotated hexagonal mesh of surface Ce atoms. The atoms that are closest to the outlines are automatically identified as step-edge atoms and their density evaluated in ML. The error in determining the density of 1 ML high steps is estimated to be ± 10 % and is marked in Fig. 2a.

Preparation of the Pt deposit. Pt was deposited on the ceria layers from a Pt wire (99.99%, MaTecK GmbH) heated by electron bombardment. Pt was deposited on the sample surface at 300 K and subsequently stabilized by increasing the sample temperature to 700 K at the rate $2\,K\,s^{-1}$. Both Pt deposition and annealing proceeded in the UHV background pressure 5×10^{-8} Pa or below. The thermal treatment supports the ionization of Pt to Pt^{2+}.

Characterization of the Pt deposit. The amount of Pt was calculated from the deposition time after calibrating the constant evaporation rate of the Pt evaporator. The evaporation rate was determined by a Quartz Crystal Microbalance and/or in a dedicated experiment from the thickness of 4-ML-thick Pt layers on $CeO_2(111)$/Cu(111) determined by attenuation of the substrate Cu $2p_{3/2}$ XPS signal measured at $hv = 1,487$ eV. This dedicated experiment was used to correlate Pt evaporation rates between the two experimental apparatuses. For calculations, we used inelastic mean free path of electrons in Pt 8 Å. The fraction of Pt^{2+} after thermal treatment was determined by fitting the ionic Pt^{2+} and neutral Pt^0 component in the PES Pt $4f$ spectrum measured at $hv = 180$ eV (cf. Fig. 1c,f,i). The error in determining the Pt and Pt^{2+} amounts on the studied samples is ± 20 % and is marked in Fig. 2a. This error represents the calibration error of the Pt evaporation rate.

Resonant PES. Reduction of the ceria surface after deposition of Pt and thermal treatment was determined with resonant PES of Ce $4f$ state. We determine the so-called resonant enhancement ratio (RER) as defined in refs 41,42 from measurements of intensities of Ce^{3+} and Ce^{4+} components of valence-band resonant PES Ce $4f$ spectra of CeO_2 measured off-resonance ($hv = 115$ eV) and on-resonance ($hv = 121.4$ eV for the Ce^{3+} component and $hv = 124.8$ eV for the Ce^{4+} component). The value of resonant enhancement ratio represents an upper estimate of the concentration of Ce^{3+} ions on the ceria surface and is plotted in Fig. 2b[17,42].

STM imaging. STM measurements were performed with commercial Pt–Ir tips (Unisoku). STM imaging of $CeO_2(111)$ and $Pt/CeO_2(111)$ films was available only via unoccupied states. We used sample voltages 2.5–3.5 V and tunnelling currents 25–75 pA.

Theory. All calculations were based on the DFT and were performed using the spin-polarized GGA + U approach[43], employing the Perdew–Burke–Ernzerhof exchange-correlation functional[44] and ultrasoft pseudopotentials[45]. The spin-polarized Kohn–Sham equations were solved with a plane-wave basis set and the Fourier representation of the charge density was limited by kinetic cutoffs of 40 and 320 Ry, respectively. The Quantum-ESPRESSO computer package was used in all the calculations[46]. In the Hubbard U term, the occupations of the f-orbitals were defined in terms of atomic wave function projectors and the value of the parameter U was set to 4.5 eV, following our previous studies[47,48].

Slab models. The ceria (111) surfaces were modelled with periodic (3 × 3) slabs being three CeO_2 ML thick and separated by more than 10 Å of vacuum in the direction perpendicular to the surface. The Brillouin zone was sampled at Gamma point. In the present work, we considered two low-energy ML-high steps that we label following ref. 25 as step I and step II. The edge of both step I and step II steps are oriented along the $[1\bar{1}0]$ direction. These surface steps were modelled with vicinal surfaces described with monoclinic periodic slabs separated by > 10 Å of vacuum in the direction perpendicular to the (111) terrace. The dimensions of the cells were 17.97 × 11.67 Å^2 along the $[1\bar{1}2]$ and $[1\bar{1}0]$ directions (step I) and 15.72 × 11.67 Å^2 along the $[1\bar{1}2]$ and $[1\bar{1}0]$ (step II). All the vicinal surfaces slabs included three CeO_2 ML. This thickness was shown to be sufficient to calculate the structural and thermodynamic properties of these steps[25]. The complete set of surface structures and systems considered in this work is listed in the Supplementary Note 1. All these systems were structurally optimized according to the Hellmann–Feynman forces. During the geometry optimization, the atomic positions of the lowermost CeO_2 ML were constrained, as well as those of the Ce atoms in the central ML, except for the Ce atoms below the step edge.

Energetics. Binding energies were computed as $1/N_{Pt} (E_{slab} + N_{Pt} E_{Pt} - E_{slab/Pt})$, where $E_{slab/Pt}$ is the total energy of the ceria slab containing N_{Pt} atoms of Pt, E_{slab} is the total energy of the corresponding relevant (stoichiometric, reduced or oxidized) Pt-free ceria slab and E_{Pt} is the total energy of a Pt atom in vacuum. The energies required to form an oxygen vacancy Ov were calculated as $(E_{slab/Ov} + \frac{1}{2} E_{O2} - E_{slab})$, where $E_{slab/Ov}$ and E_{slab} are the total energies of the ceria supercell with and without the O vacancy, respectively, whereas E_{O2} is the total energy of a gas-phase O_2 molecule compensated for the known overbinding predicted by (semi)local functionals for O_2.

References

1. Yang, X. *et al.* Single-atom catalysts: a new frontier in heterogeneous catalysis. *Acc. Chem. Res.* **46**, 1740–1748 (2013).
2. Flytzani-Stephanopoulos, M. & Gates, B. C. Atomically dispersed supported metal catalysts. *Annu. Rev. Chem. Biomol. Eng.* **3**, 545–574 (2012).
3. Fu, Q., Saltsburg, H. & Flytzani-Stephanopoulos, M. Active nonmetallic Au and Pt species on ceria-based water-gas shift catalysts. *Science* **301**, 935–938 (2003).
4. Bruix, A. *et al.* Maximum noble-metal efficiency in catalytic materials: atomically dispersed surface platinum. *Angew. Chem. Int. Ed.* **53**, 10525–10530 (2014).
5. Qiao, B. *et al.* Single-atom catalysis of CO oxidation using Pt_1/FeO_x. *Nat. Chem.* **3**, 634–641 (2011).
6. Novotný, Z. *et al.* Ordered array of single adatoms with remarkable thermal stability: $Au/Fe_3O_4(001)$. *Phys. Rev. Lett.* **108**, 216103 (2012).
7. Parkinson, G. S. *et al.* Carbon monoxide-induced adatom sintering in a $Pd-Fe_3O_4$ model catalyst. *Nat. Mater.* **12**, 724–728 (2013).
8. Bliem, R. *et al.* Subsurface cation vacancy stabilization of the magnetite (001) surface. *Science* **346**, 1215–1218 (2014).
9. Li, F., Li, Y., Zeng, X. C. & Chen, Z. Exploration of high-performance single-atom catalysts on support M_1/FeO_x for CO oxidation via computational study. *ACS Catal.* **5**, 544–552 (2015).
10. Hatanaka, M. *et al.* Ideal Pt loading for a Pt/CeO_2-based catalyst stabilized by a Pt-O-Ce bond. *Appl. Catal. B Environ.* **99**, 336–342 (2010).
11. Fiala, R. *et al.* Proton exchange membrane fuel cell made of magnetron sputtered $Pt-CeO_x$ and Pt-Co thin film catalysts. *J. Power Sources* **273**, 105–109 (2015).
12. Gong, X.-Q., Selloni, A., Batzill, M. & Diebold, U. Steps on anatase $TiO_2(101)$. *Nat. Mater.* **5**, 665–670 (2006).
13. Barth, J. V, Costantini, G. & Kern, K. Engineering atomic and molecular nanostructures at surfaces. *Nature* **437**, 671–679 (2005).
14. Vang, R. T. *et al.* Controlling the catalytic bond-breaking selectivity of Ni surfaces by step blocking. *Nat. Mater.* **4**, 160–162 (2005).
15. Gong, X., Selloni, A., Dulub, O., Jacobson, P. & Diebold, U. Small Au and Pt clusters at the anatase $TiO_2(101)$ surface: behavior at terraces, steps, and surface oxygen vacancies. *J. Am. Chem. Soc.* **130**, 370–381 (2008).
16. Dvořák, F. *et al.* Adjusting morphology and surface reduction of $CeO_2(111)$ thin films on Cu(111). *J. Phys. Chem. C* **115**, 7496–7503 (2011).
17. Duchoň, T. *et al.* Ordered phases of reduced ceria as epitaxial films on Cu(111). *J. Phys. Chem. C* **118**, 357–365 (2014).
18. Zhou, Y., Perket, J. M. & Zhou, J. Growth of Pt nanoparticles on reducible CeO_2 (111) thin films: effect of nanostructures and redox properties of ceria. *J. Phys. Chem. C* **114**, 11853–11860 (2010).
19. Shao, X., Jerratsch, J.-F., Nilius, N. & Freund, H.-J. Probing the 4f states of ceria by tunneling spectroscopy. *Phys. Chem. Chem. Phys.* **13**, 12646–12651 (2011).
20. Lu, J.-L., Gao, H.-J., Shaikhutdinov, S. & Freund, H.-J. Morphology and defect structure of the $CeO_2(111)$ films grown on Ru(0001) as studied by scanning tunneling microscopy. *Surf. Sci.* **600**, 5004–5010 (2006).
21. Zhou, Y. & Zhou, J. Interactions of Ni nanoparticles with reducible $CeO_2(111)$ thin films. *J. Phys. Chem. C* **116**, 9544–9549 (2012).
22. Mullins, D. R. *et al.* Water dissociation on $CeO_2(100)$ and $CeO_2(111)$ thin films. *J. Phys. Chem. C* **116**, 19419–19428 (2012).
23. Bruix, A. *et al.* A new type of strong metal-support interaction and the production of H_2 through the transformation of water on $Pt/CeO_2(111)$ and $Pt/CeO_x/TiO_2(110)$ catalysts. *J. Am. Chem. Soc.* **134**, 8968–8974 (2012).
24. Negreiros, F. R. & Fabris, S. Role of cluster morphology in the dynamics and reactivity of subnanometer Pt clusters supported on ceria surfaces. *J. Phys. Chem. C* **118**, 21014–21020 (2014).
25. Kozlov, S. M., Viñes, F., Nilius, N., Shaikhutdinov, S. & Neyman, K. M. Absolute surface step energies: accurate theoretical methods applied to ceria nanoislands. *J. Phys. Chem. Lett.* **3**, 1956–1961 (2012).
26. Torbrügge, S., Cranney, M. & Reichling, M. Morphology of step structures on $CeO_2(111)$. *Appl. Phys. Lett.* **93**, 073112 (2008).
27. James, T. E., Hemmingson, S. L. & Campbell, C. T. Energy of supported metal catalysts: from single atoms to large metal nanoparticles. *ACS Catal.* **5**, 5673–5678 (2015).
28. James, T. E., Hemmingson, S. L., Ito, T. & Campbell, C. T. Energetics of Cu adsorption and adhesion onto reduced $CeO_2(111)$ surfaces by calorimetry. *J. Phys. Chem. C* **119**, 17209–17217 (2015).
29. Colussi, S. *et al.* Nanofaceted Pd-O sites in Pd-Ce surface superstructures: Enhanced activity in catalytic combustion of methane. *Angew. Chem. Int. Ed.* **48**, 8481–8484 (2009).
30. Kozlov, S. M. & Neyman, K. M. O vacancies on steps on the $CeO_2(111)$ surface. *Phys. Chem. Chem. Phys.* **16**, 7823–7829 (2014).
31. Fiala, R. *et al.* $Pt-CeO_x$ thin film catalysts for PEMFC. *Catal. Today* **240**, 236–241 (2015).
32. Hatanaka, M. *et al.* Reversible changes in the Pt oxidation state and nanostructure on a ceria-based supported Pt. *J. Catal.* **266**, 182–190 (2009).
33. Nagai, Y. *et al.* Sintering inhibition mechanism of platinum supported on ceria-based oxide and Pt-oxide-support interaction. *J. Catal.* **242**, 103–109 (2006).
34. Matolín, V. *et al.* Platinum-doped CeO_2 thin film catalysts prepared by magnetron sputtering. *Langmuir* **26**, 12824–12831 (2010).
35. Zhou, J., Baddorf, A. P., Mullins, D. R. & Overbury, S. H. Growth and characterization of Rh and Pd nanoparticles on oxidized and reduced $CeO_x(111)$ thin films by scanning tunneling microscopy. *J. Phys. Chem. C* **112**, 9336–9345 (2008).
36. Zhou, Y., Perket, J. M. & Zhou, J. Growth of Pt nanoparticles on reducible $CeO_2(111)$ thin films: effect of nanostructures and redox properties of ceria. *J. Phys. Chem. C* **114**, 11853–11860 (2010).
37. Zhou, Y. & Zhou, J. Growth and sintering of Au − Pt nanoparticles on oxidized and reduced $CeO_x(111)$ thin films by scanning tunneling microscopy. *J. Phys. Chem. Lett.* **1**, 609–615 (2010).
38. Sayle, T. X. T., Parker, S. C. & Sayle, D. C. Oxidising CO to CO_2 using ceria nanoparticles. *Phys. Chem. Chem. Phys.* **7**, 2936–2941 (2005).
39. Castellani, N. J., Branda, M. M., Neyman, K. M. & Illas, F. Density functional theory study of the adsorption of Au atom on cerium oxide: effect of low-coordinated surface sites. *J. Phys. Chem. C* **113**, 4948–4954 (2009).
40. Zambelli, T., Wintterlin, J., Trost, J. & Ertl, G. Identification of the 'active sites' of a surface-catalyzed reaction. *Science* **273**, 1688–1690 (1996).
41. Matolín, V. *et al.* Water interaction with $CeO_2(111)/Cu(111)$ model catalyst surface. *Catal. Today* **181**, 124–132 (2012).
42. Mullins, D. R. The surface chemistry of cerium oxide. *Surf. Sci. Rep.* **70**, 42–85 (2015).
43. Cococcioni, M. & de Gironcoli, S. Linear response approach to the calculation of the effective interaction parameters in the LDA + U method. *Phys. Rev. B* **71**, 035105 (2005).
44. Perdew, J. P. J., Burke, K. & Ernzerhof, M. Generalized gradient approximation made simple. *Phys. Rev. Lett.* **77**, 3865–3868 (1996).
45. Vanderbilt, D. Soft self-consistent pseudopotentials in a generalized eigenvalue formalism. *Phys. Rev. B* **41**, 7892–7895 (1990).
46. Giannozzi, P. *et al.* QUANTUM ESPRESSO: a modular and open-source software project for quantum simulations of materials. *J. Phys. Condens. Matter* **21**, 395502 (2009).

47. Fabris, S., de Gironcoli, S., Baroni, S., Vicario, G. & Balducci, G. Taming multiple valency with density functionals: a case study of defective ceria. *Phys. Rev. B* **71,** 041102 (2005).
48. Fabris, S., Vicario, G., Balducci, G., De Gironcoli, S. & Baroni, S. Electronic and atomistic structures of clean and reduced ceria surfaces. *J. Phys. Chem. B* **109,** 22860–22867 (2005).

Acknowledgements

This work was supported by Czech Science Foundation (contract numbers 15-06759S and 13-10396S), and by the European Union via the FP7-NMP-2012 project chipCAT under contract number 310191 and the EU FP7 COST action CM1104. A.T. acknowledges the support of the Grant Agency of the Charles University, contract number 2048514. S.F. acknowledges the support provided by the Humboldt Foundation through a Friedrich Wilhelm Bessel Research Award. The high-performance computing resources were gratefully provided by ISCRA initiative of CINECA. CERIC-ERIC consortium is acknowledged for financial support.

Author contributions

F.D., A.T., M.V., T.S., I.M., J.M. and V.M. designed and performed the experiments. M.F.C., N.-D.T., F.R.N. and S.F. performed the DFT calculations. All authors interpreted the experimental and computational results. J.M., S.F., T.S., I.M. and V.M. wrote the manuscript. F.D., A.T., J.M., I.M., V.M. and S.F. provided funding.

Additional information

Permissions

All chapters in this book were first published in NC, by Nature Publishing Group; hereby published with permission under the Creative Commons Attribution License or equivalent. Every chapter published in this book has been scrutinized by our experts. Their significance has been extensively debated. The topics covered herein carry significant findings which will fuel the growth of the discipline. They may even be implemented as practical applications or may be referred to as a beginning point for another development.

The contributors of this book come from diverse backgrounds, making this book a truly international effort. This book will bring forth new frontiers with its revolutionizing research information and detailed analysis of the nascent developments around the world.

We would like to thank all the contributing authors for lending their expertise to make the book truly unique. They have played a crucial role in the development of this book. Without their invaluable contributions this book wouldn't have been possible. They have made vital efforts to compile up to date information on the varied aspects of this subject to make this book a valuable addition to the collection of many professionals and students.

This book was conceptualized with the vision of imparting up-to-date information and advanced data in this field. To ensure the same, a matchless editorial board was set up. Every individual on the board went through rigorous rounds of assessment to prove their worth. After which they invested a large part of their time researching and compiling the most relevant data for our readers.

The editorial board has been involved in producing this book since its inception. They have spent rigorous hours researching and exploring the diverse topics which have resulted in the successful publishing of this book. They have passed on their knowledge of decades through this book. To expedite this challenging task, the publisher supported the team at every step. A small team of assistant editors was also appointed to further simplify the editing procedure and attain best results for the readers.

Apart from the editorial board, the designing team has also invested a significant amount of their time in understanding the subject and creating the most relevant covers. They scrutinized every image to scout for the most suitable representation of the subject and create an appropriate cover for the book.

The publishing team has been an ardent support to the editorial, designing and production team. Their endless efforts to recruit the best for this project, has resulted in the accomplishment of this book. They are a veteran in the field of academics and their pool of knowledge is as vast as their experience in printing. Their expertise and guidance has proved useful at every step. Their uncompromising quality standards have made this book an exceptional effort. Their encouragement from time to time has been an inspiration for everyone.

The publisher and the editorial board hope that this book will prove to be a valuable piece of knowledge for researchers, students, practitioners and scholars across the globe.

List of Contributors

Bartosz A. Grzybowski
Department of Chemical and Biological Engineering and Department of Chemistry, Northwestern University, 2145 Sheridan Road, Evanston, Illinois 60208, USA

Thomas M. Hermans
Department of Chemical and Biological Engineering and Department of Chemistry, Northwestern University, 2145 Sheridan Road, Evanston, Illinois 60208, USA
Institut de Science et d'Ingénierie Supramoléculaires (ISIS), UMR7006, 8 allée Gaspard Monge, 67000 Strasbourg, France (T.M.H.)

Kyle J.M. Bishop
Department of Chemical Engineering, The Pennsylvania State University, 132C Fenske Lab, University Park, Pennsylvania 16802, USA

Stephen H. Davis
Engineering Sciences and Applied Mathematics, Northwestern University, 2145 Sheridan Road, Evanston, Illinois 60208, USA

Peter S. Stewart
Engineering Sciences and Applied Mathematics, Northwestern University, 2145 Sheridan Road, Evanston, Illinois 60208, USA
School of Mathematics and Statistics, University of Glasgow, University Gardens, Glasgow G12 0RB, UK (P.S.S.)

Austin P. Spencer, Boris Spokoyny, Supratim Ray, Fahad Sarvari and Elad Harel
Department of Chemistry, Northwestern University, 2145 Sheridan Road, Evanston, Illinosis 60208, USA

Haruka Sugiura, Manami Ito and Tomoya Okuaki
Department of Computational Intelligence and Systems Science, Tokyo Institute of Technology, 4259 Nagatsuta-cho, Midori-ku, Yokohama 226-8502, Japan

Yoshihito Mori
Department of Chemistry, Faculty of Science, Ochanomizu University, 2-1-1 Ohtsuka, Bunkyo-ku, Tokyo 112-8610, Japan

Hiroyuki Kitahata
Department of Physics, Graduate School of Science, Chiba University, 1-33 Yayoi-cho, Inage-ku, Chiba 263-8522, Japan

Masahiro Takinoue
Department of Computational Intelligence and Systems Science, Tokyo Institute of Technology, 4259 Nagatsuta-cho, Midori-ku, Yokohama 226-8502, Japan
PRESTO, Japan Science and Technology Agency (JST), 4-1-8 Honcho Kawaguchi, Saitama 332-0012, Japan

Chunyan Wang and Jin H. Bae
Department of Bioengineering, Rice University, Houston, Texas 77030, USA

David Yu Zhang
Department of Bioengineering, Rice University, Houston, Texas 77030, USA
Systems, Synthetic, and Physical Biology, Rice University, Houston, Texas 77030, USA

Yuki Koizumi, Naoki Shida, Masato Ohira, Hiroki Nishiyama, Ikuyoshi Tomita and Shinsuke Inagi
Department of Electronic Chemistry, Interdisciplinary Graduate School of Science and Engineering, Tokyo Institute of Technology, 4259 Nagatsuta-cho, Midori-ku, Yokohama 226-8502, Japan

Jasper H. M. van der Velde, Jochem H. Smit, Atieh Aminian Jazi, Giorgos Guoridis and Thorben Cordes
Molecular Microscopy Research Group, Zernike Institute for Advanced Materials, University of Groningen, Nijenborgh 4, 9747 AG Groningen, The Netherlands

Jens Oelerich and Gerard Roelfes
Stratingh Institute for Chemistry, University of Groningen, Nijenborgh 4, 9747 AG Groningen, The Netherlands

Andreas Herrmann and Jingyi Huang
Department of Polymer Chemistry, Zernike Institute for Advanced Materials, University of Groningen, Nijenborgh 4, 9747 AG Groningen, The Netherlands

Silvia Galiani and Christian Eggeling
MRC Human Immunology Unit, Weatherall Institute of Molecular Medicine, University of Oxford, Headley Way, Oxford OX3 9DS, UK

Kirill Kolmakov
Department NanoBiophotonics, Max-Planck-Institute of Molecular Medicine, Am Fassberg 1, 37077 Goettingen, Germany

Arkadiy Simonov and Andrew L. Goodwin
Department of Chemistry, University of Oxford, Inorganic Chemistry Laboratory, South Parks Road, Oxford OX1 3QR, UK

Alistair R. Overy
Department of Chemistry, University of Oxford, Inorganic Chemistry Laboratory, South Parks Road, Oxford OX1 3QR, UK
Diamond Light Source, Chilton, Oxfordshire, OX11 0DE, UK

Andrew B. Cairns
Department of Chemistry, University of Oxford, Inorganic Chemistry Laboratory, South Parks Road, Oxford OX1 3QR, UK
European Synchrotron Radiation Facility, 71 avenue des Martyrs, 38043 Grenoble, France

Matthew J. Cliffe
Department of Chemistry, University of Oxford, Inorganic Chemistry Laboratory, South Parks Road, Oxford OX1 3QR, UK
Department of Chemistry, University of Cambridge, Lensfield Road, Cambridge CB2 1EW, UK

Matthew G. Tucker
Diamond Light Source, Chilton, Oxfordshire, OX11 0DE, UK
ISIS Facility, Rutherford Appleton Laboratory, Harwell Oxford, Didcot, Oxfordshire OX11 0QX, UK

Simone Meloni and Ursula Rothlisberger
Laboratoire de Chimie et Biochimie Computationnelles, ISIC, FSB-BCH, École Polytechnique Fédérale de Lausanne (EPFL), Lausanne CH-1015, Switzerland
National Competence Center of Research (NCCR) MARVEL—Materials' Revolution: Computational Design and Discovery of Novel Materials, Lausanne CH-1015, Switzerland

Thomas Moehl, Michael Graetzel and Shaik Mohammed Zakeeruddin
Laboratory of Photonics and Interfaces, ISIC, Swiss Federal Institute of Technology (EPFL), Lausanne CH-1015, Switzerland

Michael Saliba, Yong Hui Lee, Peng Gao and Mohammad Khaja Nazeeruddin
Group for Molecular Engineering of Functional Materials, ISIC-Valais, Swiss Federal Institute of Technology (EPFL), Lausanne CH-1015, Switzerland

Marius Franckevičius
Laboratory of Photonics and Interfaces, ISIC, Swiss Federal Institute of Technology (EPFL), Lausanne CH-1015, Switzerland
Center for Physical Sciences and Technology, Savanoriu̧ Avenue 231, Vilnius LT-02300, Lithuania

Wolfgang Tress
Laboratory of Photonics and Interfaces, ISIC, Swiss Federal Institute of Technology (EPFL), Lausanne CH-1015, Switzerland
Group for Molecular Engineering of Functional Materials, ISIC-Valais, Swiss Federal Institute of Technology (EPFL), Lausanne CH-1015, Switzerland

Ting Xu
National Laboratory of Solid State Microstructures, College of Engineering and Applied Sciences and Collaborative Innovation Center of Advanced Microstructures, Nanjing University, 22 Hankou Road, Nanjing 210093, China
Center for Nanoscale Science and Technology, National Institute of Standards and Technology, Gaithersburg, Maryland 20899, USA
Maryland Nanocenter, University of Maryland, College Park, Maryland 20742, USA

Christopher Bohn and Henri J. Lezec
Center for Nanoscale Science and Technology, National Institute of Standards and Technology, Gaithersburg, Maryland 20899, USA

Erich C. Walter, Amit Agrawal, Jeyavel Velmurugan and Wenqi Zhu
Center for Nanoscale Science and Technology, National Institute of Standards and Technology, Gaithersburg, Maryland 20899, USA
Maryland Nanocenter, University of Maryland, College Park, Maryland 20742, USA

A. Alec Talin
Center for Nanoscale Science and Technology, National Institute of Standards and Technology, Gaithersburg, Maryland 20899, USA
Sandia National Laboratories, Livermore, California 94551, USA

Anders Kristensen, Henrik Flyvbjerg and Morten Bo Mikkelsen
Department of Micro- and Nanotechnology, Technical University of Denmark, DK-2800 Kgs. Lyngby, Denmark

Christian L. Vestergaard
Department of Micro- and Nanotechnology, Technical University of Denmark, DK-2800 Kgs. Lyngby, Denmark
Aix Marseille Université, Université de Toulon, CNRS, CPT, UMR 7332, 13288 Marseille, France (C.L.V.)

Walter Reisner
Department of Micro- and Nanotechnology, Technical University of Denmark, DK-2800 Kgs. Lyngby, Denmark
Department of Physics, McGill University, Montreal, Quebec, Canada H3A 2T8 (W.R.)

Landong Li, Junqing Yan and Naijia Guan
Collaborative Innovation Center of Chemical Science and Engineering, Tianjin 300072, China
Key Laboratory of Advanced Energy Materials Chemistry of Ministry of Education, College of Chemistry, Nankai University, Tianjin 300071, China

Tuo Wang, Zhi-Jian Zhao and Jinlong Gong
Collaborative Innovation Center of Chemical Science and Engineering, Tianjin 300072, China
Key Laboratory for Green Chemical Technology of Ministry of Education, School of Chemical Engineering and Technology, Tianjin University, 92 Weijin Road, Nankai District, Tianjin 300072, China

Jian Zhang
Department of New Energy Technology, Ningbo Institute of Materials Technology and Engineering, Chinese Academy of Sciences, Tianjin, Ningbo 315201, China

Adam Sweetman, Mohammad A. Rashid, Samuel P. Jarvis, Janette L. Dunn, Philipp Rahe and Philip Moriarty
School of Physics and Astronomy, University of Nottingham, Nottingham NG7 2RD, UK

Deming Liu, Ewa M. Goldys, James A. Piper and Chenshuo Ma
Laboratory of Advanced Cytometry, ARC Centre of Excellence for Nanoscale BioPhotonics, Department of Physics and Astronomy, Macquarie University, Sydney, New South Wales 2109, Australia

Xiaoxue Xu, Shihui Wen, Wei Ren and Dayong Jin
Laboratory of Advanced Cytometry, ARC Centre of Excellence for Nanoscale BioPhotonics, Department of Physics and Astronomy, Macquarie University, Sydney, New South Wales 2109, Australia
Faculty of Science, Institute for Biomedical Materials and Devices, University of Technology Sydney, New South Wales 2007, Australia

Shixue Dou and Yi Du
Institute for Superconducting and Electronic Materials, Innovation Campus, University of Wollongong, New South Wales 2522, Australia

Xiaogang Liu
Institute of Materials Research and Engineering, 3 Research Link, Singapore 117602, Singapore
Department of Chemistry, National University of Singapore, 3 Science Drive 3, Singapore 117543, Singapore

Xian Qin
Institute of Materials Research and Engineering, 3 Research Link, Singapore 117602, Singapore

Yuhai Zhang
Department of Chemistry, National University of Singapore, 3 Science Drive 3, Singapore 117543, Singapore

Sangtae Kim
Department of Materials Science and Engineering, Massachusetts Institute of Technology, Cambridge, Massachusetts 02139, USA

Soon Ju Choi
Department of Mechanical Engineering, Massachusetts Institute of Technology, Cambridge, Massachusetts 02139, USA

Kejie Zhao
Department of Nuclear Science and Engineering, Massachusetts Institute of Technology, Cambridge, Massachusetts 02139, USA

Sulin Zhang and Hui Yang
Department of Engineering Science and Mechanics, Pennsylvania State University, University Park, Pennsylvania 16802, USA

Giorgia Gobbi
Department of Nuclear Science and Engineering, Massachusetts Institute of Technology, Cambridge, Massachusetts 02139, USA
Politecnico di Milano, Department of Mechanical Engineering, Milan, 20156, Italy

Ju Li
Department of Materials Science and Engineering, Massachusetts Institute of Technology, Cambridge, Massachusetts 02139, USA
Department of Nuclear Science and Engineering, Massachusetts Institute of Technology, Cambridge, Massachusetts 02139, USA

Yuval Ben-Shahar and Uri Banin
The Institute of Chemistry and Center for Nanoscience and Nanotechnology, The Hebrew University of Jerusalem, Edmond Safra Campus Givat-Ram, Jerusalem 91904, Israel

Francesco Scotognella, Ilka Kriegel, Luca Moretti and Giulio Cerullo
Dipartimento di Fisica, IFN-CNR, Politecnico di Milano, 20133 Milan, Italy

Eran Rabani
Department of Chemistry, University of California and Lawrence Berkeley National Laboratory, Berkeley, California 94720-1460, USA
The Sackler Institute for Computational Molecular and Materials Science, Tel Aviv University, Tel Aviv 69978, Israel

Daniel C. Elton and Marivi Fernández-Serra
Department of Physics and Astronomy, Stony Brook University, Stony Brook, New York 11794-3800, USA
Institute for Advanced Computational Science, Stony Brook University, Stony Brook, New York 11794-3800, USA

Jiebo Li and Junrong Zheng
College of Chemistry and Molecular Engineering, Beijing National Laboratory for Molecular Sciences, Peking University, Beijing 100871, China
Department of Chemistry, Rice University, 6100 Main Street, Houston, Texas 77005, USA

Hailong Chen, Andrea Miranda and Xunmin Guo
Department of Chemistry, Rice University, 6100 Main Street, Houston, Texas 77005, USA

Zhun Zhao and Huifeng Qian
Department of Chemical and Biomolecular Engineering, Rice University, 6100 Main Street, Houston, Texas 77005, USA

Michael S. Wong
Department of Chemical and Biomolecular Engineering, Rice University, 6100 Main Street, Houston, Texas 77005, USA
Department of Chemistry, Rice University, 6100 Main Street, Houston, Texas 77005, USA

Kaijun Yuan and Yajing Chen
State Key Laboratory of Molecular Reaction Dynamics, Dalian Institute of the Chemical Physics, Chinese Academy of Sciences, Dalian, Liaoning 116023, China

Nanfeng Zheng and Guangxu Chen
College of Chemistry and Chemical Engineering, Xiamen University, Xiamen, Fujian 361005, China

Shinjiro Takano
Department of Chemistry, School of Science, The University of Tokyo, 7-3-1 Hongo, Bunkyo-ku, Tokyo 113-0033, Japan

Seiji Yamazoe and Tatsuya Tsukuda
Department of Chemistry, School of Science, The University of Tokyo, 7-3-1 Hongo, Bunkyo-ku, Tokyo 113-0033, Japan
Elements Strategy Initiative for Catalysts and Batteries (ESICB), Kyoto University, Katsura, Kyoto 615-8520, Japan

Yuichi Negishi and Wataru Kurashige
Department of Applied Chemistry, Faculty of Science, Tokyo University of Science, 1-3 Kagurazaka, Shinjuku-ku, Tokyo 162-8601, Japan

Toshihiko Yokoyama
Department of Materials Molecular Science, Institute for Molecular Science, Myodaiji, Okazaki, Aichi 444-8585, Japan

Kiyofumi Nitta
Japan Synchrotron Radiation Research Institute, SPring-8, 1-1-1 Koto, Sayo, Hyogo 679-5198, Japan

Lamuel David, Romil Bhandavat, Uriel Barrera and Gurpreet Singh
Mechanical and Nuclear Engineering Department, Kansas State University, 3002 Rathbone Hall, Kansas, Manhattan, Kansas 66506, USA

Adam D. Wright, Rebecca L. Milot, Giles E. Eperon, Henry J. Snaith, Michael B. Johnston and Laura M. Herz
Clarendon Laboratory, Department of Physics, University of Oxford, Parks Road, Oxford OX1 3PU, UK

Carla Verdi, Miguel A. Pérez-Osorio and Feliciano Giustino
Department of Materials, University of Oxford, Parks Road, Oxford OX1 3PH, UK

Biao Zhang and Romain Dugas
Chimie du Solide-Energie, FRE 3677, Collège de France, 11 Place Marcelin Berthelot, Paris, Cedex 05 75231, France
Réseau sur le Stockage Electrochimique de l'Energie (RS2E), FR CNRS 3459, Amiens 80039, France
ALISTORE-European Research Institute, Amiens 80039, France

Patrick Rozier
Réseau sur le Stockage Electrochimique de l'Energie (RS2E), FR CNRS 3459, Amiens 80039, France

ALISTORE-European Research Institute, Amiens 80039, France
University of Toulouse III Paul Sabatier, CIRIMAT CNRS UMR 5085, 118 route de Narbonne, Toulouse, Cedex 09 31062, France

Jean-Marie Tarascon and Gwenaelle Rousse
Chimie du Solide-Energie, FRE 3677, Collège de France, 11 Place Marcelin Berthelot, Paris, Cedex 05 75231, France
Réseau sur le Stockage Electrochimique de l'Energie (RS2E), FR CNRS 3459, Amiens 80039, France
ALISTORE-European Research Institute, Amiens 80039, France
Sorbonne Universités—UPMC Univ Paris 06, 4 Place Jussieu, Paris F-75005, France
University of Toulouse III Paul Sabatier, CIRIMAT CNRS UMR 5085, 118 route de Narbonne, Toulouse, Cedex 09 31062, France

Artem M. Abakumov
EMAT, University of Antwerp, Groenenborgerlaan 171, Antwerp B-2020, Belgium
Skolkovo Institute of Science and Technology, Moscow 143025, Russia

Nicola Casati
Paul Scherrer Institute, WLGA/229, CH-5232 Villigen, Switzerland

Annette Kleppe
Diamond light source Ltd., Harwell Science and innovation Campus, Didcot OX110DE, UK

Andrew P. Jephcoat
Institute for Study of the Earth's interior, Okayama University, Yamada 827, Misasa, Tottori 682-0193, Japan

Piero Macchi
Department of Chemistry and Biochemistry, University of Bern, Freiestrasse 3, Bern CH-3012, Switzerland

L. Dubrovinsky, C. McCammon and S.V. Ovsyannikov
Bayerisches Geoinstitut, University of Bayreuth, Universitaetsstrasse 30, D-95447 Bayreuth, Germany

E. Bykova and M. Bykov
Bayerisches Geoinstitut, University of Bayreuth, Universitaetsstrasse 30, D-95447 Bayreuth, Germany
Laboratory of Crystallography, University of Bayreuth, Universitaetsstrasse 30, D-95447 Bayreuth, Germany

N. Dubrovinskaia
Laboratory of Crystallography, University of Bayreuth, Universitaetsstrasse 30, D-95447 Bayreuth, Germany

H.-P. Liermann
Photon Sciences, Deutsches Elektronen-Synchrotron, Notkestrasse 85, D-22607 Hamburg, Germany

I. Kupenko
Bayerisches Geoinstitut, University of Bayreuth, Universitaetsstrasse 30, D-95447 Bayreuth, Germany
European Synchrotron Radiation Facility, 71 avenue des Martyrs, Grenoble F-38000, France

A.I. Chumakov, R. Rüffer and M. Hanfland
European Synchrotron Radiation Facility, 71 avenue des Martyrs, Grenoble F-38000, France

V. Prakapenka
Center for Advanced Radiation Sources, University of Chicago, 9700 South Cass Avenue, Illinois, Argonne 60437, USA

Jolene L. Lau and Prashant Kulshreshtha
The Molecular Foundry, Lawrence Berkeley National Laboratory, Berkeley, California 94720, USA

Kang Rae Cho
The Molecular Foundry, Lawrence Berkeley National Laboratory, Berkeley, California 94720, USA
Bioscience and Biotechnology Division, Physical and Life Sciences Directorate, Lawrence Livermore National Laboratory, Livermore, California 94550, USA

Yi-Yeoun Kim, Alexander N. Kulak and Fiona C. Meldrum
School of Chemistry, University of Leeds, Leeds LS2 9JT, UK

Pengcheng Yang and Steven P. Armes
Department of Chemistry, University of Sheffield, Brook Hill, Sheffield S3 7HF, UK

Wei Cai
Department of Mechanical Engineering, Stanford University, Stanford, California 94305, USA

Haihua Pan
The Molecular Foundry, Lawrence Berkeley National Laboratory, Berkeley, California 94720, USA
Department of Chemistry, Zhejiang University, Hangzhou 310027, China

James J. De Yoreo
The Molecular Foundry, Lawrence Berkeley National Laboratory, Berkeley, California 94720, USA
Physical Sciences Division, Pacific Northwest National Laboratory, Richland, Washington 99352, USA

J. Tyler Mefford
Department of Chemistry, The University of Texas at Austin, Austin, Texas 78712, USA

Xi Rong and Alexie M. Kolpak
Department of Mechanical Engineering, Massachusetts Institute of Technology, Cambridge, Massachusetts 02139, USA

Artem M. Abakumov
Center for Electrochemical Energy Storage, Skolkovo Institute of Science and Technology, 143026 Moscow, Russia
Electron Microscopy for Material Science, University of Antwerp, Groenenborgerlaan 171, B-2020 Antwerp, Belgium

William G. Hardin
Texas Materials Institute, The University of Texas at Austin, Austin, Texas 78712, USA

Keith P. Johnston
Texas Materials Institute, The University of Texas at Austin, Austin, Texas 78712, USA
Department of Chemical Engineering, The University of Texas at Austin, Austin, Texas 78712 , USA

Sheng Dai
Chemical Sciences Division, Oak Ridge National Laboratory, Oak Ridge, Tennessee 37831, USA

Keith J. Stevenson
Department of Chemistry, The University of Texas at Austin, Austin, Texas 78712, USA
Center for Electrochemical Energy Storage, Skolkovo Institute of Science and Technology, 143026 Moscow, Russia
Texas Materials Institute, The University of Texas at Austin, Austin, Texas 78712, USA
Center for Nano and Molecular Science and Technology, The University of Texas at Austin, Austin, Texas 78712, USA

Markus Döbbelin, Artur Ciesielski, Sébastien Haar, Paolo Samorì, Thomas Mosciatti and Fanny Richard
Nanochemistry Laboratory, ISIS & icFRC, Université de Strasbourg & CNRS, 8 allée Gaspard Monge, 67000 Strasbourg, France

Silvio Osella, Andrea Minoia, Luca Grisanti, Roberto Lazzaroni and David Beljonne
Laboratory for Chemistry of Novel Materials, University of Mons, Place du Parc 20, B-7000 Mons, Belgium

Andrea C. Ferrari and Matteo Bruna
Cambridge Graphene Centre, University of Cambridge, 9 JJ Thomson Avenue, Cambridge CB3 OFA, UK

Eko Adi Prasetyanto and Luisa De Cola
Laboratory of Supramolecular Biomaterials and Chemistry, ISIS & icFRC, Université de Strasbourg & CNRS, 8 allée Gaspard Monge, 67000 Strasbourg, France

Vincenzo Palermo
ISOF-CNR, via Gobetti 101, 40129 Bologna, Italy

Vittorio Morandi
IMM-CNR Sezione di Bologna, via Gobetti, 101, 40129 Bologna, Italy

Raffaello Mazzaro
IMM-CNR Sezione di Bologna, via Gobetti, 101, 40129 Bologna, Italy
Dipartimento di Chimica 'G. Ciamician', Università di Bologna, via Selmi 2, 40126 Bologna, Italy

Filip Dvořák, Andrii Tovt, Mykhailo Vorokhta, Tomáš Skála, Iva Matolínová, Josef Mysliveček and Vladimír Matolín
Charles University in Prague, Faculty of Mathematics and Physics, V Holešovičkách 2, Prague 18000, Czech Republic

Matteo Farnesi Camellone
CNR-IOM DEMOCRITOS, Istituto Officina dei Materiali, Consiglio Nazionale delle Ricerche, Via Bonomea 265, Trieste 34136, Italy

Fabio R. Negreiros
CNR-IOM DEMOCRITOS, Istituto Officina dei Materiali, Consiglio Nazionale delle Ricerche, Via Bonomea 265, Trieste 34136, Italy
Universidade Federal do ABC. Av. Dos Estados, 5001 Bairro Bangu, Santo André SP CEP 09210-580, Brasil

Stefano Fabris and Nguyen-Dung Tran
CNR-IOM DEMOCRITOS, Istituto Officina dei Materiali, Consiglio Nazionale delle Ricerche, Via Bonomea 265, Trieste 34136, Italy
SISSA, Scuola Internazionale Superiore di Studi Avanzati, Via Bonomea 265, Trieste 34136, Italy

Index

A
Alternating Current (ac), 35-36
Amino Acids, 41-42, 51-52, 54, 197
Aperiodic Solids, 56
Aromatic Compounds, 178
Azobenzene Composites, 210

B
Ball Milling, 169-170, 172-176, 208
Biological Homochirality, 1
Bond Stiffnesses, 143-144, 146-147

C
Chemical Bond Rupture, 79
Chemical Oscillation, 15, 19-20
Chiral Molecules, 1
Compressive Sensing, 9-11, 14
Confined Environment, 79
Controlled Growth, 105
Crystal Chemical Analysis, 186
Current Photoswitching, 210
Current-voltage Hysteresis, 64

D
Density Functional Theory, 67, 92, 98, 103, 142, 144, 149, 179, 206, 220, 223-224
Density Functional Theory (dft), 92, 98, 179
Dipole-dipole Interactions, 127-128, 131
Dynamic Force Microscopy (dfm), 98

E
Electrocatalysts, 199, 207-209
Electrochromic Polymers, 73-74, 77
Electrochromic Switching, 73-74, 76
Electron Density, 178-179, 181-182, 184
Electron-phonon Coupling, 160-162, 165, 168, 184
Electropolymerization, 35-40
Energy Migration Pathways, 135-136, 138, 140
Equilibrium, 8, 10, 15-16, 18-21, 24-28, 30, 34, 36, 59, 79, 81-86, 98, 100, 102-103, 114, 117, 133, 142, 190, 193-195, 197

F
Ferroelectric Effect, 64-65, 67, 71
Fluorophore Derivatives, 41-43, 50, 52, 54
Functional Materials, 56-57, 64

G
Glass-graphene Composite, 150
Gold Clusters, 143-144, 148

H
Heterogeneous Nanocrystals, 105, 109
Hybrid Lead Halide, 160-164, 166
Hydrogen-bond Network, 127, 132

I
Infrared Spectroscopies, 127
Inorganic Materials, 62, 97, 192
Insertion Compounds, 169-170
Intermolecular Potential, 98, 100
Interphase Formation, 169
Intramolecular Photostabilization, 41-43, 45, 50-54
Ionic Polarization, 64, 67, 70-71

L
Lift Forces, 1, 6, 8
Liquid-phase Exfoliation, 210-211, 218-219
Lithium-ion Batteries, 119, 150, 158-159
Local Lattice Strain, 192
Longitudinal Optical Phonons, 160
Low-frequency Motions, 113

M
Mechanical Energy Harvesting, 113
Metal Cluster, 143-144
Metal Nanoparticle Surfaces, 135
Micron Scale, 79, 85
Mineral-organic Composite, 192
Modulation Control, 15, 20-21
Molecular Crystal, 178, 183-184
Molecular Geometries, 178-180, 183
Monolayer Molecules, 135
Multidimensional Electronic Structure, 9

N
Nanomaterials Engineering, 105
Native Characterization, 24, 30
Non-covalent Catalysis, 24, 26-27
Nucleic Acid, 24-25, 32-34

O
Occlusion Mechanism, 192

Open-reactor System, 15-21
Optical Phonon, 127-128, 131-132, 160, 163, 166
Optimal Metal, 120, 124
Orientational Dependence, 98
Oxygen Cycling, 186-187

P
Paper Electrode, 150, 158
Perturbation Theory, 160, 165-167
Phonon Coupling, 56, 59-60, 62, 160-162, 164-165, 168, 184
Photocatalysis, 89, 96-97, 120-121, 125
Photocatalytic Hydrogen Production, 88-89, 92, 94-96, 121, 125
Plasmonics, 73, 78
Poisson's Ratio, 113, 115
Polymer Membranes, 113
Polymer Microfibre Networks, 35
Pt-ceria Catalysts, 220
Pulse-density, 15-16, 18, 20-22

R
Rutile Titanium Dioxide, 88

S
Selective Disorder, 56
Semiconductor Photocatalysts, 88

Semiconductor-metal Nanorods, 120
Silicon Oxycarbide, 150-151, 158-159
Single-element Detection, 9
Sodium-ion Batteries, 169, 177
Solar Cells, 64-66, 70-72, 111, 161, 166-168
Structural Complexity, 186
Subtractive Engineering, 105
Surface Step Decoration, 220

T
Thermodynamics, 14, 24-25, 28, 30-34, 114, 118, 198
Thiolates, 143
Titanium Dioxide, 88, 96-97
Transient Absorption, 120-122, 124-126
Transition State Theory, 79, 81

U
Ultrafast Dynamics, 9-10, 142, 219

V
Versatile Design, 41
Vibration-electron Coupling, 135
Vortex Flows, 1-2, 6

W
Water Electrolysis, 199
Wireless Electrodes, 35

www.ingramcontent.com/pod-product-compliance
Lightning Source LLC
Chambersburg PA
CBHW080516200326
41458CB00012B/4231